LABORATORY STUDIES IN INTEGRATED PRINCIPLES OF ZOOLOGY

EIGHTEENTH EDITION

Cleveland P. Hickman, Jr.
PROFESSOR EMERITUS
WASHINGTON AND LEE UNIVERSITY

Lee B. Kats
PROFESSOR
PEPPERDINE UNIVERSITY

Susan L. Keen
SENIOR LECTURER
UNIVERSITY OF CALIFORNIA, DAVIS

Original Artwork by
WILLIAM C. OBER, M.D.
WASHINGTON AND LEE UNIVERSITY
and
CLAIRE W. OBER, B.A.
WASHINGTON AND LEE UNIVERSITY

LABORATORY STUDIES IN INTEGRATED PRINCIPLES OF ZOOLOGY, EIGHTEENTH EDITION

Some ancillaries, including electronic and print components, may not be available to customers outside the United States.

This book is printed on acid-free paper.

1 2 3 4 5 6 7 8 9 LMN 21 20 19

ISBN 978-1-26-041121-8
MHID 1-260-41121-4

Portfolio Manager: *Michael R. Ivanov, PhD*
Product Developers: *Michelle Flomenhoft*
Marketing Manager: *Kelly Brown*
Content Project Managers: *Becca Gill/Ann Courtney*
Buyer: *Laura Fuller*
Design: *Jessica Cuevas*
Content Licensing Specialist: *Missy Homer*
Cover Image: © *Shutterstock/idreamphoto*
Compositor: *MPS Limited*

The Internet addresses listed in the text were accurate at the time of publication. The inclusion of a website does not indicate an endorsement by the authors or McGraw-Hill Education, and McGraw-Hill Education does not guarantee the accuracy of the information presented at these sites.

Some of the laboratory experiments included in this text may be hazardous if materials are handled improperly or if procedures are conducted incorrectly. Safety precautions are necessary when you are working with chemicals, glass test tubes, hot water baths, sharp instruments, and the like, or for any procedures that generally require caution. Your school may have set regulations regarding safety procedures that your instructor will explain to you. Should you have any problems with materials or procedures, please ask your instructor for help.

mheducation.com/highered

CONTENTS

preface *v*
laboratory safety procedures *viii*
general instructions *viii*
statement on the use of living and preserved animals in the zoology laboratory *x*
chief characteristics of the animal phyla *xi*

PART ONE

INTRODUCTION TO THE LIVING ANIMAL

EXERCISE 1

The Microscope 3

Exercise 1A: Compound Light Microscope 3
Exercise 1B: Stereoscopic Dissecting Microscope 8
Exercise 1C: Electron Microscope 9

EXERCISE 2

Cell Structure and Division 13

Exercise 2A: The Cell—Unit of Protoplasmic Organization 13
Exercise 2B: Cell Division—Mitosis and Cytokinesis 16

EXERCISE 3

Gametogenesis and Embryology 25

Exercise 3A: Meiosis—Maturation Division of Germ Cells 25
Exercise 3B: Cleavage Patterns—Spiral and Radial Cleavage 36
Exercise 3C: Frog Development 44

EXERCISE 4

Tissue Structure and Function 47

Exercise 4: Tissues Combined into Organs 54

PART TWO

THE DIVERSITY OF ANIMAL LIFE

EXERCISE 5

Ecological Relationships of Animals 63

Exercise 5A: A Study of Population Growth, with Application of the Scientific Method 63
Exercise 5B: Ecology of a Freshwater Habitat 67

EXERCISE 6

Introduction to Animal Taxonomy 77

Exercise 6A: Phylogeny Reconstruction—How to Read and Compare Cladograms 78
Exercise 6B: Use of a Taxonomic Key for Organism Identification 83

EXERCISE 7

Unicellular Eukaryotes 89

Exercise 7A: Phylum Amoebozoa—*Amoeba* and Others 89
Exercise 7B: Phyla Euglenozoa and Viridiplantae—*Euglena, Volvox,* and *Trypanosoma* 97
Exercise 7C: Phylum Apicomplexa—*Plasmodium* and *Gregarina* 106
Exercise 7D: Phylum Ciliophora—*Paramecium* and Other Ciliates 109
Experimenting in Zoology: Effect of Temperature on the Locomotor Activity of *Stentor* 115
Experimenting in Zoology: Genetic Polymorphism in *Tetrahymena* 117

EXERCISE 8

The Sponges 121

Exercise 8: Class Calcispongiae—*Sycon* 121

EXERCISE 9

The Radiate Animals 131

Exercise 9A: Class Hydrozoa—*Hydra, Obelia,* and *Gonionemus* 132
Exercise 9B: Class Scyphozoa—*Aurelia,* a "True" Jelly 138
Exercise 9C: Class Anthozoa—*Metridium* and *Astrangia* 140
Experimenting in Zoology: Predator Functional Response: Feeding Rate in *Hydra* 147

EXERCISE 10

The Flatworms 149

Exercise 10A: Class Turbellaria—Planarians 150
Exercise 10B: Class Trematoda—Digenetic Flukes 153
Exercise 10C: Class Cestoda—Tapeworms 158
Experimenting in Zoology: Planaria Regeneration Experiment 167

EXERCISE 11

Nematodes and Four Small Protostome Phyla 171

Exercise 11A: Phylum Nematoda—*Ascaris* and Others 172
Exercise 11B: A Brief Look at Some Other Protostomes 179

EXERCISE 12

The Molluscs 183

Exercise 12A: Class Bivalvia (= Pelecypoda)—Freshwater Clam 184
Exercise 12B: Class Gastropoda—Pulmonate Land Snail 191
Exercise 12C: Class Polyplacophora—Chitons 194
Exercise 12D: Class Cephalopoda—*Loligo,* the Squid 195

EXERCISE 13

The Annelids 201

Exercise 13A: Errantia (errant polychaetes)—Clamworm 202
Exercise 13B: Sedentaria—Earthworm 204
Exercise 13C: Class Hirudinida—Leech 215
Experimenting in Zoology: Behavior of Medicinal Leeches, Hirudo medicinalis 217

EXERCISE 14

The Chelicerate Arthropods 219

Exercise 14: Chelicerate Arthropods—Horseshoe Crab and Garden Spider 220

EXERCISE 15

The Crustacean Arthropods 227

Exercise 15A: Subphylum Crustacea—Crayfish, Lobsters, and Other Crustaceans 227
Experimenting in Zoology: The Phototactic Behavior of Daphnia 239

EXERCISE 16

The Arthropods 241

Exercise 16A: Myriapods—Centipedes and Millipedes 241
Exercise 16B: Insects—Grasshopper and Honeybee 243
Exercise 16C: Insects—House Cricket 251
Exercise 16D: Metamorphosis of *Drosophila* 254
Exercise 16E: Collection and Classification of Insects 255

EXERCISE 17

The Echinoderms 265

Exercise 17A: Class Asteroidea—Sea Stars 265
Exercise 17B: Class Ophiuroidea—Brittle Stars 270
Exercise 17C: Class Echinoidea—Sea Urchins 273
Exercise 17D: Class Holothuroidea—Sea Cucumbers 275
Exercise 17E: Class Crinoidea—Feather Stars and Sea Lilies 278

EXERCISE 18

Phylum Chordata: A Deuterostome Group 281

Exercise 18A: Subphylum Urochordata—*Ciona,* an Ascidian 283
Exercise 18B: Subphylum Cephalochordata—*Amphioxus* 286

EXERCISE 19

The Fishes—Lampreys, Sharks, and Bony Fishes 291

Exercise 19A: Class Petromyzontida—Lampreys (Ammocoete Larva and Adult) 291
Exercise 19B: Class Chondrichthyes—Cartilaginous Fishes 296
Exercise 19C: Class Actinopterygii—Bony Fishes 301
Experimenting in Zoology: Agonistic Behavior in Paradise Fish, *Macropodus opercularis* 307
Experimenting in Zoology: Analysis of the Multiple Hemoglobin System in *Carassius auratus,* Common Goldfish 309

EXERCISE 20

The Amphibians: Frogs 313

Exercise 20A: Behavior and Adaptations 314
Exercise 20B: Skeleton 318
Exercise 20C: Skeletal Muscles 320
Exercise 20D: Digestive, Respiratory, and Urogenital Systems 325
Exercise 20E: Circulatory System 328
Exercise 20F: Nervous System 334

EXERCISE 21

The Reptiles 337

Exercise 21: Painted Turtle 337

EXERCISE 22

The Birds 343

Exercise 22: Pigeon 343

EXERCISE 23

The Mammals: Fetal Pig 349

Exercise 23A: Skeleton 350
Exercise 23B: Muscular System 354
Exercise 23C: Digestive System 363
Exercise 23D: Urogenital System 368
Exercise 23E: Circulatory System 372
Exercise 23F: Nervous System 379
Exercise 23G: Respiratory System 384

Appendix A: Instructor's Resources for Implementing Exercises 387

Appendix B: Sources of Living Material and Prepared Microslides 420

Index 423

Definitions 436

PREFACE

Laboratory Studies in Integrated Principles of Zoology offers students hands-on experience in learning about the diversity of life. It provides students the opportunity to become acquainted with the principal groups of animals and to recognize the unique anatomical features that characterize each group as well as the patterns that link animal groups to each other. Although this manual was written to accompany *Integrated Principles of Zoology,* it can easily be adapted for use with any other introductory zoology text and with a variety of course plans. Every effort has been made to provide clear instructions and enough background material to create interest and an understanding of the subject matter. Many illustrations complement the written word.

Distinctive Features

Experimenting in Zoology. Project exercises accompany certain chapters, within sections entitled "Experimenting in Zoology." Two project exercises use molecular techniques to explore questions important to our understanding of zoology and evolution. Some of these exercises can be completed within a single laboratory period; others are followed for a longer period. In all project exercises the student follows experimental procedures, records and analyzes quantitative data, and draws conclusions from the results. Many instructors will want their students to gain additional experience by writing a laboratory report in which the student states the objectives, methods followed, results obtained, and conclusions that can be drawn from the results. The "Experimenting in Zoology" exercises are Effect of Temperature on the Locomotor Activity of *Stentor* (Exercise 7); Genetic Polymorphism in *Tetrahymena* (Exercise 7); Predator Functional Response: Feeding Rate in *Hydra* (Exercise 9); Planaria Regeneration Experiment (Exercise 10); Behavior of Medicinal Leeches, *Hirudo medicinalis* (Exercise 13); The Phototactic Behavior of *Daphnia* (Exercise 15); Agonistic Behavior in Paradise Fish, *Macropodus opercularis* (Exercise 19); Analysis of the Multiple Hemoglobin System in *Carassius auratus,* Common Goldfish (Exercise 19).

Exercises have been made more interactive, with questions placed throughout the text and with fill-in blanks provided for students to write down their responses and observations. This "active learning" approach involves students in the exercise and encourages them to think about the information as they read. Some questions may require students to consult their textbook for the answers. In some exercises, we have placed questions within the figure legends, to be answered in the blanks provided when the student consults the figure.

To help students summarize, compare, and contrast organisms within and among phyla, a table has been provided (see Chief Characteristics of Animal Phyla, pp. xi–xii). After each of the Exercises 8 to 23 has been completed, students should record developmental and morphological features of each group in the table. Instructors may ask students to draw comparisons among phyla for each feature in the table, or to consider the features of each phylum in relation to habitat or lifestyle.

An interactive exercise on how to make a cladogram forms the core study in Exercise 6, Introduction to Animal Taxonomy. The exercise explains how to map characters onto trees and how to choose between different possible evolutionary pathways using the principle of parsimony. An additional interactive exercise on cladistics called "Taxonomic Classification and Phylogenetic Trees" is found under the Instructor Resources in Connect for *Integrated Principles of Zoology.* This exercise can also be used as a supplement to Exercise 6.

Up-to-Date Coverage. We have made updates throughout that parallel the accompanying textbook, *Integrated Principles of Zoology.* Birds are now recognized as a clade within the class Reptilia. Birds and reptiles share several derived characters, including distinct skull and ankle characteristics and the presence of beta-keratin in the skin. This unites birds and nonavian reptiles as a monophyletic group. However, to apply Linnaean ranking for birds and reptiles we retain the traditional classification of class Aves and class Reptilia. All modern nonavian reptiles are now placed in the diapsid group. Turtles, formerly considered anapsids, are treated (controversially) as derived diapsids. Another change is inclusion of homopteran insects in the order Hemiptera; the order Homoptera is now obsolete.

Terminology. As with the previous editions, we have included the derivation of specialized biological terms and genera where first introduced. This will help students become familiar with the Latin and Greek roots from which technical terms are built. We repeatedly emphasize that the species is a binomial by spelling the complete species binomial in the classification breakdown for each representative species in the exercises.

Helpful Tools for the Student. There are many aids for the student in this laboratory manual. Throughout the exercises, working instructions are clearly set off from the descriptive material. Classifications, where appropriate, follow the text at the end of exercises. A revised cladogram showing the position of the group in the Eucarya and a "pie" diagram showing the relative sizes of the classes in a phylum introduce each diversity exercise. Function is explained along with anatomy. Topic headings help the student mentally organize the material. Metric tables and definitions are placed on the inside front and back covers for convenient use. Much of the artwork was designed to assist the student with difficult dissections.

Directions for preparing the exercises in this manual are found in Appendix A. For each exercise we have listed the materials required, directions for preparing solutions, suggestions for maintaining and working with living materials, suggestions for demonstrations, and a listing of appropriate references, most of which are annotated. This information is convenient to the instructor as well as to students who may later wish to consult or implement an exercise.

Exercises are divided into "Core Study" and "Further Study" sections are divided to make them more manageable for students and instructors having labs of varying lengths. The "Core Study" makes the critical concepts of each topic accessible to students in short lab periods or when only a single lab is provided for each topic. The "Further Study" section of each exercise permits in-depth exploration of materials when time permits.

In many exercises, labels for illustrations have been expanded to provide functional information for the identified structure. Where appropriate, transfer of text explanation to an illustration allows more effective understanding of structural and functional relationships.

Introductory cladograms were revised for this edition, include shared derived characters and are presented in tree format.

New to this Edition

Changes in Exercise 8 include small improvements to illustrations and expansion of the lab report with a new question. We revised the introduction to Exercise 9, changed the common name applied to scyphozoans from "jellyfish" to "jellies" (or "sea jellies"), and replaced Figure 9-5 with a new photograph of swimming *Aurelia aurita.* For Exercise 10 we updated statistics on the *Schistosoma* global health problem, made small corrections in Figures 10-2 and 10-7, and added full color to Figure 10-5. We added a new exercise on population growth in vinegar eels to the Projects and Demonstrations section of Exercise 11—students count worm populations weekly for a month to determine the effects of different concentrations of apple cider vinegar on population growth. For Exercise 12, Figure 12-7 has been improved to better illustrate the movement and oxygenation of circulation in a freshwater clam. More information on choosing between living or freshly killed earthworms for the study of structure and function has been added to Exercise 13. We added color to Figure 13.5 to bring the image closer to what students will see on their slides.

For Exercise 14, we added a photograph of a technician drawing blood from horseshoe crabs; the accompanying text describes the importance of horseshoe crab blood in the detection of human pathogens in patients, drugs, and intravenous devices. The introduction to Exercise 15 was written to better describe the morphological and developmental characteristics that distinguish the crustacean arthropods. Under the section on Other Crustaceans, we have expanded the description of barnacles (Class Thecostraca) to describe the remarkable metamorphosis they undergo during the course of their development. Color has been added by our artist to all the illustrations in Figure 15-5. We revised several of the illustrations in Exercise 16. The key to principal orders of insects was revised with the addition of several new insect illustrations; placement of illustrations was adjusted so that all now appear directly adjacent to their descriptions in the key. In Exercise 17, new photographs appear in Figure 17-5 and the legend to Figure 17-1 emphasizes that the pentaradial body plan of adult echinoderms is secondarily derived from bilateral larvae. For the vertebrate chapters, Exercises 18–23, many of the illustrations where improved and color was added to line drawings. Several interactive questions were added to the directions.

McGraw-Hill Create™

Your Book, Your Way: Create is a self-service website that allows you to build custom course materials by drawing upon McGraw-Hill Education's comprehensive, cross-disciplinary content and rights-secured third-party sources. Simply choose your content, align the content to your syllabus, and personalize your color print, black-and-white print, or eBook for your students.

1. Improve course outcomes by providing targeted content matched to your teaching style and the learning objectives for your course.
2. Increase student engagement by providing course materials that are tailored to your syllabus with integrated learning resources customized by you.
3. Enhance productivity by saving time to build a course solution that reflects the content you cover in the order in which you cover it! Receive your review password-protected PDF in minutes, review and approve, and you're ready to submit your bookstore order.

Build Your Book, Your Way at: http://create.mheducation.com.

Acknowledgments

The authors express their appreciation to the editors and support staff at McGraw-Hill Education who guided this revision and were a pleasure to work with. People who played key roles and to whom we express our gratitude were: Michael Ivanov, Senior Portfolio Manager; Michelle Flomenhoft, Senior Product Developer; Erin DeHeck, Product Developer; Kelly Brown; Senior Marketing Manager; Becca Gill, Content Project Manager; Ann Courtney, Senior Content Project Manager; Jessica Cuevas, Designer; and Laura Fuller, Buyer. We are indebted to them for their talents and dedication.

Although we make every effort to bring you an error-free manual, errors of many kinds inevitably find their way into a book of this scope and complexity. We will be grateful if readers who have comments or suggestions concerning content will send their remarks to your McGraw-Hill sales representative. To find your McGraw-Hill representative, go to www.mheducation.com and click "Get Support," select "Higher Ed," and then click the "Get Started" button under the "Find Your Sales Rep" section.

Reviewer Acknowledgments

Suggestions have been received from faculty and students throughout the country. This is vital feedback that is relied on with each edition. Each person who has offered comments and suggestions has our thanks. The efforts of many people are needed to develop and improve a product. Among these people are the reviewers and consultants who point out areas of concern, cite areas of strength, and make recommendations for change. The following reviewers helped review the 17e to help with development of the 18e:

Natalie Reynolds, MsEd, *Carl Albert State College*
Richard S. Grippo, Ph.D., *Arkansas State University*
Melissa Gutierrez, *The University of Southern Mississippi*
Dr. Eric C. Lovely, *Arkansas Tech University*
Matthew Nusnbaum, PhD., *Georgia State University*
Amy Reber, Ph.D., *Georgia State University*
Rita A. Thrasher, MS, *Pensacola State College*
Travis J. Vail, MS, *Golden West College*
Jeff Wooters, *Pensacola State College*

LABORATORY SAFETY PROCEDURES

1. Keep your work area uncluttered. Place unnecessary books, backpacks, purses, and so on somewhere other than on your desktop.
2. Avoid contact with embalming fluids. Wear rubber or disposable plastic gloves when working with preserved specimens.
3. Wear eyeglasses or safety glasses to protect your eyes from splattered embalming fluid.
4. Keep your hands away from your mouth and face while in the laboratory. Moisten labels with tap water, not your tongue.
5. Sponge down your work area and wash all laboratory instruments at the end of the period.
6. Wash your hands with soap and water at the end of the period.

GENERAL INSTRUCTIONS

Equipment

Each student will need to supply the following equipment:

Laboratory manual and textbook
Dissecting kit containing scissors, forceps, scalpel, dissecting needles, pipette (medicine dropper), probe, and ruler, graduated in millimeters
Drawing pencils, 3H or 4H
Eraser, preferably kneaded rubber
Colored pencils—red, yellow, blue, and green
Box of cleansing tissues
Loose-leaf notebook for notes and corrected drawings

The department will furnish each student with all other supplies and equipment needed during the course.

Aim and Purpose of Laboratory Work

The zoology laboratory will provide your "hands-on" experience in zoology. It is the place where you will see, touch, hear, smell—but perhaps not taste—living organisms. You will become acquainted with the major animal groups, make dissections of preserved or anesthetized specimens to study how animals are constructed, ask questions about how animals and their parts function, and gain an appreciation of some of the architectural themes and adaptations that emphasize the unity of life.

General Instructions for Laboratory Work

Prepare for the Laboratory. Before coming to the laboratory, read the entire exercise to familiarize yourself with the subject matter and procedures. Read also the appropriate sections in your textbook. Good preparation can make the difference between a frustrating afternoon of confusion and mistakes and an experience that is pleasant, meaningful, and interesting.

Follow the Manual Instructions Carefully. It is your guide to exploring and understanding the organisms or functions you are investigating. Its instructions have been written with care and with you in mind, to help you do the work (1) in logical sequence, (2) with economy of time, and (3) with a questioning attitude that will stimulate interest and curiosity.

Use Care in Making Animal Dissections. A glossary of directional terms used in dissections can be found inside the back cover. The object in dissections is to separate or expose parts or organs so as to see their relationships. Working blindly without the manual instructions may result in the destruction of parts before you have had an opportunity to identify them. **Learn the functions** of all the organs you dissect.

Record Your Observations. Keep a personal record in a notebook of everything that is pertinent, including the laboratory instructor's preliminary instruction and all experimental observations. Do not record data on scraps of paper with the intention of recopying later; record directly into a notebook. The notes are for your own use in preparing the laboratory report later.

Take Care of Equipment. Glassware and other apparatus should be washed and dried after use. Metal instruments in particular should be thoroughly dried to prevent rust or corrosion. Put away all materials and equipment in their proper places at the end of the period.

Tips on Making Drawings

You need not be an artist to make laboratory drawings. You do, however, need to be **observant.** Study your specimen carefully. Your simple line drawing is a record of your observations.

Before you draw, locate on the specimen all the structures or parts indicated in the manual instructions. Study their relationships to each other. Measure the specimen. Decide where the drawing should be placed and how much it must be enlarged or reduced to fit the page (read further for estimation of magnification). Leave ample space for labels.

When ready to draw, you may want first to rule in faint lines to represent the main axes and then sketch the general outlines lightly. When you have the outlines you want, draw them with firm, dark lines, erasing unnecessary sketch lines. Then fill in details. Do not make overlapping, fuzzy, indistinct, or unnecessary lines. Indicate differences in texture and color by stippling. Stipple deliberately, holding the pencil vertically and making a neat, round dot each time you touch the paper. Placing the dots close together or farther apart will give a variety of shading. Avoid line shading unless you are very skilled. Use color only when asked for it in the directions.

Label the Drawing Completely. Print labels neatly in lowercase letters and align them vertically and horizontally. Plan the labels so that there will be no crossed label lines. If there are to be many labels, center the drawing and label on both sides.

Indicate the magnification in size beneath the drawing—for instance, "×3" if the drawing is three times the length and width of the specimen. In the case of objects viewed through a microscope, indicate also the magnification at which you viewed the subject—for example, 430× (43× objective used with a 10× ocular).

Estimating the Magnification of a Drawing

A simple method for determining the magnification of a drawing is to find the ratio between the size of the drawing and the actual size of the object you have drawn. The magnification of the drawing can be expressed in the following formula:

$$x = \frac{\text{size of drawing}}{\text{size of object}}$$

If your drawing of the specimen is 12 cm (120 mm) long, and you have estimated the specimen to be 0.8 mm long, then $\times = 120 \div 0.8$, or 150. The drawing, then, is $\times$ 150, or 150 times the length of the object drawn.

The same formula will hold whether the drawing is an enlargement or a reduction. If, for example, the specimen is 480 mm long, and the drawing is 120 mm, then $\times = 120/480$, or 1/4.

STATEMENT ON THE USE OF LIVING AND PRESERVED ANIMALS IN THE ZOOLOGY LABORATORY

Congress has probably received more mail on the topic of animal research in universities and business firms than on any other subject. Do humans have the right to experiment on other living creatures to support their own medical, pharmaceutical, and commercial needs? A few years ago, Congress passed a series of amendments to the Federal Animal Welfare Act, a body of laws covering animal care in laboratories and other facilities. These amendments have become known as the three Rs: **r**eduction in the number of animals needed for research; **r**efinement of techniques that might cause stress or suffering; and **r**eplacement of live animals with simulations or cell cultures whenever possible. As a result, the total number of animals used each year in research and in commercial product testing has declined steadily as scientists and businesses have become more concerned and more accountable. The animal rights movement, largely comprising vocal antivivisectionists, has helped create an awareness of the needs of laboratory research animals and has stretched the resources and creativity of the researchers to discover cheaper and more humane alternatives to animal experimentation.

However, computers and cell cultures—the alternatives—can only simulate the effects on organismal systems of, for instance, drugs, when the principles are well acknowledged. When the principles are themselves being scrutinized and tested, computer modeling is insufficient. Nor can a movie or computer simulation match the visual and tactile comprehension of anatomical relationships provided by direct dissection of preserved or anesthetized animals. Medical and veterinarian progress depends on animal research. Every drug and every vaccine that you and your family have ever taken has first been tested on an animal. Animal research has wiped out smallpox and polio; has provided immunization against diseases previously common and often deadly, such as diphtheria, mumps, and rubella; has helped create treatments for cancer, diabetes, heart disease, and manic depression; and has been used in the development of surgical procedures such as heart surgery, blood transfusions, and cataract removal.

Animal research has also benefited other animals for veterinary cures. The vaccine for feline leukemia that could threaten the life of your cat, as well as the parvo vaccine given to your puppy, were first introduced to other cats and dogs. Many other vaccinations for serious animal diseases were developed through animal research—for example, rabies, distemper, anthrax, hepatitis, and tetanus.

The animal models used by the artist for the illustrations in the exercises of this laboratory manual, and the animals you will dissect in this laboratory course, were prepared for educational use following strict humane procedures. No endangered species have been used. No living vertebrate organisms will be harmed in this laboratory setting. Invertebrate animals that are to be dissected while alive are anesthetized before the procedure. The experiments selected are inoffensive, are respectful of the integrity of the animal's evolutionary contributions, and often require only close observation. The experiments closely follow the tenets of the scientific method, which cannot dictate ethical decisions but can provide the structure for common sense. Do not be wasteful. Share the animals with the other students as often as possible. At the same time, you are encouraged to observe the live animal in its natural setting and its relationships to other species, for only in this manner will you gain a full appreciation of the unique evolutionary position and special structure and systems of each animal.

Chief Characteristics of the Animal Phyla

Character/ Taxon	Embryonic Tissue Layers: none, diploblastic, triploblastic	Cephalzation: present/ absent	Adult Symmetry: radial, bilateral pentaradial	Coelom: absent, pseudo, schizo, entero	Respiratory Structures: skin, gills, lung, trachea, book lung/gill	Digestive System: absent, blind, complete	Skeletal Structure: spicule, exoskeleton, endoskeleton, shell, bone, ossicles	Circulatory System: absent, open, closed	Body Segments: absent, present	Nervous Systems: net-like, ladder-like, brain, eyes	Reproduction: monoecious, dioecious, asexual
Porifera											
Cnidaria Hydrozoa											
Cnidaria Scyphozoa											
Cnidaria Anthozoa											
Platyhelminthes Turbellaria											
Platyhelminthes Cestoda											
Platyhelminthes Trematoda											
Nematoda											
Rotifera											
Gastrotricha											
Nematomorpha											
Acanthocephala											
Mollusca Bivalvia											
Mollusca Gastropoda											

Chief Characteristics of the Animal Phyla (*Continued*)

Character/ Taxon	Embryonic Tissue Layers: none, diploblastic, triploblastic	Cephalzation: present/ absent	Adult Symmetry: radial, bilateral pentaradial	Coelom: absent, pseudo, schizo, entero	Respiratory Structures: skin, gills, lung, trachea, book lung/gill	Digestive System: absent, blind, complete	Skeletal Structure: spicule, exoskeleton, endoskeleton, shell, bone, ossicles	Circulatory System: absent, open, closed	Body Segments: absent, present	Nervous Systems: net-like, ladder-like, brain, eyes	Reproduction: monoecious, dioecious, asexual
Mollusca Polyplacophora											
Mollusca Cephalopoda											
Annelida Polychaeta											
Annelida Oligochaeta											
Annelida Hirudinida											
Arthropoda Chelicerata											
Arthropoda Crustacea											
Arthropoda Hexapoda											
Echinodermata Ophiuroidea											
Echinodermata Holothuroidea											
Echinodermata Crinoidea											
Chordata Urochordata											
Chordata Cephalochordata											
Chordata Vertebrata											

Introduction to the Living Animal

Arctic ground squirrel.

1 The Microscope 3

2 Cell Structure and Division 13

3 Gametogenesis and Embryology 25

4 Tissue Structure and Function 47

The Microscope

1 EXERCISE

EXERCISE 1A
Compound Light Microscope
Understanding the Parts and Operation of a Microscope
Getting Acquainted with Your Microscope
Taking Control of Your Microscope
Magnification in a Microscope
How to Measure Size of Microscopic Objects

EXERCISE 1B
Stereoscopic Dissecting Microscope
Exercises with the Dissecting Microscope

EXERCISE 1C
Electron Microscope
Demonstrations

For a biologist the compound microscope is probably the most important tool ever invented. It is indispensable not only in biology but also in the fields of medicine, biochemistry, and geology; in industry; and even in crime detection and many hobbies. Even though the microscope is one of the most common tools in the biologist's laboratory, too frequently it is used without any effective understanding of its construction and operation. The results may be poor illumination, badly focused optics, and misleading interpretations of what is (barely) seen.

EXERCISE 1A

Compound Light Microscope

The compound light microscope may be either monocular or binocular, with either vertical or inclined oculars.

Use both hands to carry a microscope. Grasp it firmly by the arm with one hand and support the **base** with the other. Carry it in a fully upright position.

Core Study

Understanding the Parts and Operation of a Microscope

If you are not familiar with the parts of a microscope, please study Figures 1-1 and 1-2.

The **image-forming optics** consist of (1) a set of **objectives** screwed into a **revolving nosepiece** and (2) a **body tube,** or head, with one or two **oculars (eyepieces).**

Each **objective** is a complex set of tiny lenses that provide most of the magnification. Your microscope may have two, three, or four objectives, each with its magnification, or power, engraved on the side. For example, if the objective magnifies an object 10 times, the magnification is said to be 10 diameters and is commonly written simply as 10×. Most microscopes include a **scanning objective** (3.5× or 4.5×), a **low-power objective** (10×), and a **high-power objective** (40×, 43×, or 45×). Some microscopes also carry an **oil-immersion objective** (95×, 97×, or 100×), which must always be used with a drop of oil to form a liquid bridge between the lens and the surface of the slide being viewed.

Rotate the nosepiece, noting the clicking sound when an objective swings into place under the tube. Enter here the magnifications of the objectives on your microscope: __________ __________ __________
Does your microscope have an oil-immersion lens? If so, what is its magnification? __________

Lenses in the **ocular** further magnify the image formed by the objective. The ocular most often used is the 10×. Often a pointer is mounted into the ocular. Enter here the magnification of the ocular on your microscope: __________

If there is a pointer in your microscope, rotate the ocular and note movement of the pointer. Remember that if you look through a binocular scope with only one eye, you may miss the pointer. Always use both eyes.

Directly beneath the **stage** of most microscopes is a **substage condenser,** a system of enclosed lenses that concentrates the light on the specimen above. The condenser may have a knob that allows you to move the condenser up and down.

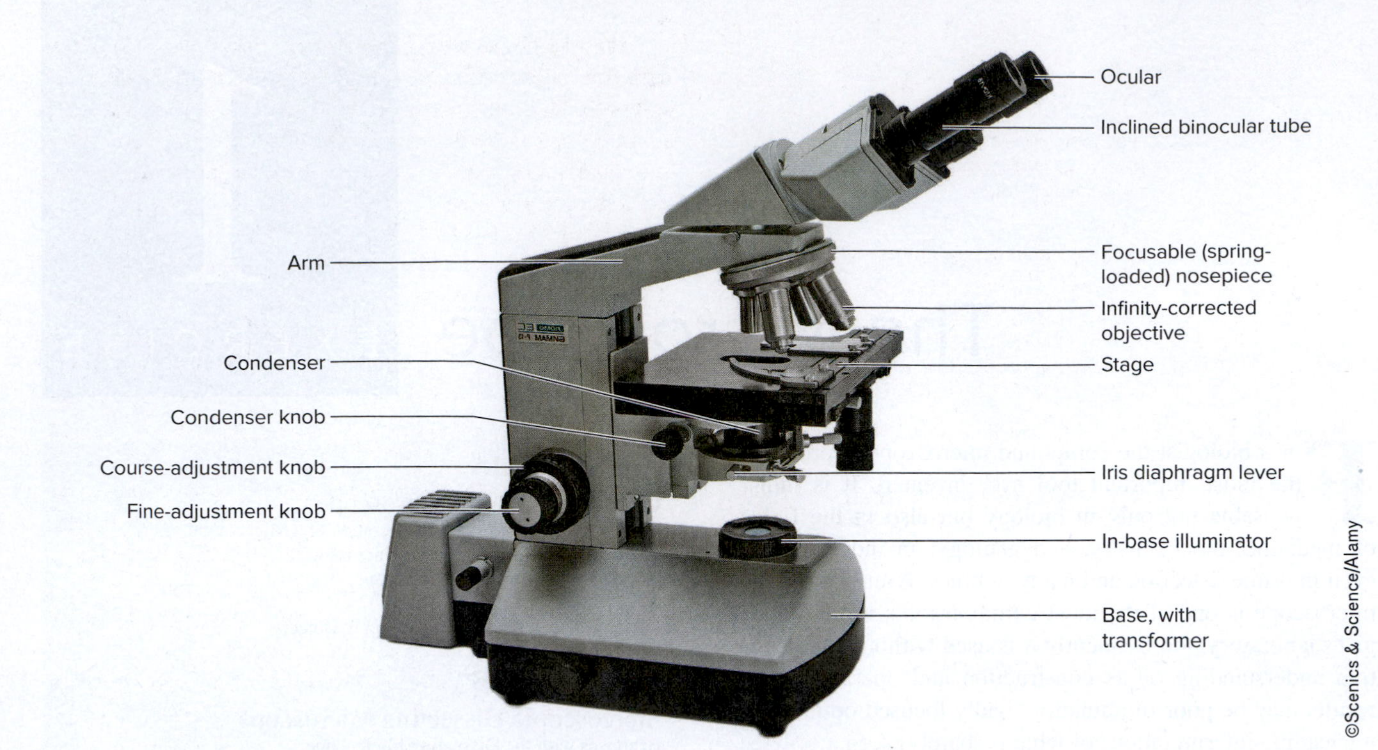

Figure 1-1
Optical and mechanical features of a compound microscope.

Beneath the condenser, many microscopes have a built-in, low-voltage **substage illuminator.** Microscopes lacking a substage illuminator employ an adjustable **reflecting mirror** that reflects natural light or light from a microscope lamp up into the optical system. The **concave surface** of the mirror is used with natural light or with a separate microscope lamp when there is no condenser on the microscope. The **plane (flat) surface** of the mirror is used with a substage condenser.

Turn on the substage illuminator. If your microscope lacks one, use a microscope lamp or position your microscope and mirror to take advantage of natural light. With the low-power objective in place, adjust the mirror to bring a bright, evenly distributed circle of light through the lens.

An adjustable iris diaphragm under the stage is used to reduce glare caused by stray light from the illuminator.

Raise or lower the substage condenser to near its uppermost limit. Then close down the iris diaphragm until all glare is gone (but don't close so far that dark halos appear around objects).*

On a straight microscope, you raise and lower the body tube with two sets of adjustment knobs. Use the **coarse-adjustment knob** for low-power work and for initial focusing. Use the **fine-adjustment knob** for final adjustment.

On a microscope with an inclined tube, you focus by *moving the stage* rather than moving the body tube. If you are using an inclined microscope, read "lower the stage" when the directions say "raise the objective" in the following exercise. In either case, distance between objective lens and object is increased.

Turn the coarse-adjustment knob and note how it moves the body tube (or stage). Find out which way to turn the knob to raise and lower the tube.

Never use the coarse-adjustment knob when a high-power objective is in place. Turn the fine-adjustment knob. This moves the tube so slightly that you cannot detect it unless you are examining an object through the ocular. The fine-adjustment knob works the same as the coarse-adjustment knob. To focus downward, turn the knob in the same direction that you would to focus down with the coarse adjustment. Practice this. *Always use the fine adjustment when the high-power objective is in place.*

Getting Acquainted with Your Microscope

You need not wear glasses when using a microscope unless they correct severe astigmatism. Nearsightedness and farsightedness can be corrected by adjusting the microscope.

*Correct adjustment of illumination, called Koehler illumination, is describedin more detail in Appendix A (p. 387).

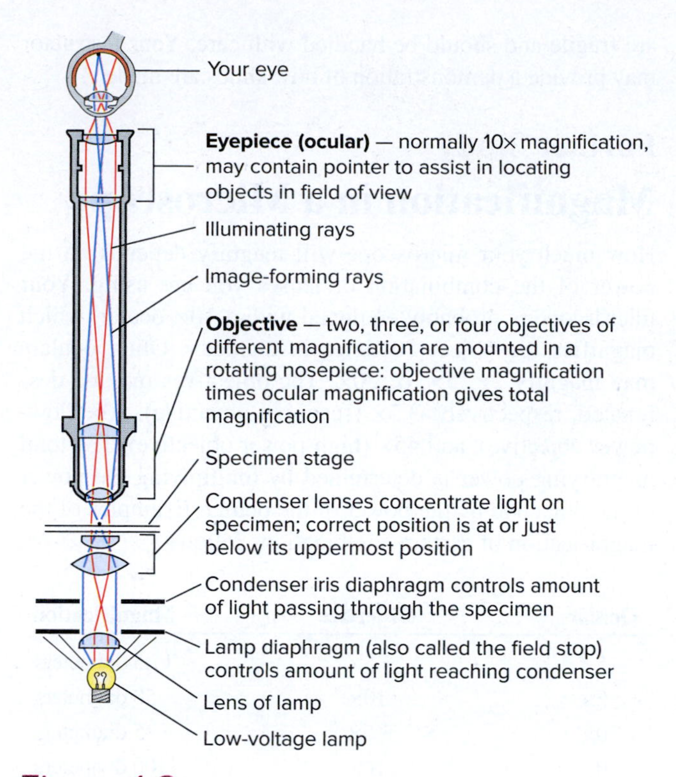

Figure 1-2
Optical path of light through a microscope.

Clean the lenses of both ocular and objectives by wiping gently with a clean sheet of lens paper. *Never touch lenses with anything except clean lens paper.*

Keep both eyes open while using a monocular microscope. If this seems difficult at first, hold a piece of paper over one eye while viewing the object with the other.

How to Focus with Low Power. Turn the low-power (10×) objective until it clicks in place over the aperture. Adjust the condenser and iris diaphragm for optimal illumination as already described.

Obtain a slide containing the letter *e* (or *a, h,* or *k*). Place it, coverslip up, on the stage with the letter centered under the objective lens. *While watching from the side,* lower the objective with the coarse adjustment until it is close to the slide surface.

While looking through the ocular, slowly raise the objective by turning the coarse adjustment toward you until the object on the slide is in sharp focus. Is the image upside down? Is it reversed; that is, does the left side of the letter appear on the right and the right side on the left? On a separate sheet of paper, draw the letter as it appears.

Shift the slide slightly to the right while viewing it through the ocular. In what direction does the image move? Move the slide away from you. What happens to the image? Turn the fine-adjustment knob toward and then away from you to observe the effect on the image. To gain experience with interpreting depth, obtain a slide with three different-colored, overlapping threads and focus with low power. Which colored thread is on top? __________ In the middle? __________ On the bottom? __________

How to Focus with High Power. Focus the object first with low power; then slowly rotate the high-power objective into position. If the microscope lenses have been constructed in a particular way by the manufacturer, the object in focus with low power will be nearly in focus under high power. Such lenses are said to be **parfocal** with respect to each other. Now turn the **fine-adjustment knob** to bring the specimen into sharp focus.

Never use the coarse adjustment while looking at an object under high power; you may drive the objective into the slide, damaging the slide or the lens of the objective.

If your microscope is not parfocal (your instructor will tell you), focus first with low power and then swing the high-power objective into place, raising the tube so that the objective lens clears the slide. Now, *still watching the high-power objective from the side*, lower it slowly to about 1 mm from the cover glass. Then, looking through the ocular, raise the tube with the fine adjustment until the object is in focus. Do this several times to acquire skill in focusing.

Because light decreases when you switch to high power, you must adjust the light with the iris diaphragm.

While viewing the object, keep your hand on the fine adjustment and constantly focus up and down. This enables you to see detail throughout the depth of the object.

How to Use an Oil-Immersion Objective. Occasionally a project or demonstration exercise requires using an oil-immersion objective. This is an objective in the 90× to 100× range. Because the resolving power of this objective is so great, you must use oil to bridge the separation between slide and objective.

To use oil immersion, bring the specimen into focus first with the low-power and then with the high-power objective. Carefully center the point of interest in the field of view; then rotate the nosepiece to move the high-power objective off to one side. Place a single drop of immersion oil on the coverslip at the point where the objective will come into position. Now move the oil-immersion objective carefully into position, *watching from the side* to be certain that the lens clears the coverslip. The oil should now form a bridge between lens and coverslip. **Carefully** adjust the fine focus to bring the specimen into focus. Adjust the iris diaphragm or substage condenser to increase light as required.

When finished, clean the lens with a lens tissue wetted with xylol and then with a lens tissue wetted with Kodak Lens Cleaner. **Never** use alcohol, which will dissolve the cement around the lens system. If the lens is to be used again soon (within a day or two), it is best not to clean the lens face. Residual oil will not harm the lens unless it is allowed to harden over a long period without use.

Taking Control of Your Microscope

1. **Proper lighting** is the first requirement for happy microscopy. Too much light is as bad as too little (beginners usually tend to use too much). Transparent objects are often

clearer in reduced light. Reduce light by closing down the iris diaphragm, *not* by lowering the substage condenser.

2. **Focus with eyes relaxed.** The image appears to your eye as though it were about 25 cm away, but the eye should be relaxed as though it were viewing an image in the distance. Look up periodically and train your eyes on something across the room. Then, if you keep your eyes relaxed, you should not have to readjust your focus very much when looking through the microscope. If you get a headache, chances are you are trying to look *into* the microscope rather than *through* it.
3. **If you are using a binocular microscope,** it is important (a) to adjust the distance between the microscope's oculars to match the distance between your own pupils and (b) to adjust the oculars for a sharp focus. If the focus is not sharp, you may have to focus each eye separately. The microscope will have one fixed and one adjustable ocular. Adjust focus for the fixed ocular first to suit that eye; then adjust the other ocular (usually by rotation) until focus is sharp for both eyes.
4. **Find the correct eye distance** from the oculars, one that affords a full view of the field. Keep relaxed, hold your head steady, and enjoy the view.
5. **Where's the dirt?** If spots or smudges appear in the field of vision, it may be dirt on the ocular, on the slide, or on the objective. To find out which, first rotate the ocular; if the spots move, the ocular needs to be cleaned (in laboratory, avoid using eye makeup, which may smear on the ocular surface). Move the slide; if the spots move with it, clean the slide. If after cleaning the ocular and slide the spots persist, it is probably a dirty objective lens. Use only special lens paper to clean lenses. The slide may be cleaned very gently with a soft, damp cloth or damp cleansing tissue. If after cleaning everything you still see spots or smudges, try moving the condenser up or down a little. Still not satisfied? You may need help from the assistant. Some people see "floaters," which drift across the field of vision while viewing a brightly illuminated object. These are defraction images of red blood cell "ghosts" in the vitreous humor of the eye. Although these are annoying at first, one can usually learn to ignore them.
6. **Keep one hand on the fine adjustment** and constantly focus up and down. This is the only way to see everything.
7. **If the fine adjustment refuses to turn,** the knob has reached its range limit. To correct this, give the fine-adjustment knob several turns in the opposite direction, and then refocus with the coarse adjustment.
8. **Be friendly to your microscope.** It must be kept dust free, so return it *carefully* to its box or cupboard when you are finished with it. Before putting it away, put the low-power objective in place and raise the tube a little. Be sure not to leave a slide on the stage.

Prepared slides you will be using in the laboratory were made at the cost of skill and patience and are expensive. They are fragile and should be handled with care. Your instructor may provide a demonstration of how slides are made.

Further Study

Magnification in a Microscope

How much your microscope will magnify depends on the power of the combination of lenses you are using. Your microscope is probably equipped with a 10× ocular, which magnifies the object 10 times in diameter. Other oculars may magnify 2×, 5×, or 20×. The objectives may be designated, respectively, 3.5× (scanning objective), 10× (low-power objective), and 45× (high-power objective). The total magnifying power is determined by multiplying the power of the objective by the power of the ocular. Examples of the magnification of certain combinations follow:

Ocular	Objective	Magnification
5×	3.5×	17.5 diameters
5×	10×	50 diameters
10×	3.5×	35 diameters
10×	10×	100 diameters
10×	45×	450 diameters
10×	90×	900 diameters

For handy reference, enter in the following table the total magnifying power for lens combinations on *your* microscope.

How to Measure Size of Microscopic Objects

It is often important to know the size of an organism or object you are viewing through the microscope. For example, if you are looking for a particular species of protozoan in a mixed culture and know that the species is usually about 400 μm long, it saves time to know just how large 400 μm will appear at either low or high power. We describe alternative methods for measuring object sizes.

Measuring Objects with Transparent Ruler Calibration. With this simple method, the viewer measures the diameter of the field of view and then estimates the proportion of the field occupied by the object. This method is not as accurate as the alternative described in the section Measuring Objects with Ocular and Stage Micrometer Calibration, but often an approximation of size is all that the viewer requires.

With scanning objective in position, place a transparent ruler on the microscope stage and focus on its edge so that you can see the scale. Move the ruler right or left so that one of the vertical millimeter lines is just visible at the edge of the circular field of view. Count the number of millimeter lines spanning the field; you will probably have to estimate the

Magnifying powers for microscope number ________						
Magnification for:	**Ocular**		**Objective**		**Magnification**	
Scanning lens	________	×	________	=	________	diameters
Low-power lens	________	×	________	=	________	diameters
High-power lens	________	×	________	=	________	diameters
Oil-immersion lens	________	×	________	=	________	diameters

last decimal fraction of a millimeter. Enter diameter in mm here __________. Now convert this figure to micrometers by multiplying by 1000. Enter this value here __________. This is the diameter in micrometers (μm) of the field for the scanning lens.

Diameters of fields of view for your low-power and high-power objectives cannot be measured directly with a transparent ruler because of the high magnification. To calculate the field diameter of the low-power objective, multiply the diameter of the scanning lens field (using the value you just determined) by the magnifying power of the scanning objective and divide by the magnifying power of the low-power objective:

$$\frac{\text{Magnification of scanning objective}}{\text{Magnification of low-power objective}} \times \text{Diameter of scanning objective field} = \text{Diameter of low-power field}$$

Make this calculation for your microscope and enter the value here __________. It usually ranges between 1.5 and 1.6 mm (1500 and 1600 μm).

Similarly, the diameter of the high-power field is determined as

$$\frac{\text{Magnification of the low-power objective}}{\text{Magnification of the high-power objective}} \times \text{Diameter of low-power field} = \text{Diameter of high-power field}$$

Diameters of fields of view for microscope number ________	
Scanning lens	________μm
Low-power lens	________μm
High-power lens	________μm

Make this calculation for your microscope and enter the value here __________. It usually ranges between 0.36 and 0.42 mm (360 and 420 μm). Now enter these values in the table above for quick reference.

Place a hair from your head and from your eyebrow on a slide. Examine with low and high power and measure their diameters. Write the diameters in micrometers here: head __________ eyebrow __________. Are they the same? Compare diameters with others in the class.

Measuring Objects with Ocular and Stage Micrometer Calibration. An **ocular micrometer** can be fitted into the microscope's ocular. It is a disc on which is engraved a scale of (usually) either 50 or 100 units. These units are arbitrary values that always appear the same distance apart no matter which objective is used in combination with the eyepiece. Therefore, the ocular micrometer cannot be used to measure objects until it has been calibrated with a **stage micrometer.**

A stage micrometer resembles an ordinary microscope slide but bears an engraved scale on its upper surface, usually 1 or 2 mm long, divided into 0.1 and 0.01 mm divisions.

Place the stage micrometer on the microscope stage and focus on the engraved scale with the low-power objective. Both scales should appear sharply defined. Rotate the eyepiece until the two scales are parallel. Now move the stage micrometer to bring the 0 line of the stage scale in exact alignment with the 0 marking of the ocular scale. The scales should be slightly superimposed as shown in Figure 1-3.

To calibrate the ocular scale for this objective, use the longest portion of the ocular scale that can be seen to coincide precisely with a line on the stage scale. For example, suppose that 70 units on the ocular scale equal 0.24 mm on the stage scale. Then

$$70 \text{ ocular units} = 0.24 \text{ mm, and}$$

$$1 \text{ ocular unit} = \frac{0.24}{70} = 0.0034 \text{ mm, and}$$

$$1 \text{ mm} = 1000 \text{ μm}$$

Therefore,

$$1 \text{ ocular unit} = 0.0034 \text{ mm} = 3.4 \text{ μm}$$

Repeat the calibration procedure for the scanning lens and the high-power lens. For handy reference, enter in the following table the micrometer values you have calculated.

To measure any object available to you in the laboratory, it is necessary only to multiply the number of divisions covered by the specimen or part thereof by the micrometer value you have determined (3.4 μm in this example). Note that the micrometer value applies only to the objective with which the calibration was made.

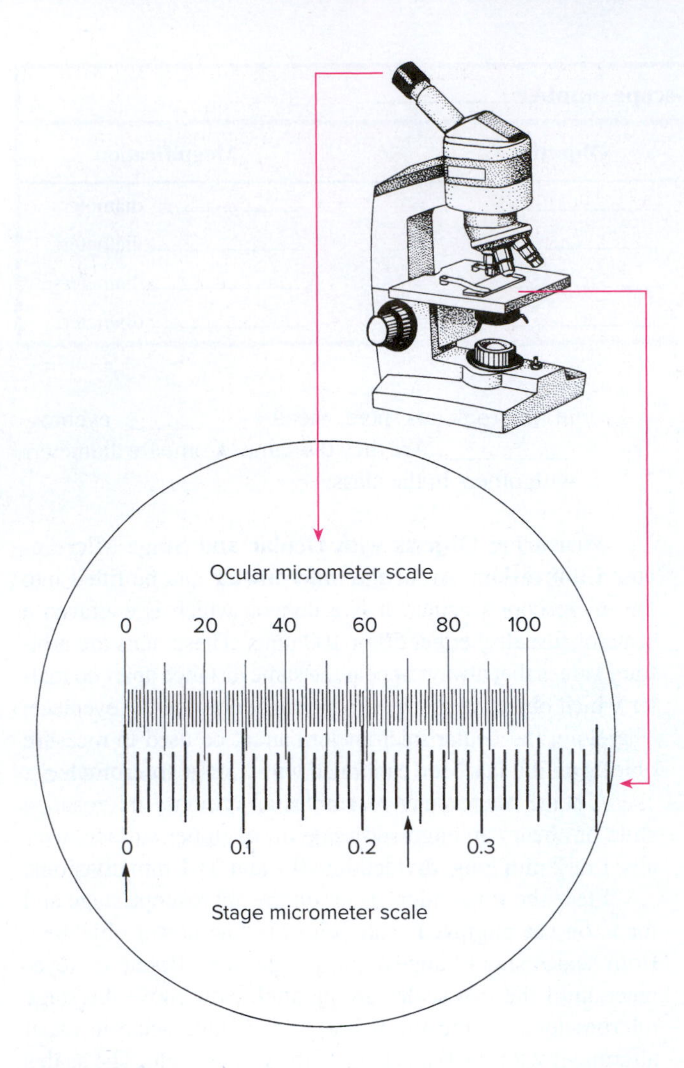

Figure 1-3
Calibrating an ocular micrometer.

Micrometer values for microscope number ________	
Scanning lens, 1 ocular unit	= __________ µm
Low-power lens, 1 ocular unit	= __________ µm
High-power lens, 1 ocular unit	= __________ µm

EXERCISE 1B

Core Study

Stereoscopic Dissecting Microscope

The stereoscopic dissecting microscope (Figure 1-4) is as indispensable to the laboratory as is the compound microscope. It enables you to study objects too large or too thick for the compound microscope. It furnishes a three-dimensional view of objects at a very low power (5× to 50×, depending on the microscope). The image is not inverted, and there is ample space for manipulation and dissection under the lens.

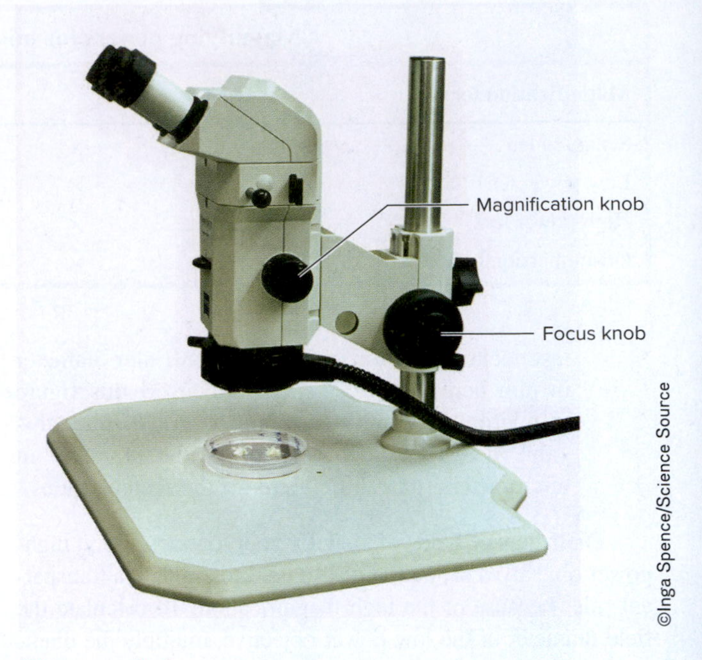

Figure 1-4
Stereoscopic dissecting microscope with illuminator.

The microscope stage can be illuminated either by **transmitted light** (light passing through the object from below) or by **reflected light** (light illuminating the object from above and being reflected by the object into the microscope). Which type of lighting is used with the compound microscope? __________ Some dissecting microscopes have a substage mirror or a substage lamp. Focusing is done with very little adjustment.

Place an object on the stage and illuminate with reflected light. Looking through the oculars, adjust them to fit the distance between your eyes so that you can look comfortably through both oculars at once. You should see a single field of vision. If focus is not sharp, you may have to focus for each eye separately. The microscope will have one fixed and one adjustable ocular. Adjust focus for the fixed ocular first to suit that eye; then adjust the other ocular (usually by rotation) until the focus is sharp for both eyes.

Move the object on the stage away from you. Which way does the image move? __________

Move the object to one side. Which way does the image move? __________ How does this compare with the compound microscope? __________

Further Study

Exercises with the Dissecting Microscope

Try the following or any similar exercises with the dissecting microscope and *keep a record* of the results of different lighting and background effects with different types of material. This will save time for you in later studies.

1. Examine some pond water. How many kinds of organisms can you see? ________ Are they more distinct on a white or a black background? ________
2. Examine a prepared slide of the whole mount of a fluke or tapeworm or a similar slide. Try it with reflected lighting, moving the light source about for the best effects. Then try transmitted lighting. If the microscope lacks substage lighting, try placing the slide over a small microscope lamp with a piece of writing paper between slide and lamp. Which method gives better illumination?
3. Examine the surface of a preserved sea star or gills of a crayfish. Study the material first without water; then study it submerged in a finger bowl or glass of water. In which preparation do you see the most detail? Why? Try the gills both with top lighting and with transmitted light. Can both methods be used with the sea star? Why?

Be sure to make good use of the dissecting microscopes that are available for your use in the laboratory. You will find them invaluable.

EXERCISE 1C

Electron Microscope

Although it is unlikely that you will have an opportunity to use an electron microscope, electron microscopy has contributed so greatly to our understanding of cell structure and function that you should have a summary understanding of its principles. In this system the visible light source of an optical microscope is replaced by a tungsten filament, which, driven by high voltage, emits a beam of electrons. The beam is shaped by magnets or electric fields, and the image is projected onto a photographic plate or a fluorescent screen for direct viewing.

The sample to be examined is treated with electron "stains" containing heavy metal ions that block the electron beam to varying degrees. It must also be cut extremely thin because the electron beam has a very low penetrating power. The sample must also be free of scratches, dirt, and other imperfections. Once the specimen is prepared (and this is the greatest challenge to successful electron microscopy), it is placed in the microscope's specimen chamber, which is evacuated. Because specimens must be dehydrated and evacuated before being placed in the specimen chamber, one obvious limitation of electron microscopy, at least at present, is that living material cannot be viewed.

The kind of electron microscope described is called a **transmission electron microscope (TEM)** because an electron beam passes through the specimen. The TEM is used to view thin sections of cells. Figure 1-5 compares a micrograph of liver cells, photographed with a compound microscope, with the greatly magnified view of a single liver cell photographed with a TEM.

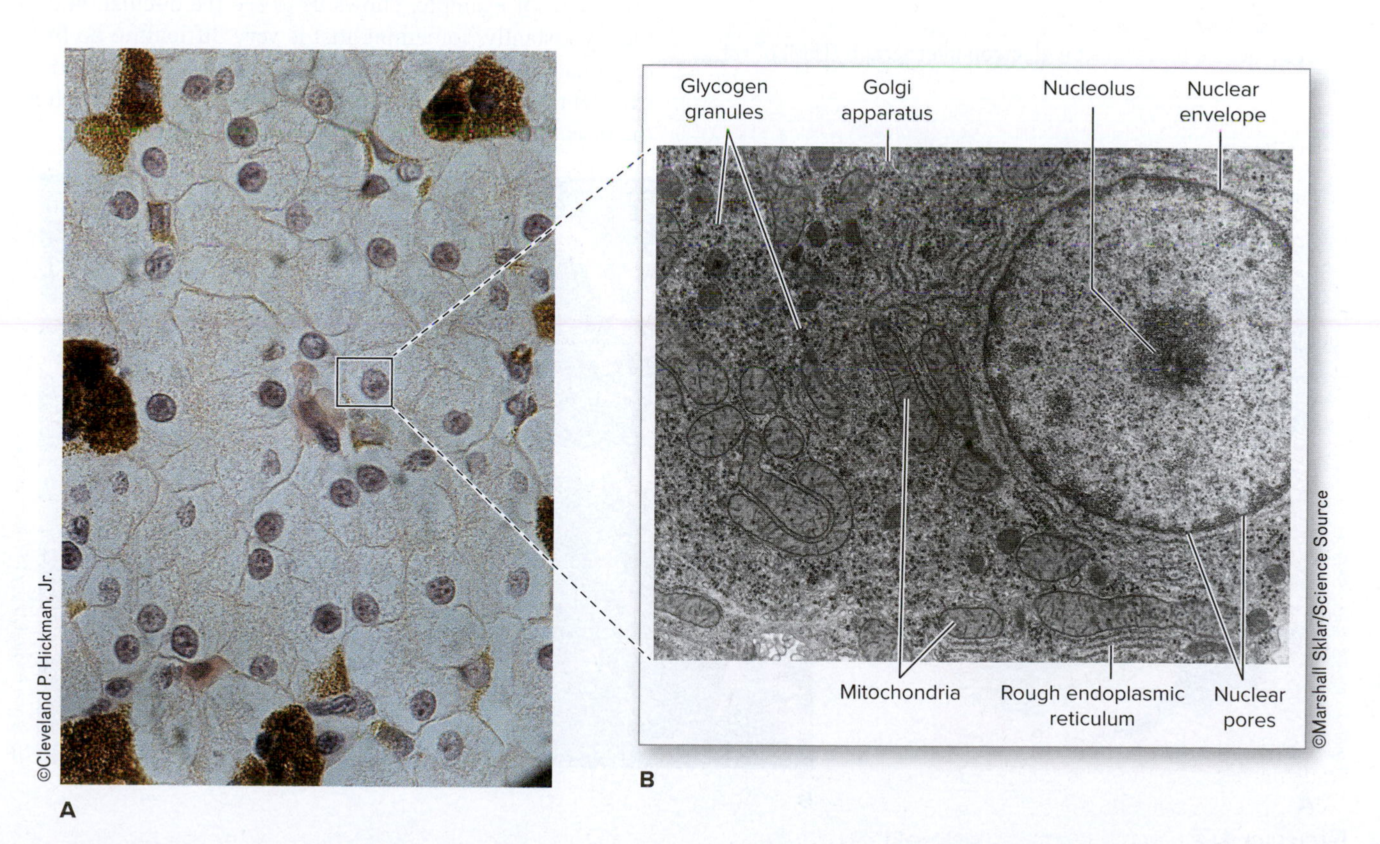

Figure 1-5
Comparison of light and electron micrographs. **A,** Light micrograph showing many liver cells, ×160. **B,** Electron micrograph showing a portion of the liver cell outlined with rectangle in B, ×7000. Electron micrograph of liver cell.

Figure 1-6
Freeze-fracture transmission electron micrograph (TEM) of rat liver.

Several specialized applications of the TEM permit shadows to be cast on the specimen at precise angles. These techniques enable the microscopist to see the general shape and profile of the object. One technique, **freeze fracturing,** is illustrated in Figure 1-6.

The **scanning electron microscope (SEM)** has become an increasingly popular biological tool. In the SEM technique, which requires an electron microscope of different design from that of the TEM, the surface topography of the specimen is revealed in great detail. The resolution of the SEM is not as great as that of the TEM (usually whole cells or tissues are viewed with the SEM), but the television-like image that is produced is of great value in understanding shape, size, and organization of biological material. Two SEM images are shown in Figure 1-7: one of single cells and the other of a tissue composed of numerous cells.

Examine Figure 1-7A, showing a macrophage (a phagocytic cell) ingesting two damaged red blood cells. Notice the ruffled surface of the macrophage; the cytoplasmic extensions behave like pseudopodia when the macrophage moves about through tissues. You can also see thin cytoplasmic extensions spreading out to envelop the red blood cells. Figure 1-7B is a SEM image of the microvasculature of a mammalian kidney. It shows several ball-like glomeruli (filtration units) and the arterioles that serve them.

SEM images are of great value in helping us understand complex relationships within tissues and organs. Figure 1-7B, for example, allows us to see the circulation of a kidney instantly, something that is very difficult to do from examination of sectioned material. The SEM also produces excellent images of the surfaces of whole animals, such as the insect in Figure 1-8.

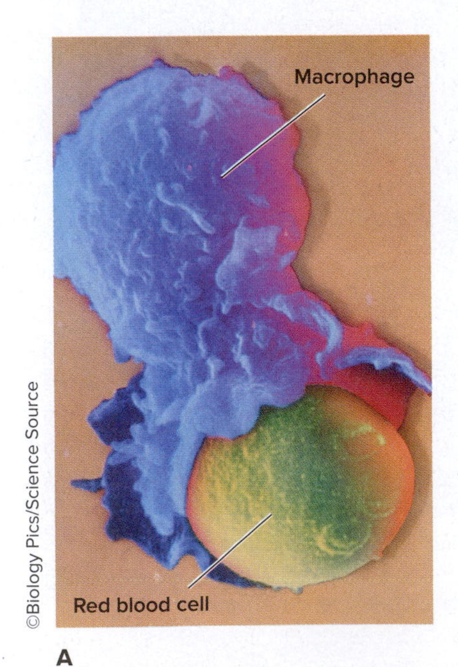

A

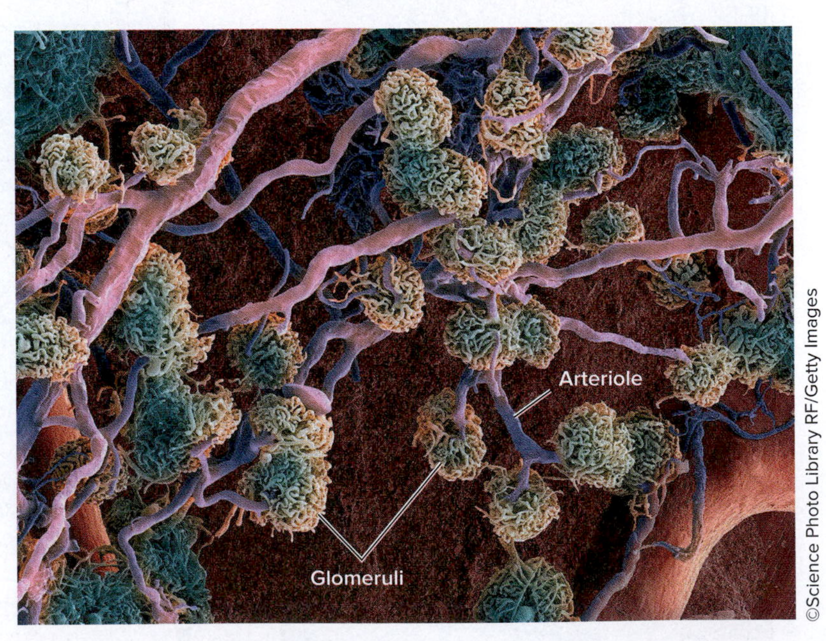

B

Figure 1-7
A, Scanning electron micrograph (SEM) of macrophage ingesting damaged red blood cells, ×5250. **B,** SEM of microcirculation of mammalian kidney. Note spherical glomeruli, the kidney's pressure filters, where urine formation begins, ×80.

From R. G. Kessel and R. H. Kardon, Tissues and Organs: A Text-Atlas of Scanning Electron Microscopy, *1979, W. H. Freeman and Co.*

©David Scharf/Science Source

Figure 1-8
Scanning electron micrograph of the head of a fruit fly, *Drosophila melanogaster*, ×70. Why do you think arthropods like this fly are said to have compound eyes? ______________________________

____

Demonstrations

Examine the photographs or books placed on display or visit one of the listed websites. The following books, atlases, and websites are especially useful.

Bubel, A. 1989. Microstructure and function of cells: electron micrographs of cell ultrastructure. New York, John Wiley & Sons. *Examines cell microarchitecture, emphasizing the diversity and structural variation of cellular elements in invertebrates, vertebrates, and plants. Drawings are used to help interpretation of electron micrograph images.*

Burgess, J., H. Martin, and R. Taylor. 1987. Microcosmos. Cambridge, U.K., Cambridge University Press. *Contains hundreds of photomicrographs of uniformly high—indeed, superb—quality. An excellent reference source of optical light, scanning, and transmission electron microscopy. The first three-fifths of the book deals with the living world. Many of the photographs are in color and most of the SEMs in Chapter 3, Animal Life, are little short of stunning. A 1990 paperback edition of this book was issued under the title* Under the Microscope: A Hidden World Revealed.

Cross, P. C., and K. L. Mercer. 1993. Cell and tissue ultrastructure: a functional perspective. New York, W. H. Freeman & Company. *Atlas of transmission electron micrographs of high quality, accompanied by text with functional explanation of ultrastructure. Insert drawings help clarify orientation of micrographs.*

Fawcett, D. W. 1966. The cell: its organelles and inclusions. Philadelphia, W. B. Saunders Company. *An excellent source of TEM images.*

Flegler, S. L., J. W. Heckman, and K. L. Klomparens. 1997. Scanning and transmission electron microscopy: an introduction. New York, Oxford University Press. *Ideal for use in the laboratory. Presents the practical and theoretical fundamentals of scanning and transmission electron microscopy with clear and concise explanations, coupled with instructive diagrams and photographs that guide you through microscope operation, image production, analytical techniques, and potential applications to various disciplines.*

Howe, J. M., and B. Fultz. 2001. Transmission electron microscopy and diffractometry of materials. New York, Springer-Verlag. *This textbook develops the concepts of transmission electron microscopy (TEM) and X-ray diffractometry (XRD) that are important for the characterization of materials. Can be used as both an introductory and an advanced-level graduate book.*

Hunter, E., P. Maloney, M. Bendayan, and M. Silver. 1993. Practical electron microscopy: a beginner's illustrated guide. Cambridge, U.K., Cambridge University Press. *Excellent beginner's guide, complete with numerous images and photographs.*

Kessel, R. G., and R. H. Kardon. 1979. Tissues and organs: a text-atlas of scanning electron microscopy. San Francisco, W. H. Freeman & Company. *More than 700 high-quality SEMs of tissues and organs, mostly mammalian.*

Kessel, R. G., and C. Y. Shih. 1974. Scanning electron microscopy in biology: a student's atlas on biological organization. New York, Springer-Verlag. *This is an excellent presentation of SEMs of protozoa, algae, cultured cells, metazoa animals, and tissues and organ systems of animals. One of the best atlases.*

Krommenhoek, W., J. Sebus, and G. J. van Esch. 1979. Biological structures. Baltimore, University Park Press. *An atlas of cells, tissues, and organs. Most of the book contains light micrographs of stained sections, reproduced in excellent color. Electron micrographs are confined to the first 25 pages, and the last section of the book presents an interesting series of X-ray photographs of the human body.*

Reimer, L. 1997. Transmission electron microscopy: physics of image formation and microanalysis. New York, Springer-Verlag. *Presents the theory of image and contrast formation and the analytical modes in transmission electron microscopy. Also discussed are the kinematic and dynamical theories of electron diffraction and their applications for crystal-structure analysis and imaging of lattices and their defects.*

Reimer, L. 1998. Scanning electron microscopy: physics of image formation and microanalysis. New York, Springer-Verlag. *Describes the physics of electron-probe formation and of electron-specimen interactions. The different imaging and analytical modes using secondary and backscattered electrons, electron-beam-induced currents, X-ray and Auger electrons, electron channeling effects, and cathodoluminescence are discussed to evaluate specific contrasts and to obtain quantitative information.*

Shih, G., and R. Kessel. 1982. Living images: biological microstructure revealed by scanning electron microscopy. Boston, Science Books International. *All SEMs, from bacteria and algae to animals and animal structures, with brief accompanying text.*

Swapp, S. Scanning electron microscopy. http://serc.carleton.edu/research_education/geochemsheets/techniques/SEM.html.

Wayne, R. O. 2009. Light and video microscopy. Burlington, Academic Press. *Prepares the reader for an accurate interpretation of the image observed through the light microscope.*

Cell Structure and Division

EXERCISE 2A
The Cell—Unit of Protoplasmic Organization
Some Examples of Cells

EXERCISE 2B
Cell Division—Mitosis and Cytokinesis
Cell Cycle and Mitosis

EXERCISE 2A

The Cell—Unit of Protoplasmic Organization

Some Examples of Cells

The cell is the basic structural and functional unit of all living organisms. In most protozoans (animal-like protists), single cells exist as individuals. In metazoans (many-celled animals), cells may be associated as tissues specialized for certain functions. Some cells are themselves highly specialized, such as nerve cells for conducting impulses, muscle cells for contraction, and eggs or spermatozoa for reproduction. But regardless of shape, size, or special function, each cell is a living, dynamic entity, capable of maintaining itself.

Core Study

Study the following preparations with both low power and higher power of the compound microscope, unless otherwise directed.

Unstained Squamous Epithelial Cell

With a clean toothpick, gently scrape the inside of your cheek. Disperse the scraping (epidermal cells that are being shed) in a drop of water on a clean slide. Add a cover glass and examine with lower power. The cells are so thin and transparent that you will need to reduce the light intensity to see them. Locate a small group of cells and switch to high power, adjusting the light intensity as necessary.

What is the shape of the **squamous epithelial cells?** _________ Each cell contains a **nucleus,** which controls the development and function of the cell. The nucleus is surrounded by translucent **cytoplasm.** Although it appears structureless in the compound microscope, it consists of numerous **organelles** and **inclusions** suspended in intracellular fluid. The metabolic and synthetic activities of a cell reside mostly in the cytoplasm. You will look more closely at a cell's organelles in your study of liver cells.

A cell is enclosed by a **plasma membrane** 8 to 10 nm* thick—too thin to be clearly seen with the compound microscope. However, you can determine its location by differences in refraction between the cell and surrounding water.

Stained Squamous Epithelial Cell

Repeat the preparation you made for an unstained squamous epithelial cell, but this time add one or two drops of a dilute stain (e.g., methylene blue, methyl green, or gentian violet) before adding a cover glass.

What part of the cell stains most intensely? _________ What advantages does a stained preparation have over unstained? _________ Look for tiny, rod-shaped bodies. These are a type of bacteria found in every mouth.

Drawings

On p. 21, sketch some squamous cells, top view, each about 2 cm wide.

Egg Cell of Sea Star

Obtain a slide of unfertilized sea star eggs (Figure 2-1). If a slide of sea star cleavage stages is used, look under low power among the various stages of cleavage development for an isolated spherical cell with a distinct nucleus. This is an unfertilized egg. Center the egg

*1 nanometer (nm) = 1/1000 micrometer (μm).

in the field of vision and focus up and down to study its general shape and structure. Then carefully turn to high power and study the parts of the cell.

The most easily differentiated parts of the cell are the central spherical **nucleus** and the surrounding **cytoplasm.** Which of these makes up most of the cell? __________ Does the cytoplasm appear to be granular or fibrillar in structure? __________ As previously mentioned, the cytoplasm contains many organelles and inclusions too small to be visible in this examination. Within the nucleus are the darkly stained **chromatin granules,** which contain DNA and form chromosomes at certain stages of the cell cycle. The chromosomes carry genes, which are responsible for hereditary qualities. Find within the nucleus a deeply stained spherical **nucleolus.** The nucleolus is rich in RNA and is the site of intensive ribosome synthesis. Ribosomes pass from the nucleus into the cytoplasm, where they serve as important sites for protein synthesis.

Further Study

Liver Cells

The liver tissue of *Amphiuma* (a large aquatic salamander from the southeastern United States) is preferred for the study of liver cells because the cells are large. However, mammalian liver tissue may also be used. In this tissue you are studying not isolated cells but cells associated in a glandular organ.

Examine a stained, prepared slide of liver tissue first with low power and then with high power.

Refer to Figure 1-5 (p. 9). In amphibians, liver cells are massed together in small groups separated by capillary spaces called **sinusoids.** The polyhedral (many-sided) cells are in direct contact with blood that circulates through the sinusoids. Examine the cells at high power to identify **nucleus, nucleolus** (a prominent, dark-staining spot within the nucleus), **nuclear envelope** that encloses the nucleus, **cytoplasm,** and **plasma membrane** that bounds the cell. In addition to the nucleolus, you will see **chromatin** (a complex of DNA and proteins) dispersed within the nucleus. Prior to cell division, the chromatin becomes organized into individual chromosomes.

Fine Structure of Liver Cell

Examine the electron micrograph of a liver cell (Figure 2-2). This photograph clearly shows some of the organelles and inclusions found in typical animal cells. These include numerous **mitochondria, endoplasmic reticulum,** and **glycogen granules.** Are these inclusions in the nucleus or the cytoplasm of the cell? __________

Nucleus. In liver cells the nucleus is a spherical structure consisting of a double-layered nuclear envelope enclosing the **nucleoplasm.** The nuclear envelope is broken at intervals by gaps, which permit continuity between the nucleoplasm and the cytoplasm surrounding the nucleus. These gaps, or **pores,** are clearly evident in Figure 1-5 (p. 9) and Figure 2-3. RNA and protein-RNA complexes pass from the nucleus to the cytoplasm through these pores.

Refer again to your slide and locate the nucleolus inside the nucleus. Do some of the nuclei have more than one

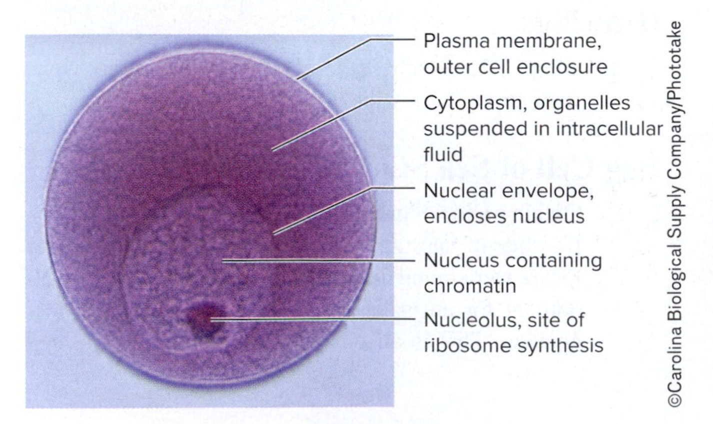

Figure 2-1
Unfertilized egg of sea star.

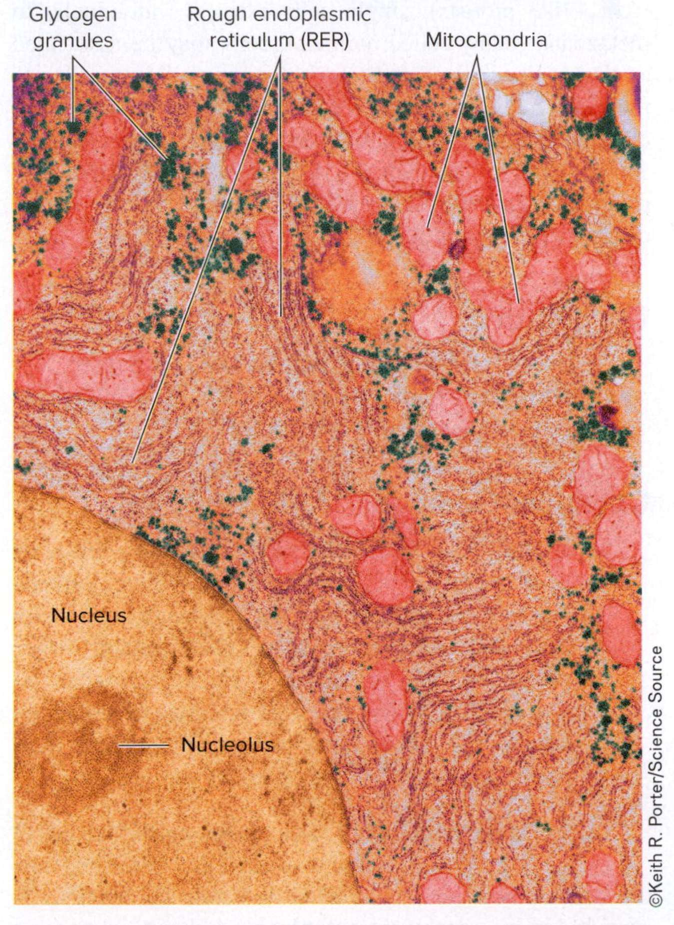

Figure 2-2
Electron micrograph of portion of liver cell, magnified about 10,000 times. Note single large nucleus containing nucleolus; mitochondria; rough endoplasmic reticulum (*RER*); and numerous glycogen granules.

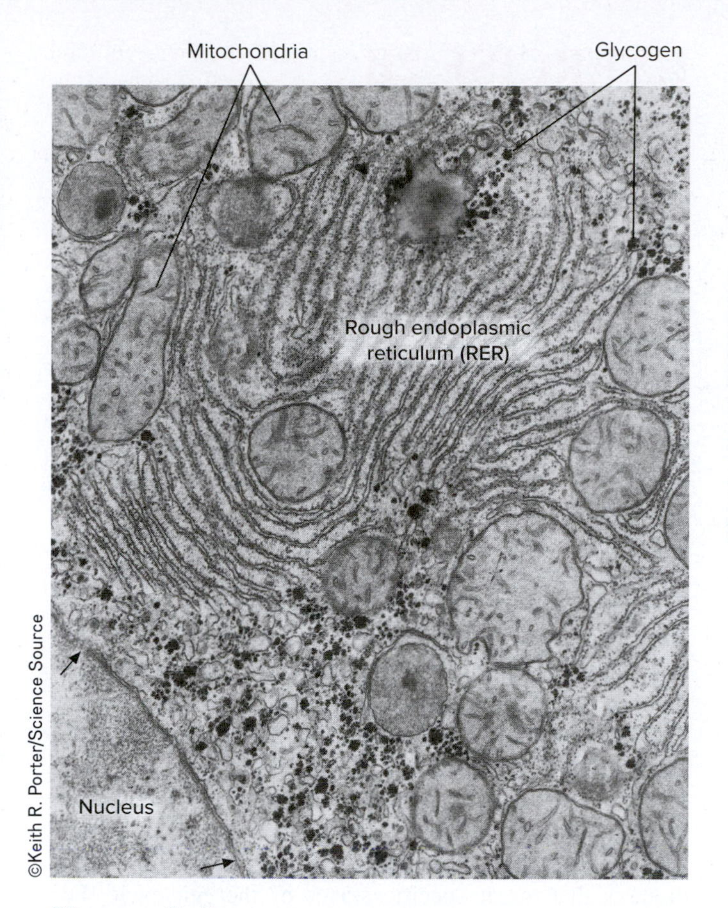

Figure 2-3

The endoplasmic reticulum is an extensive system of internal membranes of great functional importance to the eukaryotic cell. Note numerous polyribosomes, appearing as black dots along the outer surface of membranes of the endoplasmic reticulum. Polyribosomes are aggregates of protein and RNA that translate RNA copies of genes into protein. Note the nucleus with nuclear pores (*arrows*), where double membranes of the nuclear envelope pinch together.

nucleolus? __________ In cells that are actively growing and dividing, and in cells that are actively synthesizing protein, the nucleoli are large and often multiple.

Drawings

1. On p. 22 draw and label the egg cell of a sea star, about 4 cm in diameter. Label all structures from Figure 2-1 that you can see through your microscope. Under your drawing, note the magnification power at which you viewed the egg.
2. On p. 22 draw a group of liver cells, each about 2 cm in diameter, showing typical shape and nuclei. Under your drawing, note the magnification power at which you viewed the egg.

The electron micrograph shows that the nucleolus does not have a membrane (Figure 2-2) and is composed of a granular material and a fibrous network. The nucleolus forms at a particular site on a particular chromosome in the nucleus (or chromosomes if more than one nucleolus is present). The

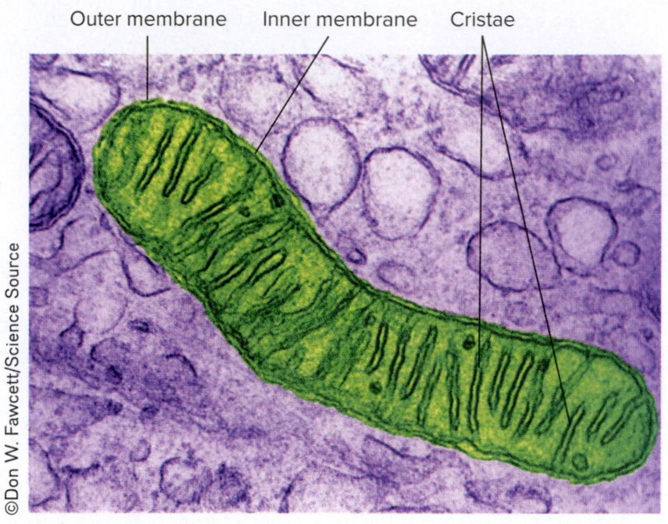

Figure 2-4

Mitochondria in longitudinal and cross section. The mitochondrion is bounded by an outer membrane and an inner membrane. The latter is folded to form numerous compartments called cristae.

DNA in the nucleolus directs synthesis of ribosomal RNA molecules, which combine into protein-RNA complexes. These move into the cytoplasm, where they become ribosomes. How are protein-RNA complexes able to pass the nuclear membrane? ________________________

Endoplasmic Reticulum and Ribosomes. The cytoplasm contains an extensive network of channels called the **endoplasmic reticulum (ER).** You cannot see these on your slide, but they are easily seen with the electron microscope. Refer again to Figures 2-2 and 2-3. The endoplasmic reticulum in these photographs is labeled *RER,* meaning "rough" endoplasmic reticulum, because at high power the membranes of the ER appear studded with numerous particles, the **ribosomes.** What important synthetic function is carried out on the ribosomes?__________

Mitochondria. Mitochondria are also abundant in Figure 2-2.

Examine a prepared slide of mitochondria* using the higher power of the compound microscope.

This slide was prepared from liver tissue of an amphibian, specially stained to reveal mitochondria. Note the prominent, dark-stained nucleus in each cell. Surrounding the nucleus is the cytoplasm, containing numerous dots and granules of different sizes and shapes; these are the mitochondria.

These organelles are seen with far more detail with the electron microscope. Refer to Figure 2-2. Note that the mitochondrion is actually composed of two membranes (more easily distinguished in Figure 2-4). The outer membrane is

*See Appendix A, pp. 387–419, for sources of prepared slides with mitochondria and Golgi.

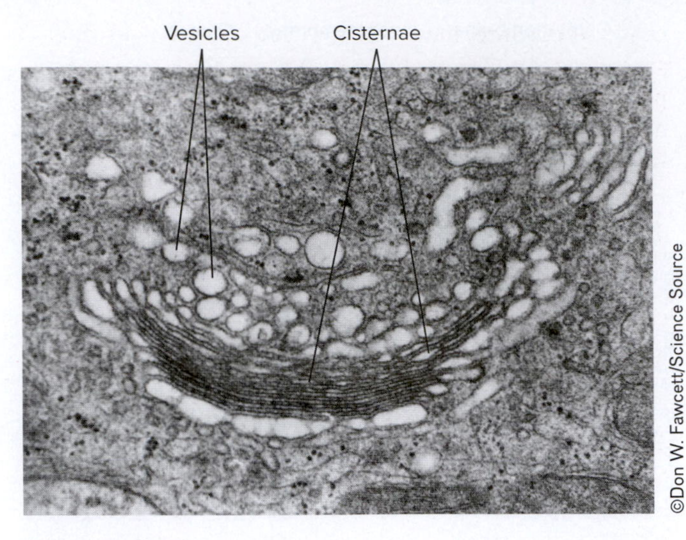

Figure 2-5
Golgi complex in cross section. Proteins formed in the endoplasmic reticulum are transported to the Golgi complex, within which they are complexed with polysaccharides. These collect in cisternae, which pinch off into vesicles; the vesicles then move to other locations in the cell.

smooth, and the inner membrane is infolded to form **cristae.** These look like partitions inside the mitochondria (Figure 2-4) and greatly increase the surface area of the inner membrane. Why are the mitochondria often called "powerhouses of the cell"? ______________________

Golgi Complex. The Golgi complex (also called Golgi apparatus) is a unique network of membrane-lined, flattened channels (called **cisternae**), usually stacked together in parallel rows.

Examine a slide of Golgi complex prepared from the stomach or small intestine of a large amphibian. The tissue was specially prepared with a silver stain that impregnates the Golgi complex, turning it black. Find the columnar epithelial cells on your slide with low power; then switch to high power.

The Golgi complex will appear as thin, darkened strands lying in parallel arrays just above the nucleus—that is, toward the free surface of the epithelial cell.

Refer now to Figure 2-5, which shows the Golgi complex at the electron microscope level. Note that it consists of parallel arrays of flattened saccules. What function or functions are carried out on the Golgi complex? ____________

____

____

The cytoplasm contains many other organelles and inclusions, such as lysosomes, microbodies, and cytoskeletal fibers that form the framework of the cell. In this exercise, you have been introduced to some of the more obvious and important organelles of a typical animal cell.

EXERCISE 2B

Core Study

Cell Division—Mitosis and Cytokinesis

All cells come from preexisting cells. Most animals start life from a fertilized egg. Beginning as a single cell, a metazoan is the result of repeated cell division. Cell division plays a role in growth, regeneration of lost parts, development, wound healing, and many other body processes. Remarkably, every day a living animal sheds about 1% or 2% of its tissue cells. To offset this loss, the tissues produce (by cell division) as many cells as they lose.

Animal and plant cells multiply in number through cell division. Each division is an orderly sequence of events in which each mother cell gives rise to two daughter cells, *each having the same number of chromosomes as the mother cells.* Chromosomes carry the hereditary factors, or genes.

Cell division consists of two distinct phases: **mitosis,** division of nuclear chromosomes, and **cytokinesis,** subsequent division of the cytoplasm. Understanding cell division and the cell cycle is important for our knowledge about disease. Cancer is probably the most well-known disease related to problems with cell division, but there are other diseases that target specific stages of the cell cycle. For instance, *Theileria* is an intracellular parasitic protozoan that is closely related to malaria. It is transferred by ticks to cattle throughout the tropics and subtropics. The parasite divides itself as the host cell divides. It attaches to the microtubules of the spindle poles during metaphase and then during anaphase it attaches to the central spindle, and as the chromatids of the host cell are pulled apart the parasite is similarly equally divided into the daughter cells of the host.

Meiosis, another type of cell division, which results in the *reduction* of chromosome number, will be studied in Exercise 3A.

Cell Cycle and Mitosis

The number of chromosomes in the cells of any particular species is generally constant, with amounts ranging from 2 to 200, but usually between 12 and 40 chromosomes. How many chromosomes are there in human cells? __________ With few exceptions, each cell of an individual organism has the species-specific number of chromosomes. With more or less than this full complement, a cell will not have the proper set of instructions for its own growth, development, and metabolism. Therefore, during cell division, the events of mitosis ensure that each daughter cell receives a complete set of chromosomes.

How long does it take a cell to divide? This depends on several things: the kind of organism, the type of tissue, the physiological condition of the organism, the temperature, and other factors. But in general the process lasts between a few minutes and 2 or 3 hours.

The stages of mitosis and the end result of cell division are quite similar in animal and plant cells. The few differences are largely correlated with differences in structure of animal and plant cells.

To see a large number of dividing animal cells, one needs to study a tissue that is growing rapidly, such as a rapidly growing embryo. Fertilized eggs from the uterus of the roundworm *Ascaris* and very young embryos of whitefish are commonly used for such studies. Epidermis of young frog tadpoles is another good source. These tissues are killed, fixed,* cut into very thin slices, mounted on slides, and stained. Each cell will naturally show only the phase at which it was fixed. But by studying many cells killed at various stages of division, you can get a good picture of the whole process. Try to think of the stages as parts of a continuous process, as though watching a film frame by frame.

Before you start to study your slide, read over the following account of the cell cycle and the various phases of mitosis in order to understand the process.

Cell Cycle

The cell cycle (Figure 2-6) is a well-ordered sequence of events that makes up the interval between one cell generation and the next. It consists of several phases. These are G_1 phase (growth phase before DNA synthesis), S phase (DNA synthesis), G_2 phase (growth phase after DNA synthesis), M phase (mitosis, or nuclear division), and C phase (cytokinesis, or cytoplasmic division). Together, G_1, S, and G_2 are called **interphase,** the period between cell divisions (mitosis and cytokinesis). Interphase once was called the "resting period" because the light microscope could not reveal DNA replication in the nucleus during this time. The nuclear envelope appears well defined during interphase, but the chromosomes themselves are difficult to see, appearing instead as an irregular network of chromatin material, rather heavily stained. One or two nucleoli may be present.

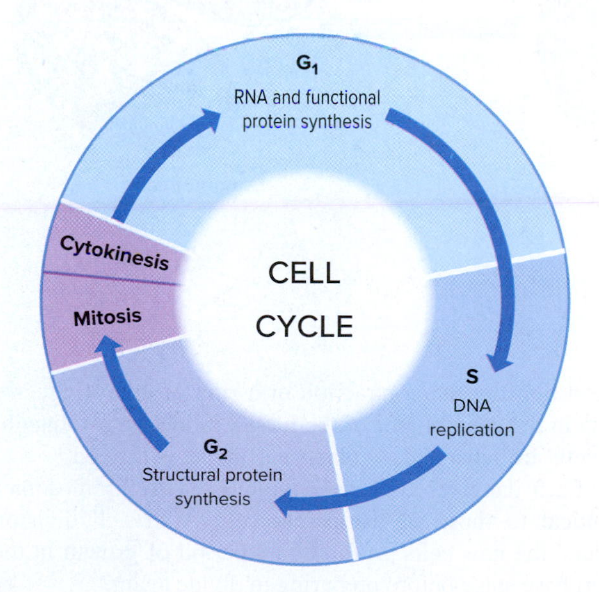

Figure 2-6
The cell cycle.

Rather than resting, a cell is growing throughout interphase. During G_1 phase, nucleotides, histones (proteins associated with DNA), and various enzymes are being synthesized in preparation for S phase. During S phase, all DNA is replicated, a process that must proceed with great accuracy, because the entire linear sequence of nucleotides in the entire genome must be reproduced. During G_2 phase, spindle and aster proteins are synthesized in preparation for mitosis. Cell organelles, such as mitochondria, are also replicated at this time. *Thus, when mitosis begins, a cell already has a duplicated set of chromosomes.*

In animal cells, there is a pair of small, dark, dot-like bodies, the **centrioles,** located near the nucleus, which also duplicate themselves (Figure 2-7). Centrioles are difficult to see except in specially prepared slides.

Mitosis

Prophase

At the beginning of prophase, the centrioles begin to separate, each forming around itself an array of microtubules called the **aster** (Figure 2-7). Even longer microtubules, the **spindle fibers,** radiate between the two centrioles to form a football-shaped **spindle.** The nuclear chromatin condenses to form **chromosomes,** which at this stage are actually two sister **chromatids** that were formed during interphase. The sister chromatids are joined together at the **centromere,** a point of constriction on the chromosome. At the centromere, each chromatid has a disc of protein, the **kinetochore,** to which the spindle fibers, now called kinetochore fibers, attach. The nucleolus and nuclear envelope begin to disappear. Plant cells have no asters, but the spindle will form in the characteristic manner.

Toward the end of prophase the chromosomes are distinct, and the characteristic number of chromosomes can be seen (Figure 2-7). In animal cells with large chromosome counts, such as human cells, special preparation is required to spread out the chromosomes so they can be counted.

Metaphase

During metaphase the chromatids arrange themselves near the center of the cell to form the **metaphase plate** (Figures 2-7 and 2-8).

Anaphase

At anaphase, the centromeres divide, freeing the sister chromatids from each other. The independent chromatids, now **daughter chromosomes,** are pulled toward opposite poles of the cell by their kinetochore fibers. As the spindle fibers shorten, the centrioles also move farther apart, dragging the daughter chromosomes closer to the poles of the cell.

*"Fixing," a term used in histology, refers to techniques used to rapidly kill and harden tissues to preserve as nearly as possible relationships present in life.

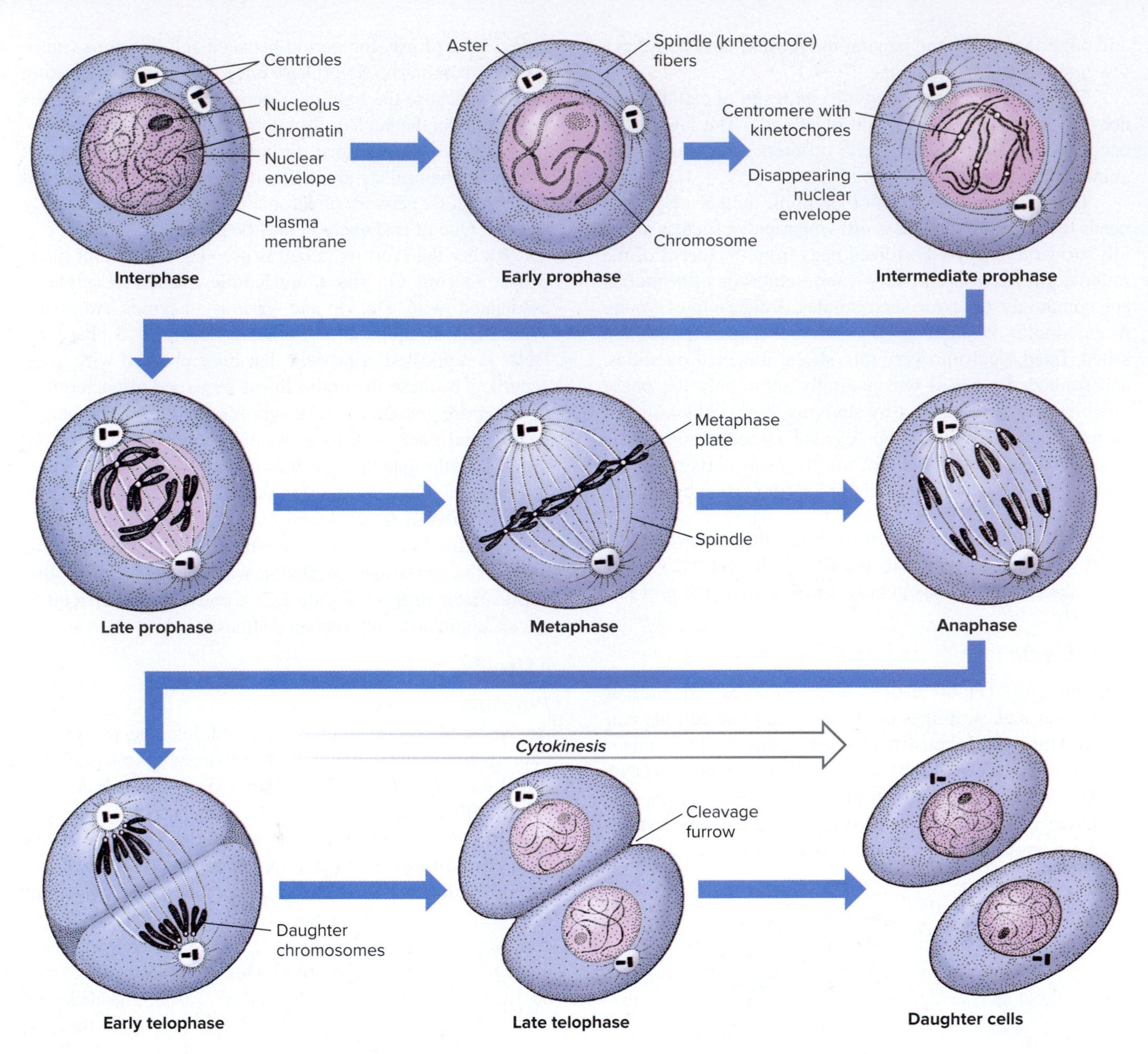

Figure 2-7
Mitosis in an animal cell, diagrammatic.

Telophase
Telophase begins as the two groups of chromosomes reach opposite poles of the cell. The chromosomes crowd together at the poles, and reorganization of a nucleus begins within each group. Telophase, the last stage of mitosis, is much like prophase in reverse. The chromosomes lose their condensed structure to again become a network of chromatin material. The spindle disappears and the nuclear envelope and nucleolus begin to re-form.

Cytokinesis
The plasma membrane indents to form a **cleavage furrow** around the cell (this actually begins at anaphase), which is drawn tight by the contraction of a ring of microfilaments lying in the deepest part of the furrow. Soon the two daughter cells are separated. In plants a cell wall is formed.

Each daughter cell has a diploid set of chromosomes identical to those of the parent cell. After cell division occurs, the new cells go through a period of growth in the interphase stage before preparing to divide again.

Slide Study of Mitosis

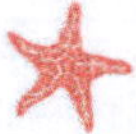

Read the following brief description of the type of slide you are to study. Then examine the slide, first with the naked eye to locate the tissue slices and then

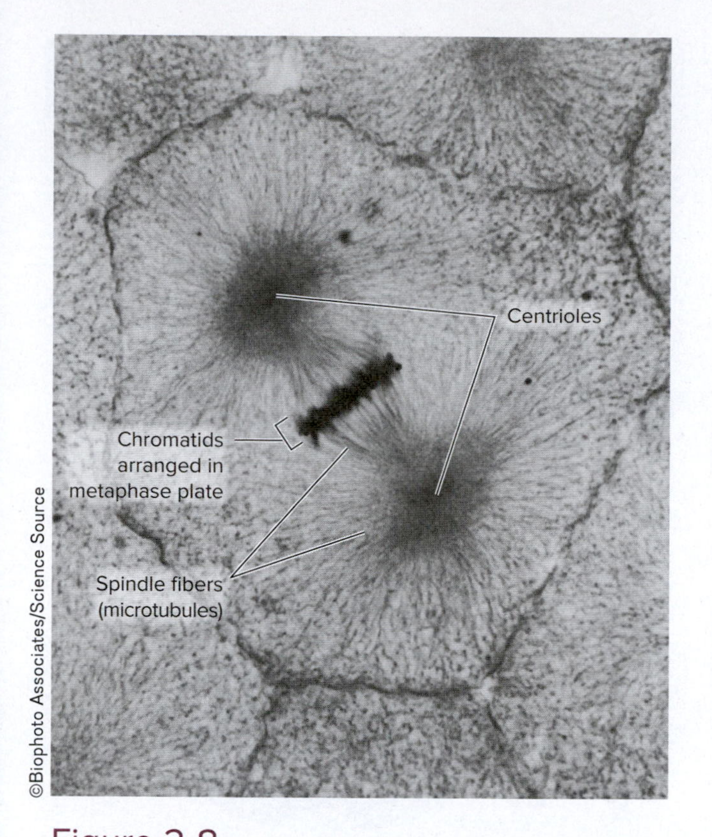

Figure 2-8
Transmission electron micrograph of metaphase, showing mitotic spindle and chromatids arranged in the metaphase plate of an animal cell, about ×15,000.

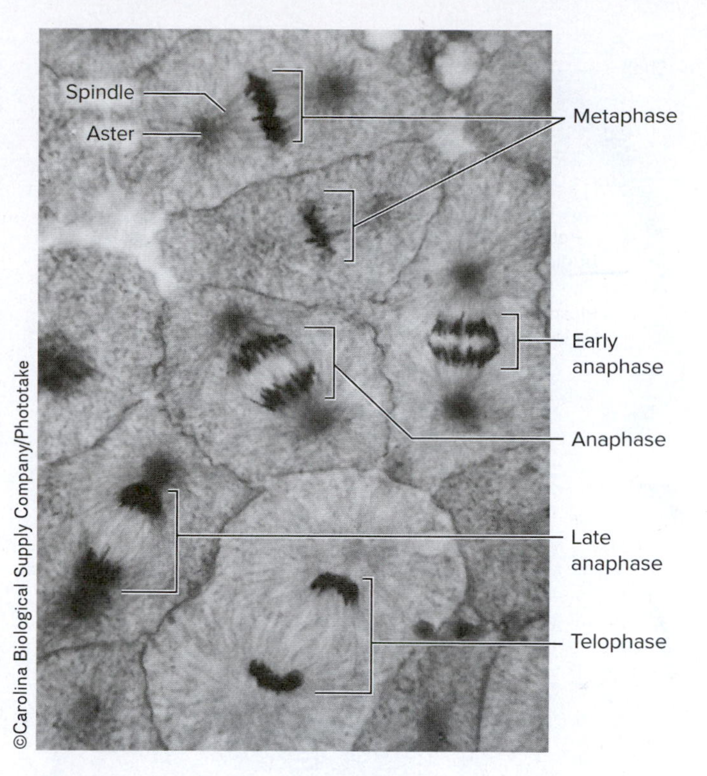

Figure 2-9
Mitosis in a whitefish embryo. This is an ideally sectioned slide. Students are unlikely to find this many ideally sectioned stages together in the same field of view.

with the low power of the microscope. Scan the tissue carefully until you are familiar with the appearance of the cells and can recognize the difference in appearance of the nuclei of various cells. When you find a phase you can recognize and identify, center it and change to high power for study. Sketch the stage you have found in the proper place on p. 23 and then find another stage.

Not all cells will clearly illustrate the characteristics of their stage of mitosis; in tissue preparation, many of the cells will be cut in such a way that all or part of the nuclear material is missing.

Mitosis in Whitefish Blastula

The blastula is a very early stage in embryonic development of whitefish* (refer to Figure 3-9G, p. 39). Each slide contains several very thin sections cut through one of these blastulas. Hold the slide over a piece of white paper and note the small spots, each of which represents one section cut through a blastula. Study a section with low power first. You will probably find all stages of mitosis represented among the cells (Figure 2-9). Note the shape of the cells and appearance of asters and spindles. Look for an ideally sectioned cell; then switch to high power to study and sketch it. Why does it appear that most cells are in interphase?

____

Mitosis in Fertilized Ascaris Eggs

Cells of the intestinal roundworm *Ascaris* have only four chromosomes, which are quite large and easy to see.

In female *Ascaris,* the eggs are fertilized in the paired uteri after copulation. Soon after fertilization, each fertilized egg (zygote) acquires a shell, secreted by the walls of the uterus. The zygote within its shell now begins to divide—the first of many cell divisions that will finally produce a new worm. To prepare these slides, longitudinal sections were made through the uterus and its eggs soon after mating had occurred, and the sections were mounted on slides and stained. Each slide section contains many small, spherical eggs in various stages of mitosis (Figure 2-10).

First examine the slide with low power. Find the walls of the uterus, which are made up of darkly stained cells. The cavity of the uterus is filled with sections of eggs. Study an egg under high power and identify its structures. Note the thick, pale **shell** on the outside. The wide, clear **perivitelline space** between shell and cytoplasm in a living egg is filled with fluid. The **cytoplasm** is stained,

*Several species of the genus *Coregonus,* a commercially valuable fish of northern lakes of North America, Europe, and Asia.

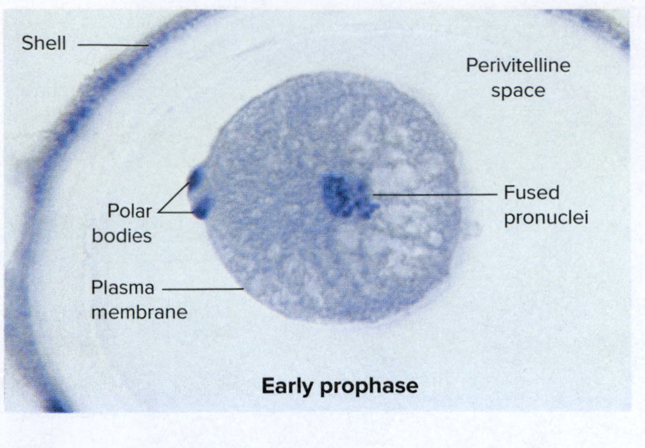

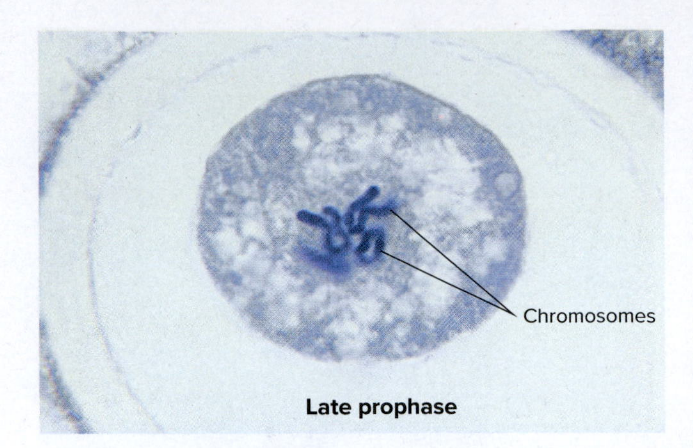

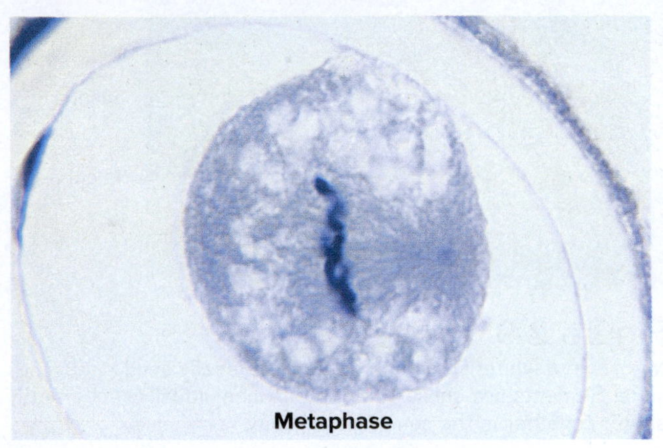

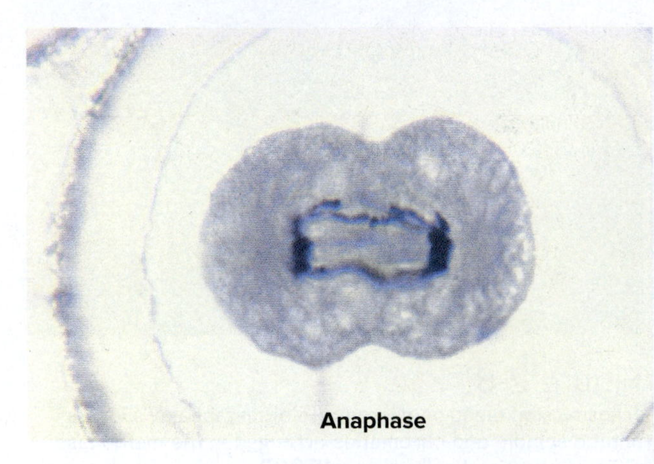

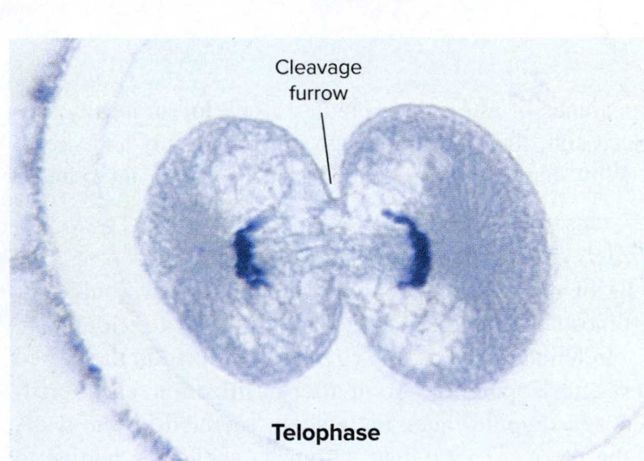

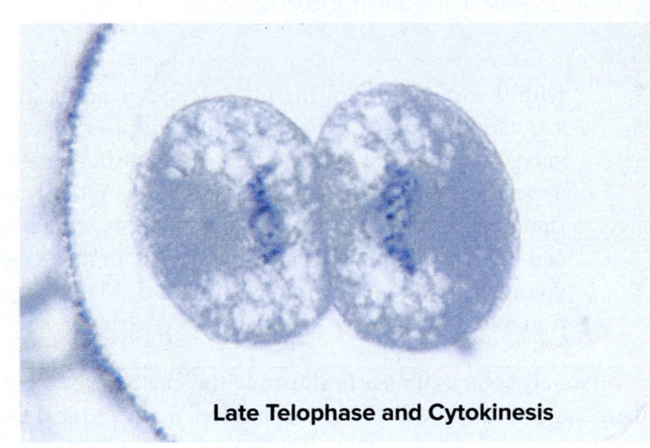

Figure 2-10
Mitosis in fertilized *Ascaris* eggs.

appears granular, and contains a number of small vacuoles. The cytoplasm is enclosed in a very thin **plasma membrane.** The **nucleus** of the fertilized egg is formed by the fusion of two **pronuclei,** one of which was contributed by the sperm at fertilization. The chromatin of the nucleus is darkly stained.

Scan the slide carefully with low power until you are familiar with the appearance of the eggs; then select an ideally sectioned cell and switch to high power for study.

Drawings

On p. 23, draw as many of the stages of mitosis as you can find, labeling the drawings adequately.

Reference

Grant, R. P. 2011. Mitotic hijacker. The New Scientist 25:22, 24.

Some Typical Cells

Name ______________________________

Date ______________________________

Section ______________________________

LAB REPORT

Squamous Cells from Cheek

LAB REPORT

Sea Star Egg

Liver Cells

Mitosis, as seen in

____

Interphase	**Early Prophase**	**Late Prophase**

Metaphase	**Early Anaphase**	**Late Anaphase**

Early Telophase	**Late Telophase**	**Daughter Cells**

LAB REPORT

NOTES

Gametogenesis and Embryology

3

EXERCISE

EXERCISE 3A

Core Study

Meiosis—Maturation Division of Germ Cells

EXERCISE 3A

Meiosis—Maturation Division of Germ Cells
Spermatogenesis
Oogenesis and Fertilization in *Ascaris*
Spermatogenesis in Grasshopper Testis

EXERCISE 3B

Cleavage Patterns—Spiral and Radial Cleavage
Spiral Cleavage: Early Embryology of the Ribbon Worm, *Cerebratulus*
Radial Cleavage: Early Embryology of the Sea Star, *Asterias*
Morula Through Larva in Both Spiral and Radial Cleavage

EXERCISE 3C

Frog Development
Study of Eggs in Frog Ovary

For animals to reproduce successfully the processes related to cell division, gamete formation, fertilization, and development must proceed without significant disruption or error. Each of these stages is susceptible to errors and these can result in genetic problems, malformed gametes, or terminated development. It is important to understand the various stages in these processes to better understand where disruptions in the reproductive cycle of animals can occur.

Mitosis (pp. 17–20) is a process for distributing the chromosomes and the DNA they contain equally at cell division. Each daughter cell receives identical copies of the genetic information. A common feature of metazoan animals is that chromosomes occur in pairs. The members of a pair are called **homologous chromosomes,** and each individual member of a pair is called a **homolog** (Gr. *homologos,* agreeing). One homolog comes from the mother and the other from the father. Human cells, for example, have 46 chromosomes (called the **diploid** number) but only 23 *different kinds* of chromosomes, each with a unique linear sequence of genes.

Meiosis is a distinctive type of nuclear division in plants and animals that differs from mitosis in two important ways. First, meiosis gives rise to *gametes*—eggs and sperm cells—for sexual reproduction. Secondly, the key events in meiosis are that a *single* replication of chromosomes is followed by *two* cell divisions. These divisions result in four daughter cells (rather than two as in mitosis), each with only half as many chromosomes as the parent cell. Consequently, mature eggs and sperm have only one member of each pair of homologous chromosomes. The gametes are now **haploid** (Gr. *haploos,* single). When the haploid gametes fuse at fertilization to form a single zygote, the full diploid number of chromosomes is restored.

The process of forming mature gametes from primary germ cells is called **gametogenesis,** and it involves both mitosis and meiosis. Early in embryonic development of a sexually reproducing animal, certain cells called **primordial germ cells** are set aside. These predestined cells migrate to the developing gonads, which later become ovaries or testes. Primordial germ cells are larger than ordinary somatic cells and are the future stock of gametes for the animal. Once in the gonad, the primordial germ cells begin to multiply by ordinary mitosis.

Thus, both body cells and cells destined to become gametes contain an identical and complete, or diploid, set of chromosomes. The **maturation process** refers to the final divisions necessary to produce functional ova and spermatozoa. Gametogenesis, then, includes many divisions by mitosis during the early multiplication stages, followed finally by a reduction in number of chromosomes by meiosis.

When an animal approaches sexual maturity, its gonads begin to produce mature eggs or sperm by meiosis. As shown in Figure 3-1, meiosis involves two divisions, called **meiosis I** and **meiosis II.** In meiosis I homologous chromosomes, called **homologs,** pair lengthwise and exchange genetic material between them by a process called **crossing over.** The homologs then separate to opposite poles of the cell. Each of the two resulting cells has half the number of chromosomes present in the original diploid cell. However, because the DNA of each chromosome has replicated before onset of meiosis, each homolog actually consists of two sister chromatids joined at their centromeres.

In meiosis II, each haploid cell divides by a process similar to normal mitosis, except that the chromosomes do not replicate between meiosis I and II. The sister chromatids simply separate to opposite poles of the cell. The result is four cells, or gametes, each with a haploid number of chromosomes (Figure 3-1). Despite the fact that eukaryotic organisms age and accumulate defects the older they get, recent studies on yeast show that older yeast cells produce gametes without any age-associated damage (Ünal et al., 2011).

Gametogenesis in the testis is called **spermatogenesis,** and in the ovary it is called **oogenesis.** The same processes are involved in both, but there is an important difference in the end result. An oocyte undergoes meiosis to produce one large, functional egg (ovum) and three small polar bodies, whereas a spermatocyte produces four functional spermatozoa (see Figure 3-3). In many animals, sterility occurs when there is a breakdown in gametogenesis.

Spermatogenesis

Consider first the development of spermatozoa (spermatogenesis) (see Figure 3-3). Early germ cells, which have migrated to the testis, are called primordial germ cells. They become **spermatogonia** as they multiply rapidly by mitosis.

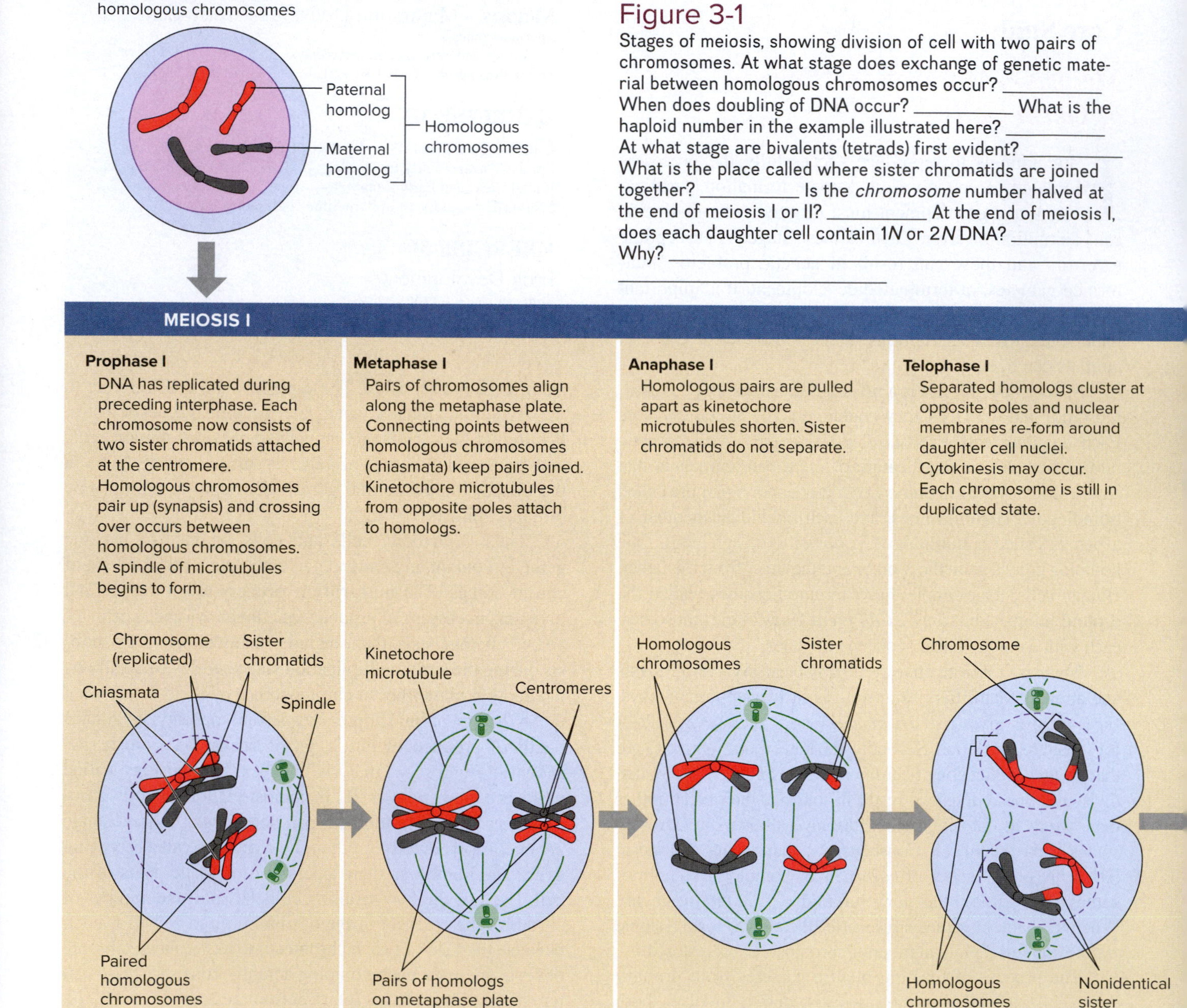

Parent cell (2n)
Parent cell with two pairs of homologous chromosomes
Paternal homolog
Maternal homolog
Homologous chromosomes
MEIOSIS I
Prophase I
DNA has replicated during preceding interphase. Each chromosome now consists of two sister chromatids attached at the centromere. Homologous chromosomes pair up (synapsis) and crossing over occurs between homologous chromosomes. A spindle of microtubules begins to form.
Chromosome (replicated)
Sister chromatids
Chiasmata
Spindle
Paired homologous chromosomes
Metaphase I
Pairs of chromosomes align along the metaphase plate. Connecting points between homologous chromosomes (chiasmata) keep pairs joined. Kinetochore microtubules from opposite poles attach to homologs.
Kinetochore microtubule
Centromeres
Pairs of homologs on metaphase plate
Anaphase I
Homologous pairs are pulled apart as kinetochore microtubules shorten. Sister chromatids do not separate.
Homologous chromosomes
Sister chromatids
Telophase I
Separated homologs cluster at opposite poles and nuclear membranes re-form around daughter cell nuclei. Cytokinesis may occur. Each chromosome is still in duplicated state.
Chromosome
Homologous chromosomes
Nonidentical sister chromatids

Figure 3-1

Stages of meiosis, showing division of cell with two pairs of chromosomes. At what stage does exchange of genetic material between homologous chromosomes occur? __________ When does doubling of DNA occur? __________ What is the haploid number in the example illustrated here? __________ At what stage are bivalents (tetrads) first evident? __________ What is the place called where sister chromatids are joined together? __________ Is the *chromosome* number halved at the end of meiosis I or II? __________ At the end of meiosis I, does each daughter cell contain 1*N* or 2*N* DNA? __________ Why? ______________________________

These early germ cells, like ordinary somatic cells, have the diploid number of chromosomes and divide by ordinary mitosis. (For a review of mitosis see pp. 17–20.)

As an organism reaches sexual maturity, some spermatogonia cease to divide; instead, they enlarge, become **primary spermatocytes,** and begin meiotic divisions. The first stage in meiosis I, prophase, is more complex than mitotic prophase because the chromosomes become rearranged by genetic recombination.

When a primary spermatocyte begins prophase of meiosis I, each chromosome has *already doubled* during the preceding interphase and consists of two chromatids so closely bound together that they cannot be distinguished with the light microscope. Next, the homologs, each consisting of two sister chromatids, line up with each other in **synapsis.** As prophase progresses, special points of contact appear, called **chiasmata** (Gr. *chiasma,* cross). At this time crossing over occurs (Figure 3-2). In this process, segments of genes are traded between homologous chromosomes. Crossing over is a profoundly important process because it results in the *recombination* of genes in chromosomes, leading to genetic variation in offspring of sexually reproducing organisms. Recombination is unique to meiosis. Neither synapsis nor recombination occurs in mitosis.

At the end of prophase I, a spindle forms and the centrioles complete their replication, as in mitosis. Metaphase begins. The homologous pairs separate and move to opposite poles (anaphase). Each chromosome is still double and is called a **dyad** (Gr. *dyas,* two) because it contains two sister chromatids. Two daughter cells **(secondary spermatocytes)** have been produced as a result of meiosis I, each containing one double chromosome from each homologous pair. Each secondary spermatocyte thus has half as many different chromosomes as any somatic cell.

The second meiotic division immediately follows. During meiosis II the sister chromatids of each chromosome separate into individual chromatids (Figure 3-1). One chromatid goes to one daughter cell and one to the other. Nuclear envelopes form around the chromatids, now full-fledged chromosomes. The daughter cells resulting from division of the secondary spermatocyte are called **spermatids** (Figure 3-3). Each spermatid has the **haploid** number of chromosomes, or just half as many as the spermatogonia or body cells. Without further division each of the four spermatids goes through an elaborate process of cytodifferentiation, producing spermatozoa. This is referred to as **spermiogenesis.**

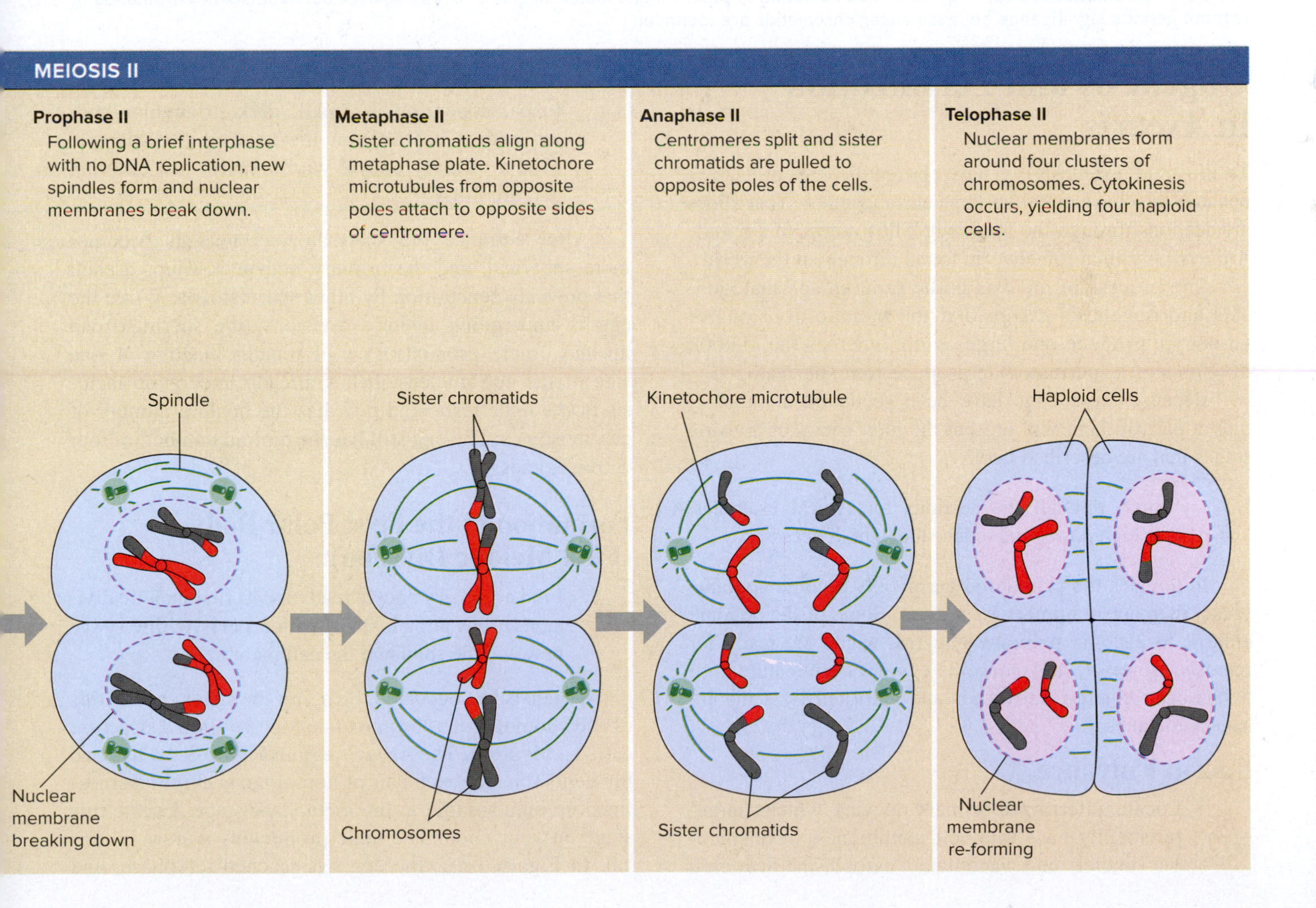

MEIOSIS II
Prophase II
Following a brief interphase with no DNA replication, new spindles form and nuclear membranes break down.
Metaphase II
Sister chromatids align along metaphase plate. Kinetochore microtubules from opposite poles attach to opposite sides of centromere.
Anaphase II
Centromeres split and sister chromatids are pulled to opposite poles of the cells.
Telophase II
Nuclear membranes form around four clusters of chromosomes. Cytokinesis occurs, yielding four haploid cells.
Spindle
Nuclear membrane breaking down
Sister chromatids
Chromosomes
Kinetochore microtubule
Sister chromatids
Haploid cells
Nuclear membrane re-forming

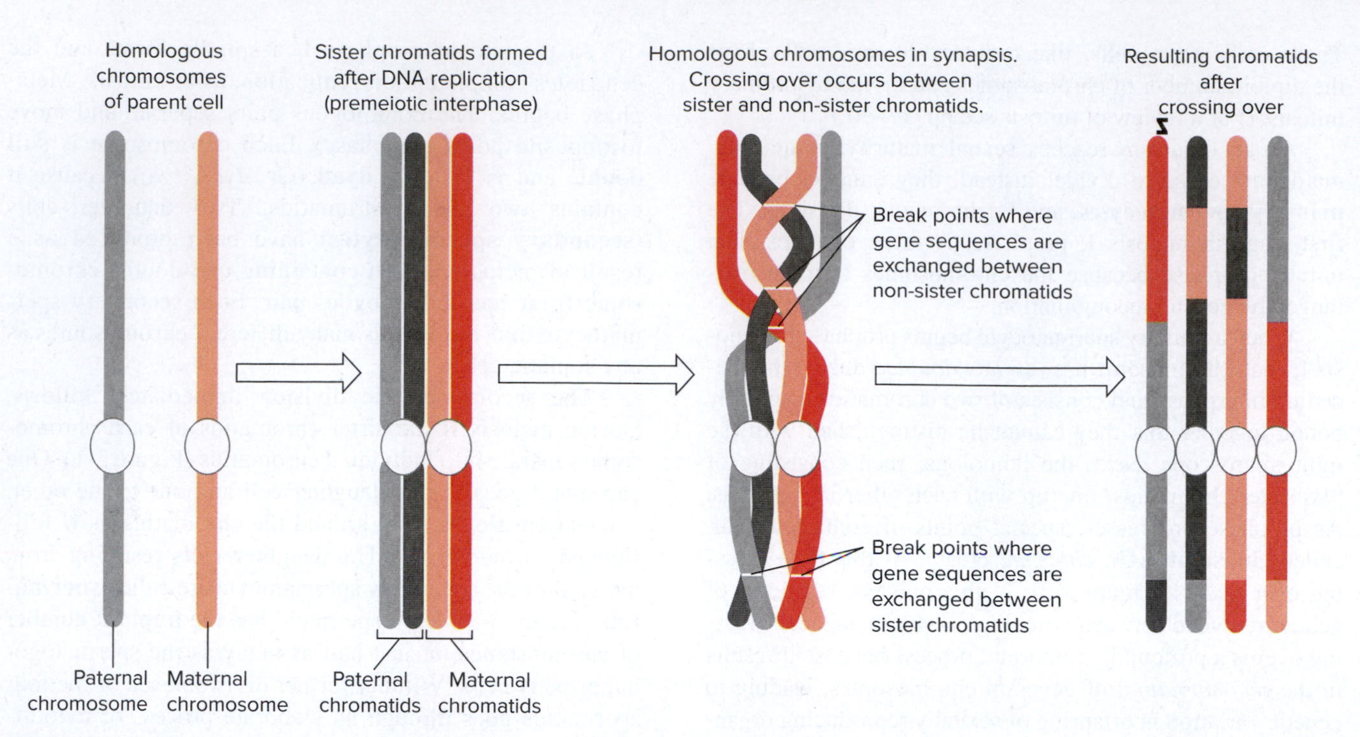

Figure 3-2
Crossing over during meiosis. Breaks and exchanges may occur on both non-sister and sister chromatids. Crossing over between non-sister chromatids results in genetic recombination of paternal and maternal genes. Crossing over between sister chromatids is without genetic significance because sister chromatids are identical.

Oogenesis and Fertilization in *Ascaris*

To illustrate oogenesis, we have chosen the eggs of *Ascaris* because they have only four large chromosomes. Your slides are sections through the long, egg-filled uterus of a female *Ascaris* (intestinal roundworm found throughout the world).

The process of meiosis is the same in spermatogenesis and oogenesis *except* that the meiotic divisions of an oocyte produce one large ovum, whereas the meiotic divisions of a spermatocyte produce four spermatozoa.

Because these eggs have been sectioned at random, only a part of them will present the ideal views of division stages that are described here.

Search through the sections for typical examples. Use high power and reduce the light, if necessary.

In females the primordial germ cells become oogonia, which increase in number by mitosis. Some of the oogonia enlarge to become **primary oocytes,** which are ready for meiosis. However, in *Ascaris* the egg does not mature until it has been entered (fertilized) and activated by a mature spermatozoon.

Sperm Entrance

Locate unfertilized primary oocytes, which characteristically have thin cell membranes, inconspicuous nuclei, and vacuolated cytoplasm. Scattered between the oocytes, find the heads of spermatozoa, appearing as small, dark, triangular bodies with a centriole at the base of the triangle. Find a primary oocyte with a sperm just entering (Figure 3-4A).

After entrance the spermatozoan nucleus becomes more spherical, and the primary oocyte develops a shell that prevents penetration by other spermatozoa. While the egg is undergoing meiotic divisions, the spermatozoan nucleus (male pronucleus) will remain inactive at one side of the egg nucleus. It has already undergone meiosis in the male testis and now has the haploid number of chromosomes. The egg still has the diploid number, or four chromosomes.

Formation of the First Polar Body (First Meiotic Division)

Find a primary oocyte ready for its first meiotic division. It has a thick **shell,** with a **perivitelline space** between the shell and the cell membrane.

Synapsis has occurred, bringing the sister chromatids of each homologous chromosome together into a structure called a **bivalent.** The *Ascaris* egg in Figure 3-4B has two bivalents (one for each pair of homologous chromosomes) lined up on a spindle in the metaphase stage. Locate this stage on your slide. The male pronucleus is now spherical. In Figure 3-4C, the egg has reached telophase, and

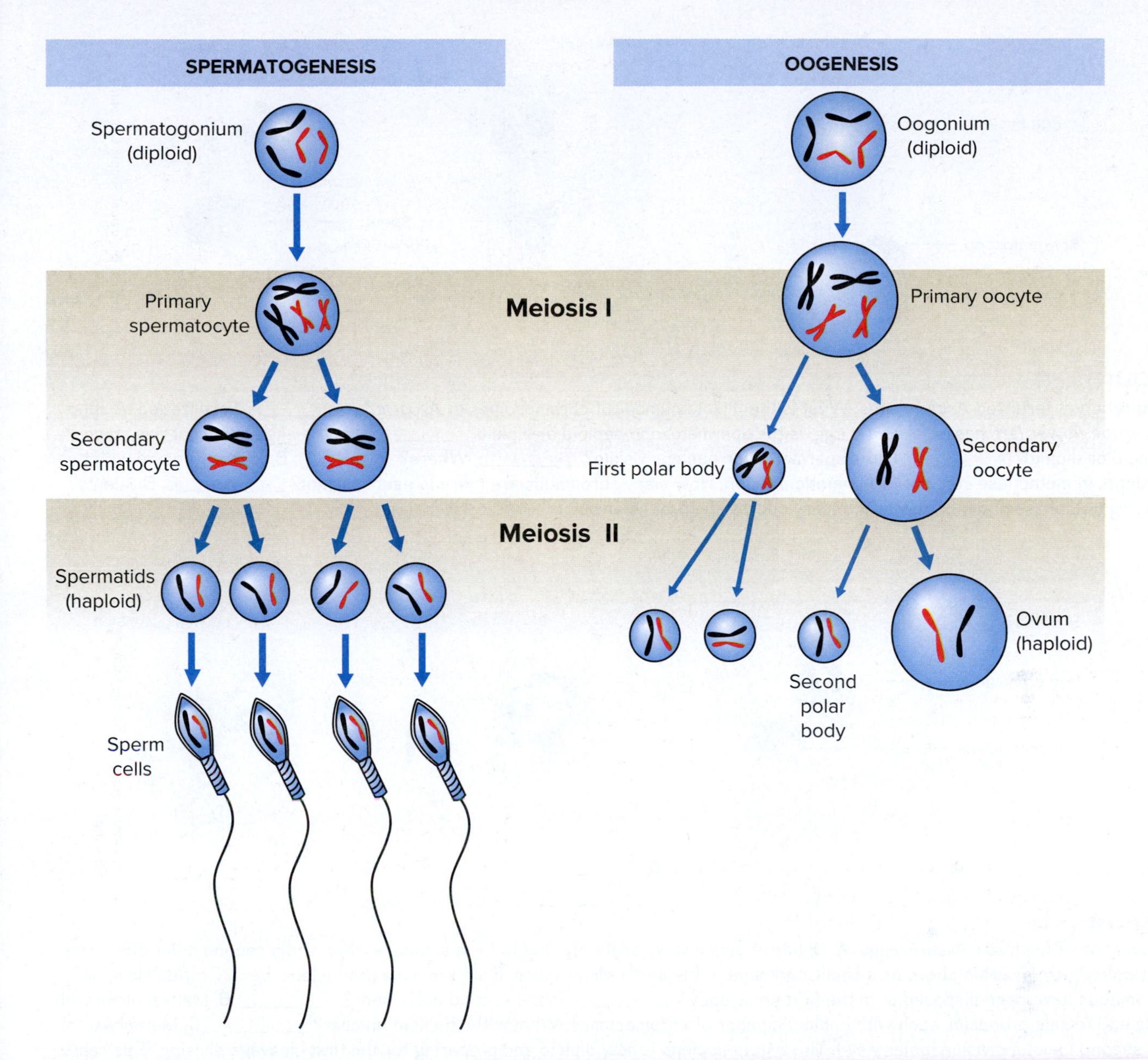

Figure 3-3
Spermatogenesis and oogenesis compared, showing division of a spermatogonium or oogonium with two pairs of chromosomes. What is the diploid number in the example illustrated here? __________ Haploid number? __________ How many chromosomes does each secondary spermatocyte contain? __________ How many chromatids? __________ If the diploid number of humans is 46, how many chromosomes are there in each spermatid? __________ In oogenesis are the polar bodies functional or nonfunctional? __________ What is the purpose of polar body formation? __________

division is almost complete. One **dyad** from each bivalent will go to each daughter cell. However, one daughter cell **(secondary oocyte)** will retain all the cytoplasm, while the dyads of the other are pushed out into the perivitelline space and become the **first polar body.** Find a secondary oocyte with two dyads in the cytoplasm and a polar body appearing as a dark spot in the perivitelline space. The spermatozoan nucleus still waits to one side (Figure 3-5A).

Formation of Second Polar Body (Second Meiotic Division)

In the secondary oocyte the two dyads divide, throwing off two chromatids in the second polar body, which, like the first, has no cytoplasm and is nonfunctional. The other two chromatids, now individual chromosomes, remain in the egg, now a **mature ovum,** and take part in the formation of the **female pronucleus.** The first polar body may or may

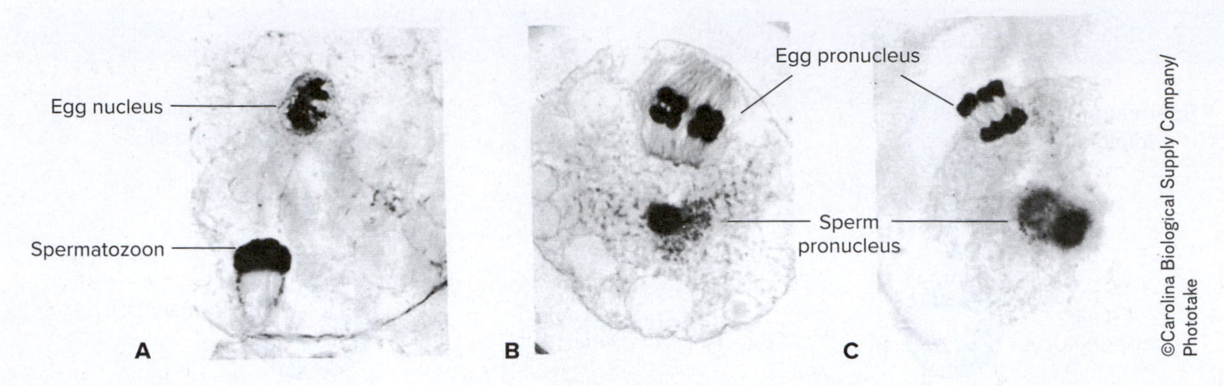

Figure 3-4

Maturation of fertilized *Ascaris* eggs. What is the diploid number of chromosomes in *Ascaris?*__________**A,** Fertilization. A spermatozoon *(lower left)* has entered the egg. Is the spermatozoon haploid or diploid? __________ Is the egg nucleus at this stage haploid or diploid? __________ Is the spermatozoan nucleus visible? __________ Where? __________**B,** Primary oocyte with two bivalents in metaphase stage of first meiotic division. How many chromatids are there in each bivalent? __________**C,** Bivalents dividing (telophase stage) to form secondary oocyte and first polar body.

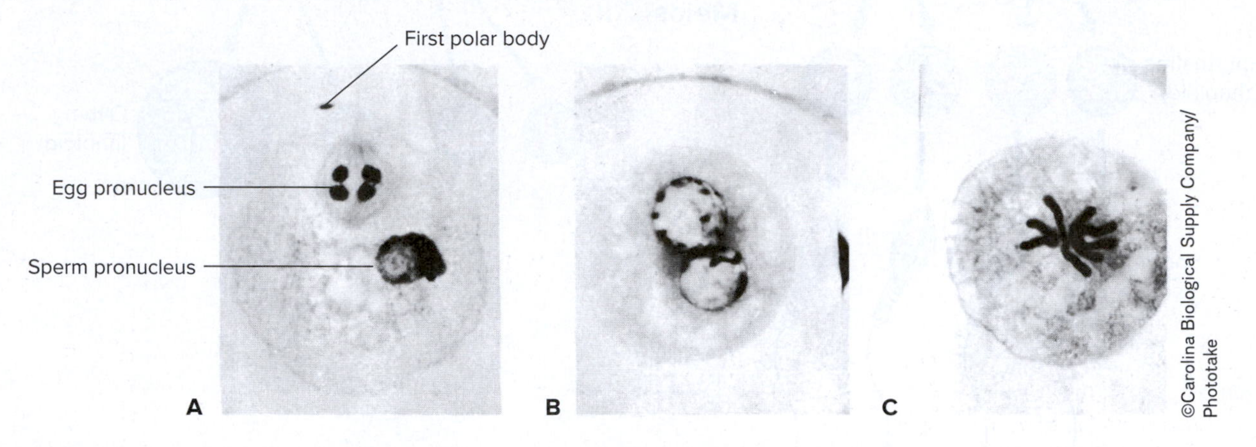

Figure 3-5

Maturation of fertilized *Ascaris* eggs. **A,** Dyads of secondary oocyte dividing to form a mature ovum and a second polar body. The first polar body is visible above as a small, dark spot in the perivitelline space. Note the male pronucleus *(center right).* How many chromatids have been disposed of in the first polar body? __________ In the second polar body? __________**B,** Mature ovum with male and female pronuclei, each with haploid number of chromosomes. What is the haploid number? __________**C,** Metaphase of the second meiotic division (polar view). The zygote nucleus is now diploid and preparing for the first cleavage division. This represents the genesis of a new embryo.

not divide, making two or three tiny, nonfunctional polar bodies in the perivitelline space, but only one functional mature ovum is produced from each oogonium. How does this compare with spermatogenesis (refer to Figure 3-3)? Note that the nucleus of each mature ovum has now the haploid number of chromosomes, the same as the mature spermatozoon. Find an egg in which the second polar body is being formed.

Mature Ovum

Look now for eggs that have completed their meiotic divisions. Note that the egg nucleus, or **female pronucleus,** is a round ball with a nuclear membrane and granular chromatin material very much like the nucleus of an interphase stage (Figure 3-5B). Note that the **male pronucleus** looks much like the female pronucleus. They are called pronuclei because they help form the **zygote nucleus.** How many chromosomes does the zygote have now? __________ Although the two pronuclei come close to each other in the egg, only rarely do they fuse together in *Ascaris.* The nuclear membrane of each pronucleus disappears, and the two chromosomes from each nucleus move on to the spindle, which forms for the first cleavage division. Figure 3-5C shows the chromosomes, **two maternal** and **two paternal,** lined up in metaphase (polar view). This will be the first of a series of mitotic (cleavage) divisions that will occur to produce the new embryo.

Drawings

Prepare drawings on p. 34 as requested by the instructor. Or you may wish to make drawings for future reference.

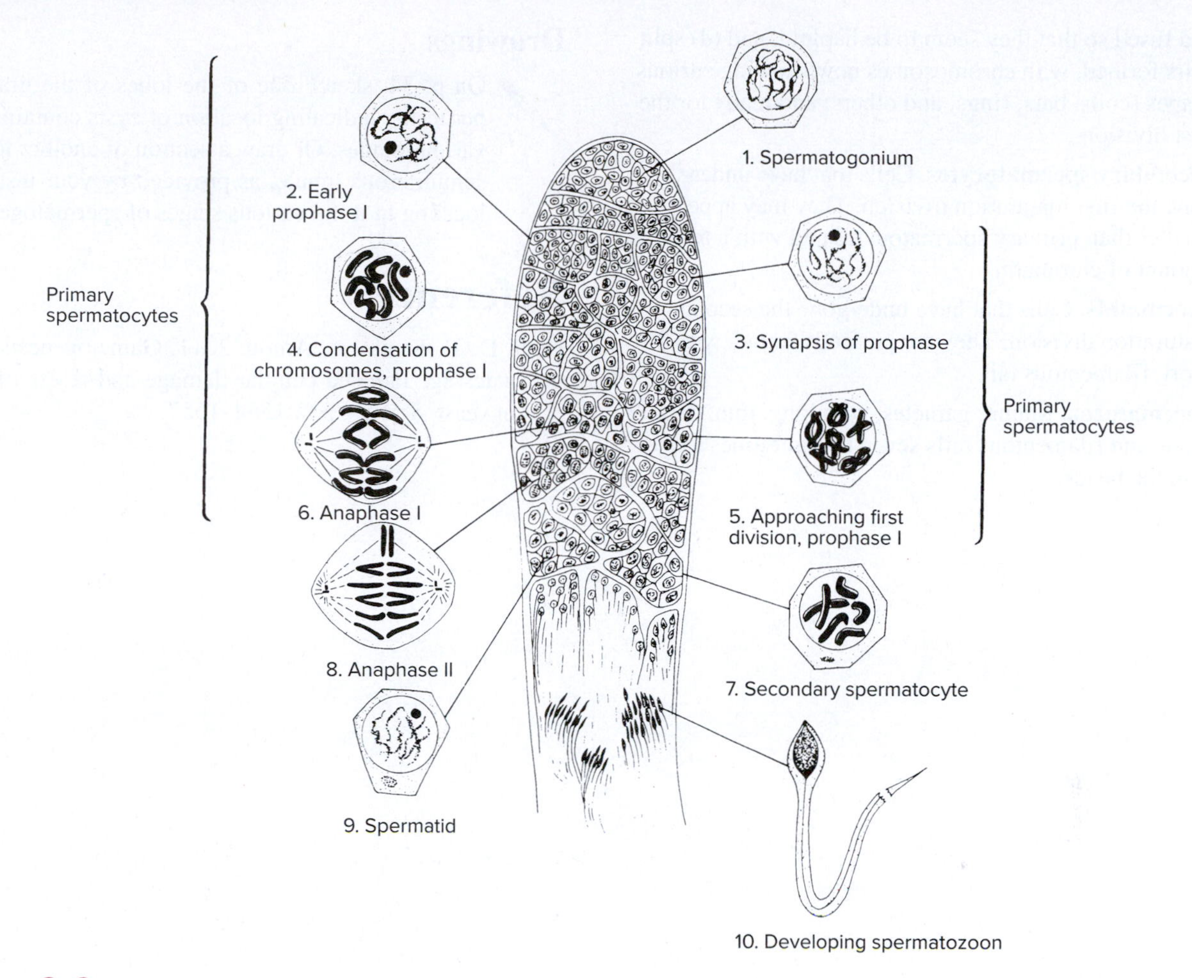

Figure 3-6
Spermatogenesis as seen in one lobe of a grasshopper testis.

Further Study

Spermatogenesis in Grasshopper Testis

The maturation stages of sperm can be seen by studying cross sections of testes of many kinds of animals. Figure 3-6 shows a section through a lobe of the testis of the lubber grasshopper, *Romalea* (native to the southeastern United States). The testis consists of a number of lobes whose pointed ends empty into the vas deferens. Each lobe contains a number of compartments (cysts) separated by tissue septa, and each cyst contains a number of cells all in the same stage of development. These cysts were formed at the blunt, or apical, end of the lobe. Here a group of primordial germ cells are continually and rapidly dividing by mitosis to form spermatogonia. As a new group of cells is formed from the primordial cells, it is pinched off to form a cyst. The spermatogonia in the new cyst begin to grow in volume, without dividing, and become known as primary spermatocytes.

Try to find on your slide, with lower power, a longitudinal section through one of the lobes. Identify the various cysts; then study with high power.

You should be able to identify cysts containing all the following stages, with those containing primary spermatocytes at the apical end and those containing mature spermatozoa at the other end. Not all the lobes will have been cut longitudinally, so not all will show all these stages. Some lobes will be cut transversely and may show only one or two stages. Search the slide for an ideal section or, if necessary, make a composite drawing from several sections.

1. **Spermatogonia.** Cells are small and crowded at the apical end.
2. **Primary spermatocytes.** These cells are larger and are found in cysts nearest those containing spermatogonia (Figure 3-6). They may be seen in several stages, with successive stages showing (a) chromatin threads, (b) chromatin threads broken into chromosomes that are pairing, (c) chromosomes thicker and in pairs (dyads)

and fused so that they seem to be haploid, and (d) split pairs formed, with chromosomes now showing curious shapes (coils, bars, rings, and others) and ready for the first division.

3. **Secondary spermatocytes.** Cells that have undergone the first maturation division. They may appear smaller than primary spermatocytes and with a reduced amount of chromatin.
4. **Spermatids.** Cells that have undergone the second maturation division. They are perfectly round, with a short, filamentous tail.
5. **Spermatozoa.** Mature gametes, with long, thin, dark heads and filamentous tails seven to eight times longer than the heads.

Drawings

On p. 34, sketch one of the lobes of the grasshopper testis, indicating location of cysts containing the various stages. Or draw a section of another testis or seminiferous tubule, as provided by your instructor, locating in it the various stages of spermatogenesis.

Reference

Ünal, E., B. Kinde, A. Amon. 2011. Gametogenesis eliminates age-induced cellular damage and resets life span in yeast. Science 332:1554–1557.

NOTES

LAB REPORT

Name ________________________________

Date ________________________________

Section ________________________________

Stages of Oogenesis in *Ascaris*

Phylum ____________________________

Genus ____________________________

Stages of Spermatogenesis in Grasshopper Testis

Name ______________________

Date ______________________

Section ______________________

LAB REPORT

Phylum ______________________

Genus ______________________

EXERCISE 3B

Core Study

Cleavage Patterns—Spiral and Radial Cleavage

Cleavage, the earliest stage in embryonic development, consists of a succession of regular mitotic cell divisions that partition the egg into a multitude of small cells clustered together like a mass of soap bubbles. In some animals, cleavage is so rapid that hundreds, sometimes thousands, of cells are produced in a matter of hours. In most animal groups, cleavage is **regular**; the egg cytoplasm is segregated into specific cells called **blastomeres** (Gr. *blastos,* bud, +*meros,* part) occupying discrete positions and having specific developmental fates. Patterns of regular cleavage depend greatly on amount and distribution of yolk in the egg. In eggs having a large amount of yolk, cleavage may be either **complete** (= holoblastic), as in amphibians, or **incomplete** (= meroblastic), as in birds and reptiles. In birds and reptiles with extreme **telolecithal** (Gr. *telos,* end, + *lekithos,* yolk) eggs, cleavage is restricted to a small disc of cytoplasm on the animal pole; this type of cleavage is called **discoidal**. The eggs of most insects follow another pattern of cleavage called **superficial.** In these, the nuclei divide mitotically into hundreds or thousands of "free" nuclei, which later migrate to the egg surface. Only then do cleavage furrows form, rapidly partitioning the cytoplasm into a superficial layer of cells.

In most invertebrates, eggs have little yolk (= **isolecithal** ["equal-yolk"]), and cleavage is complete (holoblastic) and equal. Two major kinds of complete cleavage exist: **spiral** and **radial** (Figure 3-7). The first two cleavages are the same in both kinds of eggs: the cleavage planes are along the animal-vegetal axis, producing a quartet of cells. At the third cleavage, however, these two patterns—spiral and radial—can be distinguished from each other by the geometric positioning of the cells.

In radial cleavage, the third cleavage is perpendicular to the first two, yielding two quartets of cells, with the upper quartet lying directly on top of the lower. In spiral cleavage, the third cleavage planes are oblique to the polar axis and typically produce an upper quartet of smaller cells that come to lie between the furrows of the lower quartet of larger cells.

There are other important differences between these two cleavage patterns. Spiral cleavage is typically **mosaic,** meaning that the embryo is constructed as a mosaic, with each cell fitting into its predetermined location in the larval body. If cells of the embryo are experimentally separated at this early stage, each cell will develop into partial or defective larvae because the developmental fate of each cell has already been determined. Spiral cleavage is found in several phyla, including annelids, many molluscs, some flatworms, and ribbon worms (nemerteans). All groups showing spiral cleavage belong to the grouping of animal phyla called the **Protostomia,** in which the embryonic blastopore forms the mouth.

Radial cleavage is characteristically **regulative** because cell fate does not become fixed until after the first few cleavages. While some protostomes have radial cleavage, it is found primarily in eggs of echinoderms and many chordates, especially protochordates, amphibians, and mammals. (As mentioned earlier, eggs of birds and reptiles, as well as many fishes, show discoidal cleavage.) All of these belong to the **Deuterostomia,** a group of phyla in which the mouth is formed from a secondary embryonic opening.

In this exercise you will compare spiral cleavage in a ribbon worm (a protostome) with radial cleavage in a sea star (a deuterostome).

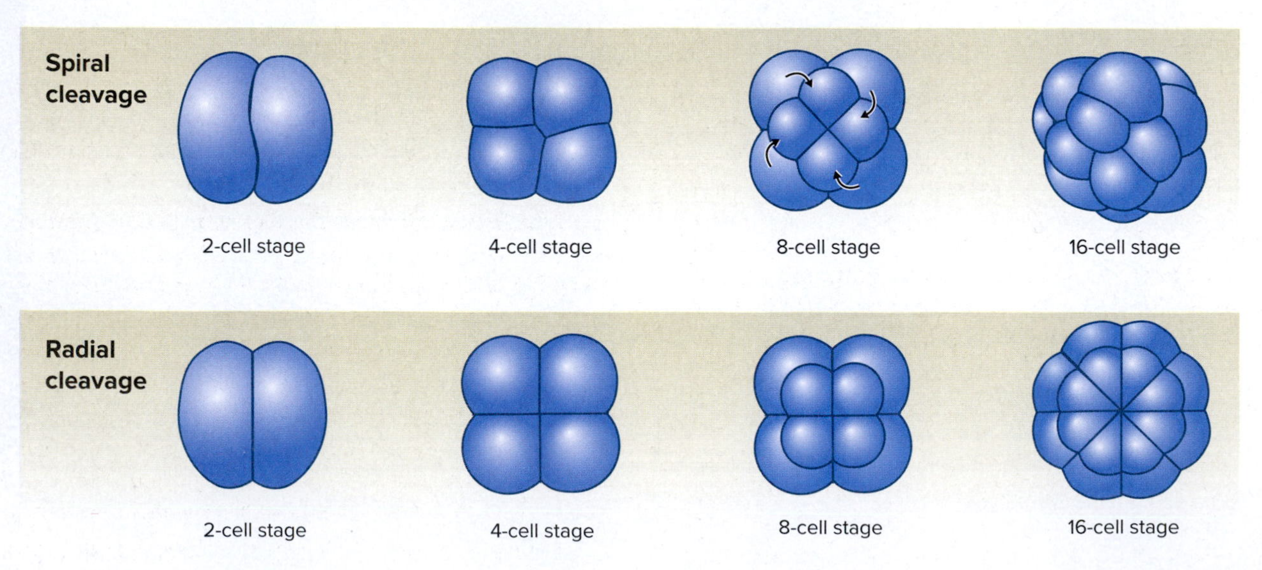

Figure 3-7
Spiral and radial cleavage compared. Which kind of cleavage is mosaic? __________ Which kind is regulative? __________ Which kind of cleavage is typical of phyla belonging to the Deuterostomia division of the animal kingdom? __________ Which kind is typical of Protostomia? __________

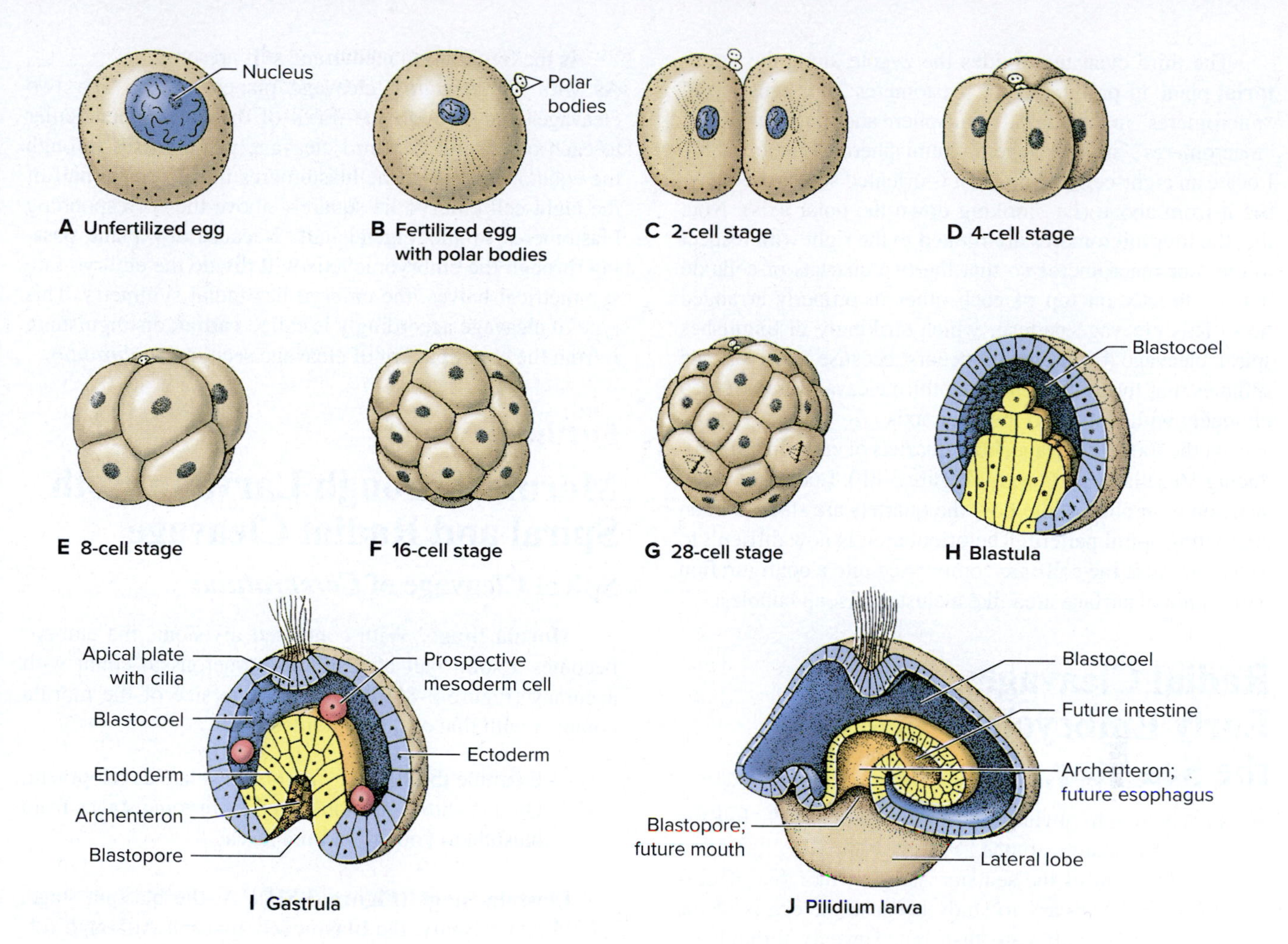

Figure 3-8

Early embryology of *Cerebratulus*, a nemertean worm. Are polar bodies formed and ejected before or after fertilization? __________ At what stage are micromeres and macromeres formed? __________ At what stage does the difference between spiral and radial cleavage become visually apparent? __________ What cavity is formed by gastrulation? __________ What is the significance of the 4d cell? __________ Is the mouth or the anus derived from the blastopore? __________

Spiral Cleavage: Early Embryology of the Ribbon Worm, *Cerebratulus*

Ribbon worms belong to the phylum **Nemertea** and often are called nemertean worms. You will study eggs of *Cerebratulus* (ser′uh-brat′u-lus), a well-known genus with species distributed in seas around the world. *Cerebratulus* and other ribbon worms follow a highly regular form of spiral cleavage that is almost diagrammatic in pattern.

On your slide of *Cerebratulus* early development,* you will see stages of development from the unfertilized or fertilized egg to about the 32-cell stage. Locate the stages with medium power and then switch to high power for study.

*See Appendix A, pp. 387–419 for sources of prepared slides of *Cerebratulus* development.

Unfertilized Ovum (Figure 3-8A). If unfertilized ova are present on your slide, they will be recognized by their irregular shape, with a conical protuberance on the vegetal pole. The nucleus is clearly visible.

Fertilized Undivided Ovum (Figure 3-8B). After sperm penetration, the egg assumes a spherical shape. First and second polar bodies are formed and constricted off from the animal pole. Focus up and down through the egg to locate the polar bodies and the large asters surrounding the female pronucleus, which is lying near the center of the egg. What is the purpose of polar body formation? __________ After expulsion of polar bodies, the haploid female pronucleus fuses with the haploid male pronucleus near the center of the egg. Fertilization of the ovum is now complete and it is called a **zygote.**

Cleavage Stages. The zygote now divides along the egg axis into two equal-sized blastomeres (Figure 3-8C). The second cleavage plane forms at right angles to the first, producing four equal-sized blastomeres (Figure 3-8D). Locate these stages on your slide.

The third cleavage divides the zygote along the **equatorial** plane to produce eight blastomeres: four equal-sized "micromeres" in the animal hemisphere and four equal-sized "macromeres" in the vegetal hemisphere (Figure 3-8E). Locate an eight-cell embryo that is oriented so you are viewing it from above (i.e., looking down the polar axis). Note that the four micromeres are rotated to the right with respect to the four macromeres so that the two quartets of cells do not lie directly on top of each other in properly arranged tiers. This cleavage pattern, which strikingly distinguishes spiral cleavage from radial, occurs because the cleavage spindles that form just before the third cleavage are oriented obliquely with respect to the polar axis.

At the fourth cleavage, both quartets of cells divide, producing 16 cells in four layers (Figure 3-8F). Locate a 16-cell stage on your slide. Note how the quartets are stacked in an alternating spiral pattern. The orientation is now difficult to follow because the cells are compressed into a configuration with minimal surface area like a cluster of soap bubbles.

Radial Cleavage: Early Embryology of the Sea Star, *Asterias*

Sea stars belong to phylum **Echinodermata,** which, like the chordates, are deuterostomes.

Development of the sea star *Asterias,* like that of *Cerebratulus,* is rather easy to study because the egg is of the **isolecithal** type, with yolk distributed evenly throughout. Cleavage is radial and complete (holoblastic).

On your slide(s) of sea star development, you will find whole mounts of embryos at all stages of development, from unfertilized ovum to bipinnaria larva. With the help of Figure 3-9, locate the best representatives of each stage, using the low-power objective. Then switch to high power to study special features of each stage.

Unfertilized Ovum (Figure 3-9A). There should be several unfertilized ova scattered on your slide. Each can be identified by its very large, conspicuous nucleus. Note the small, dark nucleolus in the nucleus.

Fertilized Undivided Ovum (Zygote) (Figure 3-9B). Soon after the eggs are spawned and activated by sperm entry, the nuclear membrane breaks down and the nucleus disappears in preparation for fusion of the male and female pronuclei. You should be able to see a **fertilization membrane,** which may appear wrinkled in your prepared whole mounts; in life it is smooth and spherical. What is the function of this membrane? ____________________________

Cleavage Stages (Figure 3-9C–E). Cleavage by mitosis begins immediately after fertilization.

Identify two-cell, four-cell, and eight-cell stages on your slide.

Is the fertilization membrane still present? __________ As with *Cerebratulus,* cleavage planes for the first two cleavages run through the poles of the egg perpendicular to each other, and the third cleavage plane passes through the equator. Note that the blastomeres in the animal half of the eight-cell embryo lie squarely above the corresponding blastomeres of the vegetal half. Because any plane passing through the embryonic axis will divide the embryo into symmetrical halves, the embryo has radial symmetry. This type of cleavage accordingly is called **radial,** distinguishing it from the spiral pattern of cleavage seen in *Cerebratulus.*

Further Study

Morula Through Larva in Both Spiral and Radial Cleavage

Spiral Cleavage of *Cerebratulus*

Morula Stage. With continued divisions, the embryo becomes a solid ball of cells too numerous to count with accuracy (Figure 3-8G). How does the size of the morula compare with that of the undivided egg?

Examine the slide of *Cerebratulus* late development. On this slide you should find embryonic stages from blastula to young pilidium larvae.

Blastula Stage (Figure 3-8H). At the blastula stage, an off-center cavity, the **blastocoel,** appears. Although difficult to see, the surface of the blastula is covered with delicate cilia.

Gastrula Stage (Figure 3-8I). In the early gastrula stage, an indentation appears at the vegetal pole. These cells continue to migrate inward by a process called **invagination** to form an internal cavity, the **archenteron.** Look for lightly stained embryos at the stage shown in Figure 3-8I. The pouch-shaped archenteron opens to the outside through the blastopore, which will become the mouth of the worm. The embryo can now swim rapidly forward in a spiraling manner with the animal pole directed forward. With favorable specimens and reduced illumination, you may be able to see the long apical tuft of cilia on the animal pole.

The gastrula is now an embryo of two **germ layers.** The outer layer is the **ectoderm,** and the inner layer that lines the archenteron is the **endoderm.** The ectoderm will give rise to the epithelium, which covers the body surface, and the nervous system. The inner endoderm gives rise to the epithelial lining of the digestive tube. The cavity between these two layers is the old blastocoel.

Locate on your slide a late gastrula lying on its side and note the mass of cells arising in the blastocoel on either side of the blastopore. These cells are giving rise to the third germ layer, the **mesoderm,** which will form the muscles, blood vessels, and reproductive system of the future adult body. Meticulous cell-lineage studies by early embryologists

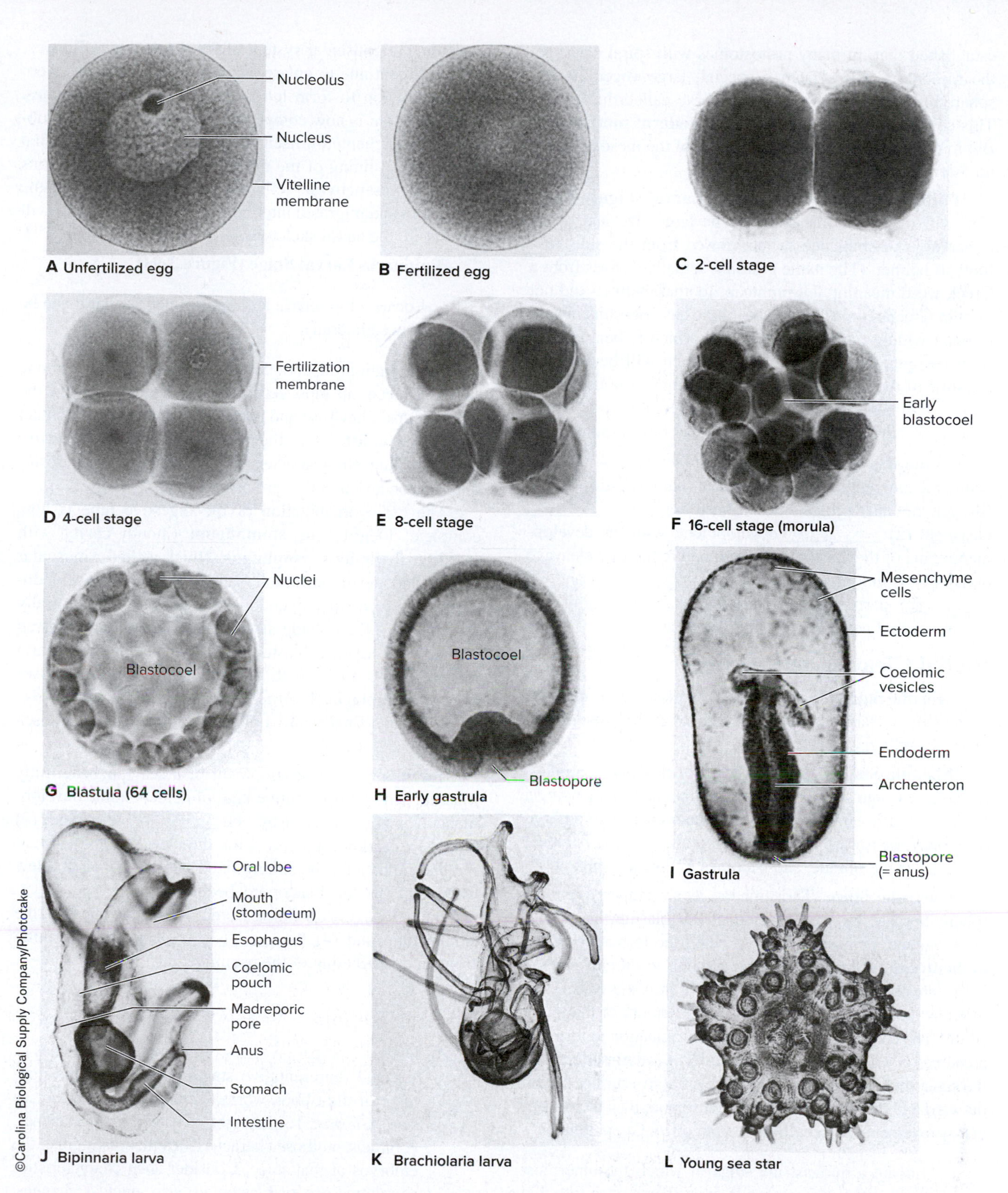

Figure 3-9

Embryology of a sea star, *Asterias* sp. With an unknown embryo, what is the earliest stage that would inform you that this is a radial-cleaving and not a spiral-cleaving embryo? __________ Is cleavage by mitosis or meiosis? __________ Is cleavage in sea stars regulative or mosaic? __________ How do these two forms of cleavage differ? __________ Into what cavity does the blastopore open? __________ What adult structure does the blastopore become? __________ Would knowing the fate of the blastopore be enough to assure you that this animal is a deuterostome? __________ What is the fate of mesenchyme cells? __________ How does the origin of mesoderm in sea stars differ from the origin of mesoderm in *Cerebratulus?*__________

established that, in many protostomes with spiral cleavage, the mesoderm arises from a special, large vegetal hemisphere blastomere of the 64-cell embryo called the 4d cell. This cell is also referred to as the **mesoderm mother cell.** But in *Cerebratulus,* the precise origin of the mesoderm has not been determined.

Pilidium Larva Stage (Helmet Larva) (Figure 3-8J). You will recognize this stage by its large size and by its remarkable resemblance, when viewed from the side, to a football helmet. (The name *pilidium* actually derives from a Greek word meaning felt nightcap; football helmets did not exist in Grecian times!) The blastopore has now sunk inward to open widely into the archenteron, which is bent over to form the gut. The large, innermost pouch will become the intestine of the worm; the tube connecting this pouch to the outside will become the esophagus of the adult body. There is no anus. Note that the pilidium is bilaterally symmetrical.

Summary. The development of *Cerebratulus* clearly shows the four features of development characteristic of many (though not all) protostomes. These are (1) __________ cleavage; (2) __________ development, with the developmental fate of the blastomeres fixed before the first cleavage division; (3) mouth derived from the __________; and (4) __________ derived from a particular blastomere (or blastomeres) located near the vegetal pole of the embryo.

Radial Cleavage of Sea Star

Morula Stage (Figure 3-9F). A mass of 16 or more cells with no large cavity in its center is the **morula** (L. *morum,* mulberry) stage.

Blastula Stage (Figure 3-9G). As cleavage continues, the cells become smaller and arrange themselves in a single layer around a central cavity, the **blastocoel.** The embryo develops cilia toward the end of this stage, escapes from the fertilization membrane, and becomes free-swimming.

Gastrula Stage. The embryo is no longer a perfect sphere because the vegetal side of the blastula has begun to invaginate, or push inward, to begin the formation of the **archenteron** (Figure 3-9H). The opening to the archenteron is the **blastopore,** which will later form the anus (recall that in *Cerebratulus* the blastopore forms the mouth of the future adult). The cells streaming inward by invagination to form the archenteron will give rise to the **endoderm** and **mesoderm.** At the same time, **ectoderm** cells on the embryo's surface spread outward and downward toward the blastopore by a morphogenetic movement called **epiboly** (ee-pib′o-lee; Gr., putting on).

Locate a midgastrula stage in which the innermost wall of the archenteron is greatly expanded into two pouchlike vesicles (Figure 3-9I).

These **coelomic vesicles** are the origin of the mesoderm that will give rise to the peritoneum (lining of the coelom), skeleton, and muscular system of the future adult. The vesicles will continue to enlarge and eventually separate from the archenteron to form left and right **coelomic pouches.** The mesoderm is now completely separated from the endoderm. The archenteron itself will give rise to the epithelial (endodermal) lining of the digestive tube. Note also at this stage the **mesenchyme** cells that have separated from the epithelium and migrated into the blastocoel. These cells will give rise to the larval skeleton.

Bipinnaria Larval Stage (Figure 3-9J).

Locate a bipinnaria larva positioned so that it can be viewed laterally.

After gastrulation the endodermal archenteron bends to one side and meets an inpocketing of the ectoderm called the **stomodeum.** They fuse and perforate at this point of contact to form the **mouth.** Thus, the mouth forms from a *secondary* larval opening, the characteristic for which the deuterostomia ("secondary mouth") are named.

The basic organization of the digestive tube is now complete. Identify the **stomodeum** (mouth cavity with ectodermal lining), **esophagus** (thick-walled, muscular tube of endoderm), **stomach** (saclike enlargement of endoderm), and **intestine,** an endodermal tube curving ventrally to connect to the outside at the **anus.** The anus is derived from the embryonic blastopore. Note that the bipinnaria is bilaterally symmetrical. The bipinnaria metamorphoses into a **brachiolaria** (L. *brachiolum,* small arm) **larva** (Figure 3-9K) and finally into a radially symmetrical young sea star (Figure 3-9L).

Summary. The development of a sea star beautifully illustrates the most primitive kind of deuterostome development. It is characterized by (1) __________ cleavage; (2) __________ development, with the development fate of the blastomeres not becoming fixed until after the first few cleavages; (3) the blastopore becoming the __________, and the __________ forming from a secondary opening, the stomodeum; and (4) the __________ germ layer forming from an outpocketing of the primitive gut.

Written Report

On p. 42 prepare a series of comparative sketches of selected representative stages in the early development of the ribbon worm, *Cerebratulus,* and the sea star, *Asterias.* To the right of each pair of sketches, comment on those characteristics that distinguish the embryos at that stage. Consider such characteristics as relative egg or blastomere size, nuclear changes, cleavage patterns, embryo shape, cell migration, origin of the germ layers, and origin and location of embryonic mouth and anus.

NOTES

LAB REPORT

Name ______________________________

Date ______________________________

Section ______________________________

Comparison of Early Development of a Protostome and a Deuterostome

	Ribbon Worm *Cerebratulus*	Sea Star *Asterias*	Distinguishing Features
Unfertilized egg			
Fertilized undivided egg			
Eight-cell stage			
Morula			

Ribbon Worm *Cerebratulus*	Sea Star *Asterias*	Distinguishing Features
Gastrula		
Free-swimming larva		

LAB REPORT

EXERCISE 3C

Core Study

Frog Development

Frog eggs are much larger than the marine invertebrate eggs studied in Exercise 3B. A frog egg is **mesolecithal** (Gr. *mesos,* middle, + *lekithos,* yolk), in which the yolk is more or less concentrated in one hemisphere (vegetal), and the cytoplasm and nucleus are found in the other hemisphere (animal). Such an egg shows definite **polarity.** The **animal pole** is the region of the egg just above the nucleus in the darkly pigmented hemisphere; the **vegetal pole** is a similar point on the light-colored yolk hemisphere diametrically opposite the animal pole.

Frog development shows a variant of radial cleavage called **bilateral** because the bilateral arrangement of the blastomeres in early cleavage is unmistakable. Cleavage is complete but unequal and begins at the animal pole. The nucleus divides by mitosis, but division of cytoplasm is retarded because of the concentration of yolk at the vegetal pole, with the result that cells of the vegetal hemisphere are larger and less numerous than those of the more rapidly dividing animal hemisphere.

You will study both preserved specimens and prepared slides of developing stages of a frog. Containers with various stages of frog egg development may be available. When you select a stage, be sure to keep it immersed in water while studying it; *do not allow it to dry.* Return your specimen to the proper container when you have finished your study of it. If the specimens have been mounted in deep-well slides, they are to be studied without removal from their containers.

Study of Eggs in Frog Ovary

Obtain a stained slide showing a section through an ovary.

Note the **ova,** or eggs, of various sizes. Smaller ova contain little yolk and have conspicuous nuclei. Note the nature and position of the **chromatin granules** in these nuclei. Compare the location of nuclei in small and large ova. Early germ cells have distinct **cell membranes,** but older ova have **follicular cell layers** surrounding them. Find an ovum with such a follicular layer. The follicular cells support the maturation of the ovum.

Eggs in Cleavage Stages

Use both whole developing eggs and stained slides showing cross sections of various stages of development.

One-Cell Stage. The one-cell stage may be represented by an unfertilized ovum or, as is more commonly the case, by a fertilized egg (zygote) that has not yet started to develop. Note the thick, jellylike layer surrounding the egg. What do you think is the function of the jelly layer? ______________ The black, or **animal, hemisphere,** which is not as heavy as the **vegetal hemisphere,** always floats uppermost in the water, where it can absorb the sun's warmth, which is necessary for development (Figure 3-10A).

Soon after fertilization, a **gray crescent** appears as a light, indistinct, crescent-shaped area on the margin of the pigmented zone (Figure 3-10B). It is formed on the side of the egg opposite that which the spermatozoon entered. It is produced by a migration of pigmented cytoplasm away from that region. Note that before fertilization the egg has radial symmetry; after fertilization it has bilateral symmetry.

Early Cleavage Stages. The frog egg divides repeatedly during cleavage. However, the presence of yolk slows down cell division so that cells in the animal hemisphere divide more rapidly than those in the vegetal hemisphere. Consequently cells of the two poles rapidly become unequal in size and number. Is cleavage by mitosis or meiosis? ________________ Do the blastomeres contain the haploid or diploid number of chromosomes? ________________

The first cleavage plane, which results in a **two-cell stage,** is vertical (meridional), beginning at the animal pole and passing through the vegetal pole. The second cleavage plane, resulting in a **four-cell stage,** is also vertical but at right angles to the first cleavage plane. The third cleavage plane is equatorial, or horizontal, but passes closer to the animal pole. The four blastomeres **(micromeres)** of the animal pole of this **eight-cell stage** are much smaller than the four **macromeres** at the vegetal pole (Figure 3-10C).

The **morula stage** contains 16 to 32 cells. There are two fourth-cleavage planes; both are vertical and appear simultaneously. Because of the greater concentration of yolk in the vegetal pole, the cells of the animal pole complete their division before those of the vegetal pole. The fifth cleavage, which results in 32 cells, is also made up of two furrows, both of which are horizontal (Figure 3-10D). Note that in these later stages the faster rate of cell division in the animal hemisphere results in a larger number of micromeres than of macromeres. As cleavage continues, the cells and cleavage planes are difficult to recognize.

Blastula Stage. The blastula stage begins at about 32 cells. A **segmentation cavity,** or **blastocoel,** has begun near the animal pole. The blastula stage is not easily recognized in the unsectioned embryo; look for it in a cross-section slide (Figure 3-10E).

Gastrula Stage. The gastrula stage is recognized by the crescent-shaped slit at the margin of the pigmented animal hemisphere. Gastrulation in a frog is more complicated than in a sea star. In the sea star eggs studied earlier, you saw that the gastrula stage was brought about by a simple folding in of one side. In a frog embryo, the heavy yolk cells prevent such a simple movement. Instead, pigmented cells of the animal hemisphere begin growing down over the vegetal cells (a process called **epiboly**), folding in at the equator

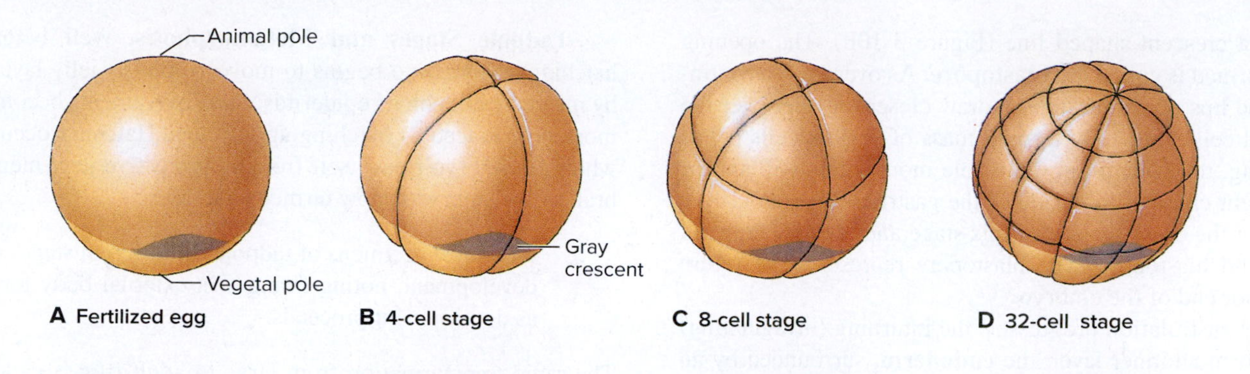

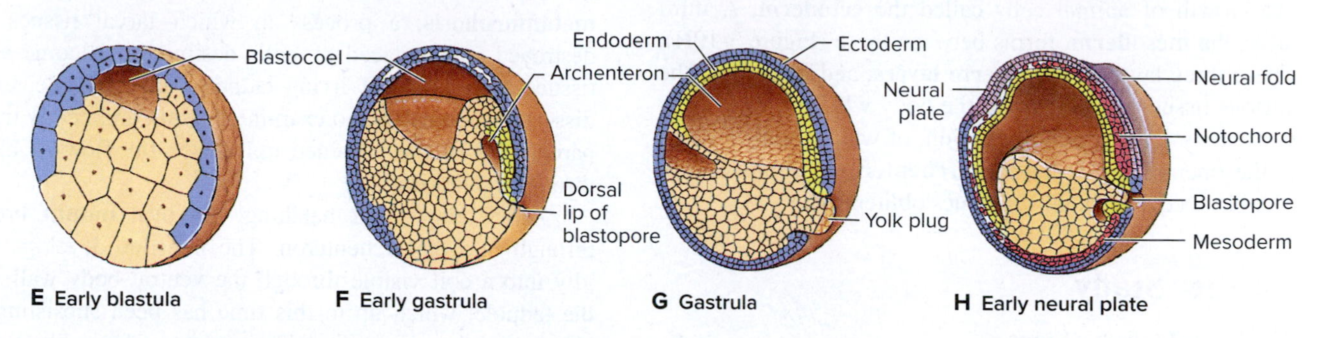

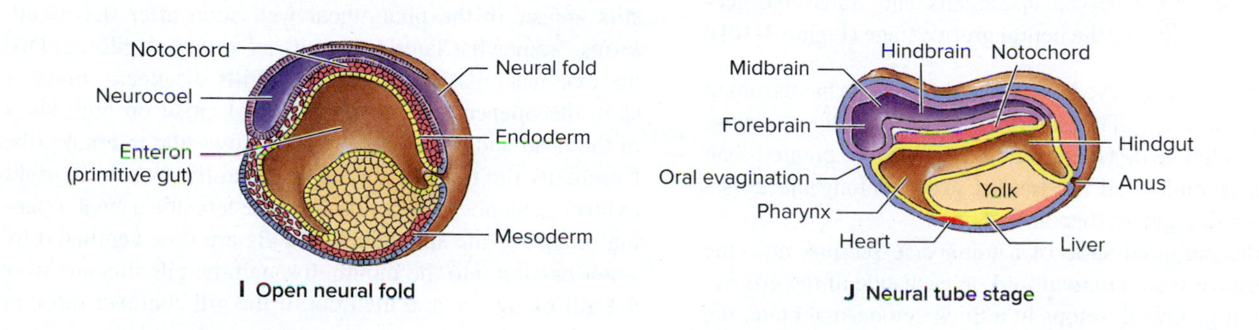

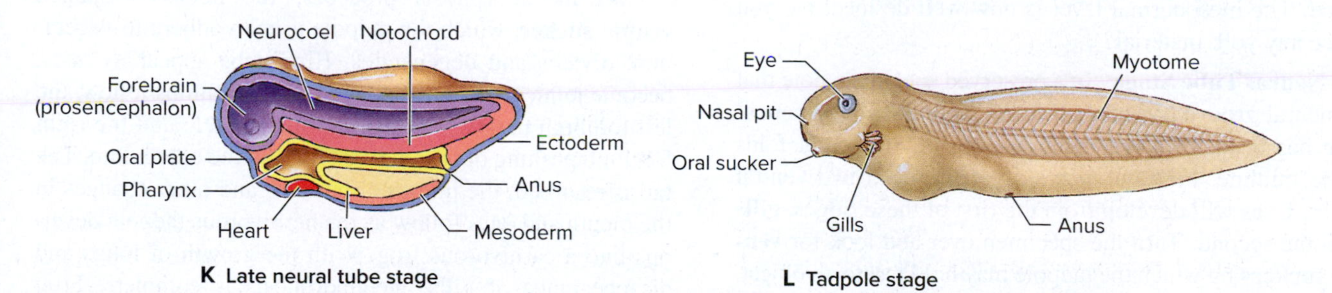

Figure 3-10

Early embryology of a frog from egg to the tadpole stage. Is cleavage of a frog egg radial, spiral, or a variant of one of these? __________ Is cleavage by mitosis or meiosis? __________ Are frogs deuterostomes or protostomes? __________ Is yolk more heavily concentrated in the animal or vegetal hemisphere? __________ What event is marked by the appearance of the gray crescent? __________ What adult structure develops from the blastopore? __________ What developmental process signals the early formation of germ layers? __________ What germ layer gives rise to the neural tube? __________ To muscles and skeleton? __________ To digestive tube? __________

along a crescent-shaped line (Figure 3-10F). The opening thus formed is called the **blastopore.** As overgrowth continues, the lips of the blastopore draw closer together over the vegetal cells until only a small mass of vegetal cells is left showing, the **yolk plug.** On whole mounts this will appear as a light circle on one side of the gastrula (Figure 3-10G). Look at the cross section of this stage and identify the yolk plug and blastopore. The blastopore represents the future posterior end of the embryo.

As gastrulation progresses, the inturning (invaginating) cells form an inner layer, the **endoderm,** surrounded by an overgrowth of animal cells called the **ectoderm.** A third layer, the **mesoderm,** forms between these (Figure 3-10H). These three layers are the **germ layers,** and from them the various tissues and organs of the body will be formed. The new cavity formed by gastrulation, of which the blastopore is the opening, is called the **archenteron** (primitive gut). The blastocoel gradually becomes obliterated.

Further Study

Neural Groove Stage.

Examine preserved specimens and transverse sections (slide) of the neural groove stage (Figure 3-10I).

Note that in the preserved specimen the embryo has assumed an elongated form. The embryo, still in its jelly covering, has not yet hatched. The posterior end is more pointed than the anterior end. Find the **neural groove** along the dorsal side. Later this groove becomes a tube.

In the prepared slide of a transverse section, note the neural groove with a **neural fold** on each side of the groove. The neural groove develops in a thick ectodermal plate, the **neural plate.** Just ventral to the neural groove is the **notochord.** The mesodermal layer is now well defined. Do you notice any yolk material?

Neural Tube Stage. In a preserved specimen, note that the neural groove has now closed (Figure 3-10J). A **neural tube** has been formed by the meeting of the neural folds in the midline. Find on each side of the head two ventral ridges. **Eyes** will develop from the first of these ridges, **gills** from the second. Turn the specimen over and look for **ventral suckers** by which the tadpole may hold fast to an object. Look for the **oral plate** at the anterior end and the **anus** at the posterior end (Figure 3-10K).

Now examine a transverse section through the embryo at this stage. Note the neural tube, notochord, and gastrocoel (archenteron). Observe the condition of the mesoderm. Masses of epidermal mesoderm lateral to the notochord give rise to dermis, muscles, and skeleton; lateral to them is the intermediate mesoderm, which forms most of the excretory system; and a thin hypomeric layer surrounds the yolk mass, which is destined to split and give rise to the coelomic cavity.

Tadpole Stages and Metamorphosis. Well before hatching, the embryo begins to move within its jelly layers by means of cilia on its epidermis. Such movements become more pronounced as hatching approaches. Hatching occurs when the embryo frees itself from its protective jelly membranes; the embryo is now termed a tadpole.

Examine specimens of tadpoles at different stages of development, noting changes in external body form as development proceeds.

The rapid transformation from larva to adult frog is called **metamorphosis,** a process in which larval tissues are destroyed and replaced by cells destined to become adult tissues and organs. If living tadpoles are available, use a dissecting microscope to examine the thin and nearly transparent skin of the flattened tail to see the flow of blood through capillaries.

A few days after hatching, the oval **mouth** breaks through into the archenteron. The intestine develops rapidly into a coil visible through the ventral body wall, and the tadpole, which up to this time has been subsisting on yolk stores, begins feeding. Three pairs of fingerlike **external gills** grow rapidly after hatching (Figure 3-10L). **Gill slits** appear in the pharyngeal wall soon after the mouth forms. Somewhat later four pairs of **internal gills** replace the external gills. As the external gills disappear, folds of skin, the **opercula** (sing. **operculum**), arise on both sides of the head and grow posteriorly to cover the internal gills. Eventually the two opercula fuse ventrally and on the right to form a chamber for the gills. On the left side a small opening remains—the **spiracle.** The gills are now ventilated by water passing into the mouth, through the gill slits and over the gill filaments, and then out of the gill chamber through the spiracle.

As metamorphosis proceeds, the horseshoe-shaped ventral **sucker,** which the tadpole uses to adhere to vegetation, divides and degenerates. Hindlimbs appear as buds, become jointed, and develop toes. The forelimbs follow, the left forelimb passing out through the spiracle and the right forelimb pushing out through the wall of the operculum. The tail is resorbed, the intestine shortens, and other changes in the mouth and jaws follow as the herbivorous tadpole develops into a carnivorous frog. With the growth of lungs and disappearance of gills, metamorphosis is complete. Frog deformities have been identified in many North American populations of frogs. Scientists have identified the cause of these deformities (missing legs or extra legs) as a parasitic flatworm that burrows into the embryonic limb bud just as limbs are emerging from the developing tadpole.

Drawings (Optional)

On separate paper, sketch such stages of frog development as you may wish to have for future reference.

Tissue Structure and Function

General Description of Basic Tissue Types

EXERCISE 4

Tissues Combined into Organs

A **tissue** is an aggregation of cells and cell products of similar structure and embryonic origin that perform a common function. Tissues represent specializations of the properties that all protoplasm possesses, such as irritability, contractility, conductivity, absorption, and excretion. The study of tissues, especially their structure and arrangement, is called **histology.**

However complex an animal may be, its cells fall into one of four major groups of tissues. These basic tissues are named **epithelial** tissue, **connective** tissue, **muscle** tissue, and **nervous** tissue.

An **organ** is an aggregation of tissues organized into a larger functional unit, such as heart or kidney. Organs work together as functional units called **systems.**

The following study is made on vertebrate tissue, but invertebrate tissues are built similarly and may be substituted at the discretion of the instructor.

Read the description of the tissues (pp. 47–51) to familiarize yourself with the general types, their functions, and where they are found. As you do, look at slides illustrating the various tissues and familiarize yourself with their appearance. Later, because tissues are usually found working together with other tissues in organs, you will study sections through certain organs, each of which will contain several types of tissues.

Drawings

On pp. 59–60, you will find places to sketch the various types of tissues that you study as assigned by your instructor. A number of the tissues can be identified by studying sections of skin, intestine, artery, nerve, and trachea (pp. 54–58). You will find the rest on special slides.

Draw only tissues that you actually see. Do not copy photographs or drawings. Where possible, show the shape of the cells and size and location of the nuclei. Indicate under each drawing (1) in what organ the tissue was seen; (2) from what animal the tissue was taken, if that information is indicated on the slide; and (3) the magnification of the specimen.

General Description of Basic Tissue Types

Epithelial Tissue

An epithelium is a sheetlike layer of cells with close cell-to-cell contact that covers surfaces and lines cavities. Its chief function is protection, but the cells are also variously specialized for secretion, excretion, absorption, lubrication, and sensory perception. Lacking its own blood supply, it receives its nourishment from the blood supply of underlying connective tissue. See Figures 4-1 through 4-6 for examples of epithelial tissue.

Connective Tissue

Connective tissues (Figures 4-7 through 4-13) are tissues of mesodermal origin that provide structural and metabolic support for the body. **Loose connective tissue** serves as a sort of "fabric" that surrounds specialized cells, underlies epithelial tissues, and contributes in many ways to all other tissues and organs in the body. Examples are areolar and adipose tissue. **Dense connective tissue** serves major supportive functions, such as bones, sheaths, ligaments, tendons, and cartilage. Despite the diversity of connective tissues, all are composed of cells and extracellular fibers embedded in a structureless ground substance (also called matrix).

Muscle Tissue

Muscle cells (Figures 4-14 through 4-16) are highly specialized contractile cells called **fibers.** The muscle fibers of both vertebrates and invertebrates are of two basic kinds: striated muscle and smooth muscle. This classification is based on the presence or absence of regular cross-striations of cells. In vertebrates, striated muscle is subdivided into skeletal and cardiac muscle.

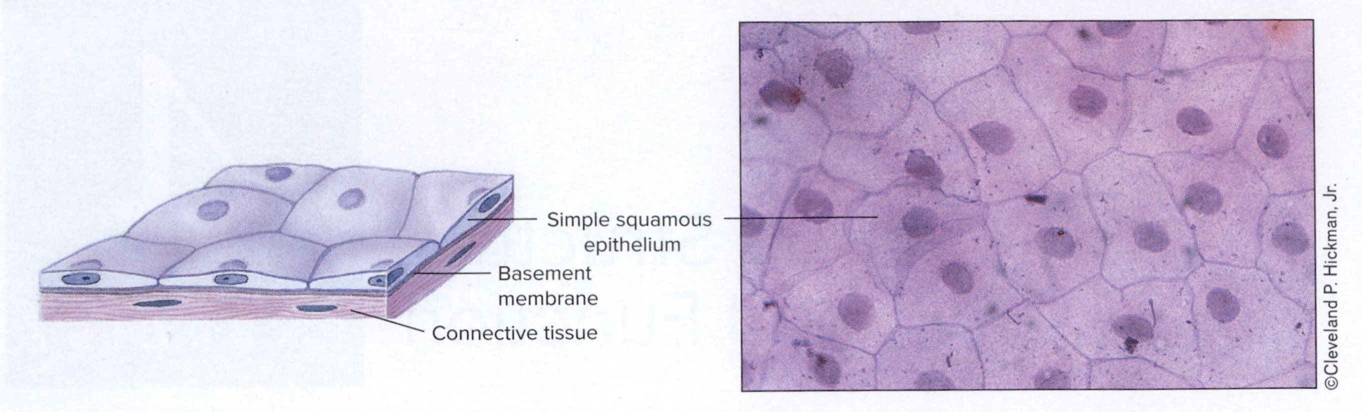

Figure 4-1

Simple squamous epithelium is composed of a single layer of flattened, irregularly shaped cells forming a continuous, pavement-like surface. It is found in areas specialized for diffusion, such as the lining of blood vessels (called endothelium), body cavity, and lungs.

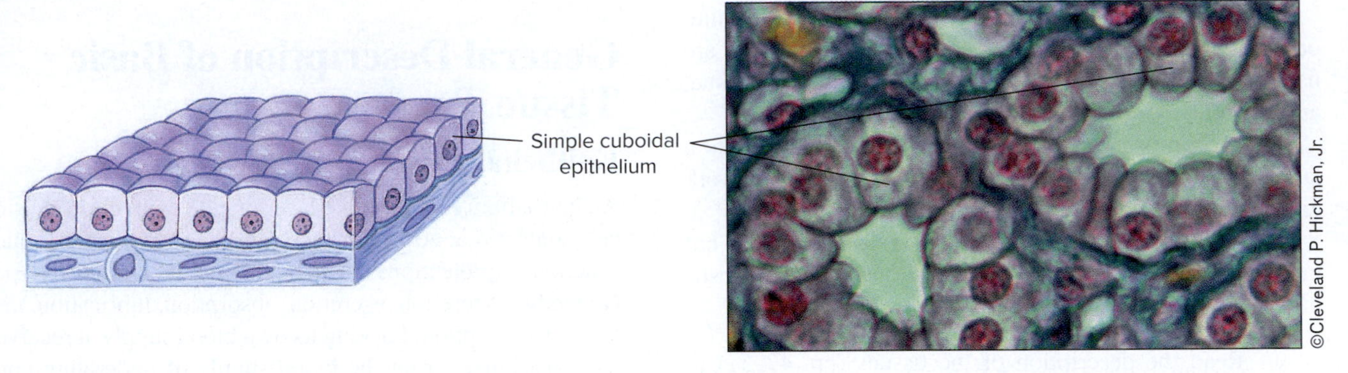

Figure 4-2

Simple cuboidal epithelium is a single layer of cubelike cells that line small ducts and tubules, such as kidney tubules, salivary glands, and mucous glands.

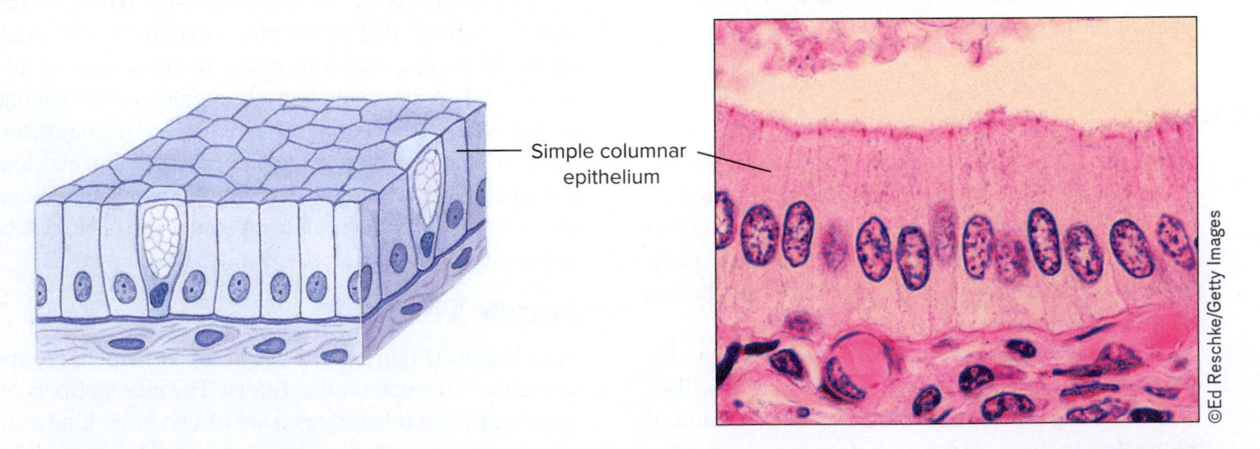

Figure 4-3

Simple columnar epithelium is similar to cuboidal, but the cells are closely packed and taller than wide. This epithelium is found on highly absorptive surfaces, such as lining of the small intestine, and on secretory surfaces, such as lining of the stomach, the oviduct, and many glands.

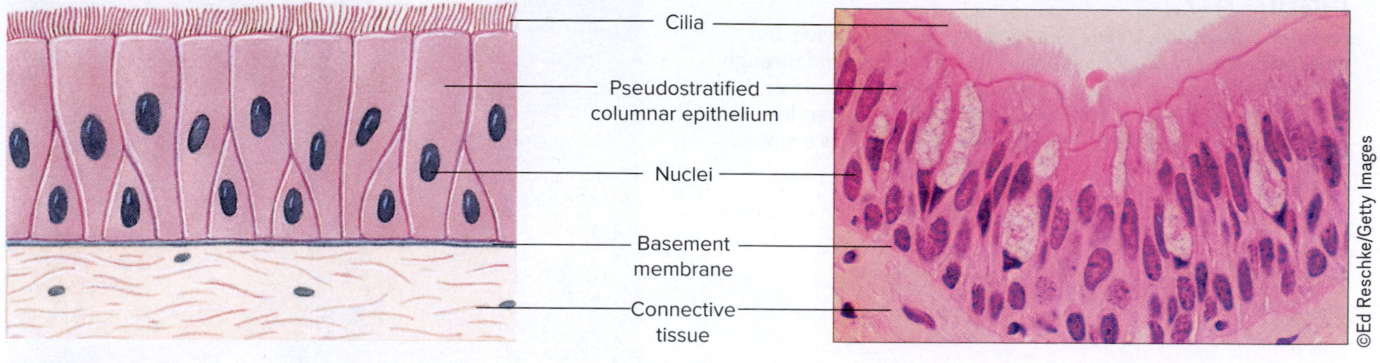

Figure 4-4

Pseudostratified columnar epithelium is actually a simple epithelium with all the cells resting on the basement membrane. They look stratified because they are not all the same height and their nuclei are located at different levels. This epithelium lines the trachea (windpipe), bronchi, and male urethra.

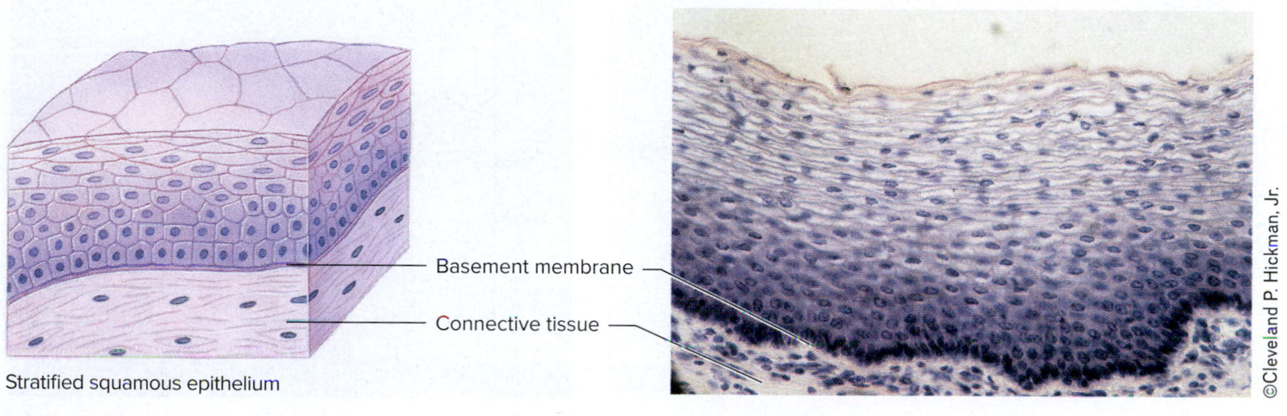

Figure 4-5

Stratified squamous epithelium consists of multiple layers of cells (hence "stratified") that are continually renewed from mitotic divisions from the basal layer of cells. It is found in areas subjected to moderate mechanical abrasion, such as the mouth, pharynx, esophagus, anal canal, and vagina. The surface of the skin is also a modified form of stratified squamous epithelium adapted to withstand constant abrasion and drying. As surface cells age and die they become keratinized, forming a tough, noncellular layer of the protein keratin.

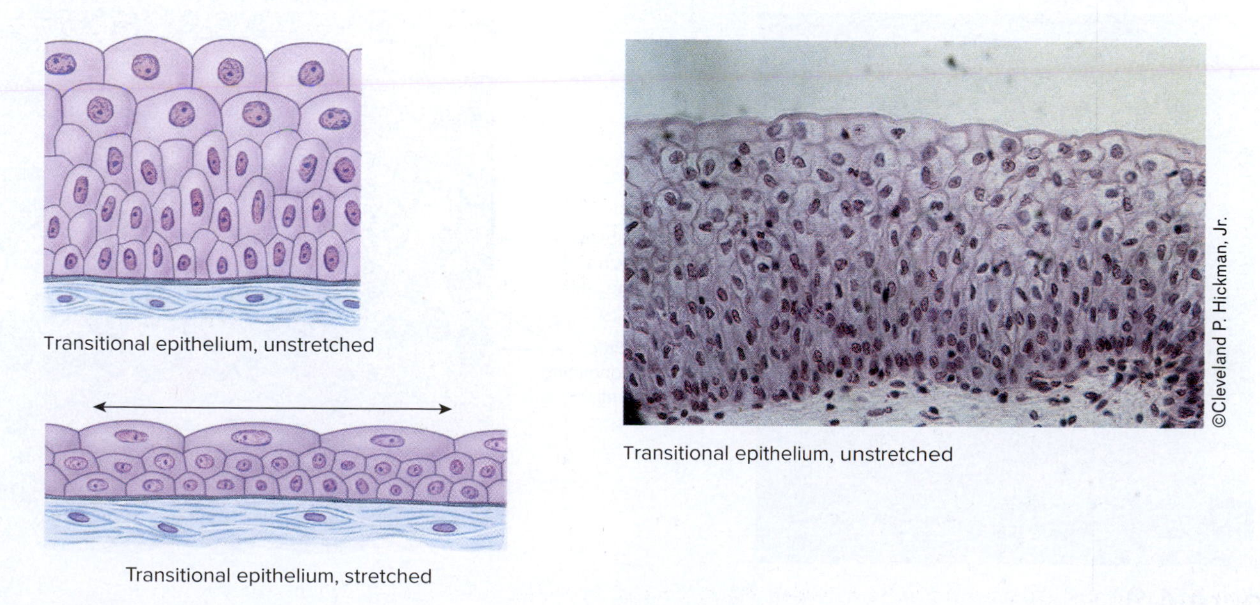

Figure 4-6

Transitional epithelium is a form of stratified epithelium that is specialized for stretching. It is found in the urinary bladder and urinary tract.

Figure 4-7

Areolar connective tissue is a loose connective tissue, the most widespread of all connective tissues. It is found throughout the body, fastening down skin, membranes, vessels, and nerves and binding muscles and other parts together. It is soft and stretchy, with a clear, jellylike matrix in which are embedded cells and three types of fibers. ×500.

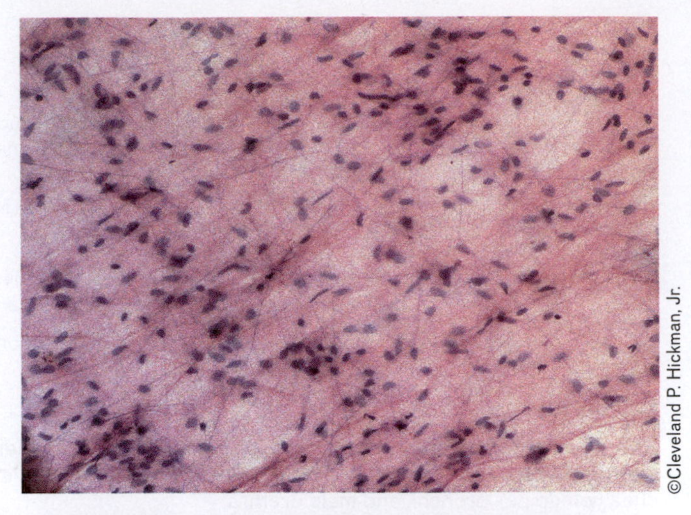

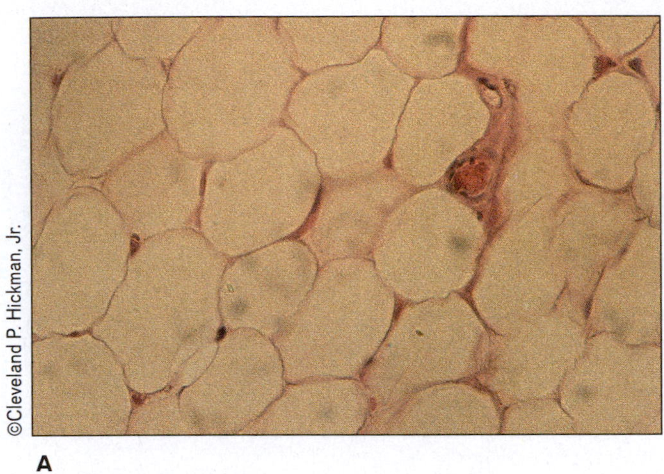

Figure 4-8

Adipose (fat) connective tissue is specialized for lipid storage. **A,** Each distended cell of white adipose tissue contains a single lipid droplet composed mostly of triglycerides. Cytoplasm and nucleus are pushed to one side. Light micrograph, ×100. **B,** Scanning electron micrograph of aggregations of human fat cells surrounded by a thin supportive network of fibers. SEM, about ×175.

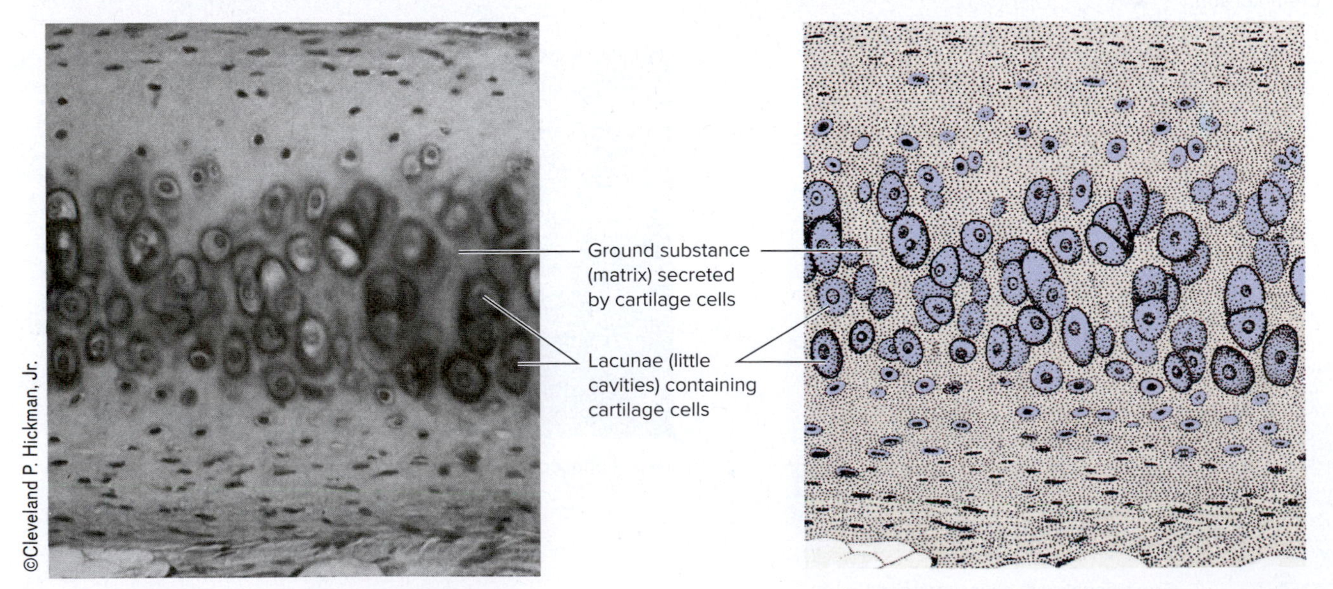

Figure 4-9

Cartilage is a type of dense connective tissue. The photomicrograph (left) with interpretive drawing (right) shows hyaline cartilage, the most common type of cartilage. It is found on the ends of long bones, and in the nose, trachea, and other places. Its ground substance, secreted by the cartilage cells, is firm but flexible.

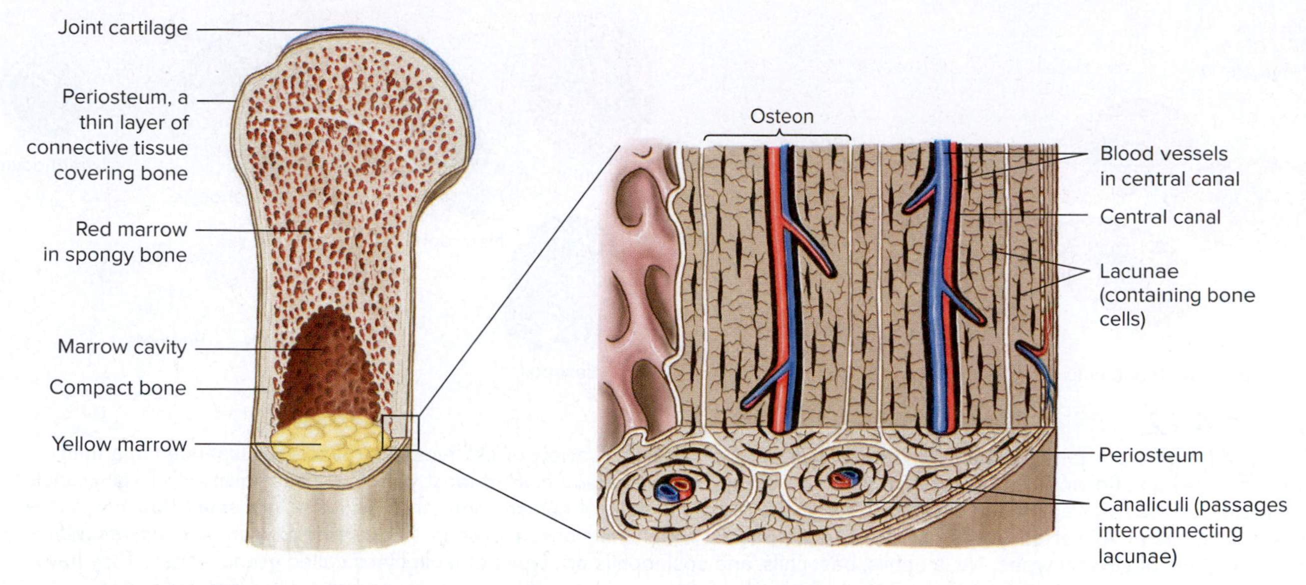

Figure 4-10

Bone is the most specialized of supportive connective tissues. In addition to its primary supportive role, bone protects vital organs with its bony framework, and it forms red blood corpuscles and most white corpuscles. Shown is the appearance of spongy and compact bone. Spongy bone consists of an interlacing framework of bony tissue with spaces filled with bone marrow. Spongy bone may develop into compact bone by additional deposition of bone matrix.

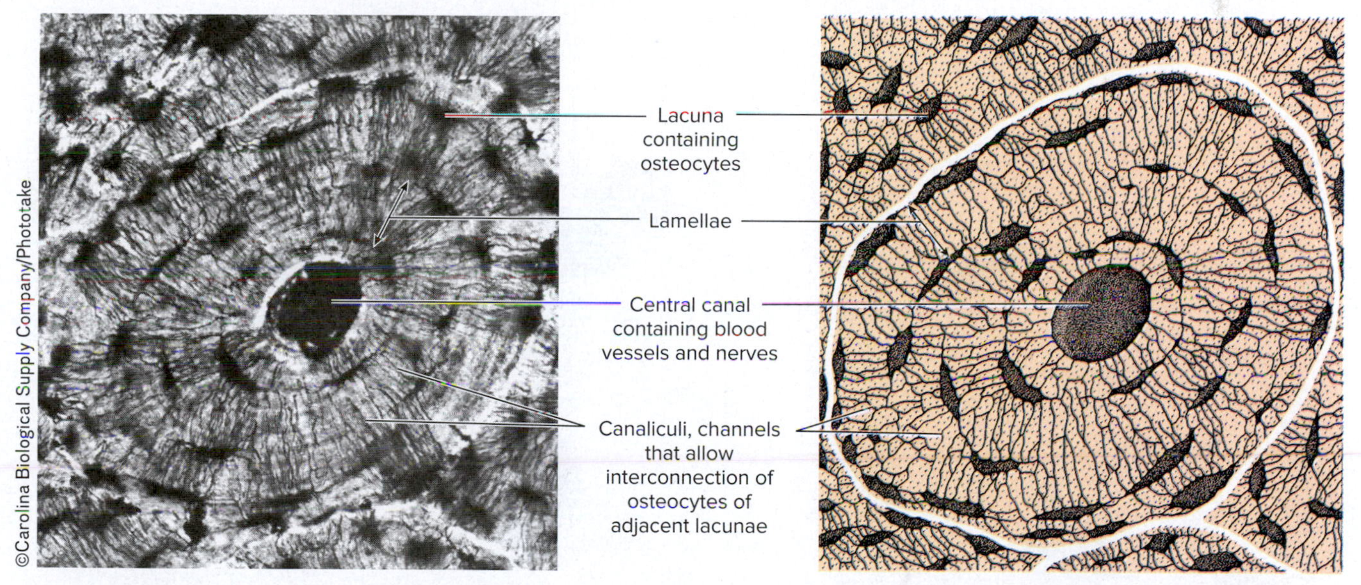

Figure 4-11

Section of compact bone, showing the cross section of a single osteon, the organizational unit of bone. Osteons are compact cylinders that are bundled and cemented together to form bone. Within each osteon are the bone-forming cells, osteoblasts, which become osteocytes.

Nervous Tissue

Nervous tissue is specialized for reception of stimuli (perception) and for conduction of nervous impulses to structures that are to act on the impulses. Nervous tissue is divided into central nervous system, consisting of brain and spinal cord, and peripheral nervous system, consisting of nerve cells and nerve cell processes that lie outside the central nervous system.

Nervous tissue consists of two distinct cell populations: nerve cells, called **neurons,** and supportive cells, called **neuroglia,** or **glial** cells. In both central and peripheral nervous systems, glial cells surround many nerve fibers with a lipid wrapping called **myelin,** and this is enclosed by a thin outer boundary, the **neurilemma** (Figure 4-17A). In the peripheral nervous system, glial cells (called Schwann cells) that wrap around nerve processes develop with gaps, or nodes, between adjacent cells (Figure 4-17A). This arrangement greatly improves speed of impulse conduction, enabling the action potential to leap from node to node (saltatory conduction).

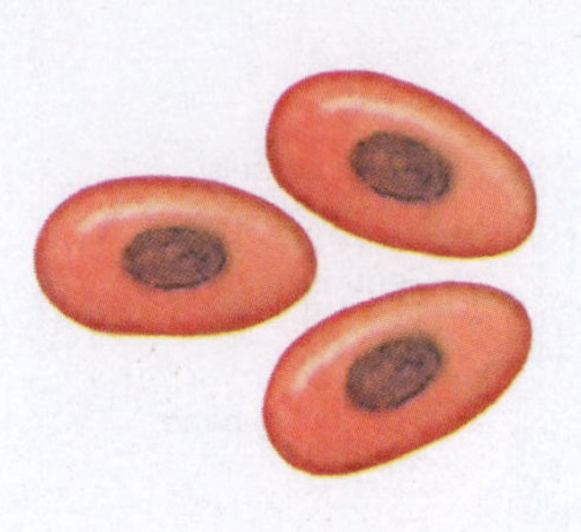

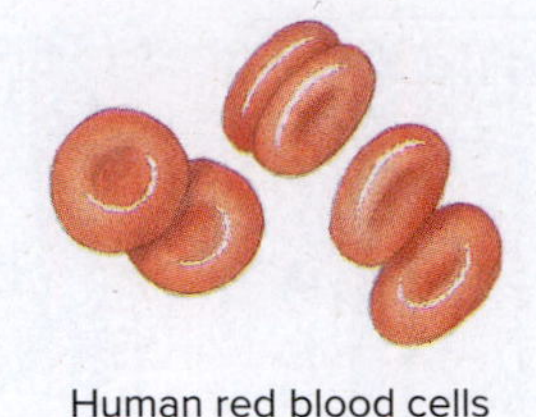

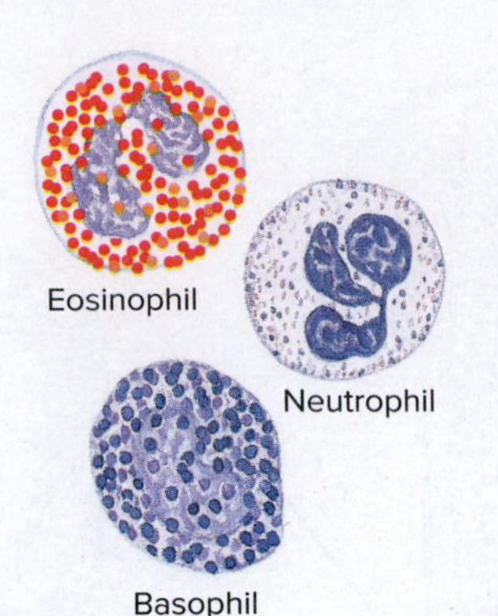

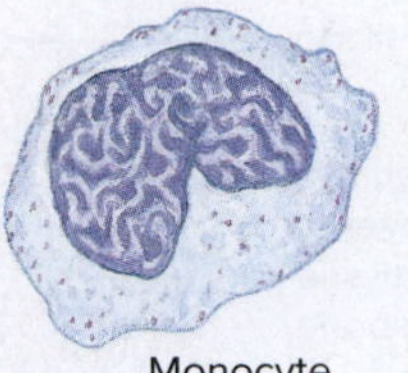

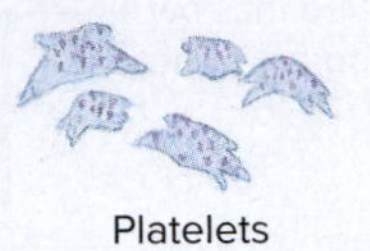

Figure 4-12

Blood and lymph are often considered to be connective tissue having a variety of cell types (corpuscles) suspended in a fluid matrix (plasma) and flowing within a system of blood vessels. The *red blood cells* of most vertebrates are oval with large granular nuclei, such as the frog red blood cells shown here. The red blood cells of humans and other mammals are small, thin, biconcave discs lacking nuclei. All vertebrate red blood cells are packed with hemoglobin, a protein that carries oxygen. *Leukocytes (white corpuscles)* are of several types. Neutrophils, basophils, and eosinophils are types of leukocytes called granulocytes. They have multilobed nuclei. Monocytes and lymphocytes lack clearly visible granules and lobed nuclei and are called agranulocytes. By ameboid movement they can pass through capillary walls and surround and ingest foreign particles and invading organisms by a process called __________. *Platelets* are fragile, disc-shaped, clot-promoting bodies present only in mammalian blood. Other vertebrates have thrombocytes, which have a similar function.

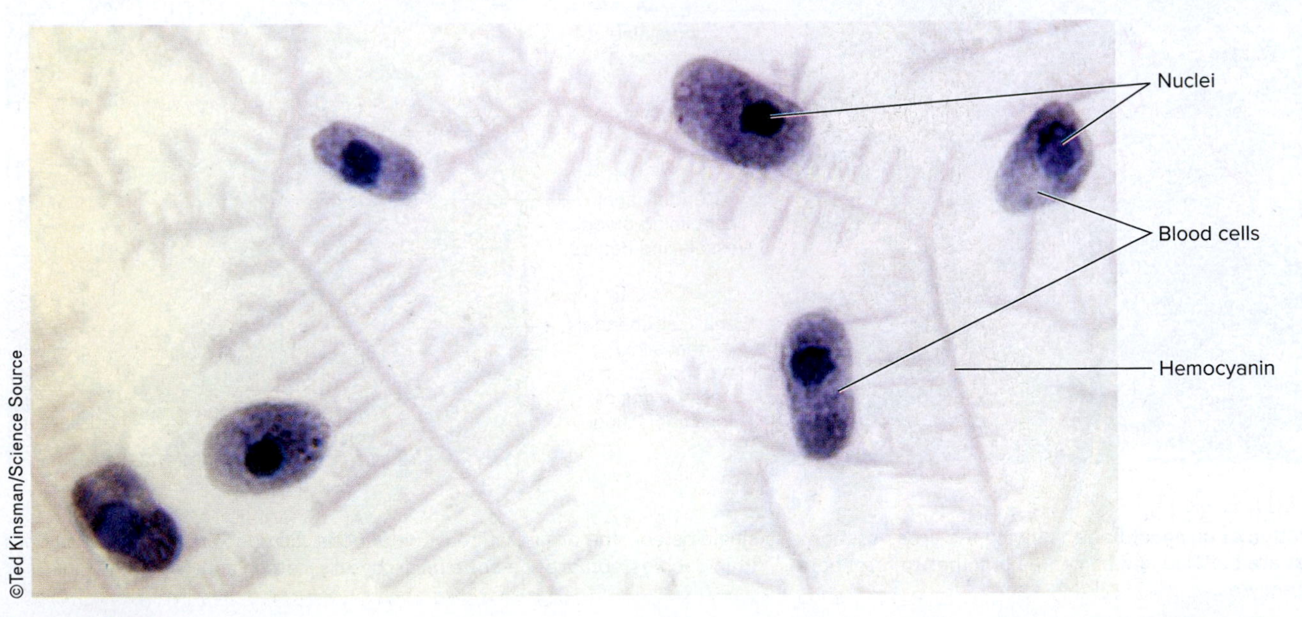

Figure 4-13

The horseshoe crab blood cells shown here also have an extracellular protein called hemocyanin. This protein contains copper and causes the blood to turn blue when in contact with oxygen. Horseshoe crab blood also has amebocyte-type cells that produce a protective gel when they contact certain bacteria. This has now become an important property used in the pharmaceutical and medical industry.

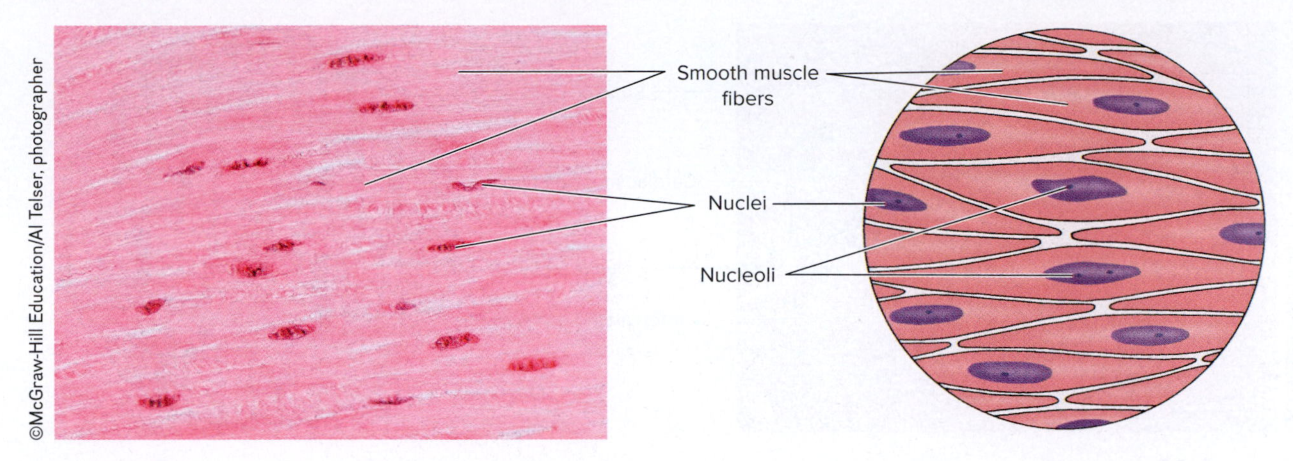

Figure 4-14

Smooth muscle is composed of long, spindle-shaped cells with centrally located nuclei. Smooth muscle is found where slow, sustained contractions are needed, such as in the digestive tract, uterus, and other visceral organs. Smooth muscle is involuntary in its action, that is, it is under unconscious control of the autonomic nervous system.

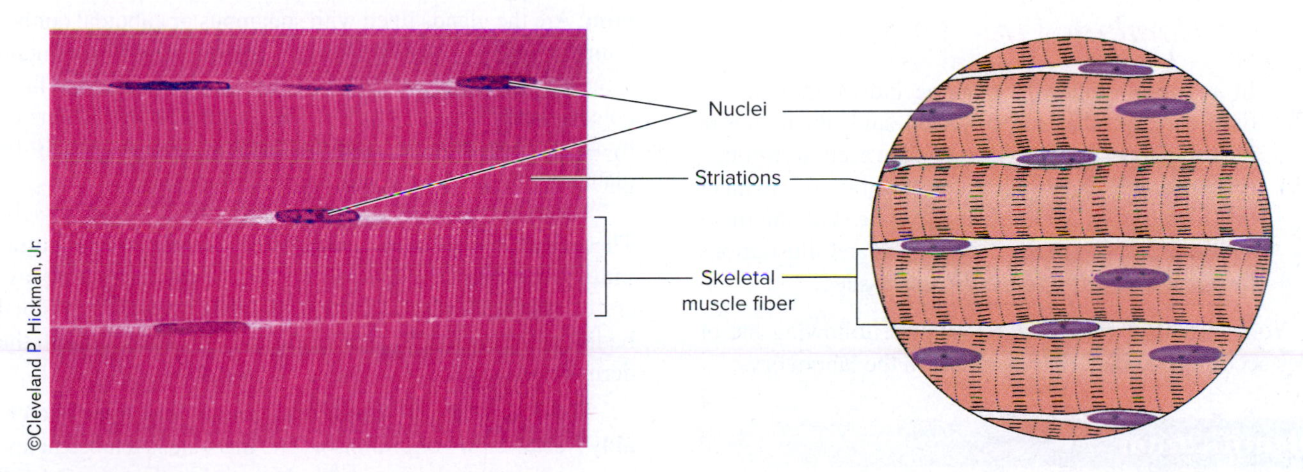

Figure 4-15

Skeletal muscle, also called striated muscle, is composed of elongate cells called fibers. Fibers are in turn composed of numerous myofibrils that extend the full length of the cell. The cross-striations, plainly visible with the microscope, are the boundaries between sarcomeres, the functional contractile units. Skeletal muscle is voluntary muscle because it is innervated and controlled by the central nervous system. Skeletal muscle contracts much more rapidly than smooth muscle but fatigues more easily and is not capable of sustained contraction.

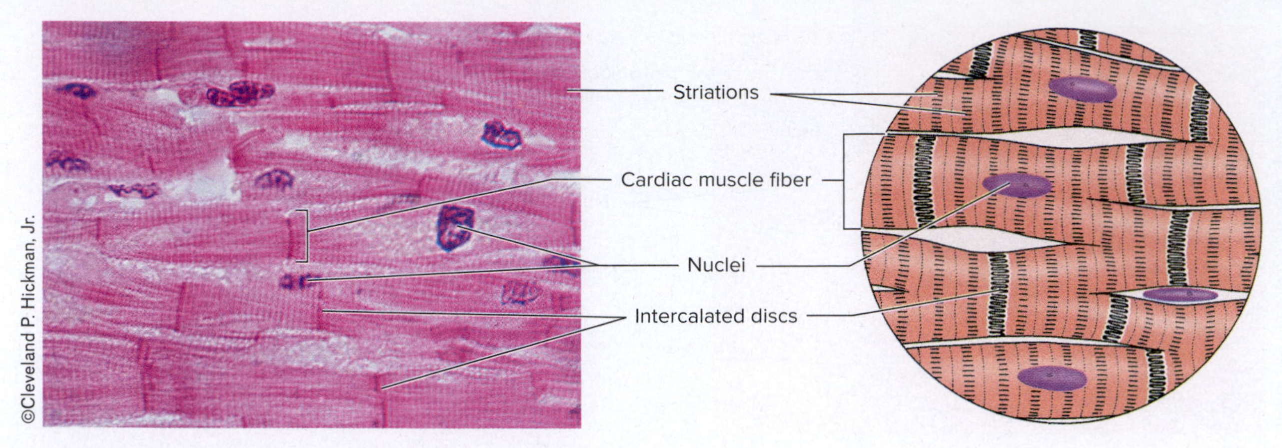

Figure 4-16

Cardiac muscle, located only in the vertebrate heart, is striated like skeletal muscle but, like smooth muscle, is involuntary. It is well adapted for rhythmic contractions. Fibers are much shorter than those of skeletal muscle, and joined end-to-end by intercalated discs. The fibers are branched, forming an interconnected network of muscle cells.

EXERCISE 4

Core Study

Tissues Combined into Organs

In addition to slides containing individual types of tissues that you have already seen, study the tissues in the following section, which contain sections through certain organs. These sections illustrate the manner in which tissues work together. While studying these slides, refer back to the descriptions and illustrations for help in identifying the various tissues.

Your instructor may wish to vary the following list of slides according to materials available in the laboratory.

Identification

Be prepared to recognize any of the various types of tissues that may later be set up by your instructor as "unknowns."

Section Through Skin of a Frog. The outer layer of the skin, the epidermis, is made up of **stratified squamous epithelium** (Figure 4-18). Note the flat surface cells that give the epithelium its name. Columnar cells at the base divide to produce new cells that push out to the surface to replace surface cells as they are worn off.

Beneath the epidermis is a thick layer of dermis, which is made up of **connective tissue** and contains glands and pigment. Connective tissue nearer the surface (spongy layer) contains loosely arranged fibers, whereas that in the deeper layer (compact layer) is much denser. Can you identify **elastic** or **collagenous fibers** in the dermis? ____________

In the spongy layer of the dermis you will find a number of mucous glands, each made up of a single layer of **epithelium.** Are the glands lined with squamous or cuboidal epithelium? ____________ The section of skin may also contain some very large **poison glands.** What kind of epithelium lines poison glands? ____________ The glands open to the outside by small ducts, but, because the ducts are narrower than the glands, not all of the cut sections will include ducts.

Scattered through the dermis are small blood vessels. These may be capillaries, made up of a single layer of **squamous epithelium,** or small arteries or veins, containing layers of **smooth muscle.** Darkly stained, irregularly shaped bodies at the base of the epidermis and scattered through the dermis are pigment cells called **chromatophores.**

Cross Section of Amphibian Small Intestine (Preferably *Necturus* or *Amphiuma*). The intestine is a tube enclosing a cavity called the lumen. The lumen of the intestine is lined with a mucous membrane that lies in many folds (Figure 4-19). The mucous membrane is made up of **columnar epithelium** in which the nuclei are located near the base of the tall cells. Is the columnar epithelium simple or pseudostratified? ________ Two cell types can be identified: **columnar cells** and **goblet cells.** Columnar cells contain digestive enzymes that are added to the intestinal contents as the cells are continually eroded away. New columnar cells are constantly being formed at the base of the mucosal folds by stem cell mitosis, then migrate upward to replace cells shed at the surface. Goblet cells, each shaped like a chalice permanently open to the lumen, continually secrete protective mucus onto the intestinal surface.

Surrounding the mucous membrane and largely conforming to its contours is a submucosal layer of **areolar**

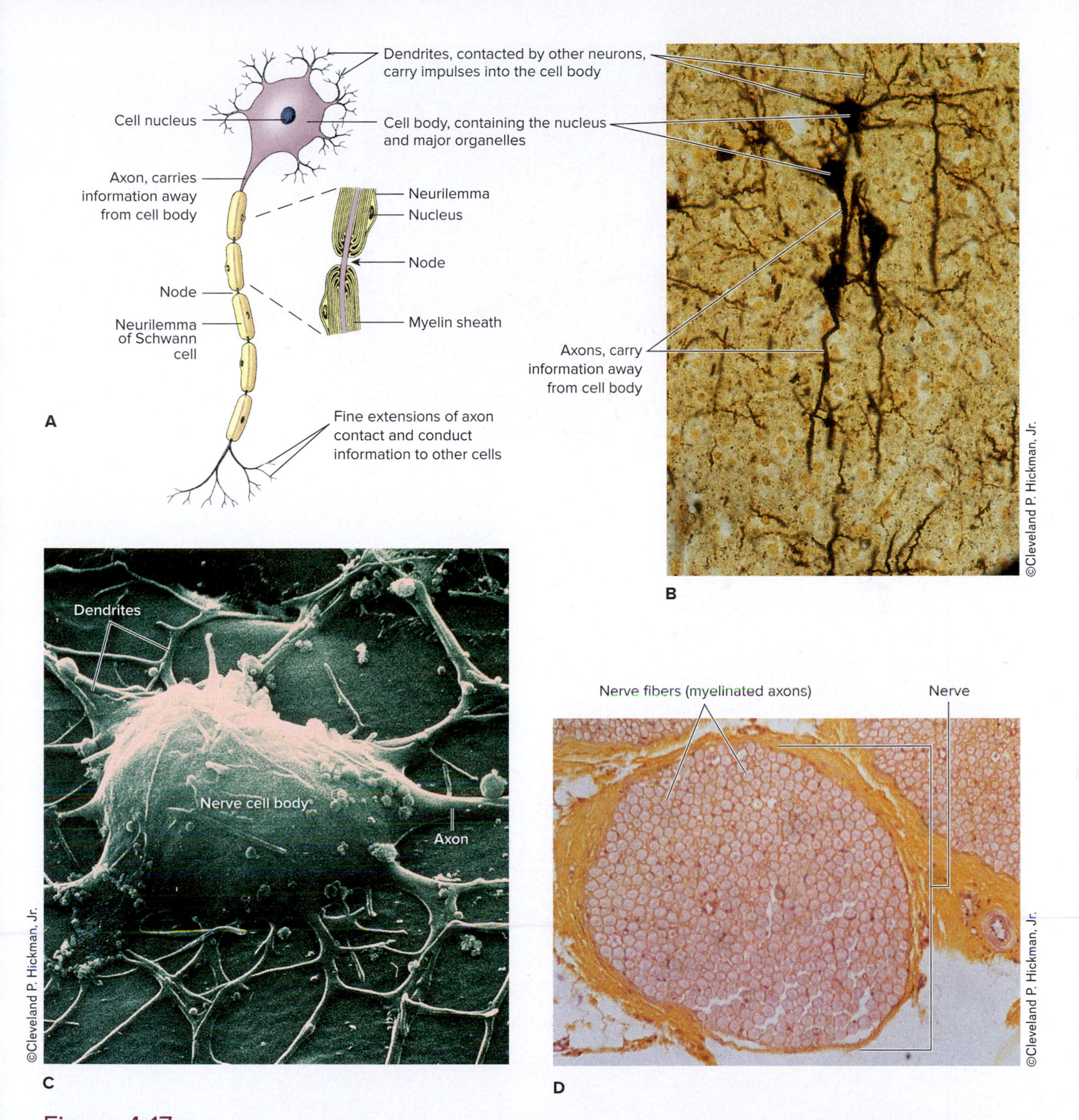

Figure 4-17

A, Diagram of a motor neuron with myelinated axon. The nodal area between insulating Schwann cells is shown enlarged. **B,** Neurons and fibers of the cerebrum, ×150. **C,** Scanning electron micrograph of a cell culture showing a large nerve cell surrounded by numerous nerve cell processes (dendrites and axons). **D,** Cross section of a nerve, showing the appearance of numerous nerve fibers (myelinated axons) in cut section, ×200.

connective tissue, containing mostly collagenous fibers. There are many blood vessels in this layer.

Outside the submucosa are layers of both circular and longitudinal **smooth muscle** responsible for the segmentation and peristaltic movements that mix the food and move it through the gut. In the circular layer, long, spindle-shaped cells can be seen fitting closely together. In the longitudinal layer, only cut ends of the fibers can be seen.

The outermost layer of the intestine is a thin layer of **squamous epithelium** that is a part of the peritoneum that

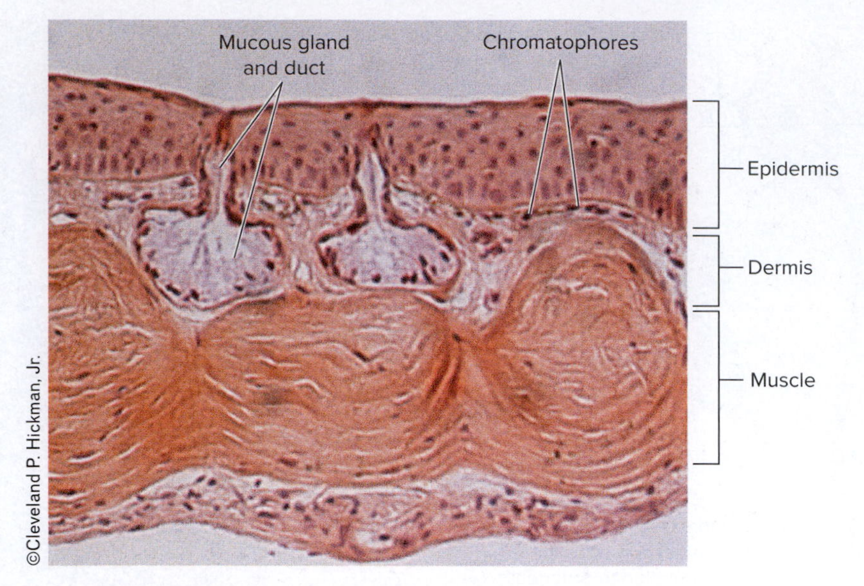

Figure 4-18
Section through the skin of a frog.

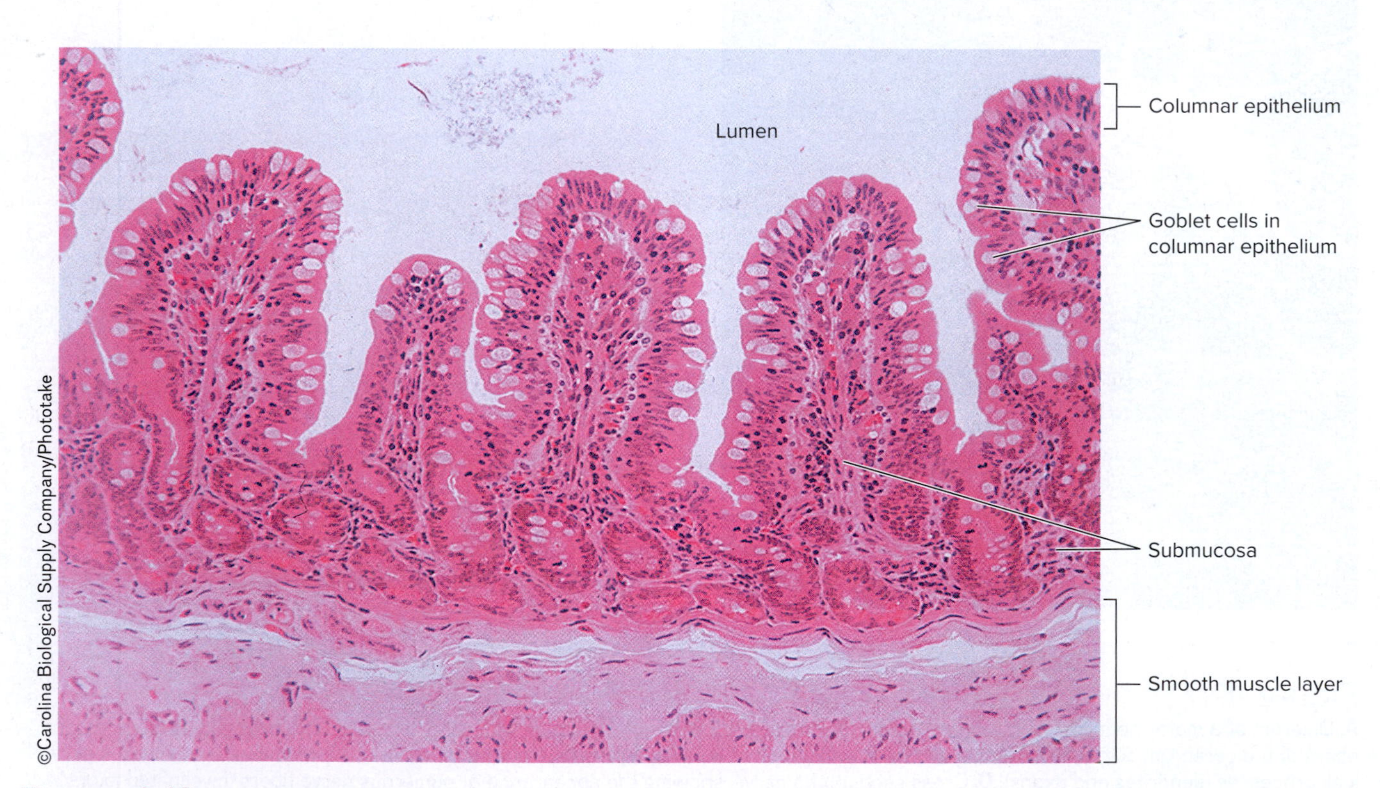

Figure 4-19
Cross section through the intestine of a monkey.

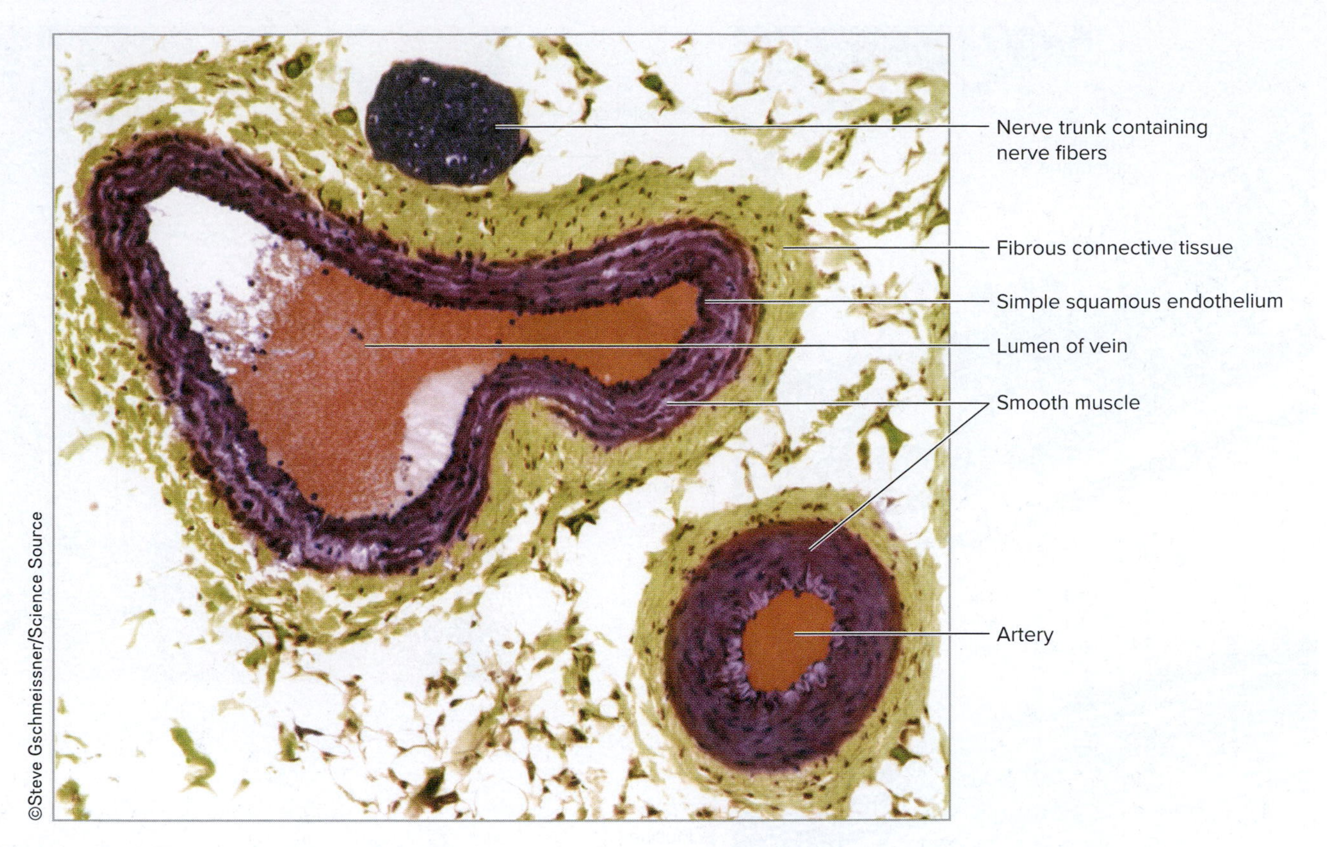

Figure 4-20
Cross section through a large artery (*center*), small vein (*left*), and nerve (*right*) containing many nerve fibers.

covers all visceral organs. The cells you see are in cut section, so they appear very thin.

Section through Artery, Vein, and Nerve. Most blood vessels are muscular organs. An artery will have a smaller diameter and thicker walls than will the vein that accompanies it, but otherwise their structures are similar. On your slide the blood vessels are probably collapsed so that the artery may appear flattened or ovoid in shape (Figure 4-20), and the thinner walls of the vein may be thrown into folds. The innermost layer of an artery or a vein is a thin **endothelium.** Are the cells of the endothelium squamous or cuboidal? __________ Outside that is a layer of **smooth muscle** circularly arranged. Is smooth muscle voluntary or involuntary muscle? __________ There are often **elastic fibers** interspersed with the muscle. The outer layer is made up of elastic **fibrous connective tissue.**

Your slide may contain one or several sections through nerves. Each nerve trunk is made up of many **nerve fibers,** each enclosed in its myelin sheath (see Figure 4-17D). Some fibers may be cut transversely and so appear circular; others may be cut diagonally or longitudinally and so appear ovoid or long.

There will probably be some **adipose tissue** scattered through the connective tissue that holds the vessels and nerves in place. This tissue will contain large, clear cells, with the small nuclei pushed over to one side (see Figure 4-8A).

Cross Section through the Trachea. The trachea, or windpipe, is a tube that leads from the larynx to the lungs. It is supported and held open by c-shaped rings of **cartilage.** What type of cartilage do you see (refer to p. 50)? __________

The innermost layer of the trachea is a mucosal lining of ciliated **pseudostratified epithelium** containing many **goblet cells** (Figure 4-21), resting on a basement membrane of **connective tissue** containing a few fibers. Do you remember the function of these goblet cells? ____________________
____

A submucosal layer contains many mucous glands composed of **cuboidal epithelium.**

Cartilage rings in the trachea do not completely surround the trachea but are open dorsally, thus providing limited expansion and constriction of tracheal diameter. **Cartilage** is easily recognizable by the large amount of clear ground substance interspersed with little lacunae (spaces), containing cartilage cells. In the space between the cartilage bands you may find some **smooth muscle** or **fibrous connective tissue.**

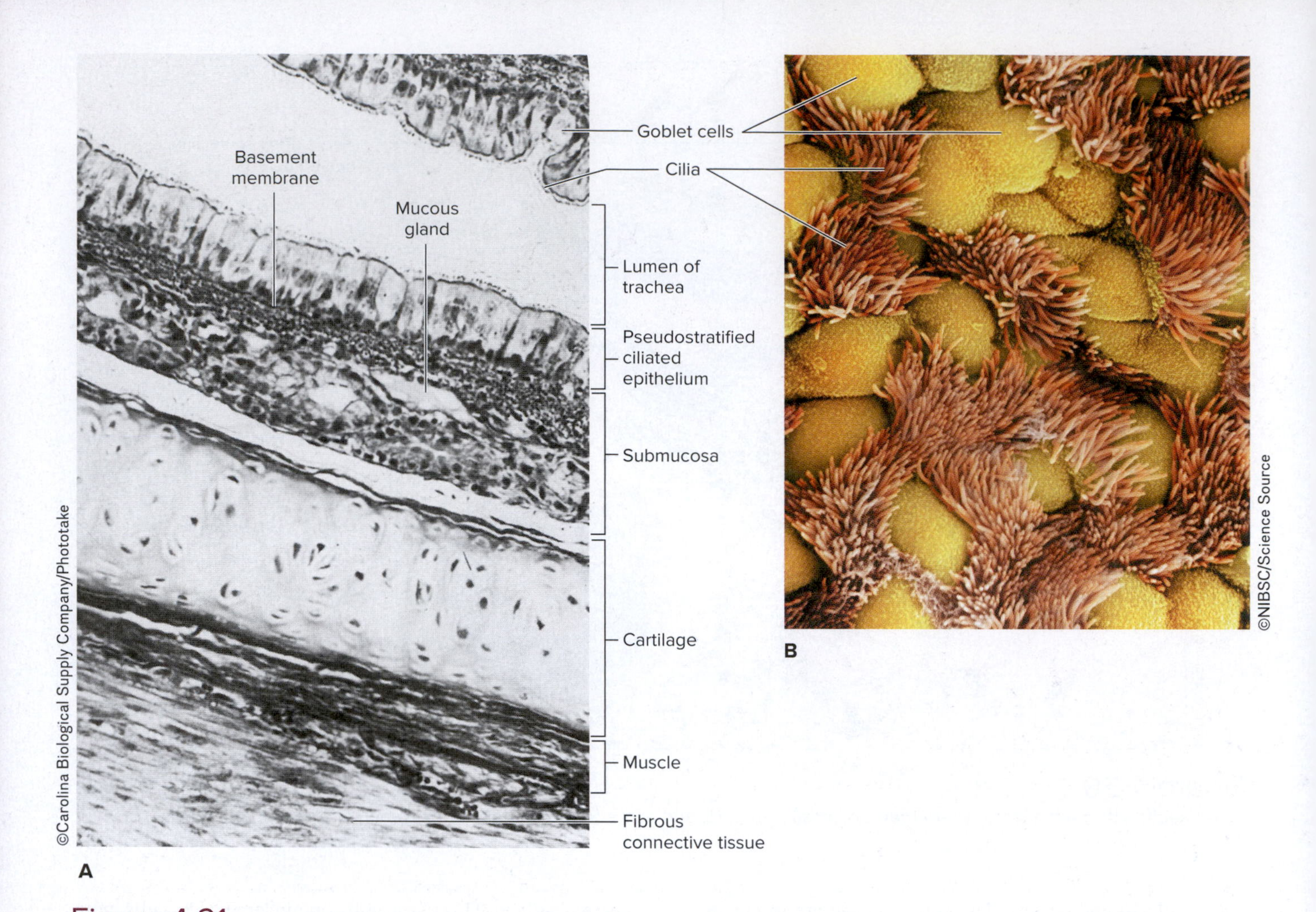

Figure 4-21

A, Photomicrograph through portion of the trachea, ×200. **B,** Scanning electron micrograph of pseudostratified epithelial surface of trachea, about ×3000. The goblet cells are carpeted with short microvilli, whereas the rest of the tracheal cells are ciliated.

Tissues

Name ____________________

Date ____________________

Section ____________________

LAB REPORT

Drawing Instructions: Where possible, show the shape of the cells and size and location of the nuclei. Indicate (1) in what organ the tissue was seen; (2) from what animal the tissue was taken; (3) the magnification of the specimen.

Simple squamous epithelium

From ____________________

Simple cuboidal epithelium

From ____________________

Simple columnar epithelium

From ____________________

Pseudostratified epithelium

From ____________________

Stratified epithelium

From ____________________

Loose connective tissue

From ____________________

Dense or fibrous connective tissue

From ____________________

LAB REPORT

Blood cells

From ___________________________

Skeletal muscle

From ___________________________

Smooth muscle

From ___________________________

Cardiac muscle

From ___________________________

Bone

From ___________________________

Cartilage

From ___________________________

PART TWO

The Diversity of Animal Life

Colony of zoanthids.

5 Ecological Relationships of Animals 63

6 Introduction to Animal Taxonomy 77

7 Unicellular Eukaryotes 89

8 The Sponges 121

9 The Radiate Animals 131

10 The Flatworms 149

11 Nematodes and Four Small Protostome Phyla 171

12 The Molluscs 183

13 The Annelids 201

14 The Chelicerate Arthropods 219

15 The Crustacean Arthropods 227

16 The Arthropods 241

17 The Echinoderms 265

18 Phylum Chordata: A Deuterostome Group 281

19 The Fishes—Lampreys, Sharks, and Bony Fishes 291

20 The Amphibians: Frogs 313

21 The Reptiles 337

22 The Birds 343

23 The Mammals 349

Ecological Relationships of Animals

EXERCISE 5A
A Study of Population Growth, with Application of the Scientific Method
The Scientific Method
Application of the Scientific Method to the Study of Populations
Experimental Procedures

EXERCISE 5B
Ecology of a Freshwater Habitat
How to Use a Taxonomic Key
Key to Common Freshwater Aquatic Invertebrates of North America

EXERCISE 5A*

A Study of Population Growth, with Application of the Scientific Method

One goal of this course is to introduce the methods scientists use to gather knowledge. In this project you will apply the "scientific method" to the problem of determining what regulates animal populations.

The Scientific Method

People, scientists included, often acquire knowledge by applying a two-stage process—**conjecture** followed by **confirmation**—although few of us think of it this way. Conjecture consists of generating a general explanation of how the world is constructed, and is often based on general observations. Confirmation tests the validity of this conjecture. We all use such a method in our everyday lives; for example, we speculate on the quality of a future music concert based on our observations of recorded music, then subsequently confirm, reject, or modify that speculation based on our experiences while attending the concert.

Scientists employ a similar method, although with a lot more rigor, in attempting to discover new facts. An idealized form of the method used by scientists is known as the **scientific method,** and is broken down into four steps.

1. **Observation.** Observations may be based on direct examination of a system, may be based on something we read, or may even be the result of discussions with others about a process or concept. Such observations frequently stimulate questions about why species exhibit certain traits, why internal organs interact the way they do, what the advantages of a particular body shape might be, or what the role of certain genes in a particular process is.

*Exercise written by James C. Munger, Department of Biology, Boise State University, Boise, Idaho, and Richard S. Inouye, Department of Biological Sciences, Idaho State University, Pocatello, Idaho.

2. **Hypothesis formulation.** Formulating a hypothesis is like saying "Let's suppose… ." Its objective is to explain, by induction, the observation. Typically several alternative hypotheses are formulated, each a possible and reasonable explanation for the observation. Weeding out these hypotheses is the role of the third and fourth steps.
3. **Prediction.** Predictions are deduced from hypotheses and often are based on some knowledge of the organisms or concepts being studied. They follow the form of "If hypothesis A is true, then I predict the following pattern." Predictions must be generated such that one set of hypotheses predicts one result, but alternative hypotheses predict another result. A prediction is worthless if it can be made for all of the hypotheses under consideration. Testing whether a prediction holds true is the way in which one or more hypotheses can be rejected, thus reducing the number of hypotheses still under consideration.
4. **Testing of predictions.** The final step is to design a test so that a prediction, if incorrect, can confidently be rejected. Tests are accomplished using observations or experimental manipulations. The confidence with which we can make such a rejection is quantified by the use of inferential statistics, which we shall not discuss. However, note that our confidence in a result increases with the use of (1) treatments known as **controls,** in which all variables except for the one manipulated are held the same, and (2) several replicates of

each treatment, to ensure that the observed result was due to conditions of the treatment and not simply due to variation among individuals. With skill (and perhaps some luck), all but one hypothesis will have been rejected. The unrejected hypothesis, however, is not proven to be true. Hypotheses can never be fully accepted; they can only be rejected (what lonely lives they must lead).

The next step is to repeat this process. With the results in hand from the tests of previous predictions, it is possible to fine-tune the hypothesis, then set about testing the new one. Following is an example of this repetitive process.

Initial observation:	Roommate breaks dish.
Generalization (hypothesis):	Roommate breaks everything.
Prediction:	Roommate will wreck borrowed car.
Test/observation:	Roommate didn't wreck car.
New generalization/ hypothesis:	Roommate breaks only dishes.
Prediction:	Roommate won't break borrowed camera.
Test:	Roommate does break camera.
New generalization/ hypothesis:	Roommate breaks small objects.

And so on ...

The unrejected hypothesis at each stage is our best guess as to how the world works. If a hypothesis withstands repeated tests and has great explanatory value, it may be elevated to the level of a scientific theory. Note that a scientific theory is not an untested hypothesis but is instead as close as scientists will come to calling a hypothesis proven. The theory of evolution is an example.

Whether a hypothesis is accepted or rejected, the observations made while testing the hypothesis frequently lead to more hypotheses, more predictions, more tests, more observations, and so on. It is often said that scientific investigation raises more questions than it answers; it is this aspect of science that many people find most exciting.

Application of the Scientific Method to the Study of Populations

If you were a scientist, you would make your own observations, formulate your own hypotheses, derive your own predictions, and perform your own tests of those predictions. A classroom situation, however, involves certain constraints, as you will see. We will apply the scientific method to a study of population growth.

Step 1—Observation

In 1798 a British economist, Thomas Malthus, published an essay in which he observed that populations do not grow indefinitely but often tend to stay at relatively stable numbers. We can make similar observations: if we look around us, we do not see populations of organisms forever growing—instead, they are relatively stable.

Step 2—Hypothesis Formulation

Why might this occur? Again we can look to Malthus, this time for one possible explanation (a hypothesis) as to why populations do not grow indefinitely. Malthus reasoned that if a population had unlimited resources, it would grow **exponentially** to infinite size. However, since no population grows to infinite size, resources must be limiting. He therefore hypothesized that a limitation of resources is the cause of limited population growth. We can depict these two possible conditions graphically. If resources are unlimited, the population should grow exponentially.

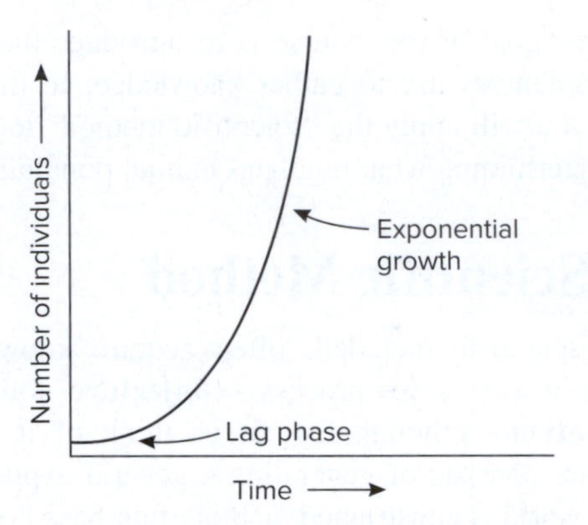

If resources are limited but in relatively constant supply, the population will experience **logistic growth,** rapidly growing at first and then eventually reaching an equilibrium, known as the **carrying capacity**.

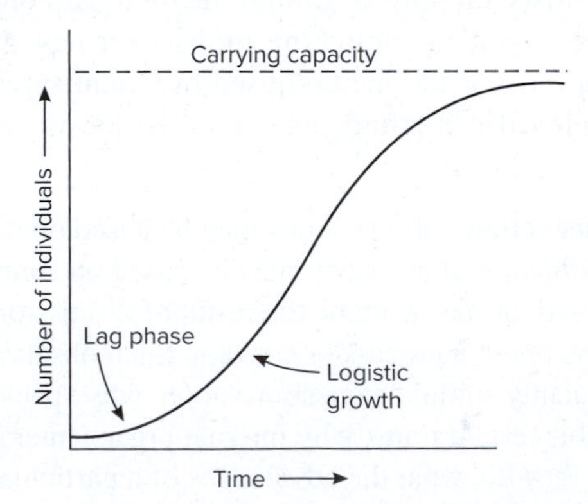

A third type of population growth occurs when resources are consumed but not renewed. This sort of population growth would occur in a test tube bacterial culture in which the population increased until all the nutrients were consumed, and then crashes.

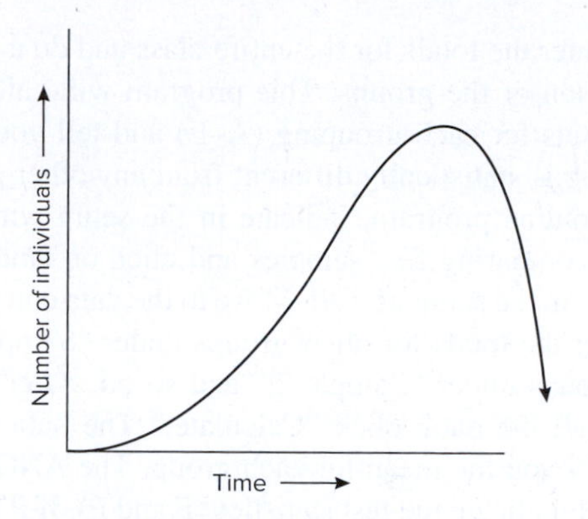

Now we can create our first hypothesis regarding what limits population growth:

H_1: Limited food limits population size. However, there are a number of factors that could limit population size, such as predation, climate, disease, limited nest sites, and intraspecific aggression, including cannibalism. Can you think of others? We can cast these factors as the following list of hypotheses:

H_1: Limited food regulates population size.
H_2: Predation regulates population size.
H_3: Disease regulates population size.
H_4: Limited nest sites regulate population size.
H_5: Intraspecific strife regulates population size.

Note that this list is not exhaustive. Also note that, in this case, the hypotheses are not mutually exclusive; that is, more than one may be true in a particular population. This is especially true if we consider a wide range of species.

Step 3—Prediction Derivation

What predictions logically follow from these hypotheses? The best predictions are those that (1) allow the investigator to decide between two or more competing hypotheses and (2) are straightforward to test.

For the purposes of discussion, we will focus on our hypotheses as they apply to the setup to be used in this exercise: a population of flour beetles eating flour, living in flour, laying eggs in flour, all contained in a small jar. In this system, we can discount the possibility of one hypothesis, predation, because we will not allow predators into the system.

A prediction that follows from H_1 is that, if we limit food availability, we expect a smaller population to result (less flour, fewer beetles; more flour, more beetles). But do the other hypotheses make different predictions? If nest sites are limited, then adding more food will increase the availability of nest sites, giving the same prediction: more flour, more beetles. If intraspecific aggression (e.g., cannibalism) is limiting, what will adding more flour do? It will give the beetles more room to hide, meaning fewer encounters; and more to eat, meaning less hunger; both mean less cannibalism. Again, more flour, more beetles. And if disease is limiting, more flour means fewer encounters among beetles and less disease transmission. Again, more flour, more beetles.

However, what if we were to vary the amount of food available while holding constant the total volume available for the beetles to roam? If H_1 were true, more food would lead to more beetles. But if H_3, H_4, or H_5 (but not H_1) were true, then more food would have no effect on beetle numbers, as long as the total volume were constant. Thus, this prediction allows us to distinguish among competing hypotheses.

Step 4—Test of Predictions

Next we need to create an experiment that will allow us to vary food without varying volume. One way to accomplish this is to put various amounts of food into jars, then add an inert filler (such as vermiculite) to maintain constant volume.

Step 5—Repeat the Process

When we look at the results from our experiment, we can consider what modifications to make to our hypotheses and what new predictions we could use to test our new hypothesis.

Experimental Procedures

We will start cultures of *Tribolium confusum* (a flour beetle) with the same initial population size but varying amounts of food and varying amounts of space. Near the end of the term, we will count the number of larvae, pupae, and adults in each container and compare age distribution and resulting densities.

T. confusum develops from egg to adult in about 28 days as follows: egg stage, 5 days; larval stages, 17 days; pupal stage, 6 days (Figure 5-1). The average life span of adult beetles is roughly 200 days.

Work in groups of four students each. Each group should prepare the following:

A. One low-density, high-food jar, containing 50 g of resource (95% whole wheat flour: 5% brewer's yeast)

B. One medium-density, medium-food jar, containing 10 g of resource

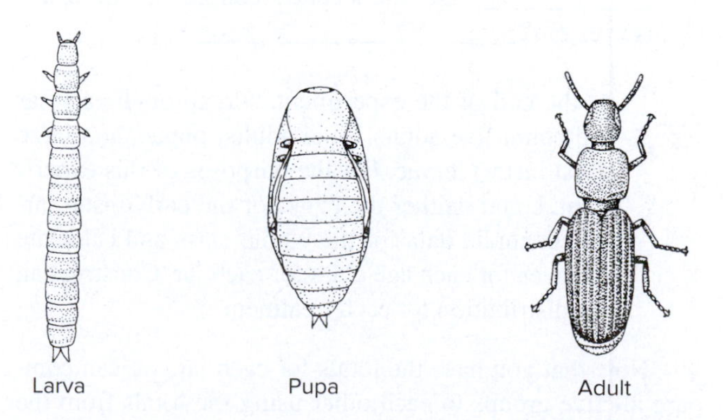

Figure 5-1
Appearance of *Tribolium confusum* at larval, pupal, and adult stages.

C. One high-density, low-food jar, containing 3 g of resource

D. One low-density, medium-food jar, containing 10 g of resource with filler (such as vermiculite screened to standardize the size) added to bring the total volume to that of jar A

E. One low-density, low-food jar, containing 3 g of resource with filler added to bring the total volume to that of jar A

Next, sort through the culture that has been provided to you. Be careful not to damage the animals. Use a fine brush or small spatula to push them around. Each group should sort and count 250 healthy-looking adults and put 50 into each of the half-pint Mason canning jars. We will assume that, within each group of 50, there are plenty of both males and females. Cover the jar with the precut wire mesh (window screen), then a piece of paper towel; then screw on the top ring. Place the jars in a cabinet with a light (for warmth—about 30° C) and an open container of water (for humidity).

You and your lab instructor may decide to try other variations on this experimental setup. For example, you might try to see if temperature is important. How would you do this? __________

Why don't we use just one jar at each density for the whole class? If there were only one jar, it would be difficult to say, because of biological variation and experimental error, that the results from that jar are representative of any jars the class might start with the same density. For example, what if only the jar your class had used for its medium-density culture had previously contained a toxic chemical? That could invalidate your results. However, if each group of 4 students prepared one jar at each density, then a class of 24 students would have a total of six jars at the same density (known as **replicates**). If all six gave approximately the same result, then we could have substantial confidence that those jars were representative of all jars at that density.

During the term, make occasional observations on beetle behavior and record those observations in your notebook. For example, do the beetles live on top of the flour or within it?_____________ Do the beetles congregate or space themselves out? ______________________

At the end of the experiment, sort through each jar and count live adults, dead adults, pupae, and large (final instar) larvae. For the purposes of this experiment, count neither the eggs nor the early instar larvae. Compile data for the whole class and calculate averages for each age class for each jar. Construct an age distribution for each treatment.

Now that you have the totals for each jar, we can compare all five groups to each other using the totals from the entire class. When we compare two or more groups statistically we use an analysis of variance (ANOVA). An excellent online program (vassarstats.net/anova1u.html) will allow you to enter the totals for the entire class and do a statistical comparison of the groups. This program will calculate the mean totals for each grouping (A–E) and tell you whether any group is statistically different from any other group. To use the online program, indicate in the setup window that you are comparing five samples and click on "independent samples" in the setup as well. Move to the data entry window and enter the totals for all A groups under "Sample 1" and all B groups under "Sample 2," and so on. After you have entered all the data, click "Calculate." The data summary will show you the mean for each group. The ANOVA summary will indicate the test statistics (F and P). If $P < 0.05$, at least one of your groups is significantly different from one of the others. Look at the summary under the Tukey test to find out which group means are different from other group means. If $P > 0.05$, then none of your groups are significantly different from one another and there will be no Tukey summary. (For a longer discussion on ANOVA, see www.statsoft.com/Textbook/ANOVA-MANOVA.)

Questions

1. Which jars will test what hypotheses? For example, what hypothesis can be tested by comparing jars A, D, and E? What would you conclude if they had the same densities of beetles?
2. What other hypotheses can you formulate to explain the observation that populations do not grow indefinitely? What predictions can you derive from these hypotheses and what tests could you perform?
3. Two life stages (eggs and pupae) are not mobile, and so are particularly vulnerable to cannibalism. Do you see evidence of this when comparing, for example, jar A with jar C?
4. An attitude commonly encountered in undergraduate science labs is that, if you did not get the result the instructor expected, then the experiment "did not work." What do you think of this view?
5. Scientists are now discussing how climate change might impact the outcome of competition (see Urban et al., 2011). How might warmer temperatures change the outcomes of an experiment like the one you conducted with *Tribolium?*

Written Report

For your report, prepare appropriate graphs and write a summary statement of the experimental approach and an explanation of the results. Answer any of the preceding questions that your instructor may assign.

Alternatively, your instructor may want you to follow the format of a scientific paper for your report: introduction, materials and methods, results, discussion, and literature cited. The questions listed above are designed to bring up possible topics that might be included in the report. Be sure to think about your results and look at your data in original ways before writing the report.

References

Edmunds J., J. M. Cushing, R. F. Costantino, S. M. Henson. B. Dennis, and R. A. Desharnais. 2003. Park's *Tribolium* competition experiments: a non-equilibrium species coexistence hypothesis. Jour. Anim. Ecology **72:**703–712.

Hasting, A., and R. F. Costantino. 1987. Cannibalistic egg-larva interactions in *Tribolium:* an explanation for the oscillations in population numbers. American Naturalist **130:**37–52.

Ho, F., and P. Dawson. 1966. Egg cannibalism by *Tribolium* larvae. Ecology **47:**318–322.

Lloyd, M. 1968. Self regulation of adult numbers by cannibalism in two laboratory strains of flour beetles *(Tribolium castaneum).* Ecology **49:**245–259.

Lutherman, C., E. Miller, and T. Park. 1939. Studies in population physiology, IX. The effect of imago population density on the duration of larval and pupal stages of *Tribolium confusum* Duval. Ecology **20:**365–373.

Park, T. 1932. Studies in population physiology: the relation of numbers to initial population growth in the flour beetle *Tribolium confusum* Duval. Ecology **13:**172–181.

Park, T. 1933. Studies in population physiology, II. Factors regulating initial growth of *Tribolium confusum* populations. Jour. Exper. Zoology **65:**17–42.

Peters, M., and P. Barbosa. 1977. Influence of population density on size, fecundity, and developmental rate of insects in culture. Ann. Rev. Entomol. **22:**431–450.

Rich, E. R. 1956. Egg cannibalism and fecundity in *Tribolium.* Ecology **37:**109–120.

Stevens, L. 1989. The genetics and evolution of cannibalism in flour beetles. Evolution **43:**169–179.

Urban, M. C., R. D. Hot, S. E. Gilman and J. Tewksbery. 2011. Heating up relations between cold fish: competition modifies responses to climate change. Jour. Anim. Ecology **80:**505–507.

Young, A. 1970. Predation and abundance in populations of flour beetles. Ecology **51:**602–619.

EXERCISE 5B

Ecology of a Freshwater Habitat

Ponds and streams have always been a source of fascination to people, but few are aware of the extraordinary complexity and variety of life that thrives beneath the water's surface. Ponds and streams are **ecosystems** containing both living (biotic) and nonliving (abiotic) components through which nutrients are cycled and recycled. The biotic component, the interacting assemblage of plants and animals, is the **community.** Ecologists often use the community concept in more restricted ways, employing it to describe particular organism/habitat relationships within the habitat, such as the floating plant community, the muddy bottom community, or the open water community.

A pond is a **lentic** (L. *lentus,* slow) habitat, as opposed to a running-water, or **lotic** (L. *lotus,* action of washing), habitat, such as a brook or stream. Within each habitat are numerous, more specialized habitats that are characterized by the plants that live there or by some special physical feature. For your visit to a pond or stream, you should be able to recognize the following habitats:

1. **Shallow-water habitat,** a swamplike area where sedges and grasses emerge from the water, providing shelter for various amphibians, birds, and mammals. This area may dry up during hot, dry spells of summer weather.
2. **Floating life and emergent plant habitat** in somewhat deeper water or slower-moving water, where both floating vegetation, such as water lilies (*Nymphaea* sp. and *Nuphar* sp.), and emergent plants, such as cattails (*Typha* sp.) and reeds, are rooted in bottom mud. The slimy undersurface of a water lily is home to a host of animal life: hydras, planarian flatworms, rotifers, snails and their egg masses, various protozoans (especially *Vorticella*), caddisfly larvae, and bryozoans. The stems of emergent vegetation also provide a surface for larvae of dragonflies, damselflies, and mayflies. The surface may also be covered in part (or even completely) by free-floating duckweed (*Lemna* sp.).
3. **Submerged plant habitat,** which is farther from shore and may occur in ponds or large stream pools, where grow completely submerged pond-weeds, such as *Anacharis* sp. (formerly *Elodea*) and *Potamogeton.* Here we would expect to find various fishes, turtles, and a variety of water beetles and aquatic bugs.
4. **Bottom habitat,** which is usually covered with mud and where heavy deposits of decaying vegetable matter have used up the available oxygen—only animals with adaptations for living in anaerobic conditions are found. Most habitat bottoms, however, harbor a wealth of animal life that burrows through the mud to extract organic nutrients or to feed on other animals: segmented worms, several kinds of insect larvae, nematodes, bivalve and gastropod molluscs, flatworms, and crayfish.
5. **Water-surface habitat**—the surface tension of the habitat's elastic skin provides a surprisingly firm platform for many plants and small creatures that either move about on its surface (water striders, water spiders, whirligig beetles, springtails) or hang from its undersurface (mosquito and diving beetle larvae, floating pond snails, flatworms, hydras).

Preparations for Fieldwork

Materials Needed

Notebooks
Long-handled dip net of strong, fine-meshed material
White plastic handbasin, white plastic buckets, or other container for sorting animals
Forceps (for handling small, hard-bodied animals)
Small paintbrush or camel hair brush (for handling soft-bodied animals)
Disposable white plastic spoons (for picking up small, soft-bodied forms)
Medicine dropper and vials (for collecting very small animals)
Various jars and plastic vials (for transporting animals)
10× hand lens (optional but very useful for on-the-spot identifications)
Thermometer
Kitchen food strainer
Plastic or canvas sheet (to spread on ground for animal sorting)
Rubber hip boots (if water is too cold for sandals or sneakers)

Depending on class size, your class may be divided into teams, each consisting of several collectors and a recorder. The recorder should record the date and time of day of the visit, weather, and location and physical surroundings of the habitat. The recorder should also make a rough sketch of the habitat, showing principal shore features (such as areas of emergent vegetation), and then note on the map where the sampling is done. As animals are collected for return to the laboratory for later study, the recorder must label the containers, noting collection information in the notebook (exact location, depth, whether from a vegetation sweep or mud bottom, etc.).

The collectors should approach the habitat cautiously, watching for animals along the shore (frogs and birds, for example) that will escape as they near the shoreline. First examine the habitat along its edges, looking for water snakes, amphibians, and amphibian eggs. Then skim the edge of the shore vegetation with a dip net, but do not disturb the bottom at this time. Empty the contents of the dip net into a white tray or handbasin and examine for small amphibians, leeches, oligochaetes, snails, aquatic insects (both adults and larvae), bryozoans, hydras, flatworms, and small crustaceans.

Next, wade into the habitat (use sneakers, or hip boots if the water is very cold) and sweep the submerged vegetation. Uprooted water plants should be laid on a piece of plastic sheeting. Allow the water to drain away; then sort out the animals by hand, using forceps, a small brush, or a plastic spoon to pick up small creatures. Return the plants immediately to the habitat. Examine lily pads and other floating vegetation carefully on the underside for snails, sponges, insect eggs, flatworms, hydras, bryozoans, rotifers, and attached protozoans. Collect these forms into separate, labeled vials or small bottles.

Finally, scoop up small samples of bottom mud and litter. Allow excess water to drain off; then dump the contents into a plastic washbasin or tray. Many of the animals will wriggle out from the debris and can be picked out by hand. Another approach is to dump the contents onto a plastic sheet or piece of canvas spread on the ground; allow the water to drain away slowly while you sort through the mud and debris. If much heavy mud is collected, separate the animals from the mud using a kitchen strainer.

Keep herbivores and carnivores separate to prevent the carnivores from emerging victorious on the way home. Carnivorous insects especially should be collected in separate containers. Air-breathing aquatic animals are best carried in damp weed or moss rather than in water. This applies to water beetles, water scorpions, water spiders, water striders, and backswimmers.

Good Pond Stewardship

A large zoology class visiting an aquatic habitat can seriously disturb the area unless certain precautions are taken.

1. Do not collect more specimens than necessary for identification. Large animals (vertebrates) *must* be identified on the spot and returned immediately to the habitat.
2. If logs or stones are turned over while collecting, replace them as they were.
3. Avoid trampling emergent vegetation along the edge. Leave no litter.

Identification of Aquatic Animal Life

Animals not identified in the field are returned to the laboratory, where you will complete the identification. How does one identify an unknown specimen? One way is by direct comparison with specimens in a museum reference collection. However, few biologists have ready access to such collections. Even with such access, most nonspecialists would find this a tedious approach, involving working through thousands of museum specimens.

A practical alternative is to use a taxonomic key. A key is a convenient tabular device that enables us to identify a specimen by comparing it feature by feature with alternative given in key couplets. Key may be designed to identify species, genera, families, orders, or any other taxon.

Following is a brief exercise in classification that shows you how to use a taxonomic key to "run down" or "key out" the classification of an organism when neither its common nor its scientific name is known. You will use the dichotomous key on pp. 70–74 to identify invertebrates to major taxonomic groups. (The key embraces all aquatic invertebrate groups having representatives in North America.) If you want to carry the identifications to genera or to species, you will need to consult more specialized keys. Some of the most important published field guides are listed in the References section. The class should draw up an inventory of all forms collected, giving the lowest taxon to which each animal can be assigned and noting the specific habitat and relative abundance of each species.

Our key is limited to invertebrate aquatic life, although some of these habitats are, of course, home to fish, newts, frogs, toads, turtles, and various birds and mammals.

How to Use a Taxonomic Key

A two-choice system serves as the basis of a **dichotomous key.** In the dichotomous key, two contrasting alternatives are offered at once, so you can choose the one that fits your specimen. At the end of the choice, you will find a reference number to the next set of alternatives to be considered. Again make a decision and proceed in the same manner until you arrive at the scientific name of the animal or the taxon to which it belongs.

This key also has the capacity for reverse use so that you can retrace your steps if you make a mistake. In each couplet, the number in parentheses refers to the number of the couplet from which that couplet was reached.

Keep in mind that individual variations exist; keys are based on the average, or "typical," adult specimen, whereas your specimen may be immature or somewhat abnormal. It is often very helpful to examine more than one specimen of a species or group, if available, when a particular descriptive character proves troublesome.

Some people resist using dichotomous keys, preferring instead to find an illustration that resembles the specimen in question. This picture approach often works well enough for certain animal groups but can be quite tedious, if not impossible, with other groups. A well-constructed dichotomous key, on the other hand, can guide you toward the correct identity by pointing out those anatomical features most appropriate for distinguishing among similar taxa. In this aquatic study, you may be able to identify many of the animals by looking at the pictures, bypassing the key entirely. However, to understand how the key works and to gain practical experience in using it, get into the habit of proceeding through the key in an orderly manner. This is the only way to become familiar with the *distinguishing* characteristics of the different groups of animals.

Written Report

Prepare a report based on the inventory assembled by the class. Begin by preparing a profile of the habitat, indicating its location, physical conditions at the time of collection (water temperature, weather, season), and major geophysical and biotic characteristics. Then list the animals found and identified in systematic order (protozoa listed first, vertebrates last), the habitat association or community in which each is found, and, if your instructor requests it, some principal adaptations of the more common forms collected.

Key to Common Freshwater Aquatic Invertebrates of North America*

In each couplet, the number in parentheses refers to the number of the couplet from which that couplet was reached, thus making it possible to retrace if a mistake is made.

1 Unicellular. Majority are microscopic (less than 500μm), but some are macroscopic (greater than 500μm). Cytoplasm and organelles are enclosed in one unified cellular covering. No appendages. Form extremely variable, including spherical, elliptical, doughnut-shaped, or constantly changing forms 2

2 (1) Fluid movement. Projections (pseudopodia) extending in several directions. Constantly changing shape. Either naket or with a test (shell) that may be spherical, caplike, or doughnut-shaped. Ameboid protozoans 3

Outer covering distinct. Some with ornate striations and sculpturing. Clear, green, brown, or pink in color. Tumbling motion or jerky movement 5

3 (2) Naked, lacking a test. *Amoeba* or related genus.

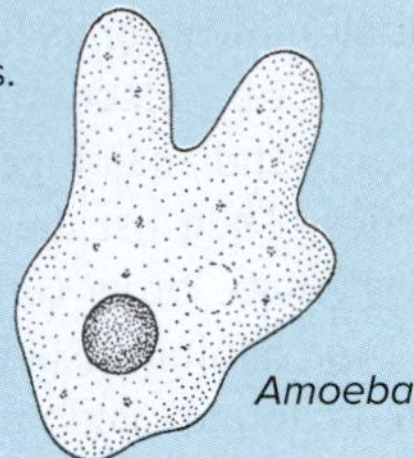

Amoeba

Possessing a test (shell) 4

4 (3) Central sphere with spines projecting outward. The heliozoan's test is doughnut-shaped from dorsal view or cap-shaped from lateral view. The testate families. *Arcella* or related genus.

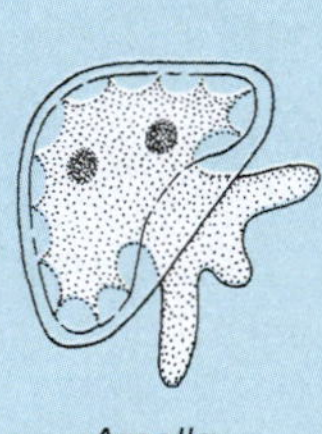

Arcella

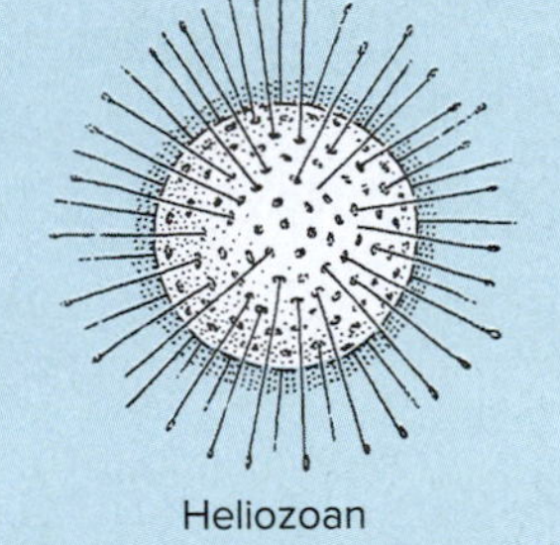

Heliozoan

5 (2) Chromoplasts (discrete packets of pigment) usually present. May be green, yellow-green, or brown. Pulled along by flagella that may not be visible. Green flagellate phyla (Viridiplantae or Euglenozoa) 6

Chromoplasts absent. Usually clear, but some are green, pink, or brown. If coloration is present, the pigmentation is uniform and not restricted to discrete packets. Cilia may cover entire surface or be restricted to localized areas. Phylum Ciliophora 7

6 (5) Green. Spherical colony consists of many cells connected by protoplasmic strands. *Volvox* or related genus.

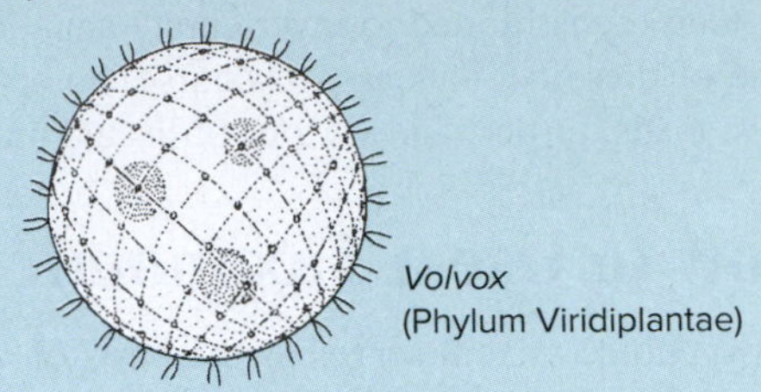

Volvox
(Phylum Viridiplantae)

Green. Solitary. *Euglena* or related genus.

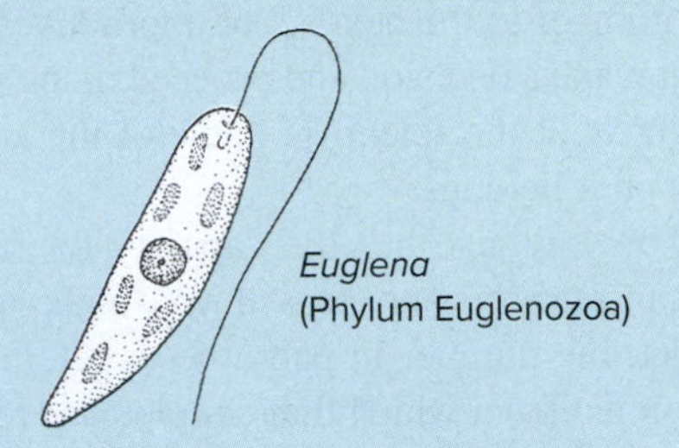

Euglena
(Phylum Euglenozoa)

7 (5) Attached to stalk to substrate. Bounces by coiling and uncoiling of stalk. *Vorticella* or related genus.

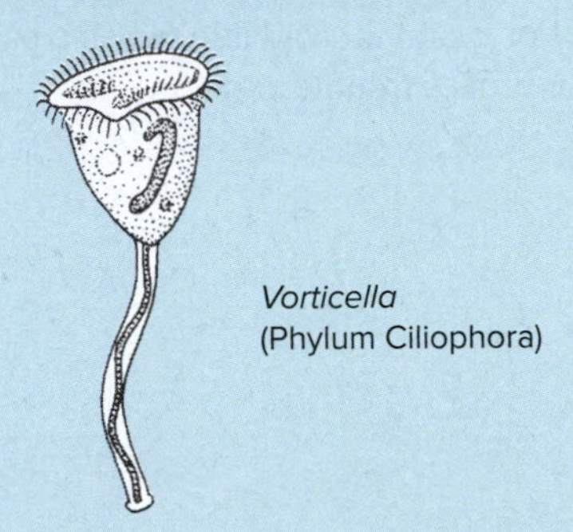

Vorticella
(Phylum Ciliophora)

Trumpet-shaped with cilia surrounding mouth. May remain attached by narrow end to substrate or can detach and swim freely. *Stentor* or related genus.

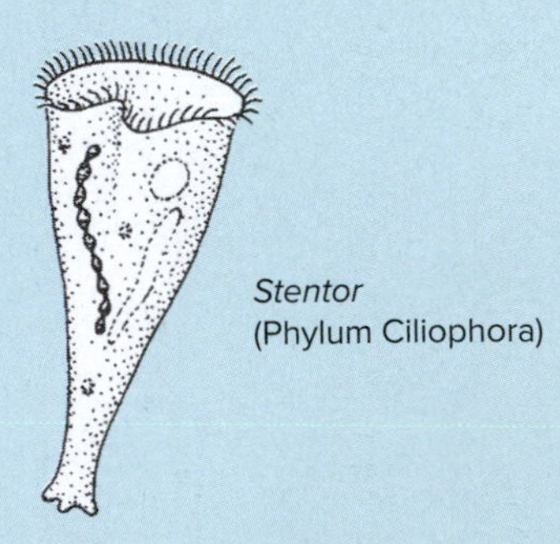

Stentor
(Phylum Ciliophora)

* Key written by Deborah Kendall, Department of Biology, Fort Lewis College, Durango, Colorado.

8 (1) Wormlike appearance. Distinct head region may or may not be present. Small appendages may be present, but segmented thoracic (second large body section in insects) legs are always absent 9

Not wormlike in appearance. Forms with many segments possess segmented legs and a distinct head region 15

9 (8) Not segmented 10

Segmented 12

10 (9) Flat with a weakly formed head region and eyespots. Phylum Platyhelminthes. Class Turbellaria. Free-living flatworms.

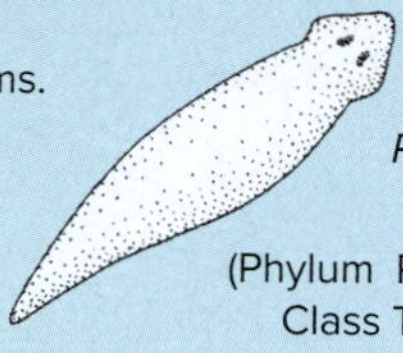

Planaria

(Phylum Platyhelminthes Class Turbellaria)

Not flat. Elliptical in cross section 11

11 (10) Body highly contractile and possessing a proboscis that may be contracted back into the head region or extended into the ambient water. Movement slow and forward-directed. Anterior end is bluntly rounded. Elliptical cross section. Locally abundant. Phylum Nemertea. Proboscis worms.

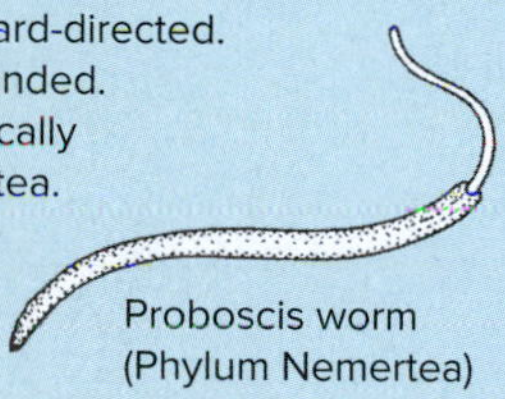

Proboscis worm (Phylum Nemertea)

Body tapers to a point at both ends. Movement is rapid and whiplike. Cylindrical cross section. Very common. Phylum Nematoda. Roundworms.

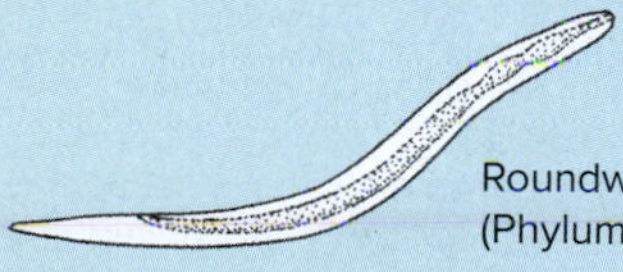

Roundworm (Phylum Nematoda)

12 (9) Nearly all body segments uniform in size. No distinct head region or appendages present. Phylum Annelida . . . 13

Body segments not uniform size. Appendages present. Phylum Arthropoda. Subphylum Uniramia. Class Insecta, in part. Larval forms 14

13 (12) Body segments telescope during crawling or swimming. Setae (hairs) extend from the body. Cylindrical cross section. Some genera, such as *Tubifex*, inhabit tubelike cases attached to the substrate. Class Oligochaeta. Freshwater earthworms.

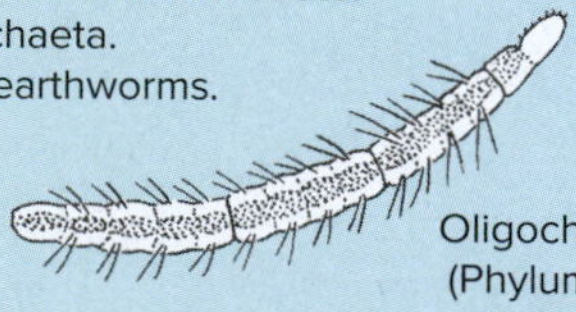

Oligochaete worm (Phylum Annelida Class Oligochaeta)

Anterior and posterior suckers. No setae. Dorsoventrally (top to bottom) flattened. May crawl by inchworm movement that consists of alternately attaching the anterior and then the posterior sucker, or may swim. Class Hirudinea. Leeches

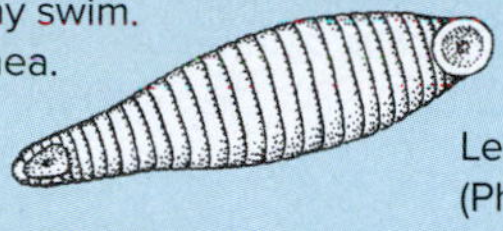

Leech (Phylum Annelida Class Hirudinea)

14 (12) Posterior appendages absent. Hardened, distinct head region present. Order Coleoptera, in part. Beetles, larval form.

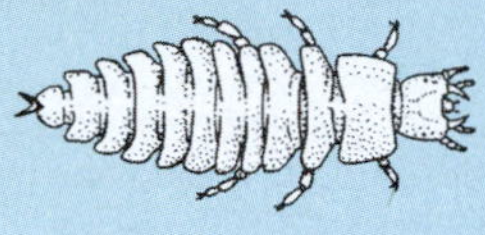

Beetle larva (Class Insecta Order Coleoptera)

Posterior appendages, such as breathing tubes, present. Prolegs (small fleshy, unsegmented appendates) may be present. Head region may or may not be present. Order Diptera, larval form.

Fly larva (Class Insecta Order Diptera)

15 (8) Attached forms 16

Not attached to substrate 17

16 (15) Solitary. Cylindrical stalk with tentacles extending from the oral region. Green or brown. Remain attached or can move by somersaulting or gliding. Phylum Cnidaria. Class Hydrozoa. *Hydra* or related genus.

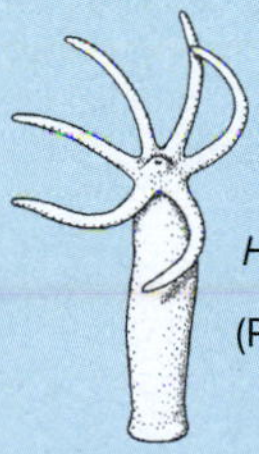

Hydra

(Phylum Cnidaria Class Hydrozoa)

Colonial. Stalks enveloped within a common thin body wall. Feathery appendages extend from oral regions. Gelatinous. Usually remain sessile but can creep along substrate. Phylum Bryozoa. Moss animals.

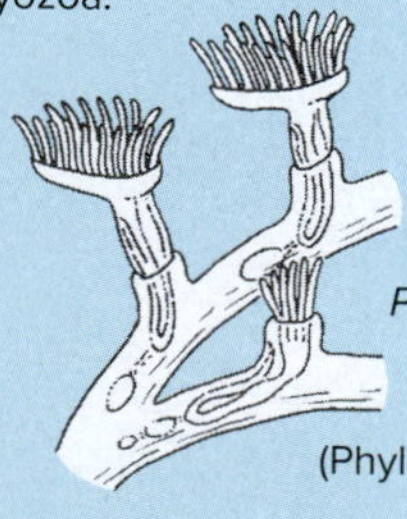

Plumatella

(Phylum Bryozoa)

17 (15) Microscopic in size 18

Macroscopic in size 19

continued

Key to Common Freshwater Aquatic Invertebrates of North America-*continued*

18 (17) Smooth body covering. Oral end is ciliated and rotating (wheel organ). Posterior end possesses small appendages (toes). May swim or telescope oral end into debris or vegetation. Phylum Rotifera. Rotifers or wheel animals.

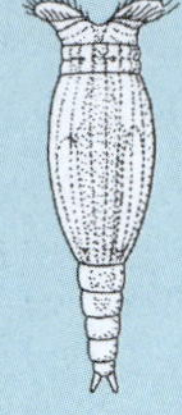

Rotifer (Phylum Rotifera)

Body covered with spines that are narrowed distally. Posterior end is forked. Smooth, gliding movement. Phylum Gastrotricha. Fork-tailed worms.

Fork-tailed worm (Phylum Gastrotricha)

19 (17) Body enclosed in an external, calcareous shell. Body sections extending from the shell are fleshy and unsegmented–for example, a foot or head. Phylum Mollusca. Freshwater shellfish 20

Body not enclosed in an external, calcareous shell. Body segmented and covered in a hard, chitinous exoskeleton. Appendages are jointed. If clam-shaped, the body appendages are segmented and covered with setae (hairs). Phylum Arthropoda. Joint-footed animals 21

20 (19) Two shells attached dorsally by a hinge. No head or tentacles present. Foot is large and muscular for digging into substrate. Class Bivalvia. Freshwater clams.

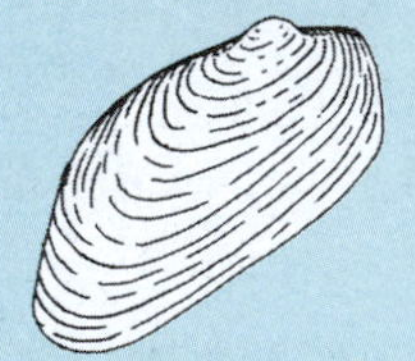

Freshwater clam (Phylum Mollusca Class Bivalvia)

One shell present. Majority possess a coiled shell (snails), but some produce a caplike shell over the body (limpets). Distinct head with tentacles. Foot is ventrally flattened for gliding over substrate. Class Gastropoda. Freshwater snails and limpets.

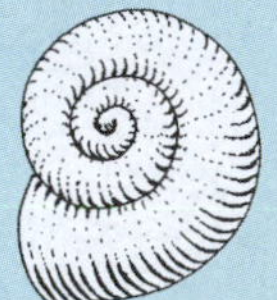

Freshwater snail (Phylum Mollusca Class Gastropoda)

Freshwater limpet (Phylum Mollusca Class Gastropoda)

21 (19) No antennae. Cephalothorax (head and thorax fused together) and abdomen present. In water mites, the cephalothorax and abdomen are fused and indistinguishable. Eight legs present in adult forms. Immatures sometimes possess six legs. Subphylum Chelicerata. Class Arachnida 22

Antennae present. Number of legs variable 23

22 (21) Chephalothorax and abdomen fused into one segment. Setae (hairs) present. Body round or ovoid and dorso-ventrally flattened. Adults possess eight legs and immatures possess six legs. Fully aquatic. Order Hydracarina. Water mites.

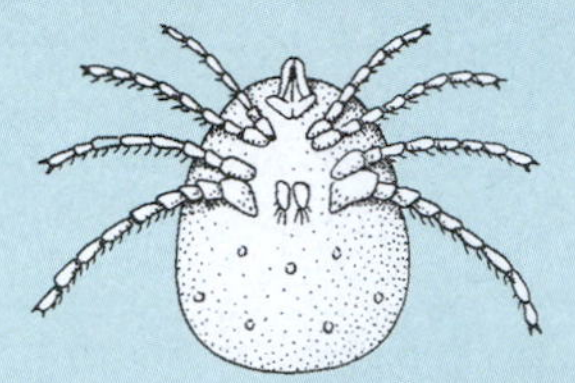

Water mite (Subphylum Chelicerata Class Arachnida Order Hydracarina)

Chephalothorax and abdomen distinct. Both adults and immatures possess eight legs. Semiaquatic. Order Araneae. Water spiders.

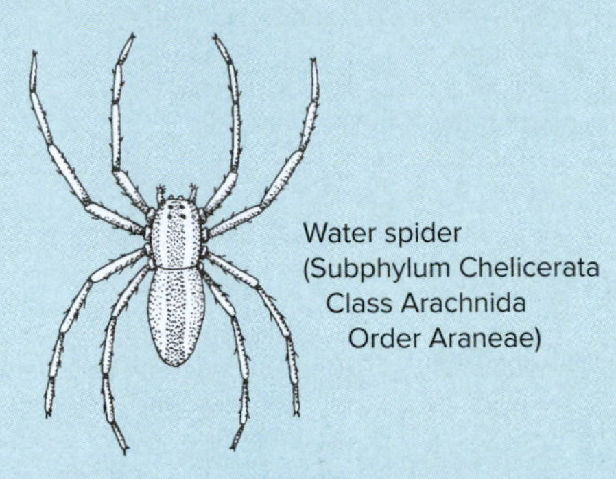

Water spider (Subphylum Chelicerata Class Arachnida Order Araneae)

23 (21) Body consists of two major body sections (cephalothorax and abdomen) that possess further segmentation. Two pairs of antennae. Subphylum Crustacea 24

Body consists of three or more major body sections that possess further segmentation. One pair of antennae and six thoracic segmented legs. Subphylum Hexapoda. Class Insecta, in part 32

24 (23) Bivalvelike (two shells) exoskeleton completely encloses the body. May extend feathery appendages to rake water for food 25

Not completely enclosed in a bivalvelike shell 26

25 (24) Shell possesses concentric rings. Order Conchostraca. Clam shrimp

Clam shrimp (Subphylum Crustacea Order Conchostraca)

Shell does not possess concentric rings. Order Ostracoda. Ostracods

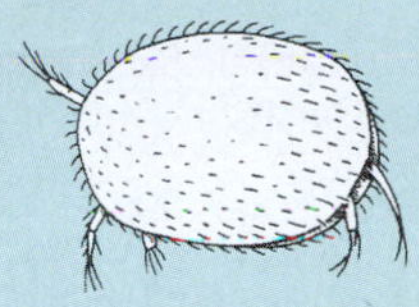

Ostracods
(Subphylum Crustacea
Class Oligostraca
Order Ostracoda)

26 (24) Large, conspicuous pair of chelate (clawlike) appendages. Eight functional walking legs. Order Decadopa. Crayfish.

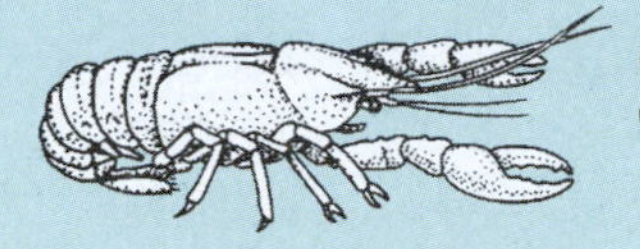

Crayfish
(Subphylum Crustacea
Class Malacostraca
Order Decapoda)

Chelate appendages absent 27

27 (26) Possessing a carapace, a single extensive shield that covers or encloses body segments 28

Not possessing a carapace. All body segments appear uniform in size 30

28 (27) Laterally compressed (side to side). Carapace covers the thorax and abdomen. Resembles a bivalvelike shell that gapes ventrally. Head distinct with one pair of large, distinct compound eyes. Order Cladocera. Water fleas.

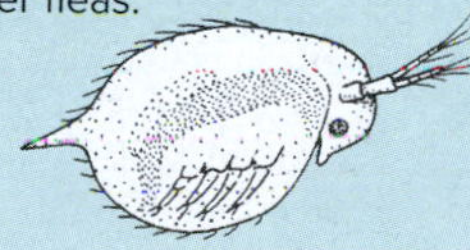

Daphnia
(Subphylum Crustacea
Class Branchiopoda
Order Cladocera)

Dorsoventrally compressed (top to bottom) 29

29 (28) Carapace covers most of the body. The visible abdominal segments appear uniform is size. Swollen area occurs behind the two dorsal median eyes. Two long, whiplike appendages extend from the posterior end. Order Notostraca. Tadpole shrimp.

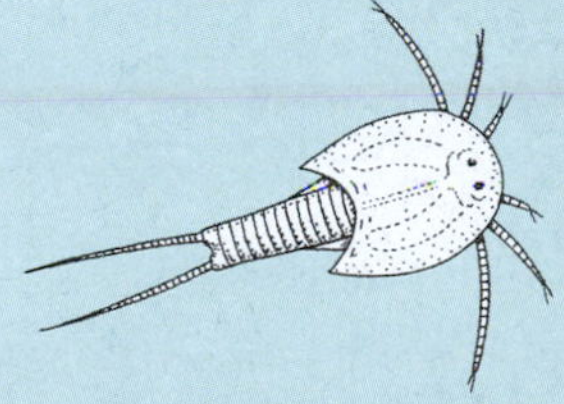

Tadpole shrimp
(Subphylum Crustacea
Class Branchiopoda
Order Notostraca)

Carapace does not cover most of the body. The visible abdominal segments do not appear uniform in size. Single dorsal median eye present. Two appendages that extend from the posterior end are short and not whip-like. In gravid female, the appendages are broad and disigned for carrying eggs. Order Copepoda.
Copepods.
Cyclops or
related genus.

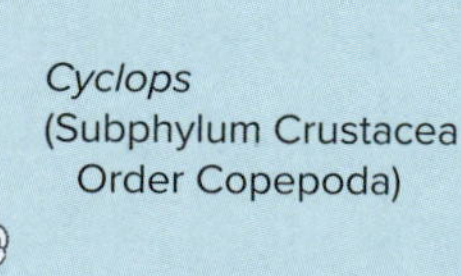

Cyclops
(Subphylum Crustacea
Order Copepoda)

30 (27) Dorsoventrally flattened (top to bottom). Resemble the terrestrial sowbug. Order Isopoda. Isopods.

Isopod
(Subphylum Crustacea
Class Malacostraca
Order Isopoda)

Laterally flattened (side to side) 31

31 (30) Body segments occur in a straight line. Appendages are flattened and leaflike. Swim with the ventral side up. Order Anostraca. Fairy shrimp. Brine shrimp are saltwater relatives.

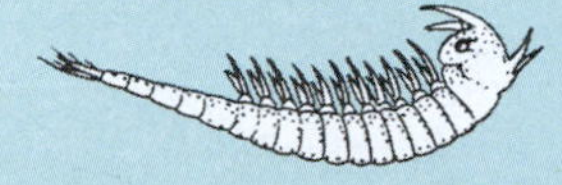

Fairy shrimp
(Subphylum Crustacea
Class Branchiopoda
Order Anostraca)

Body bent near the middle. Swim with ventral side down. Order Amphipoda. Scuds

Amphipod
(Subphylum Crustacea
Class Malacostraca
Order Amphipoda)

32 (23) Body enclosed within a case made of sand, shells, or other materials. Order Trichoptera, in part. Case building caddisflies.

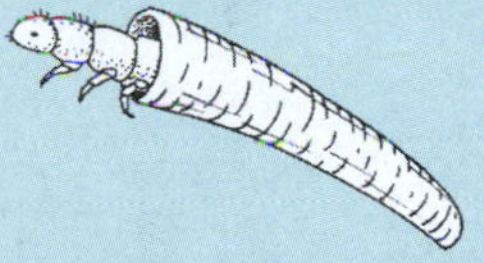

Caddisfly
(Subphylum Hexapoda
Class Insecta
Order Trichoptera)

Body not enclosed within a case 33

33 (32) Wingless 34

Possess wingpads (rudimentary wings that do not extend to the tip of the abdomen) or wings 36

34 (33) Dorsoventrally flattened. Seven to eight pairs of lateral abdominal gills. Possessing either one caudal filament (tail) or two short anal prolegs whith hooks. Order Neuroptera. Suborder Megaloptera. The alderflies, dobsonflies, and fishflies. Immature forms.

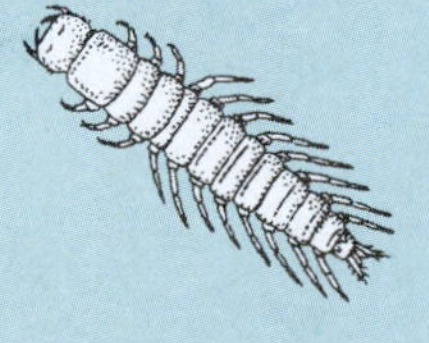

Alderfly larva
(Class Insecta
Order Neuroptera
Suborder Megaloptera)

Cylindrical cross section 35

continued

35 (34) Hooks on anal prolegs. Gills, if present, are fingerlike and branched and occur both dorsally and ventrally. Order Trichoptera, in part. Free-living caddisflies that do not build cases. Larval form.

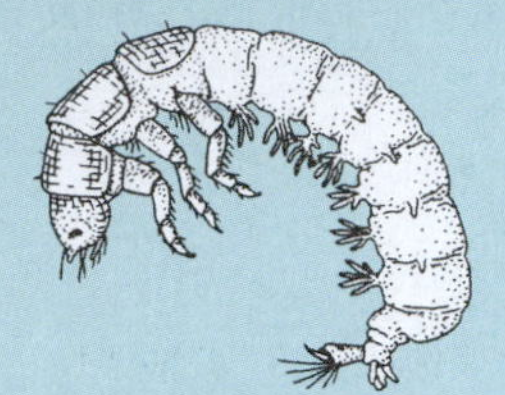

Caddisfly larva
(Class Insecta
Order Trichoptera)

Anal prolegs, if present, are not hooked. Gills, if present, arise from lateral sides of abdomen only. Order Coleoptera, in part. Beetles, larval form.

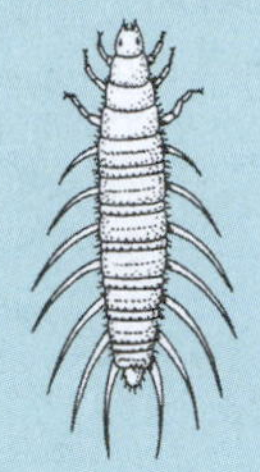

Beetle larva
(Class Insecta
Order Coleoptera)

36 (33) Possess full wings (wings that extend to tip of abdomen). Legs are variously adapted for swimming–for example, oar-like and fringed with hairs (setae). 37

Possess wing pads (wings that do not extend to tip of abdomen) 38

37 (36) Mouthparts take the form of a beak that is folded beneath the thorax. Wing tips overlap. Triangle present just behind thorax. Order Hemiptera. True bugs. Adult form.

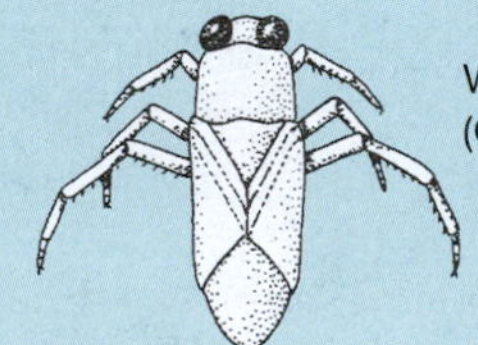

Water bug
(Class Insecta
Order Hemiptera)

Mouthparts are of the chewing type and do not take the form of a beak. Wing tips do not overlap. Order Coleoptera. Beetles. Adult form.

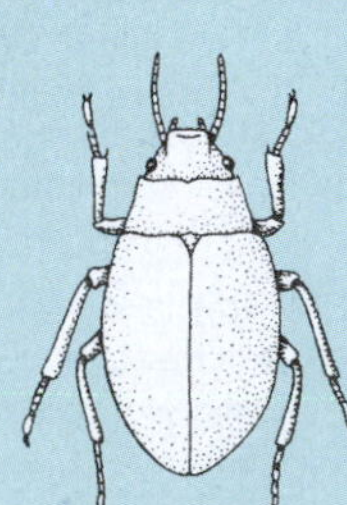

Beetle
(Class Insecta
Order Coleoptera)

38 (36) Mouthparts take the form of a beak that is folded beneath the thorax. Order Hemiptera. True bugs. Immature form.

Water bug nymph
(Class Insecta
Order Hemiptera)

Mouthparts are of the chewing type and do not take the form of a beak 39

39 (38) Mouthparts are scooplike and folded beneath the head. Order Odonata. Damselflies with three terminal leaflike gills; dragonflies with no gills.

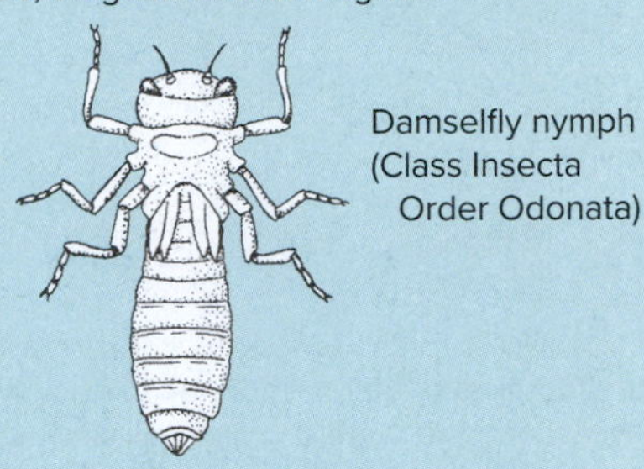

Damselfly nymph
(Class Insecta
Order Odonata)

Mouthparts are not scooplike and are present in front of the head, not folded beneath 40

40 (39) No thoracic gills. Numerous abdominal gills extend laterally. Possessing two or three long caudal appendages (tails) fringed with long setae. Tarsi (the feet, the terminal segments of the legs) with one claw. Order Ephemeroptera. Mayflies. Immature form.

Mayfly nymph
(Class Insecta
Order Ephemeroptera)

No gills or thoracic gills present (may extend to the first two abdominal segments). Tarsi with two claws. Dorsoventrally flattened. Order Plecoptera. Stoneflies. Immature form.

Stonefly nymph
(Class Insecta
Order Plecoptera)

References

The following is a selection of guidebooks and keys that may be useful for identifying animals collected on field trips.

General Field Guides and Taxonomic Keys to Pond Life

Blair, W. F., A. P. Blair, P. Brodkorb, F. R. Cagle, and G. A. Moore. 1968. Vertebrates of the United States, ed. 2. New York, McGraw-Hill Book Company.

Caduto, M. J. 1990. Pond and brook. A guide to nature in freshwater environments. Hanover, N. H., University Press of New England.

Edmondson, W. T. (editor). 1959. Ward and Whipple's freshwater biology, ed. 2. New York, John Wiley & Sons, Inc.

Klots, E. B. 1966. The new field book of freshwater life. New York, G. P. Putnam's Sons.

Needham, J. G., and P. R. Needham. 1962. A guide to the study of freshwater biology, ed. 5. San Francisco, Holden-Day, Inc.

Peckarsky, B. L., et al. 1990. Freshwater macroinvertebrates of northeastern North America. Ithaca, N. Y., Cornell University Press.

Pennak, R. W. 2001. Freshwater invertebrates of the United States, ed. 4. New York, Wiley.

Thompson, G., and J. Coldrey. 1984. The pond. Cambridge, Mass., The MIT Press (and Oxford Scientific Films).

Thorp, J. H., and A. P. Covich. 2009. Ecology and classification of North American freshwater invertebrates, ed. 3. New York, Academic Press.

Taxonomic Keys for Specific Animal Groups

Protozoa

(In addition to the listings that follow, refer also to Edmondson [1959], Needham & Needham [1962], and Pennak [2001].)

Davis, C. C. 1955. The marine and freshwater plankton. Lansing, Mich., Michigan State University.

Jahn, T. L., and F. F. Jahn. 1949. How to know the Protozoa. Dubuque, Iowa, Wm. C. Brown Group.

Kudo, R. R. 1946. Protozoology. Baltimore, Charles C. Thomas.

Annelids

Brinkhurst, R. O., and B. G. Jamieson. 1972. Aquatic oligochaetes of the world. Toronto, Toronto University Press.

Sawyer, R. T. 1972. North American freshwater leeches, exclusive of the Piscicolodae, with a key to all species. Ill. Biol. Monogr. No. 46.

Spiders and Their Kin (Chelicerate Arthropods)

Comstock, J. H. 1948. The spider book, rev. ed. Ithaca, N. Y., Comstock Publishing Company.

Gertsch, W. J. 1979. American spiders, ed. 2. New York, Van Nostrand Reinhold Company.

Headstrom, R. 1973. Spiders of the United States. New York, A. S. Barnes and Company, Inc.

Kaston, B. J. 1978. How to know the spiders, ed. 3. Dubuque, Iowa, Wm. C. Brown Group.

Levi, H. H., L. R. Levi, and H. S. Zim. 1968. A guide to spiders and their kin. Golden Nature Guide, New York, Golden Press.

Insects

Betten, C. 1934. The caddisflies or Trichoptera of New York State. New York State Museum Bulletin No. 292.

Borror, D. J., and R. E. White. 1970. A field guide to the insects of America north of Mexico. Boston, Houghton Mifflin Company.

Chu, H. F. 1992. How to know the immature insects, ed. 2. Dubuque, Iowa, McGraw-Hill.

Claassen, P. W. 1931. Plecoptera nymphs of North America. Lafayette, Ind., Thomas Say Foundation of the Entomological Society of America, Pub. No. 3.

Edmunds, G. F., Jr., S. L. Jensen, and L. Berner. 1976. The mayflies of North and Central America. New York, Dover Publications.

Harris, J. R. 1990. An angler's entomology. London, Bloomsbury.

Jewett, S. G. 1959. The stoneflies (Plecoptera) of the Pacific Northwest. Corvallis, Ore., Oregon State College Press.

LaFontaine, G. 1981. Caddisflies. New York, Lyons & Burford Publishers.

McCafferty, W. P. 1996. Aquatic entomology: the fisherman's and ecologist's illustrated guide to insects and their relatives. Boston, Science Books International.

Merritt, R. W., and K. W. Cummins. 1996. An introduction to the aquatic insects of North America, ed. 3. Dubuque, Iowa, Kendall/Hunt Publishing Company.

Miller, P. L. 1995. Dragonflies. London, Richmond Publishing.

Needham, J. G., M. J. Westfall, Jr., and M. L. May. 2000. Dragonflies of North America. Gainesville, Fla., Scientific Publishers.

Swan, L. A., and C. S. Papp. 1972. The common insects of North America. New York, Harper & Row, Publishers.

Wiggins, G. B. 1977. Larvae of the North American caddisfly genera (Trichoptera). Toronto, University of Toronto Press.

Fishes

Eddy, S., and J. C. Underhill. 1980. How to know the freshwater fishes, ed. 3. Dubuque, Iowa, Wm. C. Brown Group.

McClane, A. J. 1978. Field guide to freshwater fishes of North America. New York, Holt, Rinehart & Winston.

NOTES

Introduction to Animal Taxonomy

EXERCISE 6A
Phylogeny Reconstruction—How to Read and Compare Cladograms

EXERCISE 6B
Use of a Taxonomic Key for Organism Identification
Key to the Chief Phyla and Classes of Unicellular Eukaryotes and Animals

One goal of biology is to understand the evolutionary relationships among all living things. These relationships are indicated through taxonomy—a system of nesting where the most closely related organisms are grouped together at each of several levels. The most inclusive level is a domain. There are three domains of life: Bacteria, Archaea, and Eukarya. All members of Eukarya have a membrane-bound nucleus, among other features. There are many branches of Eukarya, most containing both unicellular and multicellular organisms. The subgroup of Eukarya that contains the animals, fungi, and some types of amebas, is called Opisthokonta. Opisthokonts are distinguished by the presence of a single posterior flagellum, when a flagellum is present. Biologists once presumed that all motile unicellular forms, such as amebas and flagellates, were closely related to animals. Thus, they assumed that there were unicellular animals (protozoans) and multicellular animals (metazoans). We know now that the unicellular forms are not closely related to animals. All animals are multicellular, so the term metazoan is not necessary, although it may still be used. Animals belong to Kingdom Animalia.

The level below kingdom is the **phylum** [plural **phyla**]. There are 32 animal phyla. In traditional Linnaean taxonomy, a phylum is subdivided into smaller groups called **classes;** classes are further subdivided into **orders;** orders into **families;** families into **genera** (sing. **genus**); and genera into **species** (sing. and pl.). In large groups, other categories, such as superclass, suborder, infraorder, and subfamily, also exist. The term **taxa** (sing. **taxon**) can refer to any of these categories.

Species is the least inclusive group. There are many definitions of a species; one defines a species as a group of actually, or potentially, interbreeding natural populations that are reproductively isolated from other such groups of populations. Because all members of a species shared a common ancestor, they also share certain morphological and molecular features inherited from that ancestor. Shared morphological features are things such as the presence of hair and the ability to produce milk, whereas shared molecular features are very similar or identical sequences of nucleotide bases in DNA or RNA, or very similar or identical sequences of amino acids that comprise certain proteins. All members of a species are not genetically identical, but they share certain features used to distinguish them from members of another species.

Some species have very restricted distributions, but others are widespread. Widespread species may look somewhat different in different geographical areas, and may be given different common names to reflect these differences in appearance. In this situation one species might have several different common names. Alternatively, a single common name may be given to different species: for example, a bird with a red breast might be called a "robin" in England and in the United States, but the "robins" might be different species in each place. To prevent confusion, each species is given a distinct latinized scientific name. The species, or scientific, name is a **binomial**—that is, it consists of two parts, a genus name and a species epithet. We call this two-name system the Linnaean system of **binomial nomenclature.** The human species is *Homo sapiens; Homo* ("a man") is the genus and *sapiens* ("mighty" or "wise") is the species epithet, actually an adjective that modifies the genus name. The genus name can be used alone when one is referring to a group of species included in that genus, such as *Rana* (a large genus of frogs) or *Felis* (a genus of cats, including wild and domestic species). The specific epithet, however, would be meaningless if used alone because the same epithet may be used in combination with different genera. The domestic cat is designated *Felis domestica; domestica* used alone is without significance because it is a commonly used epithet that identifies no particular organism. Therefore, the species epithet must always be preceded by the genus name. However, you can abbreviate the genus name when it is used in a context in which it is understood. *Felis domestica* might then be designated *F. domestica.*

In some cases, where geographical varieties or races of a species exist, three names may be used—in which case, the last name indicates the **subspecies.** When three names are thus used, the method is called **trinomial nomenclature.**

For example, northern and southern races (subspecies) of the frog *Rana clamitans* are classified as follows:

Domain Eukarya
Kingdom Animalia
Phylum Chordata
Subphylum Vertebrata
Class Amphibia
Order Anura (Salientia)
Family Ranidae
Genus *Rana*
Species *Rana clamitans*
Subspecies *Rana clamitans clamitans* green frog (northern race)
Subspecies *Rana clamitans melanota* bronze frog (southern race)

Note that all except the species and subspecies epithets are capitalized; species and subspecies epithets begin with lowercase letters. Genus, species, and subspecies names are printed in italics or are underlined when written or typed.

New species arise by speciation, also called **cladogenesis.** In cladogenesis, members of some populations of a preexisting species cease exchanging genes with individuals of other populations. For example, this may occur when two subspecies diverge from each other to the extent that they no longer select individuals of the other subspecies as mates. This process of lineage splitting could be diagrammed in an evolutionary tree where each fork in the tree represents a speciation event. Such a diagram of lineage splitting is called a **phylogeny** or a **cladogram.** Producing a cladogram that would describe the pattern of lineage splitting that led to all species, both extant and extinct, is one of the goals of evolutionary biologists. The process used to infer the structure of an evolutionary tree is called **phylogeny reconstruction.** Following is a short exercise that will help you understand the ideas underlying phylogeny reconstruction.

EXERCISE 6A

Core Study

Phylogeny Reconstruction—How to Read and Compare Cladograms

A cladogram is a branching, or tree-like, diagram that represents a hypothesis about evolutionary relationships among the taxa shown. Taxa placed on a cladogram can come from any level of a Linnaean hierarchy—one can construct a cladogram using species, genera, or phyla. Each cladogram illustrates a pattern of lineage splitting that produced the taxa we see now; we refer to these as **extant** taxa.

Figure 6-1 indicates that taxa A, B, and C shared a common ancestor at some point in the past prior to node 1. At node 1 the lineage of this ancestor split into two groups. Descendants of one group survive to the present as taxon A, whereas descendants of the other group form a lineage represented as the line that runs between node 1 and node 2. At node 2 this lineage split to form two extant taxa: B and C. Any pair of taxa resulting from a lineage split are called

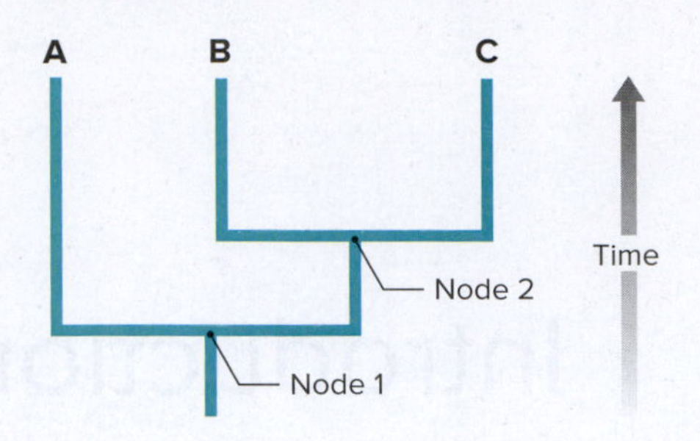

Figure 6-1

Cladogram depicting an evolutionary branching pattern for a lineage with three descendant taxa.

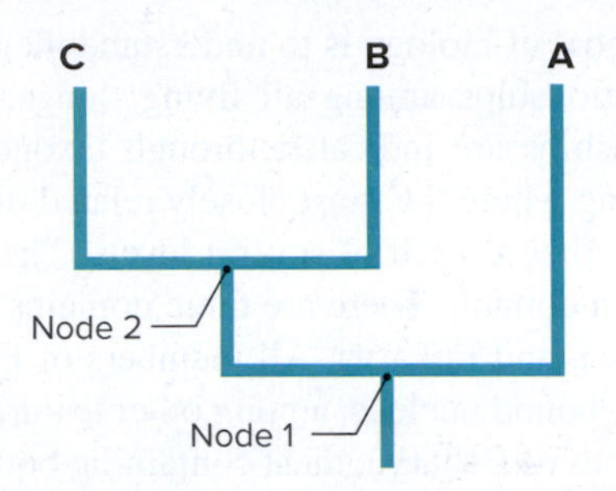

Figure 6-2

Cladogram depicting a branching pattern identical to that shown in Figure 6-1.

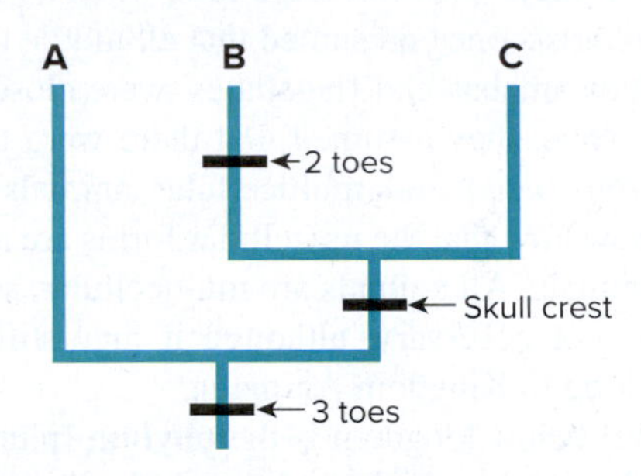

Figure 6-3

Cladogram depicting the evolution of two characters within the lineage leading to taxa A, B, and C.

sister taxa, so B and C are sister taxa. What is the sister taxon to group A? It is the other half of the lineage-splitting event that occurred at node 1, so the sister taxon to group A is the combined group B plus C.

The pattern of branching (lineage splitting) that led to particular taxa is the critical element of a cladogram, but the relative left or right positions of taxa on a cladogram are of no importance. Thus, Figure 6-2 is identical in meaning to Figure 6-1. The pattern of branching has not changed (B and C are still sister taxa); instead, each taxon pair simply rotates at each node.

If a cladogram depicts a particular pathway that evolution followed to produce the three descendant taxa that we see now, we should be able to map the evolution of characteristics present in the taxa onto the tree as well (Figure 6-3). For example, if the ancestor of all three taxa had a certain feature, such as three toes, it may have passed this feature to all of its descendants. Features of organisms are called **characters** and characters

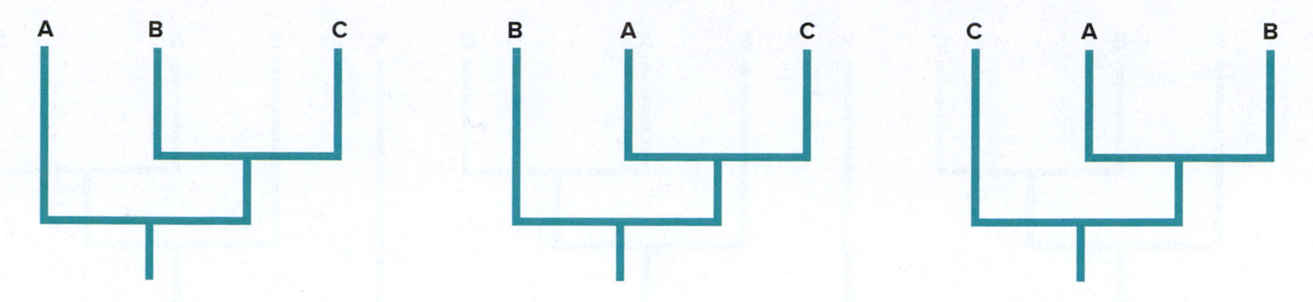

Figure 6-4

Three hypothetical branching patterns for the lineage leading to taxa A, B, and C.

may come in different versions called **character states.** In the example just given, the character is "number of toes" and the character states might be "one, two, or three toes." Another morphological character might be "wings" and the two character states might be "wings present" and "wings absent." We map characters onto a tree with a horizontal line that indicates evolution of the character. Once a character evolves it is passed to all descendants unchanged unless another horizontal line is marked to indicate a change of character state. The tree in Figure 6-3 says that the common ancestor of the three extant taxa had three toes and this state was passed to descendants A and C unchanged, but there was a loss of one toe in the lineage leading to taxon B after node 2. According to this diagram "three toes" is the **ancestral state** of the character "number of toes" and "two toes" is the **derived state** for this character. Notice where the character "skull crest" evolved on the tree. Which taxa will have a skull crest? According to this tree, skull crests are present in taxa B and C because they both inherited the crest from a common ancestor. Given that B and C both have skull crests, it is perhaps possible that each lineage independently evolved skull crests, but this explanation would require two evolutionary changes (two separate events of skull crest evolution), instead of the one event of evolution of this trait in the common ancestor. When reconstructing phylogeny, biologists began with the **Principle of Parsimony,** which states that the simplest explanation that covers all the available evidence is the one to use. The term "available evidence" refers to the character states present in the extant taxa—the tree must explain how they evolved. Thus, it is more parsimonious to assume that skull crests evolved in the common ancestor of taxa B and C than to assume that B and C independently evolved these crests.

For any group of taxa under study, one can identify characters that are likely to be passed from ancestors to descendants. Suitable characters are genetically based and show variation in character states within the taxa of interest. Now that DNA sequencing is routine, molecular data are often used to reconstruct phylogenies. The same gene is sequenced across the organisms under study, and the DNA base (A, C, G, or T) present at each position in the sequence is compared. Each position in the sequence is independent of the other positions, so a single gene sequence provides many characters for comparison. Character states are commonly recorded in a character matrix like the one shown here for two morphological characters.

Taxon	Number of Toes	Skull Crests
A	3	Absent
B	2	Present
C	3	Present

There are many potential characters that can be recorded for any group of taxa of interest. Noting each character state is called **scoring** the characters. Shortly you will see how the distribution of character states among taxa can be used to predict the most likely path of evolution, but first we must outline all possible paths of evolution.

For three taxa, there are three possible paths that evolution could have taken. These three paths, all different, are depicted in the three cladograms shown in Figure 6-4. Examine the branching pattern in each to satisfy yourself that they differ in the arrangement of sister taxa.

How will we decide which of the three hypotheses about the path of evolution is most likely? There are many different methods for reconstructing the most likely path of evolution, but the general ideas for a parsimonious reconstruction can be illustrated using an **outgroup** method. This method uses the observed distribution of character states in the taxa under study and those in an outgroup. An outgroup is a taxon distantly related to the taxa under study. For example, if the three taxa under study were members of the dog family, Canidae, we could use another four-legged vertebrate family as the outgroup. A member of the cat family or the bear family might be appropriate. If taxa A, B, and C were all canids, where would you place a line leading to the outgroup on each of the cladograms? A line leading to the outgroup (X) belongs at the base of each tree because the ancestor of all canids diverged from the ancestor it shared with bears *before* there was further subdivision of the canids into the taxa we see now (A, B, and C). A tree with an outgroup added has three nodes. An outgroup joins the tree at the lowest node, node 1 (Figure 6-5).

The need for an outgroup arises because we want to know which state for each character is ancestral and which states are derived (more recently evolved). If a character has two or three states, how can we know which appeared first in evolutionary time? If a particular character state is present in one or several outgroups—for example, in members of the bear and cat families—it is likely to have evolved in the common ancestor of all four-legged vertebrates. Thus, if the state

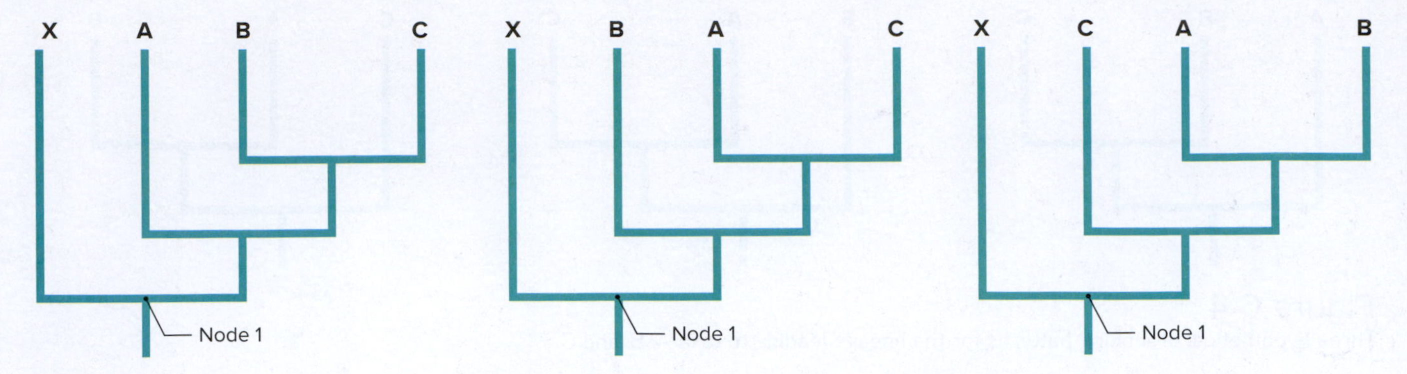

Figure 6-5

Three hypothetical branching patterns in relation to an outgroup for the lineage leading to taxa A, B, and C.

evolved in the common ancestor of all vertebrates, it must be the ancestral state for evolution within the dog family.

As shown in the table below, we add the outgroup (X) to our character matrix. We number the first two characters and add another character represented by C3.

Taxon	Character 1	Character 2	C3
A	3	Absent	Present
B	2	Present	Absent
C	3	Present	Present
X	1	Absent	Present

Notice that the character matrix can have both quantitative characters, represented here by the states 1, 2, and 3 in character 1, and qualitative characters, represented here by the presence/absence states for characters 2 and 3. We are going to code these data to make it easier to see patterns. They can be coded in different ways as long as one is consistent, but for the sake of simplicity, we will code the outgroup state as zero, and the other state as 1. If the character has more than two states, other states are coded as 2, then 3, etc. Codes are shown to the right of the states in the character matrix at the top of the next page. The last column is a summary of all the character state information for each taxon. We have condensed a variety of kinds of data on each taxon to a simple three-digit code in the summary.

We place each summary code from the table shown below above the appropriate taxon in each of the possible evolutionary pathways (Figure 6-6).

Taxon	Character 1	Character 2	Character 3	Summary
A	3 = 2	Absent = 0	Present = 0	200
B	2 = 1	Present = 1	Absent = 1	111
C	3 = 2	Present = 1	Present = 0	210
X	1 = 0	Absent = 0	Present = 0	000

Note that "absent" is coded as 0 for character 2 while "present" is coded as 0 for character 3.

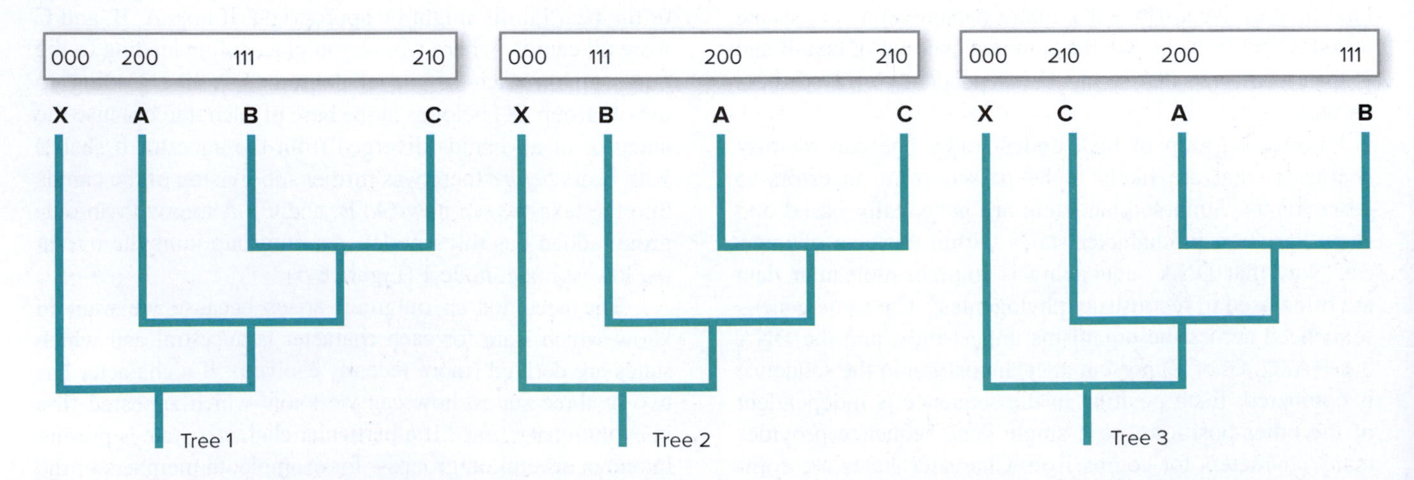

Figure 6-6

Three hypothetical branching patterns as in Figure 6-5, called trees 1 to 3, with added summary codes for each taxon.

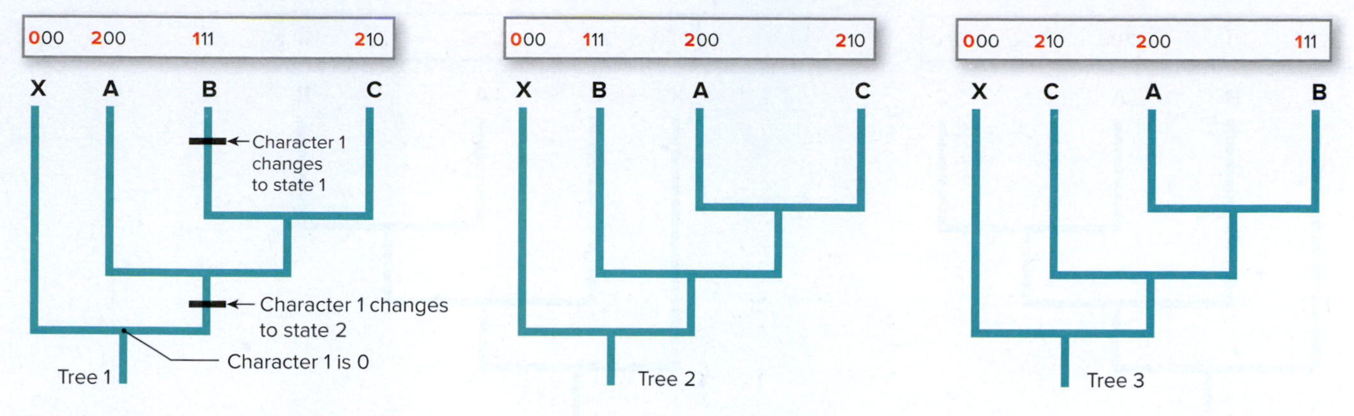

Figure 6-7

Tree 1, as in Figure 6-6, showing the evolutionary changes in character 1 that are required if this tree represents the true path of evolution. Trees 2 and 3, as in Figure 6-6, where students will map the evolutionary changes in character 1 that are required if each tree represents the true path of evolution.

For example, character 1 has state 0 in the outgroup, but it has state 1 or 2 in the taxa of interest. Assume that character 1 has state 0 at node 1 as shown on the tree in Figure 6-7. If this tree represents the true path of evolution, then character 1 changed to state 2 between nodes 1 and 2—this change allows taxa A and C to have state 2. However, taxon B has state 1 for character 1, so we must add another evolutionary step to the tree: a change to state 1 in the lineage leading to taxon B. Thus, the most parsimonious mapping procedure requires two evolutionary changes for character 1 on this tree.

To test your understanding, map character 1 onto the other two trees shown in Figure 6-7. It will take two evolutionary steps to map character 1 onto each tree, although the pattern of evolutionary change differs with each tree.

Now we map the characters onto each tree, *one character at a time,* noting where the characters must have changed to produce taxa with the character states that we observe now.

Continuing the procedure of mapping each character one at a time, we will turn to character 2 on tree 1 (Figure 6-8). From the outgroup, you know that character 2 has state 0 at node 1. Character 2 is still in state 0 at node 2 because taxon A has state 0 for this character. Character 2 requires only one change, between nodes 2 and 3, to correctly describe the pathway of evolution depicted on tree 1. Following this change, character 2 is in state 1 for taxa B and C.

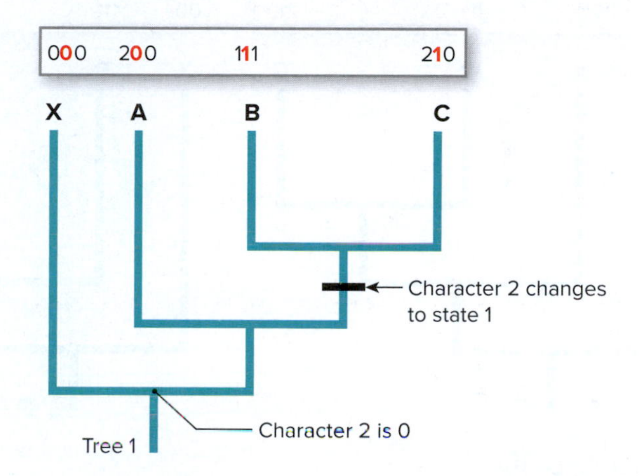

Figure 6-8

Tree 1, as in Figure 6-6, showing the evolutionary changes in character 2 that are required if this tree represents the true path of evolution.

Again, test your knowledge by mapping the changes needed to correctly depict the pathway of evolution for character 2 onto trees 2 and 3 (Figure 6-9). Did you discover that each of these trees required two changes to correctly map the evolution of character 2 onto them?

Map the changes needed for character 3 onto each tree (Figure 6-10). You have now mapped all the character changes needed for each character on each tree. Count up the total number of evolutionary changes required to map *all three characters* onto all three trees. List the total number of evolutionary changes required for:

Tree 1: ________________ Tree 2: ________________

Tree 3: ________________

Which tree represents the most likely path of evolution? We may use the Principle of Parsimony to decide which tree is the simplest possible explanation that covers all the available facts. The tree with the least number of changes is the simplest explanation for the character states that we observe in the extant taxa. This tree is our best attempt to reconstruct the phylogeny for these taxa given the data we have. Tree 1 should have the fewest changes, so check your answer.

What if two trees were to have an equivalent number of changes? Is there a way to distinguish between them? With the given data, either tree represents a likely pathway, but more characters could be studied to distinguish between two equally parsimonious trees.

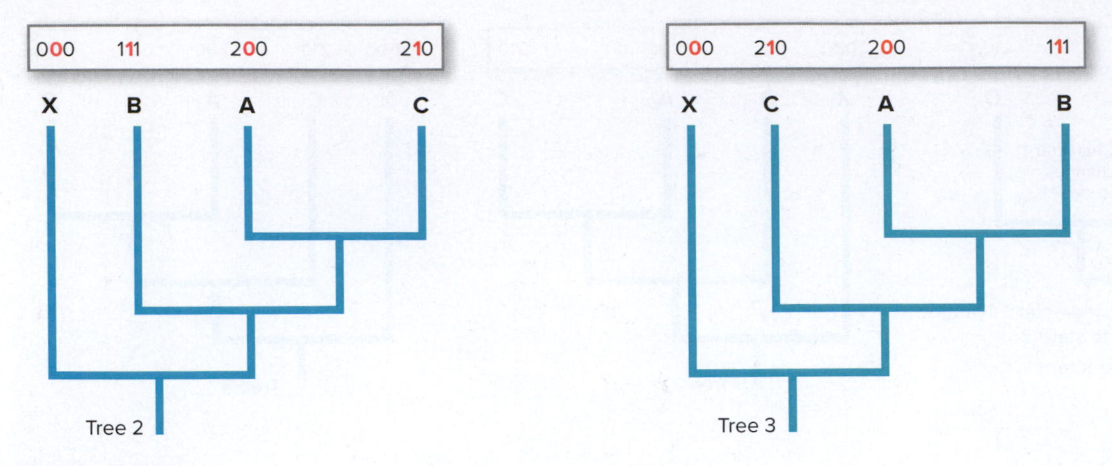

Figure 6-9

Trees 2 and 3, as in Figure 6-6, where students will map the evolutionary changes in character 2 that are required if each tree represents the true path of evolution.

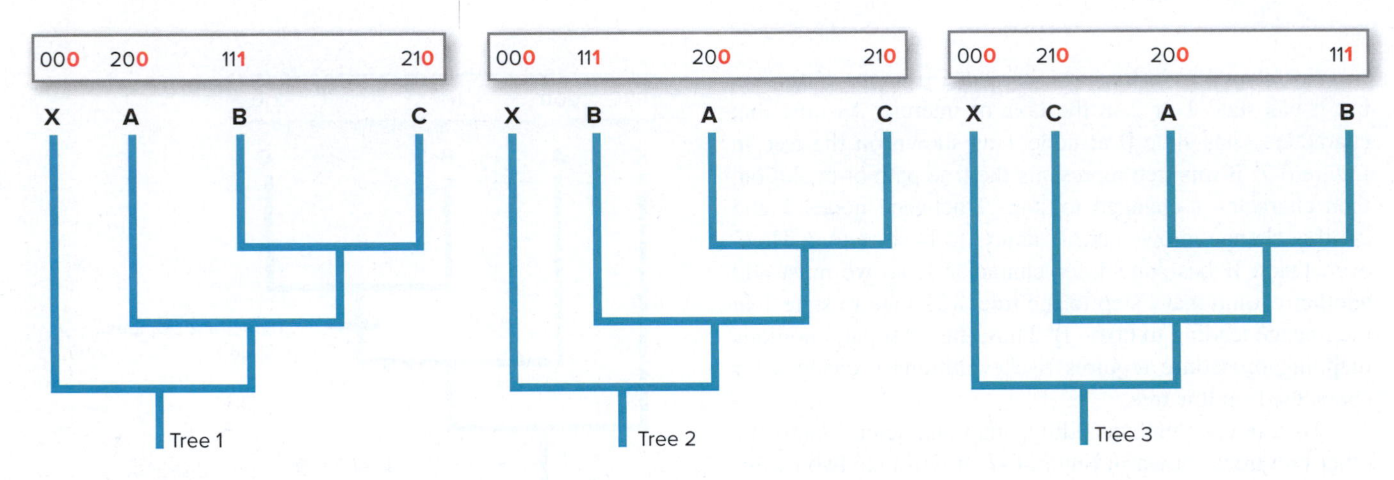

Figure 6-10

Trees 1, 2, and 3, as in Figure 6-6, where students will map the evolutionary changes in character 3 that are required if each tree represents the true path of evolution.

It may have occurred to you that when we use parsimony, we assume that all kinds of changes in characters are equally likely. Parsimony simply counts the number of changes, so evolving a skull crest is just as likely as losing a skull crest. For complex characters, such as eyes or limbs, this may not be a good assumption—it might be easier to lose the ability to make an eye than to re-evolve this ability. As biologists thought about molecular data, they also realized that not all changes in DNA bases in a gene sequence were equally likely. For example, recall that three DNA bases make a codon that specifies one amino acid in a protein. The first two bases are critical for specification of the amino acid, but the third base is not as critical—the genetic code is redundant. Because mutations in third-position bases will have fewer consequences for protein structure than mutations in bases in the first and second positions, we might expect to see more of them. When biologists have reason to assume that some changes (mutations) are more likely than others, they take this into account with a mathematical model that allows for different probabilities of changes. These model-based methods require significant computing power, especially for trees with many taxa. We saw that there were three possible trees for three taxa—there are 105 possible trees for five taxa, and over 34 million possible trees for ten taxa. To work with large data sets, you will need programming skills.

Now it is your turn to collect data on character states for a set of real organisms. We have selected four taxa in the following table, but if your instructor permits, you could choose other organisms and work your own example. We will use the sponge as the outgroup. We have filled in the first character for body symmetry. The next two characters can be described as present or absent, but the third character is quantitative. Score the remaining three characters. Code all characters using the system described previously and fill in the summary. Add more characters and make a bigger character matrix if your instructor permits.

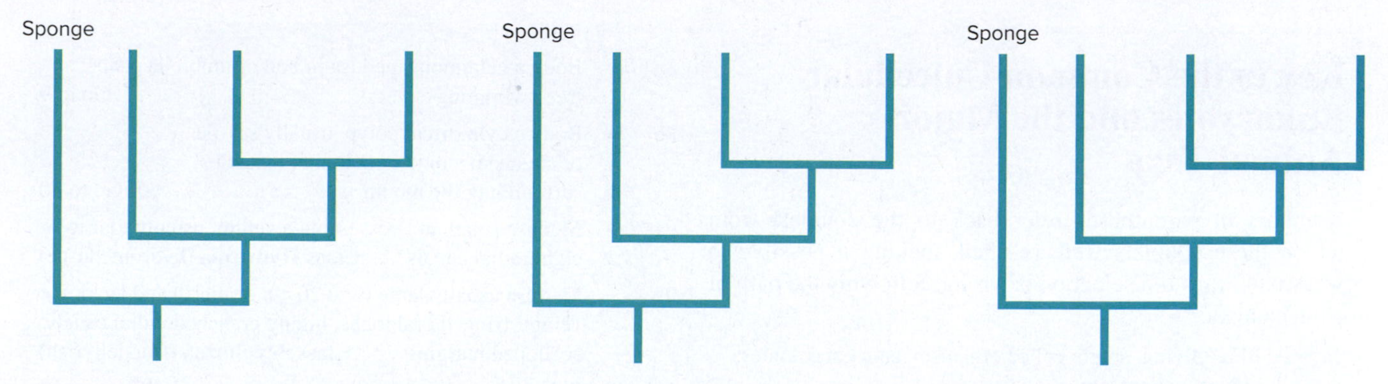

Figure 6-11
Three trees depicting a sponge as an outgroup where students will diagram the three possible branching patterns for a lineage leading to the three taxa shown in the table on the previous page. The required evolutionary changes for the four characters listed in the table are to be mapped onto each tree.

Diagram the three possible relationships for these taxa (Figure 6-11). Map the characters onto each tree one character at a time, but use the three trees for all the characters. Discover which tree is most parsimonious.

Which two taxa are most closely related to each other (sister taxa) according to your analysis? __________

Taxon	Symmetry	Head	Wings	Number of Legs	Summary
Frog	Bilateral				
Butterfly	Bilateral				
Jellyfish	Radial				
Sponge	Asymmetrical				0000

EXERCISE 6B

Further Study

Use of a Taxonomic Key for Organism Identification

In this exercise, you will identify specimens of unicellular eukaryotes and animals representing several different phyla and classes. Your instructor may assign this key or may substitute a different key based on forms common to your area. An alternative is the key to insect orders found in Exercise 16. The use of taxonomic key was explained on p. 69.

Once you have made an identification, it is important to verify its accuracy by consulting one or more references containing a text description of the distinctive characters of the species or group in question and a drawing or photograph of the species or of representatives of the group to which the species belongs.

Select a specimen and then, using the appropriate key and the following instructions, identify the specimen and record the phylum or class on pp. 87–88. Verify the identification in one or more of the reference books provided by your instructor. Identify as many of the specimens as requested by your instructor. Return all specimens to the proper trays.

Written Report

When you have identified and verified a specimen from one of the trays, give the reference source from which you verified the taxon. In giving references, always list author's surname first, followed by author's initials, year of publication, title of book or article, and publisher. Repeat for each animal identified completing the report on pp. 87–88.

Key to the Chief Phyla and Classes of Unicellular Eukaryotes and Animals

Following is a simple key to the more common phyla and classes of animals and free-living unicellular eukaryotes. The key, for the most part, uses external characters that can be visualized without dissection. It is designed for use with adult specimens. Like most keys, this key is utilitarian in the sense that the groups are not arranged in perfect phylogenetic sequence, and the characters used in the key may have no particular phylogenetic significance for the taxon. They are simply the characters that provide the best assurance of correct identification. There is one section of this key where three choices are given (see #32).

Key to the Common Unicellular Eukaryotes and the Major Animal Taxa

Numbers in parentheses refer back to the couplets from which these couplets were reached, making it possible to work backward if a selection is wrong, following the path of choices made.

1a Free-living, single-celled organism, some in colonies; chiefly microscopic, no tissues ....
Unicellular organisms .... Go to 2

1b Many-celled organisms, all heterotrophic, mostly macroscopic ....
Animals. .... Go to 4

2a (1a) Cilia or ciliary organelles in some stage; cilia used for locomotion; usually two types of nuclei; contractile vacuole and cytostome usually present ....
Phylum Ciliophora (ciliates)

2b No cilia; locomotion by pseudopodia, flagella, or both. . Go to 3

3a Locomotion by pseudopodia, body shape variable; body naked or with internal or external skeleton . . . **Amebas,** belonging to several taxa not easily distinguished by observation.

3b Flagellated, single-celled or colonial, cells containing green colored bodies (chloroplasts) ....
Phyla Viridiplantae or Euglenozoa

4a (1b) Body with numerous pores; body radially symmetrical or irregular, with one or more large openings (oscula); no mouth or digestive tract .... **Phylum Porifera** (sponges)

4b Body without numerous pores or oscula; usually symmetrical in outline; mouth and digestive tract usually present .... Go to 5

5a (4b) Body with radial, pentaradial, or biradial symmetry .... Go to 6

5b Body with bilateral symmetry .... Go to 15

6a (5a) Body mostly soft and gelatinous; cylindrical, umbrella-shaped, or somewhat spherical; body parts usually in divisions of four, six, or eight .... Go to 7

6b Body usually hard and spiny or with leathery skin; body parts in divisions of five (pentaradial); tentacles branched when present; usually with tube feet .... Go to 11

7a (6a) Body cylindrical or umbrella-shaped; mouth or rim of umbrella usually encircled with unbranched tentacles; cnidocytes (stinging cells) present; mostly marine .....
Phylum Cnidaria (jellyfish, hydroids, corals, sea anemones, and relatives)........................ Go to 8

7b Symmetry biradial; one pair of tentacles (not encircling mouth) or none; eight radial rows of ciliated comb plates ...... **Phylum Ctenophora** (comb jellies, sea walnuts)

8a (7a) Body a gelatinous medusa in bell or umbrella shape; free-swimming .... Go to 9

8b Body a cylindrical polyp, usually sessile or attached; single or colonial; tentacles surrounding the mouth.................... Go to 10

9a (8a) Medusa small and possessing a velum; usually four to eight radial canals ... **Class Hydrozoa** (hydromedusae)

9b Medusa usually large (2 to 20 cm or more) and lacking a velum; fringed oral lobes; highly branched radial canals; scalloped margins..... **Class Scyphozoa** (true jellyfish)

10a (8b) Polyps typically small, often in branching colonies with more than one type of polyp; gastrovascular cavity not divided by septa .... **Class Hydrozoa** (hydroids)

10b Polyps typically large (>0.5 cm diameter); gastrovascular cavity divided by septa .... **Class Anthozoa** (sea anemones and corals)

11a (6b) Body without arms; body with hard endoskeleton of calcareous plates, or with soft, leathery skin ..... Go to 12

11b Body with branched or unbranched arms; body with hard endoskeleton of calcareous plates .... Go to 13

12a (11a) Body with rigid endoskeleton; body globular or flattened, with movable spines .... **Class Echinoidea** (sea urchins, sea biscuits, sand dollars)

12b Saclike body with leathery skin; body elongated in mouth-to-anus axis; mouth surrounded by branching tentacles .... **Class Holothuroidea** (sea cucumbers)

13a (11b) Body with five movable, branched, and feathery arms; stalked or free-swimming; mouth and anus on oral surface, which is directed upward....... **Class Crinoidea** (sea lilies, feather stars)

13b Body with unbranched arms (except in some brittle stars); oral surface directed downward........ Go to 14

14a (13b) Arms not sharply set off from central disc; ambulacral grooves with tube feet on ventral side of each arm............... **Class Asteroidea** (sea stars)

14b Arms sharply set off from central disc; no ambulacral grooves............ **Class Ophiuroidea** (brittle stars)

15a (5b) No gill slits on pharynx, no internal skeleton (skull or vertebrae) of cartilage or bone.............. Go to 16

15b Lateral gill slits in pharynx present, or body with internal skeleton of cartilage or bone, or both; nerve cord dorsal and single .... Go to 38

16a (15a) Body slender, wormlike or leaflike, with no body segments, lateral appendages or fins, or shell..... Go to 17

16b Body not as above; if body wormlike, then having appendages or segments or both .... Go to 23

17a (16a) Body flat and soft, rarely cylindrical......... Go to 18

17b Body narrowly cylindrical with hard or tough cuticle........................... Go to 21

18 (17a) Body flattened dorsoventrally; no anus, no proboscis . . . **Phylum Platyhelminthes** (flatworms) . . . . Go to 19

18b Body long, soft, and highly contractile; long eversible proboscis present above mouth; digestive tract with anus; mostly free-living . . . . **Phylum Nemertea** (ribbon worms)

19a (18a) Mouth and digestive tract present . . . . Go to 20

19b No mouth or digestive tract; body of scolex and usually numerous boxlike pseudosegments (proglottids); body increasing in size posteriorly, usually long and ribbon-like . . . . **Class Cestoda** (tapeworms)

20a (19a) No attachment organs, no suckers around mouth; ciliated epidermis . . . . **Class Turbellaria** (planarians)

20b Attachment organs present, some with hooks at posterior end; suckers present around mouth; nonciliated epidermis; body leaflike or slender in shape . . . . . **Class Trematoda** (digenetic flukes)

21a (17b) Body round in cross-section, both ends of body pointed . . . . **Phylum Nematoda** (roundworms)

21b Body not as described above . . . . Go to 22

22a (21b) Body extremely slender and elongate . . . . **Phylum Nematomorpha** (horsehair worms)

22b Anterior retractile proboscis armed with spines . . . . **Phylum Acanthocephala** (spiny-headed worms)

23a (16b) Body slender and torpedo-shaped, with lateral and caudal fins; mouth with bristles or spines; planktonic . . . . . **Phylum Chaetognatha** (arrow worms)

23b Body not as above . . . . Go to 24

24a (23b) Body soft and unsegmented; body enclosed in shell, or body with ventral muscular foot or both; some with fleshy arms or tentacles . . . . Go to 25

24b Body segmented (often wormlike), with jointed appendages, or both . . . . Go to 30

25a (24a) Shell of two valves arranged in dorsal and ventral position to each other; ventral shell usually larger than dorsal; stalk or peduncle for attachment . . . . **Phylum Brachiopoda** (lamp shells)

25b Shell single, or of two lateral valves, or of eight dorsal plates; or shell absent or reduced and internal; ventral muscular foot or fleshy arms or tentacles present; body unsegmented . . . . **Phylum Mollusca** (molluscs) Go to 26

26a (25b) No prehensile arms with suckers; eyes small or absent . . . . Go to 27

26b Head large and well developed with two large eyes and foot modified into 8 or 10 prehensile arms with suckers . . . . **Class Cephalopoda** (nautiluses, squids, cuttlefishes, and octopuses)

27a (26a) Shell of eight dorsal plates; radula present . . . . . . **Class Polyplacophora** (chitons)

27b Single shell, or shell of two valves, or shell absent or internal . . . . Go to 28

28a (27b) Shell of two lateral valves with ligamentous hinge; muscular foot present; head reduced; no tentacles or radula. . . . . **Class Bivalvia** (bivalves)

28b Shell of one piece or absent or internal . . . . Go to 29

29a (28b) Shell tubular and open at both ends; head absent; mouth with tentacles and radula . . **Class Scaphopoda** (tooth shells)

29b Shell usually coiled or spiraled (uncoiled or absent in some); head with radula, one or two pairs of tentacles, and one pair of eyes . . **Class Gastropoda** (snails, slugs, nudibranchs, tectibranchs)

30a (24b) Body wormlike and segmented throughout; setae, parapodia, or both often present; no jointed appendages . . . . **Phylum Annelida** (segmented worms) . . . . Go to 31

30b Segmented body encased in firm exoskeleton of chitin; jointed appendages . . . . **Phylum Arthropoda** (arthropods) . . . . Go to 33

31a (30a) Setae present on each segment; segments distinct; suckers absent . . . . Go to 32

31b Setae absent; no parapodia; segments indistinct and with many annuli; clitellum present; suckers present anteriorly and posteriorly or posteriorly only . . . . . **Clitellata, Hirudinida** (leeches)

32a (31a) Many setae on each segment; parapodia or fleshy lateral appendages present (may be reduced); clitellum absent . . . . **Errantia** (marine motile segmented worms)

32b Many setae, parapodia reduced or absent, polychaete body plan, but sedentary, sessile or tube-dwelling. Clitellum absent. . . . . **Sedentaria, exclusive of Clitellata** (fan worms and other sedentary polychaetes).

32c Few setae per segment, no parapodia, clitellum present . . . . **Sedentaria, Clitellata, exclusive of Hirudinida** (earthworms and other oligochaetes).

33a (30b) Paired antennae present . . . . Go to 34

33b Antennae absent; body of cephalothorax and abdomen; segmentation often obscured . . . . Go to 37

34a (33a) Two pairs of antennae; appendages mostly biramous and specialized for different functions; many with gills; head, thorax, and abdomen present but head and at least part of thorax fused . . . . **Subphylum Crustacea** (crustaceans)

34b One pair of antennae . . . . Go to 35

35a (34b) Head, thorax, and abdomen distinct; three pairs of legs on thorax; one or two pairs of wings often present . . . . . **Subphylum Hexapoda, Class Insecta** (insects)

continued

Key to the Common Unicellular Eukaryotes and the Major Animal Taxa—*continued*

35b Body elongate, 15 or more pairs of jointed legs .... **Subphylum Myriapoda**, ... Go to 36

36a (35b) Each segment with one pair of legs; dorsoventrally flattened .... **Class Chilopoda** (centipedes)

36b Each segment with two pairs of legs; subcylindrical body .... **Class Diplopoda** (millipedes)

37a (33b) Four pairs of walking legs; head completely fused with thorax; no wings. . **Class Arachnida** (spiders, scorpions, ticks)

37b Five pairs of walking legs; lateral compound eyes present. .... **Class Merostomata** (horseshoe crabs)

38a (15b) Short and wormlike; body divided into proboscis, collar, and trunk ..... **Phylum Hemichordata** (acorn worms)

38b Body not wormlike. .... Go to 39

39a (38b) Cranium and brain absent .... Go to 40

39b Cranium and brain present. .... Go to 41

40a (39a) Adults saclike and sedentary; body covered with a test and with two siphons at one end .... **Subphylum Urochordata** (tunicates)

40b Body lance-shaped; lateral musculature in conspicuous V-shaped segments; notochord and dorsal nerve cord extend length of body. . . **Subphylum Cephalochordata** (amphioxus)

41a (39b) Body fishlike .... Go to 42

41b Body not fishlike .... Go to 45

42a (41a) Without true fins, scales, or paired fins . . . **Vertebrata, Agnatha, Cyclostomata** (hagfishes, lampreys) .... Go to 43

42b With jaws and (usually) paired appendages; notochord replaced by vertebrae . . . . . **Vertebrata, Gnathostomata** (jawed fishes, all tetrapods) .... Go to 44

43a (42a) Suctorial mouth with horny teeth; seven pairs of gill pouches. .... **Petromyzontida** (lampreys)

43b Terminal mouth with 4 pairs of tentacles; 5 to 15 pairs of gill pouches .... **Myxini** (hagfishes)

44a (42b) Skeleton cartilaginous; ventral mouth; placoid scales or no scales; five to seven pairs of gills, each in a separate pharyngeal cleft . . . . . **Chondrichthyes** (sharks, rays, skates)

44b Skeleton mostly bony; body primarily fusiform but variously modified; cycloid, ganoid, or ctenoid scales; terminal mouth with many teeth .... **Actinopterygii** (ray-finned fishes)

45a (41b) Skin with horny epidermal scales, sometimes with bony plates; paired limbs, usually with five toes, or limbs absent; no gills . . . . **Nonavian Reptiles.** (snakes, turtles, lizards, crocodilians)

45b Skin without scales .... Go to 46

46a (45b) Epidermis with feathers on body and scales on legs; forelimbs (wings) adapted for flying; toothless horny beak .... **Avian Reptiles** (Aves, birds)

46b No feathers, wings, or beak .... Go to 47

47a (46b) Skin naked, often moist, and sometimes warty . . . **Class Amphibia** (frogs, toads, salamanders, caecilians)

47b Body covered with hair (reduced in some); integument with sweat, sebaceous, and mammary glands . . . . **Class Mammalia** (mammals)

Use of a Key for Animal Identification

Name ______________________

Date ______________________

Section ______________________

LAB REPORT

In each box, print the number of a specimen and taxon (phylum or class) or, if other animal keys are used, the species (genus name and species epithet) as required by the instructor. Enter the complete reference consulted to verify the identification.

A __________ ______________________

Reference

B __________ ______________________

Reference

C __________ ______________________

Reference

D __________ ______________________

Reference

E __________ ______________________

Reference

F __________ ______________________

Reference

LAB REPORT

G ________ ______________________

Reference

H ________ ______________________

Reference

I ________ ______________________

Reference

J ________ ______________________

Reference

K ________ ______________________

Reference

L ________ ______________________

Reference

Unicellular Eukaryotes Phylum Amoebozoa, Phyla Euglenozoa and Viridiplantae, Phylum Apicomplexa, Phylum Ciliophora

The unicellular eukaryotes discussed here are a diverse assemblage characterized by two things: (1) absence of a cell wall and (2) presence of at least one motile stage in the life cycle. All functions of life are performed within the limits of a single plasma membrane. There are no organs or tissues; instead organelles function as skeletons, locomotory systems, sensory systems, conduction mechanisms, defense mechanisms, and contractile systems.

Unicellular eukaryotes are widespread ecologically, being found in fresh, marine, and brackish water and in moist soils. Some are free-living; others live as parasites or in another symbiotic relationship. See Figure 7.1 for a phylogeny of these organisms, and pp. 113–114 for a classification of unicellular eukaryotes.

EXERCISE 7A
Phylum Amoebozoa—*Amoeba* and Others
Amoeba
Other Amebas
Projects and Demonstrations

EXERCISE 7B
Phyla Euglenozoa and Viridiplantae—*Euglena, Volvox,* and *Trypanosoma*
Euglena
Volvox
Trypanosoma
Projects and Demonstrations

EXERCISE 7C
Phylum Apicomplexa—*Plasmodium* and *Gregarina*
Plasmodium
Gregarina

EXERCISE 7D
Phylum Ciliophora—*Paramecium* and Other Ciliates
Paramecium
Other Ciliates
Project
Classification of Unicellular Eukaryotes

EXPERIMENTING IN ZOOLOGY
Effect of Temperature on the Locomotor Activity of *Stentor*

EXPERIMENTING IN ZOOLOGY
Genetic Polymorphism in *Tetrahymena*

EXERCISE 7A*

*Phylum Amoebozoa—*Amoeba *and Others*

Core Study

Amoeba

Phylum Amoebozoa
Species *Amoeba proteus*

Where Found

An ameba is a unicell with a flexible plasma membrane that can be extended into a wide variety of shapes. The amebas** may be naked or enclosed in a shell. Shelled amebas belong to several different clades—find Foraminifera, "Radiolaria," and Centrohelida in Figure 7-1. Naked amebas also occur in multiple clades, including Heterolobosea and Ameobozoa,

*An engaging initial exploration of unicellular eukaryotic diversity that may be implemented before beginning Exercise 7A is described in Projects and Demonstrations, p. 94; see also p. 113.

**The term "ameba" is used by many authors as a common name for any of the naked or shelled amebas. When referring to a specific genus, the name is latinized and italicized, such as *Amoeba*, *Pelomyxa*, and *Entamoeba.*

***The genus name (*Amoeba*; from the Greek, meaning "change") may be abbreviated when used in a context in which it is understood. Recall, however, that the species is binomial (*Amoeba proteus*) and that the species epithet (*proteus* , Gr. *Proteus* , sea god that could change shape at will) has no taxonomic meaning by itself (see p. 77).

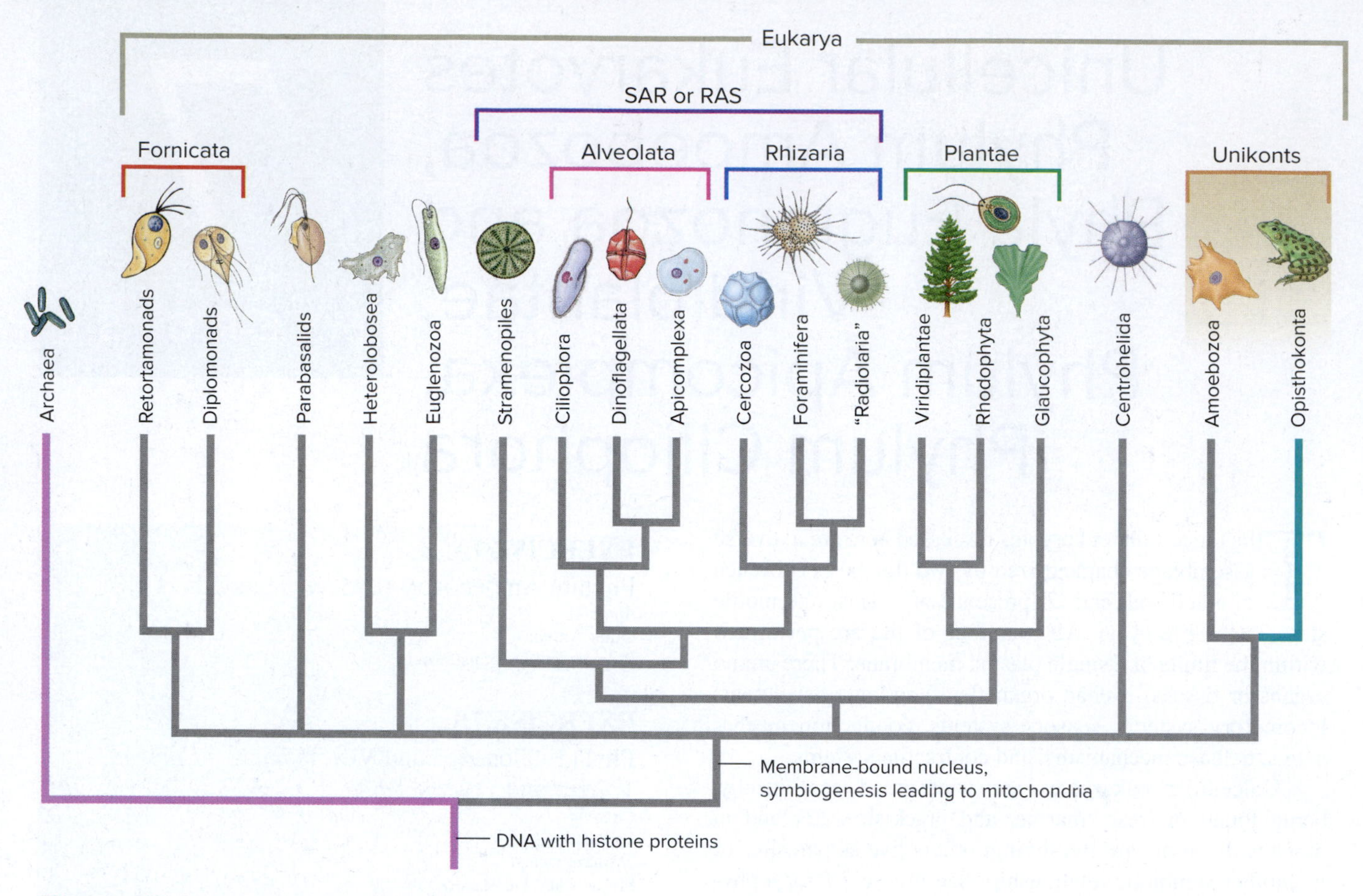

Figure 7-1
This cladogram illustrates relationships among the major eukaryotic clades. Many of these clades contain unicellular forms, but some, such as Stramenopiles, Plantae, and Unikonts, comprise unicellular and multicellular taxa.

as illustrated in Figure 7-1. Naked amebas, which include the genera *Amoeba* and *Pelomyxa,* live in both freshwater and seawater and in soil. They are bottom dwellers and must have a substratum on which to glide. *A. proteus**** is a freshwater species that is usually found in slow-moving or still-water ponds. These amebas are often found on the underside of lily pads and other water plants. They feed on algae, bacteria, unicellular eukaryotes, rotifers, and other microscopic organisms.

Study of Live Specimens

In the center of a clean slide, the instructor will place a drop of culture drawn from the bottom of the undisturbed culture bottle.* Adjust the iris diaphragm of the microscope to provide *subdued light.* With low power, explore the contents of the slide, at first *without a coverslip.*

*Amebas will be found at the bottom of the shipping container or petri dish into which they are transferred for the laboratory. Amebas must be removed with a pipette without agitating the container by squirting the contents of the pipette back into the culture. It helps to pour off most of the culture water before the lab, leaving only a few millimeters in the container. Many instructors find it best to pipette the amebas for the students.

You may see some masses of brownish or greenish plant matter, and there will probably be some small ciliates or flagellates moving about. The ameba, in contrast, is gray, rather transparent, irregularly shaped, and finely granular in appearance (Figure 7-2). If it is not apparently moving, watch it a moment to see if the granules are in motion or if the shape is slowly changing.

After you have found specimens and observed their locomotion, carefully cover with a coverslip. You will need to support the coverslip to prevent crushing the amebas as the water evaporates. Place two short lengths of hair or thread or two pieces of broken coverslip, one on each side of the drop of culture, before adding the coverslip. *Never use high power without a coverslip.* From time to time add a drop of culture water to the edge of the coverslip to replace water lost by evaporation.

Some types of substage microscope lamps will warm the stage enough to kill the ameba. Turn off the lamp when not actually viewing the specimen.

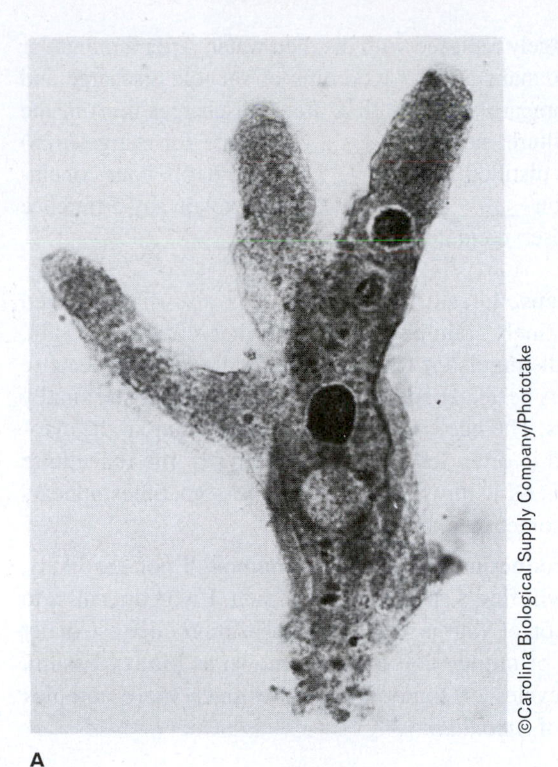

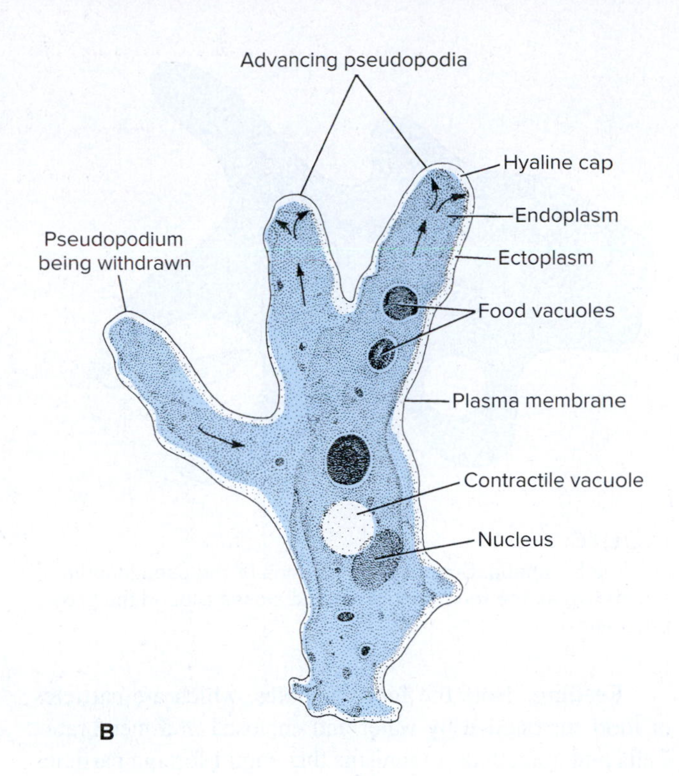

Figure 7-2
Ameba. **A**, Whole mount of a living specimen. **B**, Interpretive drawing.

General Features

The outer plasma membrane is fringed with fine, hairlike projections (too small to be seen with the light microscope), which are thought to be involved in adhesion of the cell surface to the substratum or to nutrient particles and to aid in capture and intake (ingestion) of food.

The cytoplasm enclosed by the plasma membrane is differentiated into a thin, peripheral rim of stiff **ectoplasm** and an inner, more fluid, **endoplasm.** The ectoplasm looks clear, even glassy, because it lacks the subcellular organelles present in abundance in the endoplasm (Figure 7-2B). As a result the clear ectoplasm is often referred to as a **hyaline cortex,** or "glassy rind."

Locomotion. An ameba moves and changes shape by thrusting out **pseudopodia** ("false feet"), which are extensions of the cell body. The ameba creeps forward when a protruding pseudopodium adheres to a surface and contractile elements pull the trailing end of the cell forward. The advancing end of the pseudopodium is called the **hyaline cap.** As endoplasm flows into the hyaline cap, it fountains out into the cortex and is converted into stiff ectoplasm, thus extending the sides of the pseudopodium much like a tube or sleeve. As the pseudopodium lengthens, ectoplasm at the temporary "tail end" converts again into streaming endoplasm to replenish the forward flow. At any time the action can be reversed—the endoplasm streaming back, the tube shortening, and another pseudopodium forming elsewhere.

Actin and myosin, the major proteins in animal muscles, also occur in unicells. Within the cell cytoplasm, free subunits of actin can be assembled to form fibers or a meshwork. Actin fibers push against the plasma membrane to extend it as a narrow pointed pseudopodium, called a filopodium, whereas an actin meshwork extends the plasma membrane as a broad, blunt pseudopodium, called a lobopodium. The alternate rapid assembly and disassembly of actin fibers or meshworks are responsible for the propulsive force of ameboid movement.

Observe the formation of pseudopodia. Does the ameba have a permanent anterior and posterior end? __________ Can you observe the change from endoplasm to ectoplasm, and vice versa? Does the ameba move steadily in one direction? __________ Does more than one pseudopodium ever start at once? __________ How is a pseudopodium withdrawn? __________ What happens when the ameba meets an obstruction? __________ Tap the coverslip gently with the tip of a pencil or probe. Does the ameba respond? __________ Do pseudopodia ever extend vertically as well as laterally? __________.

Drawings

On separate paper, make a series of five or six outline sketches, each about 2 cm in diameter, to show changes in shape that occur in an ameba in a period of 1 to 10 minutes. Sketch rapidly and use arrows freely to indicate direction of flow of the cytoplasm. Indicate magnification of the drawings. In the space next to your sketches, briefly explain how this ameboid movement is achieved.

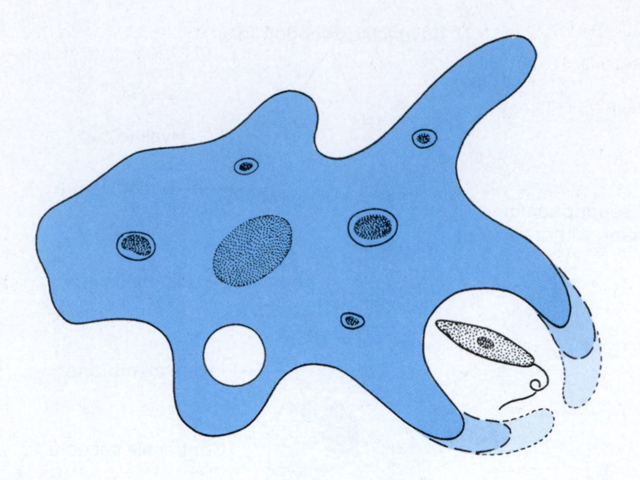

Figure 7-3
Feeding by ameba. Successive positions of the pseudopodia are shown as the food cup forms and closes around the prey organism.

Feeding. Note the **food vacuoles,** which are particles of food surrounded by water and enclosed in a membrane. Cells and unicellular organisms that engulf foreign particles are called **phagocytes,** and this type of ingestion is known as **phagocytosis** (Gr. *phagein,* to eat, + *kytos,* hollow vessel) (Figure 7-3). Phagocytosis involves encircling the prey by pseudopodia. Subsequently the prey, along with a quantity of water, becomes completely enclosed by cytoplasm to form a food vacuole. The contents are digested by hydrolytic enzymes secreted into the food vacuoles when tiny membrane-bound organelles called **lysosomes** merge with the food vacuole.

Undigested end products can be eliminated (egested) at any point along the plasma membrane.

Cells also take in fluid droplets and minute food particles by a process of channel formation called **pinocytosis** (Gr. *pinein,* to drink, + *kytos,* hollow vessel).

Osmoregulation. Look for the **contractile vacuole,** a clear bubble containing no particles. Note that this bubble gradually increases in size by accumulation of fluid and then ruptures and disappears. This organelle rids the ameba of the excess water that has been taken in along with food vacuoles or acquired by osmosis. The contractile vacuole is surrounded by a network of continuous membranous channels that are populated by proton pumps. These pumps create an osmotic gradient that moves water into the vacuole. The vacuole is then pushed against the plasma membrane, where it ruptures and empties to the outside. Note that the vacuole, as it enlarges, tends to be located in the temporary posterior end of the moving ameba.

Over a period of several minutes, time the appearance and disappearance of the contractile vacuole of an ameba in the culture medium. Then place a drop of distilled water at the edge of the coverslip and draw the water through to the other side, using a piece of absorbent paper. Repeat. The culture fluid now has been largely replaced with distilled water. After a minute or so, make another recording of vacuole discharge and compare with the first. Rate (discharges/min) in the culture medium. __________ Rate (discharges/min) in distilled water. __________ What is your conclusion? __________ How do you explain any difference observed in discharge rate? __________.

Nucleus. Locate the **nucleus.** It is disc-shaped, often indented, finely granulated, and somewhat refractive to light. You can distinguish it from the contractile vacuole because the latter is perfectly spherical, increases in size, and finally disappears. The nucleus is usually carried along in the cytoplasm and is often found near the center of the unicellular eukaryote. As it turns over and over, it sometimes appears oval and sometimes round.

Reproduction. It is possible, although not too likely, that you will find a specimen in division. If you do, call it to the attention of your instructor. Amebas reproduce asexually by a type of mitotic cell division known as **binary fission.** The life cycles of some amebas are much more complex than that of *Amoeba.*

Written Report

Answer the questions on pp. 95–96.

Further Study

Study of a Stained Slide

On a stained slide, locate the structural features that you studied in the living specimen. Note especially the **nucleus, plasma membrane, ectoplasm, endoplasm, contractile vacuole,** and **food vacuoles.**

Other Amebas

Examine stained slides or specimens from living cultures of as many of the following amebas as are available in your laboratory (Figures 7-4 and 7-5).

Parasitic Amebas

A number of species of *Entamoeba* are found in humans and other vertebrates. *E. gingivalis* lives in the mouth and feeds on bacteria around the base of the teeth. *E. histolytica* lives chiefly in the large intestine, where it causes dysentery by feeding on cells lining the gut and red blood cells (Figure 7-4). Biologists have recently discovered that this ameba nibbles on host cells, cutting away and ingesting parts of the cell in a process called trogocytosis. *E. histolytica* exists in two phases—a trophozoite (trōf′uh-zō′īt; *trophe,* food, + *zōon,* animal), or active feeding phase, and an encysted phase, in which nuclear divisions occur. The mature cyst usually contains four small nuclei. After ingestion by a suitable host, the multinucleate ameba emerges from its cyst

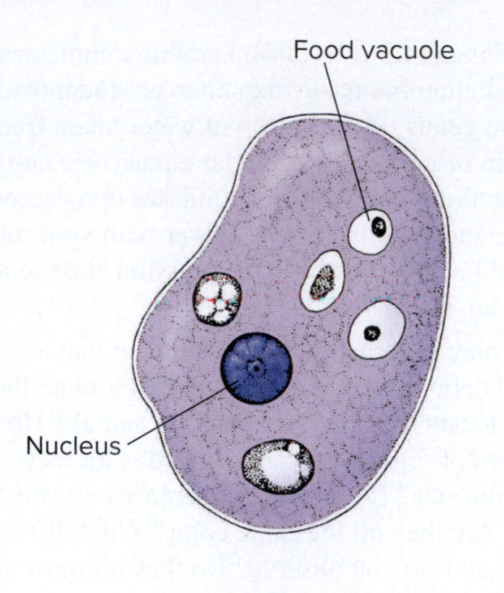

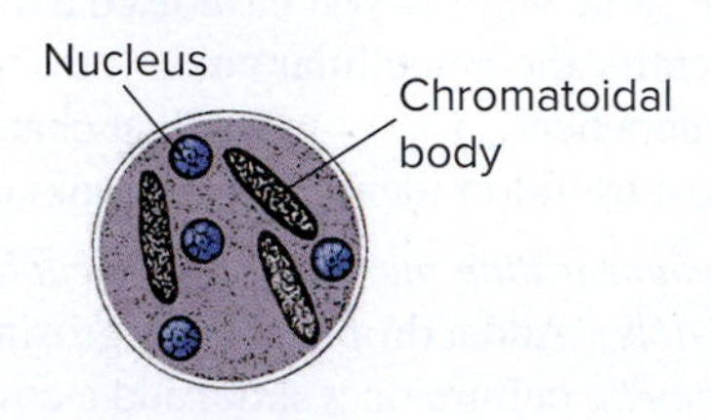

Figure 7-4

Entamoeba histolytica, the cause of amebic dysentery in humans. The trophozoite *(top image above)* is the actively moving and feeding form. The cyst *(lower image above)* can tolerate conditions outside the body and is infective to the new host. Note the four small nuclei and chromatoidal bodies.

and undergoes a series of fissions, resulting in uninucleate daughter amebas.

Shelled Amebas

Unicellular eukaryotes employ many kinds of materials for building their shells, or tests (Figure 7-5). *Arcella,* which occurs in bogs or swamps where much vegetation exists, has a hemispherical, hat-shaped skeleton (up to 260 μm in diameter) made up of siliceous or chitinous material set in a base of polymerized proteins. Small, fingerlike pseudopodia extend through an opening in the flat underside of the skeleton.

Difflugia is often found in leaf-choked puddles or in delicate aquatic vegetation. This organism has an inverted, flask-shaped skeleton (up to 500 μm long) made up of sand grains cemented together with the polymerized protein base, sometimes with the addition of diatom shells or sponge spicules. In living cultures from biological supply houses, they are often seen feeding on the single-celled filamentous algae, *Spirogyra.*

Foraminiferans

Foraminiferans, such as *Globigerina* (Figure 7-5), are marine amebas that secrete a skeleton of one or more chambers, usually calcareous and sometimes incorporating silica, sand, or sponge spicules. Long, delicate feeding pseudopodia extend through pores in the skeleton. As the foraminiferan grows, it adds new chambers to the skeleton. Upon death their skeletons join the ooze of the ocean bottom, eventually forming limestone or chalk beds. Foraminiferan skeletons made the White Cliffs of Dover and the stone quarried for the great pyramids of Egypt.

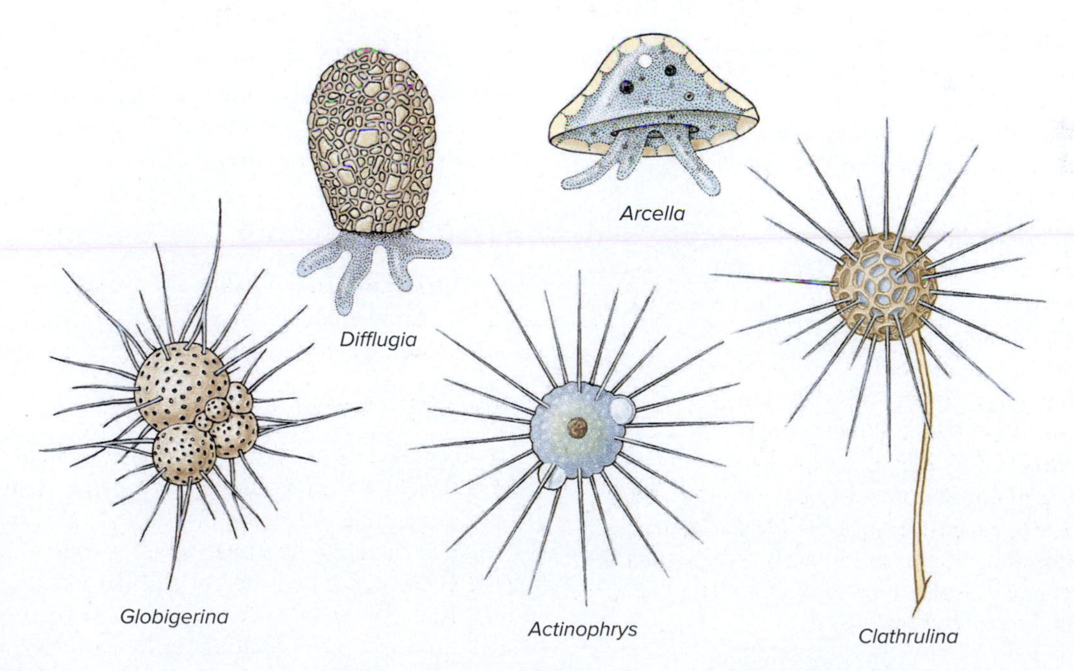

Figure 7-5

Some ameboid unicellular eukaryotes, showing different types of pseudopodia. *Arcella* and *Difflugia* are shelled amebas with blunt, lobelike pseudopodia called lobopodia. *Globigerina* is a foraminiferan with delicate pseudopodia, called filopodia, that extend through pores of a multichambered skeleton. If filopodia merge to form a network, they are called reticulopodia. *Actinophrys* and *Clathrulina* have long, slender feeding pseudopodia called axopodia that are characterized by a central array of microtubules.

Other Shelled Amebas

Centrohelids are amebas whose long, pointed axopodia typically extend through a coat of silica scales. Axopodia are stiffened by a core of microtubules. Centrohelids are predators, primarily living in freshwater.

Radiolarians are marine amebas that secrete about themselves a transparent skeleton of silica or other substances. They extend slender axopodia through pores in the skeleton to capture bacteria, microflagellates, and microalgae. They float in surface plankton; when they die, their skeletons become a part of the enormous ocean bottom ooze and often become chert. Examine a prepared slide showing a variety of shell types.

Drawings

Sketch any of the shelled amebas or their skeletons on separate paper or below.

Projects and Demonstrations

1. *Initial exploration of unicellular eukaryotes.* This exercise can introduce you to diversity, using a sample of natural pond water or a commercially available culture. For example, a Carolina culture comes with a simple dichotomous key that enables students to identify to genus. Place a drop of water taken from the bottom of a mixed unicellular eukaryotes culture sample on the center of a clean microscope depression slide and examine at low power with your microscope. Add a coverslip to the depression slide to avoid evaporation.

   You may find that you have sampled masses of algae and debris. Look carefully on and around the debris to locate the organisms in your sample. How many types of organisms can you find? Are they all the same size? If not, what differences do you observe? Are they all the same color? What differences in color do you observe? Do they all move in the same way? If you have used a dichotomous key to identify the unicellular eukaryotes to genus, list the genera here.__________ What characteristics seem most useful in identifying the unknowns?

2. *Demonstrating the contractile vacuole (water-expulsion vesicle).* Add a drop of 10% nigrosine to a drop of *Amoeba* culture on a slide, add a coverslip, and examine with high power.

Phylum ____________________

Subphylum ________________

Genus ______________________

Name ______________________________

Date ______________________________

Section ____________________________

LAB REPORT

The Ameba

1. What is the average size of specimens on your slide? __________

2. Do amebas react to stimuli? ____________________ What reactions have you observed? __________

3. What is the main function of the contractile vacuole? ____________________

4. What was the vacuole's discharge rate with the ameba in culture medium? ____________________

   Discharge rate in distilled water? ____________________ Explain any observed difference in discharge rate.

5. How does the ameba respire? ____________________

6. How does the ameba reproduce? ____________________

7. Knowing the principal mechanism of reproduction in amebas, what can you hypothesize about the genetic diversity of amebas in the culture you are examining? ____________________

8. If you observed ingestion or egestion describe what you saw. Could you identify the contents of any food vacuoles?

LAB REPORT

9. Distinguish between phagocytosis and pinocytosis.

___

___

10. How would you describe and explain ameboid movement? (Include answers to questions on p. 91.)

___

___

___

___

___

___

___

Other Amebas

1. If you found an ameba-like organism in a scraping taken from your teeth, what genus would it probably belong to? ___ What would it probably have been eating?

___

2. Drinking untreated water when on vacation in a foreign country can sometimes result in infection by *Entamoeba histolytica.* What disease does this organism cause and what are the symptoms?

___

3. How would you determine whether you have foraminiferans, centrohelids, or radiolarians in an unknown sample? Assume that you know the medium in which the organisms were living and that you can employ simple chemical tests on dried samples to distinguish them, if you think this might help.

___

___

EXERCISE 7B

Phyla Euglenozoa and Viridiplantae—*Euglena, Volvox,* and *Trypanosoma*

Core Study

Members of these phyla are unicellular eukaryotes having one or more flagella—undulating, whiplike organelles that move the cells efficiently through their fluid environment. Find the flagellated taxa in Figure 7-1. Did you notice that many clades comprise flagellated cells? We can identify two major ways that flagellates feed. The autotrophic flagellates are pigmented species, such as *Euglena* and *Volvox,* that possess chlorophyll and consequently can synthesize carbohydrates from carbon dioxide and water like any green plant.

The heterotrophic flagellates are colorless, lack photosynthetic pigments, and obtain their nutrients by absorbing them through the plasma membrane or by engulfing prey in food vacuoles. Many are free-living but some, such as *Trypanosoma,* parasitize humans and other animals. Some flagellates may switch between autotrophy and heterotrophy depending on environmental conditions.

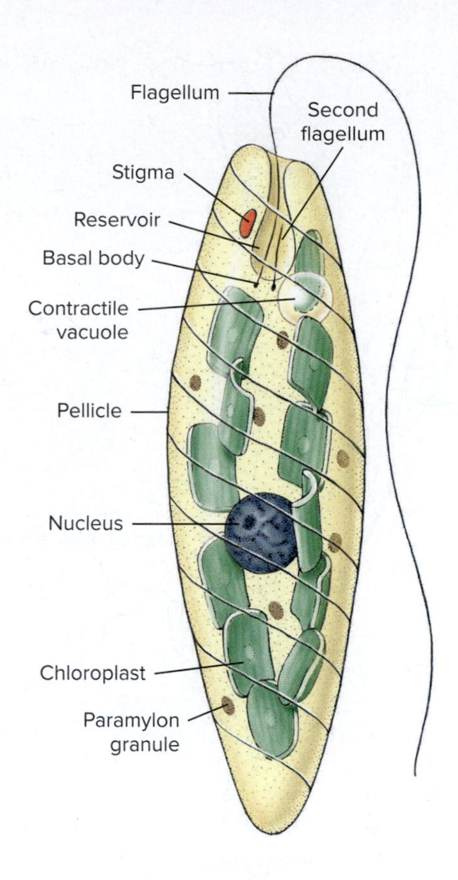

Figure 7-6
Euglena. Features shown are a combination of those visible in living and stained preparations.

Euglena

Phylum Euglenozoa
Genus *Euglena*
Species *Euglena gracilis* (or *Euglena viridis*)

Where Found

Euglenoids are most common in still pools and ponds, where they often give a greenish color to the water. Ornamental lily ponds are excellent sources. *Euglena gracilis* and *Euglena viridis* are commonly studied species.

Study of Live Specimens

Place a drop of 10% methylcellulose (or the commercial preparation Protoslo or Detain) on a clean slide, spread it thin, and add a drop of rich *Euglena* culture and a cover glass. Study with high power.

Members of *E. gracilis* range from 35 to 65 μm in length, are spindle-shaped, and are greenish. The color is caused by the presence of **chloroplasts,** which contain **chlorophyll** and are scattered through the cytoplasm (Figure 7-6). A nucleus is located centrally but is difficult to see in living specimens.

Locomotion. The blunt anterior end bears a little, whiplike **flagellum** that you may be able to see with reduced light when a specimen has slowed down considerably.* The flagellum emerges from the **reservoir,** a clear, flask-shaped space in the anterior end. In some flagellates, such as *Peranema,* the flagellum extends forward, but in *Euglena* it is directed backward along the side of the body. Movement involves generation of waves originating at the base of the flagellum and transmitted along its length to the tip (Figure 7-7). The flagellum beats at the rate of about 12 beats per second. It not only moves the organism forward but also rotates it and pushes it to one side, causing it to follow a corkscrew path, rotating about once every second as it goes.

Euglenoid Movement. When a *Euglena* stops swimming, watch how it changes shape. Keep notes on your observations. These will be used in writing your report on p. 103. These peculiar, wormlike contractions are referred to as "euglenoid movements," perhaps made possible by microtubules—tiny, hollow fibrils about 20 nm in diameter, lying just beneath the pellicle (Figure 7-8).

Body Covering. The body is covered with a protective but flexible **pellicle** secreted by the clear **ectoplasm** that surrounds the **endoplasm.** The **stigma,** or "eyespot," is a reddish pigment spot that shades a swollen basal area of the flagellum that is thought to be light-sensitive. Early microscopists, however, believed the red stigma itself was sensitive to light, so they named the genus *Euglena,* meaning "true eyeball." Why is light sensitivity important to a *Euglena?*

____.

Osmoregulation. A large contractile vacuole empties into the reservoir. It is fed by smaller vesicles around it.

Feeding. A *Euglena* takes in no solid food; the opening of the reservoir is merely for elimination of waste and excess water discharged by the contractile vacuole. Nutrition is **holophytic** (Gr. *holo,* whole, + *phyt,* plant), making use of

*Dilute methyl violet or Noland's stain may help to make the flagella easier to see. See Appendix A, pp. 393–394.

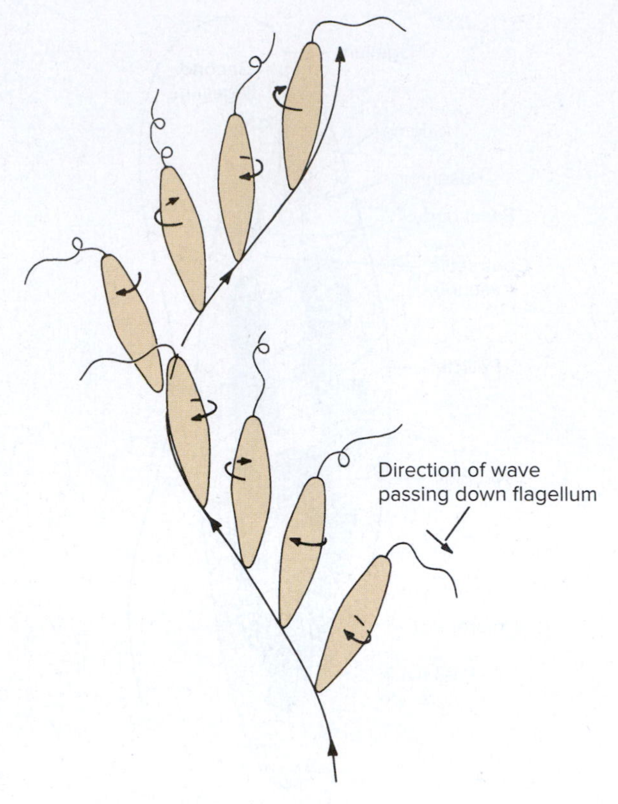

Figure 7-7
Locomotion in Euglena. Base-to-tip waves curl down the flagellum, forcing *Euglena* forward. The organism rotates as it swims, and the anterior end follows a corkscrew path. The posterior end "drags," following a straighter path.

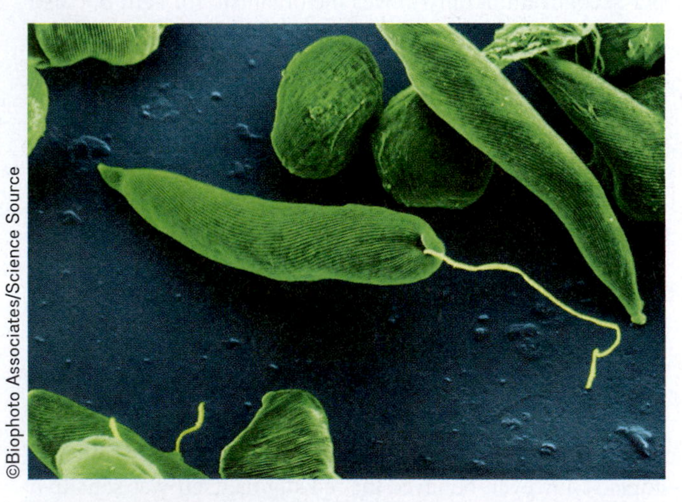

Figure 7-8
Scanning electron micrograph of *Euglena*, showing "candy-striped" ridges in the pellicle. The odd, spiraling pattern of microtubules is thought to provide the flexibility required for the twisting and contraction motions of euglenoid movement, about ×1180.

photosynthesis—the manufacture of food from carbon dioxide and water with the aid of chlorophyll and sunlight. The chlorophyll is carried within shield-shaped **chloroplasts** in the cytoplasm. Carbohydrates synthesized within the chloroplasts are stored as starch granules and **paramylon** (a carbohydrate similar to starch) concentrated in the chloroplast centers. In older cultures of *Euglena,* you may see dispersed in the cytoplasm small red, brown, or orange droplets containing lipids and pigments.

Not all photosynthetic flagellates are holophytic. Some are also saprozoic, some are holozoic, and some use a combination of methods.*

Reproduction. *Euglena* reproduces by **longitudinal fission,** which may occur when the organism is free or when it is in the encysted state. Fission begins at the anterior end and proceeds posteriorly. During encystment *Euglena* rounds up and forms a gelatinous wall around itself. In this state, it can withstand drought and other harsh conditions. A number of longitudinal divisions may occur during encystment, and many separate individuals may emerge later from a single cyst.

Stained Slide of *Euglena*

Select a good specimen on the slide. On separate paper, sketch the specimen and label the flagellum, stigma (eyespot), chloroplasts, pellicle, reservoir, and nucleus. What is the function of each of the labeled structures?

Written Report and Drawings

On p. 103, describe and explain the manner in which your specimens move about (locomotion) and alter their shape (euglenoid movement). Can you see any flagella? Record any other interesting observations. Also on p. 103, make a series of 2- to 3-cm-long drawings showing the shapes your specimen assumes in its euglenoid movement. Do *Euglena* push or pull themselves with their flagella? Can they move without using their flagella?

Further Study

Volvox

Phylum Viridiplantae
Genus *Volvox*
Species *Volvox globator*

Where Found

Volvox (L. *volvere,* to roll) is a beautiful, large, spherical organism often found with other unicellular eukaryotes in stagnant pools and ponds. It is an important component of summer plankton "blooms" that appear in nitrogen-rich ponds, pools, and ditches. The common species of *Volvox* are colonies of euglenid-like cells, but there is one species, *V. carteri,* that has evolved multicellularity.

*Saprozoic (Gr. *sapros,* rotten, + *zōon,* animal) nutrition is by absorption of dissolved salts and simple organic nutrients from the surrounding medium. Holozoic (Gr. *holo,* whole, + *zōon,* animal) nutrition involves ingestion of solid or liquid organic food particles. Thus, all major forms of nutrition—holophytic, holozoic, and saprozoic—are represented among the flagellates.

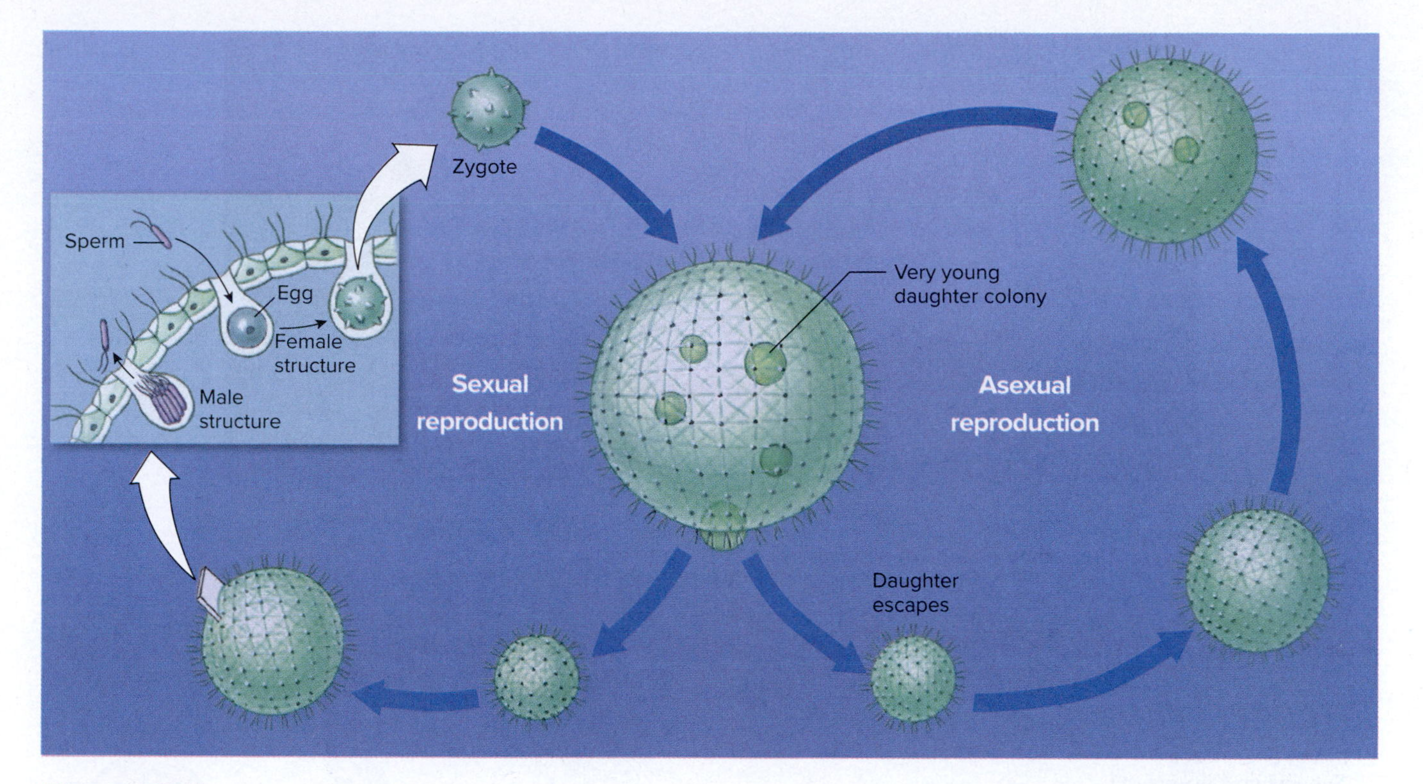

Figure 7-9
Life cycle of a colonial species of *Volvox,* phylum Viridiplantae. Notice that zygote encysts and may have a dormant phase before developing into a new colony.

General Features

Colonies of *Volvox* may reach a diameter of 2 mm (most are smaller) and are easily visible to the unaided eye. These colonies are especially interesting because they illustrate a transition between unicellular, colonial, and multicellular forms. In some species of *Volvox* we see the beginning of **cell differentiation,** resulting in the division of labor among cells—an important step toward multicellularity. One species of *Volvox, V. carteri,* limits reproduction to sex cells within the colony and is multicellular. To be defined as "multicellular," sexual reproduction must be limited to particular cells within the group. If all cells can reproduce sexually, then the organism is colonial and not multicellular, even if there are specialized cell types within the group.

Study living or preserved colonies and stained slides. For living specimens, use a pipette to transfer a colony (visible to the naked eye) to a ringed or depression slide and cover with a cover glass. Study with *low power* and focus up and down to get all the details. Be especially careful if you use high power to avoid damage to slides or objective.

Each spherical colony is composed of a variable number of one-celled individuals called **zooids.** There may be a few hundred or many thousands of cells in a colony, arranged on the surface of a gelatinous ball and connected to each other by fine **protoplasmic strands.** There are **somatic cells** and a smaller number of **reproductive cells.** The somatic cells are quite similar to those of flagellates such as *Euglena* and make up most of the colony. They handle nutrition, locomotion, and response to stimuli for the entire colony. Each somatic cell contains **chloroplasts** (which give the colony its green color), a **stigma** for light sensitivity, and a pair of **flagella** for locomotion.

Locomotion. If you have living colonies, observe locomotion. Do you notice that one end usually goes foremost? This is the anterior pole, composed of specialized cells. Some of the colonies on your slide will contain smaller **daughter colonies** revolving about inside the gelatinous center of the mother colony (Figure 7-9).

Reproduction. *Volvox* reproduces both sexually and asexually. During spring and summer, *Volvox* reproduces asexually by repeated division of the cells, which form miniature daughter colonies. These eventually escape by rupture of the mother colony. In the fall the asexual colonies develop **sex cells.** Some cells enlarge to form **macrogametes** (eggs). Other cells divide repeatedly to form packets of spindle-shaped **microgametes** (sperm). The sperm escape and fertilize the eggs, which become **zygotes.** Each zygote secretes a spiny cyst wall around itself, providing protection during winter after the mother colony dies. In spring the zygotes break out of the cysts to give rise by cell division to new asexual colonies (Figure 7-9).*

Written Report and Drawings

Answer questions on p. 104 and, on separate paper, sketch one of the colonies for your report.

*Several invertebrate groups use asexual reproduction during the warm months of the year when rapid reproduction and colonizing new habitats are more important than genetic variability furnished by sexual reproduction. Sexual reproduction occurs during the unstable conditions of winter.

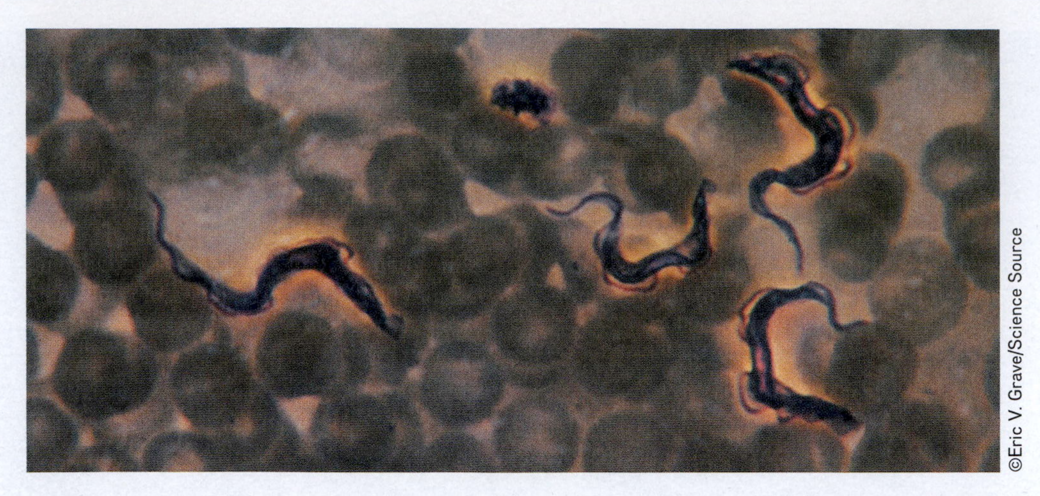

Figure 7-10
Trypanosoma brucei gambiense in the blood of a human.

Further Study

Trypanosoma

Phylum Euglenozoa
Subphylum Kinetoplasta
Class Trypanosomatidea
Genus *Trypanosoma*

Trypanosoma (Gr. *trypanon,* auger, + *soma,* body) are flagellate parasites living in blood or other tissues of all classes of vertebrates (Figures 7-10 and 7-11). Trypanosomes are **obligate heterotrophs;** that is, they lack chlorophyll and obtain their organic and inorganic requirements from the environment. Because a trypanosome's environment is the body fluid of its host, its nutrition is **osmotrophic;** it absorbs its nutrients directly from the surrounding blood or other body fluid.

Some trypanosomes are nonpathogenic but others produce severe disease, especially **sleeping sickness.** *T. brucei gambiense* and *T. brucei rhodesiense* cause African sleeping sickness in humans, and *T. brucei brucei* causes a related disease in domestic animals (note that the organisms are all subspecies of the same species, *T. brucei*). *T. b. rhodesiense,* the more virulent of the sleeping sickness trypanosomes, and *T. b. brucei* have natural reservoirs (antelope and other wild animals) that apparently are not harmed by the parasites. Trypanosomes of African sleeping sickness are transmitted by tsetse flies of the genus *Glossina.*

T. cruzi causes Chagas' disease in humans in Central and South America. It is transmitted by the bite of the "kissing bug" *(Triatoma).* Many believe that Charles Darwin contracted Chagas' disease in Chile during the voyage of the *Beagle,* resulting in his renowned ill health later in life.

Using oil immersion, examine a slide containing a stained blood smear from an infected animal. Note the numerous purple-stained trypanosomes (Figure 7-10). The body of the organism is fusiform in shape and slightly twisted. Are the two ends alike? The flagellum runs closely opposed to the body surface together with a fold of the plasma membrane; these together (fold and flagellum) constitute the **undulating membrane** so characteristic of the genus.

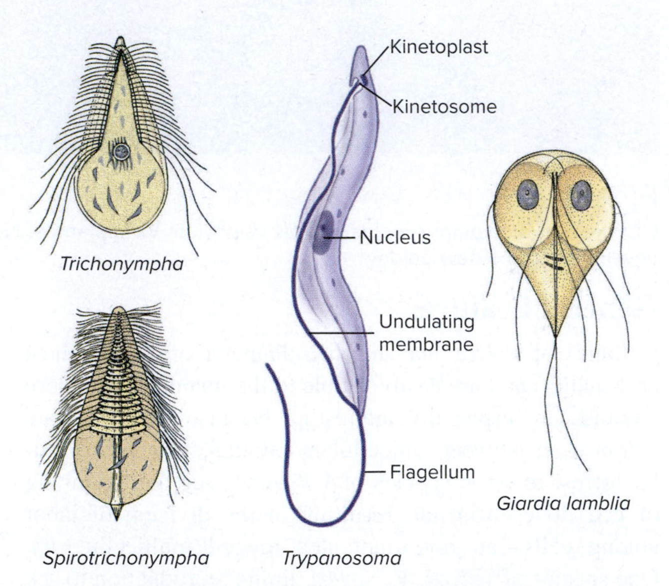

Figure 7-11
Some flagellate cells. *Trypanosoma,* a euglenozoan, is a blood parasite of various animals; some cause serious disease in humans and domestic animals. *Trichonympha* and *Spirotrichonympha* are parabasalids commonly found in the gut of termites and roaches, where they, and the bacteria they harbor, help digest cellulose in wood eaten by the insects. *Giardia lamblia* is a diplomonad that causes a highly contagious intestinal infection that is becoming increasingly common among hikers and campers who drink untreated water. Sketches are not all to the same scale.

The two subspecies of *T. brucei* (*rhodesiense* and *gambiense*) cannot be distinguished morphologically, but they differ in pathogenesis and growth rate. *T. b. rhodesiense* causes the most virulent form of sleeping sickness in central and east central Africa. *T. b. gambiense* produces the "classical" chronic form of sleeping sickness, with protracted somnambulism (sleepwalking) and nervous disorders.

Drawings

On separate paper, sketch one or more trypanosomes from your slide, paying attention to scale relative to other cells on the slide.

Written Report

Answer the questions relating to *Trypanosoma* on p. 105.

Projects and Demonstrations

1. *Demonstration of living* Trichonympha, *a symbiont of termites. Trichonympha* (Figure 7-11) belongs to an order of flagellates that have many flagella and live in the digestive tracts of roaches and termites. *T. campanula,* which lives in the gut of American termites, is a good example of **symbiotic mutualism.** Neither the termite nor the flagellate can live without the other. Most flagellates harbor bacteria in their cytoplasm that secrete the enzyme cellulase, which digests the wood (cellulose) that termites eat. There also is some recent evidence that at least some termite species secrete their own cellulase.

   Cut the head off the termite to kill it. Squeezing the posterior abdomen of the termite to extrude a drop of fluid from the anus may provide enough specimens on a slide for study. Or hold the termite body with forceps and, with another pair of forceps, pull off the posterior segment. Pull the intestine out of the body and tease it apart on a slide. Add a drop of 0.6% NaCl solution and cover. Do not wait to examine your slide because the unicellular forms will be short-lived if exposed to air. Why is *Trichonympha* important to termites? __________ In what ways does *Trichonympha* resemble *Trypanosoma?* __________ In what ways does it differ? __________

2. *Dinoflagellates.* Members of the phylum Dinoflagellata are found mostly in marine plankton. They have a peculiar arrangement of two flagella—one extends backward to propel the organism; the other encircles the body in a transverse groove and causes it to rotate. Some dinoflagellates have chloroplasts of various colors; others are colorless. Some, such as *Noctiluca* (Figure 7-12), possess luminescent granules in the cytoplasm, which are largely responsible for the phosphorescence observed in some coastal waters. Some dinoflagellates, such as *Gonyaulax,* occasionally appear in enormous numbers and cause "red tides" off the coasts of Florida *(Ptychodiscus)* and California and New England *(Gonyaulax). Ceratium,* with long spines and green chloroplasts, is found in both marine and fresh water (Figure 7-12). Why are red tides of particular concern to humans?

3. *Giardia intestinalis (lamblia).* Giardiasis occurs when one is infected with the flagellate *Giardia* of the phylum Diplomonada (Figure 7-11). The parasite is contracted from drinking water contaminated with fecal matter from infected animals such as humans, cattle, or beaver. Symptoms of infection include severe diarrhea with dehydration, intestinal pain, and weight loss. *Giardia* can swim rapidly using its flagella, but in the intestine it uses a ventral disc to attach to the intestinal surface. It also divides rapidly and can quickly build up enormous numbers; a single diarrheic stool can contain 14 billion parasites. As the unicellular eukaryotes pass into the large intestine, they encyst; the cysts transmit the infection to new hosts. When swallowed they pass through the stomach unharmed, then undergo excystation in the intestine, grow flagella, and begin their nefarious work.

   If slides of the feeding stage (trophozoite) are available, note the characteristic shape, rounded anteriorly, pointed posteriorly, and bearing several flagella. Try to find cells where you can see the ventral adhesive disc—it resembles a suction cup. How is *Giardia* similar in appearance to *Trypanosoma?* __________ In what way does it differ? __________ How might you test to determine if a person is infected with *Giardia?* __________

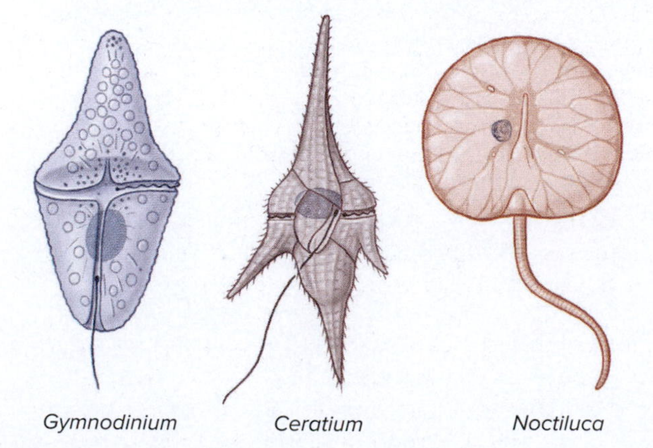

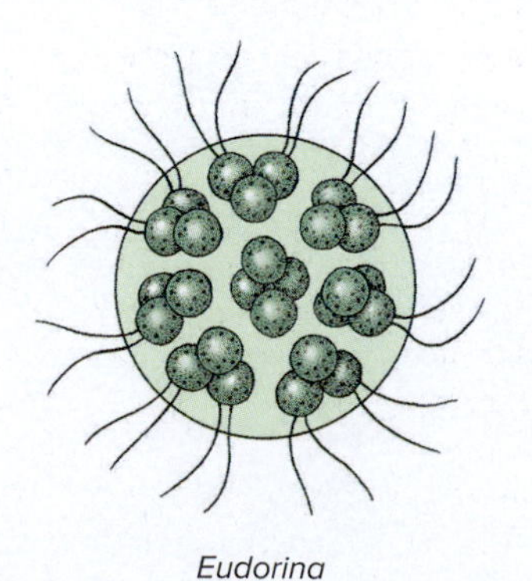

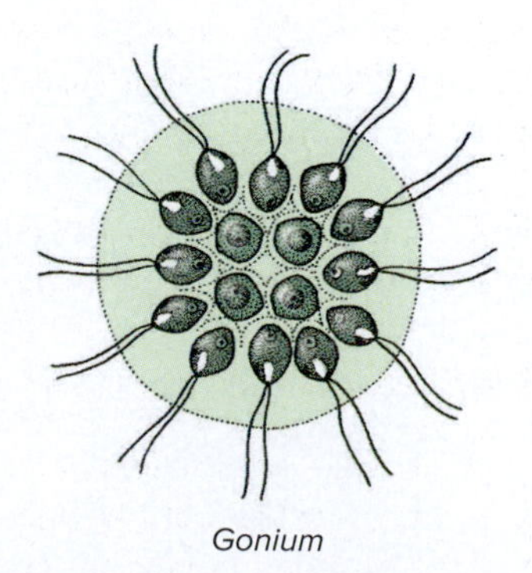

Figure 7-12
Some unicellular eukaryotes with flagella: *Noctiluca, Ceratium,* and *Gymnodinium* are dinoflagellates. *Gonium* and *Eudorina* are colonial members of the phylum Viridiplantae.

NOTES

Phylum ____________________

Subphylum ________________

Name ______________________

Date ______________________

Section ____________________

LAB REPORT

Euglena

1. What is the average length of *Euglena* specimens in your culture? ____________________
2. Describe the different forms of movement and locomotion in *Euglena*. ____________________
3. How does *Euglena* satisfy its nutritional requirements? ____________________
4. What function might the stigma of *Euglena* serve? ____________________
5. Many unicellular eukaryotes form cysts to survive harsh conditions; are the *Euglena* that emerge from a cyst genetically similar to or different from organisms of previous generations? ____________________
6. What can you infer about the habitat of *Euglena* given the fact that it forms cysts? ____________________

Observations on *Euglena* Locomotion

Volvox

1. In what ways is *Volvox* similar to *Euglena?* ______________________________

____

In what ways does it differ from Euglena? ______________________________

2. Were the colonies you examined reproducing asexually or sexually? __________ Give reasons for your answer.

____

____

____

____

3. When scientists say that *Volvox* shows beginnings of cell differentiation, what do they mean?

____

4. Why are most species of *Volvox* not considered truly multicellular? ______________________________

____

5. How do somatic and reproductive cells of *Volvox* differ in function? ______________________________

____

Which is more abundant in the colony and why? ______________________________

____

6. Why might sexual reproduction in this group be more common in the fall, whereas asexual reproduction tends to dominate in the spring? Hint: Study the product of sexual reproduction in Figure 7-9. ______________________________

____

Trypanosoma

1. Using the average diameter of the mammalian red blood cell (7.5 μm) as a guide, what do you estimate the length of *Trypanosoma* to be? ______

2. How must *Trypanosoma* cells obtain their nutritional requirements? ______

3. A doctor suspects that you might be infected with *Trypanosoma.* How might your symptoms differ depending on the part of the world where you contracted the infection?

4. What type of tissue will the doctor sample to confirm his or her diagnosis? ______

5. What key characteristic is possessed by all species of this genus? ______

6. What features of *Trypanosoma* are characteristic of the phylum Euglenozoa? ______

7. Would you expect *Trypanosoma* to have a system for digesting food like *Amoeba?* Why or why not?

LAB REPORT

EXERCISE 7C

Phylum Apicomplexa—*Plasmodium* and *Gregarina*

Core Study

Apicomplexans are almost entirely endoparasitic. On the whole, they lack special locomotor organelles, although some can move by gliding or by changing body shape, similar to euglenoid movement, and some species have flagellated gametes.

Apicomplexans are distributed via resistant spore-like stages. The process of **sporogony** (spōr-ahg′uh-nē; *spores,* seed, + *gonos,* progeny), or spore formation, follows the formation of the zygote in most apicomplexans, so that sexual reproduction leads into the production of spores.* Many apicomplexans also undergo multiple fission, or **schizogony** (ski-zahg′uh-nē; *schizein,* to split, + *gonos,* progeny), a means of spreading the population within the host, and this is followed by sporogony.

Apicomplexan nutrition is osmotrophic (nutrients are absorbed from the immediate environment).

Common apicomplexans are the gregarines (class Gregarinea), parasitic in such invertebrates as annelids, echinoderms, and ascidians, and the coccidians (class Coccidea), which are endoparasites in both invertebrates and vertebrates. Among the Coccidea are *Eimeria,* which causes coccidiosis in domestic rabbits and chickens, and *Plasmodium,* which causes malaria in humans.

Plasmodium

Phylum Apicomplexa
Class Coccidea
Family Plasmodiidae
Genus *Plasmodium*

Plasmodium (*plasma,* molded, image, + *eidos,* form) is the genus of coccidean parasites that causes malaria. Malaria ("bad air") is the most important disease in the world today in terms of lives lost and impact on global economy. Every year malaria sickens 300 million people and kills 1 million. It has always outranked war as a source of human suffering. *Plasmodium* requires two hosts—one a vertebrate, the other an invertebrate. It is transmitted to a human by a female *Anopheles* mosquito that has had a blood meal from a malaria-infected human. The mosquito, as it feeds, also injects into the human bloodstream some of its salivary juice, which contains infective **sporozoites** (Figures 7-13 and 7-14).

In humans the sporozoites leave the bloodstream and penetrate liver cells, where they undergo **schizogony** (asexual cleavage multiplication) and produce **merozoites** (mer-uh-zō′īts; *meros,* part, + *zōon,* animal). These may either infect other liver cells or be released into the blood to enter red blood cells (erythrocytes). In either case, they undergo further multiplication.

In red blood cells, they become **trophozoites** (adult stage). Cytoplasm of the trophozoite becomes nucleated and takes on a ringlike appearance, called the "signet-ring stage" (Figures 7-13 and 7-14).

Examine a slide containing a stained blood smear with various infective stages. Find a red blood cell containing a signet-ring stage. Slides stained with blood-differentiating dyes usually show the cytoplasm of the trophozoite blue and the nucleus reddish pink.

The signet-ring trophozoite now develops into a **schizont** (Gr. *schizō,* cleave, + *ontos,* a being) (Figure 7-13). The nucleus divides repeatedly, producing 8 to 32 merozoites. The host red blood cell ruptures, releasing merozoites as well as metabolic wastes that are largely responsible for the characteristic symptoms of malaria: recurring episodes of fever with intense shivering, nausea, vomiting, pain, and delirium. Many merozoites enter and infect more red blood cells, with the cycle repeating until an enormous number of host cells have become parasitized.

After several such asexual generations, some of the merozoites enter red blood cells to become sexual forms called **macrogametocytes** and **microgametocytes** (Figures 7-13 and 7-14). To which gamete (egg or sperm) in the human life cycle is each of these analogous? Macrogamete = __________. Microgamete = __________. The gametocytes may be picked up by a feeding *Anopheles* female. Gametocytes give rise within the mosquito to gametes, which unite to become zygotes. The zygotes divide to produce sporozoites, which migrate to the salivary glands, from which they may be injected along with the saliva into the blood of a human, perhaps to cause another malarial infection.

Written Report

Prepare a sketch of your own version of the life history of this organism, using words to indicate stages and arrows to show events such as reproduction, growth, and release of stages. Be sure to note in which host the parts of the cycle occur. Answer the following questions in your report, referring to your textbook when necessary. When does meiosis occur in the life cycle of this organism? In which host do the gametes come together?

Further Study

Gregarina

Phylum Apicomplexa
Class Gregarinea
Family Gregarinidae
Genus *Gregarina*

*The term "spore" is unfortunate, because these spores are not homologous to the spores produced by plants or fungi, and confusing, because the term refers to a variety of propagating, infective, and resistant stages in the life cycles of many different taxa. In *Plasmodium,* the "spore" is the **oocyst** (Gr. *oon,* egg, + *kystis,* bladder), a cystlike stage in which numerous daughter spores, called **sporozoites,** are produced by sporogony.

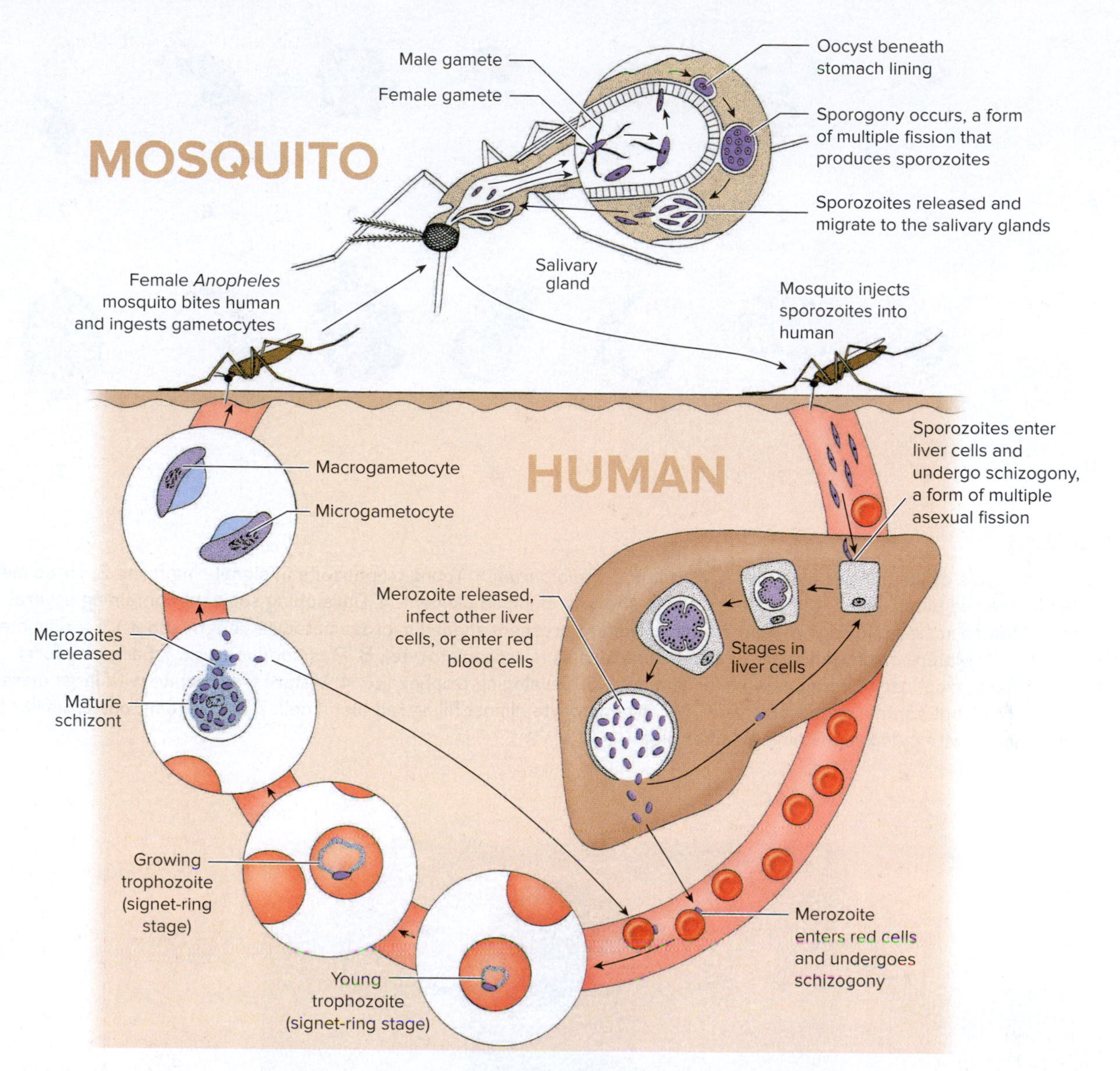

Figure 7-13

Plasmodium falciparum life cycle.

Various species of *Gregarina* (L. *gregarius,* belonging to a herd or flock) can be found in the guts of cockroaches (such as *Blatta* and *Periplaneta*), mealworms (the larvae of *Tenebrio*), and grasshoppers. Some species are surprisingly large for a unicellular organism and at one point were placed among worms by nineteenth-century zoologists.

Cut the head from a live cockroach or mealworm to kill it and quickly cut the anus free from the body wall with scissors. Remove the complete digestive tract by holding the head end of the animal and pulling the attached gut with forceps. Place the gut in a watch glass and cover with a little insect saline solution. Tease into small bits; then transfer a little of the contents in a drop of the saline to a slide for examination. Add a coverslip.

If adult feeding stages, called **trophozoites** (trōf′uh-zō′its; *trophe,* food, + *zōon,* animal), are present, they will appear as large, distinctly shaped forms ranging from nearly round to elongate and wormlike, depending on the species (Figure 7-15). In general, the body is constricted into two unequal parts: a small anterior compartment (the protomerite), which bears a holdfast device (but which usually becomes detached during transfer of the gut and its contents to the slide), and a much larger posterior compartment (the deutomerite), which contains the nucleus.

Did you find trophozoites of gregarines in your sample? If so, describe them in your notebook.

The entire life cycle takes place within the midgut of a cockroach or mealworm. Two trophozoites join together, encyst, and produce gametes, which, after fertilization, divide into **sporozoites** (spō′ruh-zō′īts; *sporos,* seed, + *zōon,* animal). These pass out with the feces. When ingested by another cockroach or mealworm, the sporozoites develop into trophozoites—the adult stage.

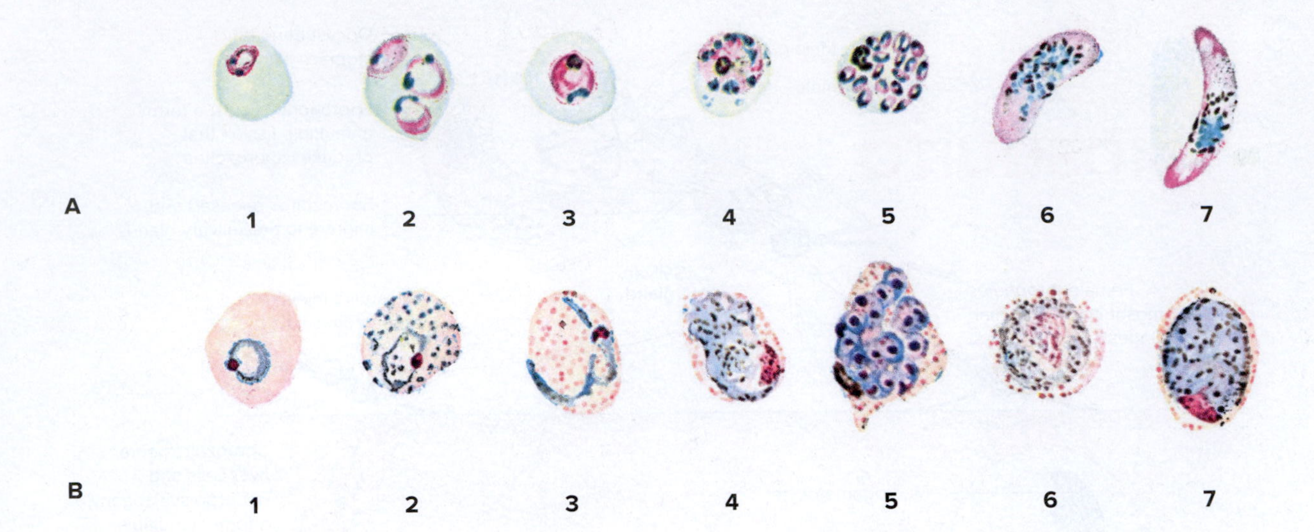

Figure 7-14

Stages in the life cycle of the malaria parasite. **A,** *Plasmodium falciparum.* 1, Young trophozoite in signet-ring form. 2, Three medium trophozoites in one red blood cell. 3, Mature trophozoite showing clumped pigment. 4, Developing schizont containing several merozoites. 5, Mature schizont filled with merozoites. 6, Mature crescent-shaped microgametocyte (length about 1.5 × diameter of red blood cell). 7, Mature macrogametocyte (about same size as microgametocyte). **B,** *Plasmodium vivax.* 1, Early ring-form trophozoite. 2, Young trophozoite with heavy chromatin dots. 3, Developing trophozoite. 4, Mature trophozoite with large mass of chromatin.5, Schizont containing merozoites. 6, Microgametocyte almost filling red blood cell. 7, Macrogametocyte, similar to microgametocyte but with cytoplasm staining darker blue.

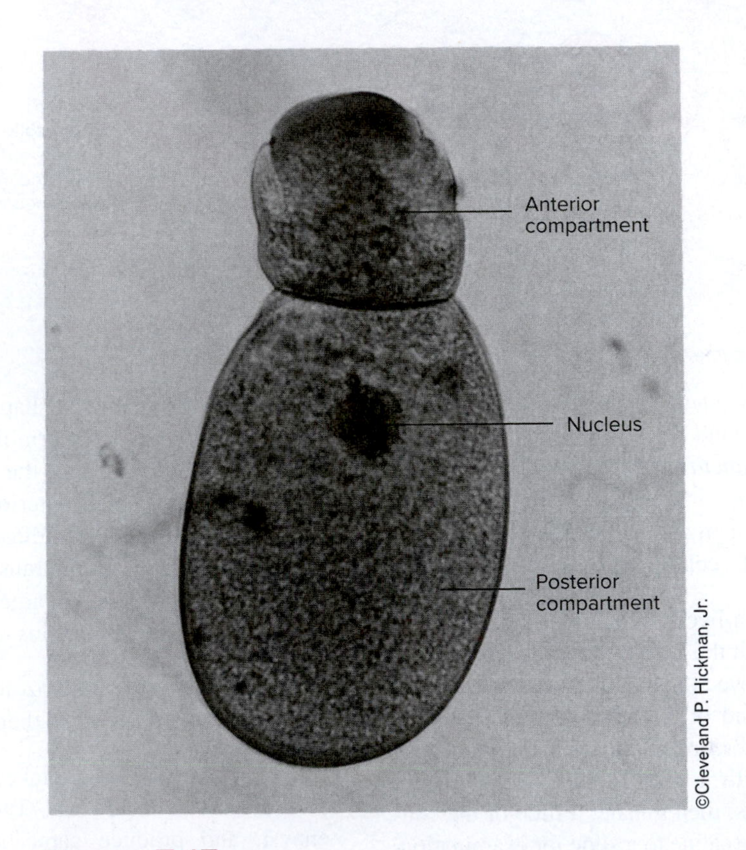

Figure 7-15

Gregarina, a parasitic apicomplexan that lives in the gut of the cockroach or mealworm. The body is divided into two compartments.

EXERCISE 7D
Phylum Ciliophora—*Paramecium* and Other Ciliates

Core Study

Paramecium

Phylum Ciliophora
Genus *Paramecium*

Where Found

Paramecium is an active ciliate common in most freshwater that contains vegetation and decayed organic matter. Find the Ciliophora in Figure 7-1. Unlike unicells with ameboid bodies or those with flagella, ciliated unicells occur in only one clade. Pond scum and even cesspools are also good sources for ciliates of many types, including *Paramecium.* Some of the most commonly studied forms are *P. multimicronucleatum, P. caudatum, P. aurelia,* and *P. bursaria.*

Study of Live Specimens

Locomotion and Behavior.

Spread out a *very few* fibers of absorbent cotton* on a slide and add a drop or two of *Paramecium* culture. Cover with a coverslip. Examine first with a scanning lens to locate the paramecia; then switch to low power for study. Add water as necessary to prevent drying out.

A paramecium is slipper-shaped, rather transparent and colorless, and very active. It can swim at the rate of 1 to 3 mm per second. Watch its swimming habits. Does it swim in a straight line, in a circle, or zigzag? __________ Does it keep one side uppermost or revolve on an axis? __________ Does it have a definite anterior end? __________ Can it reverse its direction? __________ Does it seem to be contractile, as *Euglena* is? __________ Why is the term "spiral movements" used to describe its swimming habits? __________

____

What does a paramecium do when it encounters a barrier? __________ How does it go about finding an opening? __________ Is its body flexible enough to bend or to squeeze through tight places? __________

Written Report

Describe the locomotion of *Paramecium.* Use the preceding questions to guide your description, answering as many as you can.

Describe the attempts of *Paramecium* to avoid or to pass under or around a cotton fiber barrier. Use diagrams and arrows if you wish.

What characteristics possessed by *Paramecium* tell you it belongs in phylum Ciliophora?

*Strands of cotton fibers serve as barriers to slow movement. Other methods for slowing *Paramecium* for study are given in Appendix A (p. 395).

General Structure and Function

Locate a *Paramecium* that has been stopped by a cotton fiber and study its structure with both low and high power.

Note the **oral groove** that extends obliquely from the anterior end to about the middle of the body. At the posterior end of the groove is the **mouth (cytostome).** The oral groove extends from the mouth into the body as a little canal, the **gullet (cytopharynx).** The groove and gullet are lined with strong cilia that are used in drawing in food (Figure 7-16).

Pellicle. The cytoplasm is made up of two zones, a clear outer **ectoplasm** and an inner **endoplasm.** Are there any granules in the endoplasm? Outside the ectoplasm is a complex, living **pellicle.** A delicate **plasma membrane** lies just underneath the pellicle. The pellicle and plasma membrane are more easily seen in stained preparations.

Osmoregulation. A contractile vacuole (water-expulsion vesicle) is usually located in each end of the body. *P. multimicronucleatum* may have more than two.

Observe pulsations of the vacuoles caused by their alternate filling and emptying. After one empties, note the starlike **radiating canals** that appear. These radiating, or nephridial, canals collect liquid from a network of minute tubules and empty into the vacuoles, which rupture when filled, thus expelling fluid to the outside.

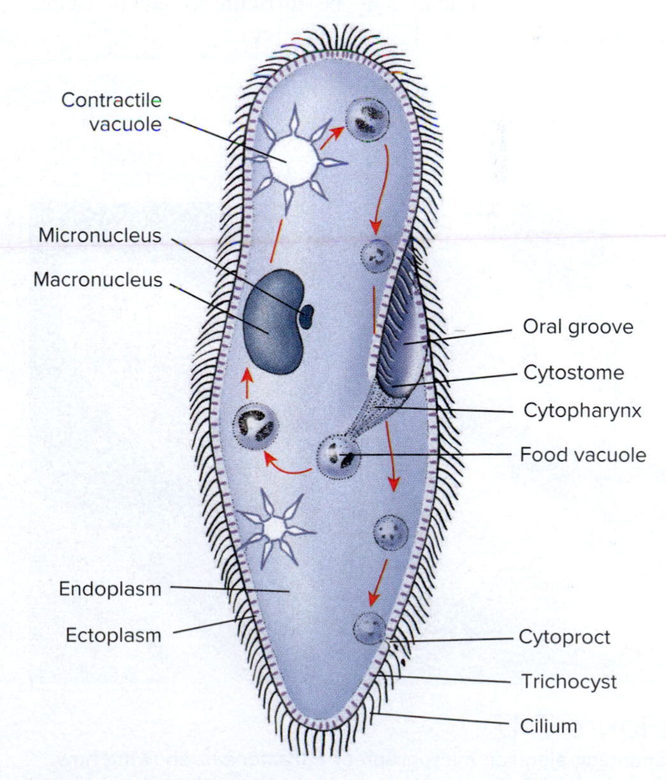

Figure 7-16
Paramecium.

Do the two vacuoles empty at the same time or alternately? __________ How much time (in seconds) is there between vacuole discharges? __________ In normal pond water, a vacuole may discharge once every 6 to 10 seconds. How does this unicell's rate compare (faster, slower, about the same)? __________

The pulsation period depends on the temperature and osmotic pressure of the medium. As the temperature increases, the pulsation rate increases. The pulsation rate can be slowed by adding drops of 0.25% solution of sodium chloride to the slide. Knowing this, how do you think the rate in marine ciliates would compare with that of freshwater ciliates? Why?

____

Written Report

Report the results of your observations with contractile vacuoles, answering the questions in the section "Osmoregulation."

Ciliary Action. Cilia perform very much as the oars of a boat; that is, they have an effective stroke that propels the unicell forward and a recovery stroke that offers little resistance (Figure 7-17). For a discussion of ciliary action, see your text.

Find a quiet specimen, cut down the light, and look closely with high power at the margin of a paramecium. Study the action of the **cilia.**

Nuclei. The nuclei may be difficult to see in living paramecia.

Figure 7-17
Scanning electron micrograph of *Paramecium,* showing how the cilia beat in waves. Look closely at the surface to see patterns. The cytostome is visible at the center, ×400.

Add a drop of acidified methyl green to the slide at the edge of the coverslip, so that the stain is drawn under by capillary action.

This will kill the organisms but should give the cytoplasm a bluish tinge and stain the nuclei green. Note the presence of two nuclei: a large macronucleus that regulates metabolism of the cell (controlling cellular functions, such as feeding and digestion), and a small micronucleus that participates in a complex form of sexual reproduction. The micronucleus is sometimes obscured by the macronucleus.

Trichocysts. If you reduce your light properly, you can, with careful focusing, observe the many small, spindle-shaped **trichocysts** lying in the ectoplasm just under the pellicle and perpendicular to the surface. Under certain kinds of stimulation the trichocysts explode, each releasing a liquid that hardens in water to form a long, slender, thread-like filament. The tangle of filaments is believed to have some protective function. In some, discharged trichocysts may be used to anchor the unicell while feeding. Trichocysts are absent in some ciliates.

Perhaps the acidified methyl green you just used to demonstrate the nuclei has caused explosion of trichocysts on your slide. Alternatively, trichocysts may be made to discharge in live paramecia by adding a drop of dilute picric acid to the slide at the side of the coverslip and drawing the acid through with filter paper touched to the opposite side of the coverslip. As the fluid reaches the paramecia, you may be able to observe the discharged trichocysts by use of subdued light and a high-power lens. How does the length of discharged trichocysts compare with the length of cilia?

Written Report

Report the results of your observations, including a sketch of a *Paramecium* with discharged trichocysts.

Further Study

Feeding and Response to Stimuli

Feeding. Nutrition in ciliates is holozoic. A paramecium is a particulate feeder; that is, it feeds on small particles, such as bacteria, which it moves toward its cytostome by action of cilia in the oral groove.

On a clean slide, spread out a few cotton fibers and add a drop or two of *Paramecium* culture. Dip a toothpick into a preparation of yeast stained with Congo red, and transfer a *very small* quantity to the culture on the slide. Mix gently with the toothpick. The mixture should be light pink. If too much yeast is added, the paramecia will be obscured. Carefully apply a coverslip.

Do not allow the culture to dry out during your observation.

Find a specimen trapped by cotton fibers. Note currents created by cilia in its **oral groove.** Watch passage of yeast particles into the groove and through the **cytostome,** or cell mouth, into a passageway called the **cytopharynx.** Watch formation of a **food vacuole,** which is a membranous sac containing water and suspended food particles. When it reaches a certain size, the vacuole breaks away and another forms. In what direction are food vacuoles carried by streaming endoplasm? ___________________

Follow the course of a food vacuole. Does it vary in size during its trip? __________ If so, how? __________ Congo red, which is red in weak acid to alkaline solutions (pH 5.0 or above), turns blue in stronger acid solutions (pH 3.0 or below). Can you observe any changes in color in the vacuoles that might indicate a change in the condition of vacuole contents? __________ Is there any subsequent color change as vacuoles near the anal pore? __________ How might this be explained? __________ Are vacuoles any smaller as they complete their circuit than when first formed? __________

An **anal pore (cytoproct)** is found between the mouth and posterior end of the body. It is a temporary opening where indigestible food is discharged, and it is seen only at that time. Have you noticed such a discharge of material in one of your specimens?

Response to Stimuli. The following simple experiments are designed to determine the type of response **(taxis)** paramecia make to selected types of stimuli. A **taxis** (pl. **taxes**) is a directed reaction and orientation of the body to a specific stimulus. A movement toward a source is a positive taxis; a movement away from a source is a negative taxis. In these experiments, the responses are **chemotactic.** Perhaps you can design and perform other experiments to determine responses to other types of stimuli, such as **phototaxis** (response to light rays), **thigmotaxis** (to contact), or **geotaxis** (to gravity).

1. Place a drop of culture on a clean slide with no coverslip. Use a hand lens, scanning lens, or dissecting microscope. (Some of you will be able to see paramecia with the naked eye.) Place a drop of **weak** acetic acid on the slide **near** the culture but **not touching it.** Locate the specimens; while observing them, draw a line with the point of a pin or toothpick from the acid to the culture.

Describe the reaction of the unicellular eukaryotes to the approach of the acid. Are they positively or negatively chemotactic to weak acid? __________ Do they move to the area where the acid is strongest, weakest, or in between? ______________________________

2. Place a drop of culture on a clean slide. Place a few grains of salt on the slide near the culture but not touching it. While observing the unicells, draw a grain or two of salt into the side of the culture drop.

Describe the reaction of the paramecia. Do they choose an area nearest the salt, farthest from the salt, or in between? ____________________

Study of Stained Slides

With both low and high power, study a paramecium on a stained slide.

Focus up and down on the body and note that its entire surface is covered with **cilia.** Examine the **pellicle.** Specially prepared slides will show the peculiar pattern of hexagonal areas. Look especially for features difficult to see in the living unstained specimens, such as the **macronucleus,** one or more **micronuclei, trichocysts, oral groove, cytostome, cytopharynx,** and **contractile vacuoles.**

Drawings

Draw and label a paramecium showing the features listed in the preceeding paragraph. Do most paramecia have one, several, or many of each of the features you have labeled? What is the function of each of the features used here?

Binary Fission.

Study a stained slide of paramecia undergoing binary fission.

Note the constriction across the middle of the body. What is happening to the macronucleus and micronucleus? At the end of the process, the halves produced by the constriction will be separate daughter unicells. How does this fission process compare with that of *Trypanosoma* (Exercise 7B)?

Conjugation.

Study also a stained slide of conjugation. Look for paired individuals lying *with oral grooves attached.*

In this position, they exchange micronuclear material. Consult your textbook for details of the process of conjugation.

Drawings

Sketch a paramecium in the process of binary fission and a pair of paramecia in the process of conjugation. If you can find a live specimen dividing, watch it over a period of time and make a series of sketches to illustrate the process.

Written Report

On separate paper, summarize your observations and describe asexual reproduction in paramecia. On what axis do they divide? What is happening to the macro and micronuclei? How would you recognize when two paramecia are in the process of sexual reproduction (conjugation)? Which nucleus is involved in sexual reproduction? Explain what happens in this process.

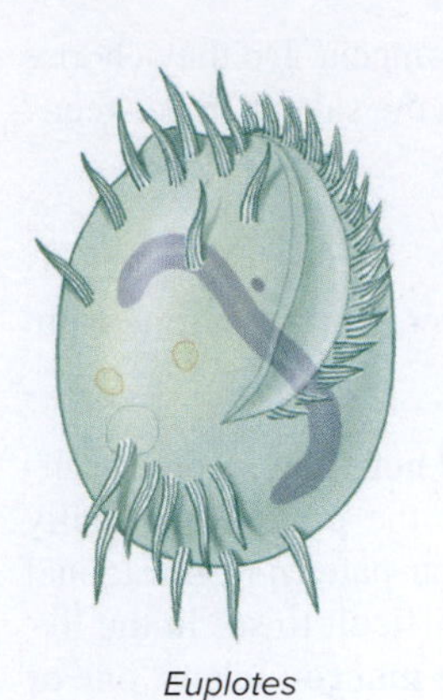
Euplotes

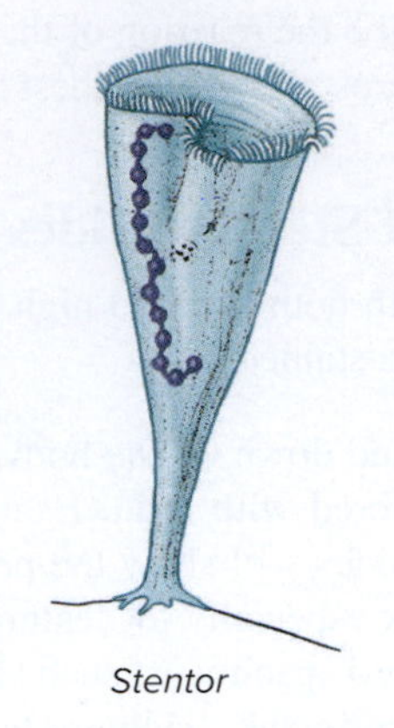
Stentor

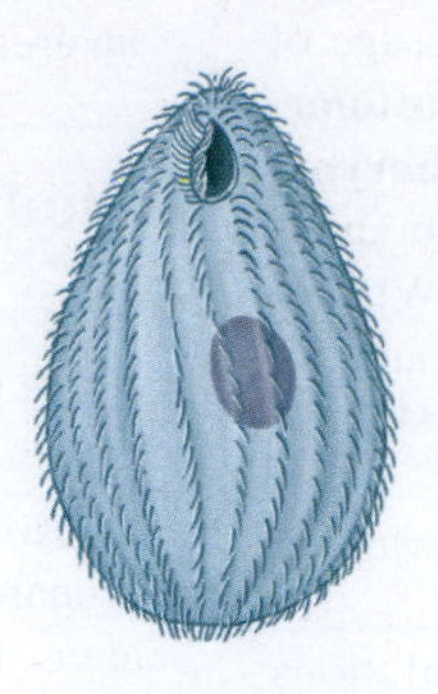
Tetrahymena

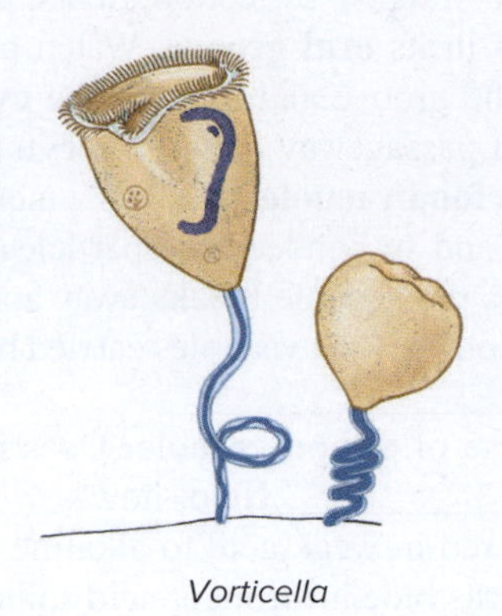
Vorticella

Zoothamnium

Figure 7-18
Some representative ciliates. *Euplotes* has stiff cirri used for crawling. Contractile myonemes in ectoplasm of *Stentor* and in stalks of *Vorticella* allow great expansion and contraction. Note the macronuclei, long and curved in *Euplotes* and *Vorticella,* shaped like a string of beads in *Stentor.*

Further Study

Other Ciliates

Vorticella

Vorticella (Figures 7-18 and 7-19) is a solitary sessile ciliate. It clings to aquatic vegetation of stagnant ponds and streams. The algae that blanket the edges of ponds and small lakes offer a good site for finding this unicellular eukaryote.

Study both living specimens and the stained slides.

What is the color of the living organism? __________ The vorticellid is attached by a long, slender **stalk** that can contract into a spiral spring shape when it is disturbed. The **body** is bell-shaped with a flaring rim, the **peristome,** at its distal end. Within the peristome is a circular **oral disc.** Note the **cilia** on the edges of the peristome and the oral disc. Note the beating of the cilia. What is their function? __________ The **cytostome** (mouth) is found between the peristome and the oral disc. From the cytostome, a short tube, the **cytopharynx,** leads into the interior. Food particles are swept into the cytopharynx by the action of the cilia, and **food vacuoles** are formed as they are in other unicellular eukaryotes. Note that the **nucleus** is made up of an elongated, U-shaped body, the **macronucleus,** and a much smaller **micronucleus.** Does the unicell have a **contractile vacuole**? __________ *Vorticella* has a surrounding **pellicle,** which helps maintain the shape of the body. Reproduction is mostly by longitudinal binary fission, but **budding** also occurs.

Sketch a specimen in your notebook, and annotate your sketch with descriptions of the behavior of the specimens you observed.

Figure 7-19
Group of living *Vorticella,* a solitary sessile ciliate.

Zoothamnium (Figure 7-18) is a stalked ciliate similar to *Vorticella* but is colonial (*Vorticella* is solitary) and has noncontractile stalks.

Stentor

Spread methylcellulose thinly on the slide before adding *Stentor* culture and coverslip.

Stentor (Figures 7-18 and 7-20) is a large ciliate with many of the same characteristics as *Paramecium* and *Vorticella.* How long is it when it is expanded? __________ When it is contracted? __________ How would you describe its shape? __________ See how many structures of this ciliate

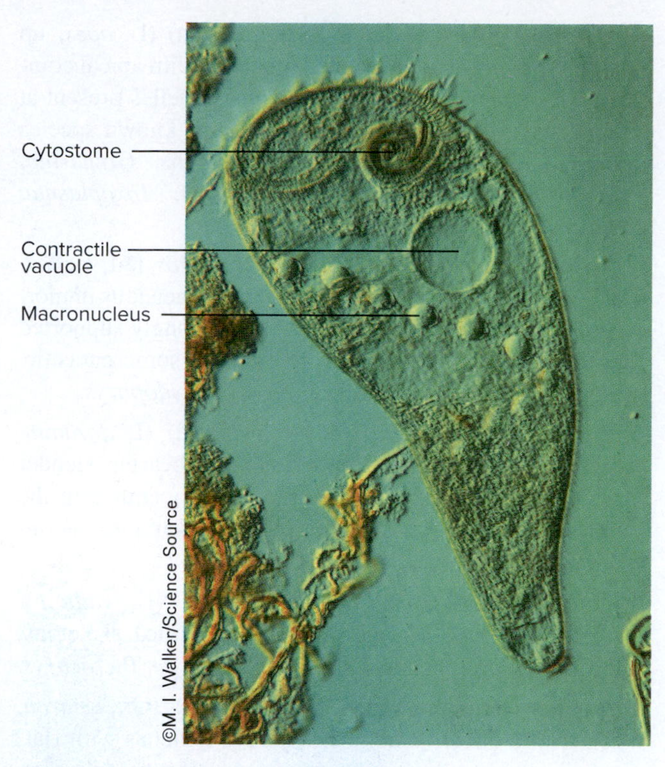

Figure 7-20
Living *Stentor* specimen.

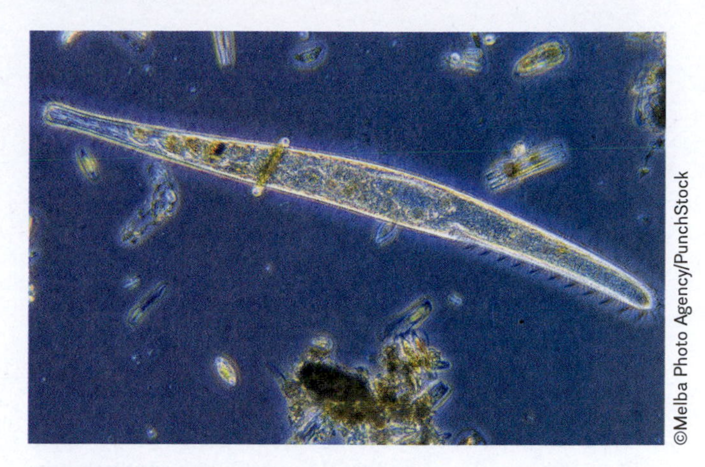

Figure 7-21
Spirostomum, an unusually large ciliate

you can identify from your knowledge of paramecia. Can you locate cytostome, cytopharynx, and contractile vacuole? __________ Is there an oral groove? __________ Can you see food vacuoles? __________ Are the cilia uniform in length? __________ Observe its striped appearance, caused by longitudinal bands of pigmentation. The blue pigment stentorin causes the blue-green color of *S. coeruleus.* Note the large **macronucleus,** stretched out like a string of beads. Small dots nearby are the micronuclei.

In your notebook, draw and label an extended specimen. Sketch its shape when the unicell is swimming. Estimate its size.

Spirostomum

Spirostomum (Figure 7-21) is one of the largest common freshwater unicellular eukaryotes. Estimate its length. __________ Different species range from 50 μm to 3 mm long. Can you see why it might be mistaken for a worm? Locate the large contractile vacuole posteriorly. It is fed by a long canal. The macronucleus is long and beadlike. Locate the cytostome and long oral groove. Focus on the surface and see the arrangement of cilia and trichocysts. Describe its swimming or other movements. Do you think it has myonemes?

In your notebook draw and label an extended specimen. Outline a contracted one.

Project

Microaquariums

A fascinating study of microorganisms and their relations to each other may be carried out in the laboratory at no expense and with only a few minutes of time at each laboratory period. Use a series of small, clean, empty jars, such as those that may have held jelly or mayonnaise. Fill each jar two-thirds full with water from the tap, a pond, a ditch, or any other source. Add to each a teaspoonful of some source material: rich soil, plants (dry grass, leaves, hay, water plants, moss, or rotting leaf mold), pond scum, sludge from a sewage plant, or any other source. Label each jar. To retard evaporation, cover with a piece of glass, plastic wrap, or the jar lid placed on loosely. Examine weekly—or more often, if you like—for several weeks or months. Each jar becomes a community that provides an interesting variety and an ever-changing cycle of life. In addition to unicellular eukaryotes, you may find a variety of crustaceans, flatworms, rotifers, gastropods, annelids, hydras, and algae. Keep a weekly record of what you find in each jar.

Classification of Unicellular Eukaryotes

Molecular analyses of unicellular eukaryotes have revolutionized our understanding of relationships among all eukaryotes and led to a proliferation of eukaryote clades, many containing both unicellular and multicellular forms. The ameboid body evolved independently many times, resulting in the reassignment of the traditional single grouping of amebas into at least eight separate phyla. This classification is not exhaustive but rather includes only phyla having examples treated in this exercise.

Phylum Diplomonada (di′plo-mon′a-da) (Gr. *diploos*, double, + L. *monas*, unit). One or two karyomastigonts (group of kinetosomes with a nucleus); individual organisms with one to four flagella; mitotic spindle within nucleus; cysts present; free-living or parasitic. Example: *Giardia*.

Phylum Parabasala (par′a-bas′a-la) (Gr. *para*, beside, + *basis,* base). With a stiffening rod, the axostyle, along longitudinal axis of body; modified region of Golgi apparatus, the parabasal body; all parasitic or endosymbiotic. Examples: *Dientamoeba, Trichomonus, Trichonympha, Spirotrichonympha.*

Phylum Euglenozoa (yu-glen-a-zo′a) (Gr. *eu-*, good, true, + *glēnē*, cavity, socket, + *zōon*, animal). Flagellates with chloroplasts with chlorophyll; flagella often with paraxial rod; mitochondria with discoid cristae; nucleoli persist during mitosis. Examples: *Euglena, Leishmania, Trypanosoma, Peranema.*

Phylum Stramenopiles (stra-men′o-piles) (L. *stramen*, straw, + *pile,* hair). Flagellates with two different flagella, one long and one short; mitochondria with tubular cristae; free-living and parasitic autotrophic and heterotrophic forms. Examples: *Actinosphaerium, Actinophyrs, Phytophthora infestans.*

Phylum Ciliophora (sil-i-of′or-a) (L. *cilium*, eyelash, + Gr. *phora*, bearing). Cilia or ciliary organelles in at least one stage of life cycle; binary fission across rows of cilia; most species free-living, many commensal, some parasitic. Examples: *Paramecium, Tetrahymena, Euplotes, Vorticella, Stentor, Epidinium, Carchesium, Trichodina, Podophrya, Zoothamnium, Spirostomum.*

Phylum Dinoflagellata (dy′no-fla-jel-at′a) (Gr. *dinos*, whirling, + *flagellum*, little whip). Typically with two flagella, one transverse and one trailing, each borne at least partially in grooves in body; about half are autotrophic with chromoplasts bearing chlorophyll, the rest heterotrophic without chromoplasts. Most are free-living, many are endosymbionts, a very few are parasitic. Examples: *Zooxanthella, Ceratium, Noctiluca, Ptychodiscus.*

Phylum Apicomplexa (ap′i-com-pleks′a) (L. *apex,* up or summit, + *complex,* twisted around). With apical complex, a characteristic set of anterior organelles present at some stage in the life cycle. All but one known species are endoparasites. Examples: *Monocystis, Gregarina, Cryptosporidium, Cyclospora, Eimeria, Toxoplasma, Plasmodium, Babesia.*

Phylum Cercozoa (ser-ko-zo′a) (Gr. *kerkos,* tail, + *zōon,* animal). Diverse group of amebas, heterogeneous in morphology and lifestyle, but monophyly strongly supported by genetic studies. Most are free-living, some parasitic. Examples: *Euglypha, Clathrulina, Chlamydophrys.*

Phylum Foraminifera (for′a-min-if′er-a) (L. *foramin,* hole, + *fero*, to bear). Shelled amebas bearing slender pseudopodia that extend through many openings in the test, forming a net that ensnares prey. Examples: *Vertebralina, Globigerina.*

Phylum "Radiolaria" (ra′de-a-la′re-a) (L. *radiolus,* small sunbeam). Marine amebas with silica skeletons, forming beautiful tests. Examples: *Tetrapyle, Pterocarys.*

Phylum Centrohelida (cen-tro-hel′i-da) (Gr. *kentron,* center of a circle, + *he-lios*, the sun). Amebas with flattened mitochondrial cristae; axoneme of axopodia that, in most, extends through a coat of variously shaped silica scales. Mostly freshwater, some marine. Examples: *Acanthocystis, Pterocystis, Heterophyrs.*

Phylum Viridiplantae (vir′i-di-plan′tee) (L. *viridis,* green, + *planto,* to set, plant). Unicellular, colonial, and multicellular green algae; all photoautotrophs with chlorophyll *a* and *b*. Examples: *Chlamydomonas, Volvox, Zea mays.*

Phylum Amoebozoa (uh-mee′bo-zo′a) (Gr. *amoibe-*, to change, + *zōon,* animal, living thing). Naked and shelled amebas, many with flagellated stages in the life cycle; mitochondria, when present, have tubular and branched cristae. Free-living and parasitic. Examples: *Entamoeba, Dictyostelium, Chaos, Arcella, Amoeba, Difflugia.*

Opisthokonta (o-pis′tho-kon′ta) (Gr. *opisthen*, behind, at the back, + *kontos*, a pole, referring to a flagellum). Many are flagellates with one posterior flagellum; the group includes nucleariid amebas, choanoflagellates, fungi, and the animals. Examples: *Codonosiga, Penicillium*, and all animals.

EXPERIMENTING IN ZOOLOGY

Effect of Temperature on the Locomotor Activity of Stentor

In this simple experiment, you will measure the effect of temperature on biological activity of the large ciliate *Stentor coeruleus*. *Stentor,* like most other unicellular eukaryotes, has a body temperature that is passively decided by the temperature of its surroundings. Most people would call such organisms "cold-blooded," but biologists try to avoid this term (as well as its counterpart, "warm-blooded") because they find it hopelessly subjective. For one thing, *Stentor* does not have blood. For another, "cold-blooded" organisms living in warm environments are not cold. We call such organisms **poikilotherms** (meaning "variable temperature") or, alternatively, **ectotherms** (meaning the body temperature is determined by the temperature of the environment). We and the rest of the mammals, and the birds, are among the few animals that have developed some orderly (and rather complicated) ways for keeping a stable body temperature. We are **homeotherms** ("same temperature") and **endotherms** (a term meaning that internally generated heat is used to elevate the body temperature).

Because biological activity is based on chemical reactions that are much influenced by temperature, unicellular eukaryotes such as *Stentor* find that everything they do is paced with water temperature. In cold water, everything slows down: swimming speed, metabolism, rate of food digestion, and so on. In warm water, it all speeds up. You might suspect that, in cold water, *Stentor* would have trouble catching its prey, but its prey have slowed down, also, as have its potential predators.

Getting Ready*

Work with a partner for this experiment. First, with cellophane tape, tape a 2 cm square piece of finely divided graph paper to the underside of a depression slide (Figure 7-22). Press the cellophane tape firmly to the slide so that it will not come loose when it gets wet. Place two or three *Stentor* in a drop of water into the depression of the slide. Apply petroleum jelly in a neat ring around the depression. The ring must be complete to prevent water from leaking in (if tap water containing chlorine gets in, the *Stentor* will die). Lower a coverslip carefully over the depression to avoid trapping bubbles beneath the coverslip. Place the slide in the bottom of a dish half-filled with cold water, place the dish on the dissecting microscope stage, and position the slide so that you can see the *Stentor.* Now carefully add crushed ice to bring the temperature down to 5°C. Stir the water with a laboratory thermometer to speed cooling. *Stentor* will be seen swimming over the graph paper grid—at this temperature, rather slowly.

*Additional notes for implementing this exercise are found in Appendix A, pp. 396–397.

How to Proceed

Choose a *Stentor* that looks active. Count the number of times this *Stentor* crosses a line in 1 minute. Your partner watches time and temperature while you watch *Stentor.* Your partner should also stir the bathwater gently with the thermometer to keep the water isothermal and add bits of crushed ice as necessary to keep the temperature steady at 5°C.

After making three separate counts at 5°C, raise the water temperature to 15°C by adding a little warm water to the bath. Hold the temperature steady at 15°C for several minutes while measuring the swimming rate three times.

Figure 7-22
Measuring the effect of temperature on the locomotor activity of *Stentor.*

Biologists often express the effect of temperature on a biological process with the temperature coefficient Q_{10}:

$$Q_{10} = \frac{k_t + 10}{k_t}$$

where k_t is the velocity constant at temperature t and $k_t + 10$ the velocity constant at 10° higher. Velocity constants are lines crossed per minute. The Q_{10} is an expression of temperature sensitivity of a biological function over a 10°C temperature range. For temperature intervals of *precisely* 10°C, you can use the following simple equation:

$$Q_{10} = SS_{(t+10)}/SS_t,$$

where SS_t is the swimming speed (lines crossed per minute) at the lower temperature, and $SS_{(t+10)}$ is the swimming speed at the higher temperature.

In general, chemical reactions have Q_{10} of about 2 to 3, while purely physical processes (such as diffusion) have much lower Q_{10} values, usually closer to 1.0.

Stentor's True Swimming Speed

How fast is *Stentor* actually swimming? The swimming speeds you have measured are "lines crossed per minute"; this is an expression, or index, of swimming speed but not a *true* swimming speed, which is distance swum per unit of time. If the graph paper grid beneath the depression slide is a square millimeter rule, you can easily determine *Stentor*'s swimming speed in millimeters per second. To get an accurate measurement, stabilize the water temperature at 20°C. Time *Stentor* with a stopwatch only when it is moving straight across or straight up the grid. Take an average of several timings. Dividing number of lines crossed by elapsed seconds gives swimming speed in millimeters per second.

Now, make an estimate of *Stentor*'s body length by comparing it with the millimeter grid. Calculate how many body lengths per second *Stentor* is swimming at 20°C. Many fish (trout, for example) swim maximally at 10 body lengths per second. Humans (practiced swimmers) can swim about 1 body length per second. How does *Stentor* compare?

Written Report

Summarize your findings in a short paragraph on separate paper. Attach a graph of swimming speed plotted against temperature.

Questions for Independent Investigation

1. At what temperature does *Stentor* reach top swimming speed? To determine this, measure swimming speed at 5° intervals, from 5° to 25°C; then plot the number of lines crossed per minute on the Y-axis of graph paper against water temperature on the X-axis.
2. Calculate Q_{10} for three different temperature increments (5° to 15°, 10° to 20°, and 15° to 25°). Are the values constant over the entire temperature range?
3. Does the temperature of acclimation affect *Stentor* swimming speeds? Compare, for example, the swimming speed at 10°C of *Stentor* from two cultures, one held for at least 3 days at room temperature and another held for 3 days at 10°C.

References

Jennings, H. S. 1904. Contributions to the study of the behavior of lower organisms. Carnegie Institution of Washington, Pub. No. 16. Experiments on the reactions of Stentor to light are described on pages 31–48 of Jennings's work.

Jennings, H. S. 1915. Behavior of lower organisms. New York, Columbia University Press.

EXPERIMENTING IN ZOOLOGY

Genetic Polymorphism in Tetrahymena*

Tetrahymena is a common unicellular, freshwater ciliate that has a teardrop shape (Figure 7-23). It is a common representative of the fauna in ponds, lakes, and streams. *Tetrahymena* has contributed greatly to our knowledge of molecular biology. It is excellent as a research organism because it is grown readily in axenic (sterile) culture and has an asexual generation time of about 2 hours or less. Prepare a wet mount slide of *Tetrahymena* from the available culture and view the unicell under high power on the compound microscope. Like most ciliates, *Tetrahymena* contains a complex set of organelles including a **macronucleus** and a **micronucleus.** The micronucleus carries the complete genetic information for the organism (five pairs of chromosomes) and is involved in sexual reproduction. The macronucleus is a somatic nucleus that carries multiple copies of DNA fragments from the chromosomes. The macronucleus appears to enhance the levels of gene expression in the rapidly growing organism by having hundreds of copies of essential genes available for transcription. The macronucleus is not involved in sexual reproduction, and it splits by a fission process during asexual reproduction.

The life cycle in *Tetrahymena* is an alternation of generations from haploid to diploid cells. Asexual reproduction occurs by normal binary fission in the diploid phase only. Sexual reproduction occurs by conjugation between two genetically different strains (known as mating types). There are seven known mating types in *Tetrahymena thermophila.* Mating type differences result from differences in genetic makeup of the organisms. *Tetrahymena,* as in other diploid forms, is often genetically **polymorphic** by possessing more than one type of allele (genetic trait) for particular genes in the genome. Genetic polymorphism also occurs in the genome at repetitive DNA sites known as **minisatellite** and **microsatellite** sites. These are regions with multiple tandem repeats of short base sequences randomly distributed in the genome. This kind of repetitive DNA exists in all eukaryotic organisms but its function is not clearly understood. Because individual organisms and clones of organisms are genetically different, DNA polymorphism may be used as markers for genetic inheritance and population analyses. Development of molecular tools has allowed for isolation and analysis of highly variable regions of DNA in organisms. Specifically important are the mini- and microsatellite regions of DNA containing variable numbers of tandem repeats (known as **VNTRs**). VNTRs have been critical in recent forensic analyses for genetically fingerprinting and identifying suspects in criminal investigations.

A fast, efficient method for genetic analysis utilizes randomly amplified polymorphic DNA markers known as **RAPDs.** In this procedure, single short primers of arbitrary sequence (10 bases in length) are used in a polymerase chain reaction (PCR)** to amplify regions between the variable repeats in a genome. PCR is a technique whereby temperature cycles of denaturation, annealing with primers, and extension with DNA polymerase are used to rapidly amplify short fragments of DNA between two regions of known sequence. The amplified fragment from a specific RAPD primer is termed a RAPD marker and is unique for a particular individual or strain of organism (e.g., a mating type in *Tetrahymena*). Figure 7-24 shows examples of RAPD bands on a DNA separation gel. RAPD markers are separated by size with agarose gel electrophoresis (Figure 7-25) and visualized by ethidium bromide*** DNA staining.

*Materials and method for implementing this exercise are found in Appendix A, pp. 397–398.

**PCR is a registered trademark of Hoffmann-LaRoche Molecular Systems, Inc.

***Caution: Ethidium bromide is a known carcinogen. Wear gloves when loading and analyzing the gel.

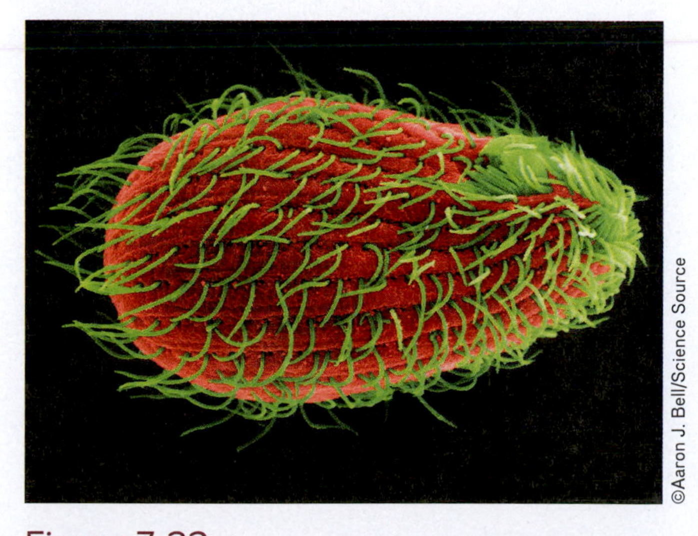

©Aaron J. Bell/Science Source

Figure 7-23
Color-enhanced photograph of *Tetrahymena piriformis* taken at ×400 on a compound microscope.

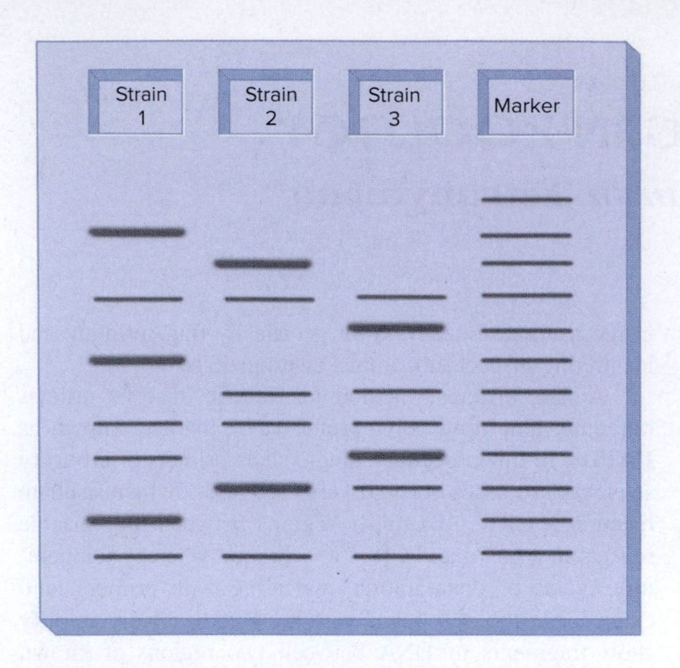

Figure 7-24
Diagram of the gel, showing RAPD bands and DNA marker bands.

You will be given samples of RAPD marker DNA prepared from two or three different mating types of *Tetrahymena.* Apply 12 µl of each sample into separate wells in a 1.4% agarose gel in a 1X Tris-Acetate EDTA buffer (TAE) containing ethidium bromide to stain the DNA. You will also load 12 µl of DNA size markers in a separate well to allow comparison of the resulting RAPD bands. The samples will be electrophoresed for 30 minutes to 1 hour at 80 volts and the bands visualized by placing the gel on a UV transilluminator.* Note the unique banding patterns that result from the genetic polymorphism between the different mating types (see Figure 7-24 for an example). A photograph or an image of the gel may be taken if equipment is available. Can you think of other uses for this technique in understanding population biology or genetic inheritance of unicellular eukaryotes, animals, or other organisms? Develop a hypothesis that you might test using the RAPD technique.

*Caution: Ultraviolet radiation from the transilluminator may cause eye or skin damage. Wear protective face and eye shields while viewing gel.

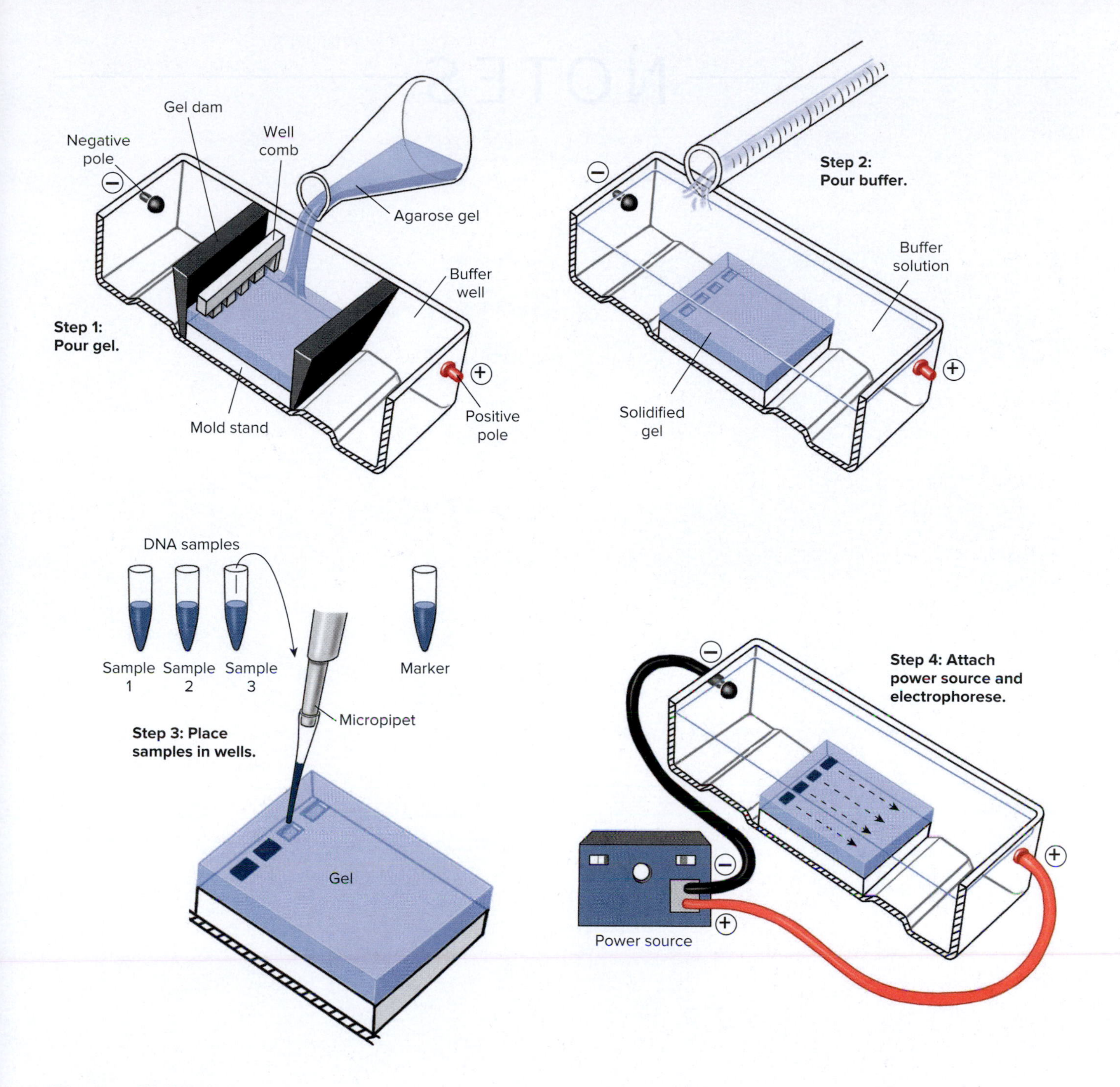

Figure 7-25

Diagram showing the horizontal submarine gel electrophoresis apparatus, power supply, and other equipment.

NOTES

The Sponges Phylum Porifera

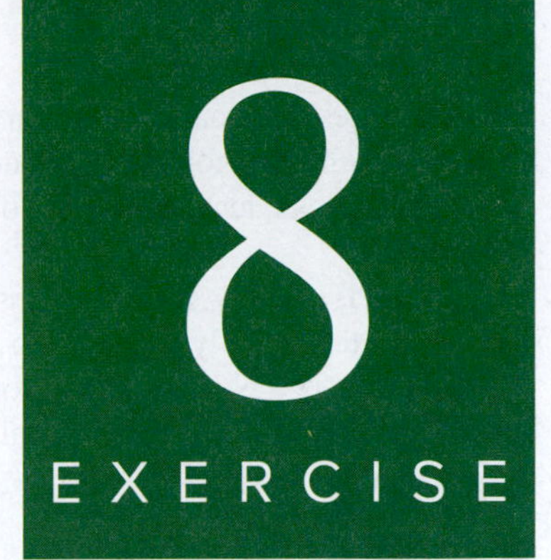

Members of phylum Porifera are considered the simplest animals—although their larvae have cell layers, adults are aggregations of cells with little or no tissue organization. They show the **cellular level of organization.** There is division of labor among their cells, but there are no organs, no systems, no mouth or digestive tract, and only very rudimentary nervous integration. Because adult sponges have no germ layers, sponges are neither diploblastic nor triploblastic. Adult sponges are all sessile in form. Some have no regular form or symmetry; others have a characteristic shape and radial symmetry. They may be either solitary or colonial.

Chief characteristics of sponges are their **pores** and **canal systems;** the flagellated sponge feeding cells, called **choanocytes,** which line their cavities and create currents of water; and their peculiar internal skeletons of **spicules** or organic fibers **(spongin).** They also have some form of internal cavity **(spongocoel)** that opens to the outside by an **osculum.**

EXERCISE 8
Class Calcispongiae—*Sycon*
Sycon, a Syconoid Sponge
Other Types of Sponge Structure

CLASSIFICATION
Phylum Porifera

EXERCISE 8
Class Calcispongiae—Sycon

Core Study

Sycon, a Syconoid Sponge

Phylum Porifera
Class Calcispongiae
Order Heterocoela
Genus *Sycon (= Scypha, Grantia)*

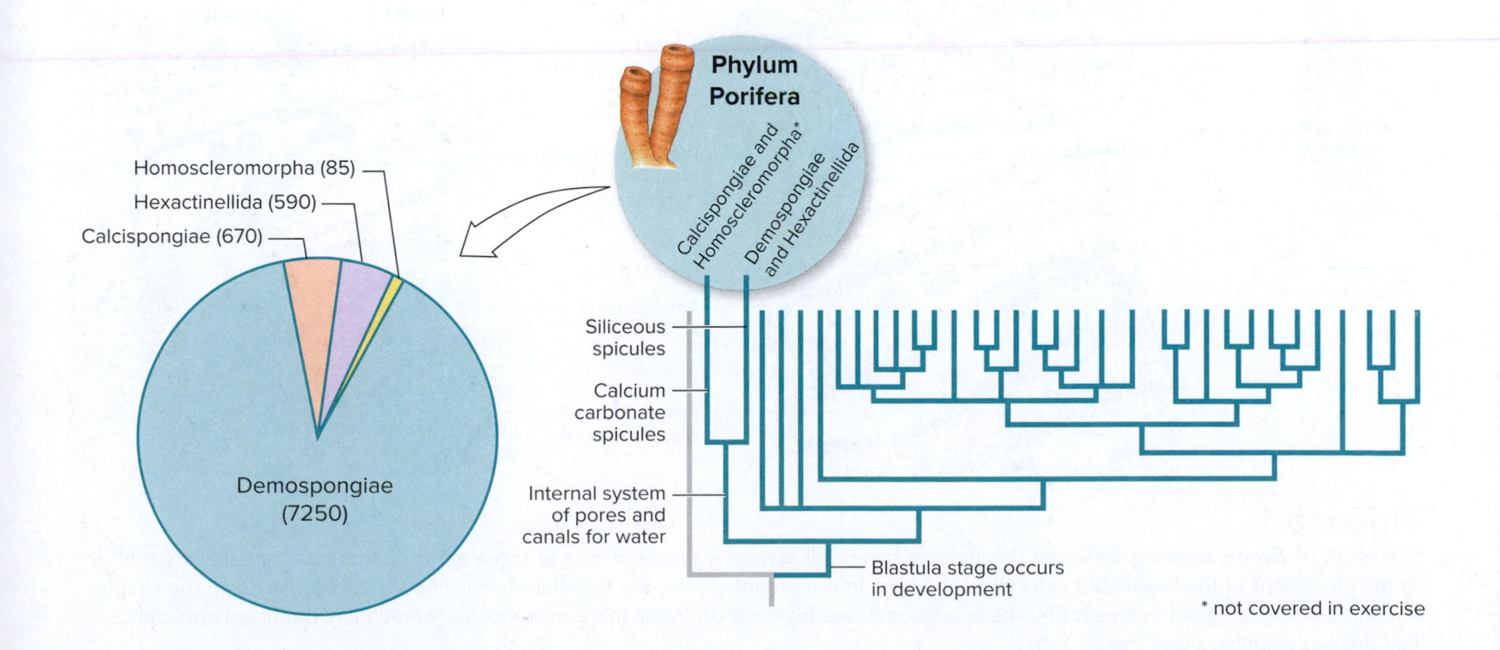

Where Found

Most sponges are marine, but there are a few freshwater species. Freshwater forms are found in small, slimy masses attached to sticks, leaves, or other objects in quiet ponds and streams.

Sycon is strictly a marine form, living in clusters in shallow water, usually attached to rocks, pilings, or shells. *Sycon* is chiefly a North Atlantic form. *Rhabdodermella* is a somewhat similar Pacific intertidal form, also belonging to class Calcispongiae.

Gross Structure

Place a preserved specimen in a small dish and cover with water. Examine with a hand lens or dissecting microscope.

Sycon (Gr., like a fig) is a **syconoid** type of sponge (Figure 8-1). What is the shape of the sponge?_________ The body wall is made up of a system of tiny, interconnected, dead-end canals whose flagellated cells draw in water from the outside through minute pores, take from it the necessary food particles and oxygen, and then empty it into a large central cavity for exit to the outside. What is the name of this central cavity? _________ All sponges have some variation of this general theme of canals and pores on which they depend for a constant flow of water.

External Structure. Is the base of the sponge open or closed? _________ The opening at the other end is the **osculum** (L., a little mouth), surrounded by a fringe of stiff, rodlike **spicules.** The external surface appears bristly when examined under magnification. Why? _________________

Note that the body wall seems to be made up of innumerable fingerlike processes pointing outward (Figure 8-1). Inside each of these processes is a **radial canal,** lined with flagellated cells, which is closed at the outer end but which opens into a central cavity, the **spongocoel** (Gr. *spongos,* sponge, + *koilos,* hollow). External spaces between these enclosed canals are **incurrent canals,** which open to the outside but end blindly at the inner end. What is the name of the openings, or pores, to the outside of the sponge? _________

Water enters the incurrent canals and passes through minute openings called **prosopyles** (Gr. *prosō,* forward, + *pylē,* gate) into radial canals and then to the spongocoel and out through the osculum. There is no mouth, anus, or digestive system. What kind of symmetry does this sponge have?
____

Spongocoel. To study the spongocoel, do the following:

Make a longitudinal cut through the midline of the body from osculum to base with a sharp razor blade. Place the two halves in a small dish and cover with water.

Find the small pores, called **apopyles** (Gr. *apo,* away from, + *pylē,* gate), that open from the radial canals into the spongocoel (Figure 8-1). Can you distinguish the tiny canals in the cut edge of the sponge wall? _________ In which direction does the water move through these canals? _________

Study of Prepared Slide

Cross sections of sponge are difficult to prepare for slides because the spicules prevent cutting sections thin enough for studying the cells. Therefore, the spicules have been dissolved away for slide preparation.

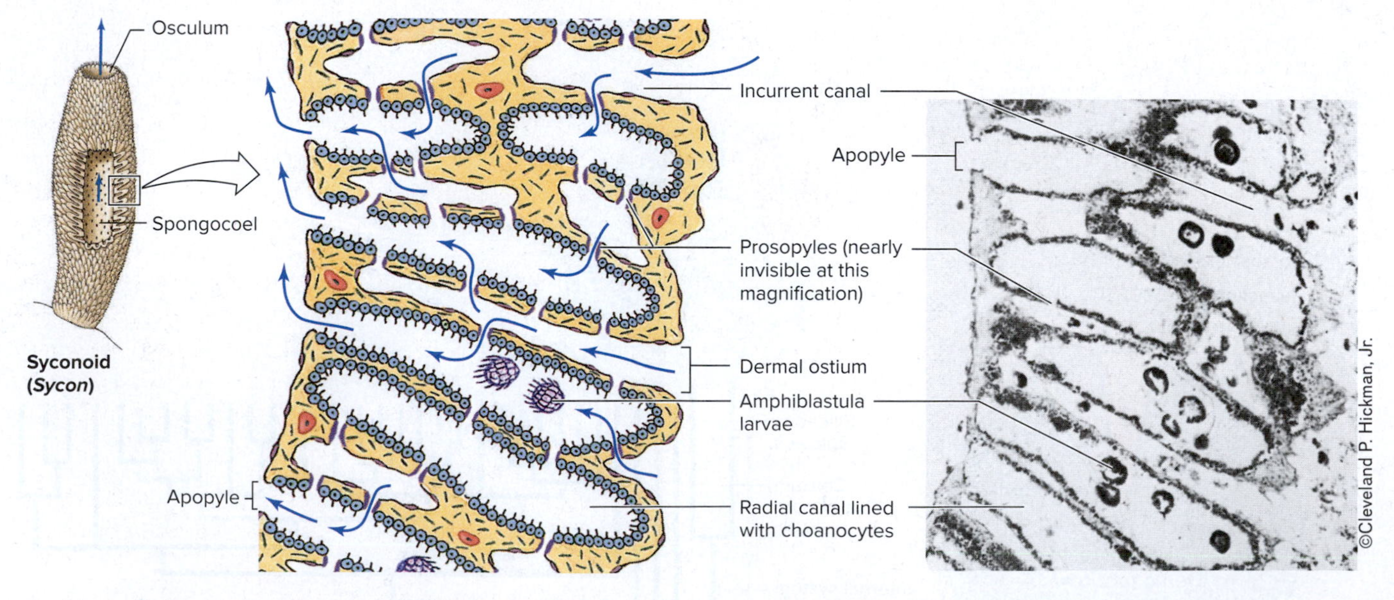

Figure 8-1

Structure of *Sycon,* showing a section through the body wall. *Sycon* is syconoid, one of three types of sponge structures that differ in the placement of the flagellated cells (choanocytes). In the syconoid plan, the flagellated cells line the radial canals. In the simpler asconoid plan, the flagellated cells line the spongocoel (see Figure 8-3). In the more complex leuconoid plan, the flagellated cells line distinct chambers (see Figure 8-5).

On a prepared slide of a cross section of *Sycon,* examine the entire section with low power to get an idea of its general relations.

Note the **spongocoel** in the middle of the section (Figure 8-1). Study the canal system. Find the **radial canals,** which open into the spongocoel by way of the **apopyles.** Are apopyle openings smaller or larger in diameter than the radial canals? __________ Some apopyle openings will be lacking in this section, and some of the radial canals will appear closed at the inner end. Follow the radial canals outward. Do they open to the outside or end blindly? __________ The radial canals may contain young larvae, called **amphiblastula larvae.** Identify the **incurrent canals,** which open to the exterior by the **dermal ostia.** Follow these canals inward and note that they also end blindly. Water passes from incurrent canals into radial canals through a number of tiny pores, or **prosopyles,** which will not be evident on the slides.

Drawings and Report for Core Study

On p. 129, draw (1) an external view of *Sycon*, showing its gross structure, and (2) a longitudinal section through the sponge showing spongocoel and internal ostia. Use arrows to show direction of water flow. Answer the questions in the laboratory report.

Further Study

Reproduction

In **sexual reproduction,** most sponges are monoecious, having both male and female sex cells in the same individual. Sperm are shed into the water and carried to other sponges, where eggs are fertilized internally. The zygotes typically are retained within the sponge to develop into free-swimming, flagellated larvae. These leave the parent through the osculum to soon settle down on a substrate where they begin development into sessile adults. What is the advantage to a sessile animal of producing free-swimming larvae? __________

Many sponges continue to grow almost indefinitely by budding and branching. They also bud off new individuals that become detached and carried away to form new sponges, a form of **asexual reproduction** in which the genotype of the parent sponge is copied to new offspring. What would be the disadvantage if this were the sole means of reproduction? __________ Is there a bud on your specimen? __________

Freshwater sponges and some marine Demospongiae reproduce asexually by means of **gemmules,** made up of clusters of amebocytes. Gemmules (L. *gemma,* bud, + *ula,* dim.) of freshwater sponges are enclosed in hard shells (Figure 8-2) and can withstand adverse conditions that would kill an adult sponge. In spring, cells in gemmules escape through the micropyle and develop into young sponges.* Marine gemmules give rise to flagellated larvae.

*An optional study of gemmules is found on p. 399.

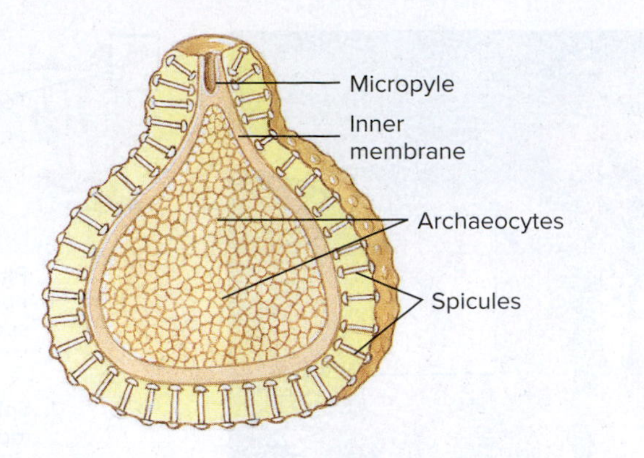

Figure 8-2
Gemmule of freshwater sponge.

Other Types of Sponge Structure

Asconoid Type of Canal System

The asconoid canal system, simplest of the three sponge body designs, is best seen in *Leucosolenia,* another marine sponge of the class Calcispongiae. *Leucosolenia* grows in a cluster, or colony (Figure 8-3), of tubular individuals in varying stages of growth. Large individuals may carry one or more buds.

After observing the external structure of a submerged specimen, cut it in half longitudinally, place it on a slide with a little water, and cover. Study with low power or use a prepared slide.

The body wall is covered with pinacocytes on the outside and filled with a gelatinous matrix (mesohyl) that contains amebocytes and spicules. Incurrent pores extend from the external surface directly to the spongocoel, which is lined with flagellated choanocytes. On living specimens, you may be able to see some flagellar activity in the spongocoel.

Leuconoid Type of Canal System

Most sponges are of the leuconoid type, the most complex of the three sponge body designs. Most leuconoids belong to the class Demospongiae (Figure 8-4). Leuconoid sponges have clusters of flagellated chambers lined with choanocytes, and water enters and leaves the chambers by systems of incurrent and excurrent canals (Figure 8-5). Water from the excurrent canals is collected into spongocoels and emptied through the oscula. In large sponges, there may be many oscula. Many marine sponges, such as *Halichondria, Microciona, Cliona,* and *Haliclona*—all belonging to Demospongiae—are of the leuconoid type.

Examine any such sponges available, in both external view and cut sections, to see this type of canal system.

Cellular Structure

The "connective tissue" of sponges is the **mesohyl** (Gr. *mesos,* middle, + *hyle,* wood) (also called mesenchyme). It holds together the various types of ameboid cells, skeletal elements, and fibrils that make up the sponge body.

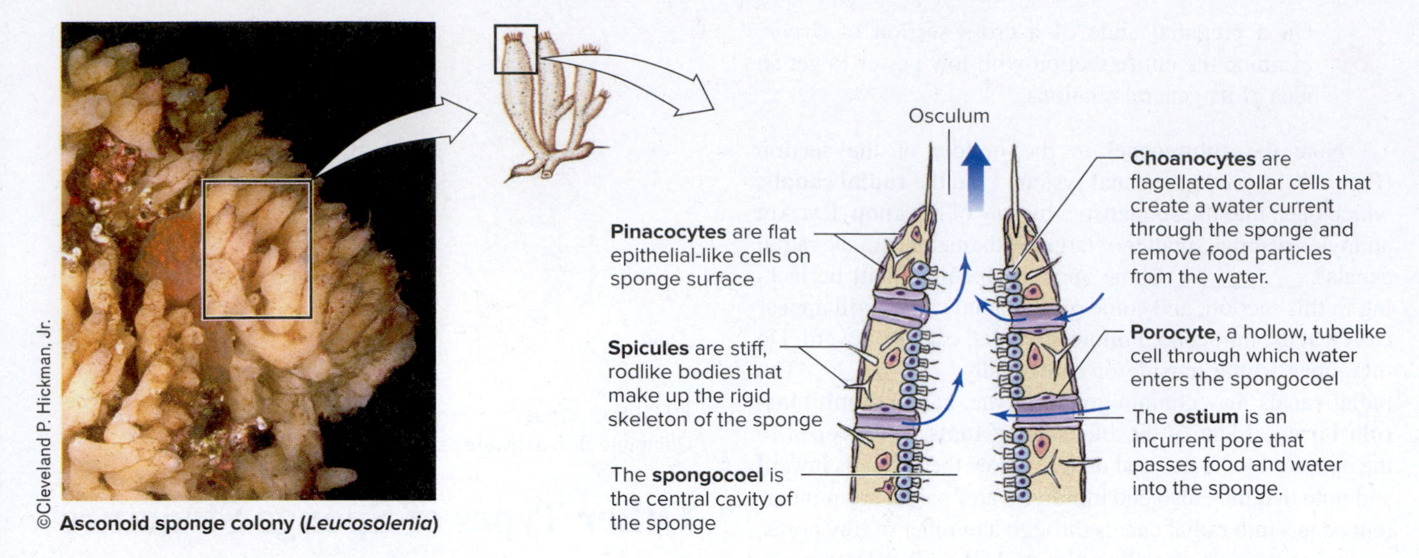
©Cleveland P. Hickman, Jr.
Asconoid sponge colony (*Leucosolenia*)
Osculum
Choanocytes are flagellated collar cells that create a water current through the sponge and remove food particles from the water.
Pinacocytes are flat epithelial-like cells on sponge surface
Porocyte, a hollow, tubelike cell through which water enters the spongocoel
Spicules are stiff, rodlike bodies that make up the rigid skeleton of the sponge
The ostium is an incurrent pore that passes food and water into the sponge.
The spongocoel is the central cavity of the sponge

Figure 8-3

Asconoid type sponge *(Leucosolenia).* Asconoid sponges have the simplest organization, with water drawn directly into the spongocoel and out the osculum. Because choanocytes are limited to capturing food from water passing adjacent to the spongocoel wall, asconoid sponges are all small and tubular. All live in dense colonies.

©William C. Ober
A
©B. Anthony Stewart/Contributor/Getty Images
B

Figure 8-4

Leuconoid type sponge, the most complex of the three main sponge body designs. **A,** *Aplysina lacunose*, globular sponge, each with a single osculum. **B,** Sponge farmers on Andros Island, Bahamas, trim a sponge harvest for shipment. The sponges, collected by divers, already have been cleaned and bleached. Most are sold as bath sponges, the most common use of natural sponges. The development of synthetic sponges has vastly reduced the natural sponge trade.

Choanocytes.

With high power, observe the "collar cells," or choanocytes (Gr. *choanē,* funnel, + *kytos,* hollow vessel), that line the radial canals (Figure 8-5B).

Although they are flagellated, you probably will not see the flagella. What is the function of the choanocytes? __________

Pinacocytes. Dermal pinacocytes (Gr. *pinax,* tablet, + *kytos,* hollow vessel) may be seen as extremely thin (squamous) cells lining the incurrent canals and spongocoel and covering the outer surface (Figure 8-5B). What is their function? __________

Amebocytes. In the jellylike **mesohyl** that lies in the wall between the pinacocytes and choanocytes, look

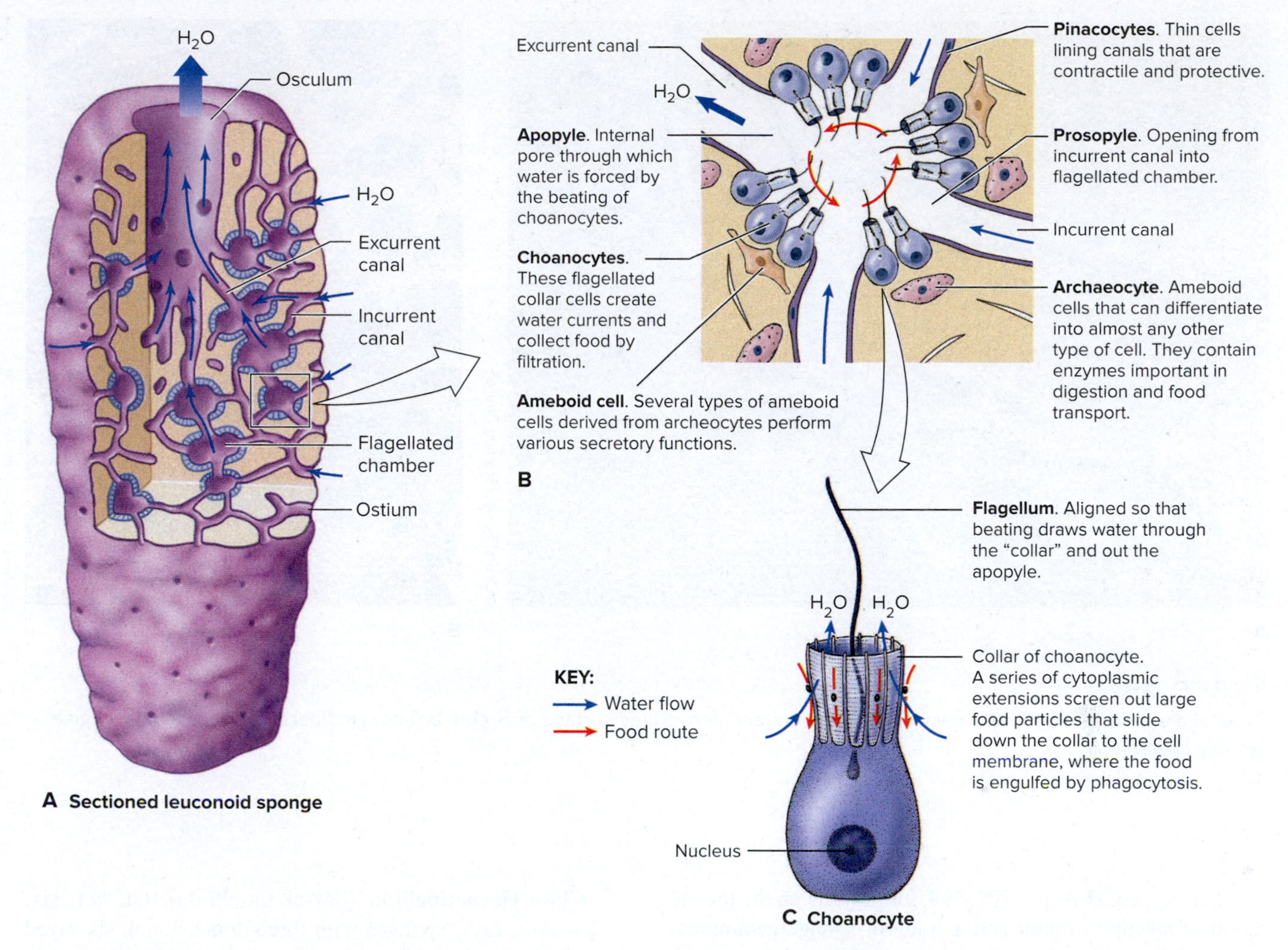

H2O
Osculum
H2O
Excurrent canal
Incurrent canal
Flagellated chamber
Ostium
A Sectioned leuconoid sponge
Excurrent canal
H2O
Apopyle. Internal pore through which water is forced by the beating of choanocytes.
Choanocytes. These flagellated collar cells create water currents and collect food by filtration.
Ameboid cell. Several types of ameboid cells derived from archeocytes perform various secretory functions.
B
Pinacocytes. Thin cells lining canals that are contractile and protective.
Prosopyle. Opening from incurrent canal into flagellated chamber.
Incurrent canal
Archaeocyte. Ameboid cells that can differentiate into almost any other type of cell. They contain enzymes important in digestion and food transport.
Flagellum. Aligned so that beating draws water through the "collar" and out the apopyle.
H2O
H2O
Collar of choanocyte. A series of cytoplasmic extensions screen out large food particles that slide down the collar to the cell membrane, where the food is engulfed by phagocytosis.
KEY:
Water flow
Food route
Nucleus
C Choanocyte

Figure 8-5

A, Leuconoid type sponge, showing flagellated chambers. **B,** Section through a flagellated chamber of a leuconoid type sponge. **C,** Choanocyte as a food-capturing cell.

for large, wandering amebocytes of various functions (Figure 8-5B). Some may differentiate into spicule-forming cells, some form sex cells, and some secrete spongin or spicules, serve as contractile cells, or aid in digestion.

On p. 129, draw a pie-shaped segment of a cross section through *Sycon,* showing a few canals and some of the cellular details of their structure.

Skeleton

Place a small bit of the sponge on a clean microscope slide; add a drop of commercial chlorine bleach, such as Clorox (sodium hypochlorite); and set aside for a few minutes to allow the cellular matter to dissolve. Break up the piece with dissecting needles, if necessary. Add a coverslip and examine under the microscope.

Look for **short monaxons** (short and pointed at both ends), **long monaxons** (long and pointed), **triradiates** (Y-shaped with three prongs) (Figure 8-6A), and **polyaxons** (T-shaped). These spicules of crystalline calcium carbonate ($CaCO_3$) form a sort of network in the walls of the animal. What is the advantage of spicules to a loosely constructed animal such as *Sycon?* ______________________________

____

Spicule types are used in the classification of sponges, along with the types of canal system. Demospongiae have siliceous (mainly $H_2Si_3O_7$) spicules, spongin fibers (composed of an insoluble scleroprotein that is resistant to protein-digesting enzymes) (Figure 8-6B), or a combination of both. Their spicules are either straight or curved monaxons or tetraxons, but never six-rayed. The glass sponges (Figure 8-7) have siliceous triaxon (six-rayed) spicules.

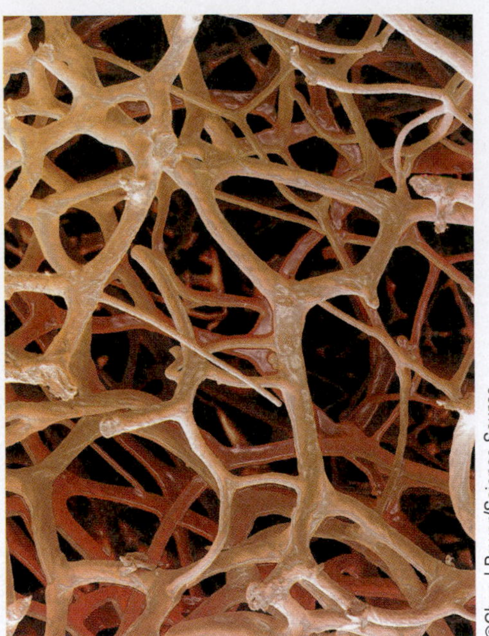

Figure 8-6
Skeletal elements. **A**, Triradiate spicules from the cut and dehydrated surface of Sycon. **B**, Spongin fibers found in Demospongiae (greatly enlarged).

See Appendix A, pp. 398–399, for projects on the preparation of spicule samples and a study of sponge gemmules, including the preparation of permanent mounts of gemmules.

Build Your Knowledge of Poriferan Features

Test your understanding of the main features of the Porifera covered in the exercise by filling out the summary table on pp. xi–xii. Try to complete the table without looking at your notes.

Classification

Phylum Porifera

Class Calcispongiae (kal-se-spun′je-e) (Gr. *calcis,* limy). Sponges with spicules of calcium carbonate, needle-shaped or three-rayed or four-rayed; canal systems asconoid, syconoid, or leuconoid; all marine. Examples: *Sycon, Leucosolenia.*

Class Hexactinellida (hex-ak-tin-el′i-da) (Gr. *hex,* six, + *aktis,* ray). Sponges with three-dimensional, six-rayed siliceous spicules; spicules often united to form network; body often cylindrical or funnel-shaped; canal systems, syconoid or leuconoid cells merge early in development to create a syncytial tissue; all marine, mostly deep-water glass sponges. Examples: *Euplectella* (Venus' flower basket), *Hyalonema.*

Class Demospongiae (de-mo-spun′je-e) (Gr. *demos,* people, + *spongos,* sponge). Sponges with siliceous spicules (not six-rayed), spongin, or both; canal systems leuconoid; one family freshwater, all others marine. Examples: *Spongilla* (freshwater sponge), *Spongia* (commercial bath sponge), *Cliona* (a boring sponge). Most sponges belong to this class.

Class Homoscleromorpha (ho-mo-skle′-ro-mor′-fa; Gr. *homos,* same, + *skleros*, hard, + *morphe*, form). Previously a subgroup of Demospongiae; skeleton reduced, spicules lacking in some; pinacocyte cells unique among sponges in having true basement membrane. Cryptic marine habitats. Example: *Plakina.*

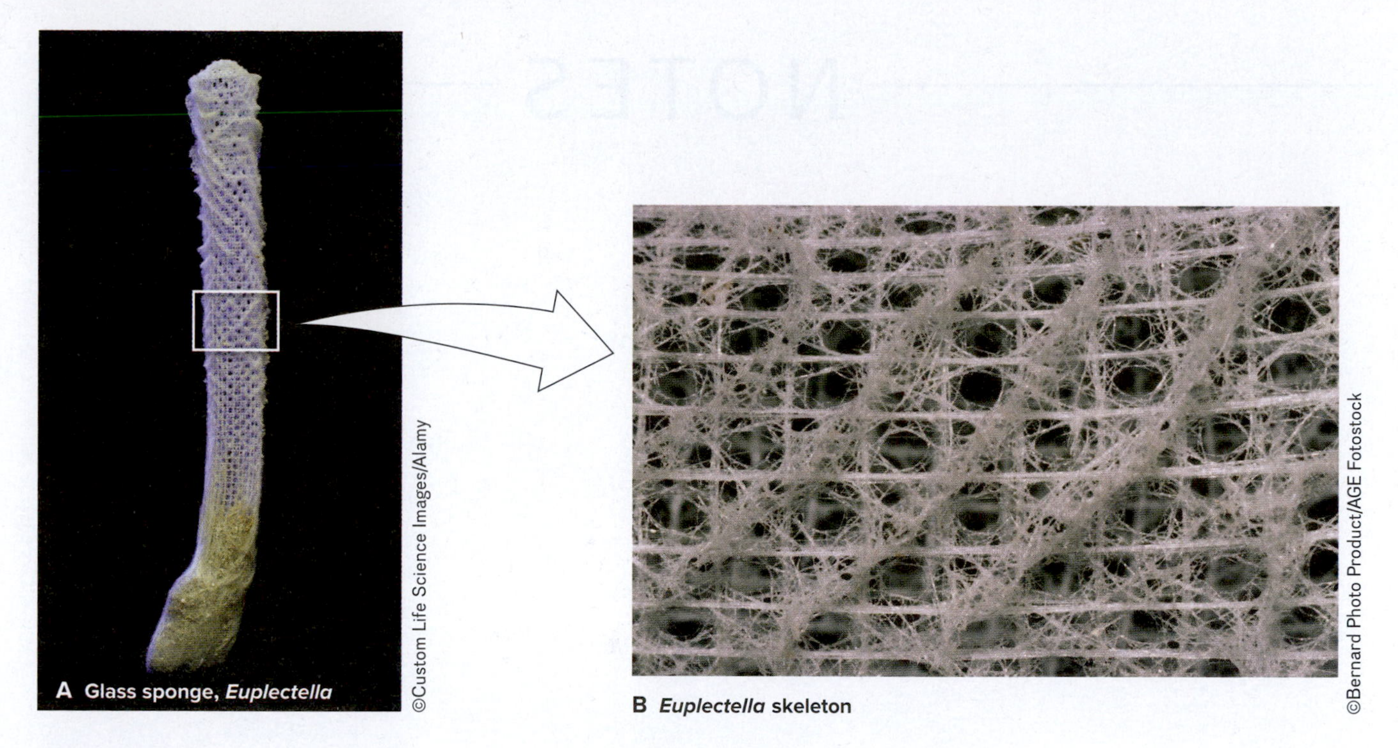

Figure 8-7

A, Beautiful skeleton of a glass sponge, *Euplectella*, of the class Hexactinellida. *Euplectella,* known as Venus' flower basket, like all glass sponges, is a deep-water form collected by dredging at depths of 1000 m or more in the western Pacific. **B**, Portion of wall of *Euplectella* with six-rayed spicules fused into a rigid latticework (about natural size).

NOTES

LAB REPORT

Name ______________________

Date ______________________

Section ______________________

Core Study

Phylum __________

Subphylum __________

Genus __________

Sycon

External View
Label the osculum.

Longitudinal Section Internal View
Label the osculum, spongocoel, and radial canal.

Sycon

1. Describe the pathway of water through *Sycon,* naming all canals and openings through which water passes from entrance to exit. ______________________

____

2. What drives the flow of water through a sponge? ______________________

3. To what level of organization do sponges belong? ______________________

4. In what way(s) does a sponge show evolutionary advancement, as compared with a colonial unicellular eukaryote, such as *Volvox*? ______________________

____

____

5. How has a syconoid sponge body plan increased the relative surface area covered with choanocytes, as compared with an asconoid body plan? ______________________

____

Further Study

A Segment of a Cross Section of Sycon

Label the osculum, incurrent canal, radial canal, and apopyle.

1. Explain how the two forms of reproduction in sponges, sexual and asexual, differ from each other.

____

____

2. What is the advantage to a sessile animal of producing free-swimming larvae? ____ What would be the disadvantage to asexual budding being the sole means of reproduction? ____

3. To what class of sponges do most marine sponges belong? ____

4. What is (are) the function(s) of choanocytes? ____

5. Explain how the skeletons of the three classes of sponges differ. ____

____

____

____

	Characteristics of the Classes of Sponges		
Characteristic	**Calcispongiae**	**Hexactinellida**	**Demospongia**
Common name(s)			
Spicule composition			
Type of canal system			
Habitat			
Representative genus			

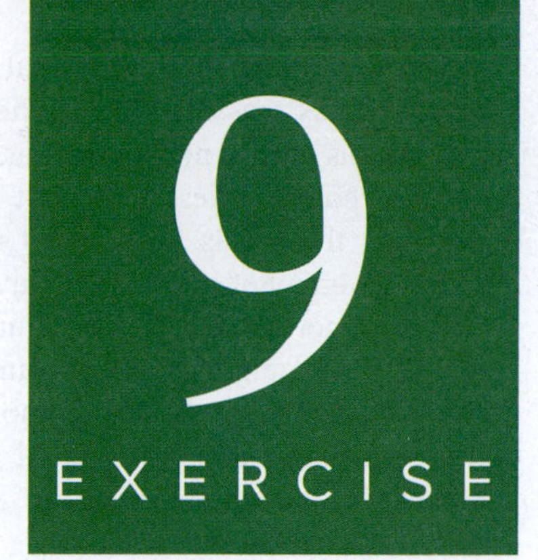

The Radiate Animals Phylum Cnidaria

The two phyla, Cnidaria (ny-dar′e-a) (= Coelenterata) and Ctenophora (te-nof′o-ra), are the earliest diverging eumetazoans—true multicellular animals. They are called radiates because all are radially (or biradially) symmetrical, a form of symmetry in which body parts are arranged concentrically around an oral-aboral axis.

Both radiate phyla are considered to be **diploblastic,** because the epidermis and gastrodermis develop from the two embryonic germ layers, ectoderm and endoderm, respectively. A third germ layer, mesoderm, appeared with the evolution of bilaterally symmetrical animals. Animals with three embryonic germ layers are called triploblastic; these will be studied in subsequent exercises.

The radiate animals are the least derived animals having a **tissue level of organization,** in which similar cells become aggregated into definite patterns or layers. With some exceptions, however, tissues of radiates are not organized into organs having specialized functions and thus lack a feature characteristic of all the more complex metazoa.

EXERCISE 9A
Class Hydrozoa—*Hydra, Obelia,* and *Gonionemus*
Hydra, a Solitary Hydroid
Obelia, a Colonial Hydroid
Gonionemus, a Hydromedusa
Projects and Demonstrations

EXERCISE 9B
Class Scyphozoa—*Aurelia,* a "True" Jelly
Aurelia
Demonstrations

EXERCISE 9C
Class Anthozoa—*Metridium* and *Astrangia*
Metridium, a Sea Anemone
Astrangia, a Stony Coral
Projects and Demonstrations

CLASSIFICATION
Phylum Cnidaria

EXPERIMENTING IN ZOOLOGY
Predator Functional Response: Feeding Rate in Hydra

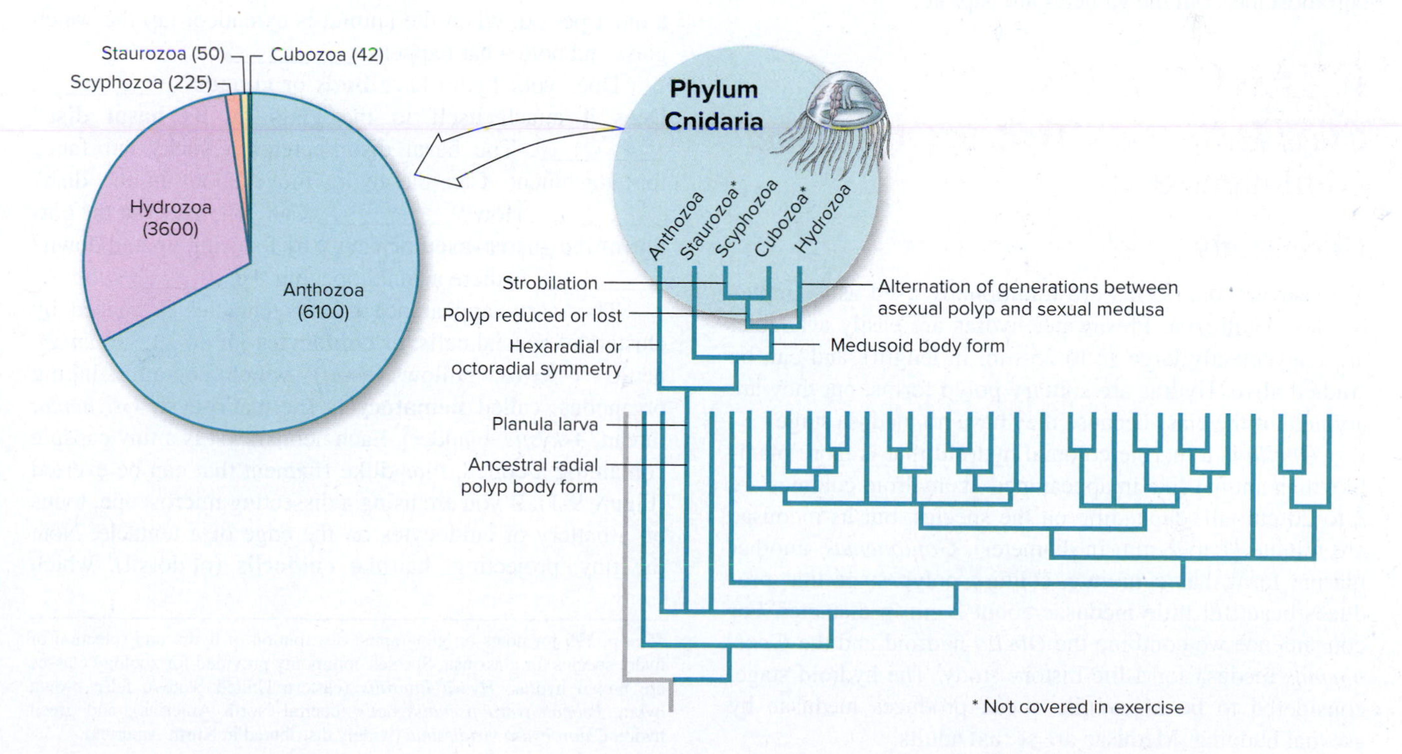

The Cnidarians include several groups familiar to most people, such as sea anemones, jellies ("jellyfish"), and corals, as well as groups not so familiar, such as hydroids, zoanthids, and comb jellies. Many are brilliantly colored, and one group, the corals, form great tropical coral reefs that harbor a diversity of life rivaled only by tropical rain forests.

Two important metazoan features shared by all cnidarians are (1) two embryological primary **germ layers** (ectoderm and endoderm) that are homologous to those of more complex metazoa and (2) internal space for digestion, the **gastrovascular cavity,** which lies along the polar axis and opens to the outside by a mouth.

Some cnidarians have a skeleton (coral, for example), but in most radiates, fluid in the gastrovascular cavity serves as a simple form of **hydrostatic skeleton.**

Although both cnidarians and their sister phylum the Ctenophora (comb jellies, not covered in this exercise), are grouped together as radiate phyla, they differ in important ways. Cnidarians have characteristic stinging organelles (cnidocytes), which are usually absent in ctenophores. **Dimorphism**—the presence in a species of more than one morphological kind of individual—is common in cnidarians but absent in ctenophores.

A distinctive characteristic of Cnidarians is that there are two main types of body form—the **polyp** (hydroid) form, which is often sessile, and the **medusa** (jelly) form, which is free-swimming. In some groups of cnidarians, both polyp and medusa stages occur in the life cycle; this is the dimorphic condition. In others, such as sea anemones and corals, there is no medusa; and in still others, such as the scyphozoans, or "true" jellies, the polyp stage is reduced or absent. In life cycles having both polyps and medusa, the juvenile polyp stage gives rise asexually to the medusa, which reproduces sexually. Both polyp and medusa have the diploid number of chromosomes, but the gametes are haploid.

EXERCISE 9A

Class Hydrozoa—Hydra, Obelia, *and* Gonionemus

Core Study

Three hydrozoan forms are traditionally used as examples of class Hydrozoa. Freshwater hydras are easily available, are conveniently large (2 to 25 mm in length), and can be studied alive. Hydras are solitary polyp forms, but they are atypical of the class because they have no medusa stage.

Obelia is a marine colonial hydroid that is more plantlike than animal-like in appearance. Its hydroid colonies are 2 to 20 cm tall, depending on the species, but its medusae are minute (1 to 2 mm in diameter). *Gonionemus,* another marine form, has a minute, solitary polyp stage that produces beautiful little medusae about 2 cm in diameter. For convenience we combine the *Obelia* hydroid and the *Gonionemus* medusa for a life-history study. The hydroid stage, considered to be a juvenile stage, produces medusae by asexual budding. Medusae are sexual adults.

Hydra, a Solitary Hydroid

Phylum Cnidaria
Class Hydrozoa
Order Hydroida
Genus *Hydra, Pelmatohydra,* or *Chlorohydra**

Where Found

Hydra (Gr., a mythical nine-headed monster slain by Hercules) is the common name applied to any of about 16 species known to occur in North America. Hydras are found in pools, quiet streams, and spring ponds, usually on the underside of leaves of aquatic vegetation, especially lily pads. To collect hydras, bring in some aquatic plants with plenty of pond water and place in a clean jar or an aquarium. In a day or so, hydras, if present, may be seen attached to the plants or to sides of the jar.

Discover Hydra Behavior and Structure

Place a live hydra in a drop of culture water on a depression slide or watch glass. Examine with a hand lens or dissecting microscope. Answer the questions in this section.

The hydra may be contracted at first. Watch it as it recovers from the shock of transfer. Does it have great powers of contraction and expansion? __________ . How long is it when fully extended? __________ . Note the cylindrical body. A conical **hypostome** is at the oral end and bears a **mouth** surrounded by **tentacles**. How many tentacles does it have? __________ . Compare with other specimens at your table. Touch a tentacle with the tip of a dissecting needle or fine artist's brush and note its reaction. What parts of the animal are most sensitive to touch? __________. After a quiet period, when the animal is extended, tap the watch glass and note what happens __________.

Does your hydra have **buds** or **gonads?** __________. Does it attach itself to the glass by its **basal disc?** __________. The basal disc secretes a sticky substance for attachment. Can the hydra move about in the dish? __________. How? __________. Can you make out the outline of the **gastrovascular** cavity by focusing up and down? __________ Is there an anal opening? __________.

The warty appearance of the tentacles is caused by clusters of special cells, or **cnidocytes** (ni′dō-sīt; Gr. *knidē,* nettle, + *kytos,* hollow vessel), which contain stinging organelles, called **nematocysts** (ne-mat′o-cyst; Gr. *nēma,* thread, + *kystis,* bladder). Each nematocyst is a tiny capsule containing a coiled, threadlike filament that can be everted (Figure 9-1). If you are using a dissecting microscope, focus on a battery of cnidocytes on the edge of a tentacle. Note the tiny, projecting, hairlike **cnidocils** (nī′dō-sil), which

*See p. 399 for notes on geographic distribution of hydra and selection of hydra species for class use. Species commonly provided for zoology classes are brown hydras, *Hydra littoralis* (eastern United States); false brown hybra, *Pelmatohydra pseudoligactis* (central North America); and green hydra, *Chlorohydra viridissima* (widely distributed in North America).

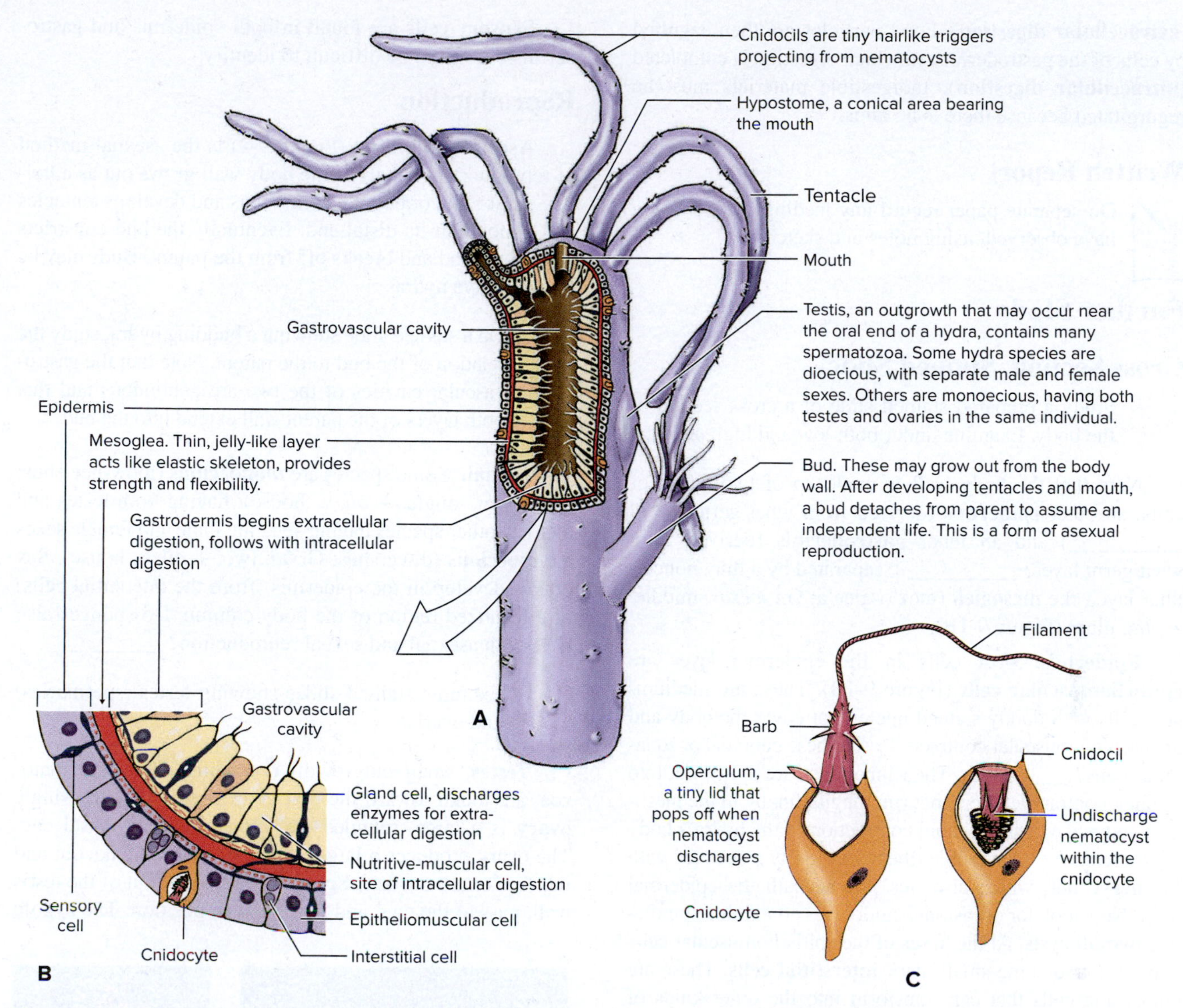

Figure 9-1

A, Structure of hydra. Both bud and developing gonad are shown, but in early life they rarely develop simultaneously. **B,** Diagrammatic cross section of a portion of the body wall. **C,** Cnidocytes with discharged *(left)* and undischarged *(right)* nematocysts.

are involved in the discharge of the nematocyst. However, chemical stimulation (as from food) is necessary to lower the threshold before a cnidocil can be stimulated by contact. A hydra may discharge a quarter of its nematocysts during food capture. But within about two days, all are replaced by interstitial cells that develop into new cnidocytes.

Feeding and Digestion. To observe the feeding reaction of hydra, do the following:

Add to the hydra culture on your slide a drop of water containing *Artemia* larvae (thoroughly washed to remove salt; see p. 400) or other suitable food organisms, such as *Daphnia* or enchytreid worms.

How does the hydra react to the presence of food? __________. How does it capture prey? __________. Does the prey struggle to escape? __________. When does the prey stop moving—before it is eaten or after? __________. What is the reaction of the hypostome? __________. How long does the feeding reaction take? __________. How does the hydra act when its appetite has been satisfied? __________.

If food is not available, or if the hydra will not eat, your instructor may want to demonstrate the feeding reaction by adding a little reduced **glutathione** (0.03 g/l) to a small dish containing some hydras. Glutathione is found in living cells. It is released, in certain of the hydra's prey, from the wounds made by nematocysts. It stimulates the hydra to open its mouth and secrete mucus to aid in the swallowing process.

Some digestion occurs within the **gastrovascular** cavity, into which gland cells secrete digestive enzymes

(extracellular digestion). Food particles are then engulfed by cells of the gastrodermis, in which digestion is completed **(intracellular digestion).** Indigestible materials must be regurgitated because there is no anus.

Written Report

On separate paper record any feeding reactions you have observed, using notes and sketches.

Further Study

Cross Section, Stained Slide

Study a prepared stained slide of a cross section of the body. Examine under both low and high power.

Note that the body wall is made up of two layers of cells, an outer **epidermis** (derived from what germ layer? __________) and an inner **gastrodermis** (derived from what germ layer? __________), separated by a thin, noncellular layer, the **mesoglea** (mez′o-glee′a; Gr. *mesos,* middle, + *glia,* glue) (Figure 9-1B).

Epidermis. Most cells in the epidermal layer are **epitheliomuscular cells** (Figure 9-1B). These are medium-size cells with darkly stained nuclei that cover the body and are used for muscular contraction. Are these cuboidal or squamous cells? __________ Their inner ends are drawn out into slender, contractile fibers that run longitudinally in the mesoglea and make possible rapid contraction of the hydra's body and/or tentacles. Contractile fibers are closely associated with the **nerve net,** which also lies just beneath the epidermal layer. Now look for occasional cnidocytes containing spindle-like **nematocysts.** At the bases of the epitheliomuscular cells may be found some small, dark **interstitial cells.** These are embryonic cells that can transform into the other kinds of cells when needed. **Gland cells** secrete mucus onto the body surface, particularly around the mouth and basal disc.

Mesoglea. This noncellular layer, lying between epidermis and gastrodermis, extends over both body and tentacles of hydra as an elastic "skeleton," providing increased flexibility to the animal. It is very thin in hydroid polyps.

Gastrodermis. The gastrodermis layer is composed principally of ciliated columnar **nutritive-muscular cells** (Figure 9-1B). These cells perform several functions. Being muscular, they enable the hydra to change shape (become longer and thinner). They also engulf food (by phagocytosis), forming food vacuoles in which food is digested intracellularly. They additionally discharge enzymes into the gastrovascular cavity, where extracellular digestion occurs. Finally, their cilia create currents that keep food particles constantly suspended and circulating. What types of cells are common to both layers of the body wall? __________________

How does digestion in the hydra compare with that in the sponge? __________ Does the sponge have intracellular or extracellular digestion? __________

Sensory cells are found in both epidermis and gastrodermis but would be difficult to identify.

Reproduction

Asexual. Budding (Figure 9-2) is the asexual method of reproduction. A part of the body wall grows out as a hollow outgrowth, or bud, that lengthens and develops tentacles and a mouth at its distal end. Eventually the bud constricts at its basal end and breaks off from the parent. Buds may be found on live hydras.

On a stained slide showing a budding hydra, study the relation of the bud to the parent. Note that the gastrovascular cavities of the two are continuous and that both layers of the parent wall extend into the bud.

Sexual. Some species are **monoecious** (mow-nee′shus; Gr. *monos,* single, + *oikos,* house), having both testes and ovaries; other species having separate male and female sexes are **dioecious** (di-ee′shus; Gr. *di,* two, + *oikos,* house). Sex organs develop in the epidermis (from the interstitial cells) of a localized region of the body column. Do sponges also have both asexual and sexual reproduction? __________

Examine stained slides showing testes (spermaries) and ovaries.

Testes, small outgrowths containing many spermatozoa, are found toward the oral end; (Figure 9-2) the single **ovary** is a large, rounded elevation nearer the basal end. The ovary produces a large, ripe **egg,** which breaks out and lies free on the surface. **Spermatozoa** break out of the testis wall, pass to the egg, and fertilize it in position. The **zygote**

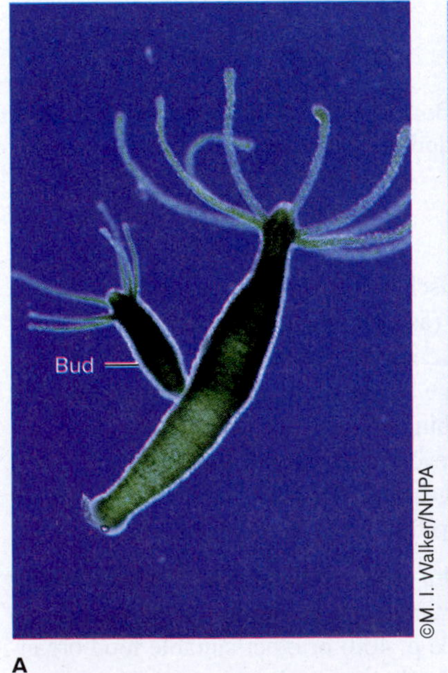

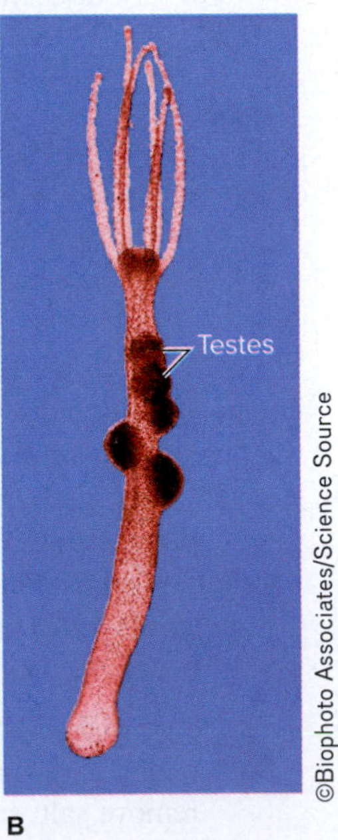

Figure 9-2
A, Hydra with one bud. **B**, Hydra with several testes.

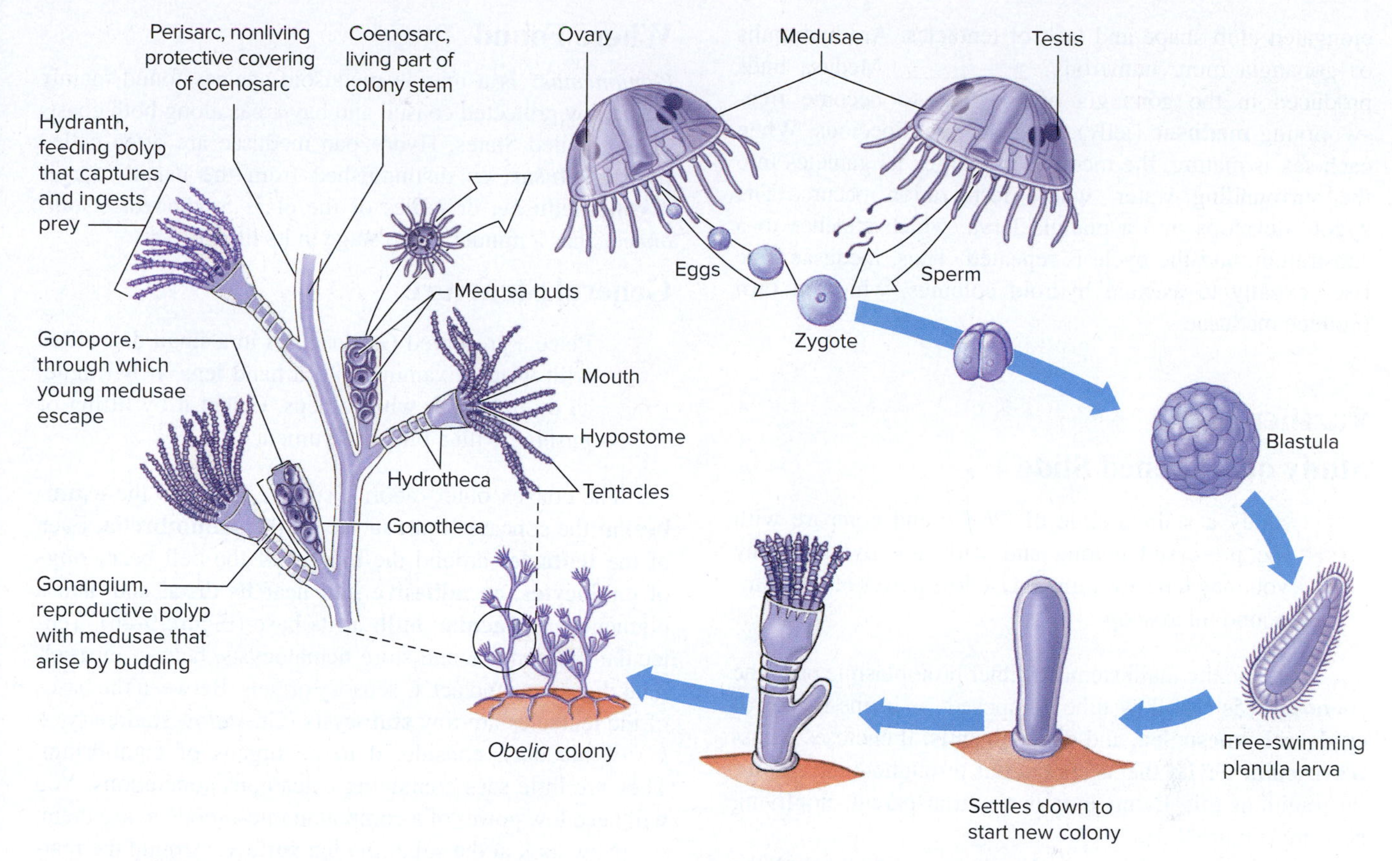

Figure 9-3

Obelia life cycle, showing alternation of polyp (asexual) and medusa (sexual) stages. In *Obelia* both its polyps and its stems are protected by continuations of the perisarc. In some hydroids, only the stems are so protected.

so formed undergoes several stages of development before dropping off the parent.

Drawings

On separate paper, sketch a budding hydra and specimens that show testes (spermaries) and ovaries.

Core Study

Obelia, a Colonial Hydroid

Phylum Cnidaria
Class Hydrozoa
Order Hydroida
Genus *Obelia*

Where Found

Obelia (Gr. *obelias,* round cake) is one of many colonial hydroids found in marine waters and attached to seaweeds, rocks, shells, and other objects. Its minute medusae make up part of the marine plankton.

Life History and General Study

The life history of *Obelia* is dimorphic, having both polyp (hydroid) and medusa (jelly) stages (Figure 9-3). A hydroid colony arises from a free-swimming **planula** larva, which settles and attaches to a substratum. Then, by budding, a colony is formed.

For study of a preserved colony,* place a small piece of marine algae with attached *Obelia* colonies in a small dish of water and observe with dissecting microscope.

Colonies resemble tiny plants attached to the seaweed by rootlike stolons. From the stolon arises a main stem, which gives rise to many lateral branches (Figure 9-3). The living part of the colony is the **coenosarc,** (sē′nō-sark; Gr. *koinos,* shared, + *sarx,* flesh), which is encased in a thin, transparent, protective **perisarc,** secreted by the epidermis. On the branches are two kinds of polyps (also called zooids): nutritive polyps called **hydranths** (Gr. *hydōr,* water, + *anthos,* flower) and reproductive polyps called **gonangia** (sing. **gonangium;** N.L. *gonas,* primary sex organ, + *angeion,* dim. of vessel). Hydranths can be recognized by their vase shape and the tentacles at their free ends; gonangia, by their

*If living *Obelia* colonies are preferred for this exercise, see p. 400 for sources of living material

elongated club shape and lack of tentacles. Are hydranths or gonangia more numerous? __________ Medusa buds produced in the gonangia break away to become free-swimming **medusae** (jelly). Medusae are dioecious. When each sex is mature, the medusa discharges its gametes into the surrounding water, where fertilization occurs. The zygote develops into a planula larva, which attaches to a substratum, and the cycle is repeated. Thus, medusae give rise sexually to asexual hydroid colonies, which in turn produce medusae.

Further Study

Study of a Stained Slide

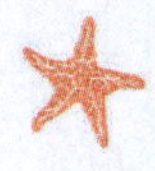

Study a stained slide of *Obelia* and compare with the preserved colony and with any living colony you may have examined. Use low power of the compound microscope.

Examine the main stem. Its inner protoplasmic part, the **coenosarc,** is a hollow tube composed, as in the hydra, of **epidermis, mesoglea,** and **gastrodermis.** It encloses a **gastrovascular cavity** that is continuous throughout the colony. Surrounding this living part is the transparent, nonliving perisarc (Figure 9-3).

Each feeding polyp, or hydranth, is continuous with the coenosarc. A transparent extension of the perisarc forms a protective cup around the hydranth. Each hydranth has an elevated **hypostome** terminating in a **mouth** and bearing a circle of **tentacles** around the base. Each tentacle has in its epidermis rings of swellings caused by clusters of **cnidocytes,** which bear the **nematocysts.** Trace the continuous **gastrovascular cavity** from the hydranth through the branches and main stem of the colony. Food taken by the hydranths can thus pass to every part of the colony.

Club-shaped reproductive **gonangia** arise at the junction of the hydranth and the coenosarc. A number of saucer-shaped **medusae buds** grow from a central stalk within the gonangium; these will develop into mature medusae. Where are the mature buds located? __________ Young medusae escape through the opening, or **gonopore,** at the distal end.

Core Study

Gonionemus, a Hydromedusa

Phylum Cnidaria
Class Hydrozoa
Order Hydroida
Genus *Gonionemus*

Obelia has a macroscopic hydroid stage, but its medusa stage is microscopic. Another hydrozoan form, *Gonionemus,* has a similar life history, but its medusa is fairly large and suitable for continued study of the hydroid life cycle.

Where Found

*Gonionemus** is a marine medusoid species found mainly in shallow protected coastal and bay areas along both coasts of the United States. Hydrozoan medusae are often called **hydromedusae,** as distinguished from the usually larger **scyphomedusae,** or jellies of the class Scyphozoa. *Gonionemus* has a minute polyp stage in its life history.

General Structure

Place a preserved *Gonionemus* in a small dish filled with water. Examine with a hand lens. It is fragile, so do not grasp with forceps. Orient it by lifting or pushing with a blunt instrument.

The convex outer (aboral) surface is called the **exumbrella;** the concave (oral) surface is the **subumbrella.** Each of the **tentacles** around the margin of the bell bears rings of **cnidocytes,** an **adhesive pad** near its distal end, and a pigmented **tentacular bulb** at its base (Figure 9-4b). Tentacular bulbs make and store nematocysts, help in intracellular digestion, and act as sensory organs. Between the bases of the tentacles are tiny **statocysts** (Gr. *statos,* stationary, + *kystis,* bladder), considered to be organs of equilibrium. They are little sacs containing calcareous concretions. You will need low power of a compound microscope to see them.

Now look at the subumbrellar surface. Around the margin, find a circular, shelflike membrane, the **velum** (L., veil, covering), that aids in swimming movement. Medusae move by a form of jet propulsion, forcing water out of the subumbrellar cavity by muscular contractions, thus propelling the animals in the opposite direction.

Note the **manubrium** (L., handle), a tubular extension bearing the **mouth,** with four liplike **oral lobes** around it.

The **gastrovascular cavity** includes the **gullet,** the **stomach** at the base of the manubrium, four **radial canals** extending to the margin, and a **ring canal** around the margin.

Note the convoluted **gonads** suspended under each of the radial canals. The sexes look alike and can be determined only by microscopic inspection of the gonadal contents.

The medusa has the same cell layers as the hydroid or polyp form. All surface areas are covered with **epidermis,** the **gastrodermis** lines the entire gastrovascular cavity, and between these two layers is the jellylike **mesoglea,** which is much thicker in the medusa than in the hydroid form.

Drawings

On p. 145, label the diagrammatic drawing of an oral-aboral section through ***Gonionemus***.

---

*The name *Gonionemus,* meaning "angled thread," may refer to a bend near the end of the slender tentacles where a tiny adhesive pad the animal uses to attach itself to seaweed is located (see Figure 9-4B).

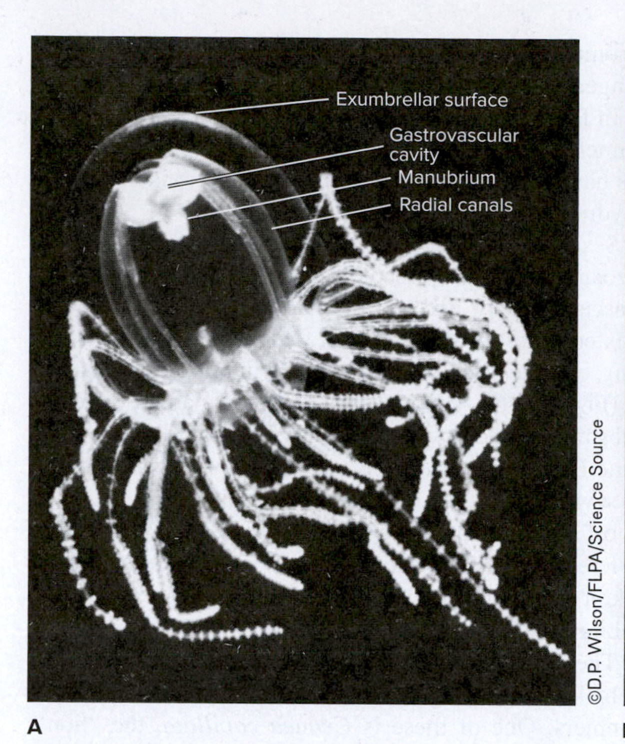

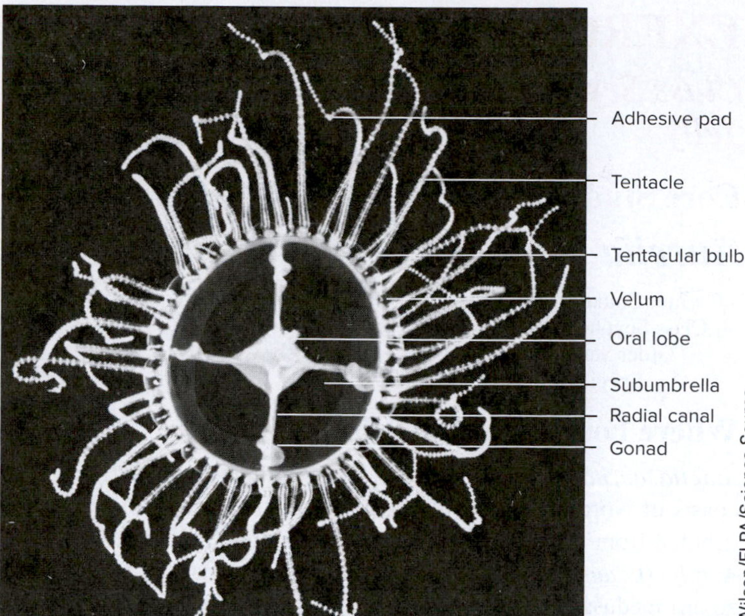

Figure 9-4
Gonionemus. **A,** Swimming. **B,** Oral view.

Further Study

Behavior

If living hydromedusae (freshwater *Craspedacusta,* marine *Gonionemus,* or others) are available, watch their movements.

Note the pulsating contractions that force water from the underside of the umbrella (subumbrella) and so propel the animal by a feeble "jet propulsion" (Figure 9-4A). Can they change direction? __________ How? __________ Does the animal swim continuously, or does it float sometimes? __________ What happens if a tentacle is touched? __________ Feeding reactions differ with different species. Some apparently depend on chance contact with food; others, such as *Gonionemus,* swim to the surface and then turn over and float downward with tentacles spread in search of food. *Gonionemus* is often found moving slowly about among marine algae and sea grass. If algae is present in the aquarium with *Gonionemus,* note how it uses its tentacles to move about. How does it use the adhesive pads on the tentacles? ______________________________

Written Report

Record your observations on separate paper.

Build Your Knowledge of Hydrozoan Features

Test your understanding of the main features of the hydrozoans covered in the exercise by filling out the summary table on pp. xi–xii. Try to complete the table without looking at your notes.

Projects and Demonstrations

1. *Portuguese man-of-war (Physalia pelagica).* The Portuguese man-of-war illustrates the highest degree of **polymorphism** among cnidarians, for several types of individuals are found in the same colony. Study a preserved specimen. Note the **pneumatophore** (Gr. *pneuma,* air, + *pherein,* to bear), or bladder. What is its function? Suspended from the pneumatophore are many zooids (polyps) that have budded from it. Two kinds of polyp individuals are the feeding polyps (gastrozooids) and fishing polyps (dactylozooids), which are equipped with long, stinging tentacles. Male and female gonophores are modified medusoid individuals that remain attached as buds, producing eggs and sperm.
2. *Obelia.* Examine some medusae of *Obelia.*
3. *Various colonial hydroids.* Examine forms of colonial hydroids, such as *Eudendrium, Tubularia, Pennaria, Sertularia,* or *Bougainvillia,* either preserved or on slides.

EXERCISE 9B

Class Scyphozoa—Aurelia, *a "True" Jelly*

Core Study

Aurelia

Phylum Cnidaria
Class Scyphozoa
Order Semaeostomeae
Species *Aurelia aurita*

Where Found and Life History

Aurelia aurita, the "moon jelly," is common along both coasts of North America. It is a cosmopolitan species distributed from temperate and tropical to subpolar latitudes.* *Aurelia* (L. *aurum,* gold) (Figure 9-5) is a scyphozoan (sy-fo-zo′an) medusa (often referred to as a scyphomedusa). Scyphomedusae are generally larger than hydrozoan medusae (hydromedusae); most of them range from 2 to 40 cm, but some reach as much as 2 m or more in diameter. The jelly layers (mesoglea) are thicker and contain cellular materials, giving scyphomedusae a firmer consistency than hydromedusae. Nevertheless, all jellies are largely water (94% to 96% water in marine species, such as *Aurelia,* and up to 99% water in some freshwater hydromedusae) and active tissues are mostly epithelial.

Scyphozoans are often called "true" jellies. Scyphomedusae are constructed along a plan similar to that of hydromedusae, but they lack a velum (the velum of hydromedusae is described on p. 136). Their parts are arranged symmetrically around the oral-aboral axis, usually in fours or multiples of four, so they are said to have tetramerous radial symmetry. Their gastrovascular systems have more canals and more modifications than those of hydrozoans.

Figure 9-5
Swimming *Aurelia aurita.*

Sexes are separate in *Aurelia,* as they are in all scyphozoans. Sex cells are shed from the gonads into the gastrovascular cavity and are discharged through the mouth for external fertilization. Within folds of the oral arms, the young embryos develop into free-swimming **planula larvae** (Figure 9-6). These escape from the parent, attach to a substratum, and develop into tiny polyps called **scyphistomae** (sy-fis′to-mee; Gr. *skyphos,* cup, + *stoma,* mouth). The scyphistoma later becomes a **strobila,** which begins to bud off young medusae (**ephyrae** [Gr. *Ephyra,* Greek city, in reference to castlelike appearance]) in layers resembling a stack of saucers (Figure 9-6). This budding process is called **strobilation** (Gr. *strobilos,* pinecone).

The large size and fiery nematocysts of many jellies make them disagreeable and sometimes dangerous to swimmers. One of these is *Cyanea capillata,* the "lion's mane jelly" of the North Atlantic.** Even more dangerous is the cubomedusan *Chironex fleckeri,* sea wasp, of the Australian region. This jelly has caused numerous fatalities in Australia; deaths occur rapidly from anaphylactic shock.

Discover Behavior of Jellies

If living *Aurelia* or other scyphozoan genera are available, observe their swimming movements (Figure 9-5).

How is movement achieved? __________ Are they strong swimmers? __________ How many times per minute does the medusa pulsate? __________ Does it swim horizontally or vertically? __________ Use a gentle touch with a small artist's brush to test reaction to touch. To test response to food chemicals, dip the brush into glucose solution, clam or oyster juice, or other food substances; small crustaceans or other small food organisms may be placed near their tentacles.

Written Report

Record your observations on separate paper.

*In recent years moon jelly populations in many habitats have increased into massive, destructive swarms that are disabling power plants by clogging cooling equipment, clogging ship engines, bursting fishing nets, and destroying fisheries. Such swarms of moon jellies, other jelly species, and comb jellies, are changing the balance of marine ecosystems.

**The lion's mane jelly appears in a Sherlock Holmes story "The Adventure of the Lion's Mane," in which a school professor becomes a victim of the jelly while swimming.

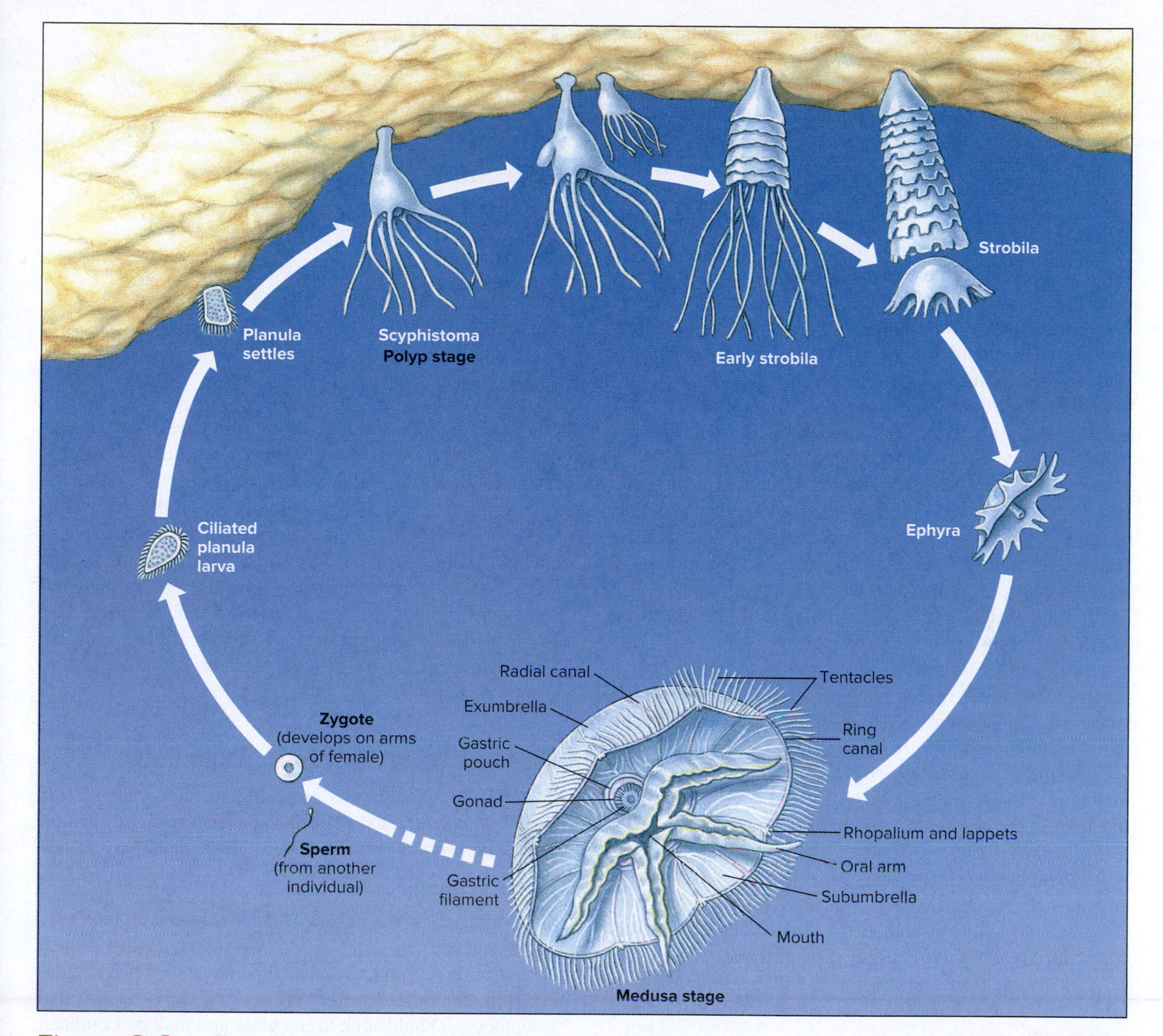

Figure 9-6

Life cycle of *Aurelia,* a marine scyphozoan medusa. This species is dimorphic with alternating sexual (medusa) and asexual (polyp) stages in the life cycle. In true jellies (scyphozoans) the sexual medusa is clearly the dominant stage. Fertilized eggs develop into planula larvae; each settles to grow into a scyphistoma, the polyp stage. The polyp gives rise to a strobila, which buds off ephyrae (immature medusae), a process called strobilation.

Further Study

General Structure

Using a ladle (the medusa is too fragile to be handled with a forceps), transfer a preserved specimen of *Aurelia* to a finger bowl of water and spread out flat.

Note that *Aurelia* is more discoidal and less cup-shaped than *Gonionemus.* When spread flat, the jelly shows a circular shape broken at eight regular intervals by marginal notches (Figures 9-5 and 9-6). Each marginal notch contains a **rhopalium** (ro-pay'li-um; N.L. from Gr. *rhopalon,* a club), a sense organ consisting of a statocyst and an ocellus. This is flanked on each side by a marginal extension, the **lappet** (Figure 9-6). What is the function of a statocyst? __________ An ocellus? __________

Snip out a rhopalium* with scissors and examine under the higher power of a dissecting microscope.

*The specific epithet *aurita* of the species *Aurelia aurita* means "eared," probably in reference to the lobelike rhopalia.

In living medusae, the excision of all rhopalia would interfere with swimming, either slowing down the contractions or stopping them altogether.

Note the short **tentacles** that form a fringe around the animal's margin. Compare these tentacles with those of *Gonionemus* medusae. How do they differ? ____________

Gastrovascular System. In the center of the oral side are four long, troughlike **oral arms.** These are modifications of the manubrium. Note that the oral arms converge toward the center of the animal, where the square **mouth** is located. The mouth opens into a short **gullet,** which leads to the **stomach.** From the stomach, four **gastric pouches** extend. They can be identified by the horseshoe-shaped **gonads** that lie within them. Near the inner edges of the gonads are numerous thin processes, the **gastric filaments,** which are provided with **nematocysts.** What would be the use of stinging cells here? __________ A complicated system of radiating canals runs from the gastric pouches to the **ring canal,** which follows the outer margin (Figure 9-6). This system of stomach and canals, resembling hub, spokes, and rim of a wheel, forms the medusoid gut, or gastrovascular cavity. Contrast this with the simple, saclike gut of polyp individuals such as hydra or *Obelia* polyps.

Jellies are carnivorous, most feeding on fish and a variety of marine invertebrates. *Aurelia,* however, is a suspension feeder that feeds largely on zooplankton. The tentacles are not used in food capture. Food organisms are caught in mucus secreted on the subumbrella and moved by cilia to the bell margin. Food is collected from the margin by the oral arms and is transferred by cilia to the stomach. Gastric filaments with nematocysts help pull in and subdue larger organisms and secrete digestive enzymes. Partially digested food and seawater are then circulated by cilia through the system of canals. In this way, nutrients and oxygen are carried to all parts of the body. The canals are lined with cells that complete digestion of the food. Thus, digestion is both extracellular and intracellular.

Build Your Knowledge of Scyphozoan Features

Test your understanding of the main features of the Scyphozoa covered in the exercise by filling out the summary table on pp. xi–xii. Try to complete the table without looking at your notes.

Demonstrations

1. *Slides.* Examine slides showing stages in the life cycle of *Aurelia*—scyphistoma, strobila, and ephyra.
2. *Scyphozoan jelly.* Examine various species of preserved scyphozoans or, if available, living specimens of the "upside-down" jelly *Cassiopeia* (see p. 401 for additional information).

EXERCISE 9C

Class Anthozoa—Metridium *and* Astrangia

Core Study

Metridium, a Sea Anemone

Phylum Cnidaria
Class Anthozoa
Subclass Hexacorallia
Species *Metridium senile*

Where Found

Metridium senile, a name from the Greek that means "ancient womb," is the most common sea anemone on the Atlantic Coast from Delaware north to the Arctic and on the Pacific Coast from Santa Catalina Island, California, north to the polar seas. It occurs from low intertidal, where it is commonly seen on rocks and pilings, to depths of perhaps 75 m. Most sea anemones are solitary sessile animals but some make clonal patches or colonies. Members of the class Anthozoa are all polyps in form; there are no medusae. There is a great variety in size, structure, and color among the sea anemones. All are marine.

Discover Behavior of Sea Anemones*

If living anemones are available, allow one to relax completely and then touch a tentacle lightly with a new (untouched) coverslip held in clean forceps.

Do the nematocysts discharge? __________ Will nematocysts discharge if the coverslip is touched more vigorously to the tentacle? __________ Now put some saliva (which contains protein) on a dry coverslip and again touch a tentacle. Is the response stronger? Why? __________ Discharged nematocysts should stick to the glass and may be examined under the microscope.

Using another relaxed anemone, drop bits of clean filter paper on the tentacles and time the type and speed of the response. Now test with bits of filter paper soaked in shrimp, clam, or mussel juice and compare the reactions. Test again with bits of clam or other sea-food. What conclusions can you draw from this simple experiment? __________

Is response to food similar to response to touch? __________ Is the reaction of these animals a part of the normal feeding reaction? __________ If live sea stars are available, try touching an anemone with the arm of a star. What happens? Is this a feeding response or a defense reaction? __________

*See p. 401 for suggestions for implementing behavior studies with sea anemones.

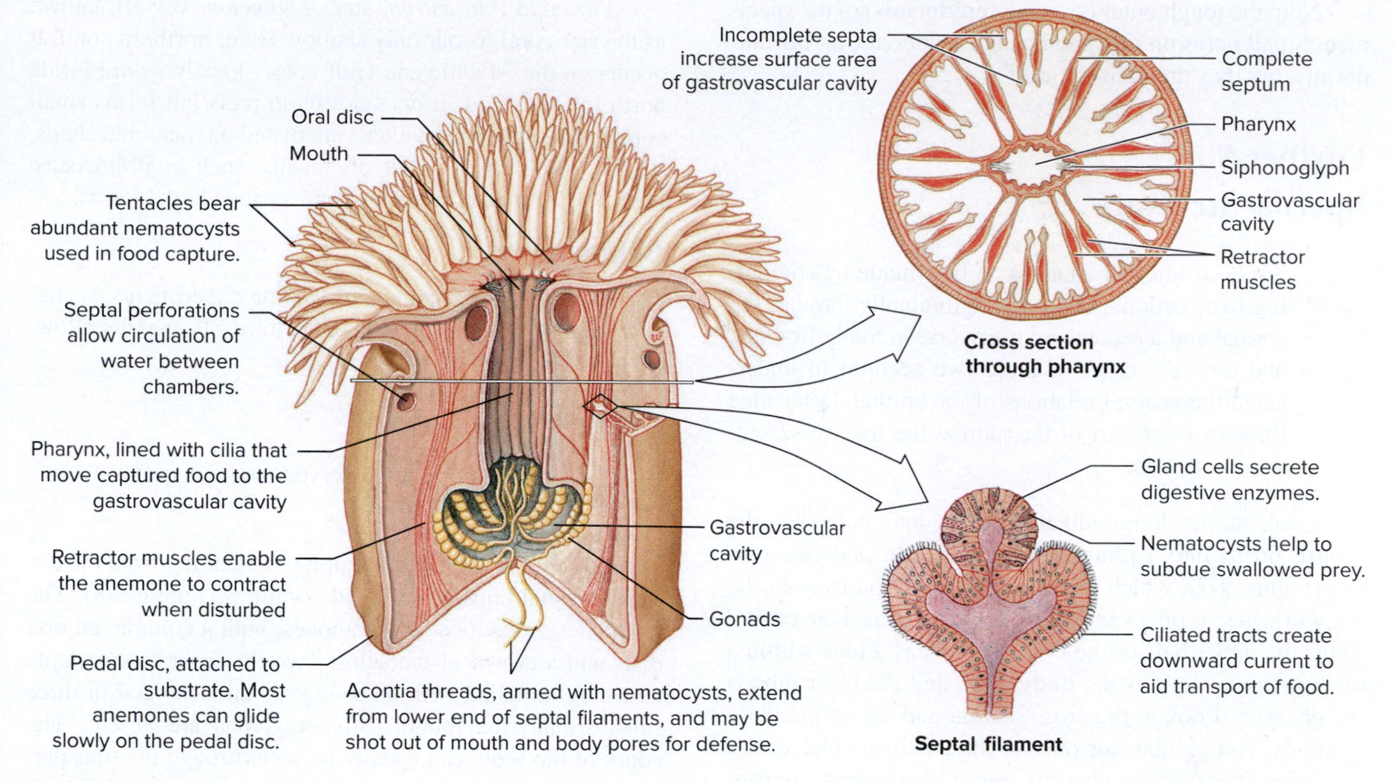
Incomplete septa increase surface area of gastrovascular cavity
Complete septum
Pharynx
Siphonoglyph
Gastrovascular cavity
Retractor muscles
Cross section through pharynx
Oral disc
Mouth
Tentacles bear abundant nematocysts used in food capture.
Septal perforations allow circulation of water between chambers.
Pharynx, lined with cilia that move captured food to the gastrovascular cavity
Retractor muscles enable the anemone to contract when disturbed
Pedal disc, attached to substrate. Most anemones can glide slowly on the pedal disc.
Acontia threads, armed with nematocysts, extend from lower end of septal filaments, and may be shot out of mouth and body pores for defense.
Gastrovascular cavity
Gonads
Gland cells secrete digestive enzymes.
Nematocysts help to subdue swallowed prey.
Ciliated tracts create downward current to aid transport of food.
Septal filament

Figure 9-7

Structure of a sea anemone. The free edges of the septa and the acontia threads are equipped with nematocysts to complete the paralysis begun by the tentacles.

Some anemones react to certain predatory stars by detaching their pedal discs and moving away from the star.

Use a glass rod to probe the pedal disc. What happens? __________ If you prod the animal vigorously, it will shoot out white, threadlike **acontia** (a-con′she-a; Gr. *akontion,* dart), which are filled with nematocysts used for defense.

Place an acontium thread on a slide, cover with a coverslip, and examine with a microscope. Do the acontia move? __________ Examine the edge of an acontium with high power. What do you see that might explain acontia movement? __________

For a dramatic demonstration of nematocysts in action, draw some methyl green under the coverslip using a piece of filter paper touched to the opposite edge of the coverslip. Viewed at high power, the undischarged nematocysts in the acontia look like long grains of rice. What happens when they discharge?

Written Report

Record your observations on separate paper.

External Structure

Place a preserved specimen in a dissecting pan. Note the sturdy nature of its body structures compared with those of other cnidarians you have studied.

The **body** is cylindrical (Figure 9-7), but in preserved specimens it may be somewhat wrinkled. Note that the body of the animal can be divided into three main regions: (1) **oral disc,** or free end, with numerous conical **tentacles** and **mouth;** (2) cylindrical **column,** forming the main body of the organism; and (3) **basal disc** (aboral end), by which during life the animal attaches itself to a solid object by means of its glandular secretions. Although it is called a sessile animal, a sea anemone can glide slowly on its basal disc.

Is there more than one row of tentacles? __________ Note that the inner surface of the mouth is lined with ridges and that a smooth-surfaced, ciliated groove, the **siphonoglyph** (sy-fun′o-glif; Gr. *siphōn,* tube, siphon, + *glyphē,* carving), is found at one side of the mouth. (In some specimens, there are two of these grooves.) Siphonoglyphs, aided by cilia, circulate water throughout the gastrovascular cavity. The mouth is separated from the nearest tentacles by a smooth space, the **peristome** (Gr. *peri,* around, + *stoma,* mouth).

Note the tough outer covering **(epidermis)** of the specimen. Small pores on tiny papillae are scattered over the epidermis, but they are hard to find.

Further Study

Internal Structure

Study of internal anatomy is best made by comparing two sections, one cut longitudinally through the animal and a second cut transversely. Study first one and then the other of these two sections to understand the general relations of the animal. Determine through what part of the animal the transverse section was made.

Look at the longitudinal sections and note that the **mouth** opens into a **pharynx** (fair′inks; Gr. *pharynx,* gullet) (Figure 9-7), which extends down only partway in the body to where it opens into the large **gastrovascular cavity.** Thus, the upper half of the body appears as a tube within a tube; the outer tube is the **body wall,** and the inner tube is the pharynx. Look at the cross section and determine these relations. Notice that not only is the gastrovascular cavity the space aboral to the pharynx, but it also extends upward to surround the pharynx.

The gastrovascular cavity is subdivided into six chambers by six vertical pairs of **primary** (complete) **septa** (Figure 9-7). These six chambers are partially subdivided by smaller pairs of **incomplete septa** with free edges expanded into **septal filaments** armed with nematocysts and bearing gland cells that secrete digestive enzymes. The septa greatly increase the digestive surface of the gastrovascular cavity. The nematocysts may help to subdue struggling prey and, by driving enzymes deep into the prey, also speed digestion.

The **gonads** are thickened bands resembling stacks of coins, often orange-red in color, that lie in the septa just peripheral to the septal filaments. Anemones are dioecious.

Astrangia, a Stony Coral

Phylum Cnidaria
Class Anthozoa
Subclass Hexacorallia
Order Scleractinia
Genus *Astrangia*
Species *Astrangia danae*

Stony corals resemble small anemones but are usually colonial. Each polyp secretes a protective calcareous cup, into which the polyp partly withdraws when disturbed. Some corals form colonies consisting of millions of individuals, each new individual building its skeleton upon the skeletons of dead ones, thus forming, over many years, great coral reefs. Reef-building corals build only in tropical or subtropical waters, where the water temperature stays at or above 21° C.

Astrangia (Gr. *astron,* star, + *angeion,* vessel), known as the star coral, is our only shallow-water northern coral. It occurs on the Atlantic and Gulf coasts locally from Florida north to Cape Cod. It does not build reefs but forms small colonies of 5 to 30 individuals encrusted on rocks and shells. Its food consists of small organisms, such as protozoans, hydroids, worms, crustaceans, and various larval forms.

Behavior

The same sort of touching and feeding experiments as suggested for *Metridium* are applicable to corals, making allowance for the smaller size.

Structure

Examine living or preserved coral polyps.

Note the delicate, transparent polyps extending from the circular skeletal cups, also called corallites (Figure 9-8). The polyps resemble those of anemones, with a column, an oral disc, and a crown of tentacles. Two dozen or more simple tentacles, supplied with nematocysts, are arranged in three rings around the mouth. Siphonoglyphs are absent. The edges of the septa can usually be seen through the transparent polyp walls. As in sea anemones, stony corals are built on a plan of six or multiples of six (hexamerous). Digestion in the gastrovascular cavity is similar to that of anemones.

The coelenterons of adjoining polyps are interconnected by pores, allowing nutrients to be distributed throughout the colony. Colonies usually arise by budding or division from a single polyp that has been sexually produced. The surface of the colony between skeletal cups is covered by a sheet of living tissue, which is an extension of the polyp walls. This tissue connects all members of the colony.

Now study a piece of skeleton from which the polyps have been removed.

The stony cup, the corallite, was the home of a polyp, which secreted it (Figure 9-8). Adjacent polyps are often connected by a sheet of living tissue, which is responsible for depositing skeletal material between the corallites. The wall of the corallite is the theca (L. *theca,* box). Rising up from the floor of the corallite are radial partitions, the sclerosepta. Overlying the sclerosepta are the true, living septa, which form the wall of the coelenteron where food is digested. As the coral polyp grows, the cup fills with calcium carbonate, so that theca and sclerosepta are continually extended upward.

Nearly all reef-building corals depend upon minute plants, called zooxanthellae, packed in their tissues. Because these symbiotic plants require sunlight, the most luxuriant coral reefs thrive in well-lit, relatively shallow waters. Using photosynthesis to fix carbon dioxide, the zooxanthellae furnish food molecules for their hosts, recycle phosphorus and nitrogenous waste compounds that otherwise would be

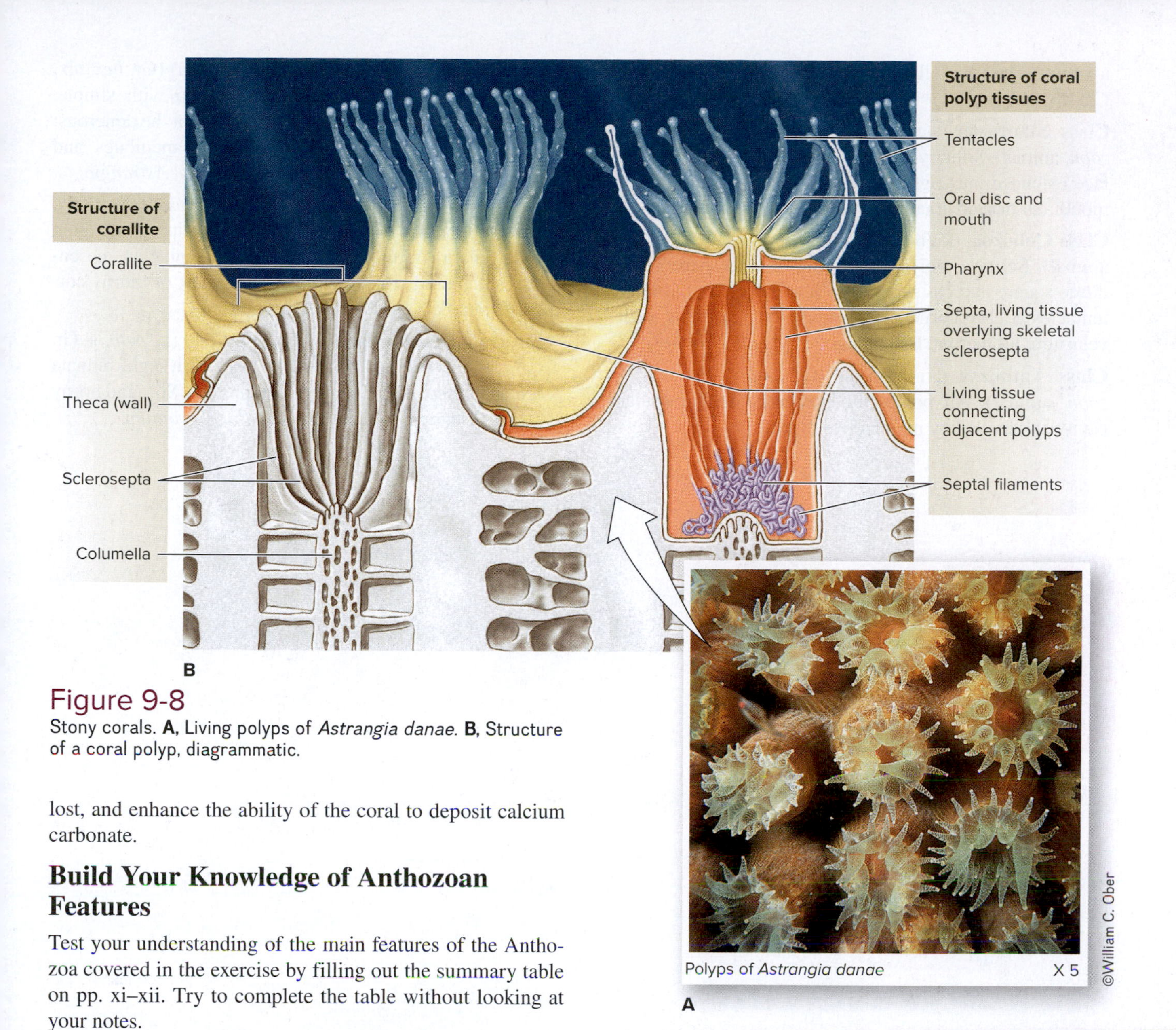

Figure 9-8
Stony corals. **A**, Living polyps of *Astrangia danae.* **B**, Structure of a coral polyp, diagrammatic.

lost, and enhance the ability of the coral to deposit calcium carbonate.

Build Your Knowledge of Anthozoan Features

Test your understanding of the main features of the Anthozoa covered in the exercise by filling out the summary table on pp. xi–xii. Try to complete the table without looking at your notes.

Projects and Demonstrations

1. *Various hard coral skeletons.* Examine such skeletons as white corals; red coral, *Corallium;* staghorn coral, *Acropora;* organpipe coral, *Tubipora;* or brain coral, *Meandrina*, looking for the structures just described.
2. *Nematocysts.* Nematocysts may be collected from various live cnidarians. Zoantharian corals (soft corals) usually have fairly large nematocysts, and the little red colonial anemone *Corynactis* is good for this purpose. Wiping a clean coverslip across the oral disc is likely to attract both discharged and undischarged nematocysts, which can be studied by inverting the coverslip over a drop of seawater on a slide. A good-quality microscope is necessary.

   Collect and compare the nematocysts of various hydrozoans, scyphozoans, and anthozoans.

Classification

Phylum Cnidaria (ny-dar′e-a)

Class Hydrozoa (hy-dro-zo′a) (Gr. *hydra,* water serpent, + *zōon,* animal). Both polyp and medusa stages represented, although one type may be suppressed; medusa with a velum; found in fresh and marine water. The hydroids. Examples: *Hydra, Obelia, Gonionemus, Tubularia, Physalia.*

Class Scyphozoa (sy-fo-zo′a) (Gr. *skyphos,* cup, + *zōon,* animal). Solitary; medusa stage emphasized; polyp reduced or absent; enlarged mesoglea; medusa without a velum. The true jelly. Examples: *Aurelia, Rhizostoma, Cassiopeia.*

Class Staurozoa (sta′ro-zo′a) Gr. *stauros,* a cross, + *zōon,* animal). Solitary, polyps only, no medusa; polyp surface extended into eight clusters of tentacles surrounding mouth; all marine. Examples: *Haliclystis, Lucernaria.*

Class Cubozoa (ku′bo-zo′a) Gr. *kybos,* a cube, + *zōon,* animal). Solitary; polyp stage reduced; bell-shaped medusae square in cross section, with a tentacle or group of tentacles at each corner; margin without velum but with velarium; all marine. Examples: *Carybdea, Chironex.*

Class Anthozoa (an-tho-zo′a) (Gr. *anthos,* flower, + *zōon,* animal). All polyps, no medusae; gastrovascular cavity subdivided by mesenteries (septa).

Subclass Hexacorallia (hek-sa-ko-ral′e-a) (Gr. *hex,* six, + *korallion,* coral) **(Zoantharia).** Polyp with simple, unbranched tentacles; septal arrangement hexamerous; skeleton, when present, external. Sea anemones and stony corals. Examples: *Metridium, Tealia, Astrangia.*

Subclass Ceriantipatharia (se-re-an-tip′a-tha′ri-a) (N. L. combination of Ceriantharia and Antipatharia, from type genera). With simple, unbranched tentacles; mesenteries unpaired. Tube anemones and black or thorny corals. Examples: *Cerianthus, Antipathes.*

Subclass Octocorallia (ok′to-ko-ral′e-a) (L. *octo,* + Gr. *korallion,* coral) **(Alcyonaria).** Polyp with eight pinnate tentacles; septal arrangement octamerous. Soft and horny corals. Examples: *Gorgonia, Renilla, Alcyonium.*

Phylum ___________

Subphylum ___________

Genus ___________

Name ___________

Date ___________

Section ___________

LAB REPORT

Hydra

1. What is meant by the term "dimorphism"? ___________

   Is hydra dimorphic?

2. Describe feeding and digestion in hydras, explaining how extracellular digestion differs from intracellular.

   ____

   ____

   What is the function of nutritive-muscular cells? ___________

3. How does hydra respire? ___________

   How does hydra excrete nitrogenous wastes? ___________

4. Cnidarians are "diploblastic." What does this term mean? ___________

5. Describe and contrast sexual and asexual reproduction in hydra. ___________

   ____

   Is the species of hydra you used for behavioral observations monoecious or dioecious? ___________

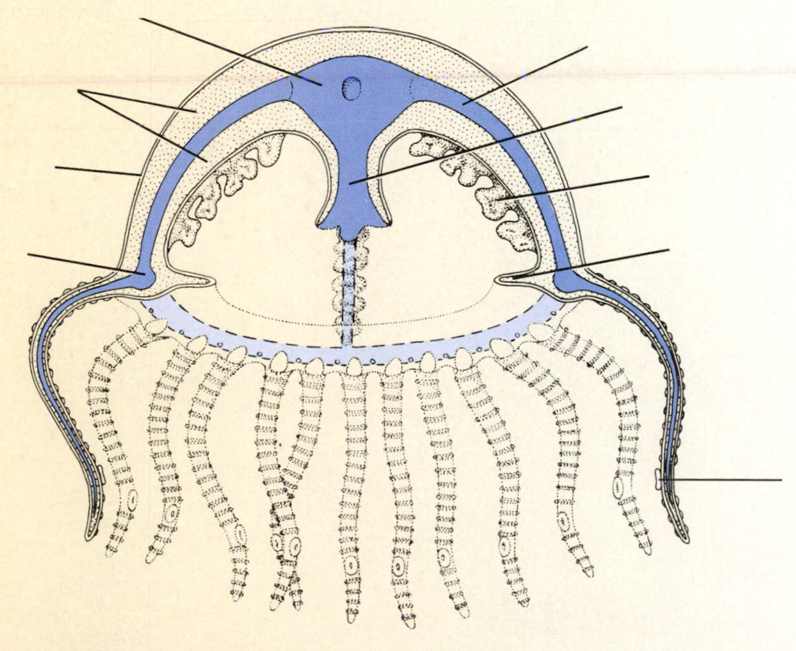

Gonionemus. Diagrammatic oral-aboral section—to be labeled.

LAB REPORT

Aurelia, a "true" jelly

1. What is the sexual stage in the life cycle of *Aurelia*? ______________________ The asexual stage? ______________________ Which stage is dominant in the *Aurelia* life cycle? __________

2. Describe the process by which young medusae are produced. ______________________

3. Does *Aurelia* have special organs for respiration and excretion? ______________________
   How does gas exchange occur in *Aurelia*? ______________________

4. What is the single anatomical feature whose presence or absence will distinguish hydromedusae from scyphomedusae? ______________________

5. On what does *Aurelia* feed? ______________________ Describe how food is captured and conveyed to the mouth. ______________________

Metridium, a sea anemone

1. Is the dominant form of reproduction in *Metridium* sexual or asexual? ______________________

2. Would the formation of clonal patches by some sea anemone species be a form of sexual or asexual reproduction? ______________________

3. What is the siphonoglyph and what is its function? ______________________

4. What are acontia threads and what is their function? ______________________

5. By what means does *Metridium* capture prey? ______________________

6. Why is the body of a sea anemone described as a tube within a tube? ______________________ What parts of the body form the outer and inner tubes? ______________________

EXPERIMENTING IN ZOOLOGY*,**

Predator Functional Response: Feeding Rate in Hydra

An important property of the predator-prey relationship is that the behavior of predators often changes as the availability of prey in the environment changes. One aspect of such behavior is known as the **functional response**, in which a predator changes its feeding rate as prey density changes (Hurd and Rathet, 1986). This measure of predation efficiency is important not only to understanding the behavioral ecology of predators but also to evaluating the ability of predators to control pest populations in biological control programs. The cnidarian, *Hydra,* is unlikely to be a satisfactory candidate for pest control, but it is a good model predator with which to measure functional response.

The obvious null hypothesis to be tested here is the following: Functional response is not affected by changing prey density. What is the most likely alternative hypothesis?

The most convenient prey to use for this experiment is *Artemia,* the brine shrimp. Brown hydra *(Hydra littoralis)* readily feed on the first stage nauplii of brine shrimp but may have trouble capturing later developmental stages as the shrimp increase in size. Therefore, brine shrimp should be hatched no earlier than 1–2 days before they are needed. Functional response rate will be recorded as the number of prey eaten during a 10-minute period as prey density increases.

How to Proceed

Assemble the materials you will need for this experiment: a dissecting microscope, depression slides, disposable pipettes, a Petri dish, a clock or stopwatch, pond water, brown hydra, and *Artemia* nauplii.

Fill a depression slide 2/3 with pond water and place a living hydra in the depression. Do not use a coverslip. With a pipette, place a few brine shrimp nauplii in the Petri dish. Put in only enough to count accurately. When the hydra appears adjusted to its new environment and is fully extended, use a pipette to transfer an exact number of shrimp (1, 2, 3, 4, 6, 8, 10, or 12) from the Petri dish to the depression slide and note the time. At the end of 10 minutes, record the number of shrimp that have been killed by the hydra. Begin with one shrimp for the first trial. Then, with a new hydra, repeat the procedure with two shrimp, then with three, and so on through the series of prey densities, ending with 12 shrimp. *Use a new hydra with each trial.* Add your data to a master sheet for the whole class.

Results and Discussion

The class data represent replicates, so you can calculate the mean and standard deviation for each prey density. Plot x = prey density and y = mean predation rate. How does the functional response of hydra change with increasing prey density? Does the curve level off or continue increasing for each successive prey density? Does the functional response variability, as reflected in the standard deviation, change with changing prey density? Why would results be variable among replicates (different students, different hydra, etc.)? What is the value of replication in such experiments?

Extending the Experiment (Optional)

Other experimental variables to consider might be increased volume of water, which might be expected to slow predation rate by reducing spatial density of brine shrimp and making it easier for the shrimp to avoid being stung. Another variable is the size of the hydra. If budding hydra are available, test them against nonbudding specimens to see if feeding rate is influenced by asexual reproduction. Why might you expect this to affect the results? What if the temperature at which you ran the experiment had been higher or lower? Can you design an experiment to test this? Can you think of other variables that might be worth testing?

Written Report

Summarize your findings in a written report organized into introduction, results, and discussion. The introduction should explain the purpose of the study and state the null hypothesis. The results should summarize the data and tell the reader what the data show. Include a table with the class data and a graph showing prey density plotted against mean predation rate. Note that tables and graphs have legends and each table and graph is placed on a separate sheet of paper. In the discussion consider the following: Was your null hypothesis rejected? What does rejection of (or failure to reject) a null hypothesis mean biologically? How does your finding add to what was known?

Reference

Hurd, L. E., and I. H. Rathet. 1986. Functional response and success in juvenile mantids. *Ecology* 7:163–167.

*Materials and method for implementing this exercise are found in Appendix A, p. 401.

**This exercise was contributed by Lawrence Hurd, Washington & Lee University.

NOTES

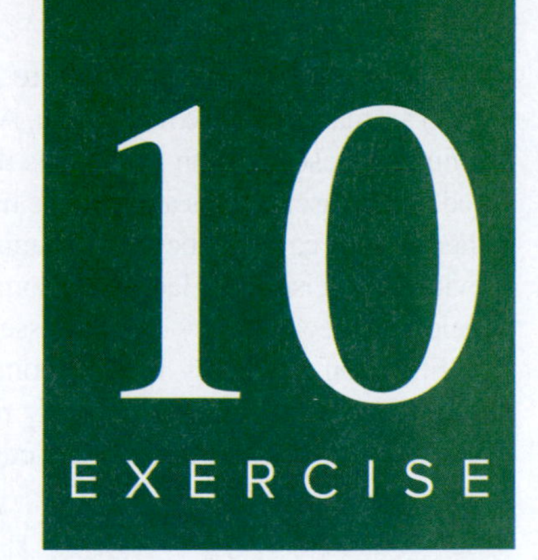

The Flatworms

Phylum Platyhelminthes

The flatworms are members of the phylum Platyhelminthes (Gr. *platys,* flat, + *helmins,* worm). This large and economically important group contains both free-living forms, such as the common planarians, and parasitic forms, such as tapeworms and flukes. They are more complex in organization than the radiate animals (cnidarians) in several ways: (1) They exhibit **bilateral symmetry,** with distinct head and associated sense organs, which allows forward, directed movement. (2) They have a **third germ layer—mesoderm**—that arises between the ectoderm and endoderm. This creates the triploblastic condition as distinguished from the two–germ layer, or diploblastic, condition of the radiate animals. The mesoderm serves as a source for many tissues, organs, and systems. (3) Their **excretory system** is made up of specialized flame cells and tubules for the removal of nitrogenous wastes. (4) They possess a highly organized nervous system and concentration of nervous tissue and sense organs in the anterior end (cephalization).

EXERCISE 10A
Class Turbellaria—Planarians
Dugesia
Projects and Demonstrations

EXERCISE 10B
Class Trematoda—Digenetic Flukes
Clonorchis, Liver Fluke of Humans
Schistosoma, Human Blood Fluke
Observations of Living Flukes
Projects and Demonstrations

EXERCISE 10C
Class Cestoda—Tapeworms
Taenia or *Dipylidium*
Projects and Demonstrations

CLASSIFICATION
Phylum Platyhelminthes

EXPERIMENTING IN ZOOLOGY
Planaria regeneration experiment

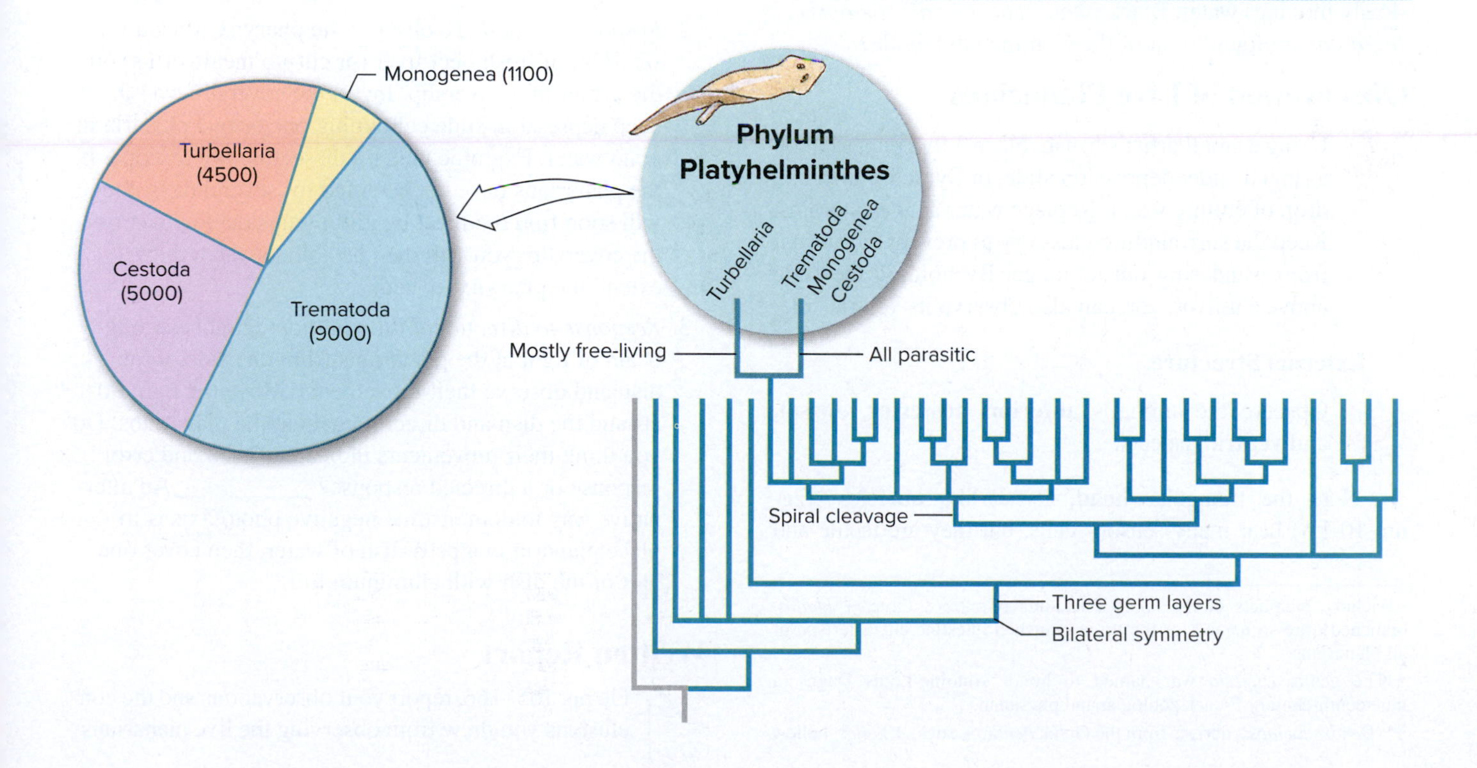

The flatworms are **acoelomate** (Gr. *a*, not, + *koilōma*, cavity) animals that have no coelom. A coelom is a cavity lying within the mesoderm. In flatworms the mesodermal space is filled with muscle fibers and a loose tissue called parenchyma, rather than a cavity. The radiate animals, considered in the previous exercise, also lack a coelom but are not considered acoelomate because they do not possess mesoderm.

Flatworms recently have become attractive models for stem cell research because of their remarkable regenerative properties, having bodies that are continually renewed from large pools of somatic stem cells.

EXERCISE 10A

Class Turbellaria—Planarians

Core Study

Dugesia

Phylum Platyhelminthes
Class Turbellaria
Order Tricladida
Genus *Dugesia*

Where Found

Freshwater triclads,* or planarians, are found on the underside of stones or submerged leaves or sticks in freshwater springs, ponds, and streams. They are often confused with leeches, which they resemble in color and somewhat in shape. There are 9 genera and more than 30 species of planarians in North America. Common and widely distributed species are the brown planarians, *Dugesia*** *tigrina* and *Dugesia dorotocephala,* adapted to warm and standing (or slowly moving) water. Black planarians, *Dendrocoelopsis**** *vaginata,* are found west of the Continental Divide.

Observation of Live Planarians

Using a small artist's brush, place a live planarian on a ringed slide, depression slide, or Syracuse dish in a drop of culture water. Replace water as it evaporates. Keep the surrounding glass dry to prevent the animal from wandering out of range. By holding the slide above a mirror, you can also observe its ventral side.

External Structure.

Observe the animal's **anterior, posterior, dorsal,** and **ventral** aspects.

Note the triangular **head.** Its earlike **auricles** (Figure 10-1A) bear many sensory cells, but they are tactile and olfactory, not auditory, in function. Are the **eyes** (more correctly called **ocelli**) movable? __________ Do they have lenses? __________ Note the pigmented **skin.** Is it the same color on the underside? __________ Is its coloring protective? __________ Are the length and breadth of the worm constant? __________ Holding the slide for a moment over a bright light, can you locate the muscular **pharynx** along the midline? When the animal feeds, the pharynx can be protruded through a ventral **mouth** opening. Can you verify this by use of the mirror?

Use a hand lens, a dissecting microscope, or the low power of a compound microscope for further examination of the eyes and body surface.

Locomotion. Observe the animal's gliding movement. Glands in its ciliated epidermis secrete a path of mucus on which the planarian propels itself by means of its cilia. Do you think cilia alone are responsible for its movement? __________ What do you think causes the waves of contractions along its body? __________ Does the animal ever leave the drop of water and travel on the dry glass? __________ Why? __________ How does it use the head and auricles? __________ Does it ever move backward? __________

Reactions to Stimuli.

Observe the responses of planarians to touch **(thigmotaxis),** food **(chemotaxis),** and light **(phototaxis)** by doing some of the following simple experiments.

1. *Response to touch. Very gently* touch the outer edges of the worm with a piece of lens paper or a soft brush. What parts of its body are most sensitive to touch? Are its reactions more localized or less localized than those of the hydra?
2. *Response to food.* To observe the pharynx, stick a *very small* bit of fresh beef liver (or cut-up mealworms) on the center of a coverslip. Invert the coverslip over a deep depression slide containing one or two planaria in pond water. Examine under a dissecting microscope. If the planarians have not been fed for several days, they will soon find the meat by gliding upside down across the coverslip; you will then be able to watch them extend the pharynx to feed.
3. *Response to directional illumination.* Direct a strong beam of light at the planarians from one side of the dish and observe their movements. Move the light 90° around the dish and direct it again at the planarians. Do you think their movements indicate a "trial and error" response or a directed response? __________ An alternative way to demonstrate negative phototaxis is to place planaria in a petri dish of water, then cover one half of the dish with aluminum foil.

Written Report

On pp. 165–166, report your observations and the conclusions you drew from observing the live planarians.

---

*Triclads, members of the order Tricladida (Gr. *treis,* three, + *klados,* branched), are so named for the three-branched intestine characteristic of all planarians.

**The genus *Dugesia* was named to honor Antoine-Louis Dugès, a nineteenth-century French zoologist and physician.

****Dendrocoelopsis* derives from the Greek *dendron,* stick, + *koilos,* hollow.

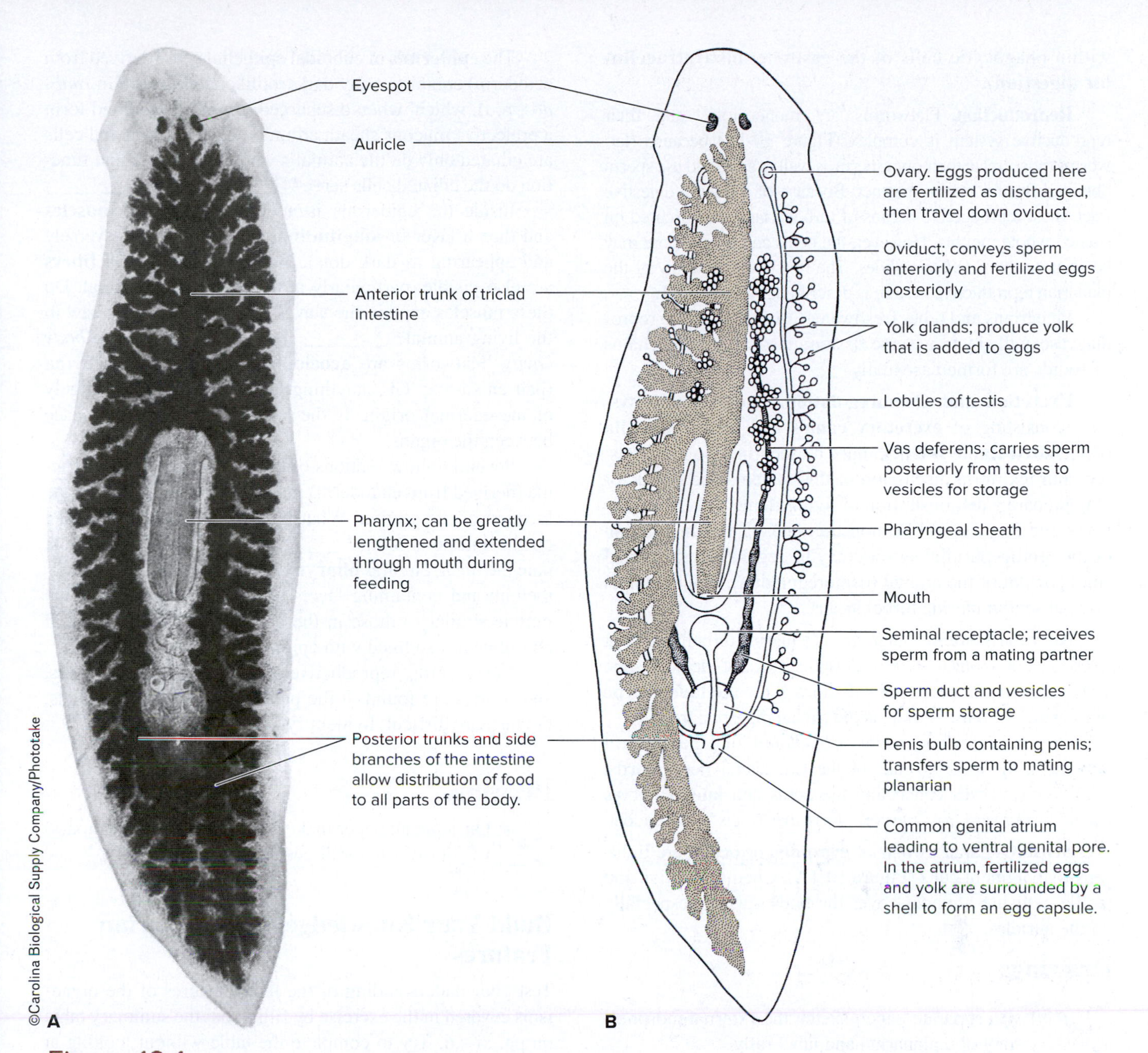

Figure 10-1

Planaria, a freshwater turbellarian. **A,** Stained whole mount. **B,** Internal anatomy; one half the intestinal trunk is removed to reveal the reproductive organs.

Observation of Stained Whole Mounts

The stained whole mount of a planarian shows an animal that was fed food mixed with India ink, carmine, or some other suitable stain before killing and fixing, resulting in a darkly stained gastrovascular tract.

Using a dissecting microscope or the low power of a compound microscope, study stained whole mounts of a freshwater planarian and the marine *Bdelloura* (de-lur'a), identifying the structures listed in the following sections.

Digestive System. As in cnidarians, the digestive tract of a turbellarian is a **gastrovascular cavity,** the branches of which fill most of the body (Figure 10-1B). Because there is no anus, undigested food is ejected through the mouth.

The muscular **pharynx** is enclosed in a **pharyngeal sheath,** but its free end can be extended through the ventral **mouth.** Ingestion occurs through the muscular sucking action of the pharynx. The pharynx opens into the intestine, which has one **anterior trunk** and two **posterior trunks,** one on each side of the pharynx. What might be an advantage to the digestive process of the branching diverticula that extend laterally from the intestinal trunks? ______ Some digestion may occur within the lumen of the digestive cavity by means of enzymes secreted by intestinal gland cells **(extracellular digestion).** As in the cnidarians, digestion is completed within phagocytic cells of the gastrodermis **(intracellular digestion).**

Reproduction. Flatworms are monoecious, and their reproductive system is complex (Figure 10-1B) because flatworms must create their own sperm as well as store and use sperm obtained from a mating partner. Because of the large digestive tract, most reproductive organs of turbellarians are obscured on stained whole mounts. However, the penis and genital pore may be seen on the *Bdelloura* slides. The functional anatomy of the planarian reproductive system is described in Figure 10-1.

Planarians and other freshwater turbellarians also reproduce asexually by transverse fission; in some species, chains of zooids are formed asexually.

Excretion and Osmoregulation. The excretory system, consisting of **excretory canals** and **protonephridia (flagellated flame cells),** cannot be seen in whole mounts; see your text for a discussion of this system. The instructor may prepare a demonstration of living flame cells from planaria (see Projects and Demonstrations). The main function of the protonephridial system may be regulation of internal fluid content of the animal (osmoregulation). The system is often absent in marine turbellarians.

Nervous System. *Bdelloura* is a marine turbellarian that lives as a commensal on the external surface of the horseshoe crab, *Limulus.* Properly stained slides show the **ladder type of nervous system,** as well as the triclad digestive tract.

On a stained whole mount of *Bdelloura,* find the **cerebral ganglia** at the anterior ends of the two lateral **nerve cords. Transverse nerves** connecting the cords and **lateral nerves** extending outward from the cords form the "rungs" of the ladder.

Sense Organs. A pair of **eyespots,** or **ocelli,** are light-sensitive pigment cups (Figure 10-1A). Chemoreceptive and tactile cells are abundant over the body surface, especially on the auricles.

Drawings

1. On separate paper, sketch the external morphology of a planarian and label fully.
2. The outline of *Bdelloura* on p. 166 contains an outline of the digestive system. Draw in and label the cerebral ganglia, nerve cords, and transverse and lateral nerves.

Written Report

Answer the questions for planarians on p. 163.

Further Study

Cross Sections of Planarian

The appearance of a cross section will depend on whether it is cut from the anterior, middle, or posterior part of the planarian (Figure 10-2).

The **epidermis** of cuboidal epithelial cells (derived from ectoderm) contains many dark, rodlike **rhabdites** (Gr. *rhabdos,* rod), which, when discharged in water, swell and form a protective mucous sheath around the body. Epithelial cells are ciliated only on the animal's ventral surface. What function do the ciliated cells serve? ____________________

Inside the epidermis is a layer of **circular muscles** and then a layer of **longitudinal muscles** (cut transversely and appearing as dark dots). **Dorsoventral muscle fibers** are also visible, particularly at the sides of the animal. Do these muscles explain the waves of contractions you saw in the living animal? __________ Note that there is *no body cavity.* Flatworms are **acoelomate** animals. **Parenchyma** (pair-en′ka-ma; Gr., anything poured in beside), largely of mesodermal origin, is the loose tissue filling up space between the organs.

Several hollow sections of the intestine and its diverticula (derived from endoderm) may be seen, depending on the location of the section. What kind of epithelial cells make up the intestinal walls? __________ In a middle section, note the thick, circular **pharynx,** covered and lined with epithelium and containing layers of circular and longitudinal muscle similar to those in the body wall. The **pharyngeal chamber** is also lined with epithelium.

Nerve cords, reproductive and excretory ducts, testes, and ovaries are found in the parenchyma of adult animals, but they are difficult to identify.

Drawings

On separate paper make a drawing of such section(s) as your instructor directs. Label completely.

Build Your Knowledge of Turbellarian Features

Test your understanding of the main features of the organisms covered in the exercise by filling out the summary table on pp. xi–xii. Try to complete the table without looking at your notes.

Projects and Demonstrations

Observing flame cells. Flame cells of the excretory system are difficult to see in living planarians because they are usually obscured by pigment in the integument. To demonstrate flame cells, compress the body (or sectioned portion of the body) of a planarian between a slide and a coverslip to crush and partially disperse the tissues. Search the tissue debris with the high power of a compound microscope. Flame cells appear as a rapidly flickering movement of the ciliated tuft, resembling a wavering current of water. Once located (often several flame cells are found close together), switch to oil immersion for closer study.

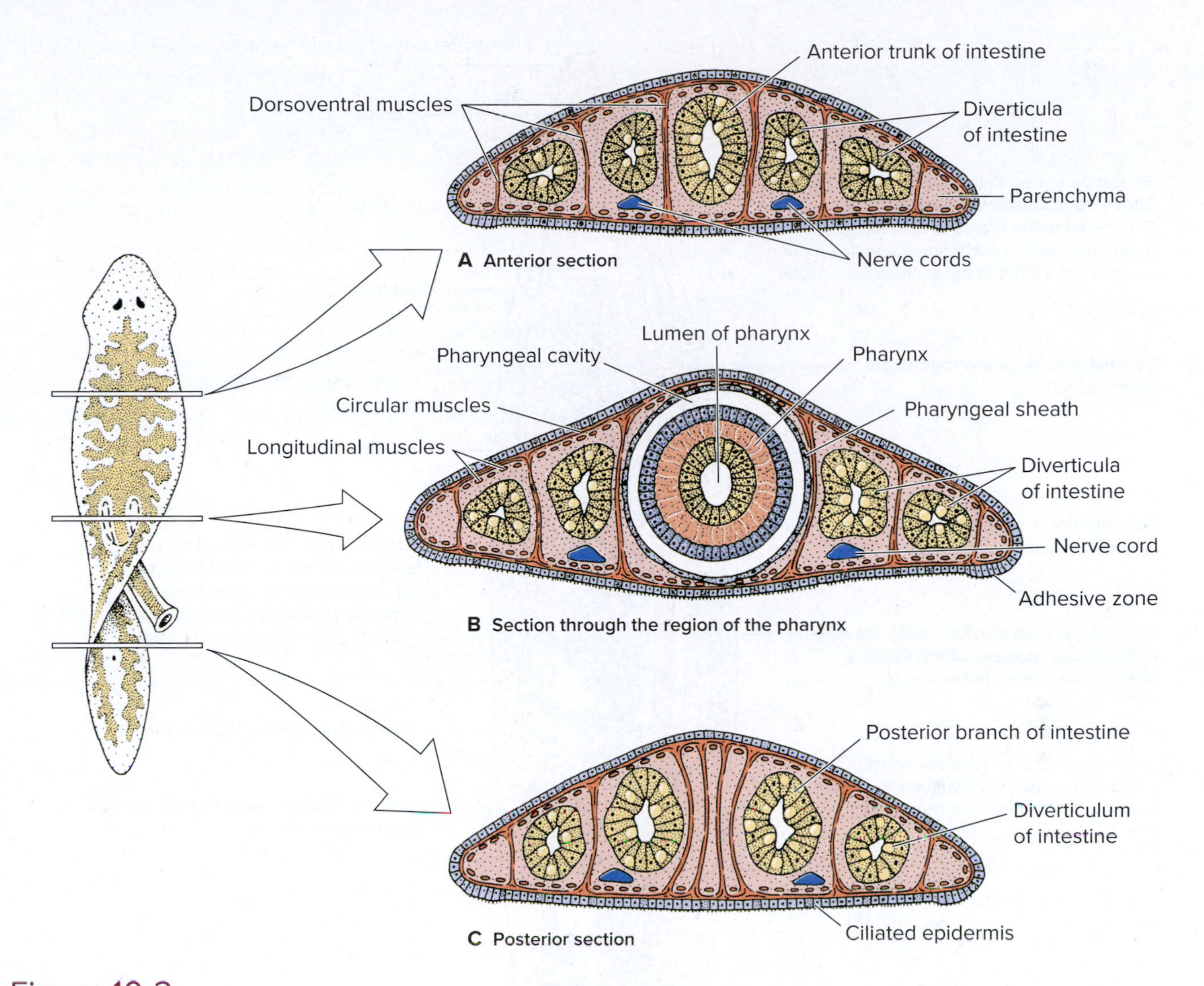

Figure 10-2
Cross section of a planarian (diagrammatic). **A,** Anterior section. **B,** Section through the region of the pharynx. **C,** Posterior section.

EXERCISE 10B

Class Trematoda—Digenetic Flukes

Core Study

Clonorchis, Liver Fluke of Humans

Phylum Platyhelminthes
Class Trematoda
Subclass Digenea
Order Opisthorchiformis
Species *Clonorchis sinensis* (human liver fluke)

Where Found

All trematodes are parasitic, harbored in or on a great variety of animals. Many of them have three different hosts in their life cycle.

The adult, or sexual, stage of *Clonorchis* lives in the human bile duct, where it feeds on bile and lacerated cells from the inflamed bile duct. It is widely distributed in Southeast Asia, especially China, Korea, Vietnam, and eastern areas of the former Soviet Union, where it causes widespread suffering and economic loss. It was estimated that 35 million people were infected worldwide in 2012. In China prevalence rates vary widely from less than 1% to 57%.

Study of a Stained Whole Mount

Study a stained slide of an adult fluke—first with a hand lens, then with the low power of the microscope.

How long is the specimen? __________ Compare it with the planarian studied in Exercise 10A in size and shape. Note the **oral sucker** at the anterior end and the **ventral sucker (acetabulum)** on the ventral surface (Figure 10-3). What is their function? ______________________

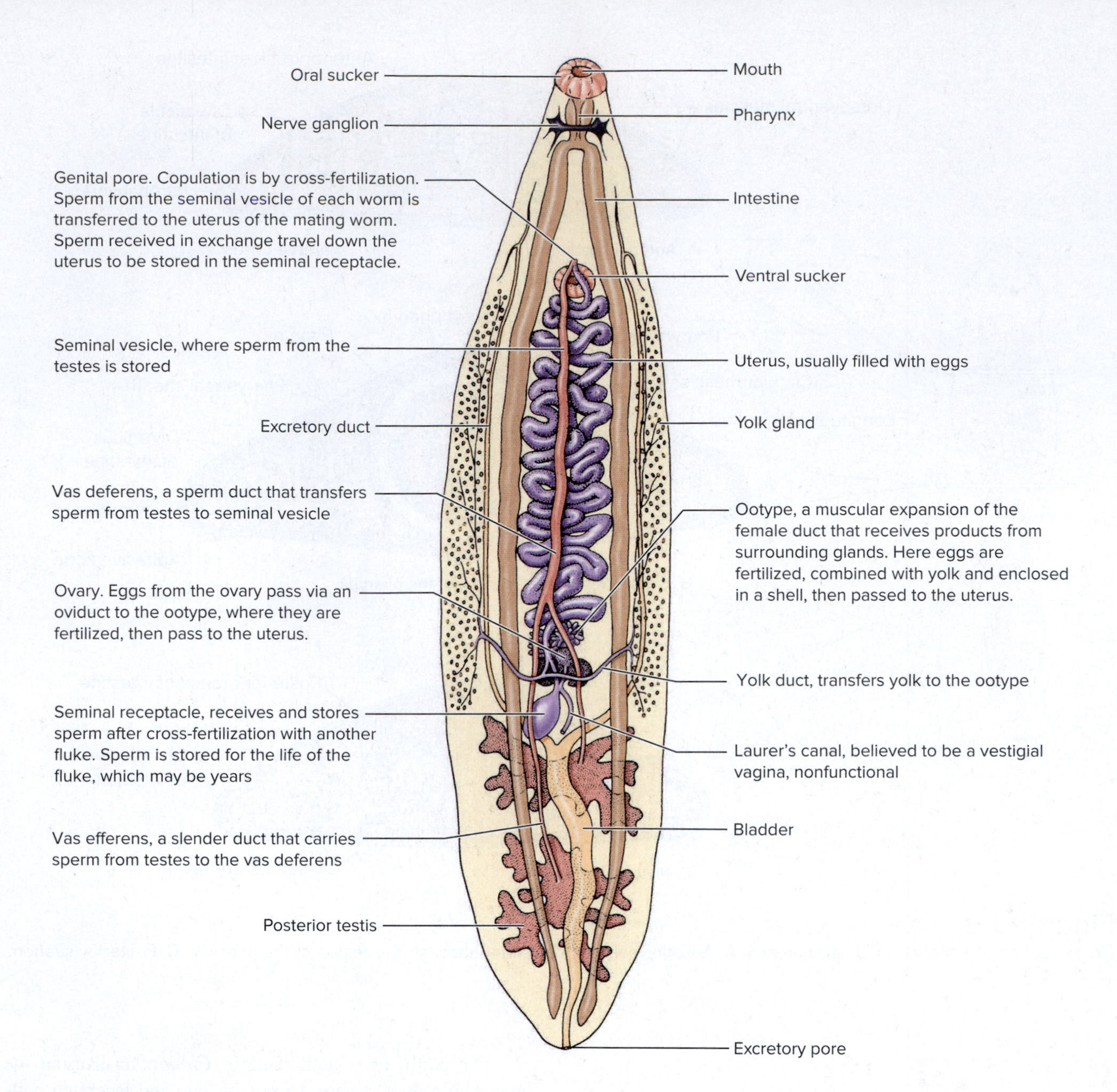

Figure 10-3
Clonorchis, the liver fluke of humans.

The body covering is a syncytial **tegument** (L. *tegumentum,* to cover). Beneath the tegument are circular and longitudinal muscle layers. The body space is filled with a spongy mass of mesenchyme cells, the **parenchyma.** The **mouth** of *Clonorchis* lies anteriorly in the oral sucker. It leads to a muscular **pharynx** and short **esophagus** that divides into two lateral branches of the digestive tract. The sucker serves for attachment and for the abrasion of the bile duct in which the fluke lives.

An excretory pore at the posterior end is the outlet of the bladder. Follow the bladder forward to see where it divides into two long tubules. The tubules collect from flame cells, which you will not be able to see in the stained specimen.

Reproduction and Life Cycle. Flukes are monoecious, each animal having both male and female reproductive systems (Figure 10-3). Locate the testes and male reproductive organs and follow the passage of sperm from testes to genital pore as shown in Figure 10-3. Similarly, while referring to Figure 10-3, follow the movement of eggs from ovary to ootype, where the eggs are fertilized and surrounded with a shell, then passed to the uterus.

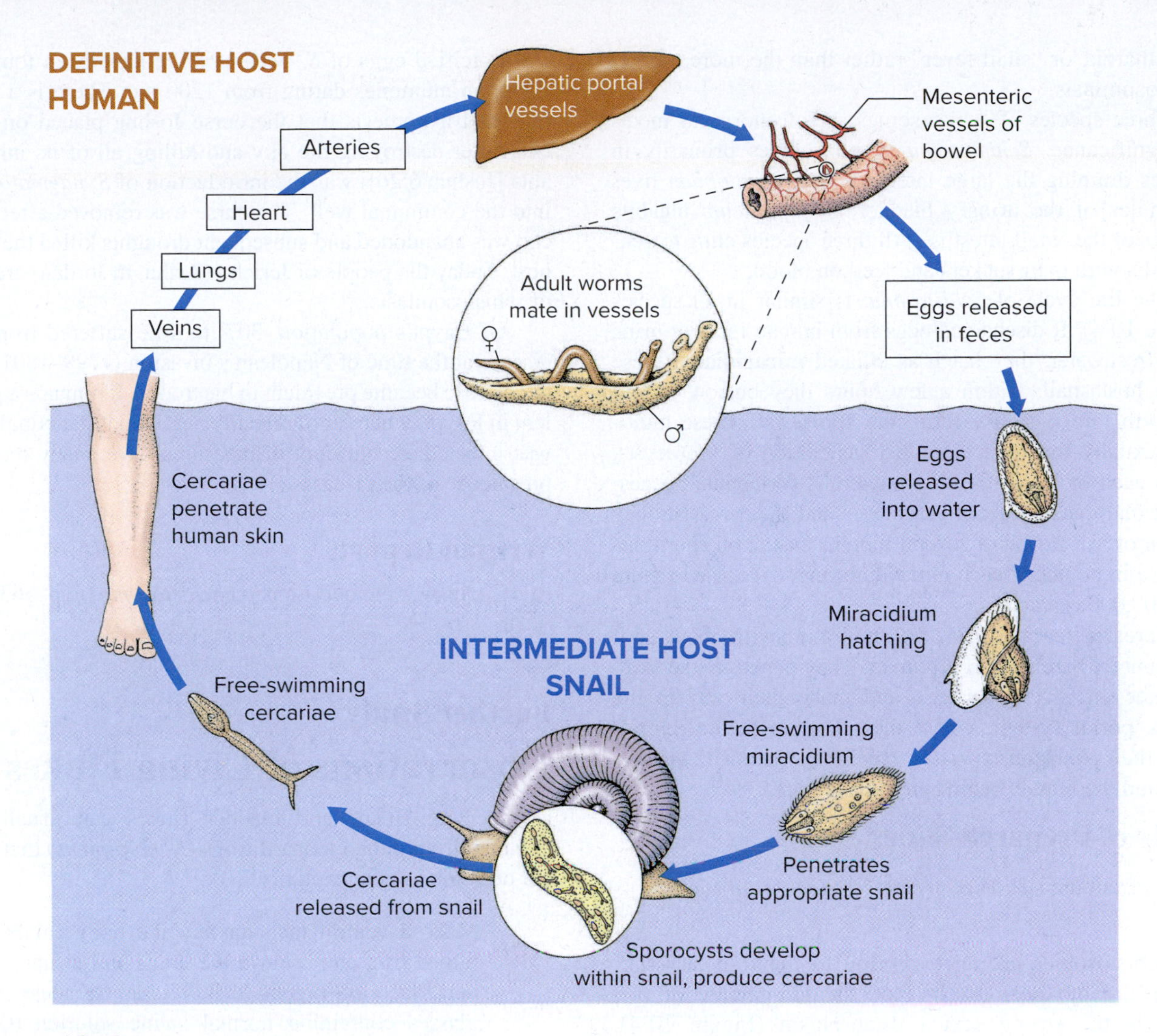

Figure 10-4
Life cycle of *Schistosoma mansoni.*

Eggs containing ciliated larvae are shed into the bile, carried to the host's intestine, and voided with feces. If feces pass into water, the larvae hatch and are eaten by snails, in which they multiply asexually through additional larval stages before leaving the snails as tadpolelike **cercariae.** If successful in finding a fish host, the cercariae bore into the muscle where they encyst. Humans (and other mammals) are infected when they eat raw or undercooked fish bearing the cysts. Young flukes then emerge, travel to the bile ducts, and mature into adults. Consult your textbook for details of this animal's life cycle. Why do you think this parasite is prevalent in Southeast Asia?

Written Report

Answer the questions on *Clonorchis* on p. 163.

Schistosoma, Human Blood Fluke

Phylum Platyhelminthes
Class Trematoda
Subclass Digenea
Order Strigeiformes
Family Schistosomatidae
Genus *Schistosoma*
Species *Schistosoma mansoni* (or *Schistosoma haematobium*)

Where Found

Schistosomes of the genus *Schistosoma* are blood flukes of humans that infect an estimated 252 million people in Asia, Africa, the Caribbean (including Puerto Rico), and northeastern South America (2016 study). It does not occur in the United States or Canada. As a major global health problem, **schistosomiasis** is exceeded only by malaria. In many areas, the disease is commonly known as "bilharzia"or "snail fever" rather than the more correct schistosomiasis.

Three species of schistosomes are of enormous medical significance. *Schistosoma mansoni* lives primarily in venules draining the large intestine; *S. haematobium* lives in venules of the urinary bladder; *S. japonicum* inhabits venules of the small intestine. All three species cling to venule walls with their suckers and feed on blood.

The life cycle of *Schistosoma* is similar in all species (Figure 10-4). If discharged eggs from human feces or urine reach freshwater, they hatch as ciliated **miracidia.** If these find a host snail, within a few hours they burrow in and transform into a saclike form, the **sporocyst.** These multiply asexually to produce another generation of sporocysts, within each of which develop numerous **cercariae.** Successive sexual generations of sporocysts and the cercariae they contain over a period of several months ensure an enormous increase in numbers: each miracidium may give rise to more than 200,000 cercariae.

Cercariae escape from the snail and swim about until they contact bare skin of a human. They penetrate the skin, enter the circulatory system, and make their way to the hepatic portal system, where they develop before migrating to their characteristic sites. Here they mate with another fluke and the female begins producing eggs.*

Study of Prepared Slides

Examine a slide of *Schistosoma mansoni* adults.

Schistosomes are an exception to the rule that flatworms are monoecious. In fact, an odd feature of this genus is the strong sexual dimorphism (Figure 10-4). Males are stouter than females and have a ventral longitudinal groove, the **gynecophoric canal** (gi′ne-ka-fore′ik; Gr. *gyne,* woman, + *pherein,* to carry). (The name *Schistosoma,* meaning "split body," refers to this canal.) Even more oddly, the thinner female normally resides there, permanently embraced by the male. Note the strong oral sucker and a secondary sucker near the anterior end called the **acetabulum.**

Examine a slide of schistosome eggs.

Note the elliptical shape of the eggs, each of which bears a sharp spine (the spine is terminal in *S. haematobium* eggs and lateral in *S. mansoni* eggs; the eggs of *S. japonicum* lack a spine). The main ill effects of schistosomiasis result from the eggs, which, as they work their way out of the venules where the adults live, cause ulceration, abscesses, bloody diarrhea, and abdominal pain.

*Although the life cycle is similar in all species of *Schistosoma,* it differs from that of other digeneans in that the cercariae penetrate the definitive host directly from a single intermediate host, a snail. There is no second intermediate host (unlike *Clonorchis,* for example, in which there are two intermediate hosts, a snail and a fish).

Calcified eggs of *S. haematobium* have been found in Egyptian mummies dating from 1200 B.C. There is a well-reasoned hypothesis that the curse Joshua placed on Jericho (after destroying the city and killing all of its inhabitants [Joshua 6:26]) was the introduction of *S. haematobium* into the communal well. The curse was removed after Jericho was abandoned and subsequent droughts killed the snail host. Today the people of Jericho (Ariha, in Jordan) are free of schistosomiasis.

Of Egypt's population, 30% to 40% suffered from the disease at the time of Napolean's invasion (1799–1801) and the disease became prevalent in his troops. It remains a problem in Egypt, where until recently nearly half the population was infected; recent control measures have vastly reduced prevalence of the disease.

Written Report

Answer the questions on *Schistosoma* on p. 163.

Further Study

Observations of Living Flukes

Living lung flukes and bladder flukes can usually be obtained from pithed leopard frogs *(Rana pipiens)* that have not been too long in captivity.

Make a ventral incision into the body cavity of a pithed frog and remove the lungs and urinary bladder. Place the organs in individual Syracuse watch glasses containing normal saline solution (0.65% NaCl). Tease each organ apart with teasing needles, looking for living flukes.

Lung Flukes

You may find lung flukes, nematodes, or both in frog lungs. Nematodes (phylum Nematoda) are slender, transparent roundworms that thrash about by bending in **S** and **C** shapes.

Flukes are mottled dark and light flatworms with slow, writhing movements. More than 40 species of *Haematoloechus* (Figure 10-5) have been found in the lungs of amphibians all over the world. How does their movement compare with that of planarians? Do they have eyes? __________ The uterus of *Haematoloechus* is probably so filled with thousands of eggs that most internal organs are obscured.

Place a specimen on a slide with some tap water and a cover glass.

The tap water and pressure of the glass may cause the animal to discharge most of its small brown eggs and thus leave the animal transparent enough to study.

You may now be able to see the intestinal tract filled with ingested blood and the uterus with its remaining eggs.

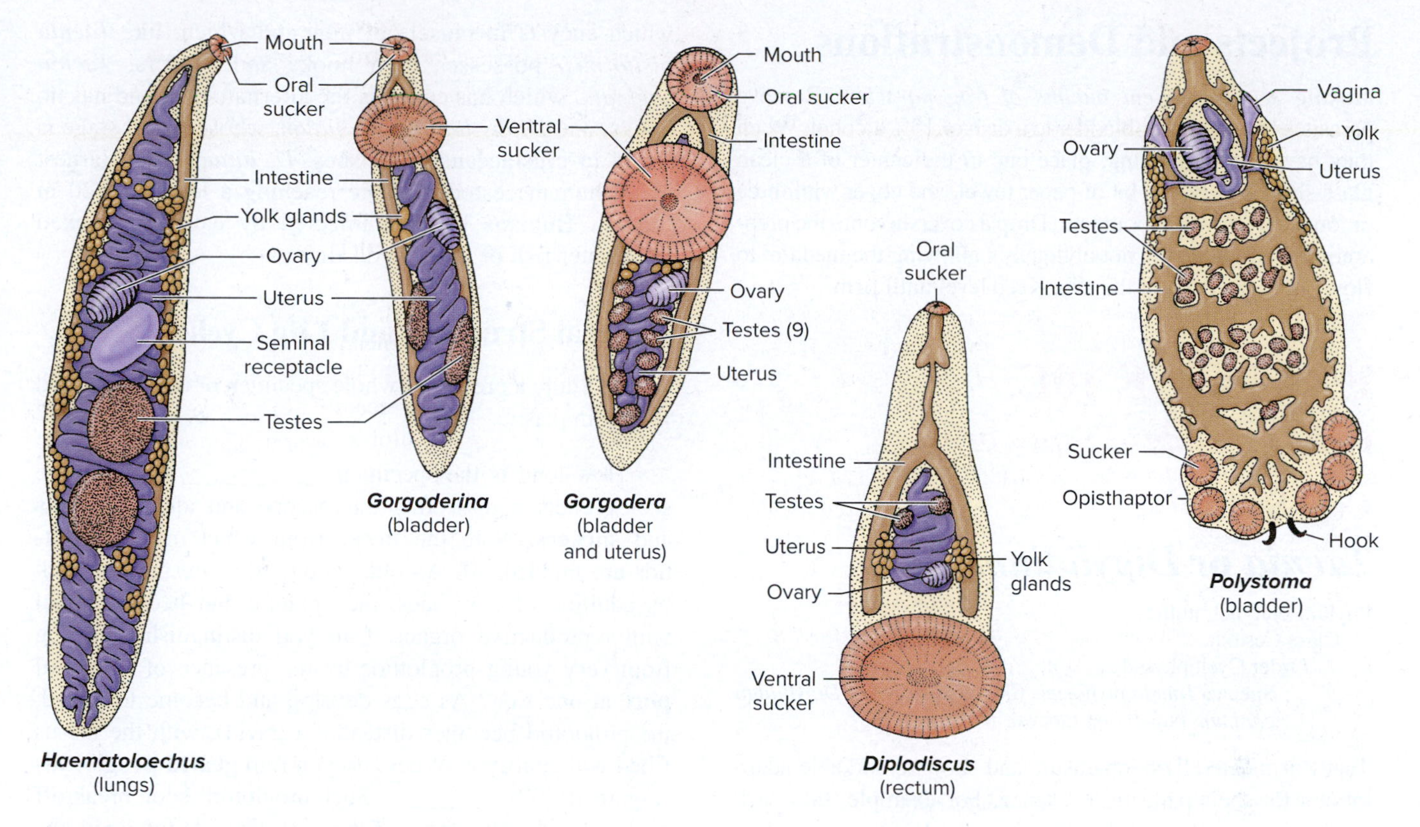

Figure 10-5
Some common trematode parasites of frogs.

The eggs of *Haematoloechus* are ingested by a certain aquatic snail in which the cercariae develop. Free-swimming cercariae encyst in the gills of the aquatic nymph of a dragonfly. Frogs become infected by ingesting either the nymph or the adult dragonfly.

Bladder Flukes

In the urinary bladder, look for small flukes with large suckers, or acetabula. Study details of the flukes on a slide with a drop of saline solution. Several species of *Gorgodera* and *Gorgoderina* occur in the bladder (Figure 10-5). These have a very large ventral sucker at the end of the anterior third of the body. In their life cycle, miracidia enter small bivalves (family Sphaeriidae), and the cercariae that develop there are usually ingested by a damselfly nymph or another aquatic insect larva that serves as food for frogs.

Polystoma (Figure 10-5) is a monogenetic fluke that has at its posterior end a large holdfast, or **opisthaptor,** consisting of six suckers and two hooks located on a muscular disc. Reproduction in *Polystoma* is associated with the frog's breeding season, and there is only one host. Its larval stage is parasitic on the gills of tadpoles.

Miracidia. If bladder flukes were found in the frog bladder, examine the saline solution in which the bladder was teased. It may contain eggs or the ciliated miracidia. If not present, this stage may be seen on prepared slides.

Living Cercariae. Cercariae may usually be found by crushing the common pond snail *Physa* in a watch glass and removing the broken shell. Just visible to the naked eye, the cercariae show up well under a dissecting microscope, characterized by their erratic swimming movements.

Or you may use a common mud snail (e.g., *Cerithidea californica*) by cracking the tip of the shell, removing the liver to a watch glass, and examining under a dissecting microscope. Uninfested livers are usually brown or green, whereas infested livers appear orange, tan, or mottled.

Place some of the cercariae on a microscope slide, cover, and examine with a compound microscope.

Drawings

On separate paper, sketch any living flukes, miracidia, or cercariae you found. Did you find any nematodes?

Build Your Knowledge of Trematode Features

Test your understanding of the main features of the Trematodes covered in the exercise by filling out the summary table on pp. xi–xii. Try to complete the table without looking at your notes.

Projects and Demonstrations

Making semipermanent mounts of frog parasites. Remove parasites from lungs or bladder to a dish of 15% alcohol. When they have ceased moving, place one in the center of a clean glass slide, blot with a bit of paper towel, and cover with three or four drops of stain mountant. Drop a coverslip onto the preparation (straight down, not obliquely), allowing the medium to flow out to the edge of the slip. Keep level until firm.

EXERCISE 10C

Class Cestoda—Tapeworms

Core Study

Taenia or *Dipylidium*

Phylum Platyhelminthes
Class Cestoda
Order Cyclophyllidea
Species *Taenia pisiformis* (dog tapeworm) or *Dipylidium caninum* (small dog tapeworm)

Tapeworms are all endoparasitic and show remarkable adaptations for their parasitic existence. For example, they lack a digestive system and there is great emphasis on reproduction. Indeed, tapeworms are first and foremost egg factories, producing enormous numbers of eggs to counter the staggering odds against a single egg completing the life cycle to become an adult worm. As their name implies, tapeworms are ribbonlike, and their long bodies (the **strobila**) are usually composed of units called **proglottids** (Gr. *proglottis,* tongue tip). (Proglottids are not to be confused with segmentation in higher forms; proglottids are formed by a continuous process of **budding** behind the scolex [Gr. *skōlēx,* worm, grub].) As new proglottids are formed, the older ones are pushed backward. The scolex, which serves as a holdfast, is usually equipped with **suckers, hooks,** or both. The body covering, or **tegument,** is an epidermis specialized for living in a hostile environment in which it must absorb nutrients but reject toxins and digestive enzymes.

Many tapeworms can fertilize their own eggs, an extreme form of inbreeding, but a strategy that guarantees offspring, especially handy if the host does not carry multiple parasites of the same species. Different proglottids of the same worm also can fertilize each other when the worm is folded back on itself, or separate worms may cross-fertilize.

The forms described here are *Taenia pisiformis* (Gr. *tainia,* ribbon; L. *pisum,* pea, + *forma,* shape), a dog or cat tapeworm whose larval stage is found in the liver of rabbits (Figure 10-6), and *Dipylidium caninum* (Gr. *di,* two, + *pyle,* entrance, + *idion,* dim. suffix; L. *caninus,* belonging to a dog), a small tapeworm of dogs or cats (Figure 10-7), with fleas as the alternate host. Other forms may be substituted in the laboratory. Among those that parasitize humans are *Taenia solium,* the larval stage of which encysts in muscle of pigs and which, like *Taenia pisiformis,* possesses both hooks and suckers; *Taenia saginata,* which has cattle as the alternate host and has no hooks; and *Diphyllobothrium latum,* whose larval stage is found in crustaceans and fishes. *D. latum* is the largest of all human cestodes, some reaching a length of 20 m (65 ft). Humans become infected by eating uncooked freshwater fish or uncooked salmon.

General Structure and Life Cycle

Study a preserved whole specimen or one embedded in plastic.

How long is the specimen? __________ Examine the **scolex** under a dissecting microscope and identify **hooks** and **suckers.** Note the **neck,** from which new proglottids are budded off. As older proglottids are pushed back by addition of new ones, they mature and become filled with reproductive organs. Can you distinguish maturing from very young proglottids by the presence of a **genital pore** at one side? As eggs develop and become fertilized, the proglottid becomes distended (gravid), with the uterus filled with embryos. Where do you find **gravid proglottids** (Figure 10-7)? __________ Such proglottids soon break off and are shed in the feces of the host. Outside the body, the proglottids of *T. pisiformis* rupture, releasing thousands of infective eggs. If eaten by a rabbit, the intermediate host, the larva penetrates the liver, where it encysts to become a bladder worm (cysticercus). If the rabbit is eaten by a dog (or other carnivore), the bladder worm everts its scolex in the dog's intestine and begins production of proglottids, thus completing the cycle.

Dogs are the principal hosts of *D. caninum* but cats and humans (mainly children) are potential hosts and fleas are the alternate hosts. Gravid proglottids passed in the feces release eggs in the perianal region of the dog, which are eaten by larval fleas, which, when mature, are nipped or licked out of the fur by the dog or cat. Encapsulated embryos hatch in the host's intestine to grow into an adult worm, completing the life cycle.

Microscopic Study

Examine with low power a prepared slide of a tapeworm whole mount containing (1) a scolex with neck and a few immature proglottids, (2) a mature proglottid, and (3) a gravid proglottid.

Scolex. Note the **suckers** (how many? __________) and a **rostellum** bearing a circle of **hooks** (Figure 10-8). What is the function of hooks and suckers? __________ Is there a mouth? __________ Is the **neck** segmented? __________ Two lightly stained tubes, one on each side, are **excretory canals,** which extend the entire length of the animal. Some of the **immature proglottids** have developed **genital pores.**

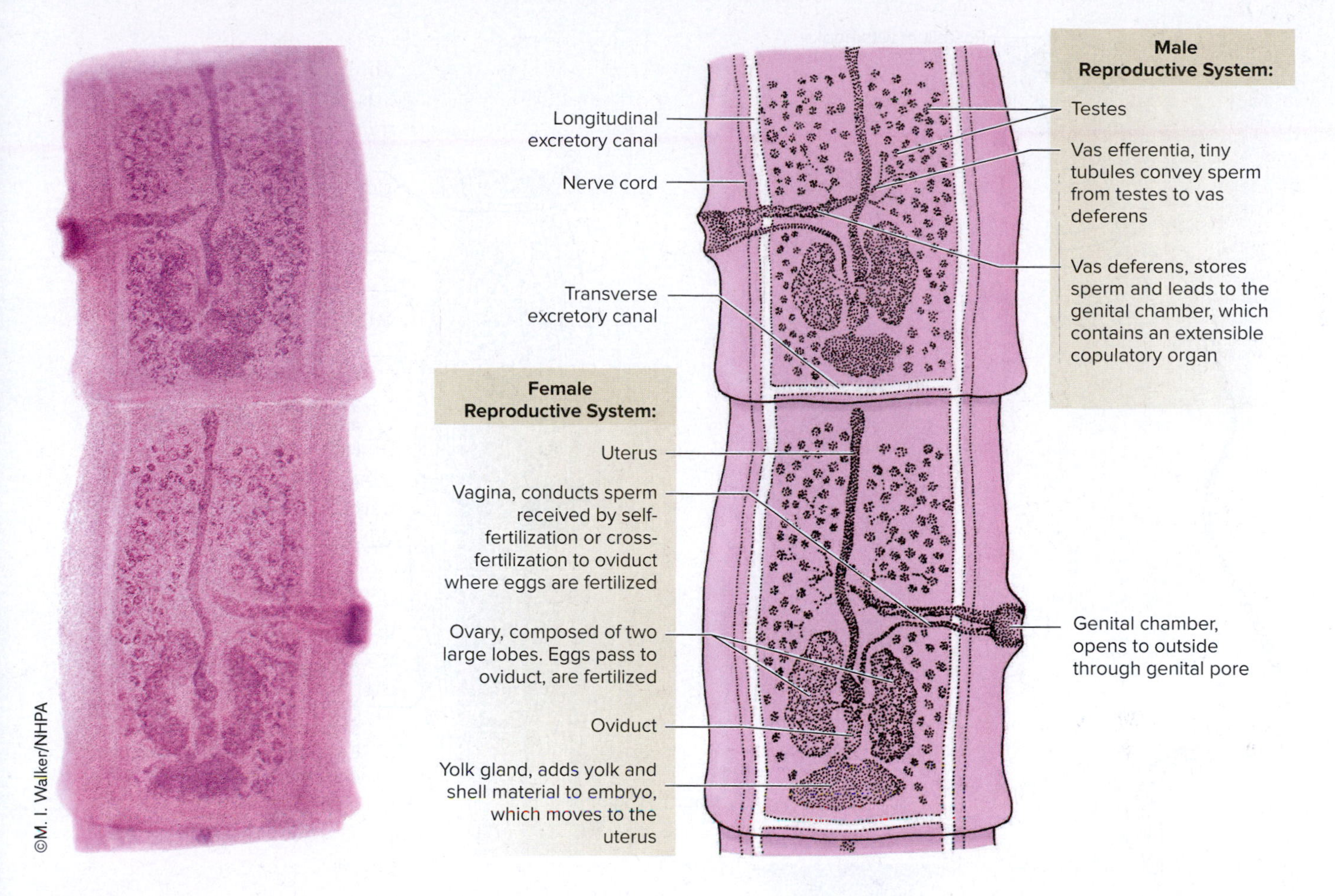

Figure 10-6

Photomicrograph and interpretive drawing of *Taenia pisiformis,* the dog tapeworm. The intermediate host is a rabbit. Genital pores may be oriented toward either right or left side.

Mature Proglottid

A mature proglottid contains complete hermaphroditic reproductive organs (both male and female). From your slide locate the genital pore (two pores in *D. caninum*), testes, vas deferens, vagina, ovaries, oviducts, yolk gland, and uterus. Study the functional anatomy of the reproductive systems of *T. pisiformis* and *D. caninum* as explained in Figures 10-6 and 10-7.

Note **excretory canals** on each side of the proglottids of both tapeworm species. They are part of the protonephridial system and empty to the outside at the posterior end of the worm. They are connected by a **transverse canal** across the posterior margin of each proglottid.

Nerve cords run just lateral to the excretory ducts. The tapeworm nervous system resembles that of planarians and flukes, although it is less well developed.

Do you find any digestive organs in the tapeworm? __________ Why? __________ What is a tapeworm's food? __________ Where is it digested? __________ How does the worm obtain it? __________ How is the structure of tapeworms adaptive for their environment? __________ What control methods can be used against pork and beef tapeworms? __________

Gravid Proglottid

Examine a **gravid proglottid.** What structures can you identify besides the distended **uterus?** If possible, study the **eggs** with high power. (CAUTION! Be very careful when changing to high power because whole-mount slides are thicker than average.)

Can you distinguish in any eggs a six-hooked larval form called an **onchosphere** (Gr. *onkinos,* hook, + *sphaira,* globe)? __________

Written Report

Answer the questions on Tapeworms on p. 164.

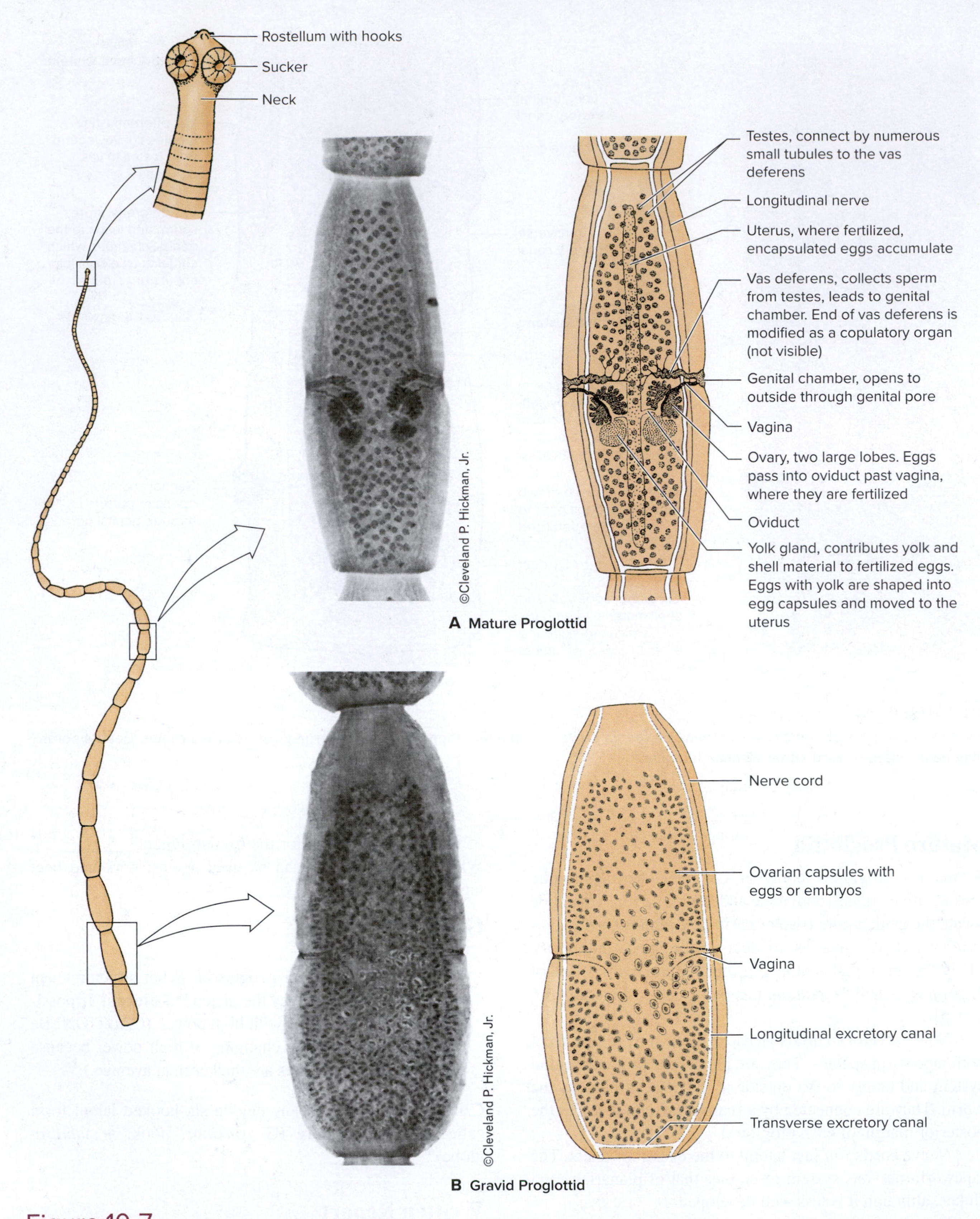

Figure 10-7

Dipylidium caninum, the small tapeworm of dogs and cats. The alternate host is a flea.

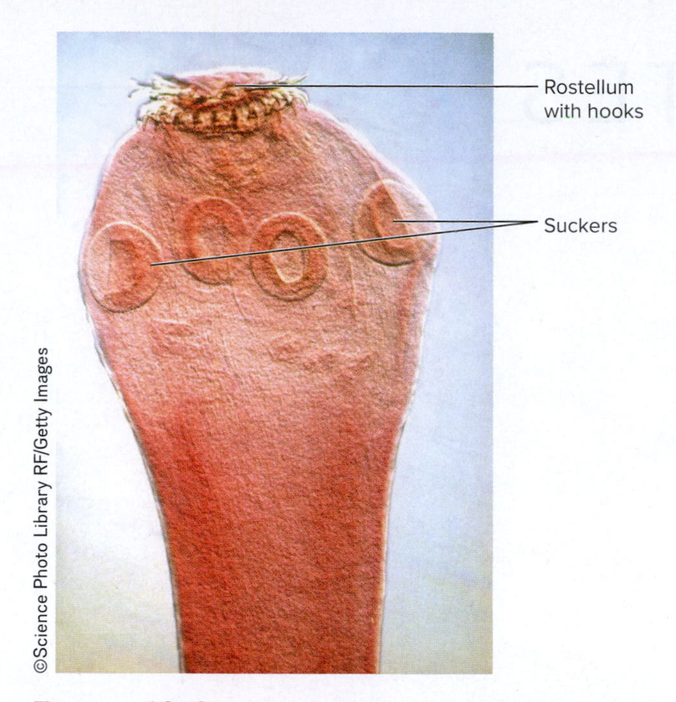

Figure 10-8
Scolex of *Taenia* sp., showing four suckers and a circlet of hooks on the rostellum.

Drawings

On separate paper, draw and label the scolex and neck, a mature proglottid, and a gravid proglottid. Include the scale of the drawings, and be sure to identify the species you are drawing.

Build Your Knowledge of Cestode Features

Test your understanding of the main features of the cestodes covered in the exercise by filling out the summary table on pp. xi–xii. Try to complete the table without looking at your notes.

Projects and Demonstrations

1. *Slides of cross sections of tapeworm.* Identify the tegument, parenchyma, excretory ducts, nerve cords, uterus, and other reproductive organs that appear in section.
2. *Preserved specimens or prepared slides.* Examine (a) preserved specimens or prepared slides of various species of tapeworms and (b) prepared slides of ova and cysts. Look for features described in this exercise.

Classification

Phylum Platyhelminthes

Class Turbellaria (tur′bel-lar′e-a) (L. *turbellae* [pl.], stir, bustle, + *aria,* like or connected with). Turbellarians. Mostly free-living, with a ciliated epidermis. A paraphyletic grouping. Example: *Dugesia tigrina.*

Class Monogenea (mon′o-gen′e-a) (Gr. *mono,* single, + *gene,* origin, birth). Monogenetic flukes. Adult body covered with syncytial tegument without cilia; leaflike to cylindrical in shape; posterior attachment organ with hooks, suckers, or clamps, usually in combination; all parasitic, mostly on skin or gills of fishes; single host; monoecious; usually free-swimming ciliated larva. Examples: *Polystoma, Gyrodactylus.*

Class Trematoda (trem′a-to′da) (Gr. *trematodes,* with holes, + *eidos,* form). Digenetic flukes. Adult body covered with nonciliated syncytial tegument; leaflike or cylindrical in shape; usually with oral and ventral suckers, no hooks; development indirect, first host a mollusc, final host usually a vertebrate; parasitic in all classes of vertebrates. Examples: *Fasciola, Clonorchis, Schistosoma.*

Class Cestoda (ses-to′da) (Gr. *Kestos,* girdle, + *eidos,* form). Tapeworms. Adult body covered with nonciliated, syncytial tegument; scolex with suckers or hooks, sometimes both, for attachment; long, ribbonlike body, usually divided into series of proglottids; no digestive organs; parasitic in digestive tract of all classes of vertebrates; first host may be invertebrate or vertebrate. Examples: *Taenia, Diphyllobothrium.*

NOTES

Name ______________________

Date ______________________

Section ______________________

LAB REPORT

Planarians

1. Why is bilateral symmetry of adaptive value for forward-moving animals? ______________________

2. What do planarians eat and how is the food digested? ______________________

3. What are rhabdites and what function do they serve? ______________________

4. What are flame cells and what is their function? ______________________

5. Flatworms are acoelomate. What fills the mesodermal space in flatworms that in coelomate animals would be a cavity?

6. Why are flatworms termed triploblastic? ______________________

7. What is meant by the term monoecious? Dioecious? ______________________

   Are planarians monoecious or dioecious? ______________________

8. What mechanism(s) do planarians use to glide across a wet surface? ______________________

Clonorchis and *Schistosoma*

1. Is *Clonorchis* monoecious or dioecious? ______________________

2. Explain how copulation and fertilization occur in *Clonorchis.* ______________________

3. Explain how an intermediate host is important in the life cycle of *Clonorchis.* ______________________

4. Is *Schistosoma* monoecious or dioecious? ______________________

5. How do the three species of *Schistosoma* differ in where the adult worms inhabit the body of their human hosts? Is there an intermediate host in the life cycle? ______________________

6. What and where is the gyneophoric canal and what is its role in the reproduction of *Schistosoma?*

LAB REPORT

Tapeworms

1. In the life cycle of *Dipylidium caninum,* what are the principal host and the intermediate host? ______

2. In the life cycle of *Taenia pisiformis,* what are the principal host and the intermediate host? ______

3. Tapeworms have no digestive tract. How do they obtain the necessary nutrients for life? ______

4. Explain the life cycle of either *D. caninum* or *T. pisiformis* after a gravid proglottid is shed in the feces of the host. ______

5. Are tapeworms monoecious or dioecious? What is meant when a mature proglottid is described as hermaphroditic? Explain how the eggs in a proglottid are fertilized. ______

6. Explain the function of the following reproductive structures in a tapeworm:

   van deferens ______

   vasa efferentia ______

   ovary and oviduct ______

   yolk gland ______

   uterus ______

Planarian Behavior

Name ______________________

Date ______________________

Section ______________________

LAB REPORT

Locomotion

Response to Touch (Thigmotaxis)

Response to Food (Chemotaxis)

Response to Directional Illumination

Turbellarian Structure

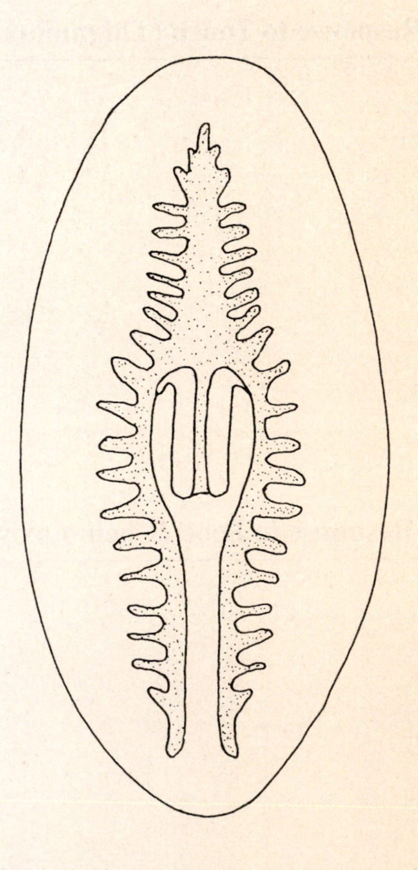

Nervous system of *Bdelloura* (digestive system is already shown). Draw in and label cerebral ganglia, nerve cords, and transverse and lateral nerves. For the digestive system, label pharynx, pharyngeal sheath, anterior trunk, and posterior trunks.

EXPERIMENTING IN ZOOLOGY

Planaria Regeneration Experiment

Planarians are easy to work with and have remarkable powers of regeneration.

Before starting, assemble your materials: sharp razor blade or scalpel blade, snap-cap vial or screw-cap specimen jar for *each* planarian, clean culture water (pond water or dechlorinated tap water), pipette, small artist's brush, lens paper, and ice cube. Decide how you want to cut the worms (Figure 10-9).

Fold a lens tissue over an ice cube and, with a camel's hair brush, transfer a planarian to the top surface of the ice cube; it will become quiescent almost immediately. Using a dissecting microscope, make the desired cut or cuts with the razor blade. If the worm is being partially split longitudinally, make certain the cut is clean and has completely separated the cut surfaces. Make a sketch in your notebook or on p. 168. With the pipette, rinse the pieces off the lens paper into the vial (previously half-filled with culture water). Label each vial with your name and the date, and prepare a sketch of the cut made. Put in a cool place.

Culture water *must* be changed every 3 or 4 days to keep the planaria healthy. *Do not* feed the animals. Remove any dead pieces, which will appear grayish or fuzzy. Additional tips for this experiment are found in Appendix A, p. 402.

Examine the worms about every 3 days for 2 weeks. To obtain two-headed or two-tailed planarians from the partial cuts shown in the second, third, and fifth sketches in Figure 10-9, you will need to recut them after 12 to 18 hours and probably again at 2 days to prevent the parts from fusing back together. Soon a blastema will form on the cut edge. Is the growth faster at the posterior end of a cephalic piece or the anterior end of a caudal piece? __________ On decapitated pieces, when do new eyes appear? __________ When can you distinguish new auricles? __________ Which is regenerated first, eyes or pharynx? __________ Do you find any evidence of anterior-posterior axis in the manner of growth of these pieces? __________

Drawings

On the record sheet, p. 168, make a sketch of the shape of each piece. Examine the specimens twice a week if possible, each time recording the date and sketching each regenerating piece. Summarize the results on p. 169. In your report, describe (1) your technique, (2) any problems you encountered, (3) potential or real sources of error in the experimental approach, and (4) your results. If you were to repeat the experiment, what would you do differently?

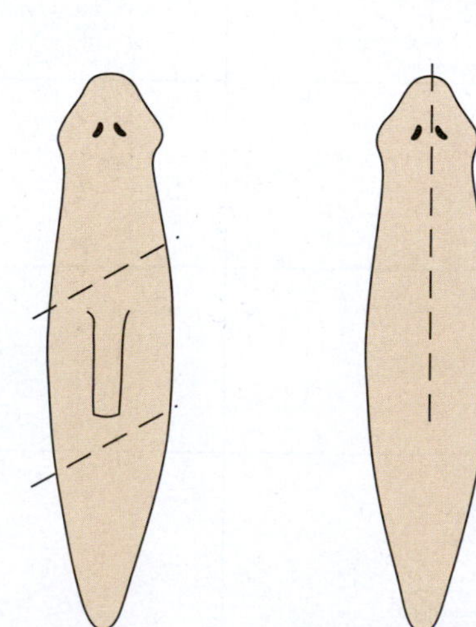

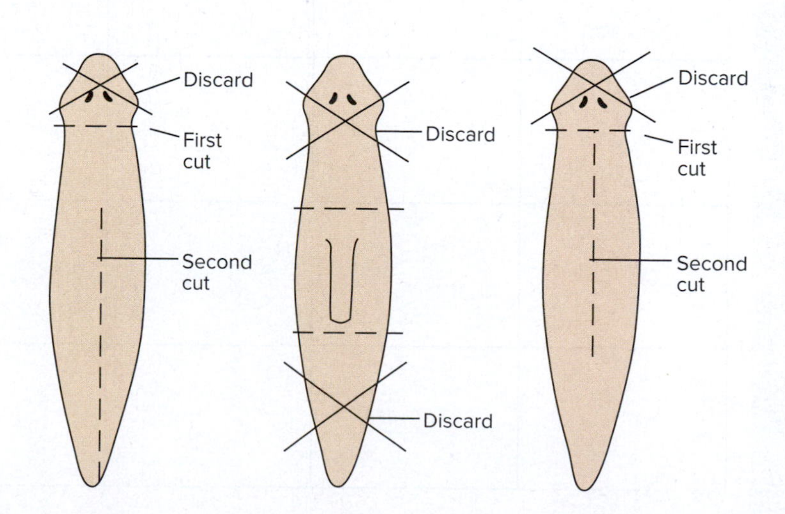

Figure 10-9
Some suggestions for regeneration experiments

LAB REPORT

Name ____________________

Date ____________________

Section ____________________

Regeneration Experiment with *Dugesia*

Sketches of original cuts of specimens:

Dates	1	2	3	4	Control

Record of growth:

Dates	1	2	3	4	Control

LAB REPORT

Methods Used

Observations and Conclusions

NOTES

Nematodes and Four Small Protostome Phyla

EXERCISE 11

Phylum Nematoda
Phylum Rotifera
Phylum Acanthocephala
Phylum Nematomorpha
Phylum Gastrotricha

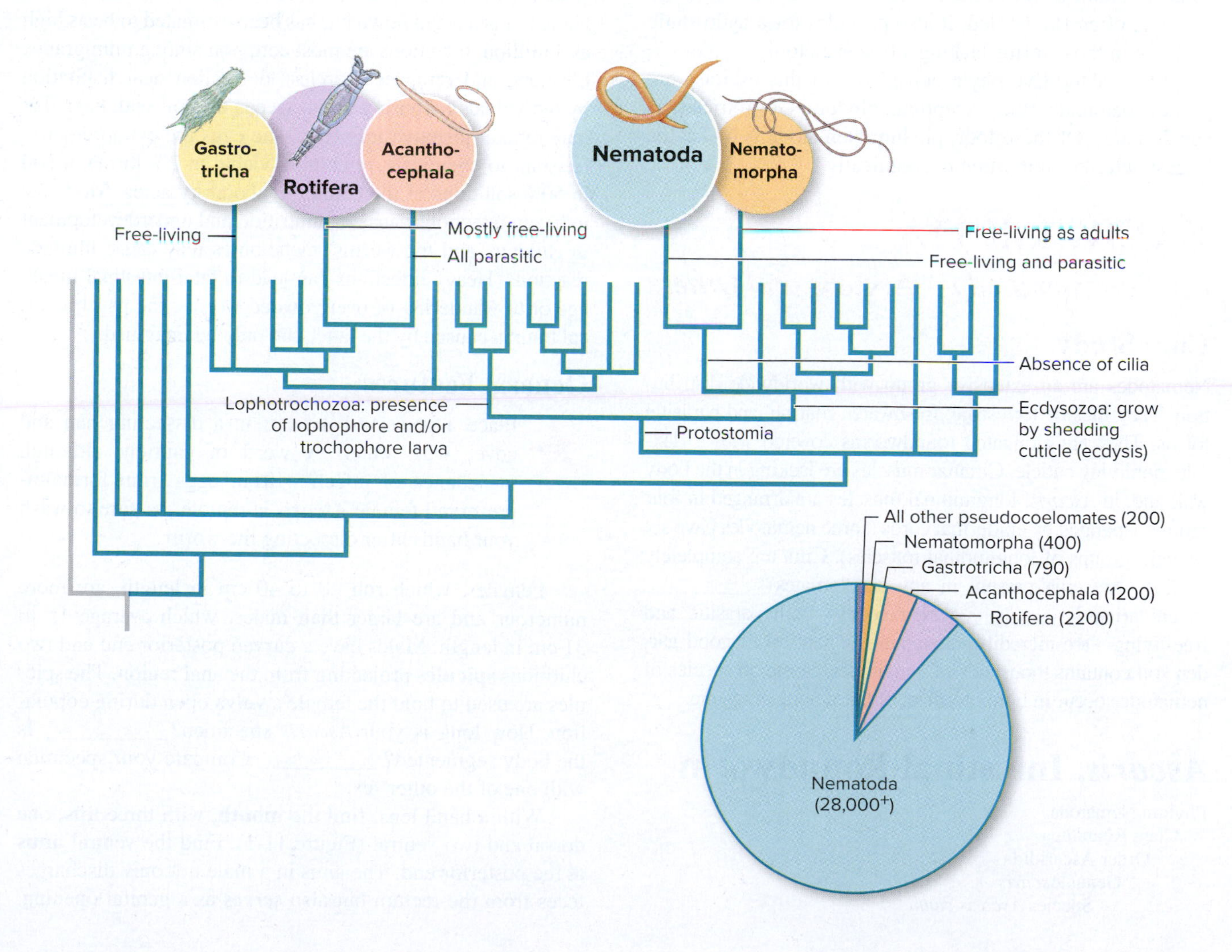

The flatworms (Platyhelminthes) considered in Exercise 10 and the phyla considered in this exercise belong to the Protostomia, a clade that embraces the great majority of invertebrate phyla (see the cladogram on the inside front cover of the manual). The Protostomia are divided in turn into two smaller clades: Lophotrochozoa and Ecdysozoa. These two clades were originally separated using molecular genetic methods (each has a distinctive molecular signature), but each also has distinctive morphological characteristics. Members of the Lophotrochozoa share either an odd horseshoe-shaped feeding structure, the lophophore, or a particular larval form, the trochophore. Members of the Ecdysozoa possess a cuticle that is shed at intervals as their bodies grow. Both lophotrochozoans and ecdysozoans are represented in the phyla considered in this exercise.

All bilateral animal phyla except the acoelomates possess a **body cavity** belonging to one of two types: (1) **true coelom,** in which a peritoneum (an epithelium of mesodermal origin) covers both the inner surface of the body wall and the outer surface of the visceral organs in the cavity, or (2) **pseudocoel,** a body cavity not entirely lined with peritoneum.

A body cavity of either type is an advantage because it provides room for organ development and storage and allows some freedom of movement within the body. Because the cavity is often fluid-filled, it also provides for a hydrostatic skeleton in those forms lacking a true skeleton.

Four of the five phyla considered in this exercise are pseudocoelomate (the exception, phylum Gastrotricha, is acoelomate). Of these four, phylum Nematoda is by far the largest and most important economically.

EXERCISE 11A

Phylum Nematoda—*Ascaris* and Others

Ascaris, Intestinal Roundworm
Some Free-Living Nematodes
Some Parasitic Nematodes
Projects and Demonstrations

EXERCISE 11B

A Brief Look at Some Other Protostomes

Phylum Rotifera
Phylum Acanthocephala
Phylum Nematomorpha
Phylum Gastrotricha

EXERCISE 11A

*Phylum Nematoda—*Ascaris *and Others*

Core Study

Nematodes are an extensive group with worldwide distribution. They include terrestrial, freshwater, marine, and parasitic forms. They are elongated roundworms covered with a flexible, nonliving cuticle. Circular muscles are lacking in the body wall and, in *Ascaris,* longitudinal muscles are arranged in four groups separated by epidermal cords (some nematodes have six or eight groups of longitudinal muscles). Cilia are completely lacking. Are cilia present in any acoelomates? __________ In cnidarians? __________ Nematodes—both parasitic and free-living—are incredibly abundant. A handful of good garden soil contains thousands of nematodes. Some 50 species of nematodes occur in humans, most of them nonpathogenic.

Ascaris, Intestinal Roundworm

Phylum Nematoda
  Class Rhabditea
    Order Ascaridida
      Genus *Ascaris*
        Species *Ascaris suum*

Where Found

Ascaris lumbricoides (Gr. *askaris,* intestinal worm) is a common intestinal parasite of humans. *A. suum,* which parasitizes pigs, is so similar to the human parasite that it was long considered merely a different strain of *Ascaris lumbricoides. Ascaris megalocephala* is common in horses.

Ascaris infections are extremely common—nearly one-quarter of the human population is infected worldwide. These infections were once common in rural areas in the southeastern United States. The number of infected persons in the United States is unknown but has been estimated to be as high as 4 million. Infections are most common among immigrants, travelers, and refugees. Careless defecation near habitation by individuals harboring worms seeds the soil with eggs that may remain infective for years. The eggs are extraordinarily resistant to chemicals, remaining viable in 2% formalin and in 50% solutions of the common laboratory acids. Moderate to heavy infections cause malnutrition and underdevelopment in children, and the worms' metabolites may cause immune reactions. Heavy infections can lead to fatal intestinal blockage or to wandering of overcrowded worms; the psychological trauma caused by the latter can only be imagined.

General Features

Place a preserved *Ascaris* in a dissecting pan and cover with water. A word of caution: although the chance of infection from eggs from formalin-preserved female *Ascaris* is remote, be sure to wash your hands after dissecting the worm.

Females, which run 20 to 40 cm in length, are more numerous and are larger than males, which average 15 to 31 cm in length. Males have a curved posterior end and two chitinous **spicules** projecting from the anal region. The spicules are used to hold the female's vulva open during copulation. How long is your *Ascaris* specimen? __________ Is the body segmented? __________ Compare your specimen with one of the other sex.

With a hand lens, find the **mouth,** with three **lips,** one dorsal and two ventral (Figure 11-1). Find the ventral **anus** at the posterior end. The anus in a male not only discharges feces from the rectum but also serves as a genital opening.

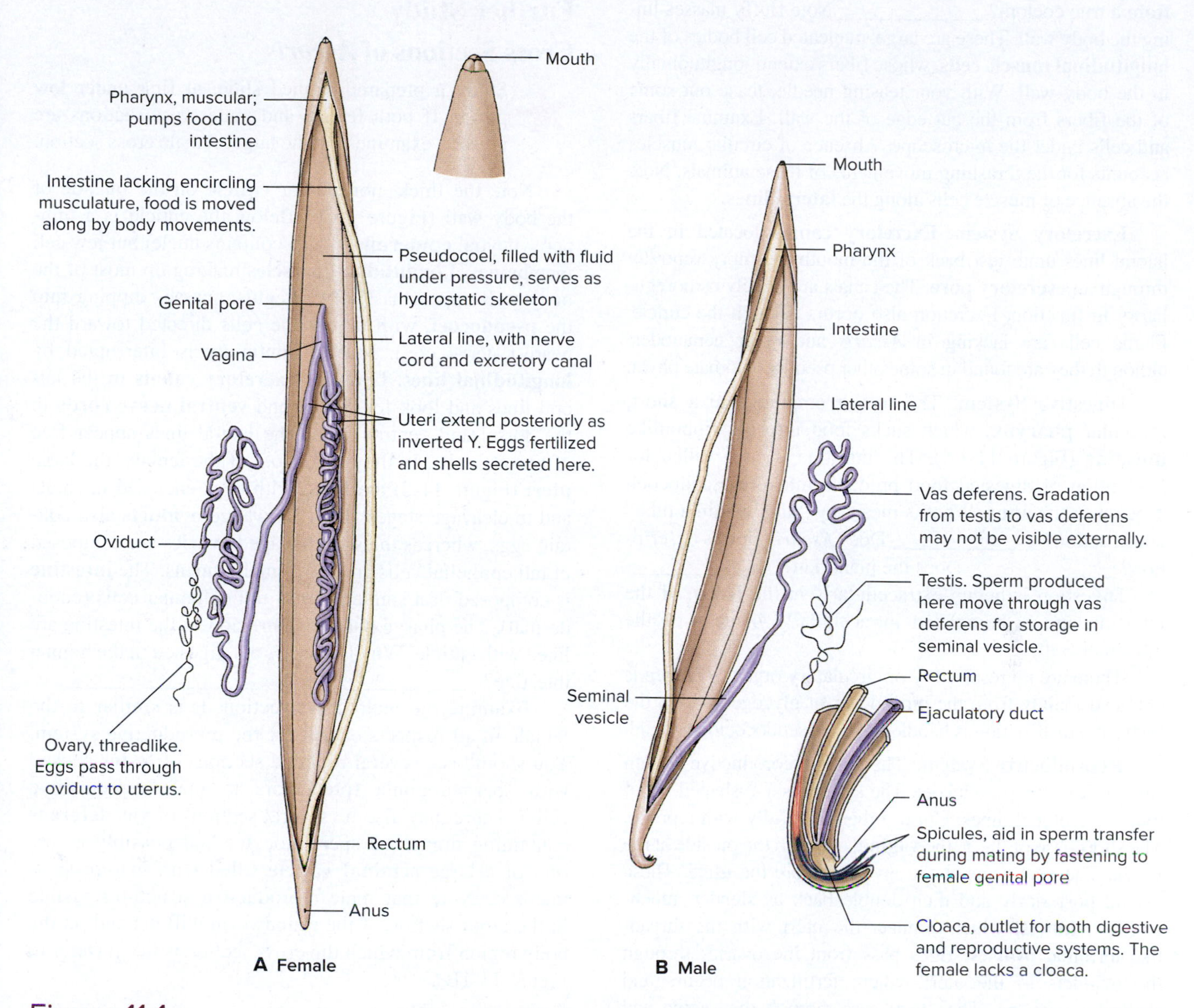

Figure 11-1

Ascaris. **A,** Internal structure of female, with insert showing detail of anterior end. **B,** Internal structure of male, with insert showing detail of cloacal region.

The female genital opening **(vulva)** is located on the ventral side about one-third the length of the body from the anterior end. It may be hard to distinguish from scars. Use a hand lens or dissecting microscope.

Note the shiny **cuticle** that covers the body wall. It is nonliving and consists primarily of **collagen,** which also is found in vertebrate connective tissue.

Four **longitudinal lines** run almost the entire length of the body—**dorsal** and **ventral median lines** and two **lateral lines.** Dorsal and ventral lines, which indicate location of bundles of nerve fibers, are very difficult to see on preserved specimens. However, along the lateral lines, the body wall is thinner, and the lines usually appear somewhat transparent. Excretory canals are located inside the lateral lines.

Internal Structure

Select a female specimen, place the worm in a dissecting pan, and cover it with water. Locate the lateral lines, where the body wall seems somewhat thinner. Now find the anus and vulva on the ventral side. This should help you identify the opposite, or middorsal, line. Now, with a razor blade, slit open the body wall *along the middorsal line,* being careful to avoid injuring the internal structures. Pin back the body wall to expose the viscera, *slanting the pins outward* to allow room for dissection.

Body Wall and Pseudocoel. Note the body cavity. Why is it called a **pseudocoel?** __________ How does it differ from a true coelom? ____________ Note fluffy masses lining the body wall. These are large, nucleated cell bodies of the **longitudinal muscle cells,** whose fibers extend longitudinally in the body wall. With your teasing needle, tease out some of the fibers from the cut edge of the wall. Examine fibers and cells under the microscope. Absence of circular muscles accounts for the thrashing movements of these animals. Note the absence of muscle cells along the **lateral lines.**

Excretory System. Excretory canals located in the lateral lines unite just back of the mouth to empty ventrally through an **excretory pore.** The canals are largely osmoregulatory in function. Excretion also occurs through the cuticle. Flame cells are lacking in *Ascaris* and other nematodes, although they are found in some other pseudocoelomate phyla.

Digestive System. The mouth empties into a short, muscular **pharynx,** which sucks food into the ribbonlike **intestine** (Figure 11-1A). The intestine is thin-walled for absorption of digested food products into the pseudocoel. Trace it to the **anus.** What is meant by "tube within a tube" construction? ____________ Does *Ascaris* fit this description? ____________ Does the planarian? ____________

Digestion is begun extracellularly in the lumen of the intestine and is completed intracellularly in cells of the intestinal wall.

There are no respiratory or circulatory organs. Oxygen is obtained mainly from the breakdown of glycogen within the body, and distribution is handled by the pseudocoelomic fluid.

Reproductive System. The female reproductive system fills most of the pseudocoel. The system is a Y-shaped set of long, convoluted tubes. Unravel them carefully with a probe. The short base of the Y, the **vagina,** opens to the outside at the **vulva.** The long arms of the inverted Y are the **uteri.** These extend posteriorly and then double back as slender, much-coiled **oviducts,** which connect the uteri with the threadlike terminal **ovaries.** Eggs pass from the ovaries through the oviducts to the uteri, where fertilization occurs and shells are secreted. Then they pass through the vagina and vulva to the outside. The uteri of an ascaris may contain up to 27 million eggs at a time, with as many as 200,000 eggs being laid per day. Study the life history of *Ascaris* from your text. Is there an intermediate host in the life cycle? ____________

The male reproductive system is essentially a single, long tube made up of a threadlike **testis,** which continues as a slightly thicker **vas deferens.** Both are much-coiled. The vas deferens connects with the wider **seminal vesicle,** which empties by a short, muscular **ejaculatory duct** into the anus. Thus, the male anus serves as an outlet for both digestive system and reproductive system and is often called a **cloaca** (L., sewer). **Spicules** secreted by and contained in spicule pouches may be extended through the anus.

What is meant by "sexual dimorphism"? ____________ How does *Ascaris* illustrate this? ____________ How is *Ascaris* transmitted from host to host? ____________ How can infestation be prevented? ____________ In what ways is *Ascaris* structurally and functionally adapted to life as a parasite in the intestine? ____________

Further Study

Cross Sections of *Ascaris*

Study a prepared stained slide, at first under low power. If both female and male cross sections are present, examine first the larger female cross section.

Note the thick, noncellular **cuticle** on the outside of the body wall (Figure 11-2). Below the cuticle is a thinner syncytial **epidermis,** which contains nuclei but few cell membranes. **Longitudinal muscles** making up most of the body wall appear as fluffy, irregular masses dipping into the **pseudocoel,** with tips of the cells directed toward the nearest nerve cord. Muscle continuity is interrupted by **longitudinal lines.** Look for **excretory canals** in the lateral lines and look for **dorsal** and **ventral nerve cords** in the dorsal and ventral lines. The lateral lines appear free of muscle cells. In the pseudocoel of the female, the large **uteri** (Figure 11-2) are filled with eggs enclosed in shells and in cleavage stages. The thin-walled **oviducts** also contain eggs, whereas the wheel-shaped **ovaries** are composed of tall epithelial cells and have small lumens. The **intestine** is composed of a single layer of tall columnar cells (endodermal). The pharynx and rectal region of the intestine are lined with cuticle. Why is *Ascaris* not digested in the human intestine? ____________

Examine the male cross section. It is similar to the female in all respects except for the reproductive system. You should see several rounded sections of **testes** packed with spermatogonia (precursors to male reproductive cells). There may also be several sections of **vas deferens** containing numerous spermatocytes and possibly a section of a large **seminal vesicle** filled with mature spermatozoa. Note that male reproductive structures visible in the cross section of the roundworm will depend on the body region from which the cross section is taken (refer to Figure 11-1B).

Table of Comparison

On pp. 181–182 is a table for comparing representatives of three of the metazoan phyla that you have studied so far. Filling in this table affords a survey of the development of these phyla and is an excellent form of review.

Alternative Core Study

Some Free-Living Nematodes

Vinegar Eels

Turbatrix aceti (tur-ba′tricks; L. female that disturbs), the vinegar eel (Figure 11-3), is often found in fermented fruit juices, particularly the sediment of nonpasteurized vinegar, where it feeds on the yeasts and bacteria found there. As in *Ascaris,* females are larger than males.

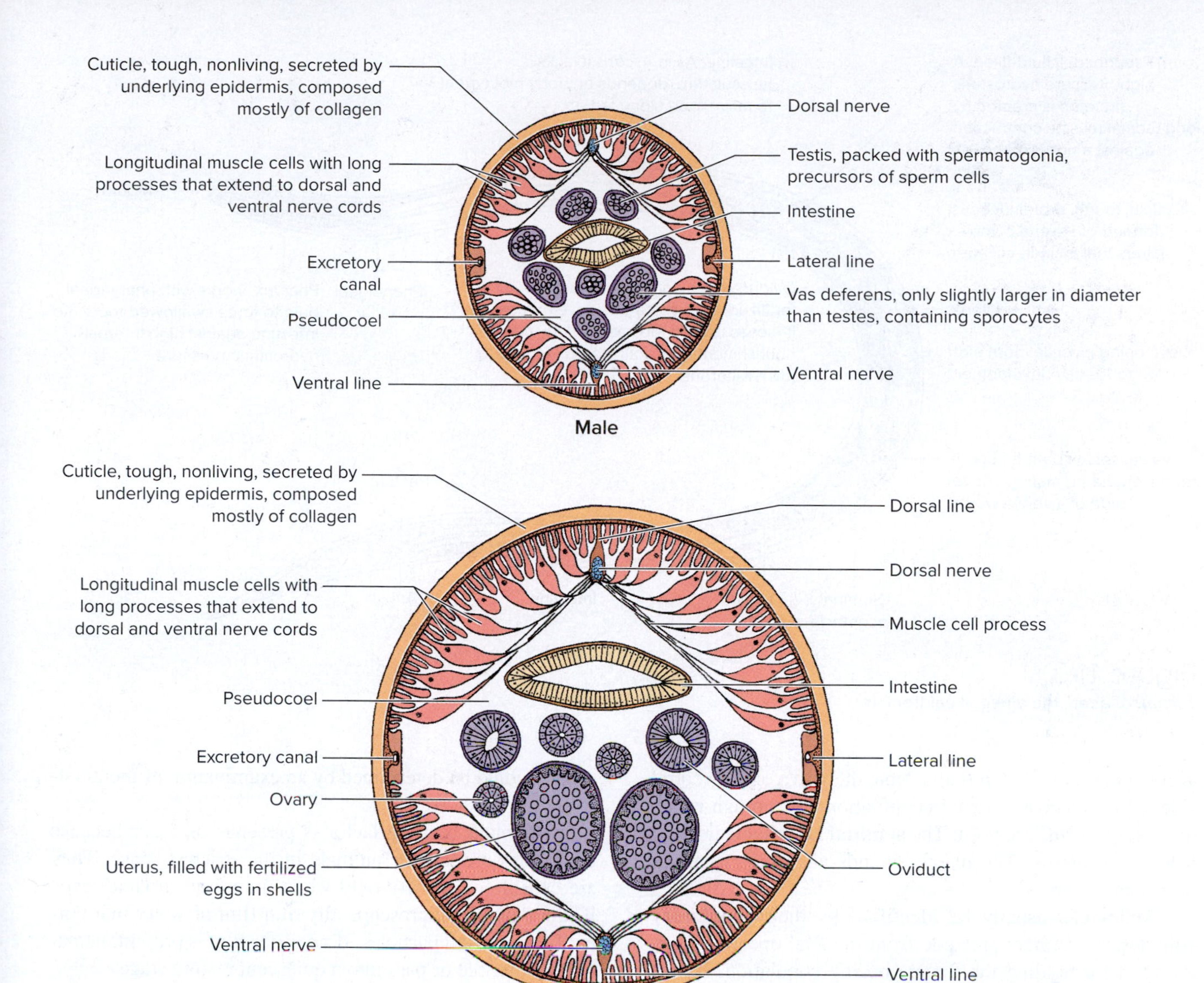

Figure 11-2

Cross sections through male and female *Ascaris* worms.

Place a drop of the culture containing worms on a clean glass slide and examine with a compound microscope.

Notice the violent thrashing movements, the body bending in S and C shapes. Because they have only longitudinal muscles in the body wall, vinegar eels, like other nematodes, can flex their body only from side to side. Locomotion is especially inefficient in free water, in which they can make little directional progress.

Add a pinch of sand to the slide and observe again. The worms are now able to use the sand grains as fulcra to make forward progress. In this manner, nematodes can readily move through the interstitial fluids of soils or between the cells of tissues. Now place another drop of culture on a clean slide and cover with a coverslip. Quiet the worms for study by *gently* warming the slide over a small flame or an incandescent lamp. Worms also may be quieted by adding a drop of 1N HCl to the culture.

Note the blunt anterior end, the **mouth, pharynx,** and **pharyngeal bulb.** This leads into a long, straight **intestine,** which ends at the ventral **anus,** located a short distance from the pointed tail end.

Select a female specimen. Large females will contain two to five developing juveniles in the **uterus.** *Turbatrix* is **ovoviviparous** (L. *ovum,* egg, + *vivere,* to live, + *parere,* to bring forth), giving birth to about 45 living young during its

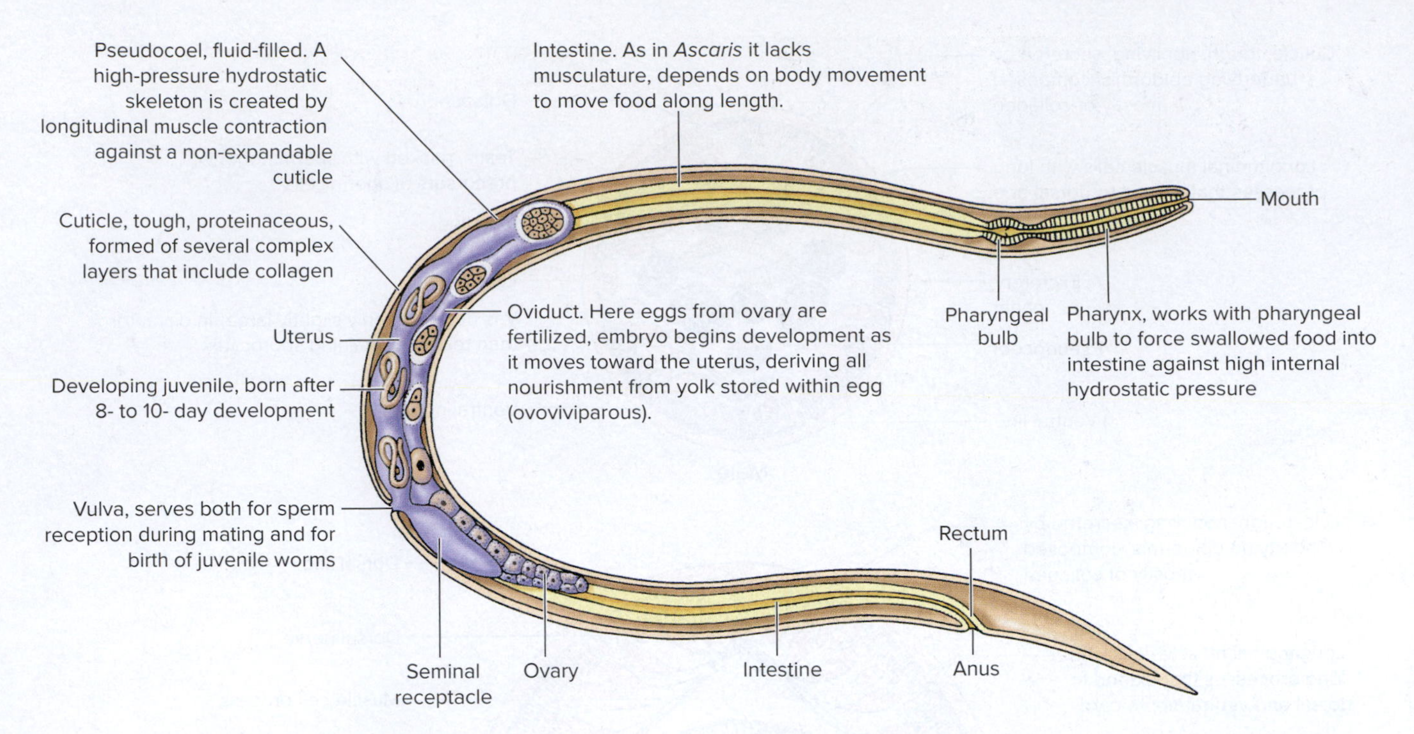
Pseudocoel, fluid-filled. A high-pressure hydrostatic skeleton is created by longitudinal muscle contraction against a non-expandable cuticle
Intestine. As in Ascaris it lacks musculature, depends on body movement to move food along length.
Mouth
Cuticle, tough, proteinaceous, formed of several complex layers that include collagen
Uterus
Oviduct. Here eggs from ovary are fertilized, embryo begins development as it moves toward the uterus, deriving all nourishment from yolk stored within egg (ovoviviparous).
Pharyngeal bulb
Pharynx, works with pharyngeal bulb to force swallowed food into intestine against high internal hydrostatic pressure
Developing juvenile, born after 8- to 10- day development
Vulva, serves both for sperm reception during mating and for birth of juvenile worms
Rectum
Seminal receptacle
Ovary
Intestine
Anus

Figure 11-3
Turbatrix aceti, the vinegar eel (female).

average life span of 10 months. Note the **vulva** on the ventral side, which receives sperm in copulation and through which the young worms are born. The **seminal receptacle** lies just behind the uterus. The **ovary** extends anteriorly from the uterus and doubles back dorsally.

Males can usually be identified by their **copulatory spicules,** which may protrude from the anal opening. They are used for holding the female during copulation. It may be possible to see the filamentous **testis** extending forward from about the middle of the body, then bending back on itself. It leads, by way of a sperm duct, to the cloaca.

Further Study

Soil Nematodes

Nematodes are present in almost every imaginable habitat. N. A. Cobb emphasized their abundance in this quotation from a 1914 U.S. Department of Agriculture yearbook: "If all the matter in the universe except the nematodes were swept away, our world would still be dimly recognizable, and if, as disembodied spirits, we could then investigate it, we should find its mountains, hills, vales, rivers, lakes, and oceans represented by a thin film of nematodes. The location of towns would be decipherable, since for every massing of human beings there would be a corresponding massing of certain nematodes. Trees would still stand in ghostly rows representing our streets and highways. The location of the various plants and animals would still be decipherable, and, had we sufficient knowledge, in many cases even their species could be determined by an examination of their erstwhile nematode parasites."

The main limiting factor is presence of water because nematodes are aquatic animals in the strictest sense. They are capable of activity only when immersed in fluid, even if it is only the microscopically thin film of water that normally covers soil particles. If no water film is present, nematodes either die or pass into a quiescent resting stage.

Soil nematodes are more abundant among roots of plants than in open soil, so the best collecting source is probably the top few centimeters of a long-established meadow turf.

Some nematodes are herbivorous, feeding on algae, fungi, and higher plants; many feed on plant roots. Carnivorous nematodes feed on other nematodes, rotifers, small oligochaetes, and so on. Some nematodes are **saprophagous** (suh-prof′uh-gus; Gr. *sapros,* rotten, + *phagos,* to eat), such as *Rhabditis,* which probably live on bacteria and decaying organic matter. Some species are omnivorous. Most nematodes are beneficial to agriculture by helping decompose organic matter. However, the few that are parasitic to plants and animals are responsible for millions of dollars in lost revenue every year in North American agriculture. They spread easily and are nearly impossible to eradicate once established.

Soil samples may be brought in from a wide variety of places. Methods that you may use to collect nematodes, as well as many annelids and arthropods, for study under the microscope are found in Appendix A, p. 404. Methods are also given for obtaining *Rhabditis* from the nephridia of earthworms.

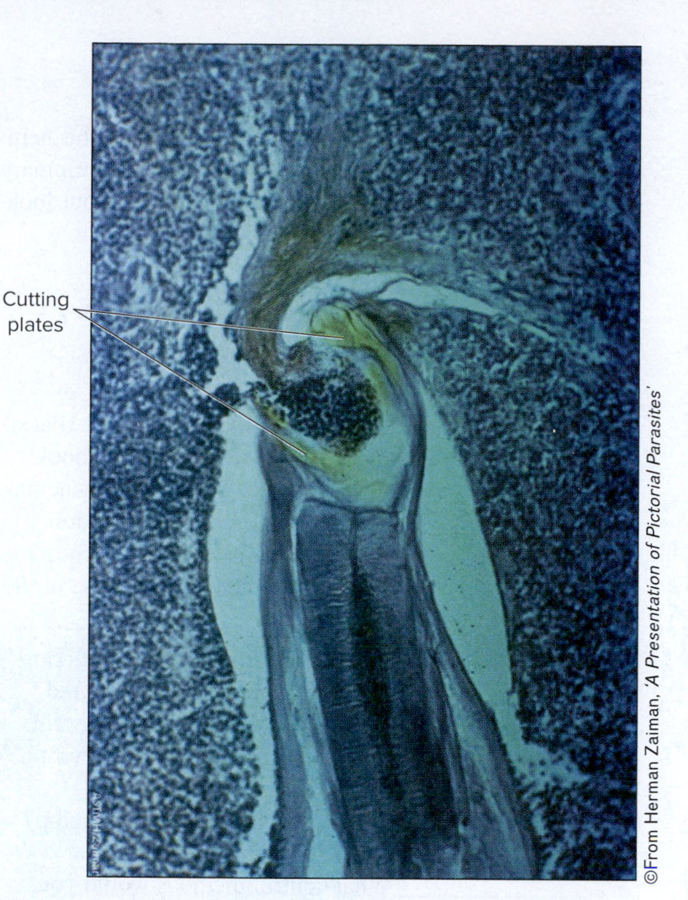

©From Herman Zaiman, 'A Presentation of Pictorial Parasites'

Figure 11-4
Section through anterior end of *Ancylostoma caninum*, the dog hookworm, in the intestine. Note the lacerating cutting plates, used for attachment to the intestinal mucosa

Some Parasitic Nematodes

Hookworms

Hookworms, *Necator americanus* ("American killer"), live in the intestines of their vertebrate hosts. They attach themselves to the mucosa and suck up the blood and tissue fluids from it. The species most important to humans are *N. americanus* and *Ancylostoma duodenale* (an-ke-los′ta-muh; Gr. *ankylos*, crooked, + *stoma*, mouth). The two species are widespread in tropical and subtropical countries and together infect an estimated 800 million people. Hookworms infect about 4% of the population in the southern United States, where 95% of the cases are *Necator* infections. *A. caninum* is a common hookworm of domestic dogs and cats (Figure 11-4).

Hookworms mature and mate in the small intestine of the host. Developing eggs are passed out in the feces. On the ground they require warmth (preferably 20° to 30° C), shade, and moisture for continued development. They hatch in 24 to 48 hours into young juveniles, which feed on fecal matter, molt their cuticles twice, and in a week or so are ready to infect a new host.

If the ground surface is dry, they migrate into the soil, but after a rain or morning dew they move to the surface, extend their bodies in a snakelike fashion, and wave back and forth. Thousands may group together, waving rhythmically in unison. Under ideal conditions, they may live for several weeks.

Infection occurs when juveniles contact the host's skin and burrow into it. Those that reach blood vessels are carried to the heart and then to the lungs. There they are carried by ciliary action up the respiratory passages to the glottis and swallowed. In the small intestine, they grow, molt, mature, and mate. Five weeks after entry, they are producing eggs.

Whether hookworm disease results from infection depends on number of worms present and nutritional condition of the infected person. Massive infections in the lungs may cause coughing, sore throat, and lung infection. In the intestinal phase, moderate infections cause an iron-deficiency anemia. Severe infections may result in severe protein deficiency. When accompanied by chronic malnutrition, as in many tropical countries, there may be irreversible damage, resulting in stunted growth and below-average intelligence.

Examine prepared slides of hookworms.

Adult males of *Necator* are typically 7 to 9 mm long; females, 9 to 11 mm long. Specimens of *Ancylostoma* are slightly longer. The anterior end curves dorsally, giving the worm a hooklike appearance. Note the large buccal capsule, which bears a pair of dorsal and a pair of ventral cutting plates surrounding its margin (Figure 11-4). A stout, muscular esophagus serves as a powerful pump.

Note on males a conspicuous copulatory bursa consisting of two lateral lobes and a smaller dorsal lobe, all supported by fleshy rays. Needlelike spicules are present, which in *Necator* are fused at the distal ends to form a characteristic hook. In females the vulva is located in about the middle of the body.

Trichina Worm

The trichina worm, *Trichinella spiralis* (Gr. *trichinos*, of hair, + *ella*, dim. suffix), is a nematode parasite in humans, hogs, rats, and other omnivorous or carnivorous mammals.

Study a slide of larvae encysted in pork muscle (Figure 11-5).

How many cysts does your slide show? __________ Would you be able to see the cysts in meat with the naked eye? ____________ The cyst wall is made of fibrous tissue that gradually becomes calcified from the host's immune reactions. How many worms are coiled in a cyst? ____________ Study the life cycle of *Trichinella* in your text. Are they oviparous or ovoviviparous? ____________ How can you prevent trichinosis? __________ Study a slide showing male and female adults. Where would the adults live in the human host? ______________________________

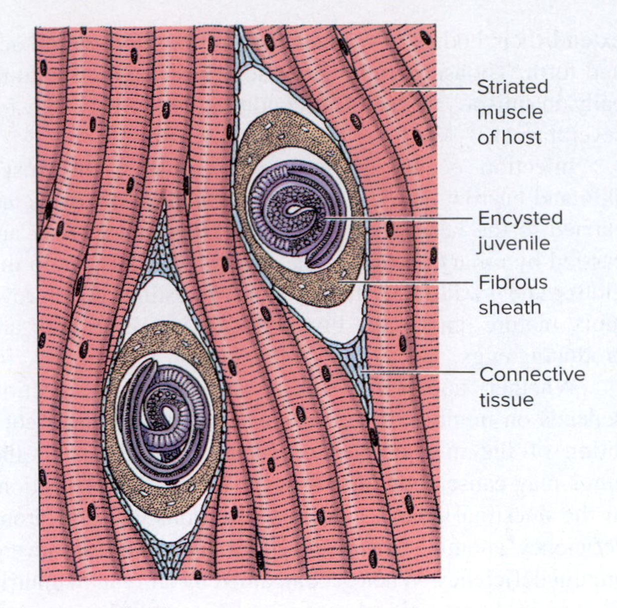

Figure 11-5
Trichinells spiralis, the trichina worm. Juveniles are shown encysted in skeletal muscle.

Pinworm

Pinworms, *Enterobius vermicularis* (en-te-robe′ee-us; Gr. *enteron,* intestine, + *bios,* life) (Figure 11-6), are the most common nematode parasite of humans in North America. They live in the large intestine and cecum. Females, up to 12 mm in length, lay their eggs at night around the anal region of their host. A single female may lay 4600 to 16,000 eggs. Scratching contaminates the hands and bedding of the host. The eggs, when swallowed, hatch in the duodenum and mature in the intestine.

Unlike hookworms, pinworms are often found at high socioeconomic levels. Although approximately one-third of all American and Canadian children will be infected with pinworms at some time in their lives, their presence is mainly an irritation and an embarrassment because pinworms cause no obvious debilitating effects. But unmeasurable is the mental stress experienced by families in their efforts to rid their households of the worms.

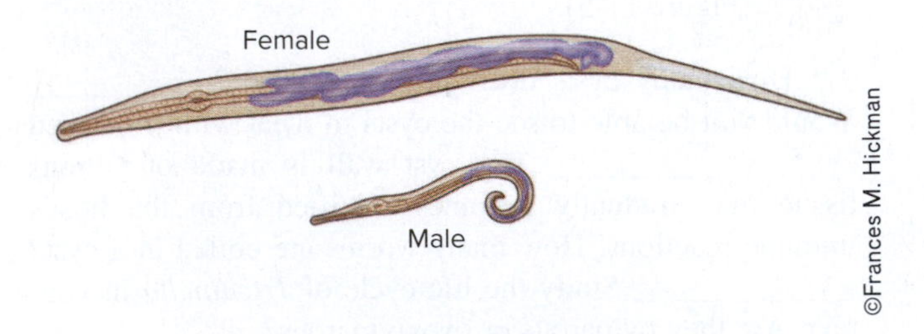

Figure 11-6
Pinworms, *Enterobius vermicularis.* The male is smaller and has a curled posterior end.

Build Your Knowledge of Nematode Features

Test your understanding of the main features of the nematodes covered in the exercise by filling out the summary table on pp. xi–xii. Try to complete the table without looking at your notes.

Projects and Demonstrations

1. *Prepared slide of* Wuchereria bancrofti *(wu-ka-rir′ ee-a; after Otto Wucherer, nineteenth-century German physician).* Examine prepared slides of the filarial worm, *Wuchereria bancrofti.* Consult your textbook for the life cycle. What disease is caused by this nematode? ____________ In what climate is infestation common? ____________ What is the alternate host? ____________ What control methods might be used? ______________________________
2. *Prepared slides of* Dracunculus medinensis *(dra-kunk′ u-les; L. dim. of* draco, *dragon).* Examine prepared slides of the guinea worm, *Dracunculus medinensis.* Consult your textbook for the life cycle of this worm. In what part of the world is this nematode common? ____________ Where do the larvae develop? ____________ How is this parasite acquired? ____________ What control methods would you use to prevent infestation? ______________________________ ______________________________
3. *Prepared slides of other nematode parasites.* Examine prepared slides of eggs, larvae, or cysts of any nematode parasites.
4. *Prepared slide of* Dirofilaria immitis *(dog heartworm).* Examine a prepared slide of microfilariae in a smear of dog blood. Dogs are infected by mosquitoes that ingest and transmit the microfilariae with their blood meals. The worms mature in the right heart and pulmonary artery. Heavy infestations cause cardiopulmonary failure. Although once confined to the southern United States, heartworm has now spread throughout the United States and warmer regions of Canada, wherever its vector, the mosquito, is found.
5. *Population growth in vinegar eels.* Students work in groups to count the growth of vinegar eels in different concentrations of apple cider vinegar. Prepare 100%, 75%, and 50% concentrations of apple cider vinegar, place in petri dishes and inoculate with a vinegar eel culture. Worms are counted twice weekly in a Sedgewick-Rafter counting chamber, and the results after a month are plotted to show population growth. Complete directions for this exercise, contributed by Eric Lovely of Arkansas Tech University, are found in Appendix A, p. 404.
6. *Methods for collecting soil nematodes.* Nematode populations in soil often reach enormous proportions

and are of great importance both in the ecology of soils and in the economic cost nematodes inflict on agriculture. Either of these procedures will give students an impression of the abundance of soil nematodes. Two procedures for collecting and demonstrating the great abundance of soil nematodes are described in Appendix A, p. 404.

7. *Methods for demonstrating* Rhabditis maupasi. The larval stages of *Rhabditis maupasi* are a common nematode parasite living in the nephridia of earthworms (other congenerics occur in freshwater and in soil). Methods for demonstrating the larval stages are found in Appendix A, p. 404–405.

EXERCISE 11B

A Brief Look at Some Other Protostomes

Core Study

Phylum Rotifera

Philodina and Others

Place a drop of rotifer culture on a depression slide, cover, and examine with subdued light under low power.

How does the animal attach itself? ____________ Is it free-swimming? ____________ Does it have a definite head end? ____________ Note the anterior discs of cilia **(corona)** that give the impression of wheels turning. Are they retractile? ____________ The cilia function both in swimming and in feeding.

The tail end (or **foot**) bears slender toes. How many? The foot contains a pedal gland that secretes a cement used for clinging to objects.

Locate the pharynx **(mastax),** which is fitted with jaws for grinding up food particles. The mastax is conspicuous in living rotifers because of its rhythmic contractions (Figure 11-7). Rotifers feed on small plankton organisms swept in by cilia. Can you identify the digestive tract?

In *Philodina* (fill-uh-dine′uh; Gr. *philos,* fond of, loving, + *dinos,* whirling) (Figure 11-7), the **cuticle** is ringed (annulated), so that it appears segmented. From watching its movements, would you say it had circular muscles? Longitudinal? Oblique? ____________ Estimate the size of the rotifers. Is there more than one variety in the culture? ____________

In many rotifers, the cuticle is thickened and rigid and is called a **lorica.** *Monostyla* and *Platyias* are examples. Some rotifers, such as *Floscularia* (Figure 11-7), live in a secreted tube. Most rotifers live in freshwater.

Phylum Acanthocephala

Spiny-Headed Worms, *Macracanthorhynchus*

The adult *Macracanthorhynchus hirudinaceus* (mak′ruh-kan-thuh-rink′us; Gr. *makros,* long, + *akantha,* thorn, + *rhynchos,* snout) (Figure 11-7) is parasitic in the small intestine of pigs, where it attaches to the intestinal lining and absorbs digested food of its host. Like tapeworms it has no digestive system at all.

The body is cylindrical and widest near the anterior end. A small, spiny proboscis on the anterior end bears six

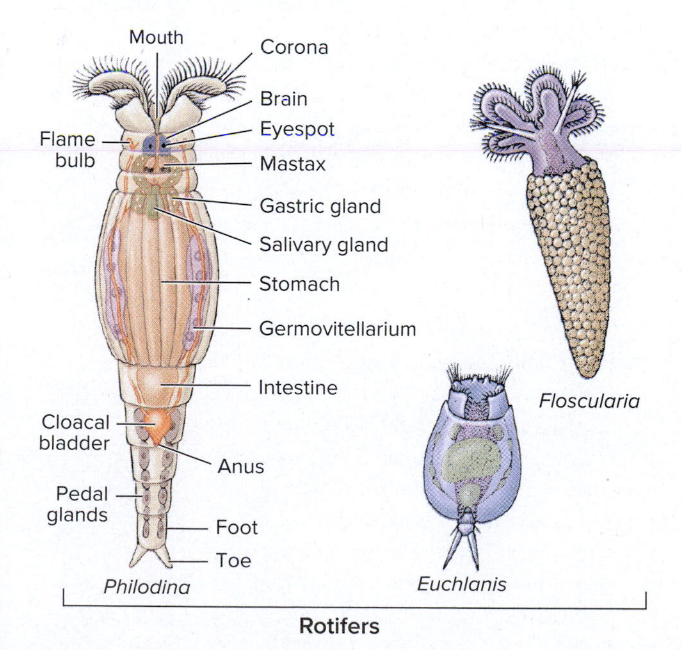

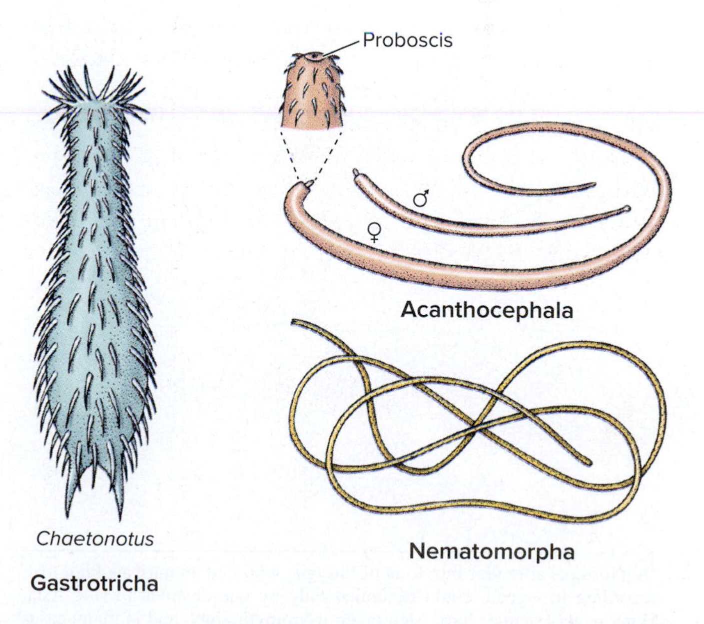

Figure 11-7
Some representative protostomes.

rows of recurved hooks for attachment to the intestinal wall. The proboscis is hollow and can be partially retracted into a proboscis sheath.

The worms are dioecious. Males are much smaller than females. Males have a genital bursa at the posterior end, which may be partly evaginated through the genital pore and is used in copulation. Eggs discharged by a female into a host feces may be eaten by white grubs (larvae of the beetle family *Scarabeidae*) in which they develop. Pigs are infected by eating the grubs or the adult beetles.

Further Study

Phylum Nematomorpha

Threadworms, or "Horsehair" Worms

Threadworms, or "horsehair" worms (Figure 11-7), such as *Paragordius* and *Gordius,** are long, cylindrical, hair-like worms often found wriggling in watering troughs, puddles, ponds, and quiet streams. Most range between 0.5 and 3 mm in diameter and from 10 to 300 mm in length, but some reach a length of 1 m.

How long are your specimens? ______________ Is the diameter uniform throughout? ______________ If you have live specimens, what would you conclude about their muscular makeup, judging from their movements? ______________

Nematomorphs have no lateral lines or excretory system, and in adults the digestive system is degenerate. They differ from nematodes in having a cloaca in both sexes. Females lay long, gelatinous strings of eggs on water plants. Larvae, which are encysted on plants, are sometimes eaten by arthropods, in which the larvae are parasitic for a while.

Phylum Gastrotricha

Chaetonotus and Others

Gastrotrichs include both freshwater and marine organisms. In size and general habits, they are similar to the rotifers and are often found in the same cultures.

Place a drop of culture on a slide, cover, and study first with low and then with high power.

Observe the manner of locomotion of gastrotrichs. They glide along on a substratum by means of ventral cilia. *Chaetonotus* (NL. *chaeta,* bristle, + Gr. *notos,* back), a common genus, is covered with short, curved dorsal spines (Figure 11-7). The rounded head bears cilia and little tufts of sensory bristles. The tail end is forked and contains cement glands similar to those of rotifers. Do gastrotrichs use the forked tail in the same manner in which rotifers use the toes?

____

Gastrotrichs have a syncytial epidermis covered with a cuticle. In feeding they use the head cilia to sweep algae, detritus, and protozoans into the mouth. How long are the specimens? ______________

Most marine gastrotrichs are hermaphroditic, but in freshwater species only parthenogenetic females are known.

Build Your Knowledge of Protostome Features

Test your understanding of the main features of the protostomes covered in the exercise by filling out the summary table on pp. xi–xii. Try to complete the table without looking at your notes.

*L. *Gordius,* after Gordius, king of Phrygia, who tied an intricate knot that, according to legend, could be untied only by one destined to rule Asia. Many generic names have been taken from mythology, and in many cases their application is quite obscure. Because horsehair worms often become inextricably entangled in knots comprising several worms, the mythological application in this case, however, is apt.

Comparing Representatives of Three Phyla

Name ____________________

Date ____________________

Section ____________________

LAB REPORT

Features	Cnidaria Hydra	Platyhelminthes Planaria	Nematoda *Ascaris*
Symmetry			
Shape			
Germ layers			
Body covering			
Cephalization (present or absent)			
Coelomic cavity (if present, state what type)			

LAB REPORT

Features	Cnidaria Hydra	Platyhelminthes Planaria	Nematoda *Ascaris*
Musculature (layers present and how arranged)			
Digestive tract and digestion			
Excretion			
Nervous system			
Sense organs			
Reproduction, sexual			
Reproduction, asexual			

Describe in what ways *Ascaris* is structurally and functionally adapted to life as a parasite in the intestine. ________

____

____

____

____

The Molluscs Phylum Mollusca

A Protostome Coelomate Group

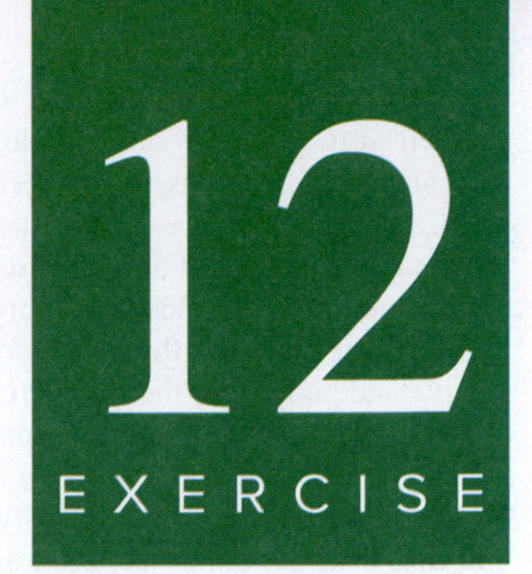

Molluscs, with over 90,000 living species, rank next to arthropods in number of named species. They include chitons, snails, slugs, clams, oysters, squids, octopuses, cuttlefish, and others. They have retained the basic features introduced by the preceding phyla, such as triploblastic structure, bilateral symmetry, cephalization, and a body cavity. The body cavity, though small and restricted to the space around the heart is now a **true coelom,** a characteristic shared by all remaining phyla. All organ systems are present. Molluscs, with spiral cleavage and an ancestral trochophore larva, belong to the Lophotrochozoa clade of animals.

EXERCISE 12A
Class Bivalvia (= Pelecypoda)—Freshwater Clam
Freshwater clam
Projects and Demonstrations

EXERCISE 12B
Class Gastropoda—Pulmonate Land Snail
Land Snail
Projects and Demonstrations

EXERCISE 12C
Class Polyplacophora—Chitons
Chitons

EXERCISE 12D
Class Cephalopoda—*Loligo,* the Squid
Loligo
Demonstrations

CLASSIFICATION
Phylum Mollusca

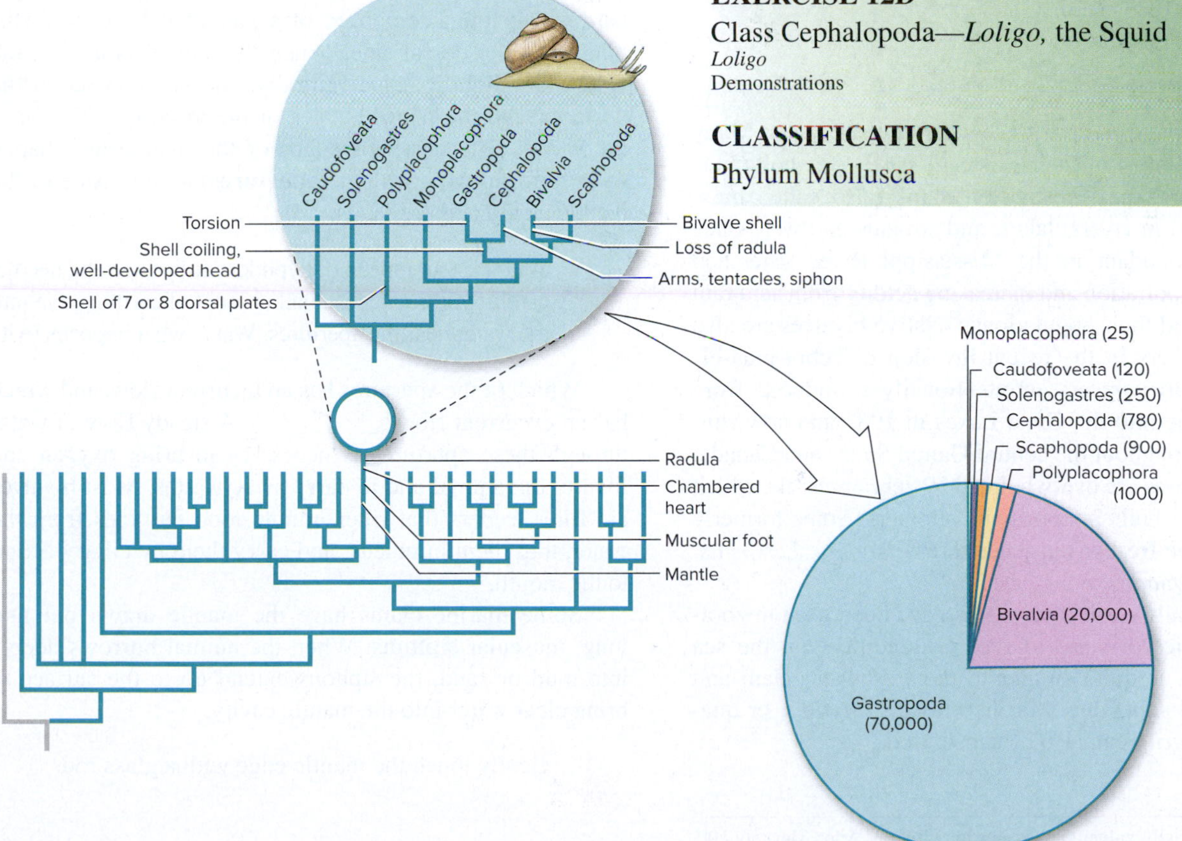

The name Mollusca means "soft-bodied" and because of their soft, plump bodies, molluscs such as clams, oysters, snails, and octopuses have been widely used for human food. Although the ancestral molluscan body has undergone enormous radiation, all have a muscular **foot,** generally used in locomotion. Two folds of skin from the dorsal body wall form a protective **mantle** that encloses a space between the mantle and body wall, the **mantle cavity.** The mantle cavity typically houses **gills** (a lung in some) and the mantle itself may secrete a protective **shell.**

Molluscs have left an extensive fossil record, indicating that their evolution has been a long one. They occupy numerous ecological niches and are found in the sea, in freshwater, and on land. They range from sedentary herbivores to fast-swimming predators.

Figure 12-1
Freshwater clam showing apertures in mantle. The dorsal, or upper, aperture is excurrent; the larger, lower aperture is incurrent.

EXERCISE 12A

Class Bivalvia (= Pelecypoda)— Freshwater Clam

Core Study

Freshwater Clam

Phylum Mollusca
Class Bivalvia
Subclass Palaeoheterodonta
Order Unionoidea
Genus *Anodonta,* others

Where Found

Bivalves are found in both freshwater and salt water. Many of them spend most of their existence partly or wholly buried in mud or sand. Freshwater clams (also called mussels) are found in rivers, lakes, and streams and were once particularly abundant in the Mississippi River watershed before stream pollution and increasing acidity from acid rain greatly depleted their populations.* Native bivalves are also the biggest losers in the recent invasion of zebra mussels *(Dreissena polymorpha),* unintentionally introduced from Eastern Europe into the Great Lakes in 1988 and now running amok throughout the central United States and Canada. Freshwater clams are overwhelmed by zebra mussels, which attach to their shells in enormous numbers. Some formerly common native freshwater genera are *Anodonta, Lampsilis, Elliptio,* and *Quadrula.*

Traditionally freshwater clams have been used in zoology laboratories because of their availability, but the sea clam, *Spisula,* is quite similar to the freshwater clam and makes a good substitute. Marine mussels, *Mytilus,* or quahogs, *Mercenaria,* can also be substituted.

*Clams are especially vulnerable to aquatic pollution. Approximately 10% of all freshwater clams became extinct in the twentieth century, and most of the rest are endangered.

Behavior and General Features

Observe living bivalves in an aquarium, if they are available.

Freshwater clams lie half-buried in the sand, and their reactions are slow.

What is the natural position of a clam at rest? ____________ When moving? ____________ Note that it leaves a furrow in the sand when it moves. The soft body is protected by a hard **exoskeleton** composed of a pair of **valves,** or shells, hinged on the dorsal side. When the animal is at rest, the valves are slightly agape ventrally, and you can see at the posterior end the fringed edges of the **mantle,** which lines the valves. The posterior edges of the mantle are shaped so as to form two openings **(apertures)** to the inside of the mantle cavity (Figure 12-1).

With Pasteur pipette or hypodermic syringe and needle, carefully introduce a small amount of carmine dye into the water near the apertures. Watch what happens to it.

Which of the apertures has an incurrent flow, and which has an excurrent flow? __________ A steady flow of water through these apertures is necessary to bring oxygen and food to the animal and to carry away wastes. Most bivalves are filter feeders that filter minute food particles from the water, trap them in mucus, and carry them by ciliary action to the mouth.

Some marine clams have the mantle drawn out into long, muscular **siphons**. When the animal burrows deeply into mud or sand, the siphons extend up to the surface to bring clear water into the mantle cavity.

Gently touch the mantle edge with a glass rod.

What happens? The mantle around the apertures is highly sensitive, not only to touch but also to chemical stimuli, a necessity if the animal is to close its valves to exclude water containing unpleasant or harmful substances.

Lift a clam out of the sand to observe hasty withdrawal of the foot. Then lay it on its side on the sand to see if it will right itself.

The foot is as soft, flexible, and sensitive as the human tongue. Mucous glands keep the foot well protected with mucus.

If there is a marine tank containing scallops, compare the method of locomotion of scallops with that of clams.

Written Report

Report your observations on clam behavior on separate paper.

External Structure

The bivalve shell protects the animal from predators, serves as a skeleton for muscle attachment, and, in burrowing forms, helps keep mud and sand out of the mantle cavity.

The two valves of the clam are attached by a **hinge ligament** on the dorsal side; the ventral side is free for the protrusion of the foot. A swollen hump, the **umbo** (pl. **umbones**), near the anterior end of the hinge, is the oldest and thickest part of the shell and the part most resistant to boring gastropod predators. Concentric **lines of growth** around the umbo indicate growth periods. Where is the youngest part of the shell? ____________

The outer, horny layer of the shell is the **periostracum,** which protects the underlying calcium carbonate from being dissolved by acid in the water. The periostracum is thin and tends to be worn away from older parts of the shell.

Now determine the correct orientation of the clam. Note that the umbones are dorsal and located toward the anterior end. Identify the right and left valves.

Internal Structure

The directions that follow apply to living clams (see Appendix A, p. 405) but may be easily adapted to preserved specimens.

Obtain a living clam that has been heated to about 40° C (104° F), causing the valves to gap slightly. *Caution! Do not overheat.* Your instructor or teaching assistant will open the clam by inserting a strong, short-bladed knife (*not* a scalpel) between the *left* valve and *left* mantle at the posterior end and cutting the posterior adductor muscle as close as possible to the left valve (Figure 12-2A). Cut the anterior adductor muscle in the same manner. Hold the clam against a firm surface when cutting, and keep your hand clear of the knife blade. The valve will gap open because of the action of the hinge ligament. Separate the mantle completely from the left valve and lift the loosened left valve. Place the clam in a dissecting pan and flood the body with pond or dechlorinated water, allowing the right valve to serve as a container.

Examine the inner surface of the left valve (Figure 12-3).

The inner, iridescent mother-of-pearl surface is the **nacreous layer,** which lies next to the mantle and is secreted continuously by the mantle surface. Between the inner nacreous layer and outer periostracum is a **prismatic layer,** made up of crystalline calcium carbonate. It is secreted by glands in the edge of the mantle.

With the aid of Figure 12-3, locate the **hinge ligament,** which acts as a spring to force the shells apart, and scars of the **anterior** and **posterior adductor muscles,** which pull the shells closed. Near the adductors are smaller muscles that retract and extend the foot (Figure 12-4). The **pallial line** marks the location of the pallial muscle that retracts the edge of the mantle (Figure 12-5). Along the dorsal margin of the shell is a ridge of lateral **hinge teeth** that interlock the valves to prevent them from slipping apart.

The Mantle

With the left valve removed, examine the thin **mantle** covering all the soft tissues of the clam.

Posteriorly the edges of the two mantles are thickened, darkly pigmented, and fused together dorsally to form the ventral **incurrent aperture** and dorsal **excurrent aperture.** The apertures permit a continuous flow of water through the mantle cavity. In many burrowing clams, the apertures are extended into siphons.

Examine the edge of the mantle, which forms three parallel folds: an outer, a middle, and an inner fold (Figure 12-5).

The outer fold (closest to the shell) secretes the horn-like periostracum as well as the prismatic layer of the shell. The middle lobe is sensory in function; in some bivalves (scallops, for example), it is drawn out into specialized sensory structures, such as tentacles and eyes. The ciliated surface of the inner fold assists in water circulation within the mantle cavity and sweeps out debris; the inner fold also seals the mantle cavity when the clam closes its shell. It is this inner fold that, on the posterior side, forms the clam's incurrent and excurrent apertures.

Sometimes a foreign object, such as a sand grain or parasite, becomes lodged between the mantle and the shell. The outer epithelium of the mantle then secretes nacre around the object, forming a pearl. Often, however, the formative pearl becomes fused with the nacreous layer of the shell. Do you find any pearls or other evidence of such irritation in the valve or mantle?

Muscles. Locate the large **adductor muscles,** which close the valves (see Figure 12-4). Slightly dorsal to them are the **foot retractor muscles.** The **foot protractor muscle** is small and will be found lying in the visceral mass just posterior to the anterior adductor.

Pericardium. On the dorsal side of the animal, locate the delicate, almost transparent **pericardial membrane** surrounding the heart. In living clams, the beating ventricle is

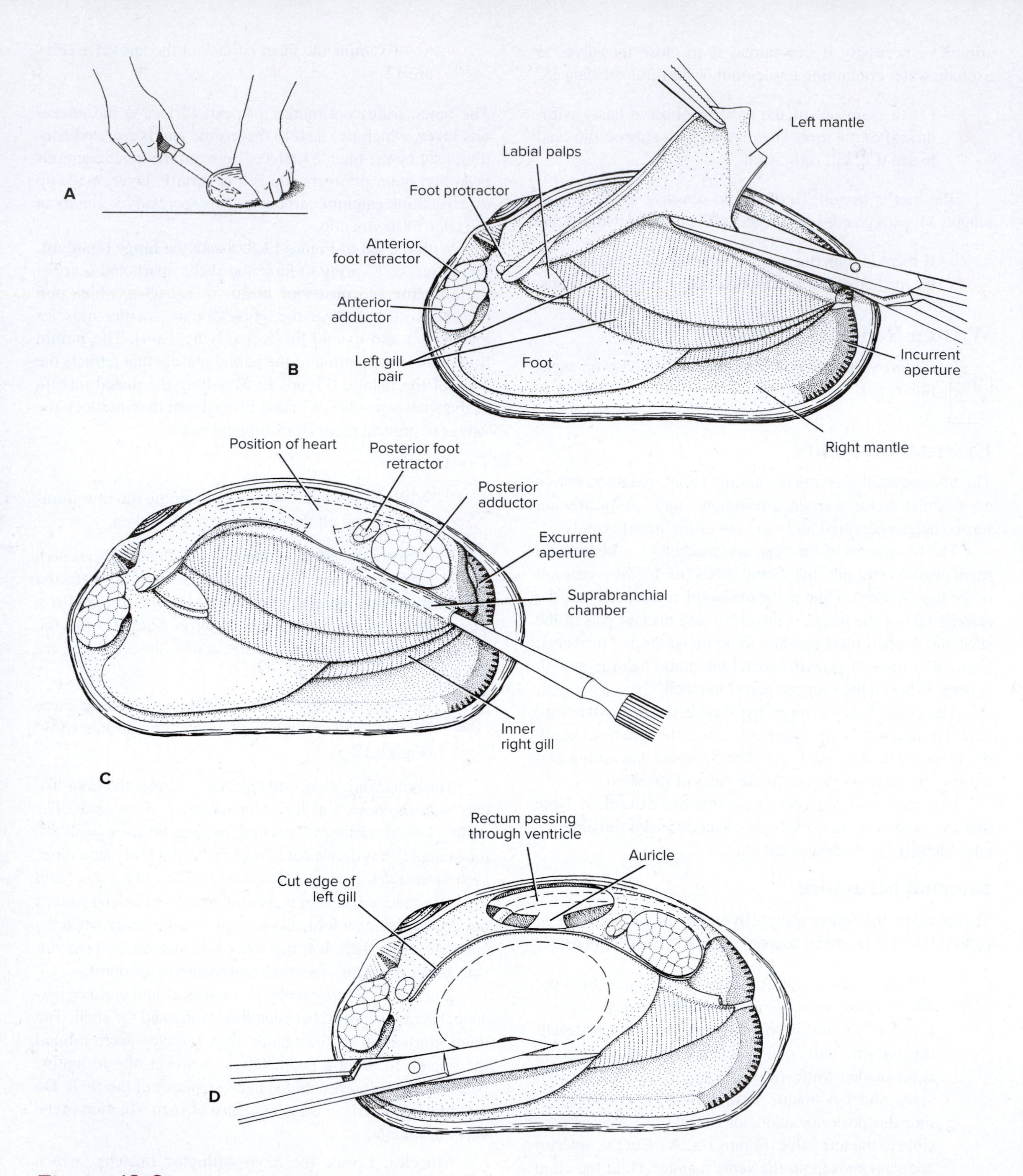

Figure 12-2

Dissection of a freshwater clam showing the clam body lying in the right valve. **A,** Cutting the posterior adductor. **B,** Trimming off the left mantle. **C,** Probing the suprabranchial chamber. **D,** Cutting to expose the visceral mass.

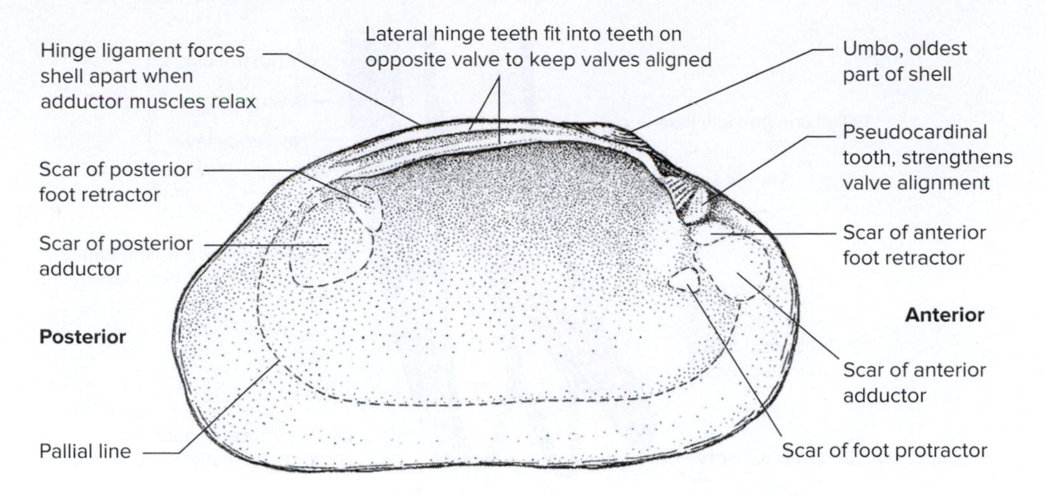

Figure 12-3
Left valve of a freshwater clam, showing the muscle scars.

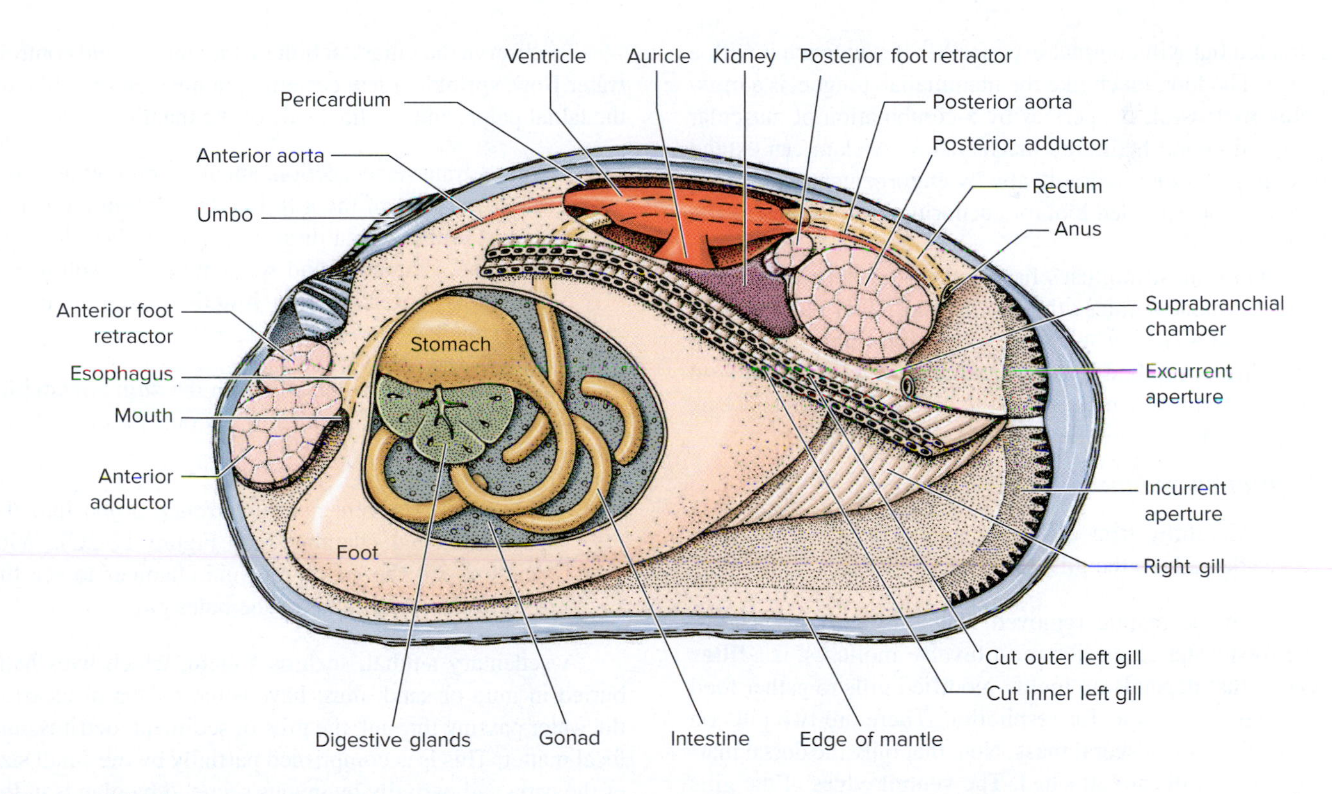

Figure 12-4
Anatomy of a unionid clam showing the clam lying in the right valve.

visible through the pericardium. Do not open the pericardium until instructed to do so later.

Mantle Cavity
Lift up the mantle to expose the outer pair of **gills** and body mass beneath. The entire space between the right and left lobes of the mantle is the **mantle cavity.** Cilia on both mantle and gills keep water flowing through the mantle cavity. If the outer gill is much thicker than the inner gill, the animal is probably a female in which the gill is serving as a **brood chamber** for developing embryos. Are clams monoecious or dioecious? ______________________________

The soft portion of the body is the **visceral mass.** The muscular **foot** lies ventral to the visceral mass; it will be

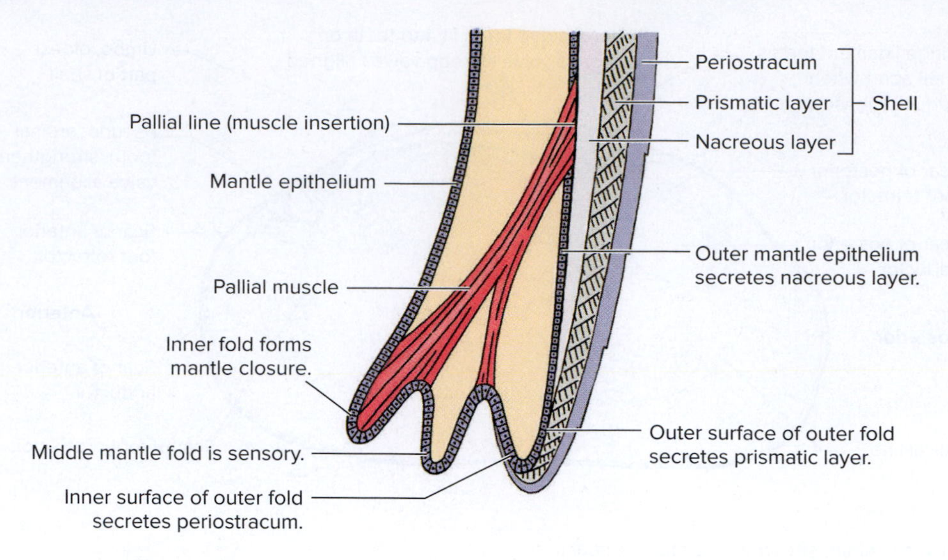

Figure 12-5

Section through margin of freshwater clam shell and mantle.

retracted but will contract even farther if you touch it with a probe. The foot, much like the mammalian tongue, is a **muscular hydrostat.** It operates by a combination of muscular contraction and hydraulic mechanisms. A clam can extend or enlarge its foot hydraulically by engorgement with blood and uses the extended foot for anchoring itself or drawing its body forward.

While most molluscs have a well-developed head, the clam has none—unless the mouth and labial palps can be considered a head. The two pairs of **labial palps,** a pair on each side of the body, are ciliated and serve to guide food particles trapped in mucus into the mouth. Bivalves also lack a radula, which would be of no use to a filter feeder.

Respiration and the Gills

Carefully trim off the mantle where it is attached dorsally to the gills (see Figure 12-2B).

With the mantle removed, the gills are conspicuous. The freshwater clam, like most bivalve molluscs, is a **filter feeder** that depends on highly modified gills to gather food materials, as well as for respiration. There are two gills on each side of the visceral mass. Note that only the dorsal margins of the gills are attached. The ventral edges of the gills hang free in the mantle cavity.

Strong currents produced by cilia on the **gill filaments** draw water into the incurrent siphon, then into the **water tubes** through numerous tiny pores **(ostia)** between the gill filaments (Figure 12-6). Food particles, mostly phytoplankton and organic debris, are carried by ciliary currents and mucus down the gill surface toward the food groove on the lower edge of the gill. Here, mucus and food particles are rolled into a food string, which passes into the mouth by way of the labial palps. Large particles are dropped to the edge of the mantle and discarded.

To observe the ciliary action that maintains and controls water flow, sprinkle a few carmine granules on the gills, on the labial palps, and on the inside of the mantle.

Cut a transverse section about 1 mm wide from the lower part of the gill. Lay the section on a slide and examine with a dissecting microscope. Identify lamellae, filaments, and water tubes, as well as the partitions **(interlamellar junctions)** that hold the two lamellae apart (Figure 12-6).

Water tubes connect dorsally with the **suprabranchial chamber,** which in turn empties to the outside through the excurrent aperture.

From the excurrent aperture, run a probe into the suprabranchial chamber (see Figure 12-2C). With scissors, slit the suprabranchial chamber to see the tops of the water tubes of the outer gill.

A sedentary animal, such as a clam, which lives half-buried in mud or sand, must have some means of clearing the water passing through the gills of sediment, detritus, and fecal matter. This is accomplished partially by the small size of the ostia and partially by mucus secreted by glands in the roof of the mantle cavity, which traps particles too large for the ostia. Larger debris drops off the gills, while smaller food particles are carried toward the mouth.

The water tubes in the female serve as brood pouches for eggs or larvae during breeding season.

Further Study

Circulatory System

Near the dorsal midline, just below the hinge, is the thin-walled **pericardial sac,** within which lies the **heart.**

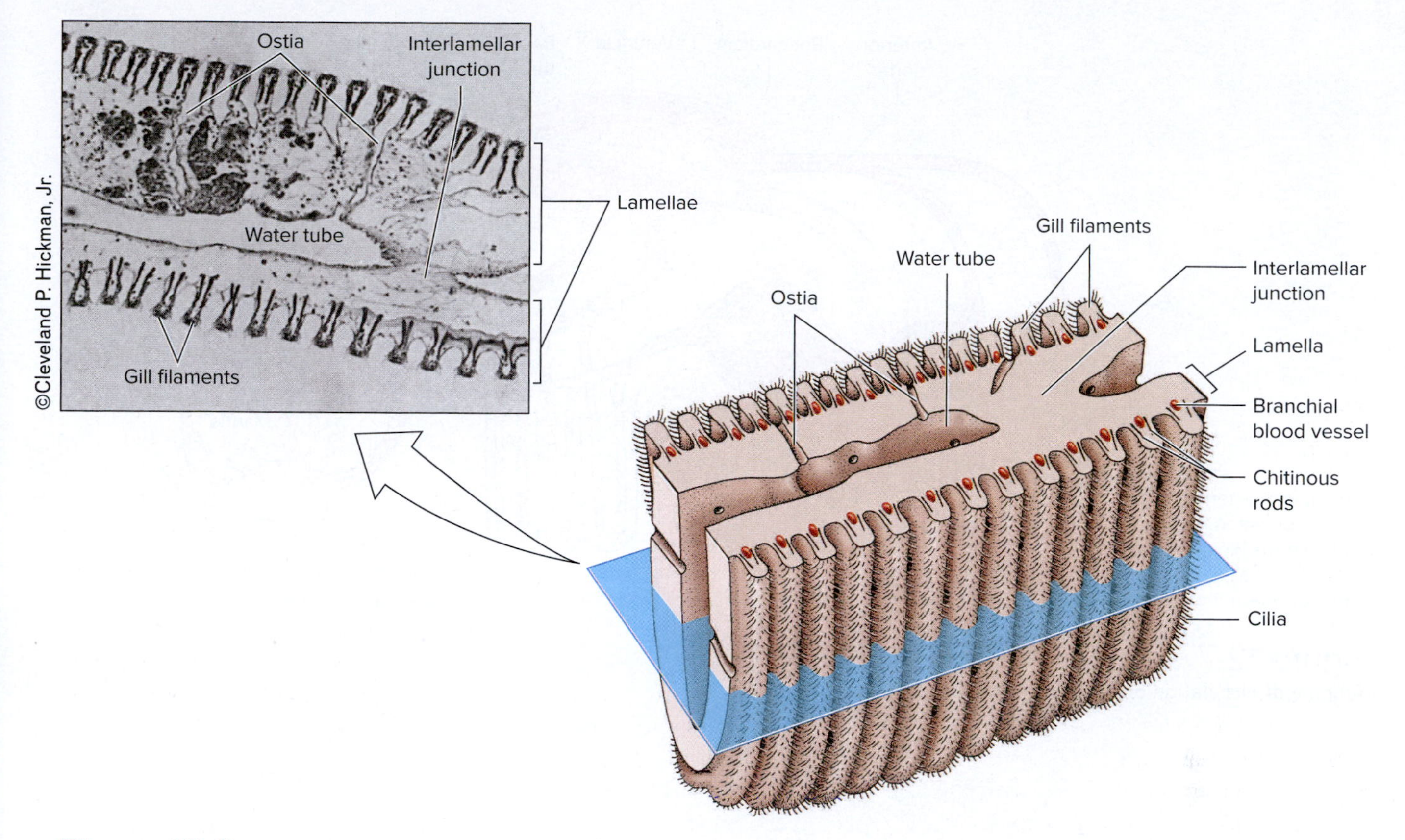

Figure 12-6

Cross section through a portion of the gill of a freshwater clam.

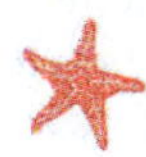

Carefully slit open the pericardium, using fine-tipped surgical scissors. Be cautious not to injure the delicate heart inside.

The three-chambered heart is composed of a single **ventricle** and a pair of **auricles** (see Figure 12-4).

If the ventricle is still beating, count the rate (beats/minute).

Note that the ventricle surrounds the intestine. Two aortae leave the ventricle; the **anterior aorta** passes to the visceral mass and intestine, and the **posterior aorta** runs along the ventral side of the rectum to the mantle.

The paired auricles are fan-shaped and very thin-walled. Pass a probe under the uppermost (left) auricle and lift carefully to see its connection to the ventricle.

At your instructor's option, you can inject a small amount of carmine solution into the ventricle using a tuberculin syringe and a small (26 or 27 gauge) hypodermic needle. The carmine solution will flow into and reveal the two aortae.

The pericardial space around the heart is a part of the **coelomic cavity,** greatly reduced in molluscs.

Almost all molluscs (most cephalopod molluscs are an exception) have an **open circulatory system** (Figure 12-7). In a clam the two aortae carry blood into smaller and smaller vessels and finally into sinus spaces that bathe the tissues directly. There are no capillaries present in a closed circulatory system. However, the sinuses are not spacious cavities; rather, they are narrow clefts and channels in connective tissue surrounding the muscular and nervous tissues. Although the system is "open," it functions much like a system of capillaries.

Coelom

Although molluscs have a true coelom, it is small. The pericardial cavity is part of the coelom, as is the small space around the gonads. A true coelom, you recall, is distinguished from other cavities by being lined with epithelium that arises from the mesoderm. What is the name of the lining of the coelom? ______________________________

Excretory System

A pair of dark kidneys lie under the floor of the pericardial sinus. They are roughly U-shaped tubes. The kidneys pick up waste from blood vessels, with which they are richly supplied, and from the pericardial sinus, with which they connect. Waste is discharged into the suprabranchial chamber and carried away with the exhalant current.

Digestive System

Locate again the **labial palps** and **mouth.**

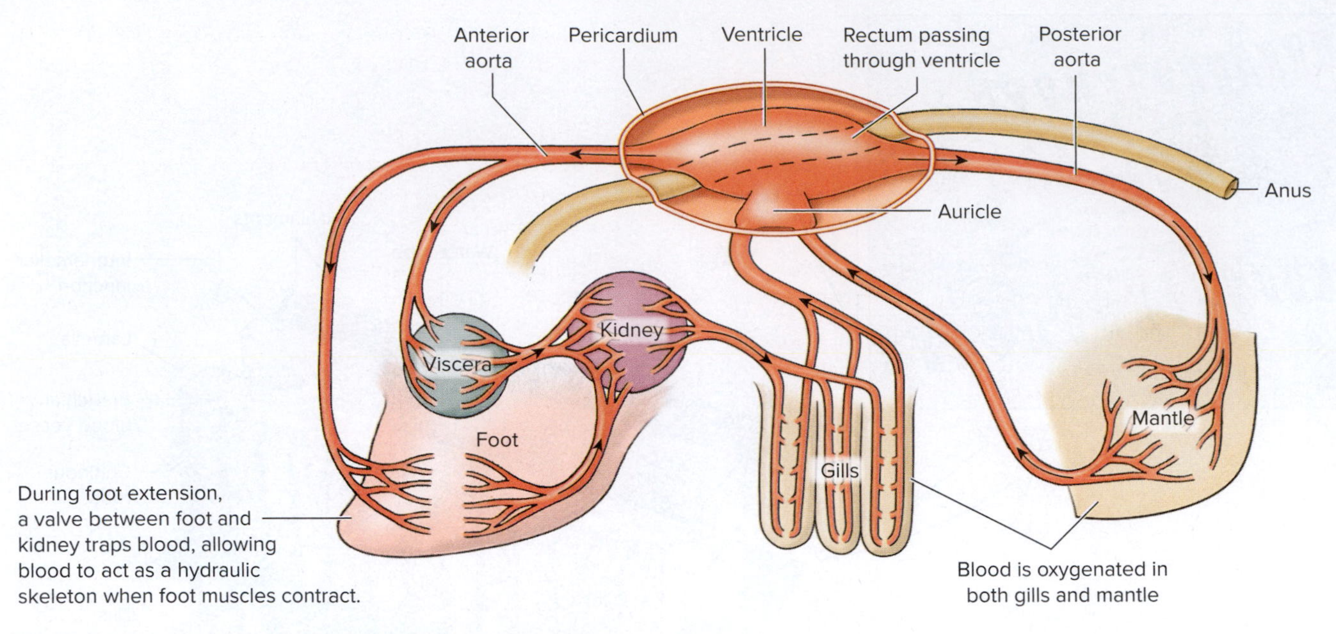

Figure 12-7
Scheme of circulation of a freshwater clam.

To reveal the alimentary canal, cut through the surface tissue on one side of the visceral mass and foot and strip away the epithelium (see Figure 12-2D).

The mouth leads into a short **esophagus** that widens into the **stomach,** surrounded by greenish brown **digestive glands.** The stomach narrows into a tubular **intestine,** which can be seen looping back and forth through the visceral mass. The intestine connects to the rectum, seen earlier passing through the ventricle. Trace the rectum as it passes dorsal to the posterior adductor muscle to its end, the **anus,** which empties feces into the exhalant current. Surrounding the intestine is light brown tissue of the **gonad** (see Figure 12-4).

In freshly collected clams, you may find a solid, gelatinous rod, the **crystalline style,** projecting into the stomach. It is composed of mucoproteins and digestive enzymes (chiefly amylase), which are released into the food. The crystalline style disappears within a few days after clams are collected and usually is absent from specimens purchased from biological supply houses. Digestion is mostly intracellular.

Reproductive System

Sexes are separate but are difficult to distinguish, except by the swollen gills of the pregnant female. The **gonads** (**ovaries** or **testes**) are a brownish mass of minute tubes filling the space between the coils of the intestine.

Make a wet mount of gonadal tissue and determine whether there are eggs or sperm in it.

The gonads discharge their products into the suprabranchial chamber. Spermatozoa pass into the surrounding water. They enter a female with the inhalant current and fertilize eggs in the suprabranchial chamber. The **zygotes** settle into the water tubes of the outer gill (brood pouch), where each zygote develops into a tiny bivalved larval form known as a **glochidium** (Gr. *glochis,* point, + *idion,* dim.) (found only in freshwater clams) (Figure 12-8). Glochidia, about the size of dust particles, escape through the excurrent siphon. Glochidia have valves bearing hooks, by which they fasten themselves to gills, fins, or skin of a passing fish. Here they encyst and live as parasites for several weeks. Carried by their fish host, the larvae are dispersed upstream into lakes and rivers, enabling bivalve mollusc populations to invade and flourish in virtually all freshwater environments. After a growth period, the young clams break loose and sink to the bottom sand, where they develop as free-living adults.

If a female with a swollen brood pouch is available, make a wet mount of some gill contents and examine with a microscope to determine whether eggs or glochidia are present.

Nervous System

The nervous system of a clam is not highly centralized. Dissection of the nervous system is difficult and often impractical. Three pairs of **ganglia** (small groups of nerve cells) are connected to each other by nerves. **Cerebropleural ganglia** are found one on each side of the esophagus on the posterior

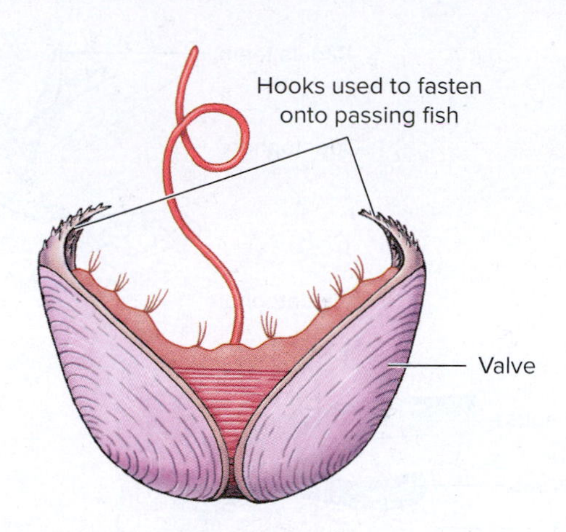

Figure 12-8
Glochidium, larval form of freshwater bivalve molluscs. Each glochidium is about 0.3 mm in diameter.

surface of the anterior adductor muscle. **Pedal ganglia** are fused and are found in the anterior part of the foot. **Visceral ganglia** are fused into a star-shaped body just ventral to the posterior adductor muscle. They are covered by a yellowish membrane and are connected to the cerebropleural ganglia by nerves.

Sense organs are poorly developed in clams. They are involved with touch, chemical sensitivity, balance, and light sensitivity. They are most numerous on the edge of the mantle, particularly around the incurrent aperture, but you will not be able to see them. However, in scallops, such as *Pecten,* the ocelli are large and numerous, forming a distinctive row of steel-blue "eyes" along the edges of each mantle.

Oral Report

Be prepared to demonstrate your dissection and explain the clam's structures and their functions.

Build Your Knowledge of Bivalve Mollusc Features

Test your understanding of the main features of the Bivalvia covered in the exercise by filling out the summary table on pp. xi–xii. Try to complete the table without looking at your notes.

Projects and Demonstrations

1. *Cross section of entire clam after removal from the shell.* This study is made from prepared slides. What you will see in it depends to some extent on the region through which the body was cut. Identify **mantle, mantle cavity, gills, lamellae, water tubes, intestine, foot, suprabranchial space, gonad,** and other structures revealed in the cross section.
2. *Oysters, scallops, and sea clams.* Examine the shells of oysters, scallops, and sea clams.
3. *Shipworm,* Teredo.* Examine a piece of wood into which the wormlike molluscs, *Teredo,* have bored. Examine a preserved specimen of the shipworm. Note the small valves, the small body, and the prolonged siphon that make up the bulk of the shipworm.
4. *Prepared stained slides of glochidia larvae.* Note especially the valves and muscles. Do they have **adhesive threads** between the valves?
5. *Female clam with glochidia in the gills.* Preserved specimens of clams with brood pouches (marsupia) may be studied. Which of the gills serve as brood pouches? How numerous are the glochidia?
6. *Effect of temperature on ciliary action of clam gill.* Remove a piece of bivalve gill and pin out on a layer of wax in a glass dish. Support the gill on two small glass rods approximately 1 cm apart. Place a 1 mm disc of aluminum foil on the surface of the gill and time its progress across a given distance. Cover with ice-cold water. Record rate of transport. As the water warms to room temperature, record rate at intervals of 5°. Judicial additions of slightly warmer water may also be tried. Make several determinations at each temperature and average. Make a diagram plotting mean rates of transport (millimeters/minute [mm/min]) against degrees of temperature.
7. *Experiments with the clam heart.* The heartbeat can be vividly demonstrated by passing a thread under the auricle and recording with a heart transducer and suitable chart recorder. 5-hydroxytryptamine is excitatory and acetylcholine is inhibitory in clam hearts. Both are applied directly on the exposed heart with a pipette as 1:10,000 solutions. Flush with saline (frog Ringers diluted 1:6 with water) after noting the effect of each drug.

EXERCISE 12B

Core Study

Class Gastropoda—Pulmonate Land Snail

Land Snail

Phylum Mollusca
Class Gastropoda
Subclass Pulmonata
Superorder Stylommatophora
Genus *Helix, Polygyra,* others

*See Lane, C. E. 1961. The teredo. Sci. Amer. (Feb.); see also interesting Wikipedia entries to shipworms on the Web.

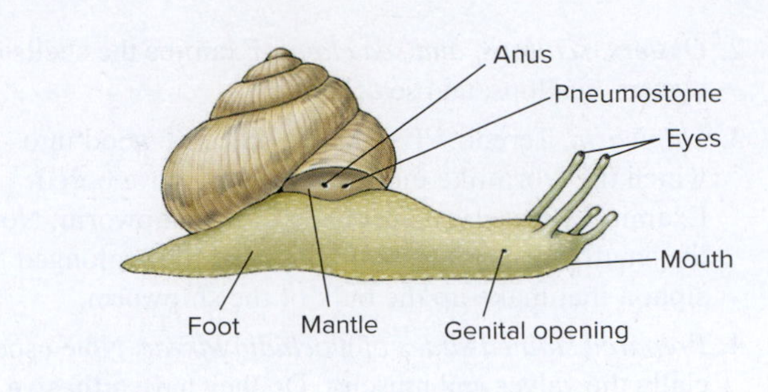

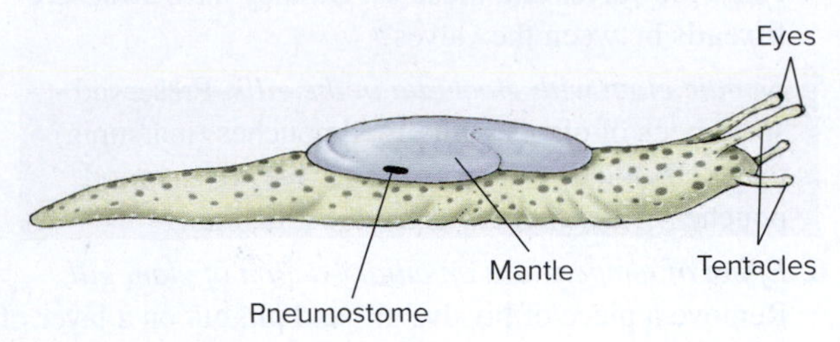

Figure 12-9
Common pulmonates. **Top**, *Helix*, a common land snail. **Bottom**, *Limax*, a common garden slug.

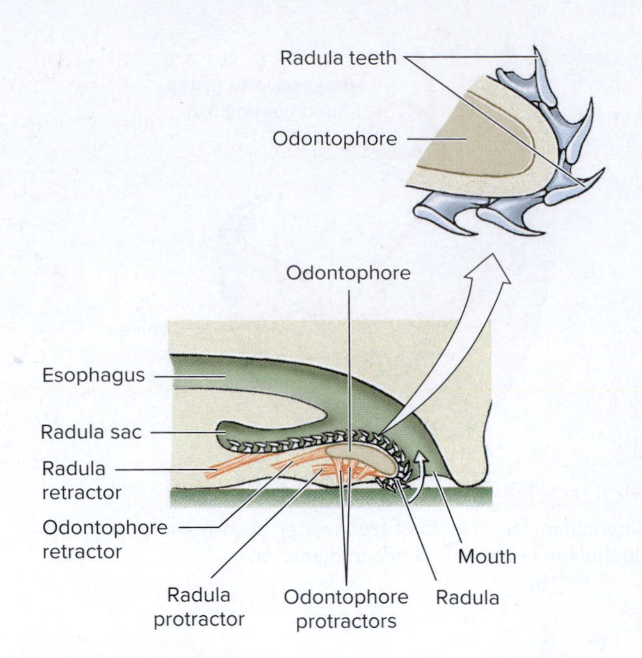

Figure 12-10
Diagrammatic longitudinal section of a gastropod head, showing the radula and radula sac. The radula moves back and forth over the odontophore cartilage. As the animal grazes, the mouth opens, the odontophore is thrust forward against the substratum, the teeth scrape food into the pharynx, the odontophore retracts, and the mouth opens. The sequence is repeated rhythmically.

Where Found

Most land snails (Figure 12-9) prefer fairly moist habitats. They are common in wooded areas, where they spend their days in the damp leaf mold on the ground, coming out to feed on the vegetation at night. *Helix* (Gr. *helix,* twisted) (Figure 12-9, top) is a large snail from southern Europe, and introduced all around the world. The most common species, *H. aspera*, has become a garden pest, especially in California. *Polygyra* (Gr. *poly*, many, + *gyros,* round) is an American land snail with approximately 30 recognized species.

Subclass Pulmonata constitutes one of three major groups of gastropods. Two other subclasses—Prosobranchia and Opisthobranchia—contain mainly marine snails and marine nudibranchs and tectibranchs.

Behavior

Place a land snail on a moistened glass plate and invert the plate over a dish, so that you can watch the ventral side of the foot under a dissecting microscope.

Do you see waves of motion? Can you see evidence of mucus? Of ciliary action? __________ If you are fortunate, you may see the action of the **radula** (L., scraper) in the mouth. The radula is a series of tiny teeth attached to a ribbonlike organ that moves rapidly back and forth with an action like that of a rasp or file (Figure 12-10).

Prop up the plate in a vertical position. In which direction does the snail move? __________ Now rotate the plate 90°, so that the snail is at right angles to its former position. When it resumes its travels, in which direction does it move? __________ Now rotate the plate again and observe.

Do you think the snail is influenced in its movements by the forces of gravity (geotaxis; refer to p. 111 for an explanation of taxes and the difference between a positive and a negative taxis)? ____________ Try several snails. Do they respond in a similar manner?

Place a snail in a finger bowl and, using the following suggestions, observe its behavior.

Place a piece of lettuce leaf near a snail and see whether it will eat. How long does it take the snail to find the food? __________ Note the **head** with its **tentacles.** How do the two pairs of tentacles differ? __________ Where are the **eyes** located? __________ Touch an anterior tentacle gently and describe what happens. __________ Touch some other parts of the body and note results. Pick up a snail very carefully from the glass on which it is moving and observe exactly what it does.

The Shell

Examine the shell of a preserved specimen. Note the nature of the spiral shell. Is it symmetrical? ____________ To what part of the clam shell does the **apex** correspond? __________ The body of the snail extends through the **aperture.** A **whorl** is one complete spiral turn of the shell. On the whorls are fine **lines of growth** running parallel to the edge of the aperture. Holding the apex of the shell upward and the aperture toward you, note whether the aperture is at the right or left. **Dextral,** or right-handed, shells will have the aperture toward the right and **sinistral,** or left-handed, shells will have it toward the left (Figure 12-11). Is your snail dextral or sinistral? __________ Examine a piece of broken shell for the characteristic three layers—the outer **periostracum,** the middle **prismatic layer,** and the shiny inner **nacre.**

Further Study

Surface Anatomy

If the snail has not contracted enough during killing and preservation* to allow you to slip it carefully from the shell, you may use scissors to cut carefully around the spiral between the whorls, removing pieces of shell as you go and leaving only the central parts, or **columella** (L., pillar). Try not to damage the coiled part of the visceral mass.

The body is made up of **head,** muscular **foot,** and coiled **visceral hump** (Figure 12-12). Identify the **tentacles, eyes,** ventral **mouth** with three lips, and **genital aperture** just above and behind the right side of the mouth. The foot bears

*See Appendix A, p. 405 for a method for relaxing snails for study.

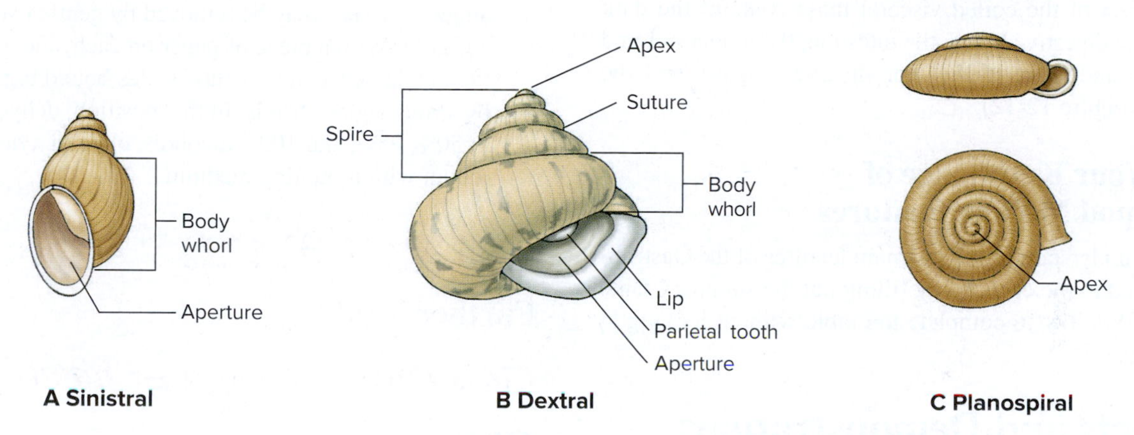

Figure 12-11

Structure of a snail shell. **A,** *Physa,* a sinistral, or left-handed, freshwater snail with lymnaeiform shell (height exceeds width). **B,** Mesodon (= *Polygyra*), a dextral (right-handed) snail with heliciform shell (width exceeds height). **C,** *Helicodiscus,* a land snail with planospiral, or flattened, spire (two views).

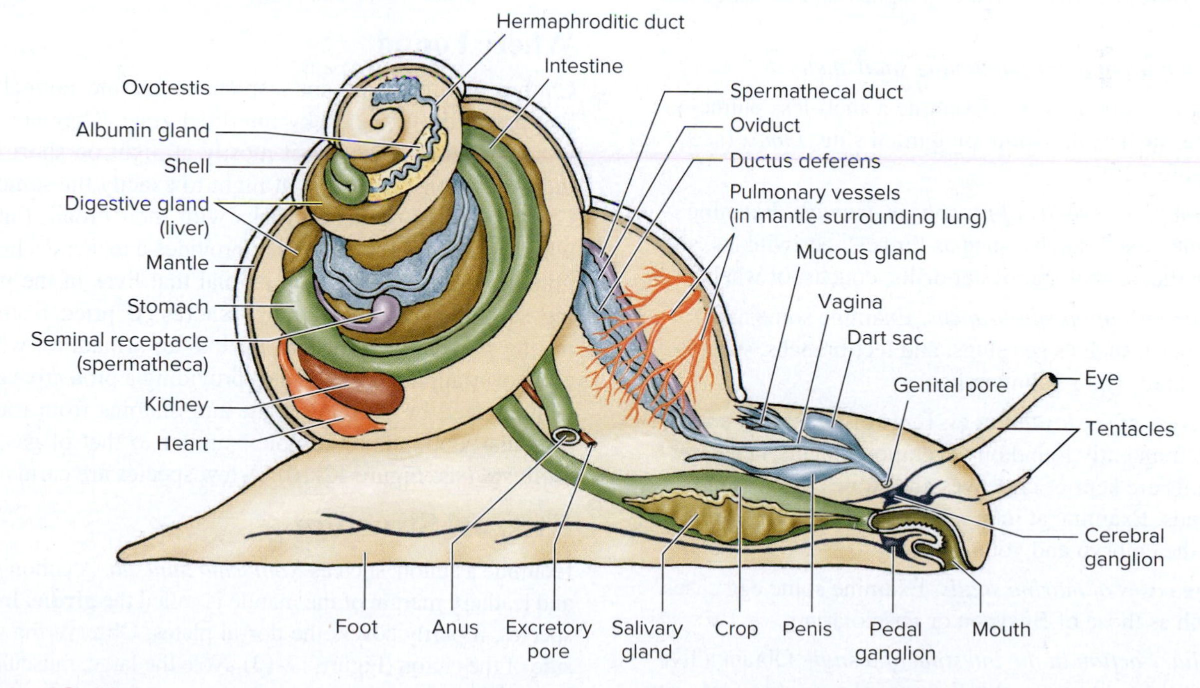

Figure 12-12

Anatomy of a pulmonate snail (diagrammatic).

a **mucous gland** just below the mouth. Mucous secretions aid in locomotion.

Note the thin **mantle** that covers the visceral hump and forms the roof of the **mantle cavity.** It is thickened anteriorly to form the **collar** that secretes the shell.

Find a small opening, the **pneumostome** (Gr. *pneuma,* air, + *stoma,* mouth), under the edge of the collar (see Figure 12-9). It opens into a highly vascular portion of the mantle cavity, located in the first half-turn of the spiral, which serves as a respiratory chamber **(lung)** in pulmonates. Here diffusion of gases occurs between air and blood. Oxygen is carried by the pigment hemocyanin. Most aquatic gastropods possess gills. The mantle cavity in the second half-turn contains the heart and a large kidney.

The rest of the coiled visceral mass contains the dark lobes of the digestive gland, the intestine, the lighter-colored albumin gland (part of the reproductive system), and the ovotestis (Figure 12-12).

Build Your Knowledge of Gastropod Mollusc Features

Test your understanding of the main features of the Gastropoda covered in the exercise by filling out the summary table on pp. xi–xii. Try to complete the table without looking at your notes.

Projects and Demonstrations

1. *Example of a pond snail.* Watch a pond snail attached to the glass side of an aquarium. Note the broad foot by which it clings to the glass. Can you see the motion of the **radula** as the animal eats algae that has settled on the glass?
2. *Examples of other pulmonate snail shells.*
3. *Shell-less pulmonate.* Examine a shell-less pulmonate, such as the common garden slug, *Limax* (see Figure 12-9).
4. *Examples of shells of prosobranch snails.* Examine some prosobranchs, such as limpets, periwinkles, slipper shells, abalone, oyster drills, conchs, or whelks.
5. *Examples of opisthobranchs.* Examine some nudibranchs, such as sea slugs, and tectibranchs, such as sea hares and sea butterflies.
6. *Masses of fresh snail eggs.* Eggs of freshwater snails are frequently found on vegetation in aquaria where snails are kept or on leaves and stones in streams and ponds. Examine at intervals to follow the development of the embryo and young snails.
7. *Egg cases of marine snails.* Examine some egg cases, such as those of Busycon or Fasciolaria.
8. *Ciliary action in the intestine of a snail.* Obtain a live aquatic snail (preferably *Lymnaea stagnalis*). Cut open the shell and remove the viscera. Carefully slit open the body and remove the intestine. Slit the intestine and place a small portion, intestinal surface up, on a slide in a drop of saline solution. Examine with high power of your microscope. Note the progressive undulations of cilia over the surface. Ciliary action is best seen some time after you have made the preparation. Are the cilia independent of nerve action? ____________ Place a drop of warm (45° C) saline solution on the preparation and note what happens. What is the function of cilia in this region? ____________________
9. *Isolation of the radula of a snail.* Cut off the head of a snail and soak it in a 10% solution of KOH for 2 or 3 days or until the soft tissues are destroyed and only the radula remains. Transfer the radula to water and wash for 1 or 2 hours in running water. Pieces of attached tissues may be removed by gentle teasing with a needle. With a piece of paper on each side, place the radula between two glass slides bound together by strong rubber bands. In this position, dehydrate it in 50%, 95%, and 100% alcohols, clear in xylol, and mount with mounting medium.

EXERCISE 12C

Further Study

Class Polyplacophora—Chitons

Chitons

Phylum Mollusca
  Class Polyplacophora
    Family Mopaliidae
      Genus *Katharina,* others

Where Found

Chitons (kyt′ens) are inconspicuous marine animals that live primarily in the rocky intertidal zone. They are "stay-at-homes" that move about mostly at night on short foraging expeditions, returning at night to exactly the same spot. Chitons adhere tightly to rocks with their broad, flat foot, which forms a suction cup that provides a tenacious hold, an important adaptation for an animal that lives in the pounding surf of high-energy, rocky shores. If pried from their resting place, they will curl up like an armadillo, with the eight overlapping dorsal plates providing a protective armor. Chitons feed by scraping algae and diatoms from rocks by a radula-odontophore complex similar to that of gastropod molluscs (see Figure 12-10). A few species are carnivorous.

External Structure

Examine a chiton, such as *Katharina tunicata.* A chiton's thick and leathery margin of the mantle is called the **girdle.** In many species, it partly covers the dorsal plates. Observe the ventral side of the chiton (Figure 12-13). Note the large, muscular foot on which the animal moves by waves of muscular activity, as in gastropods. The **head,** with its central **mouth,** is easily recognized anterior to the foot but is not highly developed. It

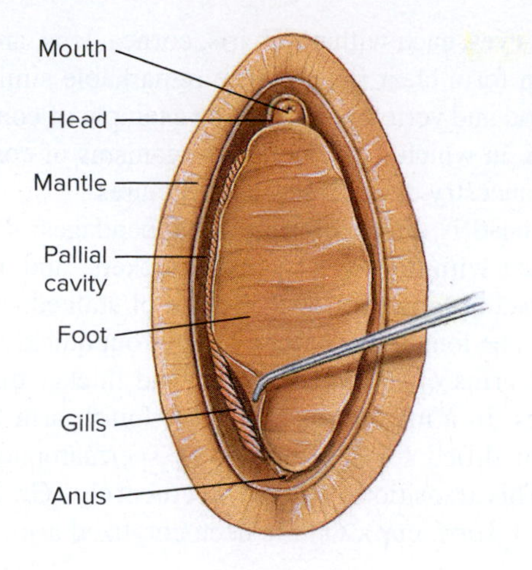

Figure 12-13
Ventral view of the chiton *Katharina tunicata.*

bears no tentacles or eyes but has a subradular organ that can be extended from the mouth and is thought to be a taste organ (chemosensory). On each side of the body is a distinct groove that separates the foot and head from the mantle; this is the **mantle cavity.** Within this groove lies a single row of many delicate **gills,** or **ctenidia.** Cilia on the gills move a stream of water in a posterior direction; the water emerges from a small median cleft in the girdle at the posterior end. Locate the **anus** at the posterior end of the mantle cavity. Just anterior to the anus are two excretory pores, one on each side of the foot. In *Katharina* these are located 1 to 2 cm anterior to the anus. Just anterior to each excretory pore is a reproductive pore.

The internal anatomy will not be examined in this exercise. The chiton has a long, coiled intestine with primarily extracellular digestion and a simple, ladderlike nervous system. Most chitons have separate sexes, external fertilization, and a trochophore larval stage that feeds on plankton until it settles and transforms into a juvenile chiton.

Build Your Knowledge of Polyplacophoran Features

Test your understanding of the main features of the Polyplacophora covered in the exercise by filling out the summary table on pp. xi–xii. Try to complete the table without looking at your notes.

EXERCISE 12D

Core Study

*Class Cephalopoda—*Loligo, *the Squid*

Loligo

Phylum Mollusca
Class Cephalopoda
Subclass Coleoidea
Order Teuthoidea
Genus *Loligo*

Where Found

Squids and octopuses are marine animals. Active squids are free-swimming and are found in offshore waters at various depths, whereas octopuses are often found in shallow tidal waters.

Squids range in size from 2 cm up. The giant squid *Architeuthis* (Gr. *archaios,* ancient, + *teuthis,* squid) may measure 15 m from tentacle tip to posterior end. *Loligo* (L., a cuttlefish) averages about 30 cm.

Behavior

Because squids ordinarily do not survive long in aquaria, they may not be available for observation. If they are, notice the swimming movements. A squid can move swiftly, either forward or backward, by using its fins and ejecting water through its **funnel.** When water is ejected forward, the squid moves backward; when water is ejected backward, the animal moves forward—movement by jet propulsion. Even when it is resting, there is a gentle, rhythmical movement of the fins, a muscular movement that also aids in bringing water to the gills in the mantle cavity.

Notice the color changes in the integument caused by contraction and expansion of many **chromatophores** (pigment cells) in the skin. Each chromatophore is a minute organ filled with pigment granules and surrounded by a series of radial muscles. Contraction of the radial muscles stretches the entire organ to form a sheet of pigmented cytoplasm. When the muscles relax, the organ with its pigment shrinks to a small sphere. The pigments, called **ommochromes,** may be black, red, brown, yellow, or orange. Colors appear suddenly when the chromatophores expand and may disappear just as quickly when the chromatophores shrink. Thus, the colors of squids (and those of most other cephalopod molluscs, such as octopuses) can change rapidly. Elaborate color changes that may sweep quickly across a squid's body serve as a complicated and highly developed means of communication. Chromatophores of cephalopod molluscs are quite different from those of vertebrates (frogs, for example), which are irregularly shaped cells containing pigment granules that can be concentrated into a small area or dispersed throughout the cell. No muscles are involved and pigment changes are much slower than those of the cephalopods.

When the animal is attacked, it can emit a cloud of ink from its ink sac through its funnel.

Food is caught by a pair of retractile tentacles extended with lightning speed, then is held and manipulated by the eight arms and killed by a poison injection. Small fishes, shrimps, and crabs are favorite foods.

External Structure

Squids are called "head-footed" because the end of the head, which bears tentacles, is homologous to the ventral side of the clam, and the tentacles represent a modification of the foot. To orient the animal morphologically, hold it with its pointed end uppermost; this is the dorsal "side" of the animal. The head, with the arms, tentacles, and funnel, is the ventral surface. The mouth is on the anterior surface, and the funnel is posterior. Functionally, however, the animal swims horizontally by moving forward and backward. Thus, the *morphological* anterior surface is the *functional* dorsal side, and the *morphological* ventral surface is the *functional* anterior side. We will use the morphological orientation in the following directions, as they are shown in Figure 12-14.

Notice the streamlined body, with the **head** and **arms** at one end and a pair of **lateral fins** at the other (Figure 12-14). The visceral mass, dorsal to the head, is covered with a thick **mantle,** the free end of which forms a loosely fitting **collar** about the neck. The head bears a pair of complex, highly advanced **eyes,** each with pupil, iris, cornea, lens, and retina, which can form clear images. The remarkable similarity of cephalopod and vertebrate eyes is an example of **convergent evolution,** in which two groups of organisms of completely different ancestry develop similar structures.

The head is drawn out into 10 appendages: 4 pairs of **arms,** each with 2 rows of stalked suckers, and 1 pair of long, retractile **tentacles,** with 2 rows of stalked suckers at the ends. The long tentacles can be shot out quickly to catch prey. The arms of males are longer and thicker than those of females. In a mature male, the left fourth arm becomes slightly modified for the transfer of spermatophores to a female. This transition is called **hectocotyly** (Gr. *hekaton,* hundred, + *kotyl,* cup). On the hectocotylized arm, some of

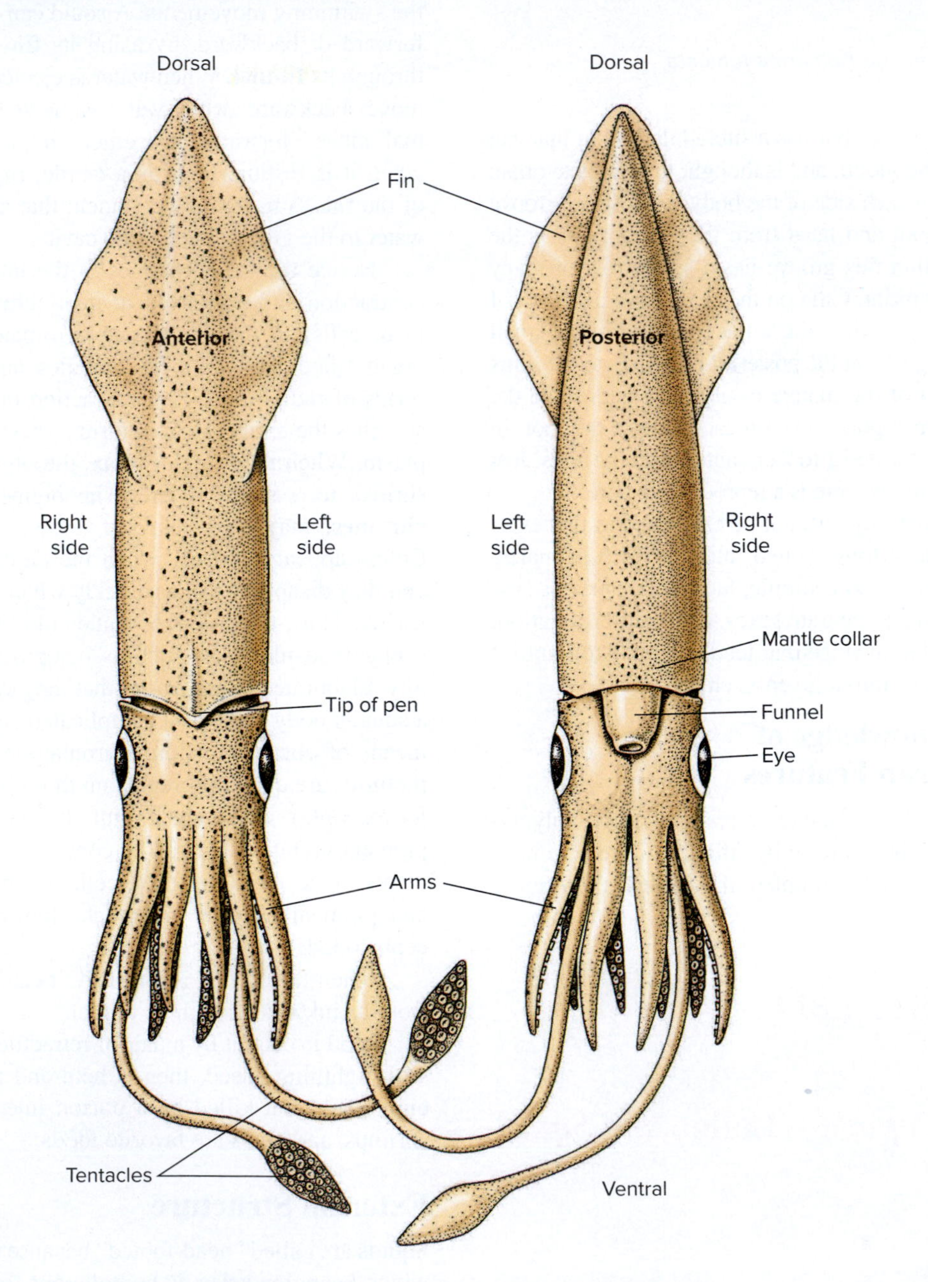

Figure 12-14

External structure of a squid. **Left,** anterior view. **Right,** posterior view.

the suckers are smaller and form an adhesion area for carrying the spermatophore.

The **mouth** lies within the circle of arms. It is surrounded by a **peristomial membrane,** around which is a **buccal membrane** with seven projections, each with suckers on the inner surface. Probe in the mouth to find two horny, beaklike **jaws.**

A muscular **funnel (siphon)** usually projects under the collar on the posterior side, but it may be partially withdrawn. Water forced through the funnel by muscular contraction of the mantle furnishes the power for jet propulsion locomotion. Wastes, sexual products, and ink are carried out by the current of water that enters through the collar and leaves through the funnel. The siphon, or funnel, of the squid is not homologous to the siphon of the clam; the clam siphon is a modification of the mantle, whereas the squid siphon, along with the arms and tentacles, is a modification of the foot.

The mottled appearance of the skin is caused by **chromatophores,** pigment cells.

Mantle Cavity

Beginning near the funnel, make a longitudinal incision through the mantle from the collar to the tip. Pin out the mantle and cover with water.

The space between mantle and visceral mass is the **mantle cavity.** The mantle itself is made up largely of circular muscles covered with integument. The funnel contains both circular and longitudinal muscles.

Locate the interlocking **pallial cartilages** (Figure 12-15). They help support the funnel and close the space between the neck and the mantle, so that water inhaled around the collar can be expelled only by way of the funnel. Lateral to the funnel, find large, saclike valves that prevent outflow of water by way of the collar.

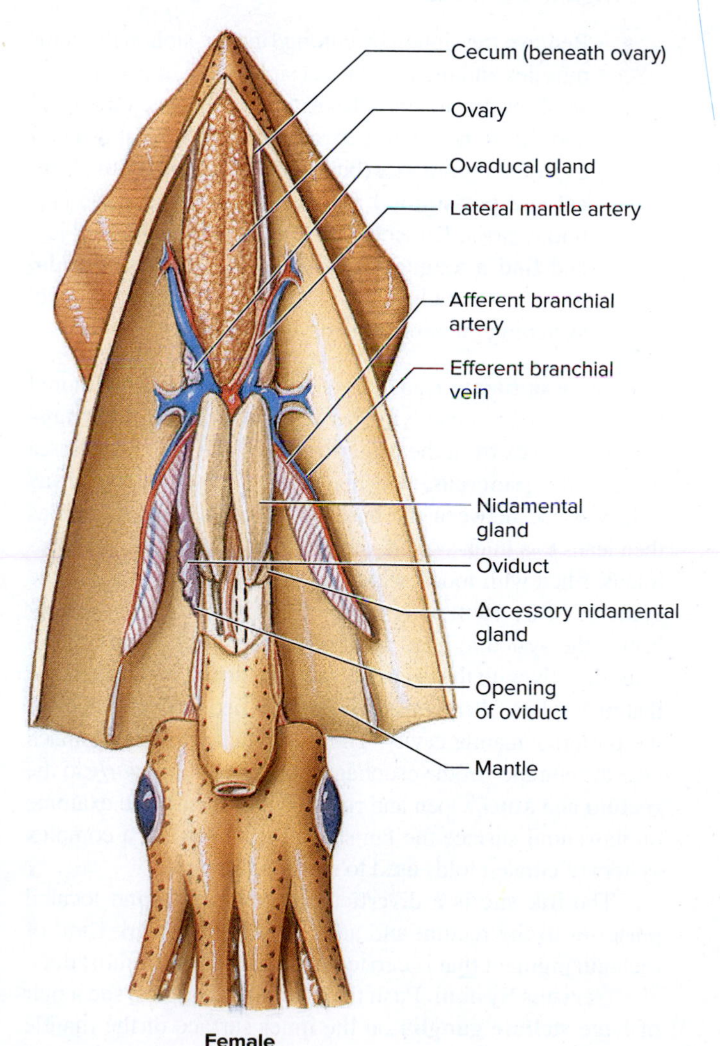

Figure 12-15
Posterior view of mantle cavity of a squid. **Left,** female. **Right,** male.

Slit open the funnel to see the muscular, tonguelike valve that prevents inflow of water through the funnel.

This valve allows a buildup of hydrostatic pressure in the mantle cavity before a jet stream of water is ejected through the funnel.

Locate a large pair of **funnel retractor muscles** and beneath them the even larger **head retractor muscles.** Locate the free end of the **rectum** with its **anus** near the inner opening of the funnel. Between it and the visceral mass is the **ink sac.** Do not puncture it. When a squid is endangered, it can send out a cloud of black ink through the funnel as it darts off in another direction.*

A pair of long **gills** is attached at one end to the visceral mass and at the other to the mantle (Figure 12-15). They are located so that water entering the mantle cavity passes directly over them. Unlike in bivalves, the gills are not ciliated. The mantle cavity is ventilated by action of the mantle itself. Contraction of the radial muscles in the body wall causes the wall to become thinner and the capacity of the mantle cavity to become greater, so that water flows in around the collar. Water is then expelled through the funnel when mantle muscles contract. This movement, also used in locomotion, permits very efficient ventilation of the gills.

In most introductory laboratories, observation of the squid will end with study of the mantle cavity. For the more advanced laboratory or for the enterprising student, a study of internal structure follows.

Further Study

Internal Structure

A thin skin covers the organs of the visceral mass and encloses the **coelom.** Remove this membrane carefully as you expose the visceral organs. If the specimen is a female, a pair of large, whitish **nidamental** (L. *nidamentum,* nesting materials) glands (which secrete the outer capsules of the egg masses) should be carefully removed (Figure 12-15). Note their location and lay them aside for later study.

Remove the ink sac by separating it carefully from the rectum using fine, sharp-pointed scissors. Do not puncture the ink sac.

Respiratory and Circulatory Systems. At the base of each **gill** is a small, whitish, bulblike **branchial heart** (gill heart) (Figure 12-15). Deoxygenated blood from the branchial heart is carried to the gill by the afferent **branchial artery,** and oxygenated blood is returned by an efferent **branchial vein** to the **systemic heart** (true heart), a larger, whitish organ lying between the branchial hearts. The systemic heart may be partially hidden by the more anterior **kidneys,** which have a spongelike texture, and the pancreas, seen later in dissection. Each branchial heart receives deoxygenated blood from a large, thin-walled, conical **posterior vena cava.** Each branchial heart also receives blood from a fork of the **anterior vena cava** (cephalic vein, not easily visible) and from a small **anterior mantle vein.** The systemic heart pumps oxygenated blood through the **cephalic aorta** (anterior, not visible in Figure 12-15) and the short **posterior aorta,** which branches immediately to form the **lateral mantle arteries** and the single **median mantle artery** (not shown in Figure 12-15; this artery extends to the mantle and will have been severed when the mantle cavity was opened).

Excretory System. A pair of **kidneys,** somewhat triangular in shape, of spongy texture, and usually white or pale in uninjected specimens, lies between and slightly anterior to the branchial hearts. The kidneys will have the color of the injection fluid, if your specimen was injected. A **renal papilla** lies at the anterior tip of each kidney.

Digestive System.

Remove the siphon by cutting first the siphon retractor muscles and then the lateral siphon valves and the two small protractor muscles extending between the head and the siphon. Cut between the two ventral arms to expose the **pharynx** (buccal bulb). Cut away the **buccal** and **peristomial membranes** to expose the chitinous **jaws.** Dissect away the overlapping lower jaw and find a tonguelike cartilage, bearing the **radula.** Remove the radula and examine under a microscope, sketching the arrangement of its minute teeth.

The **esophagus** leads down through the cream-colored **liver,** a soft, pale organ lying between the head retractor muscles. It emerges from the posterior end of the liver and passes through the **pancreas,** which is exposed by scraping away kidney tissue between the bases of the gills. The esophagus then leads to a thick-walled, muscular **stomach** (in some specimens, filled with food), which lies somewhat posterior to the systemic heart. Removing some of the pancreatic tissue now brings the systemic heart into view. The stomach communicates directly with the **cecum,** a large, delicate, thin-walled sac that may, when filled with partially liquefied food, fill much of the posterior mantle cavity. The **intestine** leaves the stomach near the entrance of the esophagus and passes *anteriorly* to the **rectum** and **anus.** Open and rinse out the cecum and examine on its ventral surface the fan-shaped **spiral valve,** a complex system of ciliated folds used to sort food particles.

The **ink sac** is a diverticulum of the intestine located posterior to the rectum and anus. It secretes a dark fluid of melanin pigment that is carried to the rectum by a short duct.

Nervous System. Push the head to one side to see a pair of large **stellate ganglia** on the inner surface of the mantle close to the neck. These ganglia function in movement of the mantle. From each ganglion, several large nerves radiate out over the inner mantle surface. Each nerve contains,

*Several videos of squids ejecting ink may be found on YouTube. See, for example, https://www.youtube.com/watch?v=aq0nt_iBOKI.

along with smaller fibers, one of the giant fibers, which are used in rapid maximal contraction of the mantle. Directions will not be given here for dissection of the brain, which is composed of ganglia lying partly above and partly below the esophagus.

Sense Organs. Sense organs of cephalopods are highly developed. The **eyes** are capable of forming an image. Remove the thin, outer, transparent integument **(false cornea)** to uncover the **true cornea.** Cut away the cornea to observe the circular **iris diaphragm.** Behind the iris is an almost spherical **lens,** suspended by a **ciliary muscle.** Remove the lens to see the darkly pigmented sensory lining **(retina)** of the optic cavity.

A fold of tissue behind each eye and somewhat covered by the edge of the mantle contains a sensory area assumed to be olfactory in function. Sensory cells are numerous in the skin, particularly in the rims of the suckers. **Statocysts** are found embedded in the cartilages on each side of the brain.

Reproductive Organs. In males, the **testis** is an elongated, light-colored organ in the posterior coelomic cavity, usually partly concealed by the cecum (Figure 12-15). Sperm are shed into the coelomic cavity from an opening in the testis and swept by ciliary currents into the enlarged free end of the **vas deferens.** Sperm pass through the convoluted vas deferens to an enlarged, convoluted tubule, the **spermatophoric gland.** Here sperm are packaged into spermatophores (capsules that enclose the sperm), which are stored in the **spermatophoric sac** (not shown in Figure 12-15). The spermatophoric sac lies posterior to the spermatophoric gland and adjacent to the vas deferens. During copulation, the fingerlike processes at the tip of the hectocotylized arm take the spermatophores from the **penis** and transfer them to the female.

In females, the **nidamental glands** (removed earlier) are conspicuous white organs filling most of the lower part of the mantle cavity. The **ovary** lies posterior and sheds eggs into the coelomic cavity. Push the ovary to one side and try to locate the **oviduct** (it may be covered by the cecum). Near the left branchial heart, the oviduct enlarges into the **oviducal gland,** which secretes the egg shell. The oviduct continues anteriorly beside the nidamental glands to its flared opening, the **ostium,** in the mantle cavity. A mature female has a small pouch, or **sperm receptacle,** on the buccal membrane in the median ventral line, where a male may place the spermatophore. In the process of mating, the male may thrust the spermatophores inside the female's mantle cavity or into the sperm receptacle.

A female uses one of her arms to pick up strings of eggs as they come from her siphon, fertilizes them with spermatozoa from the pouch, and then attaches the strings to some object in the sea. The young hatch in 2 to 3 weeks.

Skeletal System. Dissect out the chitinous **pen** that lies dorsal to the visceral organs and extends from the free edge of the collar to the apex of the mantle. There are also a number of **cartilages** in the head, near the siphon, and in the mantle.

Build Your Knowledge of Cephalopod Features

Test your understanding of the main features of the Cephalopoda covered in the exercise by filling out the summary table on pp. xi–xii. Try to complete the table without looking at your notes.

Demonstrations

1. *Microslides showing spermatophores of* Loligo
2. *Preserved octopuses and cuttlefish* (Sepia)
3. *Shells of* Nautilus
4. *Dried cuttlebone of* Sepia
5. *Dissection of an injected cephalopod to show circulatory system*
6. *Dissection of a cephalopod brain*
7. *Living cephalopod, if available*

Classification

Phylum Mollusca

This abbreviated classification is limited to classes surveyed in this exercise. Consult your text for a complete mollusc classification.

Class Polyplacophora (pol′y-pla-kof′o-ra) (Gr. *polys,* many, + *plax,* plate, + *phora,* bearing). Chitons. Elongated, dorsally flattened body with reduced head; bilaterally symmetrical; radula present; shell of seven or eight dorsal plates; foot broad and flat; gills multiple, along sides of body between foot and mantle edge; sexes usually separate. Examples: *Katharina, Mopalia.*

Class Gastropoda (gas-trop′o-da) (Gr. *gaster,* belly, + *pous, podos,* foot). Snails, slugs, conchs, whelks, and others. Body asymmetrical, usually in a coiled shell (shell uncoiled or absent in some); head well developed, with radula; foot large and flat; one or two gills or with mantle modified into secondary gills or lung; dioecious or monoecious. Examples: *Busycon, Physa, Helix, Aplysia.*

Class Bivalvia (bi-val′vi-a) (L. *bi,* two, + *valva,* valve). Bivalves. Body enclosed in a two-lobed mantle; shell of two lateral valves of variable size and form, with dorsal hinge; cephalization much reduced; no radula; foot usually wedge-shaped; gills platelike; sexes usually separate. Examples: *Anodonta, Venus, Tagelus, Teredo.*

Class Cephalopoda (sef′a-lop′o-da) (Gr. *kephalē,* head, + *pous, podos,* foot). Squids, nautiloids, and octopuses. Shell often reduced or absent; head well developed with eyes and radula; foot modified into arms or tentacles; siphon present; sexes separate. Examples: *Loligo, Octopus, Sepia.*

NOTES

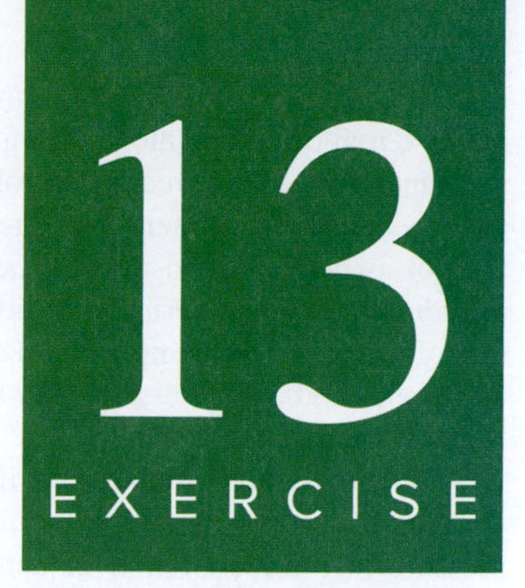

The Annelids

Phylum Annelida

The annelids include a variety of earthworms, leeches, and marine polychaetes. Their various adaptations allow for freshwater, marine, terrestrial, and parasitic living. They are typically elongate, wormlike animals; are circular in cross section; and have muscular body walls. The most distinguishing characteristic that sets them apart from other wormlike creatures is their **segmentation.** They are often referred to collectively as the "segmented worms." This repetition of body parts, also called **metamerism** (me-ta′me-ri′sum; Gr. *meta,* between, + *meros,* part), not only is external but also is seen internally in the serial repetition of body organs.

EXERCISE 13A
Errantia—Clamworm
Nereis
Projects and Demonstrations

EXERCISE 13B
Sedentaria—Earthworm
Lumbricus, Common Earthworm

EXERCISE 13C
Class Hirudinida—Leech
Hirudo, Medicinal Leech

CLASSIFICATION
Phylum Annelida

EXPERIMENTING IN ZOOLOGY
Behavior of Medicinal Leeches, Hirudo medicinalis

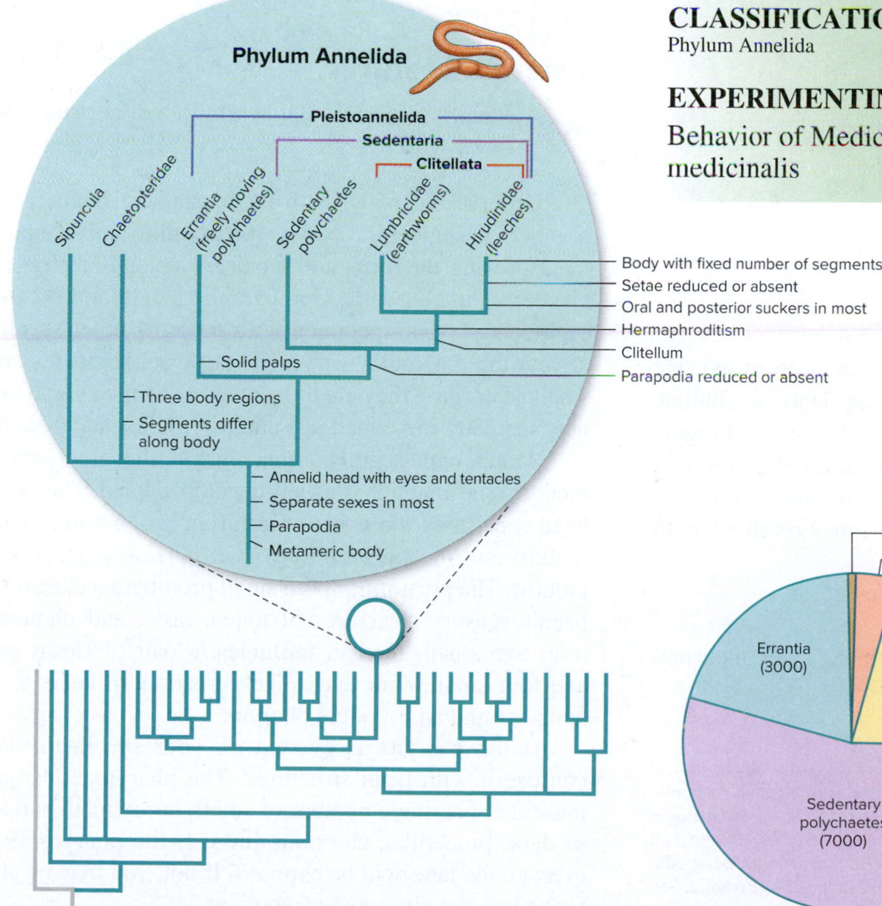

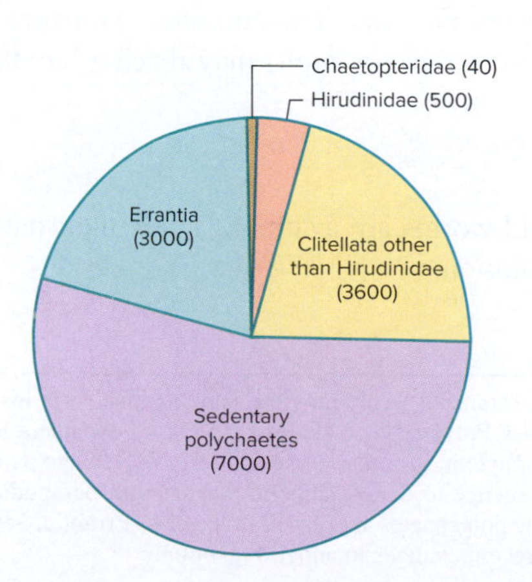

The evolution of segmentation was of great significance because it allowed heightened complexity of body structure and function, such as greater control of movement and more advanced nervous and sensory systems. Such advancements are not as noticeable in annelids as in arthropods, but the introduction of metamerism coincided with the rapid evolution of much greater complexity of structure and function in arthropods and chordates, the only other phyla emphasizing segmentation.

Division of the coelomic cavity into fluid-filled compartments also has increased the usefulness of hydrostatic pressure in the locomotion of annelids. The coordination between their well-developed neuromuscular system and more efficient hydrostatic skeleton makes annelids proficient in swimming, creeping, and burrowing.

Annelids have a complete mouth-to-anus digestive tract with muscular walls, so that digestive tract movements are independent of body movements. There is a well-developed closed circulatory system with pumping vessels, a high degree of cephalization, and an excretory system of nephridia. Some annelids have respiratory organs.

EXERCISE 13A

Core Study

Errantia (errant polychaetes)— Clamworm*

Nereis

Phylum Annelida
Class Errantia
Order Phyllodocida
Family Nereididae
Genus *Nereis*

Where Found

Clamworms (also called sandworms or ragworms) are strictly marine. They live in the mud and debris of shallow coastal waters, often in burrows lined with mucus. Largely nocturnal, they usually are concealed by day under stones, in coral crevices, or in their burrows. The common clamworm *Nereis virens* (Gr. *Nereis,* a sea nymph) may reach a length of 0.5 m.

Behavior

If living nereid worms are available, study their patterns of locomotion.

When the animal is quiescent or moving slowly, note that the **parapodia** (Gr. *para,* beside, + *pous, podos,* foot) (lateral appendages) undergo a circular motion that involves an effective stroke and a recovery stroke, each parapodium tracking an ellipse during each two-stroke cycle. In the effective stroke, the parapodium makes contact with the substratum, lifting the body slightly off the ground. The two parapodia of each segment act alternately, and successive waves of parapodial activity pass along the worm.

For more rapid locomotion, the worm uses undulatory movements of the body produced by muscular contraction and relaxation in addition to parapodial action. As the parapodia on one side move forward in the recovery stroke, longitudinal muscles on that side contract; as the parapodia sweep backward in their effective stroke, the muscles relax. Watch a worm in action and note these waves of undulatory movements. Can the worm move swiftly? Does it seek cover? Does it maintain contact with substratum, or does it swim freely?

Place near the worm a glass tube with an opening a little wider than the worm. Does the worm enter it? Why? Place a bit of fresh mollusc or fish meat near the entrance of the tube. You may be able to observe feeding reactions.

Written Report (Optional)

Record your observations on separate paper.

External Features

Place a preserved clamworm in a dissecting pan and cover with water.

Observe the body, with its specialized **head,** variable number of segments bearing **parapodia,** and caudal segment bearing the **anus** and a pair of feelers, or **cirri** (sing. **cirrus;** sir′us; L., curl). Compare the length and number of segments of your specimen with those of other specimens at your table. Do they vary? The posterior segments are the smallest because they are the youngest. As the animal grows, new segments are added just anterior to the caudal segment.

Along with segmentation, one of the great advancements of the annelids is a well-developed head. The two-part head consists of the **prostomium** (Gr. *pro,* before, + *stoma,* mouth) and the **peristomium** (Gr. *peri,* around, + *stoma,* mouth). The prostomium is a small protuberance bearing the head's sensory structures of touch, taste, and photoreception: two small, median **tentacles,** a pair of fleshy **palps,** and four small, dark **eyes.** The peristomium bears sensory tentacles and an eversible pharynx.

If the pharynx is everted on your specimen, do not confuse it with head structures. The pharynx is large and muscular, bearing a number of small, horny teeth and a pair of dark, pincerlike, chitinous **jaws.** If the pharynx is fully everted, the jaws will be exposed. If not, you may be able to probe into the pharynx to find them.

---

*The Errantia include the errant, or freely moving, annelids that were historically placed in the class Polychaeta. Although the class Polychaeta is now recognized as a paraphyletic group, we use the term polychaete as a descriptive term of convenience to denote annelid worms with parapodia and many setae. The motile polychaetes are placed in the class Errantia. See p. 216 for explanation of recent changes in annelid taxonomy.

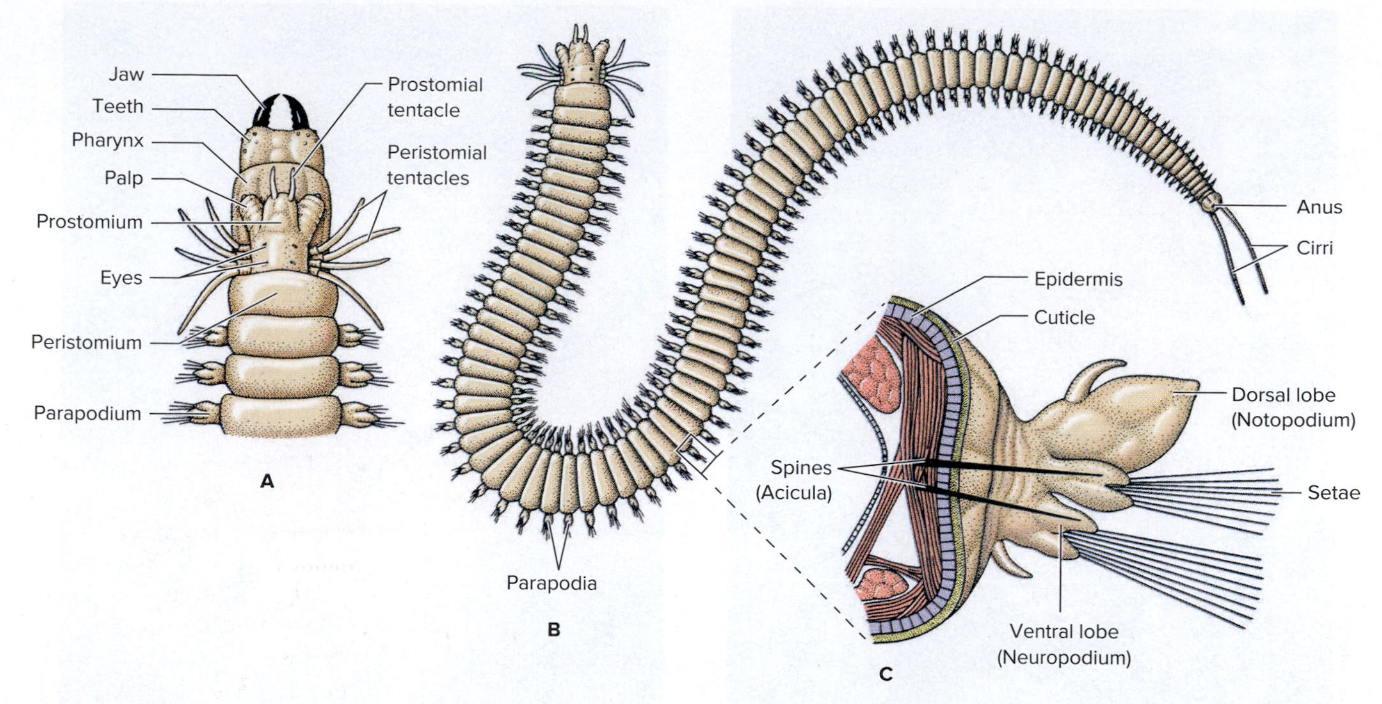

Figure 13-1

Structure of the clamworm *Nereis*. **A,** Anterior view with pharynx extended. **B,** Dorsal view of the entire worm with pharynx withdrawn. **C,** Structure of parapodium.

Clamworms remain in their burrows for safety during the day—they are an attractive food for many fishes—emerging at night to feed. They are predatory carnivores that use their strong jaws to capture and drag food back into their burrows for safety.

Cut off, close to the body, a parapodium from the posterior third of the body. Mount it in water on a slide, cover with a coverslip, and examine with a hand lens or dissecting microscope.

Parapodia are used for respiration as well as locomotion. Are they all identical? ____________ Each parapodium has a dorsal lobe and a ventral lobe (Figure 13-1). Each lobe has a bundle of bristles called **setae** (see′tee; sing. **seta;** L., bristle) and a long, chitinous, deeply embedded spine. The spines are the supporting structures of the parapodium (they are more conspicuous in the posterior parapodia). Each spine is attached by muscles that can protrude it as the parapodium goes into its effective stroke and retract it during the recovery stroke. How might the spines help in locomotion?

____

Peel off a piece of the thin cuticle that covers the animal and study it in a wet mount under a microscope.

The cuticle is fibrous, and its iridescence is caused by its cross-striations. It is full of small pores, through which the gland cells of the underlying epidermis discharge their products.

Internal Structure

Because the internal structure of polychaetes is similar to that of oligochaetes, study of the internal anatomy will be limited to that of earthworms.

Build Your Knowledge of Polychaete Features

Test your understanding of the main features of the polychaetes covered in the exercise by filling out the summary table on pp. xi–xii. Try to complete the table without looking at your notes.

Projects and Demonstrations

There are more than 10,000 species of polychaetes, most of them marine, and most included within the Pleistoannelida (see Classification on p. 216). Besides the errant, or free-moving, worms (class Errantia), such as *Nereis* and its relatives, there are many sedentary species (class Sedentaria), including burrowing and tube-dwelling forms (Figure 13-2). A polychaete excluded from the Pleistoannelida is the unusual and fascinating parchment worm *Chaetopterus*, described below.

Examine the marine aquarium for living polychaetes and examine the preserved material on the demonstration table.

1. *Chaetopterus* (Gr. *chaitē*, bristle, + *pteron*, wing) *variopedatus* is a suspension feeder often found on

Figure 13-2

Annelid tubeworms such as these fanworms secrete the tubes in which they live. They are sessile suspension feeders that have feathered crowns well adapted for the capture of food. **A**, Social featherduster worm, *Bispira brunnea*. **B**, Christmas tree worms, *Spirobranchus giganteus*.

mudflats. This species secretes a U-shaped parchment tube, through which it pumps and filters a continuous stream of water with its fanlike parapodia. The animal is profoundly adapted to its permanent tube environment and completely helpless if removed from it. Feeding currents generated by three pairs of modified parapodia are filtered through a mucous bag held open by another highly modified pair of parapodia. Plankton trapped in the mucus is gathered into a food-laden mucous pellet, which is passed forward by cilia to the mouth.

2. *Diopatra* is a tubeworm that camouflages the exposed portion of its tube with bits of shell, seaweed, and sand. Note its gills and long, sensitive antennae. Fanworms (Figure 13-2) are ciliary feeders, trapping minute food particles on mucus on the radioles of their feathery crowns and then moving the particles down toward their mouth in ciliated grooves.
3. *Amphitrite* (Gr., a mythical sea nymph) is a deposit feeder. It runs long, extensible tentacles over the mud or sand, picking up food particles. These are carried along each tentacle in a ciliated groove to the mouth. Note the cluster of gills just below the tentacles.
4. *Examine tubes of various worms.* What kinds of materials do you find? Examine an old seashell or a rock bearing calcareous worm tubes. The variety of secreted tubes and materials used to supplement them seems endless.

EXERCISE 13B

Core Study

Sedentaria—Earthworm

Lumbricus, Common Earthworm

Phylum Annelida
Class Sedentaria
Order Clitellata
Family Lumbricidae
Genus *Lumbricus*
Species *Lumbricus terrestris*

Where Found

Earthworms prefer moist, rich soil that is not too dry or sandy. They are found all over the earth. They are chiefly nocturnal and come out of their burrows at night to forage. A good way to find them is to search with a flashlight around the rich soil of lawn shrubbery. Large "night crawlers" are easily found this way, especially during warm, moist nights of spring and early summer. *Lumbricus terrestris* (L. *lumbricum,* earthworm), named by Linnaeus, is one of the most common earthworms in Europe, Asia, and North America and has been introduced all over the world.

Behavior

Wet the center of a paper towel with pond or dechlorinated water, leaving the rest of the paper dry. Place a live earthworm on the moist area. Using the following suggestions, observe its behavior.

Is the skin of the worm dry or moist? ____________ Do you find any obvious respiratory organs? ____________ Where do you think exchange of gases occurs? ____________ Would this necessitate a dry or damp environment? ____________ Does the worm respond positively or negatively to moisture? ____________

Notice the mechanics of crawling. The earthworm's body wall contains well-developed layers of circular and longitudinal muscles. As it crawls, notice the progressing peristaltic waves of contraction. These are produced by alternate contraction and relaxation of longitudinal and circular muscles in the body wall acting against noncompressible coelomic fluid. An earthworm's body is divided internally into segments by septa. When longitudinal muscles of a segment contract, the segment becomes shorter and thicker because the volume within each body segment remains constant. Conversely, when circular muscles contract, the segment elongates. Do the waves of circular muscle contraction move anteriorly or posteriorly when the worm as a whole is moving forward? ____________ How far apart (what proportion of total body length) are the waves of contraction? ____________ Watch the anterior end as the worm advances. Do short and thick or long and thin regions advance the head end forward? ____________

Run a finger along the side of the worm. Do you detect the presence of small setae (bristles)? How might these setae be used? ____________

How does the animal respond when you gently touch its anterior end? ____________ Its posterior end? ____________ Draw the towel to the edge of the desk and see what happens when the worm's head projects over the edge of the table. Is it positively or negatively thigmotactic (responsive to touch)? ____________ Turn the worm over and see if it can right itself, and how.

Can you devise a means of determining whether an earthworm is positively or negatively geotactic (responsive to gravity)?

Place the earthworm on a large plate of wet glass. Does this difference in substratum affect its locomotion? Is friction important in earthworm locomotion?

Does the earthworm have eyes or other obvious sensory organs? ____________ Can you devise a means of determining whether it responds positively or negatively to light? ____________

Written Report

On p. 212, record the responses of the earthworm. Comment on hydrotaxis, locomotion, thigmotaxis, phototaxis, and the importance of friction.

External Structure

This study may be completed using either living and anesthetized earthworms or freshly killed earthworms. To anesthetize an earthworm, immerse it for 30 to 40 minutes in 7% ethanol.* When the worm is completely limp, transfer it to a dissecting tray that has been dampened with water. Examine with a hand lens or dissecting microscope.

To kill living earthworms, immerse them in 10–15% ethanol for 30 minutes just before the laboratory. Place in a dissecting tray dampened with water and examine with hand lens or dissecting microscope.

What are the most obvious differences between the earthworm and the clamworm? List two or three here. ____________

The first four segments make up the head region. The first segment is the **peristomium.** It bears the **mouth,** which is overhung by a lobe, the **prostomium.** The earthworm's head, lacking specialized sense organs, is not a typical annelid head.

Find the **anus** in the last segment. Observe the saddle-like **clitellum** (L. *clitellae,* packsaddle), which in mature worms secretes egg capsules, into which eggs are laid. In what segments does it occur? ____________ Because leeches share with earthworms the presence of a clitellum, both are included in the clade (order) Clitellata. Unlike the earthworm's clitellum, which is always present, that of the leech is visible only during the reproductive season.

How many pairs of setae are on each segment, and where are they located? ____________ Use a hand lens or dissecting microscope to determine this.

There are many external openings other than mouth and anus. Earthworms are monoecious. Note **male pores** on the ventral surface of segment 15. These are conspicuous openings of sperm ducts, from which spermatozoa are discharged. Note two long **seminal grooves** extending between

*Prepared by diluting 74 ml of 95% ethanol with 1 L of water. See Appendix A, p. 406, for a note on the use of live, freshly killed, or preserved earthworms for this exercise.

the male pores and the clitellum. These guide the flow of spermatozoa during copulation. Use a small hand lens to look for small **female pores** on the ventral side of segment 14. Here the oviducts discharge eggs. You may not be able to see the openings of 2 pairs of **seminal receptacles** in grooves between segments 9 and 10 and between 10 and 11.

There are paired excretory openings, **nephridiopores,** located on the lateroventral surface of each segment (except the first three and the last). Look also for **dorsal pores** from the coelomic cavity located at the anterior edge of the mid-dorsal line on each segment from 8 or 9 to the last. Many earthworms eject a malodorous coelomic fluid through the dorsal pores in response to mechanical or chemical irritation or when subjected to extremes of heat or cold.

Drawings

Complete the external ventral view of the earthworm on p. 211. Draw in and label prostomium, peristomium, mouth, setae, male pores, female pores, seminal grooves, clitellum, and anus.

Internal Structure and Function

Reanesthetize the earthworm in 7% ethanol, if necessary. Place the anesthetized worm dorsal side up in a dissecting pan and straighten it by passing one pin through the fourth or fifth segment (just behind the peristomium) and another pin through any segment near the posterior end of the worm. With a razor blade or new scalpel blade, and beginning at about the fortieth segment (just behind the clitellum), cut through the body wall at a point just to one side of the dark mid-dorsal line (the **dorsal blood vessel**). Use fine-tipped scissors to complete the middorsal cut all the way to the head, pulling up on the scissors as you proceed to avoid damaging internal organs. Keep your incision slightly to one side of the dorsal blood vessel. With a pipette, squirt some isotonic salt solution on the internal organs to keep them moist. Now, starting at the posterior end of the incision, pin the animal open. You will need to break the septa (partitions between the metameres) with a needle as you proceed anteriorly. When you have finished, remove all the pins except those anchoring the worm at the anterior end. Stretch out the worm by pulling gently on the posterior end and repin, placing the pins at an oblique angle. If you are using a dissecting microscope rather than a hand lens, you may need to position the worm to one side of the dissecting tray for viewing. Now flood the tray with enough isotonic saline (0.6% NaCl) to cover the earthworm completely.

Note peristaltic movements of the **digestive tract,** which propel food posteriorly. Find 3 pairs of cream-colored **seminal vesicles** in segments 9 to 12 (Figure 13-3), 2 pairs of glistening white **seminal receptacles** in segments 9 and 10, and a pair of delicate, almost transparent, tubular **nephridia** in the coelomic cavity of each segment. Note the **dorsal vessel** riding on the digestive tract. In which direction is blood flowing in this vessel? ____________ A total of 5 pairs of pulsating **aortic arches,** sometimes called "hearts," surround the **esophagus** in segments 7 to 11 (some of these arches are covered by the seminal vesicles).

Digestive System. Identify the **mouth;** the muscular **pharynx;** the slender **esophagus** in segments 6 to 13; the thin-walled **crop;** and the muscular **gizzard, intestine,** and **anus.** Note the functions of these organs as summarized with the labels in Figure 13-3. Locate the **calciferous glands** that lie on both sides of the esophagus (usually partly concealed by the seminal vesicles). Bright yellow or green **chloragogue cells** often cover the intestine and much of the dorsal vessel. They are known to store glycogen and lipids but probably have other functions as well, similar to those of the vertebrate liver. Make an off-center longitudinal cut into the intestine in the region of the clitellum to expose the **typhlosole** (Gr. *typlos,* blind, + *sōlēn,* channel), a ridgelike structure projecting into the lumen of the intestine. The typhlosole increases the surface available for digestive enzyme production and absorption.

Circulatory System. An earthworm has a **closed** circulatory system. Note that both **dorsal vessel** and **aortic arches** (identified earlier) are contractile, with the dorsal vessel being the chief pumping organ and the arches maintaining a steady flow of blood into the **ventral vessel** beneath the digestive tract.

Retract the digestive tract. Lift up the white nerve cord in the ventral wall. Note the **subneural vessel** clinging to its lower surface and a pair of **lateroneural vessels,** with one located on each side of the nerve cord. Be able to trace blood flow from the dorsal vessel to the intestinal wall and back, to the epidermis and back, and to the nerve cord and back.

Reproductive System. An earthworm is monoecious; it has both male and female organs in the same individual, but cross-fertilization occurs during copulation. The **male organs** (Figure 13-3) consist of three pairs of **seminal vesicles** attached to segments 9, 11, and 12; two pairs of small, branched **testes** housed within reservoirs of the seminal vesicles; and two small **sperm ducts** that connect the testes to the male pores in segment 15 (Figure 13-3). Both testes and sperm ducts are too small to be found easily.

The female organs are also small. Two pairs of small, round **seminal receptacles,** easily seen in segments 9 and 10, store spermatozoa after copulation. You should be able to find the paired **ovaries** that lie ventral to the third pair of seminal vesicles. The paired **oviducts** with ciliated funnels that carry eggs to the female pores in the next segment will probably not be seen.

Earthworm Copulation. When mating, two earthworms, attracted to each other by glandular secretions, extend their anterior ends from their burrows and, with heads pointing in opposite directions, join their ventral surfaces in such a way that the seminal receptacle openings of one worm

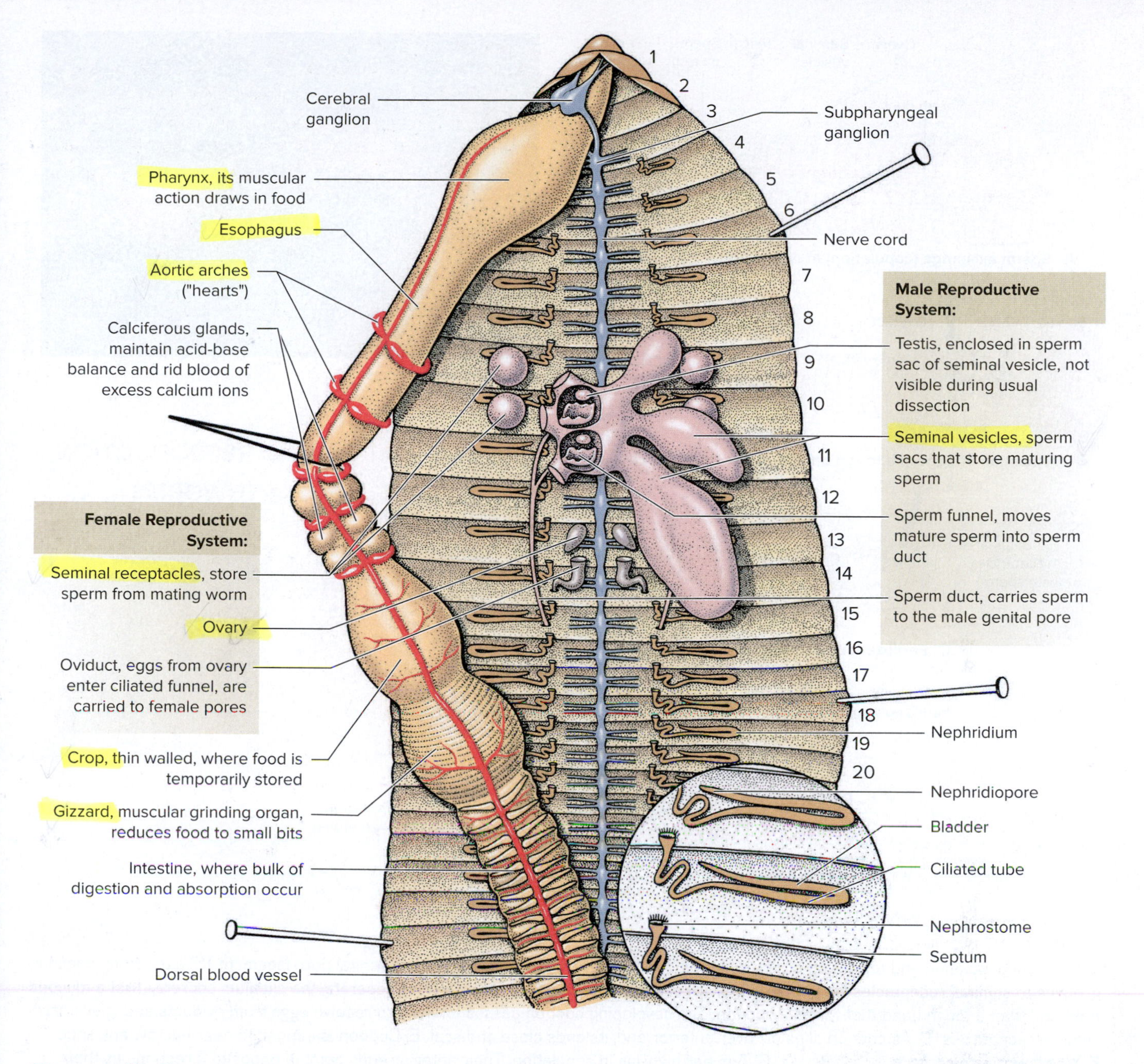

Figure 13-3
Internal structure of *Lumbricus*, dorsal view.

lie in opposition to the clitellum of the other (Figure 13-4). Each worm secretes quantities of mucus, so that each is enveloped in a slime tube extending from segment 9 to the posterior end of the clitellum. Seminal fluid discharged from the sperm ducts of each worm is carried along the seminal grooves by contraction of longitudinal muscles and enters the seminal receptacles of the mate. After copulation the worms separate, and each clitellum produces a secretion that finally hardens over its outer surface. The worm moves backward, drawing the hardened tube over its head (Figure 13-4C). As it is moved forward, the tube receives eggs from the oviducts, sperm from the seminal receptacles, and a nutritive albuminous fluid from skin glands. Fertilization occurs in the cocoon. As the worm withdraws, the cocoon closes and is deposited on the ground. Young worms hatch in 2 to 3 weeks.

Excretory System. This consists of a pair of excretory organs, called **nephridia,** in each segment except the first three and the last one. Each nephridium actually occupies two segments because it begins with a ciliated funnel (nephrostome) that projects through the anterior septum of the segment containing the body of the nephridium (see inset in Figure 13-3).

Use a dissecting microscope to examine a nephridium. They are largest in the region just posterior to the clitellum. With fine-tipped scissors, carefully remove a

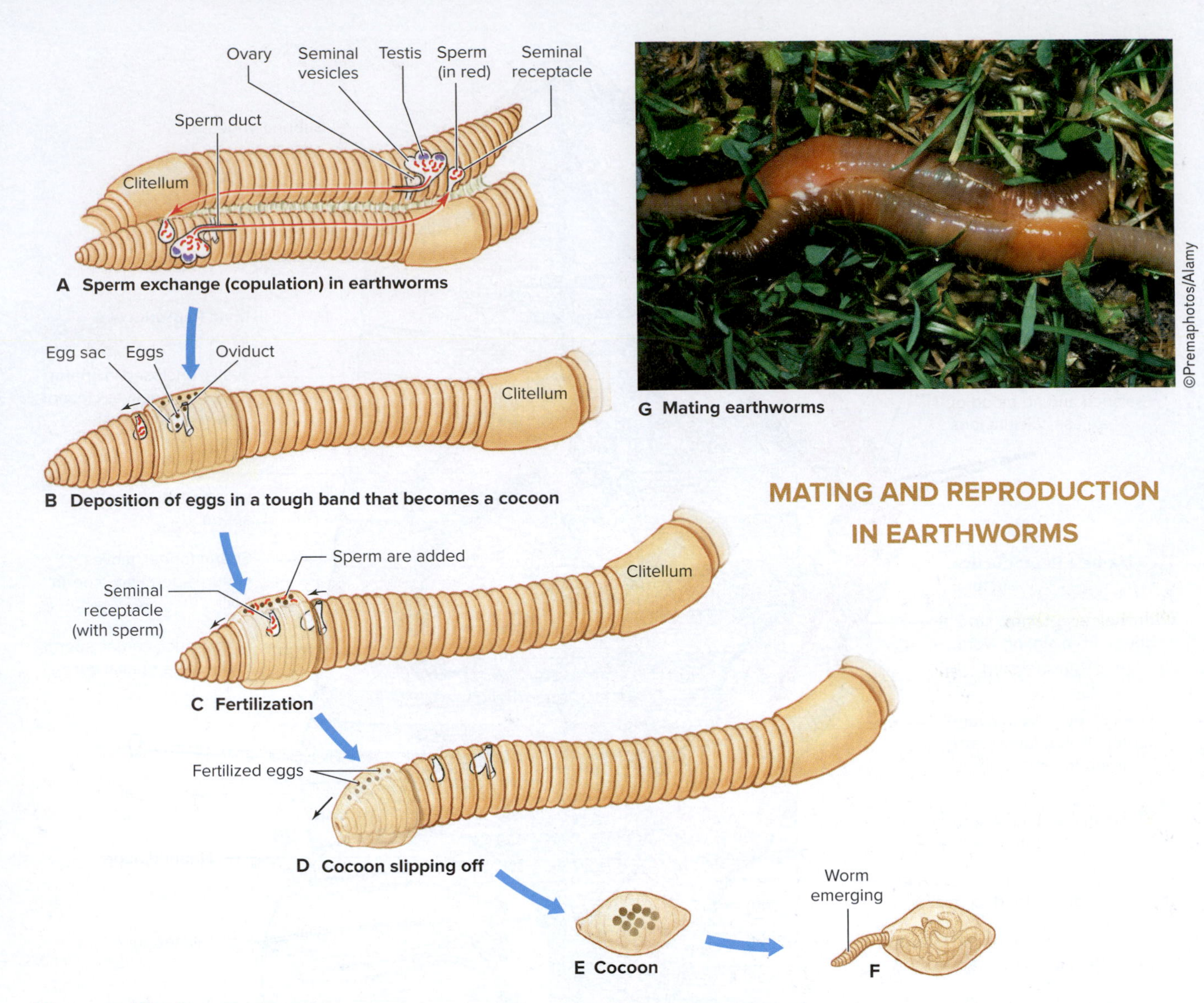

Figure 13-4

Earthworm copulation and formation of egg cocoons. **A,** Mutual insemination; sperm from genital pore (segment 15) pass along seminal grooves to seminal receptacles (segments 9 and 10) of each mate. **B** and **C,** After worms separate, the clitellum secretes first a mucous tube and then a tough band that forms a cocoon. The developing cocoon passes forward to receive eggs from oviducts and sperm from seminal receptacles. **D,** As cocoon slips off over anterior end, its ends close and seal. **E,** Cocoon is deposited near burrow entrance. **F,** Young worms emerge in 2 to 3 weeks. **G,** Two earthworms in copulation. Their anterior ends point in opposite directions as their ventral surfaces are held together by mucous bands secreted by the clitella.

nephridium, along with a small portion of the septum through which the nephrostome projects. Mount on a slide with a drop or two of saline solution, cover with a coverslip, and examine with a compound microscope.

Note the slender tubule passing from the nephrostome to the looped nephridium. Note ciliary activity in one narrow portion of the tubule. You may also see parasitic nematodes in the large bladder segment of the tubule. Coelomic fluid is drawn by ciliary activity into the nephrostome and then flows through the narrow tubule, where ions, especially sodium and chloride, are reabsorbed. Urine, containing wastes, collects in the bladder, which empties to the outside through a **nephridiopore.**

Nervous System

If you have not already done so, extend the dorsal incision to the first segment.

Find a small pair of white **cerebral ganglia** (the "brain") lying above the pharynx and partially hidden by dilator muscles. This connects by a pair of nerves to the first ganglia of the **ventral nerve cord.** The ventral nerve cord is a complex of nerves and segmental ganglia that coordinate sensory and motor impulses that control the smooth muscular waves of forward movement. The nerve cord also contains giant fibers passing the length of the cord. Giant fibers carry very rapid nerve impulses, measured at up to 45 m/sec,

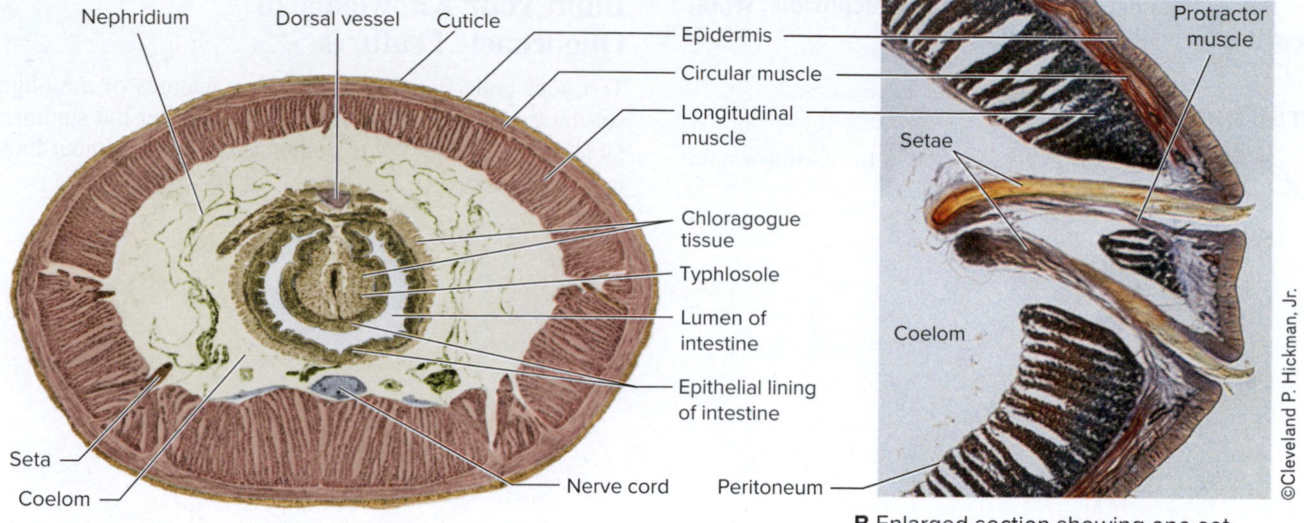

A Cross section of an earthworm through intestine

B Enlarged section showing one set of setae and their protractor muscles.

Figure 13-5

A. Cross section of an earthworm through the intestinal region. **B.** Portion of a cross section of an earthworm, showing one set of setae with their protractor muscles.

that enable the worm to instantly contract its whole body in response to strong stimulation and retire rapidly into its burrow. You can expose and examine the ventral nerve cord by removing or laying aside the digestive tract.

Although earthworms have no eyes, they do have many photoreceptors in their epidermis, enabling them to shun strong light.

Oral Report

Be prepared to (1) demonstrate your dissection to your instructor, (2) point out both external and internal structures you have studied, and (3) explain their functions.

Further Study

Histology of Cross Section (Figure 13-5)

Examine a stained slide with low power. Note the tube-within-a-tube arrangement of intestine and body wall. Identify the following.

Cuticle. Thin, noncellular, and secreted by the epidermis.

Epidermis. (Ectodermal.) Columnar epithelium containing mucous gland cells. Mucus prevents the skin from drying out.

Circular Muscle Layer. Smooth muscle fibers running around the circumference of the body. How does their contraction affect body shape? ____________________

Longitudinal Muscle Layer. Thick layer of obliquely striated fibers that run longitudinally. The muscle layers may be interrupted by the setae and dorsal pore. How does their contraction affect body shape? ____________________

Peritoneum. (Mesodermal.) The peritoneum (Gr. *peritonaios,* stretched around), a thin epithelial layer lining the body wall and covering the visceral organs (Figure 13-5). Peritoneum lining the body wall is called parietal (L. *paries,* wall) peritoneum. Peritoneum covering the digestive tract and other visceral organs is called visceral (L. *viscera,* bowels) peritoneum.

The space between the parietal and visceral peritoneum is the **coelom.**

Setae. Also called bristles, these are brownish spines moved by tiny muscles (Figure 13-5). How many setae are there on each segment? ______________ You may not see setae unless they happen to be in the plane of section. The setae are important in the worm's locomotion and also serve to anchor the worm firmly in its burrow.

Alimentary Canal. The intestine is surrounded by chloragogue tissue. Chloragogue tissue plays a role in intermediary metabolism similar to that of the liver in vertebrates. Inside the chloragogue layer is a layer of **longitudinal muscle.** Why does it appear as a circle of dots? __________ Next is a **circular muscle layer,** followed by a layer of ciliated columnar epithelium (endodermal), which lines the intestine. Intestinal contents are moved along by peristaltic movement. Is such movement possible without longitudinal and circular muscles? __________ Is peristalsis possible in the intestine of the flatworm or *Ascaris?* __________

Ventral Nerve Cord. Use high power to identify three **giant fibers** in the dorsal side of the nerve cord, as well as nerve cells and fibers in the rest of the cord.

Blood Vessels. Identify the **dorsal** vessel above the typhlosole, **ventral** vessel below the intestine, **subneural** vessel below the nerve cord, and **lateral neurals** beside the nerve cord.

Some slides may also reveal parts of **nephridia, septa, mesenteries,** and other structures.

Drawings

Sketch and label a cross section of the earthworm on p. 211 as it appears on your slide.

Build Your Knowledge of Oligochaete Features

Test your understanding of the main features of the oligochaetes covered in the exercise by filling out the summary table on pp. xi–xii. Try to complete the table without looking at your notes.

Phylum ______________________

Genus ______________________

Name ______________________

Date ______________________

Section ______________________

LAB REPORT

Earthworm

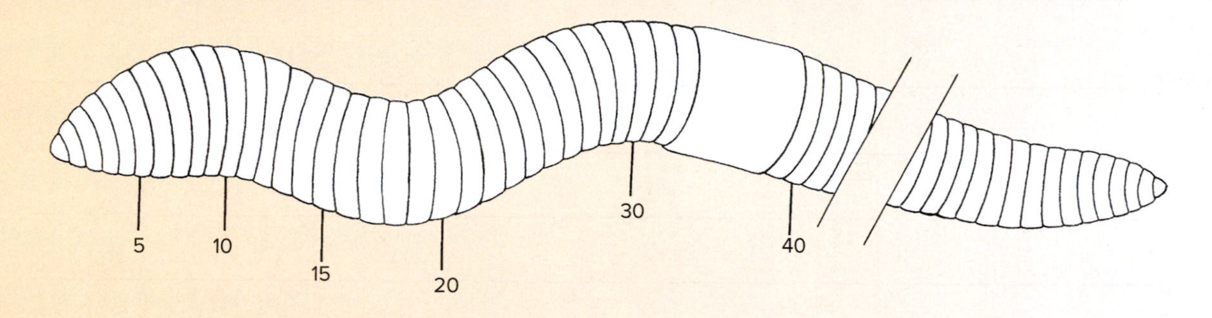

External structure of an earthworm, ventral view

Cross section through the body of an earthworm

LAB REPORT

Observations on the Behavioral Responses of an Earthworm

Hydrotaxis ______________________________

____

____

____

____

Locomotion ______________________________

____

____

____

____

Thigmotaxis (response to touch) ______________________________

____

____

____

____

Phototaxis ______________________________

____

____

____

____

Importance of friction ______________________________

____

____

____

____

Clamworm

1. Clamworms are polychaetes of the Class Errantia. What is the derivation of the polychaete name and what anatomical characteristic is indicated in the name? ____________________

2. Clamworms have a well-developed two-part head. What sensory structures are borne on each part? ____________________
____

3. What are parapodia? ____________________ What function or functions are served by the parapodia? ____________________

4. What characteristic of annelid worms is described by the term "metamerism?" ____________________
____

Earthworm

1. Is the skin of a living earthworm dry or moist? __________ Are there obvious respiratory organs on the worm's surface? __________ Where does respiratory exchange occur? __________ Does this require a dry or moist environment? __________

2. What are setae? ____________________ Does an earthworm bear setae on the body surface? __________ If so, how might they be used? ____________________

3. What is the function of the clitellum? ____________________

4. What is the function of the earthworm pharynx? __________ Function of the crop? __________ Function of the gizzard? ____________________

5. If earthworms have no eyes, how do they detect the presence or absence of light in their environment? __________
____

6. Why is the circulatory system of an earthworm described as a closed system? ____________________
How does it differ from the open circulatory system of a clam? ____________________

LAB REPORT

NOTES

EXERCISE 13C

Class Hirudinida—Leech

Hirudo, Medicinal Leech

Phylum Annelida
Class Hirudinida
Order Hirudinea
Genus *Hirudo*
Species *Hirudo medicinalis*

Where Found

Leeches are predaceous and mostly fluid feeders. Some are true bloodsuckers, attaching themselves to a host during feeding periods.

Hirudo is one of the freshwater leeches found in lakes, ponds, streams, and marshes and is native to Europe. *H. medicinalis* (Figure 13-6) is often referred to as the "medicinal leech" because it was used in bloodletting from several centuries BC to near the end of the nineteenth century. It is now occasionally used in plastic surgery and in the treatment of hematomas. It feeds on the blood of amphibians and mammals. Living leeches are readily available from biological supply houses.

Behavior

Note the use of ventral **suckers**—a smaller oral sucker at the anterior end and a larger caudal sucker at the posterior end.

Attempt to pull an attached leech free from its substratum.

Are the suckers powerful? The leech combines the use of its suckers with muscular body contractions in creeping.

Place a leech on a glass plate and watch it move.

The oral sucker attaches and the body contracts; then the caudal sucker attaches and the body extends forward before the oral sucker attaches again. How does movement of a leech on the glass plate compare with movement of the earthworm that you previously observed?

Drop a leech into the water to observe its undulating, free-swimming motions.

Medicinal leeches are attracted to heat. Why would this be? ____________________ To investigate this further, see the Experimenting in Zoology feature at the end of the chapter.

External Features

Study a preserved specimen of *Hirudo* (Figure 13-7).

How does its shape compare with that of an earthworm? Can you distinguish between the dorsal and ventral surfaces?

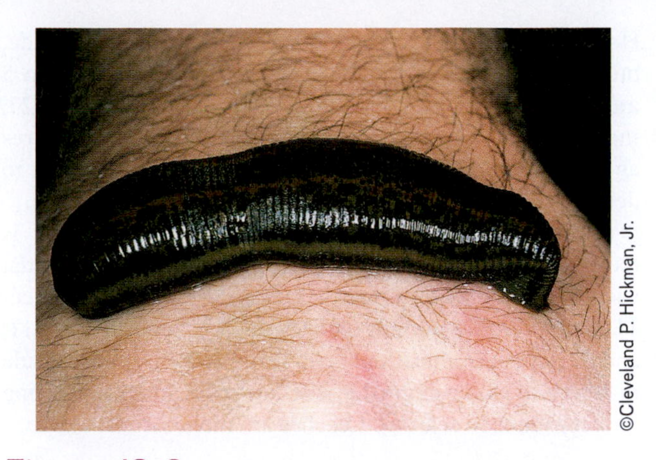

Figure 13-6
Medicinal leech, *Hirudo medicinalis,* feeding on blood from a human arm.

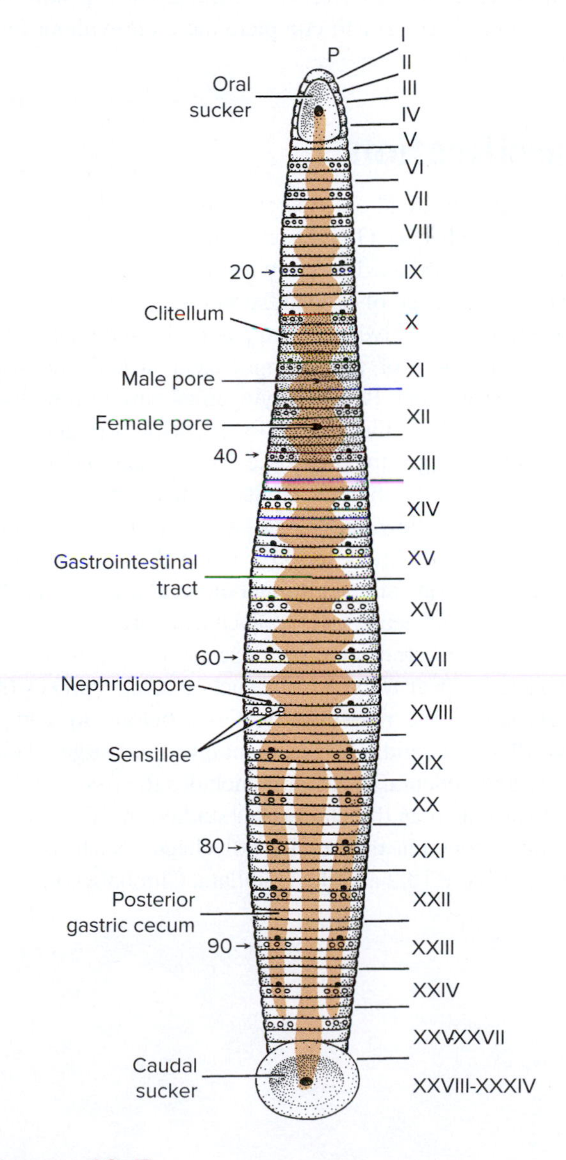

Figure 13-7
External anatomy of *Hirudo medicinalis*, showing segments (Roman numerals) and annuli (Arabic numerals). The position of the gut, not visible externally, is indicated in color.

How? __________ A leech is segmented inside and outside, but externally each segment also is marked off into 1 to 5 **annuli,** the larger number being found in segments 9 to 23, inclusive. Each true segment bears a pair of **nephridiopores** and, on one of its annuli, a row of **sensillae** (used mainly to detect water movements) or (at the anterior end) eyespots.

The oral sucker contains the **mouth.** The **anus** is located in the middorsal line at the junction with the caudal sucker. A **male genital pore** is located ventrally on segment 11, and a **female pore** is on segment 12. Segments 10, 11, and 12 serve as a functional **clitellum,** secreting the capsule in which eggs develop (the clitellum is visible only during the reproductive season). Leeches are monoecious.

Build Your Knowledge of the Hirudinida

Test your understanding of the main features of the Hirudinida covered in the exercise by filling out the summary table on pp. xi–xii. Try to complete the table without looking at your notes.

Classification

Traditional classification separated the annelids into three classes: Polychaeta, Oligochaeta, and Hirudinidae. This classification was based on body form—in particular, on the presence or absence of parapodia and the number of setae. Members of class Polychaeta had parapodia and many setae, whereas members of class Oligochaeta lacked parapodia and had fewer setae. Recent phylogenies based on molecular characters show that both polychaetes and oligochaetes are paraphyletic groups. Only the Hirudinidae, comprising the leeches, is monophyletic. Consequently, the terms oligochaete (meaning "few long hairs") and polychaete (meaning "many long hairs") are descriptive rather than taxonomic.

Classification using additional molecular characters has shown that many features of polychaetes are, in fact, features of the ancestral annelid. Descendants of the earliest polychaetes appear to be worms such as *Chaetopterus* (see pp. 203–204). The remaining annelids belong to a group called Pleistoannelida, which contains two large clades, Errantia and Sedentaria. Errantia includes the freely moving polychaetes such as the clamworm studied in this exercise. Sedentaria contains the sessile polychaetes such as those shown in Figure 13.2 and the Clitellata. Clitellata comprises the annelids with an oligochaete body plan, such as the earthworm, and the leeches.

Phylum Annelida

Basal Annelida Annelids of uncertain taxonomy that diverged from the most recent common ancestor of annelids before the ancestor of Pleistoannelida. Example: *Chaetopterus.*

Subphylum Pleistoannelida (plis′to-an-ill′a-da) (Gr. *pleistos*, most, + Fr. *anneler*, to arrange in rings <L. *annelus*, ring). Marine, freshwater, and terrestrial annelids, most with segmented bodies.

Errantia Freely moving polychaetes (pol′e-ke′ta) Gr. *polys*, many, + *chaitē,* long hair). Segmented inside and out; parapodia with many setae; distinct head with eyes, palps, and tentacles; no clitellum; separate sexes, trochophore larva usually present; mostly marine. Examples: *Nereis, Aphrodite.*

Sedentaria Sedentary annelids including tube-dwelling polychaetes and those living in burrows, as well as members of Clitellata: oligochaetes (ol′i-go-ke′ta) (Gr. *oligos*, few, + *chaitē*, long hair) and leeches with a clitellum at some phase of the life cycle. Clitellate annelids with an oligochaete body plan have conspicuous segmentation inside and out; number of segments variable; setae few or absent; no parapodia, head poorly developed; coelom spacious and usually divided by intersegmental septa; direct development; chiefly terrestrial and freshwater. Examples: *Sabella, Serpulis, Arenicola, Lumbricus, Tubifex, Hirudo.*

Clade Clitellata (cli-tel′a-ta) Annelids with an oligochaete body plan having conspicuous segmentation; number of segments variable; few setae; no parapodia; head absent; chiefly freshwater and terrestrial. Examples: *Lumbricus, Tubifex, Hirudo.*

Family Lumbricidae (lum-bri′si-dee) (L. *lumbricus*, earthworm). Oligochaetes including common terrestrial earthworms; often large with well-developed and often complex reproductive systems. Example: *Lumbricus.*

Class Hirudinidae (hir-uh-din′i-dee) (L. *hirudo*, leech). Segments 33 or 34 in number, with many annuli; clitellum present; anterior and posterior suckers; setae absent (except *Acanthobdella*); parapodia absent; terrestrial, freshwater, marine. Examples: *Hirudo, Placobdella.*

EXPERIMENTING IN ZOOLOGY

Behavior of Medicinal Leeches, Hirudo medicinalis

Medicinal leeches, *Hirudo medicinalis,* are best known because of their use in medical procedures. Historically, these leeches were used to draw blood from victims of bites or stings from insects or venomous animals. In addition, especially during the eighteenth and nineteenth centuries, leeches were used to draw out "corrupt blood" from people suffering from disease. Today these annelids are frequently used to restore venous circulation after reconstructive surgical procedures, particularly when a finger, a toe, or an ear is surgically reattached after an accident. In 2004 the Food and Drug Administration (FDA) for the first time cleared the commercial marketing of leeches for medicinal purposes. Aside from their use in modern medicine, medicinal leeches are wild animals that have evolved specific adaptations for locating their mammalian prey. Because mammals are endotherms, their body temperatures typically are above ambient water temperatures. Medicinal leeches are able to locate mammals that have entered their ponds or lakes because they have evolved the ability to detect the slightly elevated water temperatures generated by the presence of a mammal.

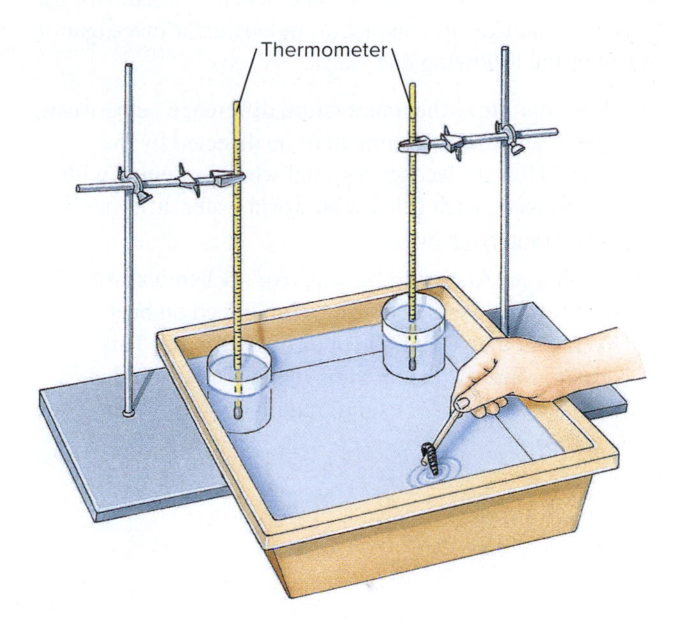

Figure 13-8
Setup for observing the behavior of leeches.

Getting Ready

Work with a partner for this exercise. Obtain a plastic tub approximately 40 cm wide by 60 cm long. Wash the tub and fill it to a depth of 8 to 10 cm using pond water, bottled water, or dechlorinated tap water. After taking the temperature of the water in the tub, fill two 500 ml beakers with water of the same temperature. Place the two filled beakers onto opposite sides of the tub about 30 cm from one end (Figure 13-8). Use ring stands and clamps to suspend thermometers for monitoring the temperatures throughout the experiment. Have a stopwatch available for timing leech behavior.

How to Proceed

Use a short wooden dowel to choose an active medicinal leech. The leech will attach itself to the dowel or balance momentarily on it while you transfer it to your experimental tub. If the leech will not stay on the dowel, gently place it on a paper towel to transfer it to the tub. Release the leech at the end of the tub. Try to release the leech so as not to bias its behavior toward either of the beakers. How does the leech behave when you release it? Does it go toward the beakers? When leeches find potential prey items or simply want to crawl on a substrate, they first attach to an object using their anterior sucker. Does the leech attach itself to the side of the tub? Does it attach itself to one of the beakers? If so, how much time expires between when you first release the leech and when it attaches itself to a beaker? Once the leech attaches itself to a beaker or the tub, gently remove the leech with the wooden dowel and repeat the trial by replacing the leech at the end of the tub opposite the beakers.

After three trials with the beakers at room temperature, replace the water in one beaker with water warmed to 40° C. Again monitor the temperatures in the beakers. Watch that the temperature in the warm beaker remains at approximately 40° C. Release the leech at the end of the tub opposite the beakers. Does the leech attach itself to one of the beakers? To which beaker does it attach? How much time elapses between release and attachment? Remove the leech from the beaker and conduct at least two more trials.

Does the leech appear to be attracted to the beaker filled with warm water? Compare your results with those of your classmates. Your instructor may want you to combine your data with your classmates' and compare control treatments with experimental treatments, using a statistical test. How many times out of the total control trials does a leech attach to a beaker? How many times in the warm beaker trials? Do the leeches attach to the warm water beakers faster than when there are only beakers filled with room temperature water?

Questions for Independent Investigation

Medicinal leeches are hearty animals that are ideal specimens for these types of experiments. While repeated trials using the same animal violates the important statistical assumption of independence (see Exercise 5A), you may be able to borrow five or six of the other leeches ordered for use by your classmates to conduct an independent investigation on one of the following questions:

1. How small may the temperature difference be between "prey" and the environment to be detected by the leech? How do leeches respond when presented with two beakers, each filled with warm water differing by only a degree or two?
2. Leeches are known to be sensitive to chemical stimuli in the water. Medicinal leeches will feed on beef liver. Can they detect the odor of beef liver? This question may be explored by using a setup similar to that described for examining their sensitivities to temperature differences. Place beef liver inside a plastic beaker that has been pierced repeatedly with a heated pin or needle.
3. How do leeches respond when presented with a choice of attaching either to a beaker filled with warm water (thermal cues) or to one that is releasing food odor (chemical cues)? Which appears to be the more attractive stimulus?

References

Dickinson, M. H., and C. M. Lent. 1984. Feeding behavior of the medicinal leech, *Hirudo medicinalis*. Jour. Comp. Phys. **154:**449–455.

Lent, C. M., and M. H. Dickinson. 1987. On the termination of ingestive behavior by the medicinal leech. Jour. Exp. Bio. **131:**1–15.

Lent, C. M., K. H. Fliegner, E. Freedman, and M. H. Dickinson. 1988. Ingestive behavior and physiology of the medicinal leech. Jour. Exp. Bio. **137:**513–527.

The Chelicerate Arthropods

Phylum Arthropoda
Subphylum Chelicerata

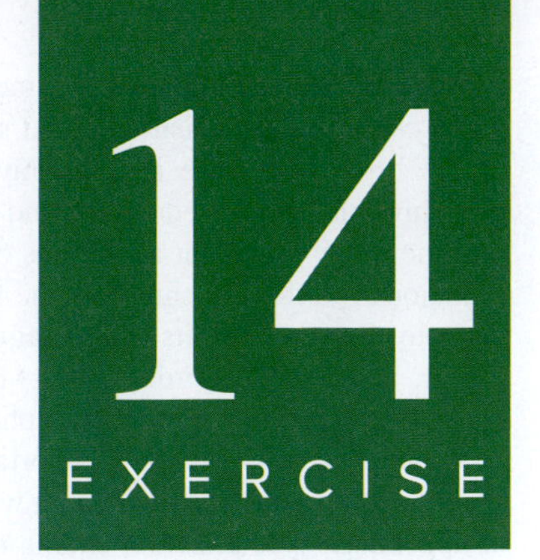

There are an estimated million named species of arthropods, a huge and diverse phylum that includes chelicerates (spiders, scorpions, and their allies), crustaceans, myriapods (millipedes, centipedes), and hexapods (insects, and their kin).

Arthropods have **jointed appendages,** a fact that gave them their name (Gr. *arthron,* joint, + *pous, podos,* foot). Like that of annelids, the arthropod body is constructed of

EXERCISE 14
Chelicerate Arthropods—Horseshoe Crab and Garden Spider
Horseshoe Crab
Garden Spider

CLASSIFICATION
Phylum Arthropoda

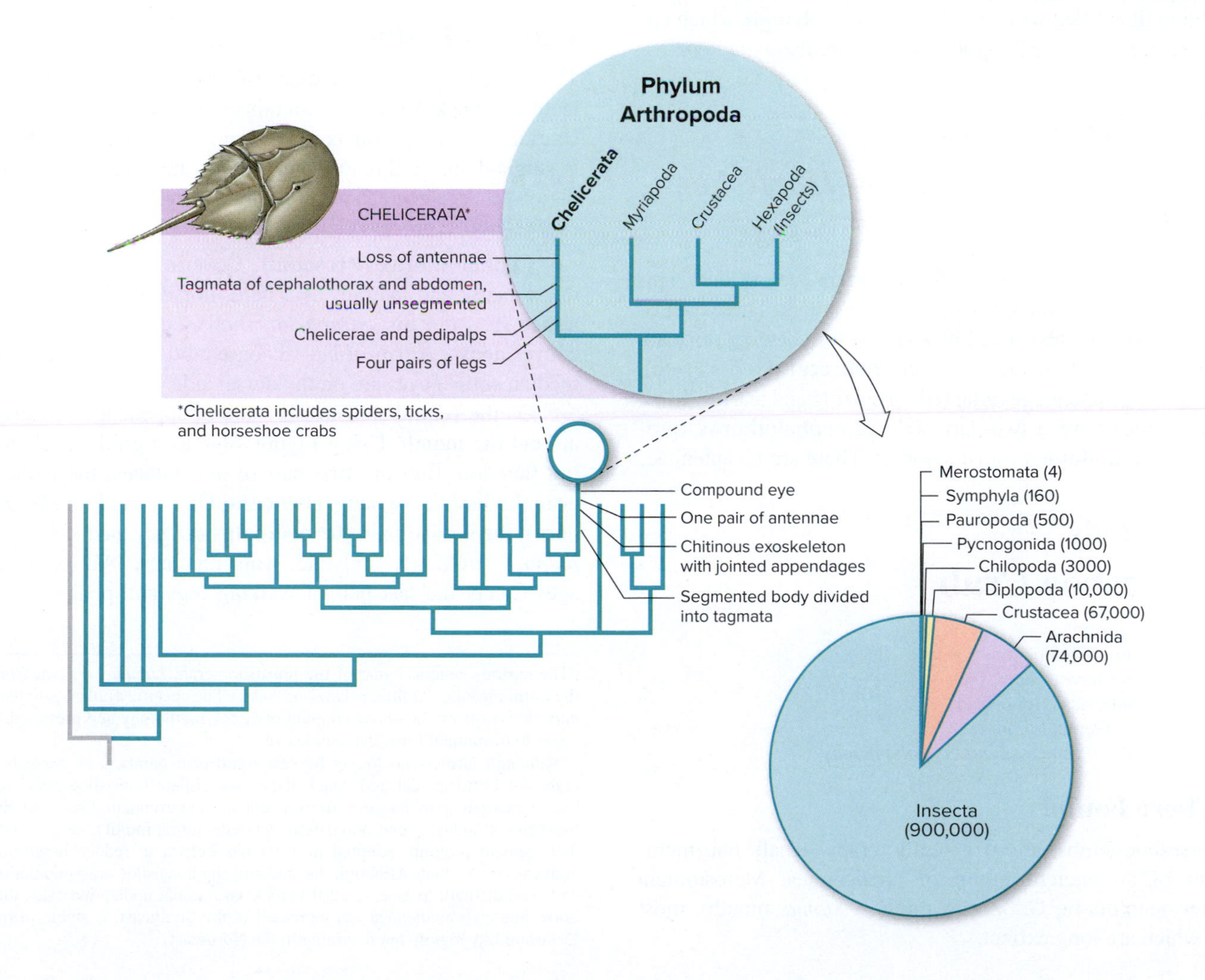

an extended series of repeated segments. This design principle is called **metamerism** = serial segmentation. However, unlike in annelids, the initial array of repeated segments has, through fusion, reduction, and specialization, evolved into the many divergent anatomies of advanced arthropods. A major evolutionary shift was the fusion of segments into discrete, functional units called **tagmata** (sing. **tagma;** Gr. *tagma,* arrangement, order, row)—for example, head, thorax, and abdomen of insects or cephalothorax and abdomen of spiders and ticks. Another important evolutionary trend is specialization of appendages. You will see several examples of appendage specialization among the three arthropod subphyla to be studied in this exercise and Exercises 15 and 16.

In addition to features shared by previous phyla, such as **triploblastic development, true coelom, bilateral symmetry, cephalization,** and **all organ systems,** arthropods have developed **striated muscle** for rapid movement; an **exoskeleton,** or cuticle, containing the tough nitrogenous polysaccharide **chitin** for support and protection; **gills** and a very efficient **tracheal system** for gaseous exchange; and **greater specialization** of body organs, especially specialization of form and function among the appendages. The coelom is much reduced. Instead, the major body space is a **hemocoel** derived from fusion of the embryonic blastocoel with the developing coelom. It is filled with blood, called **hemolymph,** which circulates through an arthropod's "open" circulatory system.

EXERCISE 14

Chelicerate Arthropods—Horseshoe Crab and Garden Spider

The chelicerates include horseshoe crabs, sea spiders, spiders, scorpions, mites, ticks, and some others. These arthropods do not possess mandibles (jaws) for chewing. Instead, the first pair of appendages, called **chelicerae** (ke-liss′er-uh), are feeding appendages adapted for seizing and tearing. Most chelicerates have a two-part body of **cephalothorax** (prosoma) and **abdomen** (opisthosoma). There are no antennae.

Core Study

Horseshoe Crab

Phylum Arthropoda
Subphylum Chelicerata
Class Merostomata
Subclass Xiphosurida
Genus *Limulus*
Species *Limulus polyphemus*

Where Found

Horseshoe crabs are not really crabs at all but members of an ancient group of chelicerates, Merostomata (mer′o-sto′ma-ta; Gr. *mēros,* thigh, + *stoma,* mouth), most of which are long extinct.

Horseshoe crabs are called "living fossils" because they are survivors of a group that was abundant during the Silurian period, 425 million years ago. The genus *Limulus,* to which the living horseshoe crab *Limulus polyphemus** belongs, arose 250 million years ago and has changed little during this vast span of time. Horseshoe crabs are marine bottom dwellers that feed on molluscs, worms, and dead fish. They live along the Atlantic Coast from Nova Scotia to the tip of Florida and the Gulf of Mexico to the Yucatan Peninsula. However, they are historically abundant along the coasts of Virginia, Delaware, and New Jersey, where they come ashore in great numbers during spring high tides to spawn. Their eggs are important food for millions of migratory shorebirds, which feast on the eggs and may more than double their weight before continuing their northward migration. It has been suggested that recent declines in horseshoe crab numbers may be impacting the success of migratory bird species. Until recently, horseshoe crabs were used by farmers as fertilizer and as feed for farm animals.** Today, horseshoe crabs are collected by the biomedical industry to support the production of a clotting agent that aids in the detection of human pathogens in patients, drugs, and intravenous devices (Figure 14-1). No other procedure has the same accuracy as a test developed from horseshoe crab blood.

External Features

The entire body of a horseshoe crab is covered with a tough, leathery **exoskeleton** that contains chitin, a tough polysaccharide. Among products made from horseshoe crab chitin is surgical suture thread that promotes healing as it slowly dissolves. As a horseshoe crab grows, it must shed (molt) the exoskeleton, a process called **ecdysis.**

Cephalothorax (Prosoma). Covering the cephalothorax dorsally and laterally is a hard, horseshoe-shaped **carapace** (F. from Sp. *carapacho,* shell), concave below and convex above. A pair of lateral compound eyes and a pair of median simple eyes are on the dorsal side.

On the ventral side are six pairs of appendages, located around the mouth. Using Figure 14-2 as a guide, and noting function, find the first pair of appendages, the **chelicerae** (ke-liss′uh-ree; sing. **chelicera;** Gr. *chēlē,* claw,+*keras,* horn); the second pair, the **pedipalps** (L. *pes, pedis,* foot, + *palpus,* stroking); and the **walking legs.** All appendages except the last pair of walking legs and pedipalps of

*The curious generic name of the horseshoe crab, *Limulus,* derives from the Latin meaning "a little askew" or "odd." The specific epithet, *polyphemus,* derives from the one-eyed giant of Greek mythology and presumably refers to the animal's median simple eye.

**Although farmers no longer harvest significant numbers of horseshoe crabs for fertilizer, eel and conch fishermen collect horseshoe crabs for bait. Responding to concern from scientists, environmentalists, and the biochemical industry that populations were declining rapidly, an interstate management program adopted in 1998 has helped to reduce horseshoe crab capture for bait. Although the status of the horseshoe crab population had been difficult to assess, a 2013 stock assessment update indicates that horseshoe crab abundance has increased in the Southeast, is stable in the Delaware Bay region, but declining in the Northeast.

Figure 14-1
Technician collecting blood from horseshoe crabs.

the male are chelate—that is, they bear pincers, or **chelae** (ke′lee; sing. **chela;** Gr. *chēlē,* claw). Lacking specialized appendages for chewing food, such as the mandibles of other arthropods, the walking legs bear spiny processes called **gnathobases** (Gr. *gnathos,* jaw, +*basis,* base) on their basal segments that serve a similar function. Move the appendages and note how these processes would tear up food and move it toward the mouth. The chelae of the appendages pick up food and pass it to the gnathobases. Note that the last pair of walking legs has no chelae; instead, each has four movable, bladelike processes, one of which is tipped with a pair of spines. These legs are used to push against the sand to help in forward movement and in burrowing. Between the last pair of walking legs is a small, rudimentary pair of appendages called **chilaria.**

Abdomen (Opisthosoma). The abdomen bears six pairs of spines along the sides and, on its ventral side, six pairs of flat, platelike appendages. The first of these forms the **genital operculum,** on the underside of which are two **genital pores.**

The other five abdominal appendages are modified as **gills.** Lift up one of these flaps to see the many (100 to 150) leaflike folds called **lamellae** (la-mel′ee; sing. **lamella;** L. dim. of *lamina,* plate). Because of this leaflike arrangement, such gills are called **book gills** (Figure 14-2). Exchange of gases between blood and surrounding water takes place in the lamellae. Movement of the gills not only circulates water over them but also pumps blood in and out of the lamellae. The blood contains **hemocyanin,** a respiratory pigment used in oxygen transport, as well as an amebocyte that kills invading bacteria. A chemical extracted from the amebocytes is now used to assure absence of harmful bacteria in intravenous drugs, vaccines, and implantable medical devices. Each summer, blood is collected from thousands of horseshoe crabs for extraction of amebocytes; the crabs are returned to the sea after bleeding.

Beating of the abdominal flaps can also be used in swimming and may aid the animal in burrowing by creating a water current that washes out mud or sand posteriorly.

Telson. The long, slender **telson** (Gr., extremity), or tail spine, is used for anchoring when the animal is burrowing or plowing through the sand or in righting itself when turned over. The **anus** is located under the proximal end of the telson.

Reproduction

During mating and egg-laying, horseshoe crabs aggregate in shallow water during high tides of full and new moons in spring and summer. A male clasps a female's carapace with his modified pedipalps and is carried around by the larger female. The female lays eggs in a depression in the

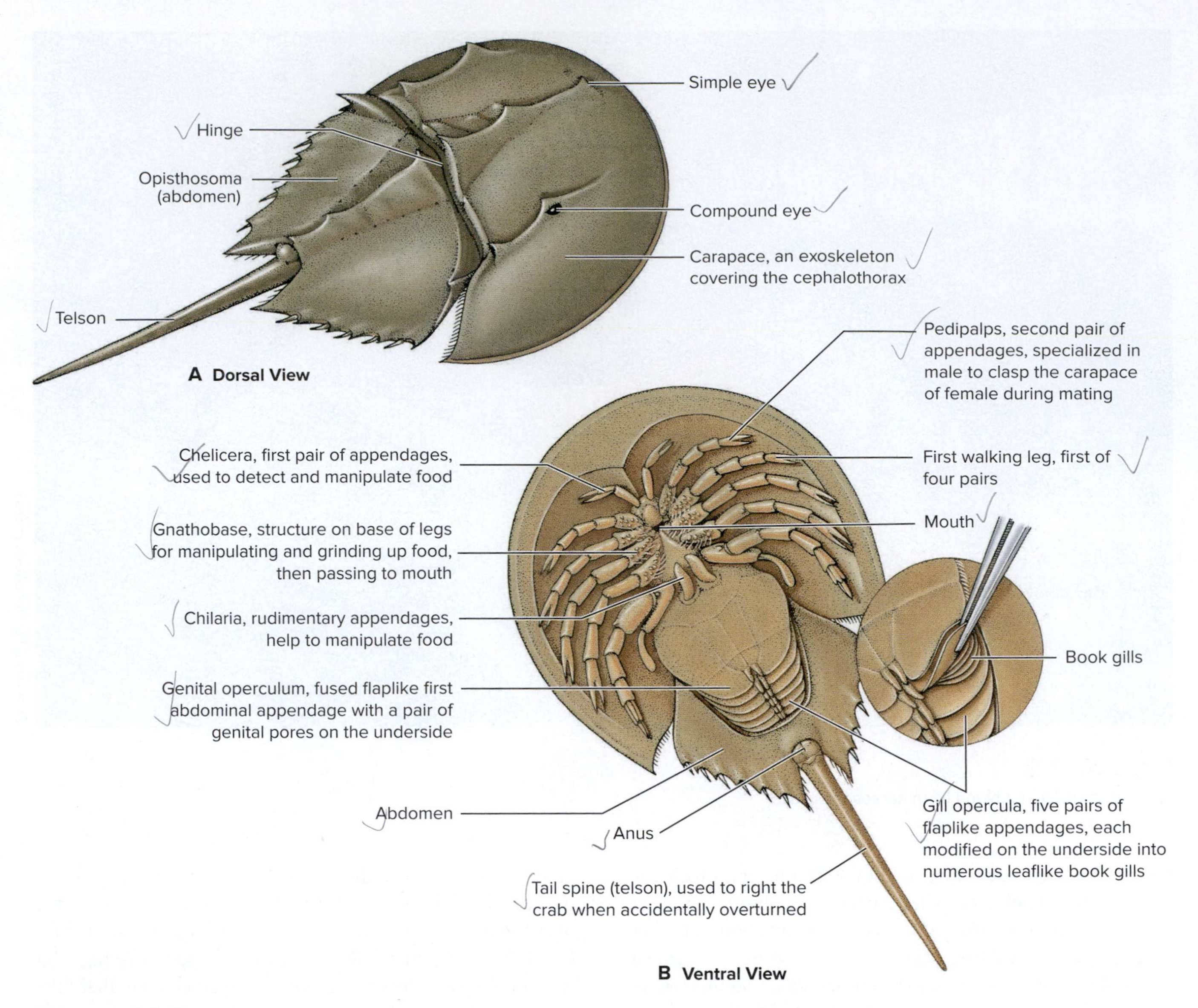

Figure 14-2

A, Dorsal view of *Limulus,* the horseshoe crab. **B,** Ventral view of a female. The inset shows an operculum lifted to reveal the location of book gills.

sand near the high-tide mark while the male sprays sperm on the eggs as they emerge. After several weeks, the eggs hatch as free-swimming **trilobite larvae** (Figure 14-3), so named because of their superficial resemblance to a trilobite. A larva looks much like an adult except that it lacks the tail spine and has only two of the five pairs of book gills. As it develops through a series of molts, segments and appendages are added until the young animal reaches adult form.

Behavior

If live horseshoe crabs are available in a marine aquarium, you may make some observations.

Where do you find the resting animals? Swimming about, resting on the sand, or covered with sand? Before disturbing them, drop some bits of fresh shrimp, oyster, or fish meat near them. Do they respond? How? (Normally they do most of their feeding at night.)

To observe their respiratory gill movements, lift an animal in the water, so that you can watch the beating of the gills. Gill movement is probably faster than when the animal is resting. Why? Each time the flaps move forward, blood flows into the gill lamellae; as the flaps move backward, the blood flows out.

Free the animal near the surface of the water and see how it swims. Does it turn over? Which appendages does it use in swimming? As it settles to the sandy bottom, watch its reactions. Does it try to burrow? How? Can you see the use of the last pair of legs? Does the animal arch its back? Does it use the telson?

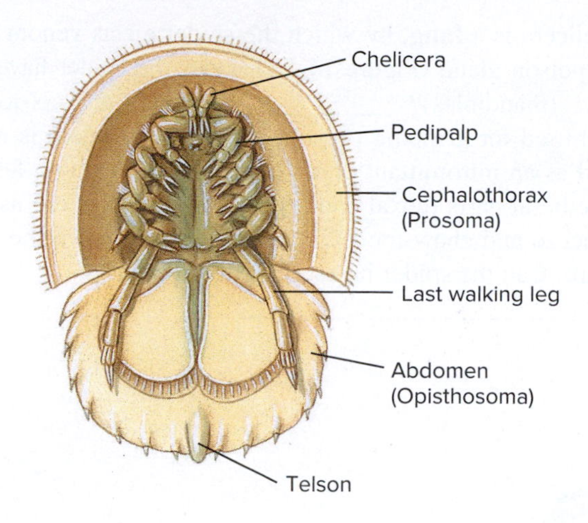

Figure 14-3
Trilobite larva of the horseshoe crab.

Written Report

Record your observations on separate paper.

Further Study

Garden Spider

Phylum Arthropoda
Subphylum Chelicerata
Class Arachnida
Order Araneae
Genus *Argiope*

Where Found

Spiders are distributed in all kinds of habitats, such as forests, deserts, mountains, swamps, land, and water. The garden spider (*Argiope,* ar-ji′uh-pee; Gr., nymph of mythology), which builds its orb webs in sunny places in gardens and tall grass, is found throughout the United States and Canada. Two common species are *A. aurantia,* with a mottled black and yellow abdomen (Figure 14-4), and *A. trifasciata,* with black and yellow bands on the abdomen. The males are about one-fourth the size of the females (4 to 8 mm) and are rarely seen. They spin smaller webs near those of the female.

Behavior

A garden spider will spin its orb web in captivity, and, if the spider is kept in a large glass container with tall grass or twigs and covered with screen or cheesecloth, the process can be watched. The spider spins a symmetric orb web and prefers to hang head downward in the center, holding its forelegs and hind legs close together (Figure 14-4). If web building can be observed, note the process—how the spider

Figure 14-4
Black and yellow garden spider, *Argiope aurantia,* which builds its orb web in gardens or tall grasses and then hangs there, head down, awaiting an unwary insect.

makes a supporting framework, then the radial spokes, and finally the spiral—first a temporary spiral beginning outside and then a permanent spiral beginning at the center. If possible, examine bits of the silk under a microscope. Is there more than one kind of silk? ______________________________

Adding insects to the terrarium containing the spider may allow you to witness a capture in the net, biting of the prey to paralyze it, and then securing of the prey with silken thread.

Spiders secrete enzymes to begin digestion outside the body. *Argiope* can apparently crush and tear the prey as well as suck out the liquid parts. How long does it take the spider to complete a meal? ______________________________

Can you see the eyes? ____________ Most spiders have poor vision (jumping spiders are an exception) but are covered with sensory hairs and are very sensitive to touch and vibration.

External Features

Study a preserved specimen and handle gently. There is always danger of breaking off appendages or the abdomen from the rest of the body. Use a hand lens or dissecting microscope to examine the parts. Keeping the specimen moist will help prevent its becoming brittle.

The chitinous **exoskeleton** is hard, thin, and somewhat flexible. **Sensory hairs** project from all parts of the body. The tagmata of the arachnid include the anterior

cephalothorax and the posterior **abdomen** joined by a slender waist, or **pedicel.**

Cephalothorax. Most spiders have six to eight **eyes** on the anterior dorsal surface, but some have fewer. Spiders do not have compound eyes; all are simple ocelli. How many eyes does your specimen have? __________ Use a dissecting microscope to find them.

Identify the paired chelicerae, which are vertically oriented on the front of the face. The terminal segment of a chelicera is a **fang,** by which the spider ejects venom from its poison gland (Figure 14-5B). Does the spider have true jaws (mandibles)? __________ Pedipalps are six-jointed and used for gripping prey. In males the pedipalp is modified as an intromittant organ to transfer sperm to a female. The basal parts (coxal endites) of the pedipalps are used to squeeze and chew food. Find the mouth between the pedipalps. Can the spider ingest only liquid food? __________
____

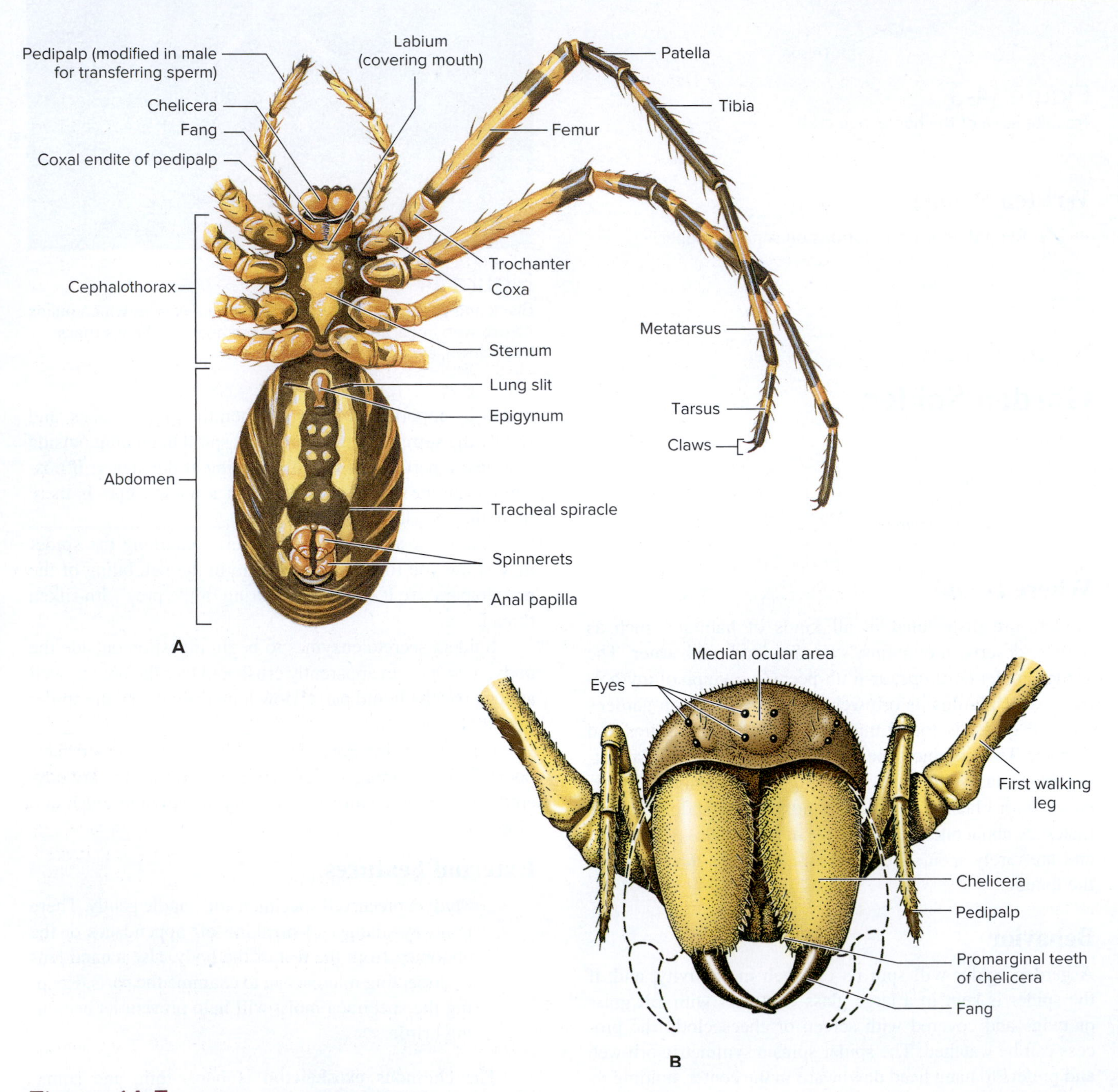

Figure 14-5
A, Ventral view of a female garden spider, *Argiope aurantia*. **B,** Anterior, or "face," view.

How many pairs of **walking legs** are there? __________ Each leg is made up of seven segments, as follows (from base to distal end): **coxa, trochanter, femur, patella, tibia, metatarsus,** and **tarsus.** The tarsus has claws and a tuft of hair at its terminal end.

Abdomen. On the ventral surface at the anterior end, find two lateral, slitlike openings that mark the location of the **book lungs.** Book lungs are fashioned very much as are the book gills of horseshoe crabs, except that they are enclosed internally in pockets. The inner walls of these pockets are folded into long, thin plates (leaves of the "book") held apart by bars, so that there are always air spaces between them. Gas exchange occurs between blood circulating inside the lamellae and air flowing in the spaces between the lamellae. These air spaces connect with a small air chamber in each lung that opens to the outside through the slitlike openings, or spiracles, already observed (Figure 14-5). Some spiders—the tarantula, for example—have two pairs of book lungs. By peeling forward the cover of the lung and examining with a dissecting microscope, you should be able to see many thin respiratory leaves.

Between the spiracles, locate the **epigynum** (e-pij′in-um; Gr. *epi,* on, + *gynē,* woman), which conceals the female genital pore. Preserved specimens are probably all females.

Posteriorly on the abdomen, just in front of the spinnerets, is a small **tracheal spiracle.** This is an opening into a small chamber from which tracheal tubes extend into the body. What is the function of the tracheal system? _______________ Arachnid tracheal systems are similar to those of insects but less extensive. Garden spiders have both book lungs and tracheae, but not all spiders do; some have only one type of respiratory organ.

There are three pairs of **spinnerets** on a raised surface. The middle pair is quite small, but the other two pairs are rather large, conical, and readily movable. The ends of the spinnerets have a variety of tiny silk spouts, each producing a particular type of silk. Silk is secreted as a fluid by the silk glands and hardens with exposure to air. Examine the spinnerets with a dissecting microscope.

A small, fleshy papilla just posterior to the spinnerets bears the **anus.**

Build Your Knowledge of Chelicerate Features

Test your understanding of the main features of the chelicerates covered in the exercise by filling out the summary table on pp. xi–xii. Try to complete the table without looking at your notes.

Classification

Phylum Arthropoda

Subphylum Trilobita (tri-lo-bi′ta) (Gr. *tri,* three, + *lobos,* lobe). Trilobites. All extinct forms; Cambrian to Carboniferous; body divided by two longitudinal furrows into three lobes; distinct head, thorax, and abdomen; biramous (two-branched) appendages.

Subphylum Chelicerata (ke-liss′uh-ra′ta) (Gr. *chēlē,* claw, + *keras,* horn, + *ata,* group suffix). Eurypterids, horseshoe crabs, spiders, and ticks. First pair of appendages modified to form chelicerae; pair of pedipalps and four pairs of legs; no antennae, no mandibles; cephalothorax and abdomen usually unsegmented.

Class Merostomata (mer′o-sto′ma-ta) (Gr. *mēros,* thigh, + *stoma,* mouth, + *ata,* group suffix). Aquatic chelicerates that include horseshoe crabs *(Limulus)* and extinct Eurypterida.

Class Pycnogonida (pik′no-gon′i-da) (Gr. *pyknos,* compact, + *gony,* knee, angle). Sea spiders.

Class Arachnida (ar-ack′ni-da) (Gr. *arachnē,* spider). Spiders, scorpions, and their allies. Segments fused into cephalothorax; head with paired chelicerae and pedipalps; four pairs of legs; abdomen segmented or unsegmented, with or without appendages; respiration by gills, tracheae, or book lungs. Example: *Argiope.*

Subphylum Crustacea (crus-ta′she-a) (L. *crusta,* shell, + *acea,* group suffix). Crustaceans. With gills; body covered with carapace; exoskeleton with limy salts; appendages currently considered biramous and variously modified for different functions; head with two pairs of antennae. Examples: *Cambarus, Homarus.*

Subphylum Myriapoda (mir-ee-ap′o-da) (Gr. *myrias,* a myriad, + *podus,* foot). Centipedes, millipedes, pauropods, and symphylans. All appendages uniramous; head appendages consisting of one pair of antennae, one pair of mandibles, and one or two pairs of maxillae.

Class Diplopoda (di-plop′o-da) (Gr. *diploos,* double, + *pous, podos,* foot). Millipedes. Subcylindric body elongated and wormlike; variable number of segments; usually two pairs of legs to a segment. Example: *Spirobolus.*

Class Chilopoda (ki-lop′o-da) (Gr. *cheilos,* lip, + *pous, podos,* foot). Centipedes. Elongated with dorsoventrally flattened body; variable number of segments, each with pair of legs; tracheae present. Example: *Lithobius.*

Class Pauropoda (pau-rop′o-da) (Gr. *pauros,* small, + *pous, podos,* foot). Pauropods. Minute, soft-bodied forms with 12 segments and 9 or 10 pairs of legs. Example: *Pauropus.*

Class Symphyla (sym′fy-la) (Gr. *syn,* together, + *phylon,* tribe). Garden centipedes. Centipede-like bodies of 15 to 22 segments and usually 12 pairs of legs. Example: garden centipede, *Scutigerella.*

Subphylum Hexapoda (hek-sap′oda) (Gr. *hex,* six + *pous, podus,* foot). Insects. Body with distinct head, thorax, and abdomen; thorax usually with two pairs of wings; three pairs of jointed legs. Example: lubber grasshopper, *Romalea.*

NOTES

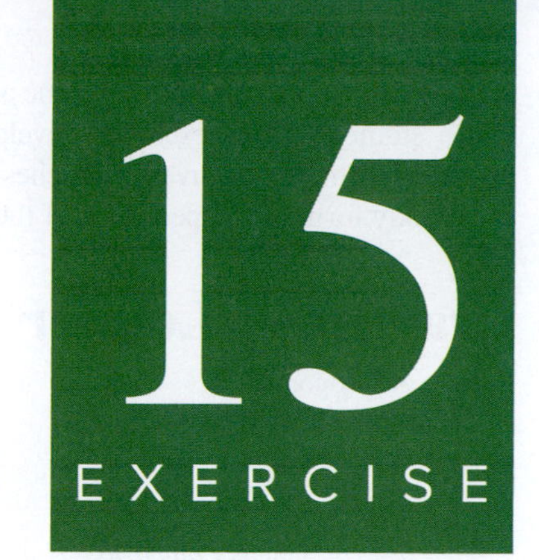

The Crustacean Arthropods

EXERCISE 15A

Subphylum Crustacea—Crayfish, Lobsters, and Other Crustaceans

EXERCISE 15A
Subphylum Crustacea—Crayfish, Lobsters, and Other Crustaceans
Crayfish or Lobster
Other Crustaceans
Crustacean Development, Exemplified by Brine Shrimp
Projects and Demonstrations

EXPERIMENTING IN ZOOLOGY
The Phototactic Behavior of Daphnia

Crustaceans are **gill-breathing arthropods,** sometimes referred to as "insects of the sea." Just as insects dominate land, crustaceans abound in the world's oceans, lakes, and rivers. Like other arthropods, crustaceans have an exoskeleton composed of chitin to which their muscles attach. Crustaceans also are distinguished by having five head segments bearing **two pairs of antennae, mandibles,** and **two pairs of maxillae.** In most crustaceans, the head segments are fused with the thoracic segments to form a **cephalothorax.** All appendages are ancestrally **biramous** (bear two branches), and some appendages of present-day adult crustaceans are biramous. Most crustaceans have a **nauplius** larva, which

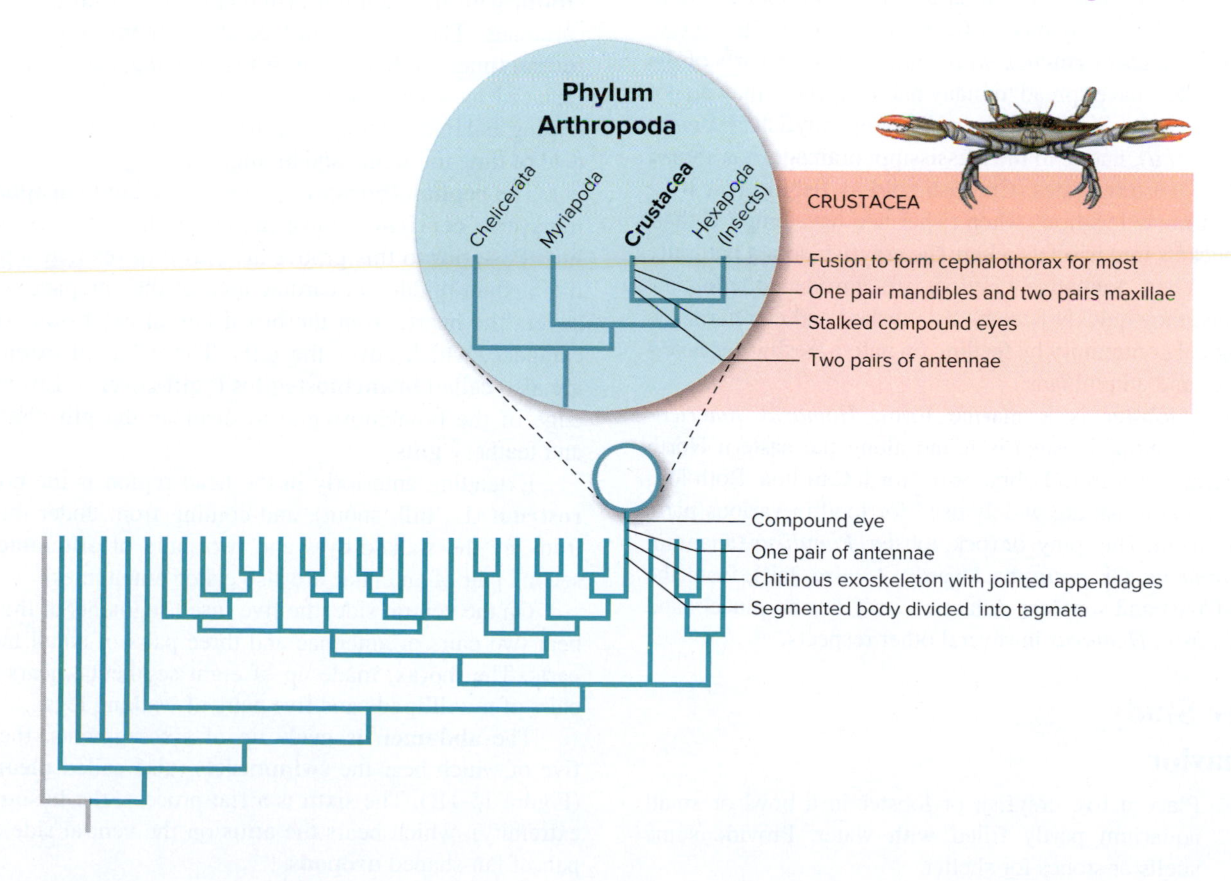

undergoes metamorphosis to become adult. However, in crayfishes studied in this exercise, development is direct, with no larval form; a tiny juvenile hatches from the egg with the same body form and appendages of the adult.

Crayfish or Lobster

Phylum Arthropoda
Subphylum Crustacea
Group Vericrustacea
Class Malacostraca
Order Decapoda
Genus *Cambarus*

The following description applies to either crayfish (*Cambarus, Procambarus, Pacifastacus,* and *Orconectes*) or lobsters (*Homarus* and others), for crayfish and lobsters are very similar, except in size.

Where Found

Crayfish, called locally and variously crawfish, crawdads, and mudbugs, are found in freshwater streams and ponds all over the world. There are about 350 species of crayfish in the United States, more species of crayfish than in all the rest of the world. The southern United States, especially the Gulf Coast, has the largest number of crayfish species. This area also produces most edible crayfish, and millions are raised each year on watery crayfish "farms" in Louisiana.

Crayfish are omnivorous, feeding on all kinds of succulent aquatic vegetation and on animals such as snails, worms, and small vertebrates. Crayfish are hearty animals, able to survive in almost any type of freshwater habitat. Rusty crayfish *(Orconectes rusticus),* while native to some parts of the Great Lakes, have spread to many habitats where they do not naturally occur. Similarly, the red swamp crayfish *(Procambarus clarkii),* native to the Mississippi drainage, has spread to many western states. Crayfish sold as fishing bait have spread to new locations when fishermen have emptied their bait buckets into habitats where the crayfish do not naturally occur. When nonnative crayfish are introduced into new habitats, they quickly become established and can harm the biological community by feeding on native freshwater invertebrates and amphibians.

The lobster is a marine form. *Homarus americanus* (F. *homard,* lobster) is found along the eastern North American coast from Labrador to North Carolina. Both lobsters and crayfish are widely used for food in various parts of the world. The spiny, or rock, lobster, *Panulirus* (anagram of *Palinurus,* Gr. *palin,* backwards, + *oura,* tail), from the West Coast and southern Atlantic Coast, has no pincers and differs from *Homarus* in several other respects.

Core Study

Behavior

Place a live crayfish or lobster in a bowl or small aquarium partly filled with water. Provide some shells or stones for shelter.

How are the antennae used? ______________ What is the response when you stroke the antennae? ______________ Notice the compound eyes on the ends of stalks. Are the eyestalks movable? ______________ Of what advantage would such eyes be to the animal? ______________ What evidence of segmentation do you see on the dorsal side of the animal?

____

Note the five pairs of legs. The first pair are called chelipeds because they bear the large claws (chelae). How are the chelipeds used? ______________ When the animal is startled, what does it do? ______________ Is the escape movement forward or backward? ______________ How is it accomplished? ______________ Is the tail fan involved? ______________

Lift up a crayfish and notice the swimmerets on the abdomen. Release it near the surface of the water. Can it swim? ______________ How? ______________ The females use the swimmerets to carry and aerate eggs during the breeding season.

Drop a bit of meat or fish near an undisturbed crayfish in an aquarium. What appendages are used to pick up and handle food? ______________ Notice the activity of the small mouth appendages as the animal feeds.

External Features

Place a preserved crayfish or lobster in a dissecting pan and add water to the pan.

The **exoskeleton** is a cuticle secreted by the epidermis and hardened with a nitrogenous polysaccharide called **chitin,** with the addition of mineral salts, such as calcium carbonate. The cuticle must be shed, or **molted** (ecdysis), several times while the crayfish is growing, each time being replaced by a new, soft exoskeleton, which soon hardens. During and immediately after molting, crayfish spend a great deal of time immobile and hiding. Why? ______________

The **cephalothorax** is covered by a hard **carapace.** A transverse **cervical groove** marks the head-thorax fusion line. Posterior to this groove are two grooves that separate the median middorsal cardiac area of the carapace, which covers the heart, from the broad lateral extensions of the carapace, which cover the gills. These lateral extensions are also called **branchiostegites** ("gill-cover"). Lift up the edge of the branchiostegite to disclose the **gill chamber** and feathery **gills.**

Extending anteriorly in the head region is the pointed **rostrum** (L., bill, snout), and coming from under the rostrum are the stalked **eyes** and two pairs of **antennae** (the second pair of antennae are also called antennules).

On the ventral side, the five fused segments of the head bear two pairs of antennae and three pairs of small mouthparts. The thorax, made up of eight segments, bears three pairs of **maxillipeds** and five pairs of walking legs.

The **abdomen** is made up of six segments, the first five of which bear the **swimmerets** (also called **pleopods**) (Figure 15-1B). The sixth is a flat process, the **telson** (Gr., extremity), which bears the **anus** on the ventral side and a pair of fan-shaped **uropods.**

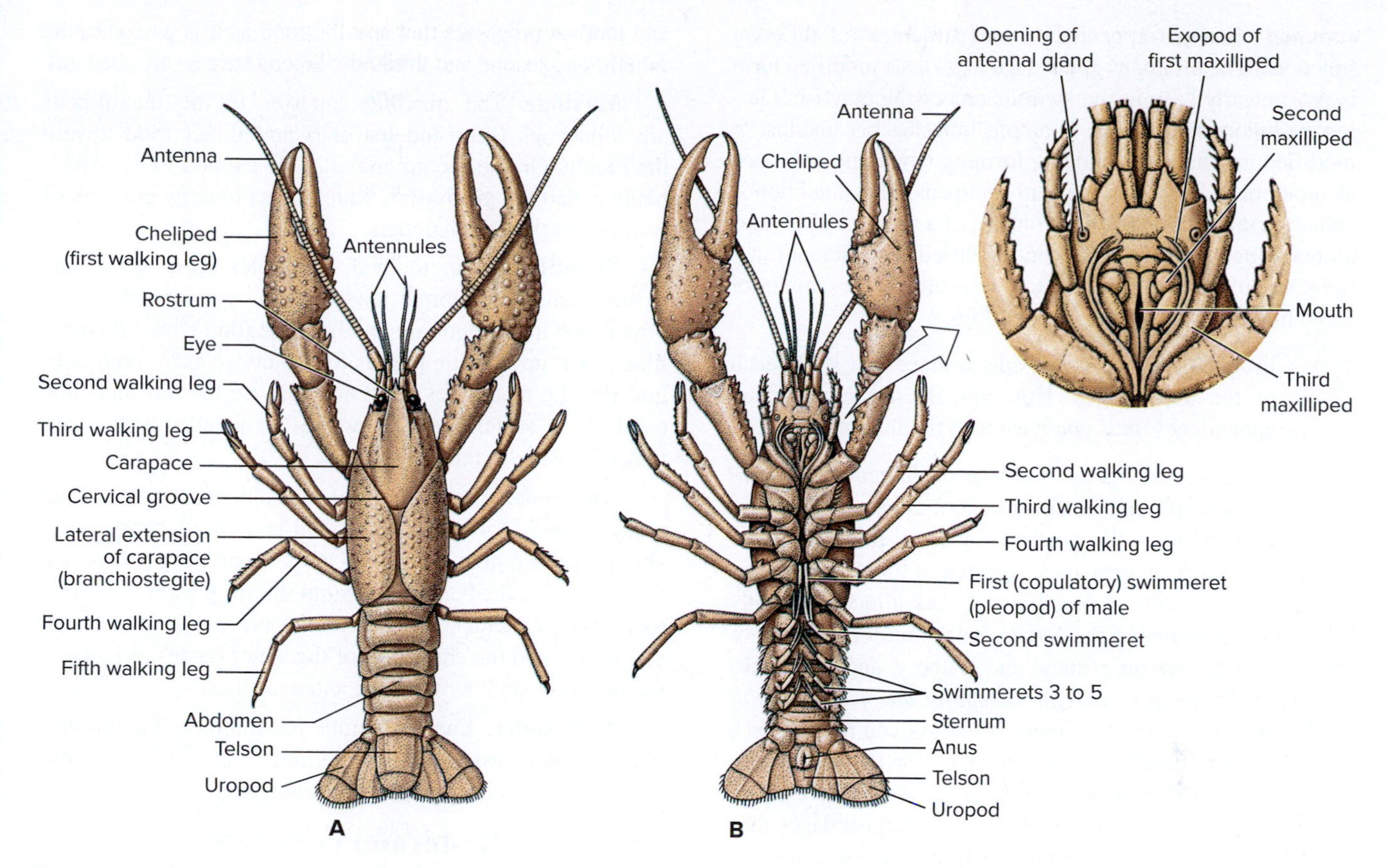

Figure 15-1

External structure of crayfish. **A,** Dorsal view. **B,** Ventral view.

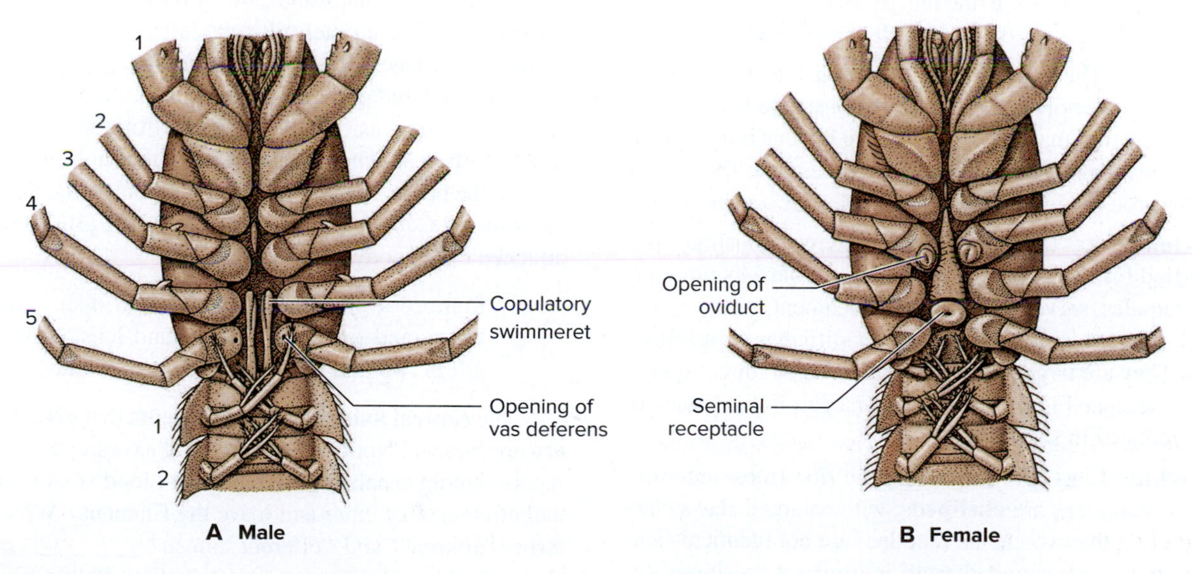

Figure 15-2

Ventral views of crayfish, male and female.

Genital Openings. In males the genital openings of the sperm ducts are located medially at the base of each of the fifth walking legs (Figure 15-2). In females the genital openings of the oviducts are located at the base of each third walking leg. Where is the opening of the seminal receptacle? ________________ During copulation the male turns the female over and presses the sperm along grooves of the specialized first (copulatory) swimmerets into the seminal receptacle of the female.

Dissection of the Appendages

As you dissect crayfish appendages, you will see how they illustrate the **principle of serial homology.** It is an exceptional example of the evolutionary modification of a basic plan—the common biramous appendage—into structures of different functions. The biramous appendage in its least modified form is most clearly seen in the swimmerets of the crayfish. During evolution, some of the biramous branches became lost or modified and new parts added, forming new structures such as mouthparts, walking legs, and chelipeds. This condition is called **serial homology,** the evolution of a series of structures all homologous to each other but modified for different functions. Crayfish and their relatives possess the best examples of serial homology in the animal kingdom.

From the animal's left side, remove and study each of the appendages. However, do not remove any appendages until you have read the instructions.

Although appendages are numbered consecutively beginning at the anterior end of the animal, it is easier to remove them by beginning at the **posterior** end and proceeding forward. To remove an appendage, grasp it at the base with forceps as near the body as possible and work it loose gradually by gently manipulating the forceps back and forth. Some of the appendages are quite small and feathery, and some have attached gills. *Be very careful to remove all of the parts of each appendage together* (Figure 15-3). As you remove each one, identify its **medial** and **lateral** sides. Pin the appendages in order on a sheet of paper in the bottom of a dissecting pan. Keep covered with water. Alternatively, the appendages may be glued (e.g., with Elmer's Glue-All) in order on a piece of white cardboard and labeled. Each biramous appendage, at least in its early growth stages, consists of a basal part, the **protopod,** which bears a medial (inner) branch called the **endopod,** and a lateral (outer) branch, the **exopod.**

Uropods. The broad uropods (Gr. *oura,* tail, + *pous,* foot) on the most posterior segment are biramous; together with the medial telson, they make up the strong **tail fan** that is used in the rapid backward movements so important in escape. The tail fan also helps protect eggs and young on the female's swimmerets.

Swimmerets (Pleopods). Swimmerets illustrate the unmodified biramous plan. They aid in forward locomotion and, in females, serve as a place of attachment for the eggs.

The first two pairs in males are modified as **copulatory organs.** They are large and grooved and used to direct sperm onto females (see Figure 15-2). The first pair in the female is usually reduced in size.

Walking Legs (Pereiopods). The first (most anterior) pair of walking legs are **chelipeds,** with enlarged claws **chelae**. Note that the two chelae (keé-lee) are not identical. The heavier chela with rounded teeth is used for crushing; the other, more slender, chela with sharp teeth is used for tearing and seizing prey. This difference is more distinctive in lobsters than in crayfish. The other four pairs are used for walking and food handling. There is no exopod, so they are called **uniramous** (single-branched). The first four pairs are attached to gills. Which of the four pairs of walking legs bear chelae? ____________________

Maxillipeds. The three paired maxillipeds (L. *maxilla,* jaw, + *pes,* foot) are feeding appendages with bristly edges and toothed processes that tear the food as it is passed to the mouth. The second and third maxillipeds have an attached gill.

Maxillae. The maxillae anterior to the maxillipeds are foliaceous (thin and leaflike) and direct food toward the mouth. On the second maxilla, the exopod forms a long blade called the **gill bailer,** which beats to draw currents of water from the gill chambers.

Mandibles. The toothed mandibles work from side to side (unlike vertebrate jaws that move up and down) to direct food into the mouth and hold the food while the maxillae tear it up. A little palp is folded above each tooth margin. Pry the mandibles apart and remove the left mandible carefully. A strong mandibular muscle attached to the base may tear off with the mandible.

Antennae. The endopod of each antenna is a very long, many-jointed filament. The exopod is a broad, sharp, movable projection near the base. On its broad protopod on the ventral side of the head is the **renal opening** from the excretory gland. Antennae and antennules are sensitive to touch, vibrations, and the chemistry of the water (taste). When you fed the crayfish, how were the antennae used? __________

Antennules. Each antennule has a three-jointed protopod as well as two long, many-jointed filaments. The antennules are also concerned with equilibrium.

Branchial (Respiratory) System

Remove part of the carapace on the animal's right side and note the feathery **gills** lying in the branchial chamber. You have already seen that some of the gills are attached to certain appendages. These outer gills are called the foot gills. How many appendages have gills attached? ____________ Move the appendages to determine this. Separate the gills carefully, laying aside the foot gills. Another row of gills underneath is attached to membranes that hold the appendages to the body. These gills are called joint gills. Some genera, but not *Cambarus,* have a third row of gills (side gills) attached to the body wall.

Remove a gill, place it in a small dish, cover with water, and examine with a hand lens. Now cut the gill in two and look at one of the cut ends.

The **central axis** bears **gill filaments** that give it a feathery appearance. Notice that the central axis and the little filaments contain canals. These represent blood vessels (afferent and efferent) that enter and leave the filaments. What do the terms "afferent" and "efferent" mean? __________ What change in blood happens as it passes through the filaments? ____________ Water enters the gill chamber by the free ventral edge of the carapace and is drawn forward over the gills by action of the gill bailer of the second maxilla, facilitated by movements of the other appendages.

Brief Preview of Internal Structure

Remove the dorsal portion of the exoskeleton by inserting the point of a scissors under the posterior edge of the lateral carapace about 1.3 cm to one side

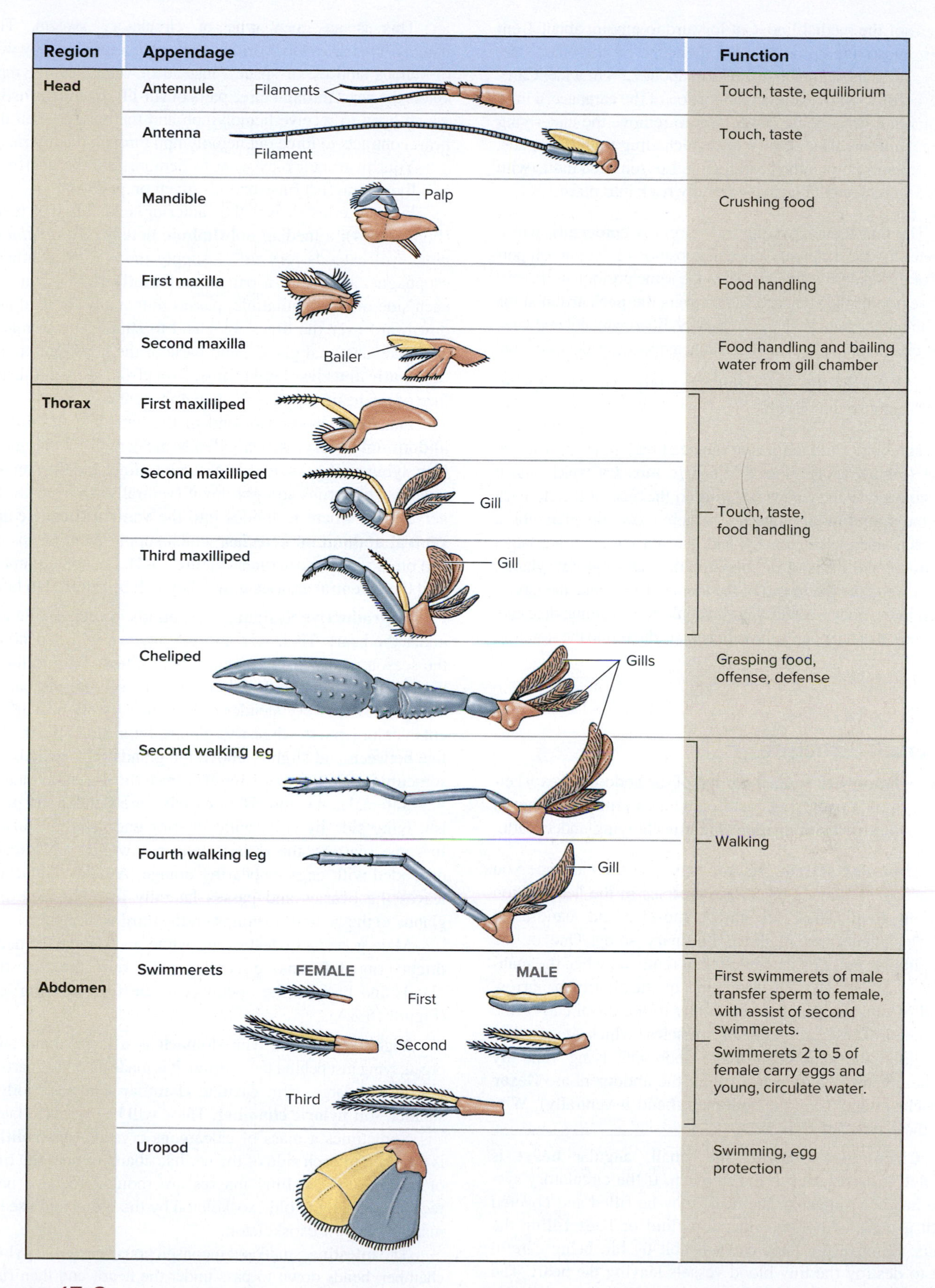

Region
Appendage
Function
Head
Antennule
Filaments
Touch, taste, equilibrium
Antenna
Filament
Touch, taste
Mandible
Palp
Crushing food
First maxilla
Food handling
Second maxilla
Bailer
Food handling and bailing water from gill chamber
Thorax
First maxilliped
Second maxilliped
Gill
Third maxilliped
Gill
Touch, taste, food handling
Cheliped
Gills
Grasping food, offense, defense
Second walking leg
Fourth walking leg
Gill
Walking
Abdomen
Swimmerets
FEMALE
MALE
First
Second
Third
First swimmerets of male transfer sperm to female, with assist of second swimmerets.
Swimmerets 2 to 5 of female carry eggs and young, circulate water.
Uropod
Swimming, egg protection

Figure 15-3

Appendages of the crayfish left side. Base (protopod) in brown; medial branch (endopod) in blue; lateral branch (exopod) in yellow.

of the medial line. Cut forward to a point about 1 cm posterior to the eye. Do the same on the other side, thus loosening a dorsal strip about 2.5 cm wide. Carefully remove this center portion of the carapace, a little at a time, being careful not to remove the underlying **epidermis** and **muscles,** which cling to the carapace, especially in the head region. Loosen such tissue with a scalpel and push it carefully back into place.

The thin tissue covering the viscera is **epidermis,** which secretes the exoskeleton. Notice the position of the pinkish portion of the epidermis, which lies in the same position as the cardiac region of the carapace. This covers the **pericardial sinus** containing the **heart.** The sinus may be filled with colored latex if the circulatory system has been injected for easy identification.

Remove the epidermis carefully with forceps to expose the viscera.

The large **stomach** lies in the head region, anterior to the heart (Figure 15-4). Note the **gastric muscles** which attach the stomach to the carapace. Lying to the side of the stomach are the **mandibular muscles,** which move the mandibles. On each side of the stomach and beneath the heart are large, cream-colored lobes of the **hepatopancreas** (digestive gland). They extend the full length of the thorax. This gland, the largest organ in the body, secretes digestive juices, containing digestive enzymes, that are poured through hepatic ducts into the stomach.

Further Study

Internal Structure

Remove the dorsal portion of the abdominal exoskeleton, uncovering each segment carefully so as not to destroy the long **extensor muscle** lying underneath.

Muscular system. As you removed the carapace you noticed the tendency of certain muscles in the head region to cling to it. These are gastric muscles and mandibular muscles mentioned in the preliminary study. One of the mandibular muscles may have been removed when the mandible was removed. On each side of the thorax, a narrow band of muscles runs longitudinally to the distal end of the abdomen. These are **extensor muscles,** which are used to straighten the abdomen. In the abdomen, lying ventrally to the extensors and nearly filling the abdomen, are **flexor muscles,** which flex the abdomen (bend it ventrally). Why are these muscles so large? ____________________

Circulatory System. The small, angular **heart** is located just posterior to the stomach. If the circulatory system has been injected, the heart may be filled and covered with a mass of colored injection fluid or latex filling the sinus. Remove the latex carefully bit by bit, being careful not to destroy the tiny blood vessels leaving the heart. The heart lies in a cavity called the **pericardial sinus,** which is enclosed in a membrane, the **pericardium.** The pericardium may have been removed with the carapace.

This is an open type of circulatory system. The hemolymph leaves the heart in arteries but returns to it by way of venous sinuses, or spaces, instead of veins. Hemolymph enters the heart through three pairs of slit-like openings, **ostia,** which open to receive hemolymph and then close when the heart contracts to force out hemolymph through the arteries.*

Your instructor may want to demonstrate the heartbeat of a living crayfish (instructions given on p. 236).

Five arteries leave the anterior end of the heart (Figure 15-4): a median **ophthalmic artery** (Gr. *ophthalamos,* eye) extends forward to supply the cardiac stomach, esophagus, and head; a pair of **antennal arteries,** one on each side of the ophthalmic, passes diagonally forward and downward over the digestive gland to supply the stomach, antennae, antennal glands, and parts of the head; and a pair of **hepatic arteries** (Gr. *hepatos,* liver) from the ventral surface of the heart supplies the hepatopancreas.

Leaving the posterior end of the heart are the **dorsal abdominal artery,** which extends the length of the abdomen, lying on the dorsal side of the intestine, and the **sternal artery,** which runs straight down (ventrally) to beneath the nerve cord, where it divides into the **ventral thoracic** and **ventral abdominal arteries,** which supply the appendages and other ventral structures (Figure 15-4). Do not attempt to find these ventral arteries now. They will be referred to later.

Reproductive System. The **gonads** in each sex lie just under the heart. Their size and prominence will depend on the season in which the animals were killed. To find them, lay aside the heart and abdominal extensor muscle bands. The gonads are very slender organs, usually slightly different in color from the digestive glands, lying along the medial line between and slightly above the glands. The gonads are sometimes difficult to distinguish from the digestive glands.

In females the gonads, or **ovaries,** are slender and pinkish, lying side by side, with anterior ends slightly raised. In some seasons, the ovaries may be swollen and greatly distended with eggs, appearing orange. A pair of **oviducts** leaves the ovaries and passes laterally over the digestive glands to the genital openings on the third walking legs.

Male gonads, or **testes,** are white and delicate. **Sperm ducts (vas deferens)** pass diagonally over the digestive glands and back to the openings in the fifth walking legs (Figure 15-4A).

Digestive System. The **stomach** is a large, thin-walled organ, lying just behind the rostrum. It is made up of two parts—anteriorly a large, firm **cardiac chamber** and posteriorly a smaller, soft **pyloric chamber.** These will be examined later.

Sometimes a mass of calcareous crystals **(gastroliths)** is attached to each side of the cardiac chamber near the time of molting. These limy masses are thought to have been recovered from the old exoskeleton by the blood and used in making the new exoskeleton.

The **intestine,** small and inconspicuous, leaves the pyloric chamber, bends down to pass under the heart, and then rises

*The ostia are best seen in uninjected specimens.

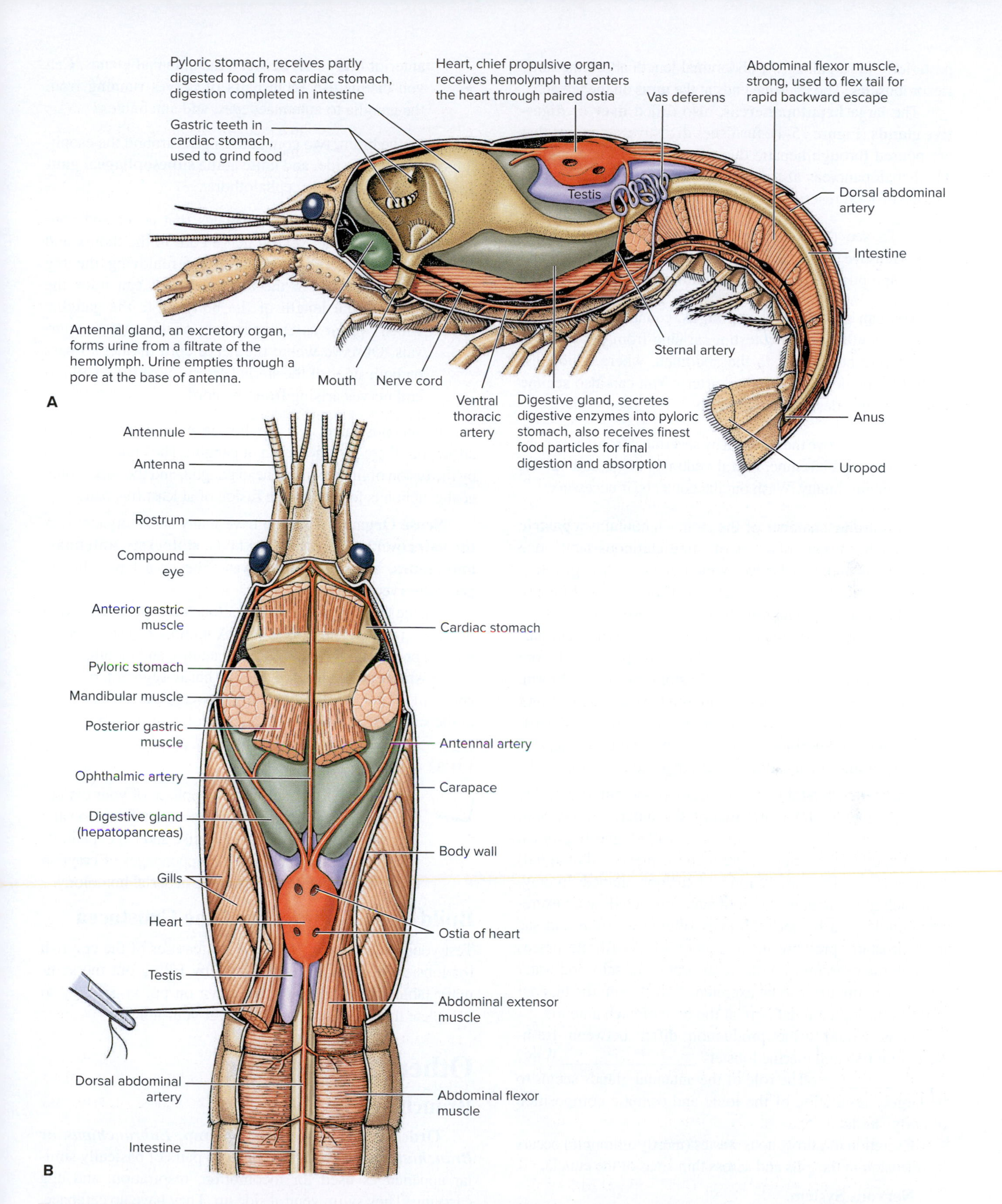

Figure 15-4

Internal structure of male crayfish. **A,** Lateral view of left side. **B,** Dorsal view. What median artery extends forward from the heart to supply the cardiac stomach, esophagus, and head? ______________________ What pair of arteries pass diagonally forward and downward over the digestive gland to supply the stomach, antennae, antennal glands, and part of the head? ______________________ What artery leaves the posterior of the heart? ______________________

posteriorly to run along the abdominal length above the large flexor muscles. The intestine ends at the **anus** on the telson.

The large **hepatopancreas,** also called **liver** or **digestive glands** (Figure 15-4), furnishes digestive secretions that are poured through hepatic ducts into the pyloric chamber. The hepatopancreas also is the chief site of absorption and storage of food reserves.

To see the whole length of the digestive system, remove the left lobe of the hepatopancreas and gonad and push aside the left mandibular muscle.

You can now see the **esophagus,** connecting the stomach and mouth, and the intestine, arising from the stomach and running posteriorly to the abdomen, where it lies just ventral to the dorsal abdominal artery. You can also see the **sternal artery** descending ventrally.

Now remove the stomach by severing it from the esophagus and intestine, turn it ventral side up, and open it longitudinally. Wash out the contents, if necessary.

The cardiac chamber of the stomach contains a **gastric mill,** which consists of a set of three chitinous teeth, one dorsomedial and two lateral, which are used for grinding food. They are held by a framework of ossicles and bars in the stomach and operated by the gastric muscles.

In the cardiac stomach, the food is ground up and partially digested by enzymes from the hepatopancreas before it is filtered into the smaller pyloric stomach in liquid form. Large particles must be egested through the esophagus. Rows of setae and folds of the stomach lining strain the finest particles and pass them from the pyloric stomach into the hepatopancreas or into the intestine, where digestion is completed.

Excretory System. In the head region anterior to the digestive glands and lying against the anterior body wall is a pair of **antennal glands** (also called **green glands,** though they will not appear green in the preserved material) (Figure 15-4). They are round and cushion-shaped. In crayfish each gland contains an **end sac,** connected by an excretory tubule to a bladder. Fluid is filtered into the end sac by hydrostatic pressure in the hemocoel. As filtrate passes through the excretory tubule, reabsorption of salts and water occurs, leaving urine to be excreted. A duct from the bladder empties through a renal pore at the base of each antenna.

How would urine production differ between freshwater crayfish and marine lobster? ________________ Why? ________________ The role of the antennal glands seems to be largely regulation of the ionic and osmotic composition of body fluids.

Excretion of nitrogenous wastes (mostly ammonia) occurs by diffusion in the gills and across thin areas of the cuticle.

Nervous System.

Carefully remove all of the viscera, leaving the esophagus and sternal artery in place. The brain is a pair of **supraesophageal ganglia** that lie against the anterior body wall between the antennal glands. Can you distinguish three pairs of nerves running from the ganglia to antennae, eyes, and antennules?

From the brain, two connectives pass around the esophagus, one on each side, and unite at the **subesophageal ganglion** on the floor of the cephalothorax.

Chip away the calcified plates that cover and conceal the double ventral nerve cord in the thorax and follow the cord posteriorly. By removing the big flexor muscles in the abdomen, you can trace the cord for the length of the body. Note the ganglia, which appear as enlargements of the cord at intervals. Observe where the nerve cord divides to pass on both sides of the sternal artery. Note the small lateral nerves arising from the cord.

In annelids there is a ganglion in each segment, but in arthropods there is some fusion of ganglia. The brain is formed by the fusion of three pairs of head ganglia, and the subesophageal ganglion is formed by the fusion of at least five pairs.

Sense Organs. Crayfish have many sense organs: **tactile hairs** over many parts of the body, **statocysts, antennae, antennules,** and **compound eyes.** The eyes have already been observed.

Tactile hairs are specialized for touch reception, detection of water currents, and orientation. A number of chemoreceptors are present on the antennae, antennules, and mouthparts.

A **statocyst** is located in the basal segment of each antennule. The pressure of sand grains against sensory hairs in the statocyst gives crayfish a sense of equilibrium.

Oral Report

Be prepared to demonstrate any phase of your dissection to your instructor. (1) Locate on the dissection any structure mentioned in the exercise and (2) explain its function. (3) Explain how the appendages of crayfish or lobster illustrate the principle of serial homology.

Build Your Knowledge of the Crustacea

Test your understanding of the main features of the crayfish (or lobster) covered in the exercise by filling out the summary table for Arthropoda Crustacea on pp. xi–xii. Try to complete the table without looking at your notes.

Other Crustaceans

Branchiopoda

Order Anostraca, Fairy Shrimp, *Eubranchipus* or *Branchinecta.* Fairy shrimp have 11 pairs of basically similar appendages used for locomotion, respiration, and egg carrying. They swim ventral side up. They have no carapace. Note the dark eyes borne on unsegmented stalks. Females carry eggs in a ventral brood sac, which can usually be seen when the animal is moving (Figure 15-5).

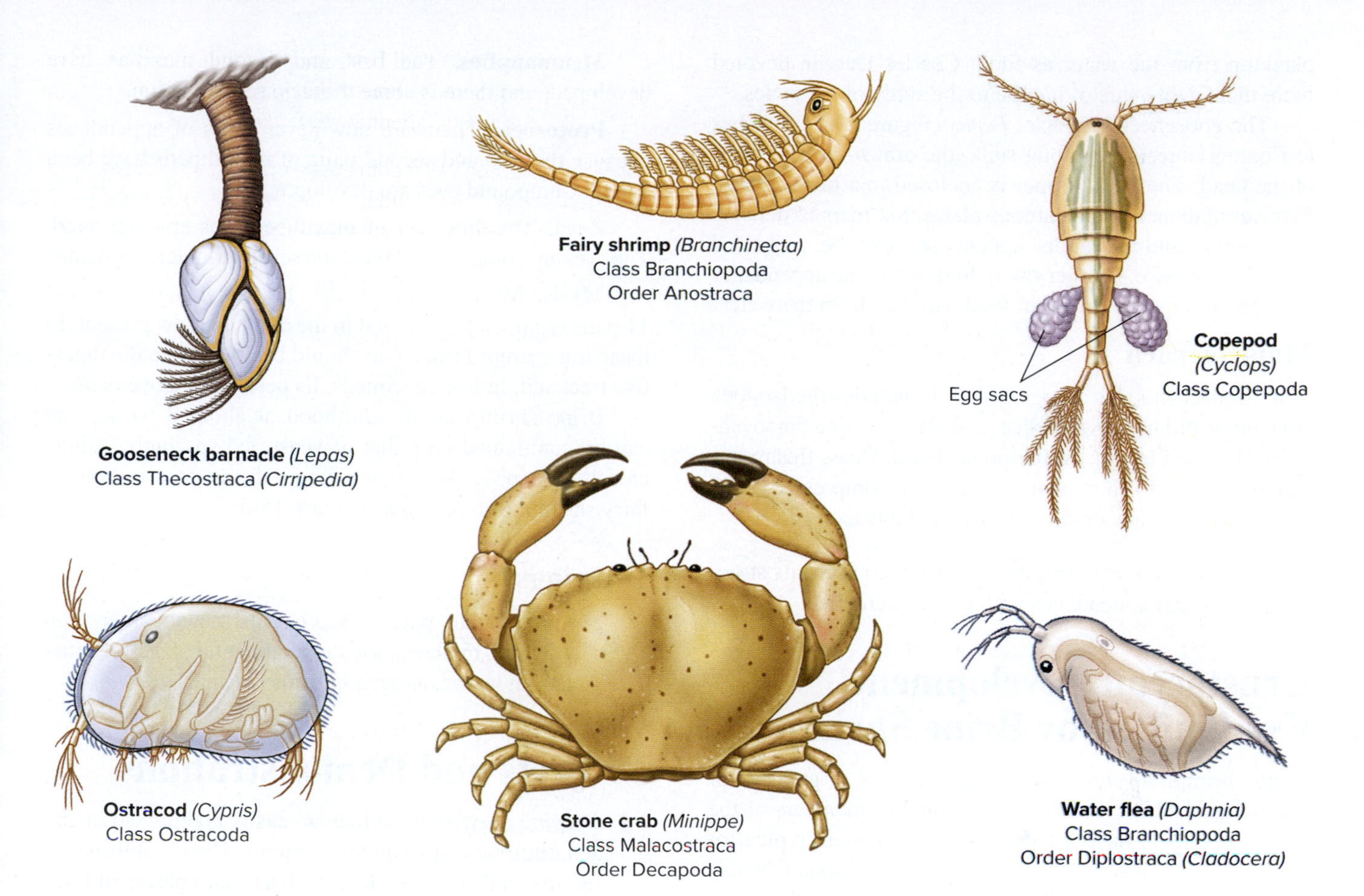

Figure 15-5
Some representative crustaceans.

Order Diplostraca (Cladocera), Water Flea, *Daphnia.* Water fleas (Figure 15-5) are common in pond water. Mount a living *Daphnia* on a slide, using enough water to prevent crushing but not enough to allow free swimming. *Daphnia* is 1 to 3 mm long and, except for the head, is covered by a thin, transparent carapace. Large, biramous second antennae are the chief organs of locomotion. There are five pairs of small, leaflike swimmerets on the thorax. These are used to filter microscopic algae from the water for food. Blood in the open system is red from the presence of hemoglobin. The paired eyes are fused. The brood pouch in females is large and posterior to the abdomen. They can reproduce parthenogenetically. (See p. 236 for demonstration of heartbeat.)

See the Experimenting in Zoology feature at the end of this chapter for a study of the phototactic behavior of *Daphnia.*

Ostracoda

Ostracods with their transparent bivalved carapace resemble tiny clams (1 to 2 mm). They have a median eye and seven pairs of appendages, including large antennules and antennae (Figure 15-5).

Copepoda

Cyclops and other copepods are found everywhere in fresh and brackish water. *Cyclops* has a median eye near the base of the rostrum and long antennae, modified in males. The cephalothorax bears appendages; the abdomen has none. The sixth thoracic segment in females carries large, pendulous egg sacs. The last abdominal segment bears a pair of caudal projections covered with setae (Figure 15-5).

Thecostraca (Cirripedia)

Barnacles (all marine) might be mistaken for molluscs because they are enclosed in a calcareous shell. *Balanus,* the acorn barnacle, has a six-piece shell that surrounds the animal like a parapet. The animal protrudes its six pairs of biramous appendages through the opening at the top to create water currents and sweep in plankton and particles of detritus.

Barnacles are among the most curious of crustaceans. Hatching as a free-swimming nauplius larva, a barnacle eventually settles, cements itself to a suitable surface, and begins producing calcareous plates around itself. The legs (cirri) now extend outward to form a comb that sieves plankton from the water as food. Charles Darwin devoted more than eight years of his life to the study of barnacles.

The gooseneck barnacle, *Lepas* (Figure 15-5), attaches to floating objects by a long stalk, the drawn-out front part of the head. The body proper is enclosed in a bivalve carapace strengthened by calcareous plates. Six pairs of delicate filamentous and biramous appendages can be protruded from the carapace. Feathery with long setae, the appendages form an effective net to strain food particles from the water.

Malacostraca

Class Malacostraca is a large group. It includes the Isopoda (sow bugs, pill bugs, wood lice, and others with a dorsoventrally flattened body); Amphipoda (beach fleas, freshwater *Gammarus,* and others that are laterally compressed); and Decapoda (shrimps, crabs, crayfish, and lobsters).

Examine a crab (Figure 15-5) and compare its structure and appendages with those of crayfish.

Crustacean Development, Exemplified by Brine Shrimp

Brine shrimp, *Artemia salina* (Gr. *Artemis,* a goddess of mythology), which are not true shrimp but members of the order Anostraca, have a pattern of development typical of most marine crustaceans. Brine shrimp are found in salt lakes around the world. They are filter feeders. Since the 1930s, adults and cysts (containing nauplius larvae in diapause) have been harvested and cultured in the laboratory. They are used as food in the larviculture of many marine organisms, including fish and shellfish. If time permits, your instructor may have prepared a series of cultures, started 10 days before the laboratory period and every 2 days thereafter. This should provide you with five larval stages for study. The larvae hatch 24 to 48 hours after the dry cysts are placed in natural or artificial seawater. The stages are classified on the basis of number of body segments and number of appendages present. Because growth cannot continue without molting (ecdysis), the larvae will go through a number of molts before reaching adult size. You may find some of the shed exoskeletons in the cultures.*

The animals can be narcotized or killed if you add a few drops of ether or chloroform per 10 ml of culture water.

To observe the larvae, mount a drop of the culture containing a few larvae on a slide and cover with a coverslip. Try to identify the stages listed in the following text.

Nauplius. A newly hatched larva, called a nauplius, has a single median ocellus and three pairs of appendages (two pairs of antennae and one pair of mandibles), and the trunk is still unsegmented.

*More complete directions for raising brine shrimp are found in Appendix A, p. 400.

Metanauplius. The first and second maxillae have developed, and there is some thoracic segmentation.

Protozoea. There are now seven pairs of appendages because the first and second pairs of maxillipeds have been added. Compound eyes are developing.

Zoea. The third pair of maxillipeds has now appeared. The eyes are complete and there are several thoracic segments.

Mysis. Most or all of the 19 body segments and 11 pairs of appendages found in the adult are now present. In these transparent forms, you should be able to see the digestive tract and, in live specimens, its peristaltic movements.

Brine shrimp attain adulthood at about 3 weeks and can be maintained on a diet of yeast, fed sparingly. Adults closely resemble their much larger freshwater cousins—fairy shrimp, *Eubranchipus* (Figure 15-5).

Drawings

On separate paper, sketch crustaceans other than crayfish, or sketch the developmental stages of crustacean larvae, as seen in brine shrimp.

Projects and Demonstrations

1. *Keeping crayfish.* Crayfish are easily maintained in an aquarium tank at room temperature. They will thrive if fed sparingly on raw meat, such as small pieces of fish, shrimp, frog, or beef. Crayfish need plenty of oxygen. Keep the water in the aquarium shallow and provide rocks for cover. Rocks will also enable the crayfish to crawl out of the water. The addition of aquatic plants enhances the appearance of the aquarium and provides food. Remember not to release nonnative crayfish into the wild when you are through observing them.
2. *Heartbeat of crayfish.* Anesthetize a crayfish in a suitably sized dish by immersing in 15% ethanol, or club soda, freshly opened. Remove from the anesthetic when the animal fails to respond to prodding; then remove the carapace and observe the beating heart. The crayfish should survive several hours under these conditions.
3. *Respiratory current of crayfish.* This can be demonstrated by gently releasing dilute methylene blue with a Pasteur pipette or syringe held under the rear of the animal's carapace. The ink or dye emerges anteriorly as two colored jets.
4. *Organization of antennal glands.* These can be identified by injecting into the hemocoel of a cray fish 0.25 ml of 0.5% aqueous solution of Congo red. After 18 to 24 hours, remove one of the glands and examine under a dissecting microscope. Cells of the end sac (coelomosac) should be red; channels of the labyrinth, blue.
5. *Heartbeat of* Daphnia. Place a small drop of petrolatum (Vaseline) in the center of a Syracuse watch glass.

Fasten *Daphnia* by one valve of its carapace to the petrolatum, but be careful that water circulates between the valves. Observe the heartbeat. Start with water at 0° C. As the water slowly rises to room temperature, record the heartbeats at the different temperatures. Determine the number of heartbeats in 15 seconds at each 2° or 3° rise in temperature.

6. *Tree hermit crabs. Coenobita clypeatus* are available from biological supply companies (e.g., Carolina Biological) and make interesting displays for the laboratory. They are easily maintained in a sand-based terrarium, where they can be fed dry toast, lettuce, bananas, and the like.
7. *Blood cells in arthropods.* Unlike vertebrate blood, arthropod blood contains no erythrocytes. The respiratory pigment, usually hemocyanin, is dissolved in the plasma rather than carried in cells. Arthropods possess a variety of leukocytes or amebocytes. Because their blood clots much more rapidly than does vertebrate blood, their observation is sometimes more difficult.

   To observe in the natural state, place a drop of paraffin oil on a clean slide and focus on the drop. Snip off the tip of an antenna with petroleum-jellied scissors, place the tip in the oil drop, and observe and sketch the cells as they emerge.

   Phagocytosis in crayfish cells may be observed by injecting the animal with a suspension of carmine or India ink. Examination of the blood an hour or so later may reveal particles ingested by leukocytes.
8. *Other suitable demonstrations*
   a. Female crayfish carrying eggs and young; preserved specimens are available from biological supply companies
   b. Various kinds of barnacles
   c. Various kinds of crabs
   d. Cross section of crayfish gill

LAB REPORT

Name ______________________

Date ______________________

Section ______________________

1. What anatomical characteristics distinguish the Crustacea from all other members of the Phylum Arthropoda?

2. What is meant by the term **serial homology?**

   Why are appendages of a crayfish considered an excellent example of serial homology?

3. What do the terms "afferent" and "efferent" mean?

4. Describe the two-part stomach of the crayfish.

   What is the function of each part?

5. What is the function of the crayfish statocyst?

   Where is it located? ______________________ In what less-advanced group studied earlier are statocysts found?

EXPERIMENTING IN ZOOLOGY

The Phototactic Behavior of Daphnia

Many species of freshwater zooplankton move toward light (positive phototaxis). In nature, movement toward light often means movement toward shallower, warmer water that is rich in algae and microorganisms. However, most zooplankton have few defense mechanisms and are very vulnerable to predators. Movement toward light would seem to make them more visible to potential predators. Thus, natural selection might favor one type of response to light when no predation risk is apparent and a different response to light when predation risk exists.

Getting Ready

Work with a partner for this exercise. Try to work in a room that can be darkened. Obtain an open-ended glass tube 2 to 3 cm in diameter and 25 to 30 cm long (Figure 15-6). Wash the tube and place a black rubber stopper tightly into one end. Use a wax pencil to mark off three equal sections of tube: left, middle, and right. Fill the tube to within 3 to 4 cm of the top with bottled water or dechlorinated tap water. Have a second rubber stopper available for the open end. Have a stopwatch available for timing the experiment.

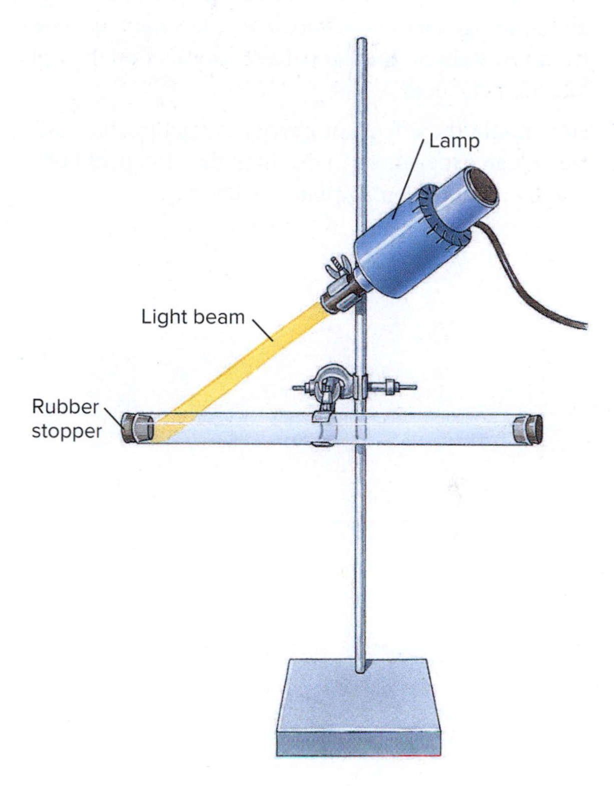

Figure 15-6
Setup for observing the phototactic behavior of *Daphnia.*

How to Proceed

Use a pipette to count out 20 *Daphnia* and add them to the tube. Place the second stopper firmly into the open end. Clamp the tube to the ring stand, so that it is horizontal. Turn out the room lights and darken the room as is possible. Allow the *Daphnia* to acclimate to the tube for the next 10 minutes. At the end of each minute for the next 10 minutes, record where the *Daphnia* are distributed in the tube (left, middle, right). If light from the windows is insufficient to see the *Daphnia,* use diffuse lighting from a lamp some distance from the tube. Turn the lamp on just long enough to count the *Daphnia* and then turn it off until the next observation. After the tenth observation, calculate the mean percentage of *Daphnia* in each section of the tube over the 10 minutes. Wash out the tube and fill it with clean water and 20 new *Daphnia* and repeat the experiment. Calculate the mean distribution of *Daphnia* for the two trials.

After two trials with *Daphnia* in the dark, repeat the experiment with a dissecting scope illuminator or fiber optic light (Figure 15-6) clamped at an angle, so that it illuminates the end of one stopper, making it visible to the *Daphnia.* Clamp it far enough away from the tube that it does not heat the water. Again record the distribution of *Daphnia* in the tube over the next 10 minutes and repeat the experiment. Calculate the mean distribution of *Daphnia* for the two trials.

Compile the data from the entire class and calculate overall means for all control trials and all lighted trials.

Do *Daphnia* appear to respond to light? Do they exhibit negative or positive phototaxis? In the light trials, do you notice any change in distribution over the 10 minutes?

Questions for Independent Investigation

Studies on vertical migrations in zooplankton have found that stimuli from predators can influence phototaxis. Aquatic organisms such as *Daphnia* are often very sensitive to chemical stimuli in the water.

1. Design an experiment to test how fish odor might affect phototaxis in *Daphnia.* Try adding a small amount of water (40 to 50 ml) from a tank that

contains a bass or sunfish to your experimental tube containing the *Daphnia.* What would be a good control for this experiment?

2. Do *Daphnia* respond to chemical cues from bass or sunfish the same way they respond to cues from frog tadpoles? Why might there be a difference?
3. *Daphnia* also have many invertebrate predators (e.g., hydra, midge larvae—*Chaoborus*). Do chemical cues from invertebrate predators have an effect on their phototactic behavior?
4. How might the effects of gravity impact phototaxis? Design an experiment to examine the effects of both gravity and light on *Daphnia* behavior.

References

DeMeester, L. 1991. An analysis of the phototactic behaviour of *Daphnia magna* clones and their sexual descendants. Hydrobiologia **225:**217–227.

DeMeester, L. 1993. Genotype, fish-mediated chemicals, and phototactic behavior in *Daphnia magna.* Ecology **74:**1467–1474.

Neill, W. 1990. Induced vertical migration in copepods as a defense against invertebrate predators. Nature **345:**524–525.

Parejko, K., and S. Dodson. 1991. The evolutionary ecology of an antipredator reaction norm: *Daphnia pulex* and *Chaoborus americanus.* Evolution **45:**1665–1674.

The Arthropods

Myriapods and Hexapods

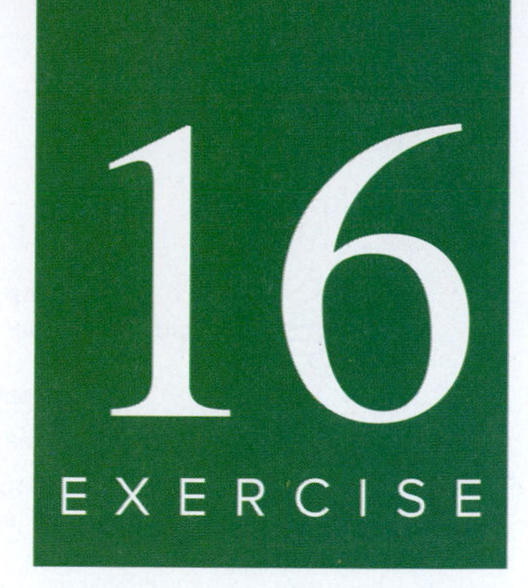

EXERCISE 16A
Myriapods—Centipedes and Millipedes
Class Chilopoda—Centipedes
Class Diplopoda—Millipedes

EXERCISE 16B
Insects—Grasshopper and Honeybee
Romalea, Lubber Grasshopper
Apis, Honeybee
Projects and Demonstrations

EXERCISE 16C
Insects—House Cricket
Acheta domesticus, House Cricket

EXERCISE 16D
Metamorphosis of *Drosophila*
Drosophila, Fruit Fly

EXERCISE 16E
Collection and Classification of Insects
Key to the Principal Orders of Insects

EXERCISE 16A

Myriapods—Centipedes and Millipedes

The subphylum **Myriapoda** (Gr. *myries,* myriad, numberless, + *podos,* foot) includes several groups (Chilopoda, Diplopoda, Pauropoda, and Symphyla). They have two tagmata—head and trunk. There is one pair of antennae, and the trunk bears paired appendages on all but the last segment (Figure 16-1).

Core Study

Class Chilopoda—Centipedes

Phylum Arthropoda
Subphylum Myriapoda
Class Chilopoda
Genus *Scolopendra* or *Lithobius*

Centipedes, or "hundred-leggers," are active predators that live in moist places under logs, stones, and bark, where they feed on worms, larvae, and insects.

If living examples are available, watch their locomotion and note the agile use of the body and legs.

Note the general shape of the body and arrangement of segments and appendages. Is the body circular or flattened in cross section? ____________

On a preserved specimen of *Lithobius* (Gr. *lithos,* stone, + *bios,* life) or *Scolopendra* (Gr., a kind of centipede) (Figures 16-1 and 16-2), examine the head.

Some species have simple **ocelli;** others have large, faceted **eyes** resembling the compound eyes of insects. Which does your specimen have? ____________

Pull aside the large poison fangs to uncover the mouth area and examine under a dissecting microscope.

Find the **antennae; labrum,** anterior to the mouth; **mandibles** and **first maxillae,** lateral to the mouth; and **second maxillae,** bearing a long palp and a short labial portion just posterior to the mouth.

The first trunk appendages are the prehensile **maxillipeds,** each bearing a terminal **poison fang.** Do the rest of the trunk appendages bear legs? ____________ The last segment bears the **gonopores,** and the **anus** is located on a short telson. Find the **spiracles.** In *Lithobius* they are located near the bases of the legs. In *Scolopendra* the spiracles are located more dorsally and are present on alternate segments.

Class Diplopoda—Millipedes

Phylum Arthropoda
Subphylum Myriapoda
Class Diplopoda
Genus *Julus* or *Spirobolus*

Millipedes, or "thousand-foot worms," are found throughout the world, usually hiding in damp woods under bark, leaves, rocks, and logs. They are herbivorous, feeding on decaying wood or leaves. Their most distinguishing feature is the presence of diplosegments, which are double trunk segments derived from the fusion of two single segments during

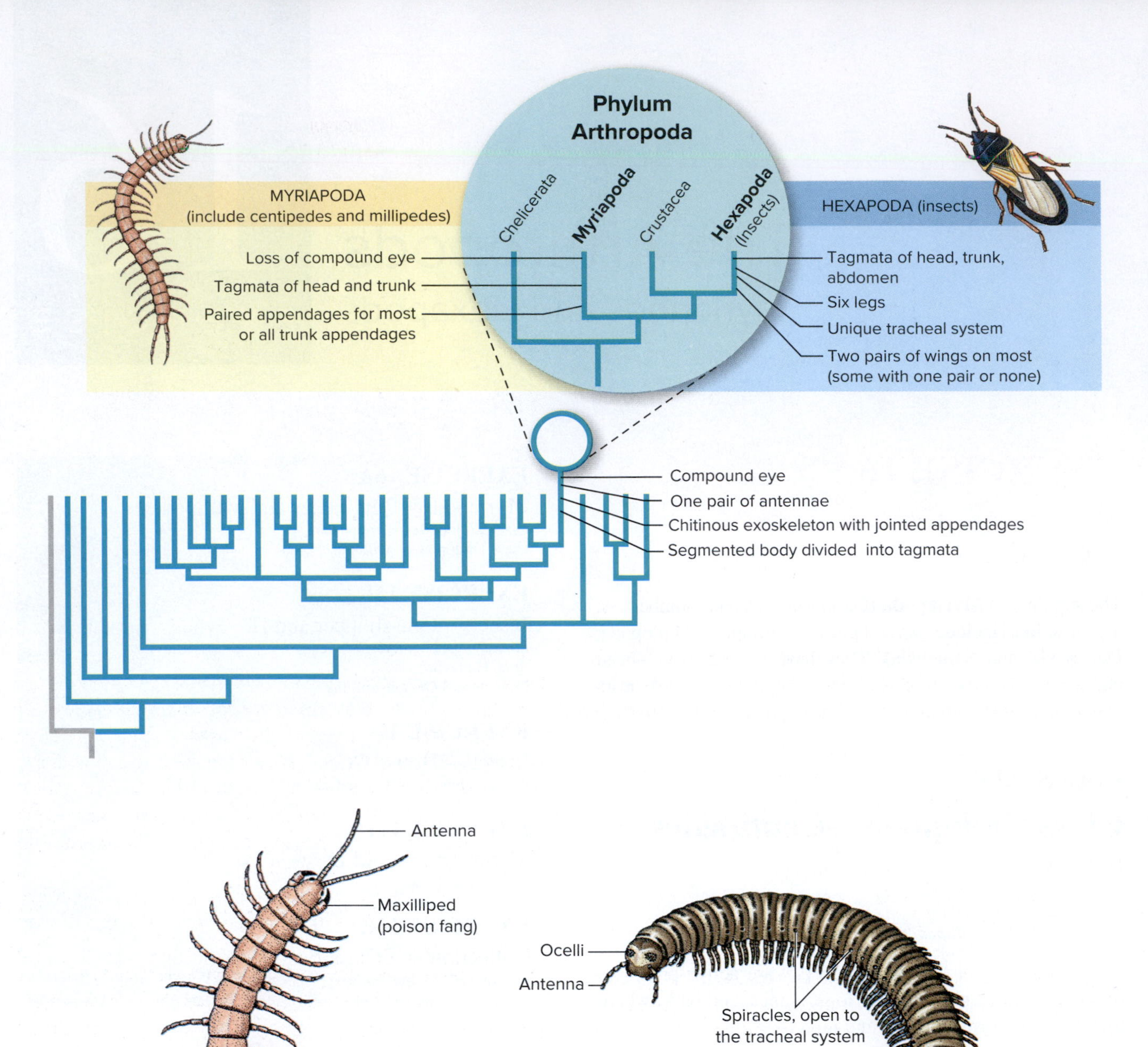

Phylum Arthropoda
MYRIAPODA
(include centipedes and millipedes)
Chelicerata
Myriapoda
Crustacea
Hexapoda
(insects)
HEXAPODA (insects)
Loss of compound eye
Tagmata of head and trunk
Paired appendages for most
or all trunk appendages
Tagmata of head, trunk,
abdomen
Six legs
Unique tracheal system
Two pairs of wings on most
(some with one pair or none)
Compound eye
One pair of antennae
Chitinous exoskeleton with jointed appendages
Segmented body divided into tagmata
Antenna
Maxilliped
(poison fang)
Last pair of legs are sensory,
not used in locomotion
Centipede *Scolopendra*
Ocelli
Antenna
Spiracles, open to
the tracheal system
Diplosegment with two
pairs of legs each segment
Millipede *Julus*

Figure 16-1

Examples of myriapods—centipede and millipede.

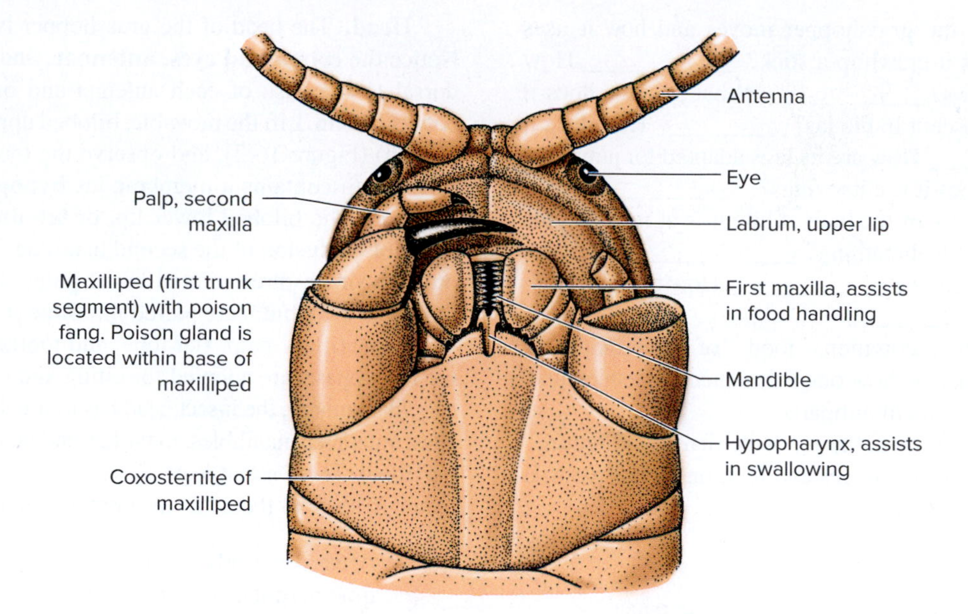

Figure 16-2
Mouthparts of *Scolopendra,* ventral view.

embryonic development and each bearing two pairs of legs. The name "Diplopoda" comes from Greek *diploos,* double, and *pous, podos,* foot (see Figure 16-1).

If living millipedes such as *Spirobolus* (L. *spira,* coil, + Gr. *bōlos,* lump) or *Julus* (Gr. *iulus,* plant-down) are available, note the shape of the body, use of the antennae, use of the legs, and ability to roll up.

Notice the rhythm of the power and recovery strokes of the legs with opposite sides of each segment exactly out of phase. It is similar to that seen in the polychaete *Nereis.* The movements of the legs pass in regular, successive waves toward the anterior end. If a specimen has been chilled, you should be able to determine the number of legs working together in each wave and the number of segments between waves.

Study a preserved specimen, and locate on the head the **ocelli, antennae, labrum, mandibles,** and **labium** (fused second maxillae).

Are there appendages on the first trunk segments? ____________ How many pairs on the next three segments? ____________ The **gonopores** open on the third trunk segment at the bases of the legs. How does this compare with the centipede? ____________ The dorsal overlapping of the exoskeletal plates provides full protection for the animal, even when rolled into a ball. Notice that each of the diplosegments has two pairs of **spiracles.**

EXERCISE 16B

Insects—Grasshopper and Honeybee

Insects are by far the largest group of animals. It is estimated that there are 1.1 million named species of insects—more than all other named animals combined—and that several million species of insects remain to be discovered and described.

Among their chief characteristics are **three pairs of walking legs; one pair of antennae;** a body typically divided into **head, thorax,** and **abdomen;** and a respiratory system of **tracheal tubes.** Most insects are also provided with one or two pairs of **wings.** Their sense organs are often specialized and perhaps account for much of their success in the competition for ecological niches.

Most insects are less than 2.5 cm long, but they range from 1 mm to 20 cm, with the largest insects usually living in tropical areas.

A grasshopper has a fairly typical insect body plan. A honeybee, on the other hand, has become specialized for particular conditions. Not only is its morphology modified for special functions and adaptations, but it is a social insect in which patterns of group organization involve different types of individuals and division of labor.

Core Study

Romalea, Lubber Grasshopper

Phylum Arthropoda
Subphylum Hexapoda
Class Insecta
Order Orthoptera
Genus *Romalea*

Behavior

If living grasshoppers are available, observe them in a terrarium or place one in a glass jar with moist paper towel in the bottom and a stick or another object to perch on.

Is the color adaptive, given its natural habitat? __________ Can it move its head? __________

Observe how the grasshopper moves and how it uses its legs. How does it crawl up a stick? ____________ How does it use its claws? ____________ What position does it assume when quiescent in the jar? __________ How does it jump? ____________ How are its legs adapted for jumping? ____________ Does it use its wings? ____________

Note movements of the body while it is at rest. Are the movements related to breathing? ____________ How does the grasshopper breathe? ____________ How does it get air into its body? ____________

What is the common food of grasshoppers? ____________ Observe how one eats a piece of lettuce leaf. Watch how it moves its mouthparts.

Take a specimen from the jar. Note that it will regurgitate its greenish digestive juices and food on a glass plate. Is this a defensive adaptation?

Written Report

Record your observations on separate paper.

External Structure

Study a preserved specimen of a grasshopper. It is a model of compactness, so use a dissecting microscope or hand lens to observe smaller structures.

Note the division of the body into three tagmata: **head, thorax,** and **abdomen.** Is the grasshopper segmented throughout, or is segmentation more apparent in certain regions of the body? ____________

The chitinous **exoskeleton** is secreted by the underlying epidermis. It is composed of hard plates, called **sclerites,** which are bounded by sutures of soft cuticle.

Head. The head of the grasshopper is freely movable. Notice the **compound eyes, antennae,** and three **ocelli,** one dorsal to the base of each antenna and one in the groove between them. Lift the movable, bilobed upper lip, or **labrum** (L., lip) (Figure 16-3), and observe the toothed **mandibles.** The mouth contains a membranous **hypopharynx** for tasting food. The bilobed lower lip, or **labium** (L., lip), is the result of the fusion of the second maxillae. The labium bears on each side a three-jointed labial palp. Between the mandible and the labium are paired **maxillae** (L., jawbone), each with a maxillary palp, flat lobe, and toothed jaw. Note how the mouthparts are adapted for biting and chewing.

In summary, the insect head has four pairs of true appendages: antennae, mandibles, maxillae, and labium (fused second maxillae). There are at least six segments in the head region, although some of them are apparent only in the insect embryo.

After you study the rest of the external features, if time permits, your instructor may ask you to use forceps and teasing needles to carefully remove all the mouthparts and arrange them in their relative positions on a sheet of paper. Permanent demonstration mounts can be made very easily by a method given in Appendix A, p. 409.

Thorax. The thorax is made up of three segments: **prothorax, mesothorax,** and **metathorax,** each bearing a pair of legs (Figure 16-4). The mesothorax and metathorax also bear a pair of wings. **Spiracles** (external openings of the insect's tracheal system) are located above the legs in the mesothorax and metathorax. Note the leathery **forewings** (on the mesothorax) and the membranous **hindwings** (on the metathorax). Which is more useful for flight? ____________ What appears to be the chief function of the

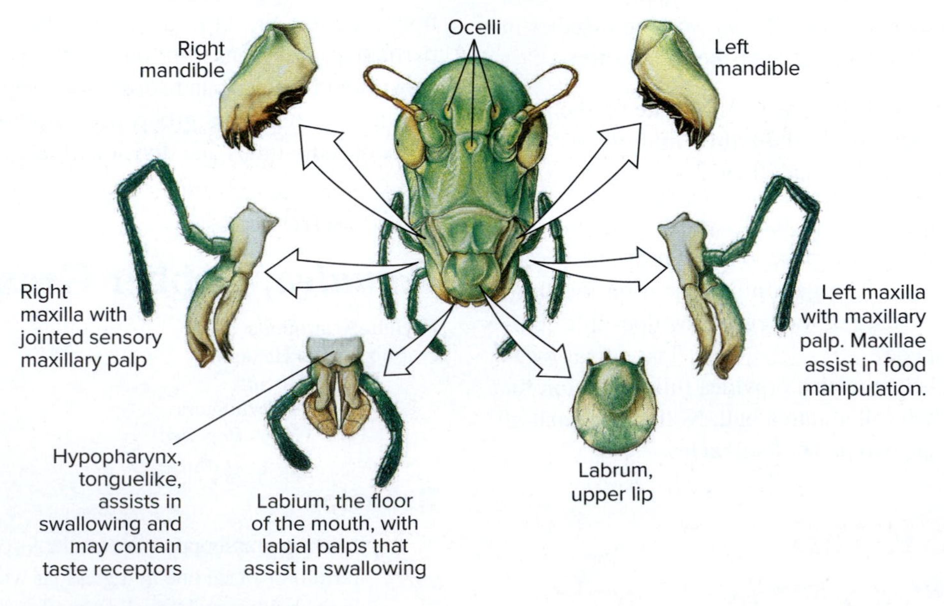

Figure 16-3
Head and mouthparts of a grasshopper.

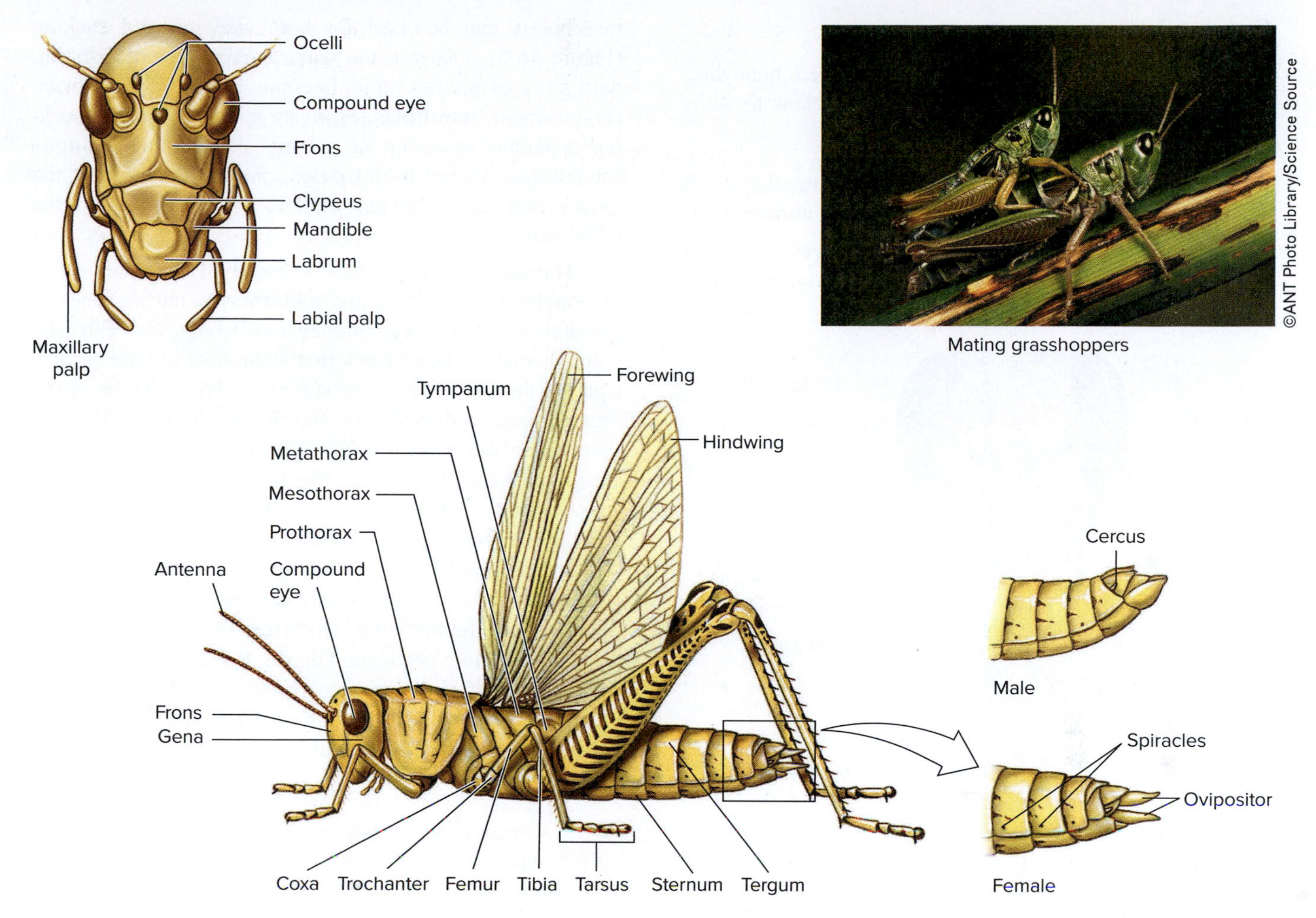

Figure 16-4
External features of a female grasshopper. The terminal segments of a male with external genitalia are shown in the inset.

forewings? ____________ The small **veins** in the wings, or tracheal tubes, are used by entomologists in the identification and classification of insects.

Examine the grasshopper's legs and identify the basal **coxa** (L., hip); small **trochanter** (Gr., ball of hip joint); large **femur;** slender, spiny **tibia;** and five-jointed **tarsus** with two **claws** and a terminal pad, the **arolium.** Which pair of legs is most specialized, and for what function? __________

Abdomen. There are 11 segments in the abdomen of the grasshopper. Notice the large **tympanum,** the organ of hearing, one located on each side of the first abdominal segment. On which of the abdominal segments are the paired spiracles located? __________ In both sexes, segments 2 to 8 are similar and unmodified, and segments 9 and 10 are partially fused.

The eleventh segment forms the genitalia (secondary sex organs). On each side behind the tenth segment is a projection, the **cercus** (pl. **cerci;** Gr. *kerkos,* tail). In females the posterior end of the abdomen is pointed and consists of two pairs of plates, with a smaller pair between, with the whole forming the **ovipositor.** Between the plates is the opening of the **oviduct.** The end of the abdomen in males is rounded. What is the sex of your specimen? __________

Internal Structure

Study of the internal structures of a grasshopper is not always satisfactory because the various organs appear somewhat poorly defined. Internal structures are more easily studied in an anesthetized cricket, instructions for which are given in Exercise 16C.

Drawings

 Label the external view of the grasshopper on p. 249.

Further Study

Apis, Honeybee

Phylum Arthropoda
Subphylum Hexapoda
Class Insecta
Order Hymenoptera
Genus *Apis*
Species *Apis mellifera*

External Structure

Examine a preserved or freshly killed honeybee. Use a dissecting microscope or hand lens to study smaller structures.

The body of a honeybee, like that of a grasshopper, is divided into three tagmata: **head, thorax,** and **abdomen.**

Head. Identify **antennae, compound eyes,** and three **ocelli,** located dorsally between the compound eyes. The mouthparts can be used for both chewing and sucking (Figure 16-5). Observe the narrow upper lip, or **labrum,** with a row of bristles on its free margin. Below the labrum are brownish **mandibles.** From its mouth projects a sucking apparatus made up of a long, slender, hairy **tongue** (or **labium**); paired **labial palps,** one on each side of the tongue; and paired, broad **maxillae,** one on each side of the labial palps.

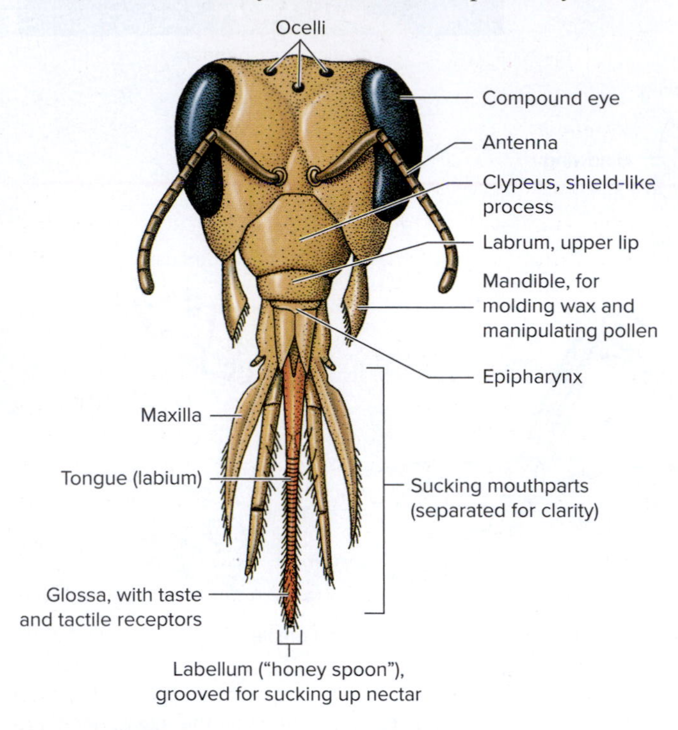

Figure 16-5
Head and mouthparts of a honeybee.

Thorax. The thorax is composed of three segments, **prothorax, mesothorax,** and **metathorax,** but the lines of division between the segments are less distinguishable than on a grasshopper. A honeybee's first abdominal segment is also a part of the thorax, lying anterior to its narrow "waist." Each segment bears a pair of legs, and the mesothorax and metathorax each bear a pair of wings. Note that both pairs of wings are thin and membranous, unlike those of a grasshopper, in which only the hindwings are membranous. How would this difference affect the wings' function? ___________

Notice small hooks on the front margin of the hindwing. During flight the hooks catch hold of a groove near the margin of the forewing. The wings of a bee may vibrate 400 times or more per second during flight.

The legs of a honeybee have the same segments as those of a grasshopper but are highly specialized with tools for collecting pollen, which provides essential proteins for honeybee larvae.

Use your hand lens, or remove the legs from one side of the bee and examine under a dissecting microscope.

Note hairs on the **foreleg.** Are they branched? _________ These hairs collect pollen. The **pollen brush** (Figure 16-6) consists of long hairs on the proximal end of the tarsus. Pollen brushes on the forelegs and middle legs

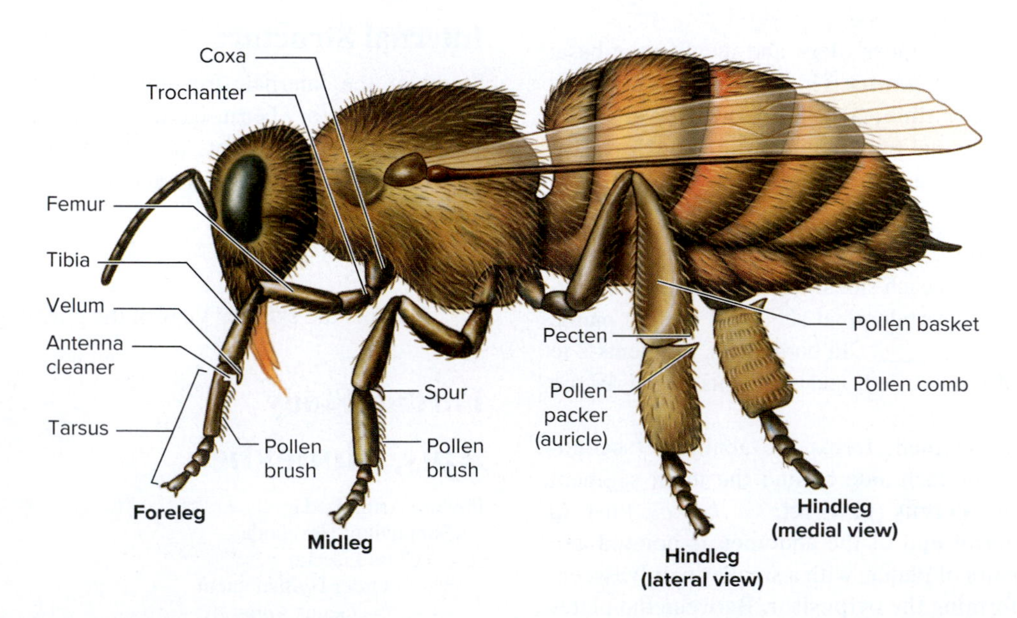

Figure 16-6
Adaptive legs from the left side of a worker honeybee.

brush pollen off the body hairs and deposit it on the pollen combs of the hindlegs (Figure 16-6).

At the distal end of the tibia is a movable spine, the **velum** (L., covering). The velum covers a **semicircular notch,** which bears a row of stiff bristles, the **antenna comb.** The velum, notch, and comb together make up the **antenna cleaner.** The antenna is freed from pollen as it is drawn through this antenna cleaner.

The **middle leg** has a long, sharp **spur** projecting from the end of the tibia that is used to remove pollen from the pollen basket of the leg behind. The middle leg also bears a pollen brush.

The **hindleg** is the largest leg and the most specialized. One of its striking adaptations is the **pollen basket,** a wide groove with bristles on the outer surface of the tibia. By keeping these bristles moist with mouth secretions, the bee can use the basket for carrying pollen. On the inner surface of the metatarsus are **pollen combs,** which are composed of rows of stout spines. Large spines found along the distal end of the tibia and the proximal end of the metatarsus make up the **pollen packer.** The pecten (pollen rake) removes pollen from the pollen brush of the opposite leg; then, when the leg is bent, the auricle packs it into the pollen basket. The bee carries her baskets full of protein-rich pollen back to the hive and pushes the pollen into a cell, where it will be cared for by other workers.

Abdomen. The abdomen of a bee has 10 segments, the first of which is actually part of the thorax, as mentioned before. The last three segments are modified and hidden within the seventh segment. Can you identify five pairs of spiracles on the abdomen? ____________________

Honeybee Sting. The amazingly intricate **sting** is a modified ovipositor that has evolved for defense. It consists of a large **poison sac** that receives secretions (a mixture of proteins, peptides, and other compounds) from two acid glands by way of a common duct (Figure 16-7). The poison sac discharges into the cavity of a sting bulb; from there the poison passes through the canal of the sting shaft and into the wound. The sting bulb also receives a short alkaline gland of uncertain function.

When a honeybee stings, it thrusts the barbs of the sting into the victim's flesh. The entire sting apparatus is torn out of the bee's abdomen, fatally wounding the bee. The sting actually consists of a stylet together with two barb lancets that are moved rapidly back and forth on the stylet by the action of protractor and retractor muscles. Therefore, even after the bee has hurriedly departed, the self-sacrificial sting continues to work in deeper and poison continues to be injected into the wound for 30 to 60 seconds.

The sting of the queen bee is used only against rival queens and can be withdrawn and reused many times. It is more firmly attached within the queen's abdomen, and the lancets have fewer and smaller barbs.

Honeybees as Social Animals.

In the hive of a honeybee, there are three **castes: workers, queen,** and **drones. Workers** are sexually inactive genetic females and make up most of the society. They do most of the work of the hive, except lay eggs. They collect food, clean out the hive, make honey and wax, care for the young, guard the hive, and ventilate the hive.

The single **queen** is a sexually mature female that lays the eggs, which may or may not be fertilized by sperm stored in her spermatheca.

Drones are males; all develop from unfertilized eggs and consequently are haploid (workers and the queen are

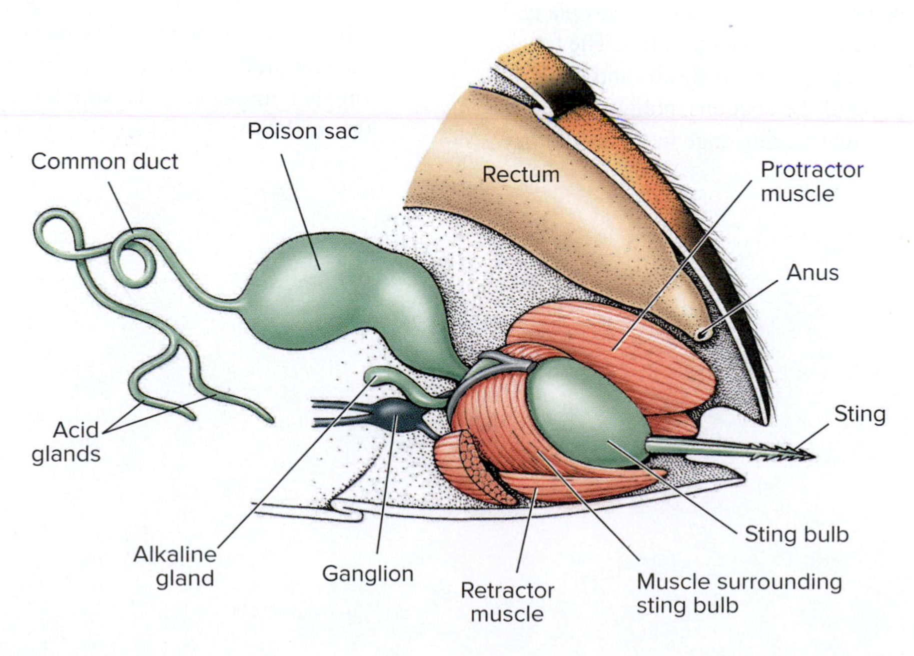

Figure 16-7
Sting of a worker honeybee.

diploid). There are usually only a few hundred drones in a hive. Their main duty is to fertilize the queen during the nuptial flight, although usually only one is required to provide the queen enough sperm to last her lifetime.

Oral Report

Be prepared to (1) identify the major external features of a grasshopper and/or honeybee, (2) explain the adaptations of worker honeybee appendages, and (3) explain the sting of a worker honeybee.

Projects and Demonstrations

1. *Capillary circulation in a cockroach wing.* Use adults of *Periplaneta* or tropical cockroach *Blaberus.* From a small piece of Styrofoam or balsa wood, hollow out a depression about the size of the cockroach. Anesthetize the roach with CO_2 and place it in the depression. Pin a piece of paper across the head and pronotum; pin a second piece of black paper across the abdomen, leaving one of the wings exposed *above* the paper in its normal resting position. Slip a piece of shiny white paper under the wing. Shine a focused beam of intense light on the wing from a low angle and observe the wing with a dissecting microscope. The circulating blood in the capillaries should now be visible. The light must be intense, and it may be necessary to adjust the angle of light for best effect. Visibility depends on reflection of light from circulating blood cells. Sometimes it is easier to see circulation in the hindwing (flow stops if the forewing is extended into the flying position).
2. *Real-life study of cecropia or polyphemus moth life cycles.* Both large larvae and cocoons of cecropia and polyphemus moths can be collected in fall. The larvae feed on leaves of many species of shrubs and trees. To observe the spinning of the cocoons, collect large larvae and place them in a rearing cage in the laboratory. The rearing cage can be an aquarium tank with 1 inch of moist sand in the bottom and a jar of water or wet sand to contain fresh branches of the tree species from which the caterpillars were collected. Place cotton around the mouth of the jar to prevent larvae from falling into it. Place the larvae on the leaves and cover the aquarium with a glass plate. Replace the leaves when they wilt or are eaten by larvae.

   Collected cocoons can be stored in a box of moss, which must always be kept slightly moist. If the box is stored in a cool cellar or unheated room, adults will emerge in spring. If kept in a warm room, they will probably emerge in midwinter. Before adults emerge, place the cocoons in a large cage or screened box containing branches to which the moths can cling while drying their wings.
3. *Demonstration of insect eggs and egg cases.* Eggs of walkingsticks and egg masses (oothecae) of praying mantids can be obtained from biological supply companies and kept in a refrigerator until a few weeks before their hatching is desired. Place in a terrarium or any glass container that can be covered with cheesecloth or fine-meshed wire screening.

   Mantids are carnivorous and have voracious appetites. If not provided with sufficient food (*Drosophila* or other insects), they will devour each other. The nymphs should not be kept in direct sun or drafts, as they require enough natural humidity to protect them during molts. If the skins they are molting are too dry, they may not be able to free themselves and consequently die during the process.

   Grasshopper eggs are sometimes available and will hatch if kept in damp soil at room temperature. The nymphs can be fed fresh lettuce.
4. *Firefly luminescence.* Dried firefly lanterns together with ATP and all necessary material for demonstrating the emission of light can be obtained from biological supply companies. ATP supplies the energy needed for the oxidation of the luciferin. Instructions are included.

Phylum ______________________

Subphylum ______________________

Genus ______________________

Name ______________________

Date ______________________

Section ______________________

LAB REPORT

Lubber Grasshopper

External view of female (male abdomen at bottom left)

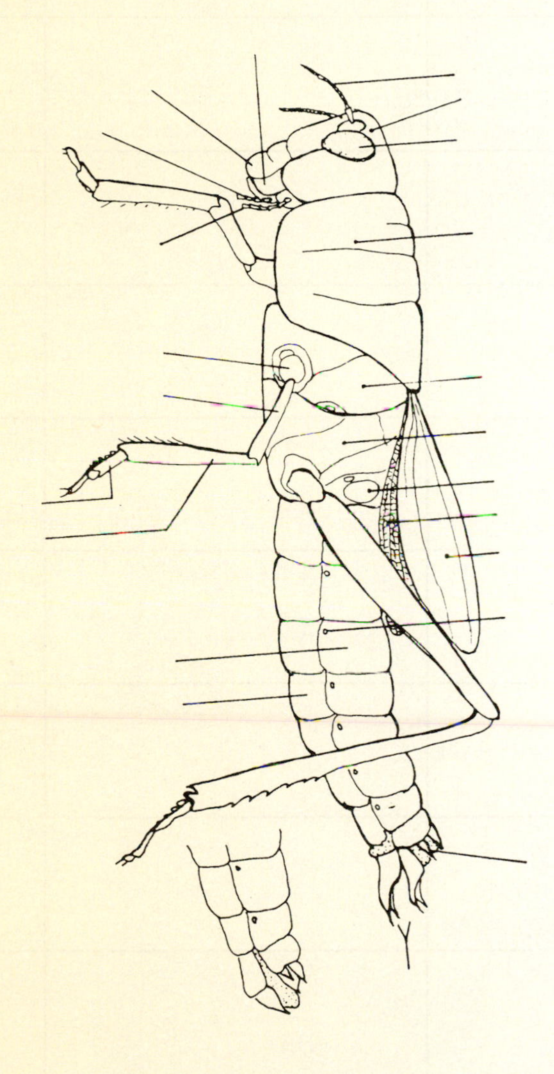

LAB REPORT

Name ______________________________

Date ______________________________

Section ______________________________

Comparing Characteristics of Annelida and Arthropoda

	Annelida	Arthropoda
Segmentation		
Body covering		
Main body cavity		
Appendages		
Circulatory system		
Respiration		
Excretory organs		
Sense organs		
Method of growth		
Division of animal kingdom (protostome/deuterostome)		

EXERCISE 16C

Insects—House Cricket[*, **]

Core Study

Acheta domesticus, House Cricket

Phylum Arthropoda
Subphylum Hexapoda
Class Insecta
Order Orthoptera
Genus *Acheta*
Species *Acheta domesticus*

House crickets, or gray crickets, commercially available as fish bait, are a European import and are not native to this country. Although crickets have wings, the flight muscles do not completely develop and crickets never fly. Adult house crickets live an average of 2 months, during which time females will deposit up to 2000 eggs. Only adult males sing (more accurately, stridulate), and one of the functions of their stridulating is to attract females. At a rearing temperature of 30°C, the eggs hatch in 13 days. There are eight larval instars (growth stages) before the final ecdysis (molt) to the adult 48 days later.

External Structure

Examine several alcohol-preserved crickets. Determine the sex of the crickets (Figures 16-8E and 16-9C), and determine which are larval instars and which are adults.

Only adult crickets have fully developed wings. The last two or three larval instars of crickets have external wing pads (buds). Crickets have **gradual metamorphosis.** The animal grows by successive molts (ecdysis), with each stage called a larval instar; after the last molt, the insect is called an adult.

Pin an adult male or female preserved cricket lateral side up in the dissecting dish.

Note that the basic 18 segments of the insect are functionally organized into three body regions, or tagmata: **head** (5 segments), **thorax** (3 segments), and **abdomen** (10 segments). Feeding and sensory organs are on the head, locomotory organs (wings and legs) are on the thorax, and digestive and reproductive organs are in the abdomen. All head segments are fused into a unit head capsule, and the mouthparts represent the modified appendages. The 3 thoracic segments are the **prothorax, mesothorax,** and **metathorax.** The abdominal segments are simply numbered 1 through 9 (the cerci represent segment 10). The entire dorsum is called the **tergum** (L., back) (or notum), and any one specific segment of the dorsal plate is called a **tergite.** The lateral body surface is the **pleuron** (Gr., side) (**pleurite** for one segment), and the ventral body surface is the **sternum** (L., breastbone) (**sternite** for one segment). Find and identify all of these parts.

Cut off a prothoracic leg and identify all of the segments and the tympanic membrane (Figure 16-8). Cut off the mesothoracic wing of a male cricket (Figure 16-8A), and determine how male crickets are able to stridulate. Place a cricket ventral side up, bend the head back, fold out the labrum, and pin the head in this position (Figure 16-8C). Identify all the mouthparts by moving each with fine forceps. Carefully remove one maxilla and the labium to see the **hypopharynx.**

The hypopharynx acts as a tongue and bears the openings of the **salivary glands.**

Functional Observations

Work in pairs. Anesthetize two crickets with carbon dioxide. Immobilize one cricket ventral side up by crossing insect pins over the insect (do not stick the pins through the cricket). When the cricket recovers from the anesthetic, feed it some colored, moistened food and observe the action of the mouthparts under a dissecting microscope.

The heart is an almost transparent tube visible through the intersegmental membranes along the middorsal line. Cut the wings off the other cricket and immobilize it *dorsal side up* with insect pins crossed over the body. When the cricket recovers from the anesthetic, determine the ventilation rate (abdominal contractions) and the heart rate in ventilations and beats per minute.

Internal Anatomy

If necessary, reanesthetize the cricket from which the wings have been removed, and pin the animal *dorsal side up* through the head and epiproct (Gr. *epi,* upon, + *proktos,* anus) (see Figure 16-9A) onto the wax in the dish. Flood the entire animal with saline and cut the cricket open with fine-tipped scissors. Pin open the cricket with insect pins (Figure 16-9A).

Observe **peristalsis** of the gut and movement of the **Malpighian tubules** (Figure 16-9B). The cricket will remain alive and the organ systems will function normally for several hours under these conditions. The chalky white **fat body,** which functions as a liver, is spread throughout the body and will vary greatly in amount and location according to age and diet of the cricket. Note the **tracheal tubes,** which appear silvery because of contained air. These branch throughout the body.

Sever the esophagus and rectum. Grasp the esophagus with forceps and gradually lift the entire digestive tract out by carefully cutting each tracheal

[*]Exercise contributed by J. P. Woodring of Louisiana State University.
[**]Materials and notes for this exercise are found on p. 409.

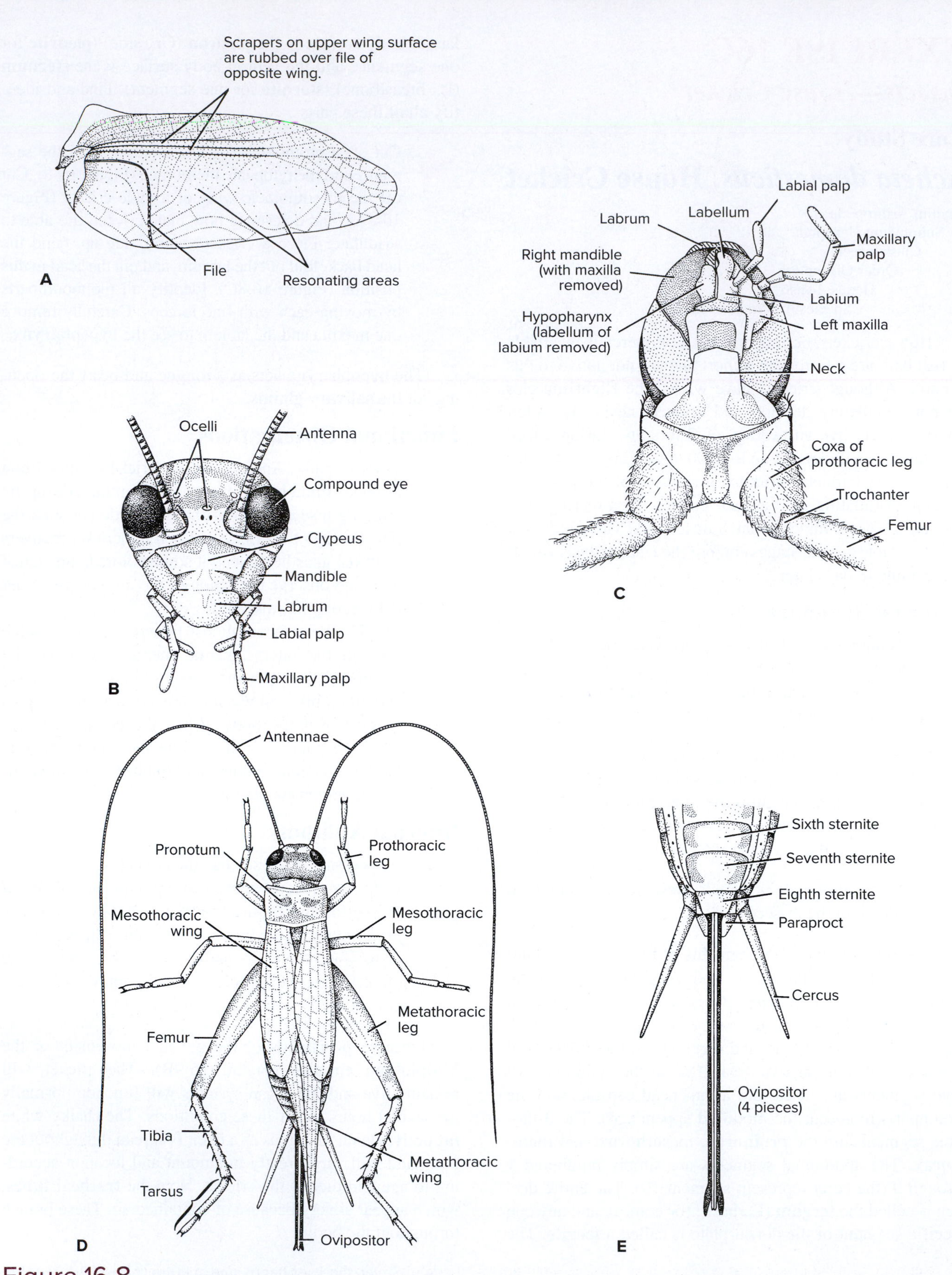

Figure 16-8

External structure of cricket. **A,** Undersurface of male mesothoracic wing. **B,** Frontal view and **C,** ventral view of head. **D,** Dorsal aspect of adult female. **E,** Ventral aspect of adult female abdomen.

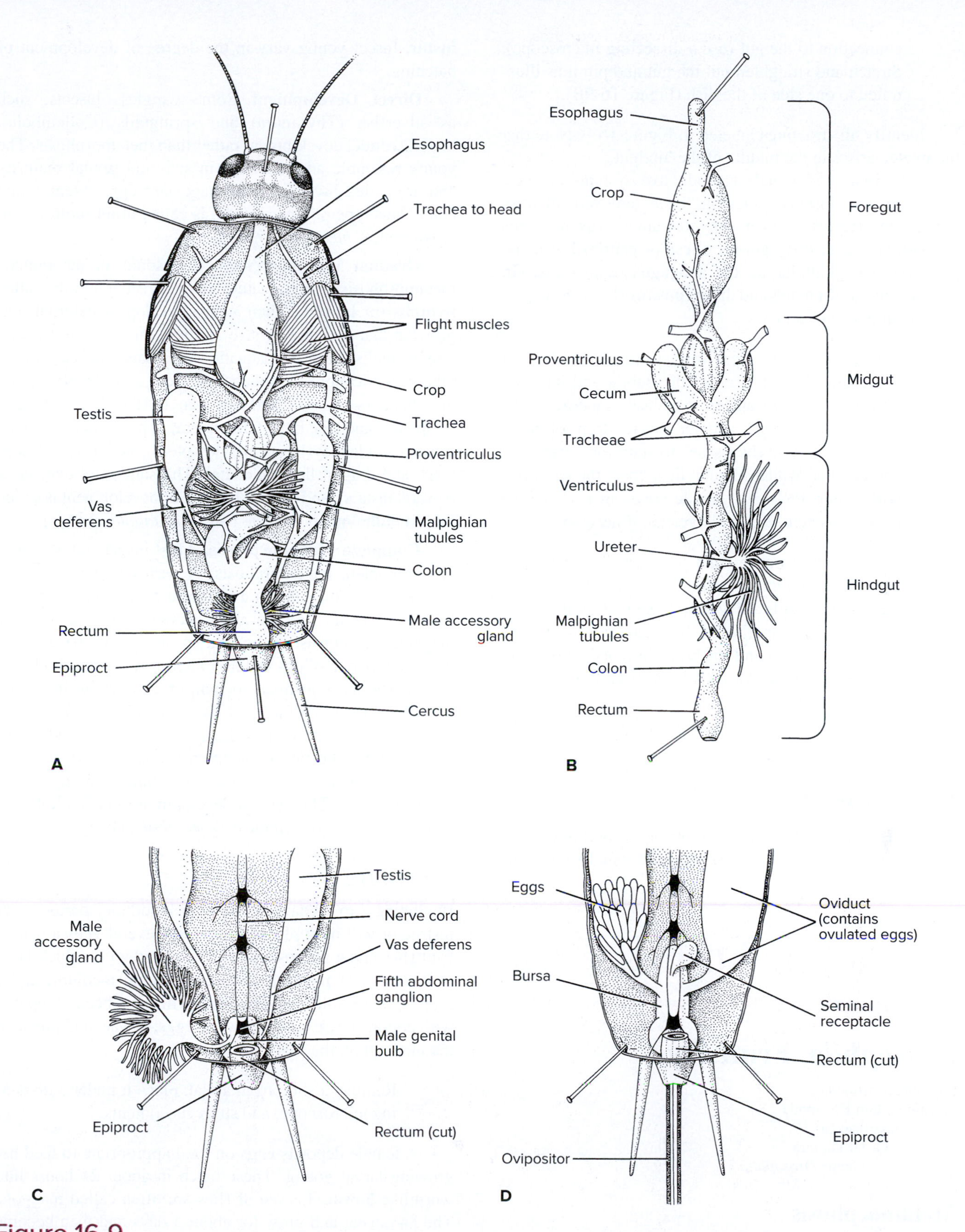

Figure 16-9

A, Internal anatomy of cricket (fat bodies not shown). **B,** Isolated gut. **C,** Male reproductive anatomy. **D,** Female reproductive anatomy.

connection to the gut (use a dissecting microscope). Stretch and straighten out the gut and pin it as illustrated to one side of the dish (Figure 16-9B).

Identify all structures labeled in Figure 16-9. Note that the **ureter** arises in the middle of the hindgut.

Both male and female reproductive systems are best seen after the digestive tract has been removed. Identify the gonads (**testes** or **ovaries**), gonoducts (**vas deferens** or **oviducts**), **accessory gland, bursa** or **genital bulb,** and genitalia. The genitalia are the **aedeagus** (e-de'a-gus; Gr. *aidoia,* genitals) of males and the **ovipositor** (L. *ovum,* egg, + *ponere,* to place) of females.

Pour off the saline from your specimen, but leave it pinned in place. Rinse out the debris and fat body with a mild jet of water from a plastic squeeze bottle. Pour off all the water and add several drops of methylene blue dye to the tissues to stain the nerve cord and ganglia. Wait 1 minute; then cover the specimen with water. Pick away tissue covering parts of the nervous system. Rinse and restain if necessary.

A double ventral **nerve cord** passes back through the body. In the thorax are three pairs of large, fused **segmental ganglia.** The nerve cords are widely separated. In the abdomen are five smaller ganglia; the fifth ganglion (overlying the genital bulb in males and the bursa in females) is larger than the rest and supplies all the posterior end of the body (Figure 16-9C and D).

Two cerebral ganglia lie in the head but are not visible without dissection of the head.

Oral Report

Be prepared to locate on your dissection any structure mentioned in the exercise and state its function.

Further Study

EXERCISE 16D

Metamorphosis of Drosophila

Drosophila, Fruit Fly

Phylum Arthropoda
Subphylum Hexapoda
Class Insecta
Order Diptera
Genus *Drosophila*

Metamorphosis

Early development of an insect occurs within the egg. Most insects change form during the growth stages after hatching. This is called **metamorphosis.** As it grows, an insect undergoes a series of molts; each stage between molts is called an **instar.** Insect young vary in the degree of development of hatching.

Direct Development. Some wingless insects, such as silverfish (Thysanura) and springtails (Collembola), undergo direct development rather than metamorphosis. The young resemble adults except in size and sexual maturity and are called juveniles. The stages are egg, juvenile, and adult. This type of development is called **ametabolous** (Gr. *a,* without, + *metabolē,* change).

Gradual Metamorphosis. In gradual, or incomplete, metamorphosis, the immature forms are commonly called **nymphs** (or **larvae**). Their wings develop as external outgrowths that increase in size as the animal grows by successive molts. Larvae of aquatic forms have tracheal gills or other modifications for aquatic life. Terrestrial forms include grasshoppers, locusts and the like (Orthoptera), termites (Isoptera), and true bugs (Hemiptera). Those having aquatic larvae include mayflies (Ephemeroptera), stoneflies (Plecoptera), and dragonflies (Odonata). The stages are egg, larva (several instars), and adult. This type of development is called **hemimetabolous** (Gr. *hemi,* half, + *metabolē,* change).

Complete Metamorphosis. Most insects (about 88%) have complete metamorphosis. A series of larval instars different from the adult are followed by a pupal stage and finally the adult. The wormlike larvae, called caterpillars, maggots, bagworms, grubs, and so on, usually have chewing mouthparts. Wings develop internally during the larval stages. The larva forms a cocoon, or case, about itself and becomes a pupa. At the final molt, the adult emerges fully grown. Flies and fruit flies (Diptera), butterflies and moths (Lepidoptera), beetles (Coleoptera), and the like undergo complete metamorphosis. Thus, the stages are egg, larva, pupa, and adult. This type of development is called **holometabolous** (Gr. *holo,* complete, + *metabolē,* change).

Procedure

Beginning 7 to 10 days before your laboratory period, your instructor will have placed in each of several vials a pair of fruit flies and a microscope slide covered with a thick layer of banana-agar medium (or a commercial medium) and a drop of yeast solution. The vials will have been stoppered with cotton. There should now be eggs, larvae, and pupae in the medium on the various slides.

Remove a slide from a vial, place it under a dissecting microscope, and study its contents.

A female deposits **eggs** on food appropriate to feed her growing larval young. These hatch in about 24 hours into wormlike **larvae.** Larvae of flies are often called maggots. The larvae eat and grow for about 5 days and then become inactive **pupae,** each enclosed in a brownish case, within which final metamorphosis occurs. About 4 days after pupae are formed, **adults** emerge. Adults are full size when they appear.

To study adults, you may kill them by placing a few drops of ether on the cotton plug in the vial.

Can you distinguish sexual differences in the adults? ____________ Compare the structure of a fruit fly with that of a honeybee and a grasshopper. What difference do you note in the wings? ____________ Fruit flies belong to the order Diptera (Gr. *di-*, two, + *pteron,* wing). Is the order well named? ____________

Drawings

On separate paper, sketch eggs, larvae, and pupa of developing *Drosophila.*

Further Study

EXERCISE 16E

Collection and Classification of Insects

Where to Collect Insects

1. Using a sweep net, sweep over grass, alfalfa, or weed patches.
2. Spread a cloth, a newspaper, or an inverted umbrella under a bush or shrub; beat or shake the plant vigorously.
3. Overturn stones, logs, bark, leaf mold, and rubbish; look under dung in pastures.
4. Watch for butterflies to alight; then drop a net over them.
5. Look around outdoor lights at night.
6. At night, suspend a sheet from a limb or clothesline, with the lower part of the sheet spread on the ground. Direct automobile headlights or a spotlight on the sheet. The insects will be attracted to the white light, hit against the sheet, and drop onto the cloth below.
7. Moths may be baited by daubing a mixture of crushed banana or peach and molasses or sugar on the bark of trees; visit the trees at night with a flashlight to collect the moths.
8. Locate soil insects by placing humus and leaf matter in a Berlese funnel.
9. Water insects may be seined with a water net. Aquatic insect larvae may be found in either quiet or running water, attached to plants and leaves, or under stones, or may be sieved out of bottom mud.
10. In early spring, the sap exuding from stumps and tree trunks attracts various insects.

Killing Insects

If an insect is to be preserved, it must be killed in such a way as not to damage it. Killing bottles of various sizes and shapes may be used. Wide-mouthed plastic jars, such as peanut butter jars with screw-top lids, are best. Each jar should be conspicuously labeled POISON.

A temporary killing bottle can be made quickly by wetting a piece of cotton with a fumigant, such as ethyl acetate (nail polish remover); placing it in the bottom of a bottle; covering it with several discs of blotting paper; and screwing on the lid. *Do not inhale* the fumigant. Half-pint or pint jars are satisfactory. The fumigant must be replenished before each collecting trip. A few loose strips of paper towel added to the bottle will help absorb moisture and protect small insects. Replace the paper as it becomes moist.

A more permanent killing jar can be made by cementing a snap-cap vial to the underside of the lid of a wide-mouthed jar, with the vial opening facing down. The vial is filled with cotton saturated with ethyl acetate before use. When the killing jar is not in use, the cap may be snapped on the vial to retard evaporation. Similar killing jars may be purchased from biological supply houses.

Hard-bodied insects may require an hour or so to ensure death. If you are collecting many kinds of insects, it is good to have several killing bottles so that large insects, such as butterflies, dragonflies, and large beetles, may be kept in separate jars, away from smaller, more fragile insects.

To transfer an insect from a net to the killing jar, fold the net over, uncover the jar, work the jar into the net, and place its mouth over the insect as it clings to the cloth; hold the lid over the jar for a few seconds until the insect is quiet before removing it from the net and placing it in the jar. Remove the insect from the killing jar when it is dead and place it between layers of paper towel or in paper envelopes to prevent damage.

Soft-bodied insects and especially larval forms should not be put into the killing jar. They are killed by dropping them into 70% to 80% ethyl alcohol or by injecting their bodies with alcohol. The alcohol should be changed after a few days because it becomes diluted by the body fluids of the animals.

Relaxing

Insects that have dried out are fragile and easily broken. If an insect cannot be mounted before it has become dry, it can be relaxed. Prepare a relaxing jar by placing a layer of wet sand or wet paper towel in the bottom of a glass jar, adding a few drops of formalin or carbolic acid to discourage mold, and covering the sand with a disc of blotting paper to protect the insect. The insect may be left in the relaxing jar for 1 to 3 days, as necessary.

Displaying Insects

1. Insects can be mounted in shallow boxes on a layer of cotton and covered with glass plates or sheets of acetate. An attractive box of this kind can be made by cutting out the main portion of the lid of the box, so that a 1.5- to 2.5-cm rim, or "frame," remains around

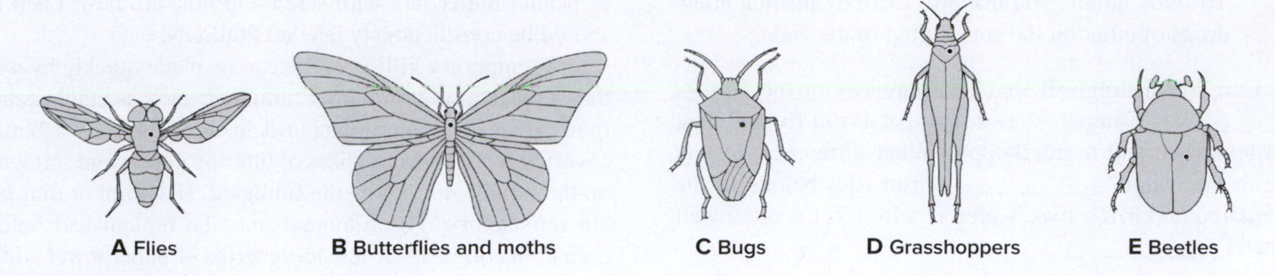

Figure 16-10

How to pin insects. The dots indicate the location of pins. **A,** Flies. **B,** Butterflies and moths. **C,** Bugs. **D,** Grasshoppers. **E,** Beetles.

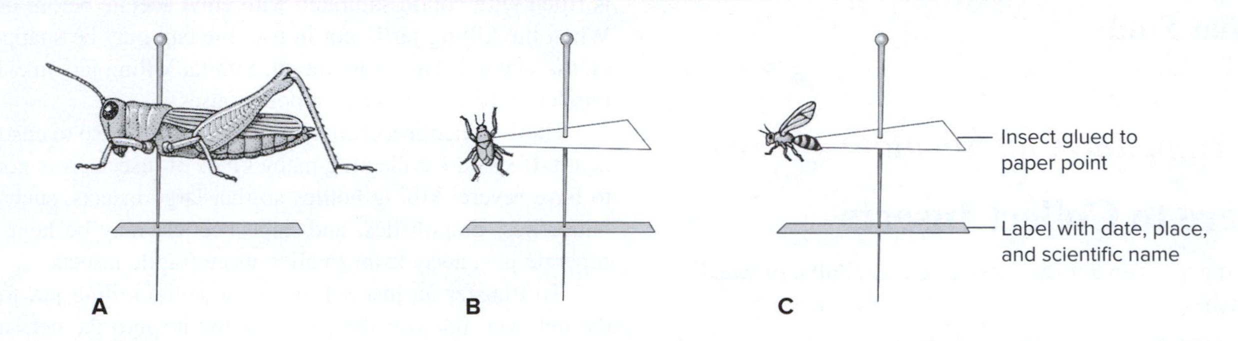

Figure 16-11

Pinning and labeling insects. **A,** Larger insects are pinned through the body. Minute ones can be glued to a paper point, either **B,** dorsal side up or **C,** laterally. Dates, places, and scientific names may be printed on slips of paper beneath the insect.

each side, then gluing a piece of glass or acetate to the inside of the lid behind the frame. The box itself is lined with cotton, and the insects are held in place on the cotton by the acetate or glass in the lid.

2. Pinned insects can be displayed in purchased Schmitt boxes, cigar boxes, or cardboard boxes deep enough to take an insect pin. Balsa wood or layers of soft corrugated cardboard should be glued to the bottom of the box.

   Pin insects with regular insect pins, which are longer and thinner than common pins. Sizes 2 and 3 are convenient sizes for general use.

   Pin grasshoppers through the posterior part of the pronotum, a little to the right of the midline (Figure 16-10D). Butterflies and moths are pinned through the center of the thorax, whereas bees, wasps, and flies are pinned through the thorax but a little to the right of midline. Pin bugs (hemipterans) through the scutellum, a bit to the right of midline, and beetles through the right elytron (forewing), halfway between the two ends of the body. To insert the pin, hold the insect between the thumb and forefinger of one hand and insert the pin with the other. Mount all specimens at a uniform height—about 2.5 cm above the point of the pin.

   A tiny label with date, place, and name of the collector can be added to the pin under the insect, with another label giving the scientific name of the insect placed under the first.

   A very small insect can be mounted on the point of an elongated, triangular piece of light cardboard about 8 to 10 mm long and 3 to 4 mm wide at the base (Figure 16-11). After putting the triangular cardboard mount on the pin, hold the pin and touch the tip of the cardboard triangle to the glue; then touch the insect with the gluey tip. Use as little glue as possible to hold the insect.

   Many winged insects, such as butterflies, moths, and mayflies, are usually mounted first on a spreading board, which has a groove into which the body is pinned. The wings are then spread out and held in the proper position to dry. Care must be taken not to damage the insect or to rub off the scales of a lepidopteran. The wings are carefully moved into position by pins and held by strips of paper pinned to the board. For butterflies, moths, and mayflies, the rear margins of the forewings should be straight across, at right angles to the body; the hindwings should be far enough forward so that there is no gap between the forewings and hindwings. The forewing should overlap the front edge of the hindwing a little. With damselflies, dragonflies, grasshoppers, and most other insects, the front margins of the hindwings should be straight across, with the front wing far enough forward so that forewings and hindwings do not touch.

Try to maneuver the wings by means of a pin held near the base of the wings—not through the wings, because that would make holes in them. Pin the wings securely with paper strips, using several strips if necessary. When completed, hold the body down with forceps and remove the pin from the thorax.

If the specimen is to be displayed on cotton under glass, the spreading board is not necessary. Pin the insect upside down (feet up) on a flat pinning surface, such as corrugated cardboard, and pin the wings with strips of paper as before.

Drying may take several days. If the abdomen, when touched gently with a pin, can be moved independently of the wings, the specimen is not dry enough; wait until the body is stiff before removing the pins and strips of paper.

Key to the Principal Orders of Insects*

The following key will enable you to place your insects in the correct order. The use of a two-choice, or dichotomous, key such as this one is explained in Exercise 5B. This key is designed for use with adult (final instar) insects and is therefore not suitable for identification of insect larvae. If you wish to carry identification to family, genus, or species, consult one of the references listed at the end of this key.

Terms Used in the Key

Cercus (pl. cerci) One of a pair of jointed anal appendages.

Tarsus (pl. tarsi) The leg segment distal to the tibia, consisting of one or more segments or subdivisions.

1 All wings functional, obvious, and membranous or covered with scales or hairs 2

Without functional wings, or with forewings thickened and concealing functional membranous hindwings 15

2 (1) Wings covered with minute scales; mouthparts usually a coiled tube (butterflies and moths) .... **Lepidoptera**

Wings usually clear, not covered with scales; mouthparts not a coiled tube 3

3 (2) With one pair of wings (true flies) .... **Diptera**

With two pairs of wings 4

4 (3) Wings long, narrow, fringed with long hairs, body length 5 mm or less (thrips) .... **Thysanoptera**

Wings narrow and fringed, body usually longer than 5 mm 5

5 (4) Abdomen with two or three threadlike "tails"; hindwings small (mayflies) .... **Ephemeroptera**

Abdomen with only short filaments or none; hindwings larger 6

6 (5) Forewings clearly longer and with greater area than hindwings 7

Forewings not longer, or only slightly longer, than hindwings, and with same or less area than hindwings 9

7 (6) Forewings noticeably hairy; antennae as long as or longer than body (caddisflies) .... **Trichoptera**

Wings transparent or translucent, not hairy; antennae shorter than body 8

8 (7) Tarsi two-segmented or three-segmented; body not wasplike or beelike 14

Tarsi five-segmented; usually wasplike or beelike (sawflies, ichneumons, winged ants, wasps, bees .... **Hymenoptera**

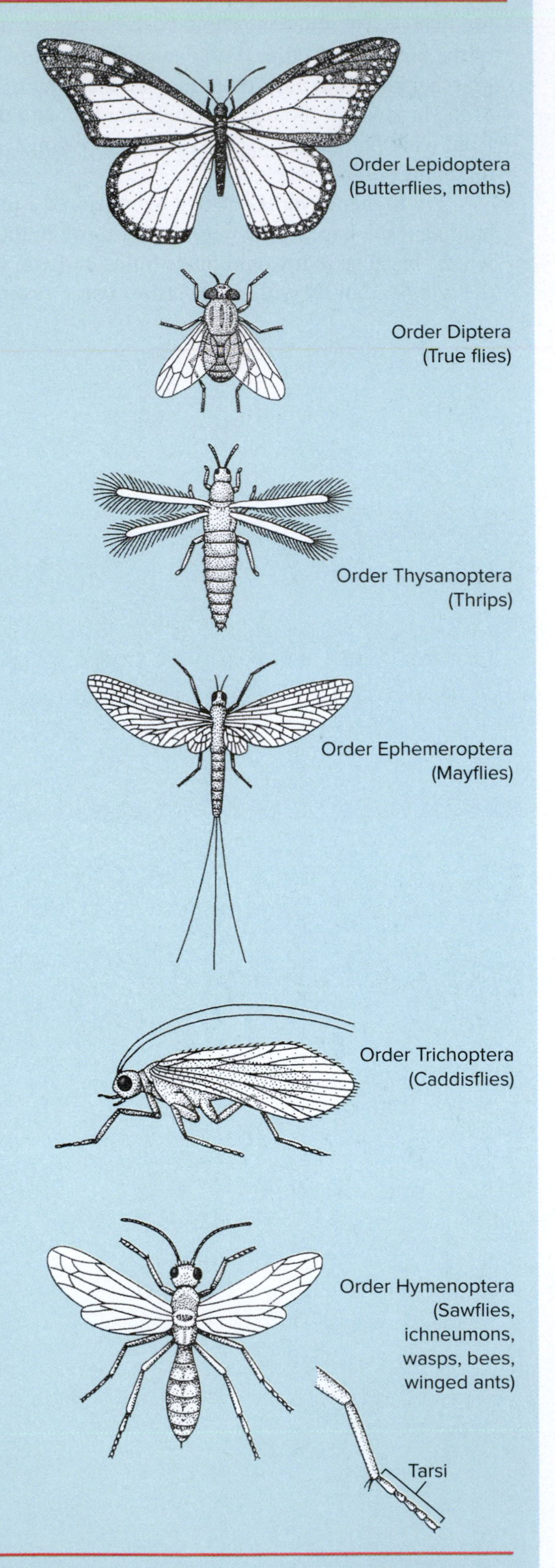

* Key adapted from Borror, D.J., and R.E. White. 1970. A field guide to the insects of America north of Mexico. Boston, Houghton Mifflin Company.

9 (6) Head prolonged ventrally into a beaklike structure (scorpionflies) . . . . **Mecoptera**

Head not prolonged vertically 10

Order Mecoptera (Scorpionflies)

10 (9) Antennae very short and bristlelike; eyes large; abdomen long and slender (dragonflies, damselflies) . . . . **Odonata**

Antennae not short and bristlelike; eyes moderate to small 11

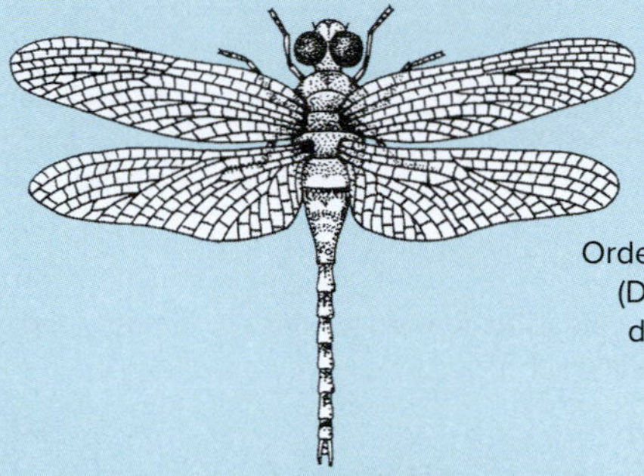

Order Odonata (Dragonflies, damselflies)

11 (10) Hindwings broader than forewings; cerci present (stoneflies) . . . . **Plecoptera**

Hindwings little if any broader than forewings; cerci absent 12

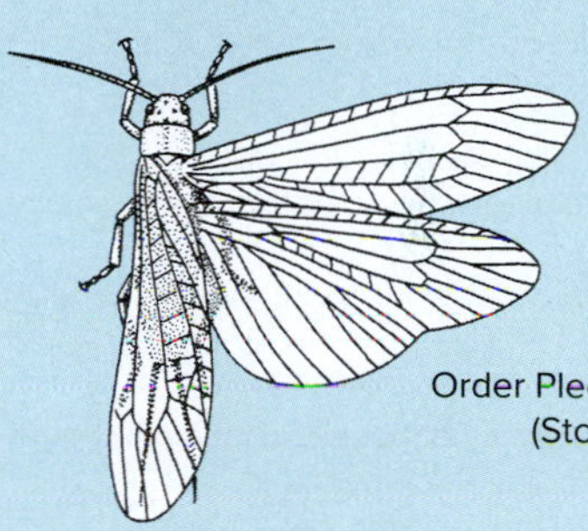

Order Plecoptera (Stoneflies)

12 (11) Mothlike; wings noticably hairy and opaque; antennae as long as or longer than body (caddisflies) . . . . **Trichoptera**

Not mothlike; wings not noticeably hairy, usually clear; antennae shorter than body 13

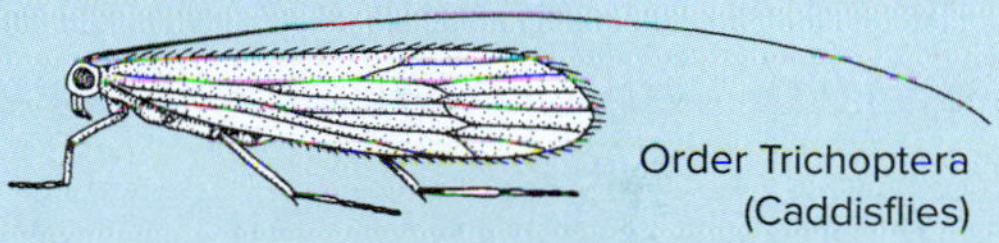

Order Trichoptera (Caddisflies)

13 (12) Wings with few crossed veins; tarsi four-segmented; length to 8 mm (termites). . . . . **Isoptera**

Order Isoptera (Termites)

Wings with numerous cross veins; tarsi five-segmented; length to 75 mm (fishflies, dobsonflies, lacewings, ant lions) . . . . . **Neuroptera**

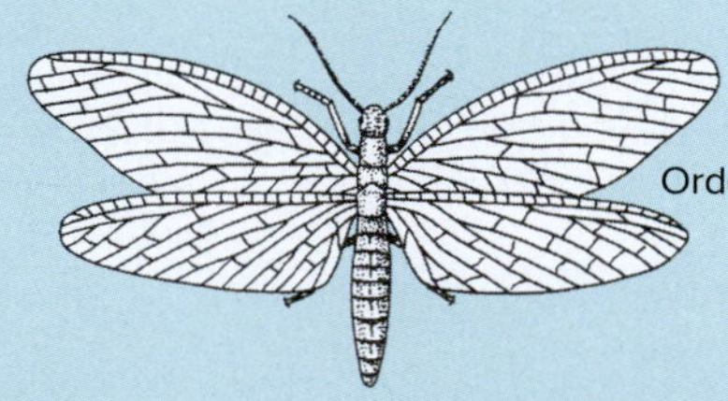

Order Neuroptera (Fishflies, Dobsonflies, lacewings, ant lions)

continued

Key to the Principal Orders of Insects (*continued*)

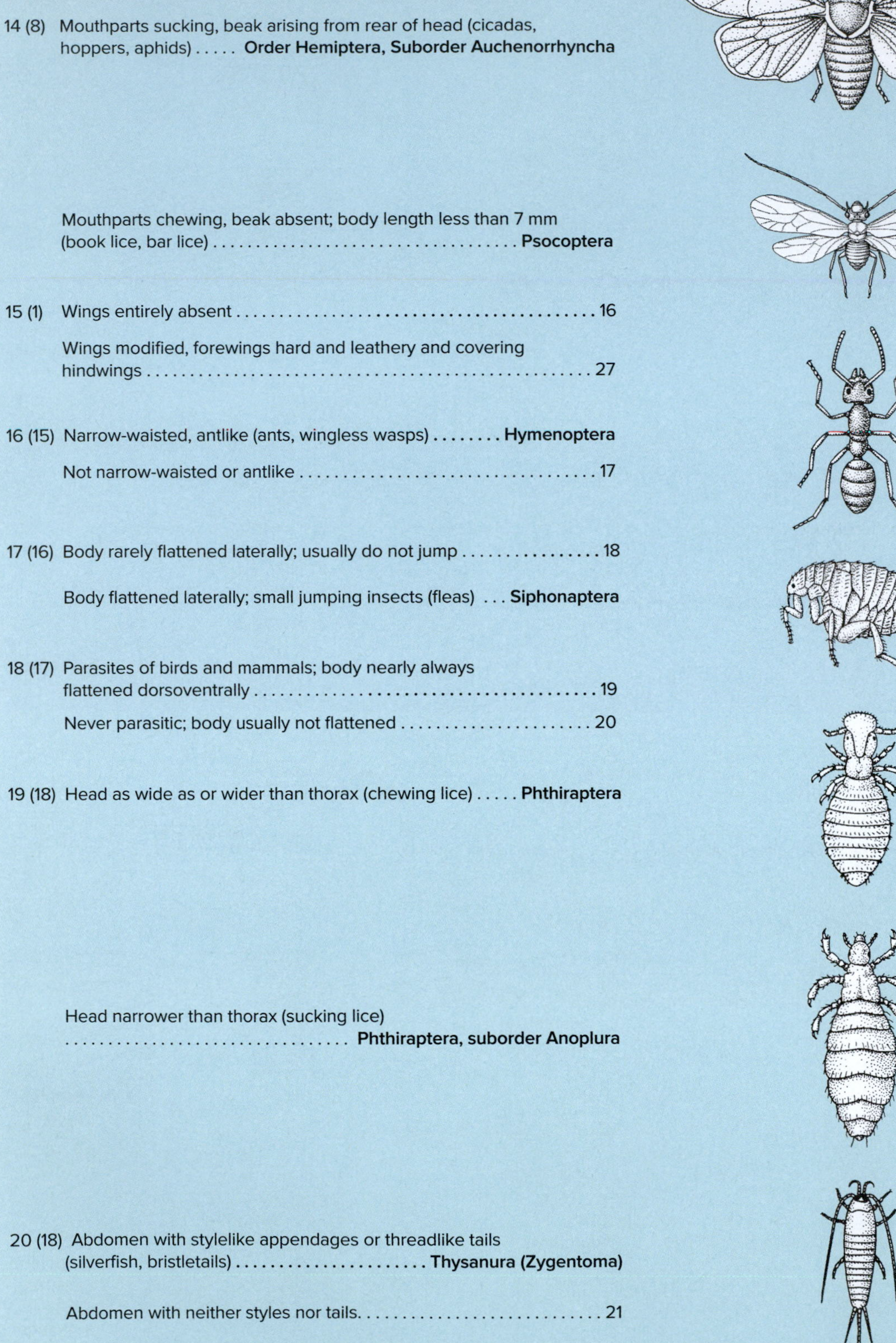

14 (8) Mouthparts sucking, beak arising from rear of head (cicadas, hoppers, aphids) . . . . . **Order Hemiptera, Suborder Auchenorrhyncha**

Mouthparts chewing, beak absent; body length less than 7 mm (book lice, bar lice) . . . . **Psocoptera**

15 (1) Wings entirely absent 16

Wings modified, forewings hard and leathery and covering hindwings 27

16 (15) Narrow-waisted, antlike (ants, wingless wasps) . . . . **Hymenoptera**

Not narrow-waisted or antlike 17

17 (16) Body rarely flattened laterally; usually do not jump 18

Body flattened laterally; small jumping insects (fleas) . . . **Siphonaptera**

18 (17) Parasites of birds and mammals; body nearly always flattened dorsoventrally 19

Never parasitic; body usually not flattened 20

19 (18) Head as wide as or wider than thorax (chewing lice) . . . . . **Phthiraptera**

Head narrower than thorax (sucking lice) . . . . **Phthiraptera, suborder Anoplura**

20 (18) Abdomen with stylelike appendages or threadlike tails (silverfish, bristletails) . . . . **Thysanura (Zygentoma)**

Abdomen with neither styles nor tails. 21

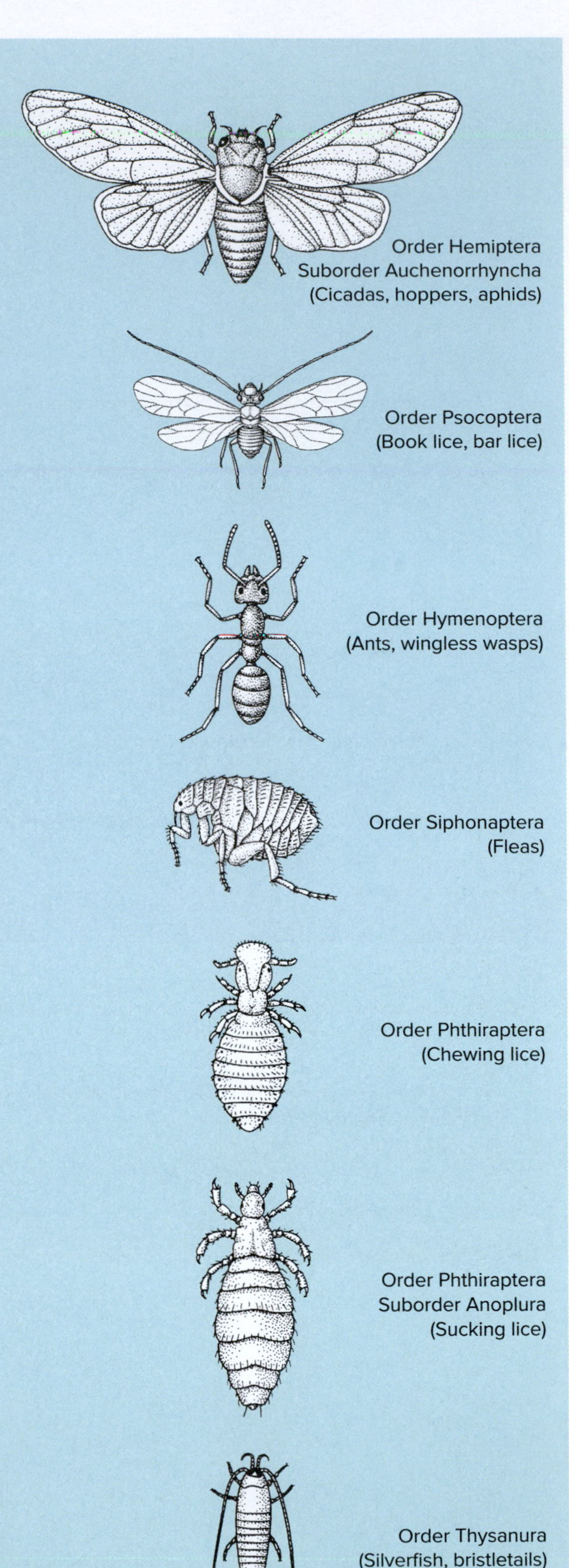

21 (20) Abdomen with a forked, tail-like jumping mechanism (springtails) .... **Collembola**

Abdomen lacking a jumping mechansim 22

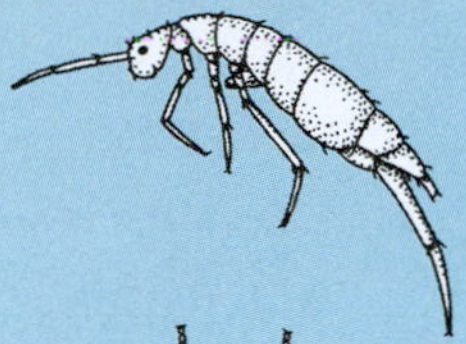

Order Collembola
(Springtails)

22 (21) Abdomen usually with two short tubes; small, plump, soft-bodied (aphids, others) . . . **Hemiptera, suborder Sternorrhyncha**

Abdomen without tubes; usually not plump and soft-bodied 23

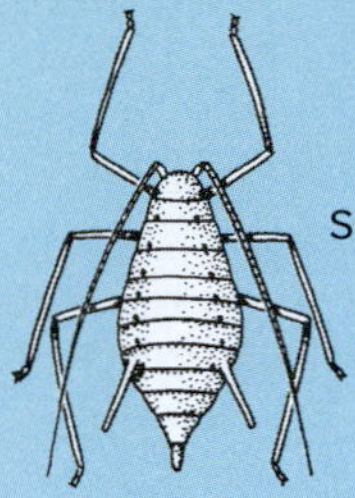

Order Hemiptera
Suborder Sternorrhyncha
(Aphids)

23 (22) Lacking pigment, whitish; soft-bodied 24

Distinctly pigmented; usually hard-bodied 25

24 (23) Antennae long, hairlike; tarsi two-segmented or three-segmented (psocids) .... **Psocoptera**

Order Psocoptera
(Book lice, bar lice)

Antennae short, beadlike; tarsi four segmented (termites) .... **Isoptera**

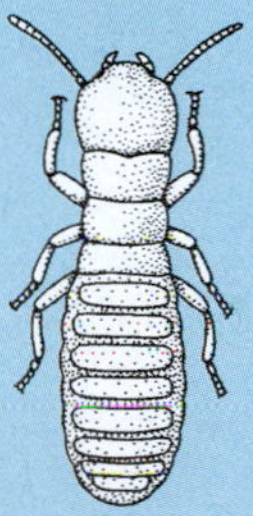

Order Isoptera
(Termites)

25 (23) Body shape variable; length over 5 mm. 26

Body narrow; length less than 5 mm (thrips) .... **Thysanoptera**

26 (25) Antennae four-segmented or five-segmented; mouthparts sucking (wingless bugs) .... **Hemiptera, suborder Heteroptera**

Antennae many-segmented; mouthparts chewing 31

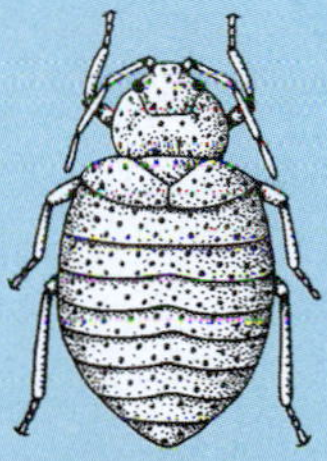

Order Hemiptera
Suborder Heteroptera
(Wingless bugs)

27 (15) Abdomen with forcepslike cerci (earwigs). .... **Dermaptera**

Abdomen lacks forcepslike cerci 28

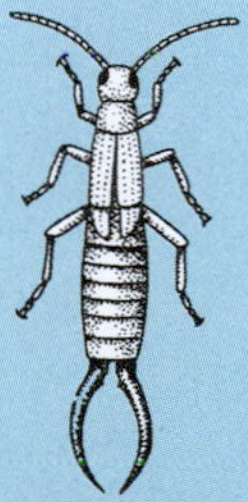

Order Dermaptera
(Earwigs)

28 (27) Mouthparts sucking; beak usually elongate. 29

Mouthparts chewing 30

29 (28) Forewings nearly always thickened at base, membranous at tip; beak rises from front or bottom of head (true bugs) .... **Hemiptera, suborder Heteroptera**

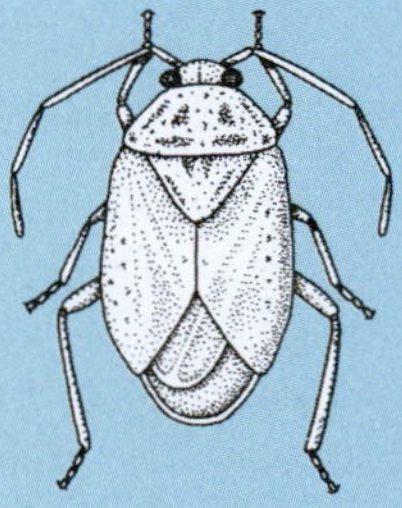

Order Hemiptera
Suborder Heteroptera
(True bugs)

continued

Key to the Principal Orders of Insects (*continued*)

Forewings of uniform texture throughout; beak arises from hind part of head (hoppers) .... **Hemiptera, suborder Auchenorrhyncha**

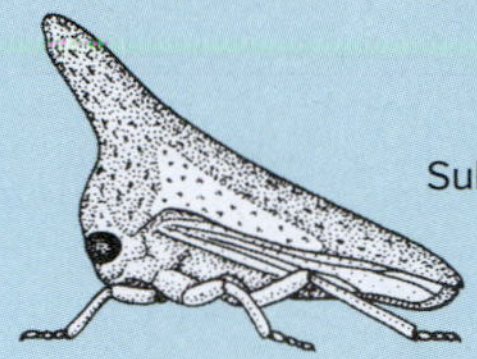

Order Hemiptera
Suborder Auchenorrhyncha
(Hoppers)

30 (28) Forewings with veins, at rest held rooflike over abdomen or overlapping; hindlegs modified for jumping (grasshoppers, crickets, katydids) .... **Orthoptera**

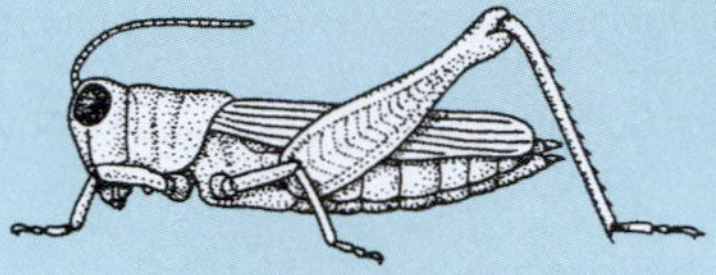

Order Orthoptera
(Grasshoppers, crickets, katydids)

Forewings without veins, meeting in a straight line down back (beetles) .... **Coleoptera**

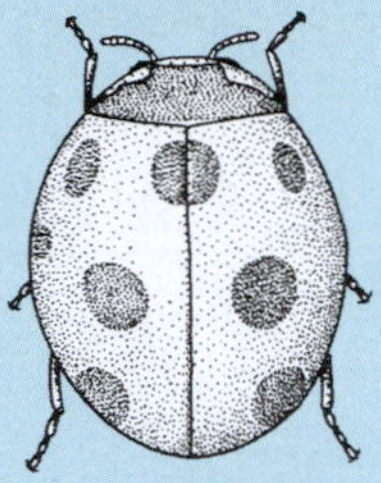

Order Coleoptera
(Beetles)

31 (26) Body oval and flattened (cockroaches) .... **Blattodea**

Body slender and elongate 32

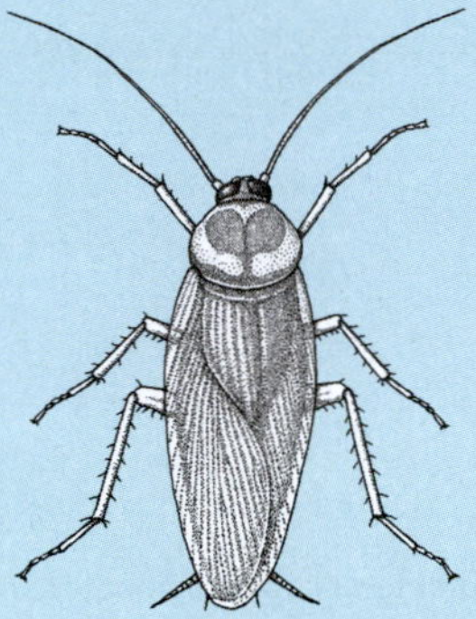

Order Blattodea
(Cockroaches)

32 (31) Sticklike body (walking sticks) .... **Phasmatodea**

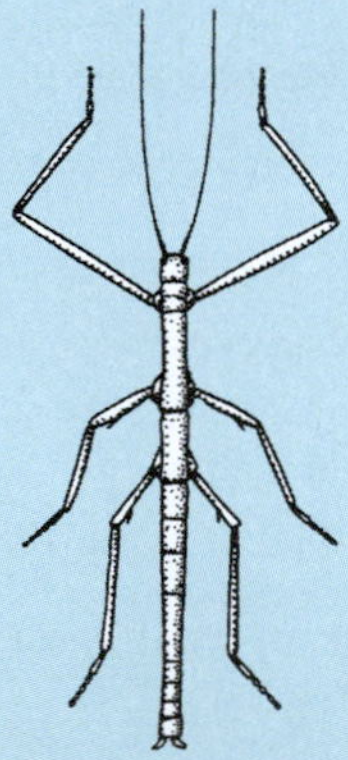

Order Phasmatodea
(Walking sticks)

Elongate body with raptorial front legs (mantids) .... **Mantodea**

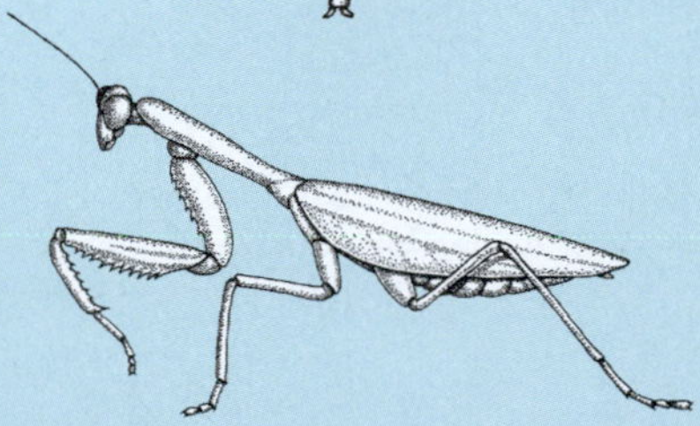

Order Mandodea
(Mantids)

References

Note that, in addition to the following general guidebooks, regional guides are often useful for surrounding areas.

Arnett, R. H. 2001. American insects: a handbook of the insects of America and Mexico, ed. 2. Boca Raton, Florida, CRC Press.

Arnett, R. H., Jr., and M. C. Thomas, eds. 2000. American beetles, vol. 1. Boca Raton, Florida, CRC Press.

Borror, D. J., and R. E. White. 1998. A field guide to the insects: America north of Mexico. Boston, Houghton Mifflin Harcourt.

Chu, H. F. 1949. How to know the immature insects. Dubuque, Iowa, Wm. C. Brown Company.

Claassen, P. W. 1931. Plecoptera nymphs of America (north of Mexico). Springfield, Ill., Pub. of the Thomas Say Foundation, by Charles C Thomas.

Covell, C. V., Jr. 2005. A field guide to the moths of eastern North America, ed. 2. Martinsville, VA, Virginia Museum of Natural History.

Dillon, E. S., and L. S. Dillon. 1972. A manual of common beetles of eastern North America. New York, Dover Publications, Inc.

Edmunds, G. F., Jr., S. L. Jensen, and L. Berner. 1976. The mayflies of North and Central America. Minneapolis, University of Minnesota Press.

Ehrlich, P. R., and A. H. Ehrlich. 1961. How to know the butterflies. Dubuque, Iowa, Wm. C. Brown Company.

Fisher, B. L., and Stepfan P. Cover. 2007. Ants of North America: A guide to the genera. Berkeley, University of California Press.

Harris, J. R. 1952. An angler's entomology. New York, F. A. Praeger.

Hogue, C. L. 1993. Insects of the Los Angeles Basin. Los Angeles, Natural History Museum of L. A. County.

Holland, W. J. 1968. The moth book: a popular guide to a knowledge of the moths of North America. New York, Dover Publications, Inc.

Jaques, H. E. 1947. How to know the insects, ed. 2. Dubuque, Iowa, Wm. C. Brown Company.

Jaques, H. E. 1951. How to know the beetles. Dubuque, Iowa, Wm. C. Brown Company.

Jewett, S. G. 1959. The stoneflies (Plecoptera) of the Pacific Northwest. Corvallis, Oregon State College.

Kaufman, K., and F. P. Brock. 2006. Butterflies of North America (Kaufman Field Guides). Boston, Houghton Mifflin Harcourt.

Klots, A. B. 1951. A field guide to the butterflies of North America east of the great plains. Boston, Houghton Mifflin Company.

LaFontaine, G. 1981. Caddisflies. New York, Lyons & Burford.

Lehmkahl, D. M. 1979. How to know the aquatic insects. The pictured key nature series. Dubuque, Iowa, Wm. C. Brown Company.

McPherson, J. E. 1982. The Pentatomoidea (Hemiptera) of northeastern North America. Carbondale, Ill., Southern Illinois University Press.

Merritt, R. W., and K. W. Cummins. 1996. An introduction to the aquatic insects of North America, ed. 3. Dubuque, Iowa, Kendall/Hunt Publishing Company.

Miller, P. L. 1984. Dragonflies. New York, Cambridge University Press.

National Audubon Society. 1980. National Audubon Society field guide to North American insects and spiders. New York, Alfred A Knopf, Inc.

Needham, J. G., and M. J. Westfall, Jr. 1975. A manual of the dragonflies of North America (Anisoptera): including the Greater Antilles and the provinces of the Mexican border. Berkeley, University of California Press.

Opler, P. A. 1998. A field guide to eastern butterflies, ed. 2. The Peterson field guide series. Boston, Houghton Mifflin Harcourt.

Otte, D. 1981–84. The North American grasshoppers. Vol. 1: Acrididae (Gomphocerinae and Acridinae); Vol. 2: Acrididae (Oedipodinae). Cambridge, Mass., Harvard University Press.

Pyle, R. M. 1981. The Audubon Society field guide to North American butterflies. New York, Alfred A. Knopf, Inc.

Sborboni, V., and S. Forestiero. 1998. Butterflies of the world. Buffalo, N.Y., Firefly Books Inc.

Schuh, R. T., and J. A. Slater. 1995. True bugs of the world. Ithaca, N.Y., Cornell University Press.

Scott, J. A. 1986. The butterflies of North America: a natural history and field guide. Stanford, Calif., Stanford University Press.

Smart, P. 1975. The international butterfly book. New York, Crowell Company.

Stehr, F. W. (ed.). 1991. Immature insects. Vols. 1 & 2. Dubuque, Iowa, Kendall/Hunt Publishing Company.

Swan, L. A., and C. S. Papp. 1972. The common insects of North America. New York, Harper & Row, Publishers.

Triplehorn, C. A., and N. F. Johnson. 2005. Borror and DeLong's introduction to the study of insects, ed. 7. Belmont, Calif., Brooks/Cole.

Westfall, M. J., Jr., and J. L. May. 1996. Damselflies of North America. Jodhpur, India, Scientific Publishers.

White, R. E. 1983. A field guide to the beetles of North America. Boston, Houghton Mifflin Company.

Wiggins, G. B. 1977. Larvae of the North American caddisfly genera (Trichoptera). Toronto, University of Toronto Press.

NOTES

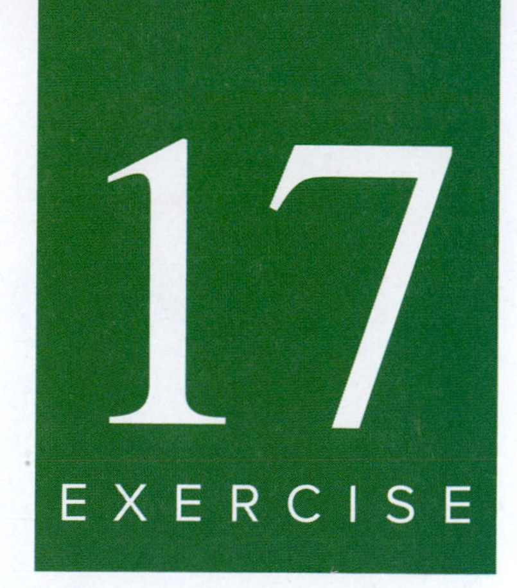

The Echinoderms

Phylum Echinodermata

A Deuterostome Group

Echinoderms are an all-marine phylum comprising of sea lilies, sea stars, brittle stars, sea urchins, sand dollars, and sea cucumbers. They are a strange group, strikingly different from any other animal phylum, a group that abandoned bilateral symmetry to become radial.

Echinoderms are deuterostomes, thus sharing with Hemichordata and Chordata several embryological features that set them apart from all the rest of the animal kingdom: anus developing from or near the blastopore and mouth developing elsewhere, enterocoelous coelom, radial and regulative cleavage, and mesoderm derived from enterocoelous pouches. Thus, all three phyla are presumably derived from a common ancestor.

It is an echinoderm's **dermal endoskeleton** of calcareous plates and spines, often fused into an investing armor, that provides the group with its name (L. *echinatus,* prickly, + *derma,* skin). They have a **water-vascular system** that powers a multitude of tiny tube feet used for locomotion and food gathering. Many are invested with pincerlike **pedicellariae** that snap at creatures that would settle on them. To breathe, many echinoderms rely on numerous **dermal branchiae** (skin gills) that project delicately through spaces in their skeletal armor. As if to emphasize their uniqueness, echinoderms lack a definite head, their nervous system and sense organs are poorly developed, locomotion is slow, and they lack segmentation. Of all their distinguishing features, however, none delineates the group more conspicuously than its **pentaradial symmetry:** body parts always arranged radially in five or multiples of five, no matter how the body plan has become adapted to different feeding strategies. This radial symmetry is, however, secondarily acquired because their larvae are unmistakably bilaterally symmetrical (Figure 17-1).

EXERCISE 17A
Class Asteroidea—Sea Stars
Asterias

EXERCISE 17B
Class Ophiuroidea—Brittle Stars
Brittle Stars

EXERCISE 17C
Class Echinoidea—Sea Urchins
Arbacia

EXERCISE 17D
Class Holothuroidea—Sea Cucumbers
Sea Cucumbers

EXERCISE 17E
Class Crinoidea—Feather Stars and Sea Lilies
Feather Stars

CLASSIFICATION
Phylum Echinodermata

EXERCISE 17A

Class Asteroidea—Sea Stars

Core Study

Asterias

Phylum Echinodermata
Subphylum Eleutherozoa
Class Asteroidea
Genus *Asterias*

Sea stars are the "prima donnas" of echinoderms, familiar to many people as beautifully symmetrical symbols of marine life. Those commonly seen are intertidal species, especially along rocky coastlines. Many live on high-energy beaches that receive full force of the surf. Other species inhabit a variety of benthic habitats, often at great depths in the ocean. Sea stars (also called starfish), like other echinoderms, are strictly bottom dwellers. Some are particle feeders, but most are predators of slow-moving prey, such as molluscs (their favorite food), crabs, corals, and worms, since sea stars are themselves slow-moving animals. Adult sea stars seem to have few enemies, suggesting that they produce something that discourages potential predators.

Behavior

Examine a living sea star in a dish of seawater. Using the following suggestions, observe its behavior.*

*Suggestions for demonstrations with living sea stars are found in Appendix A, p. 411.

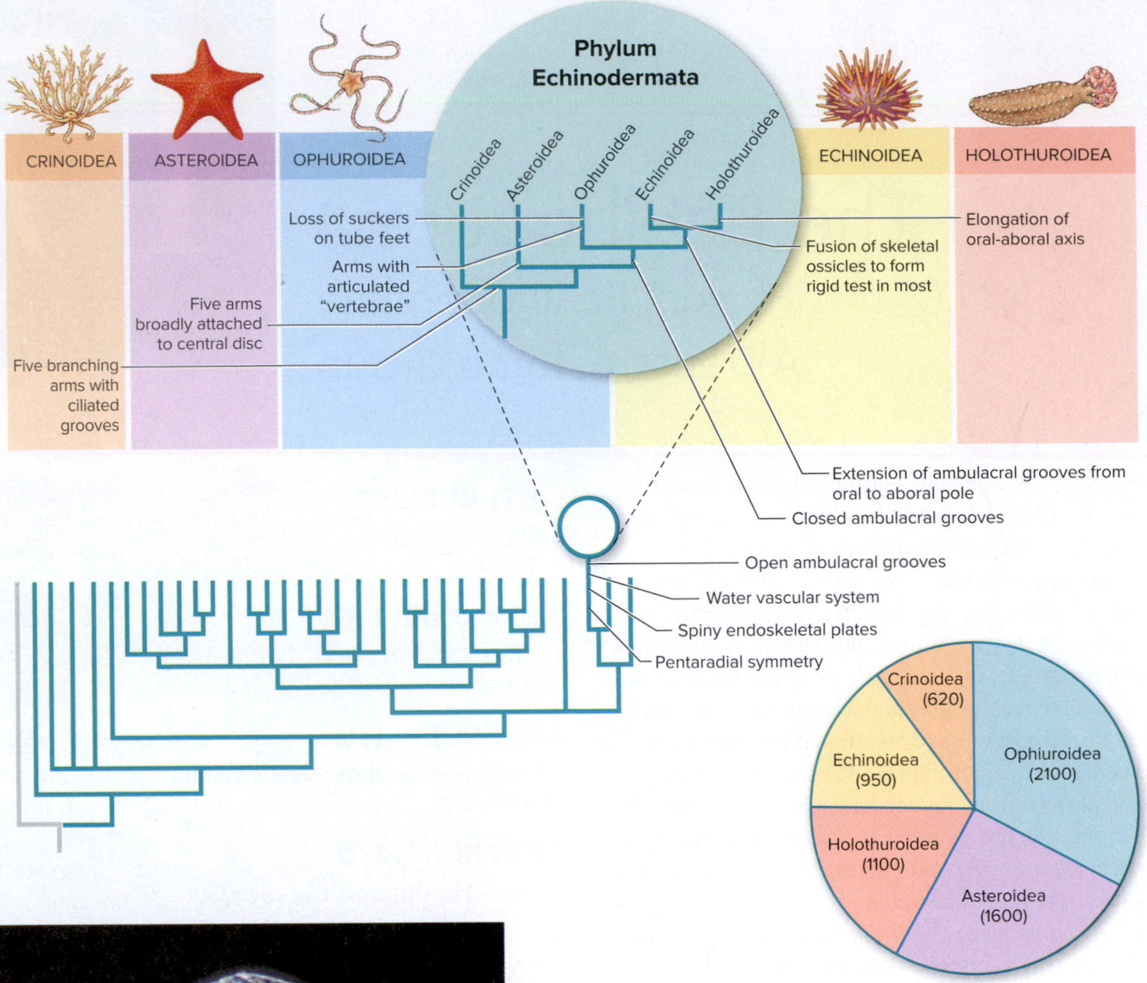

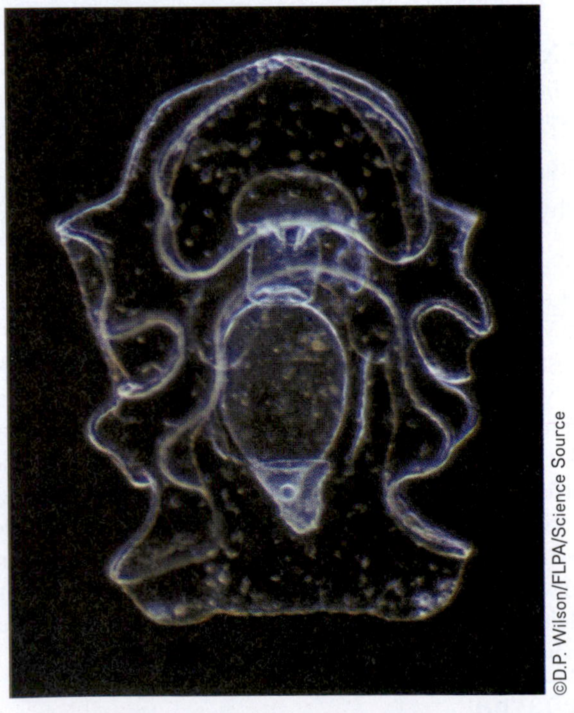

Figure 17-1
Bilaterally symmetrical auricularia larva of the sea cucumber, ×100. Because all echinoderms have bilateral larvae such as this one, the pentaradial (five-part) body plan of adult echinoderms is secondarily derived, one of several unique characteristics that distinguish the echinoderms.

Note the general body plan of the star with its **pentaradial** (Gr. *pente,* five, + L. *radius,* ray) symmetry; its five **arms,** or **rays;** its **oral-aboral flattening;** and its **mouth** on the underside. Lift up the dish and look at this oral surface. Notice the rows of **tube feet.** How are they used? _____________ How are the ends of the tube feet shaped? _____________ What is the sequence of action of a single tube foot as the animal is moving? _____________ Tube feet are filled with fluid and are muscular, providing the necessary components for a hydraulic skeleton. When a foot contracts, water flows into a bulblike **ampulla** inside the arm. Tube feet and ampullae are parts of the **water-vascular system.**

Tilt the dish to one side (pour out a little water, if necessary) and watch the animal's reaction. Does it move up or down the inclined plane? _____________ Now tilt the dish in the opposite direction. Does the animal change direction? _____________ Is it positively geotactic (moves toward the earth) or negatively geotactic? _____________

Place a piece of fresh seafood (oyster, fish, or shrimp) near one of the arms. Is there any reaction? _____________ Are the arms flexible? _____________ If the sea star makes no move toward the food, touch the tip of an arm with it, or slip it under the end of an arm. Hold the dish up and look underneath. How is the food grasped? _____________

Figure 17-2
Distal portion of the ray of a living sea star *(Asterias).*

How is it moved toward the mouth? ____________ What position does the animal assume when feeding? ________________________

Examine the aboral surface with a hand lens or dissecting microscope. The **epidermis** is ciliated. Place a drop of carmine suspension (in seawater) on the exposed surface and note the direction of the ciliary currents. Notice the calcareous **spines** protruding through the skin (Figure 17-2). These are extensions of skeletal ossicles. Do the spines move? ____________ Small, fingerlike bulges in the epidermis are **skin gills** (also called **dermal branchiae**) and **papulae** (sing. **papula;** L., pimple), concerned with gaseous exchange. Around the spines, you will see small **pedicellariae** (L. *pediculus,* little foot, + *aria,* like) (Figure 17-2). These calcareous, two-jawed pincers are modified spines concerned with capturing tiny prey and protecting the dermal branchiae from collecting sediment and small parasites. Touch with a small camel's hair brush to observe the pincer action.

Look at the tip of each arm to see a small, red **eyespot** and elongate tube feet modified as **sensory tentacles** (Figure 17-2).

Written Report

Record your observations on separate paper.

External Structure

Place a preserved sea star in a dissecting pan and cover with water. If your observations of external structure are on a living sea star, place the animal in a clean pan or culture dish and cover with seawater.

The star-shaped body is composed of a **central disc** and five **arms** (rays). Are all the arms alike? ____________ What would account for some of the arms being shorter than others? ____________ Compare your specimen with those of your neighbors. Preserved specimens may seem rigid but live stars can bend their arms by means of muscles.

Aboral Surface. The central disc bears a small, porous **madreporite plate** (Fr. *madrépore,* reef-building coral, + *ite,* suffix for body part) composed of calcium carbonate (Figure 17-3). It allows seawater to seep into an intricate **water-vascular system,** which provides the means of locomotion. The arms on each side of the madreporite are called the **bivium;** the other three are the **trivium.** The **anus** opens in the center of the central disc, but it is probably too small to see.

Submerge one of the arms in water and examine the dorsal body wall under a dissecting microscope. Compare with a piece of dried body wall.

The body is covered with a thin, ciliated **epidermis,** through which white calcareous **spines** extend from the endoskeleton beneath. Are the spines movable? ____________ Surrounding the base of each spine is a raised ring of skin bearing tiny, calcareous, pincerlike **pedicellariae.** Some are also found between the spines. Sometimes pedicellariae can be seen more easily on a dried piece of body wall. Some pedicellariae have straight jaws and others have curved jaws. They are moved by tiny muscles. What is the function of the pedicellariae? ____________ Between the spines are soft, transparent, fingerlike projections, the **skin gills,** or **dermal branchiae.** These are hollow evaginations of the coelomic cavity. What is the function of the dermal branchiae? ____________

Drawings

Sketch some pedicellariae in your notebook.

A small, pigmented **eyespot** is located at the tip of each arm.

Oral Surface. The **ambulacral** (L. *ambulare,* to walk) **groove** of each arm contains rows of **tube feet (podia).** The **ambulacral spines** bordering the groove are movable and can interlock when the groove is contracted to protect the tube feet. Note the size, shape, number, and arrangement of the tube feet. How many rows of tube feet are there?

Scrape away some tube feet and note arrangement of the pores through which they extend. If you are making observations on a living sea star, do not scrape off tube feet but instead brush them with a camel's hair brush, noting the response of both tube feet and ambulacral ossicles.

Tube feet are a part of the water-vascular system. They are hollow, and their tips form suction discs for attachment (Figures 17-3 and 17-4). These are effective not only in locomotion but also in opening of bivalves for food.

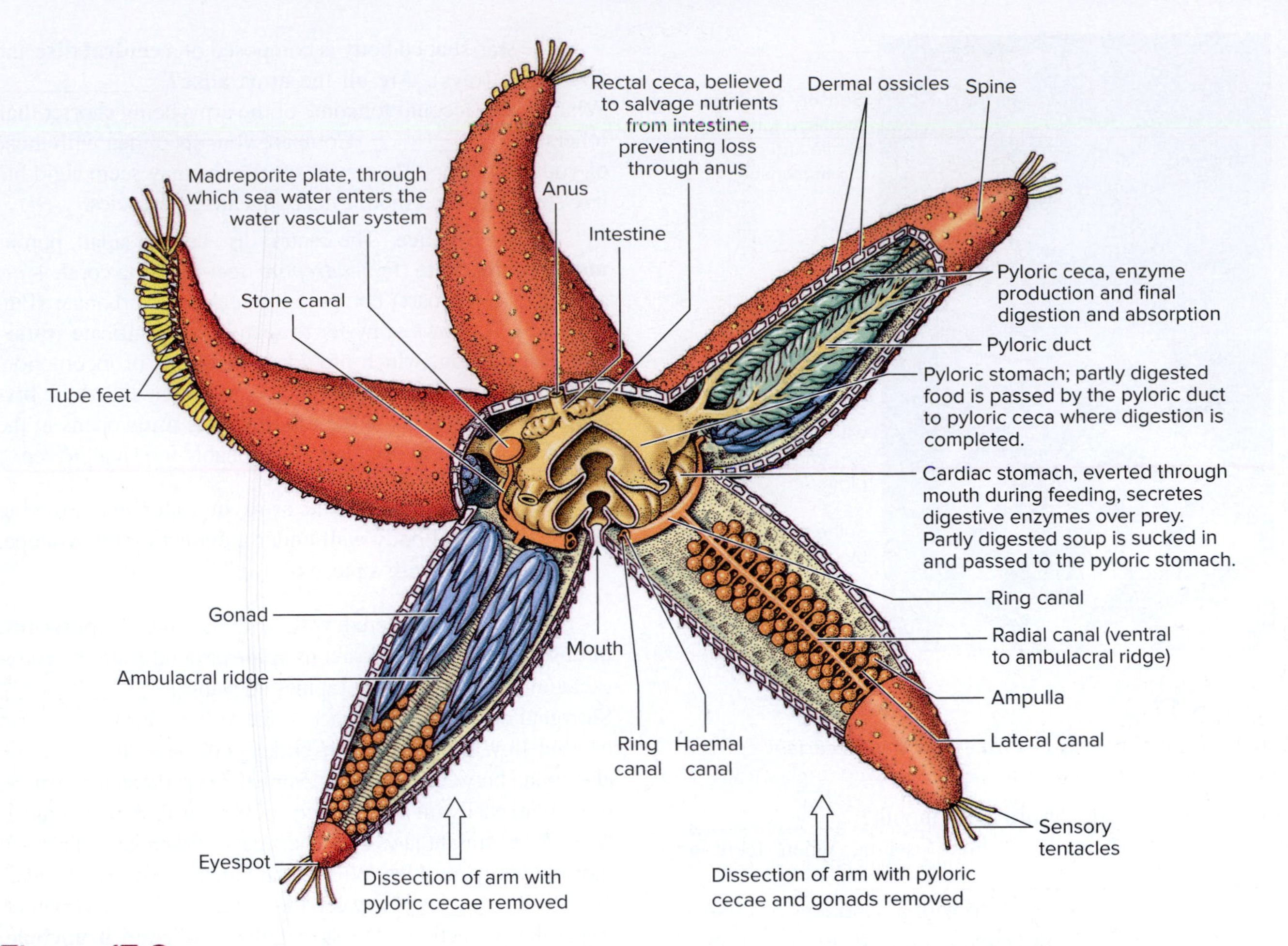

Figure 17-3

Anatomy of a sea star. Gonad size varies greatly with season when specimen was collected.

The central **mouth** is surrounded by five pairs of movable spines. Push the spines outward, bend the arms back slightly, and note the thin **peristomial membrane,** which surrounds the mouth. Sometimes the mouth is filled with part of the everted **stomach.**

Endoskeleton

Echinoderms are the first invertebrates to have a mesodermal endoskeleton. It is formed of calcareous plates, or **ossicles,** bound together by connective tissue.

If you are studying a preserved sea star, cut off part of one of the arms of the bivium, remove the aboral wall from the severed piece, and study its inner surface. Compare with a piece of dried body wall, which is excellent for examining the arrangement of ossicles.

Note the skeletal network of irregular ossicles. Now look at the cut edge of the body wall and identify the outer layer of **epidermis;** the thicker layer of **dermis,** or connective tissue in which the ossicles are embedded; and the thin inner layer of ciliated **peritoneum.** Do sea stars have a true coelom? ____________ The dermis and peritoneum are mesodermal in origin. Are the spines part of the endoskeleton? ____________ Hold the piece up to the light. The thin places you see between the ossicles are where the dermal branchiae extend through the connective tissue.

Look at the oral surface of the severed arm. Note how the **tube feet** extend up between the **ambulacral ossicles,** emerging on the inside surface of the wall as bulblike **ampullae** (am-pool'ee; sing. **ampulla;** L., flask). Compress some of the ampullae and note the effect on the tube feet. Press on the tube feet and see the effect on the ampullae. Both are muscular and can regulate the water pressure by contraction. Scrape away some of the ampullae and tube feet and examine the shape of the ossicles in the ambulacral groove. How do they differ from ossicles elsewhere? ____________ Note the alternating arrangement of **ambulacral pores,** through which the tube feet extend. Although the ambulacral ossicles form a groove on the oral surface of each arm, they form an **ambulacral ridge** on the inner surface.

Drawings

In your notebook, sketch a series of ambulacral ossicles, showing the arrangement of the openings for the tube feet.

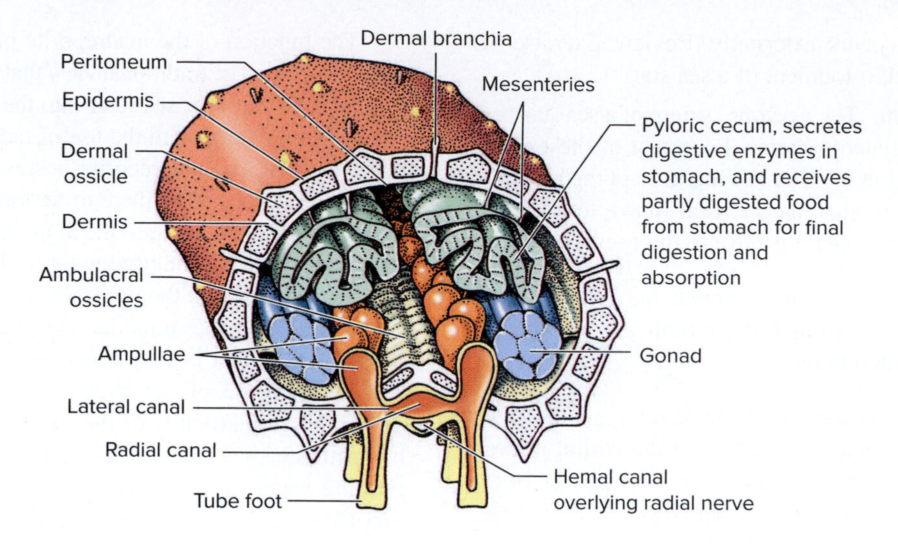

Figure 17-4
Cross section of a sea star arm.

Further Study

Internal Structure (Dissection)*

Place the specimen aboral side up in a dissecting pan and cover with water. Select the three arms of the trivium and snip off their distal ends. Insert a scissors point under the body wall at the cut end of one of the arms. Carefully cut along the dorsolateral margins of each arm to the central disc. Lift up the loosened wall and carefully free any clinging organs. Uncover the central disc, but cut around thc madreporite plate, leaving it in place. Be careful not to injure the delicate tissue underneath. The tissue around the anal opening in the center of the disc will cling. Cut the very short intestine close to the aboral wall before lifting off the body wall.

The **coelomic cavity** inside the arms and disc contains **coelomic fluid,** which bathes the visceral organs.

Digestive System. A pentagonal **pyloric stomach** (*Gr. pylōros,* gatekeeper) (Figure 17-3) lies in the central disc, and from it a **pyloric duct** extends into each arm, where it divides to connect with a pair of large, much-lobulated **pyloric ceca** (digestive glands) (Figure 17-3). A very short **intestine** leads up from the center of the stomach to the anus in the center of the disc. Attached to the intestine are two small, branched **rectal ceca.** Below (ventral to) the pyloric stomach is the larger, five-lobed **cardiac stomach,** which fills most of the central disc (Figure 17-3). Each lobe of the stomach is attached to the ambulacral ridge of an arm by a pair of **gastric ligaments,** which prevent too much eversion of the stomach.

*Directions for anesthetizing living stars for this study are given in Appendix A, p. 411.

When a sea star feeds on a bivalve, it folds itself around the animal, attaches its tube feet to the valves, and exerts enough pull to cause the shell to gape slightly. Then, by contracting its body walls to increase coelomic fluid pressure, it everts its stomach and inserts it into the slightly opened clam shell. There it digests the soft parts of the clam with digestive juices from the pyloric ceca. Partly digested material is drawn into the stomach and pyloric ceca, where digestion is completed. There is little waste fecal matter. When the sea star is finished feeding, the stomach withdraws into the coelom by contraction of thc stomach muscles and relaxation of the body wall, which allows coelomic fluid to flow back into the arms.

Many stars feed on small bivalves by engulfing the entire animal, digesting its contents, then casting the shell out through the mouth.

Reproductive System. Sexes are separate (dioecious) in sea stars. Remove the pyloric ceca from one arm to find paired **gonads** attached to the sides of the ray where the ray joins the disc (Figure 17-3). During the breeding season, the gonads are much larger than at other times, when they may shrink to a small fraction of their breeding size. Each gonad opens aborally to the exterior at the point of attachment by a very small **reproductive duct** and **genital pore.** The sex can seldom be determined by simple inspection of the gonads, although the female gonads may be a little coarser in texture and more orange than the male gonads.

Make a wet mount of a mashed bit of gonad and examine with a microscope.

In the ovary, eggs with large nuclei can be found; in the testes, many small sperm are to be found. In early summer, large streams of eggs and sperm are shed into the water, where fertilization occurs externally. Review Exercise 3B (pp. 36–40) for the development of a sea star.

Nervous System. The nervous system of a sea star consists of three interrelated systems. Foremost of these is the **oral** system, consisting of a **nerve ring** around the mouth in the peristomial membrane and a **radial nerve** to each arm running along the ambulacral groove to the eyespot.

To find the nerve ring, remove the tube feet and movable spines around the mouth and expose the peristomial membrane.

The nerve ring is a whitish thickening on the outer margin of this membrane. To see one of the **radial nerves,** bend an arm aborally and look along the oral surface of the ambulacral groove for a whitish cord (Figure 17-4). Trace the nerve from the ring to its termination in the arm.

The other two systems are an aboral system lying near the upper surface of the sea star and a deep system positioned between the oral and aboral systems. You will be unable to see these. Freely connected to these three systems is the **epidermal nerve plexus.** This consists of a network of nerve cell bodies and their processes lying just beneath the epidermis that coordinate responses of the dermal branchiae to tactile stimulation.

Sense Organs. Chemoreceptors and cells sensitive to touch are found all over the surface.

Each pigmented **eyespot** consists of a number of light-sensitive ocelli.

Water-Vascular System. A water-vascular system is found only in echinoderms, which use it for locomotion and, in the case of sea stars, for opening of clam shells. If this system in your specimen has been injected with a colored injection medium, its features can be studied to greater advantage.

Carefully remove the stomach from the central disc.

The **madreporite plate** (orange in life) on the aboral surface (Figure 17-3) contains ciliated grooves and pores. From it a somewhat curved **stone canal** (yellow in life and named for calcareous deposits in its wall) leads to a **ring canal** (Figure 17-3), which is found around the outer edge of the peristomial membrane next to the skeletal region of the central disc. The ring canal may be difficult to find if not injected.

Five **radial canals,** one in each arm, radiate out from the ring canal, running along the apex of the ambulacral groove just below the ambulacral ossicles and above the radial nerve. The position of the radial canal is best seen in a cross section of one of the arms (Figure 17-4). Short **lateral canals,** each with a valve, connect the radial canal with each tube foot. Now look on the inside of an arm and study the alternating arrangement of the ampullae. Note how each ampulla connects with a tube foot through a **pore** between the ambulacral plates.

The function of the madreporite plate has been a subject of controversy. The traditional view that the madreporite serves as a point of entry for seawater into the water-vascular system has been confirmed with the use of radioactive isotopes. Seawater entering the madreporite passes down the stone canal to the ring canals, from there to the radial canals, and finally through the lateral canals to the ampullae and tube feet.

Tube feet have longitudinal muscles; ampullae have circular muscles. When the tube feet are contracted, most of the water is held in the ampullae (Figure 17-4). When ampullae contract, water is forced into the elastic tube feet, which elongate because of the hydrostatic pressure within them. When the cuplike ends of the extended tube feet contact a hard surface, they attach with suction force and then contract, pushing water back into the ampullae and pulling the animal forward. Valves in the lateral canals prevent backflow of water into the radial canals. Although a single foot is not very strong, hundreds of them working together can move the animal along slowly and can create a tremendous pull on the shell of a bivalve mollusc. Suckers are of little use on a sandy surface, where the tube feet serve as tiny legs. Some species have no suckers but use the stiff podia like little legs to "walk." A large sea star can travel about 15 cm per minute.

Cross Section of an Arm of a Sea Star (Microslide)

Identify the **epidermis** covering the entire animal, **dermis** containing the **ossicles,** muscular tissue, **peritoneum,** and **coelomic cavity.** The spines have not yet erupted because these slides were prepared from young sea stars. Observe the **dermal branchiae** projecting from the coelom through the body wall. **Pyloric ceca** hang from the aboral wall by **mesenteries.** Notice the ampullae in the coelom and their connection to the **tube feet** and **lateral canals.** The canals may not always be seen. Why? _______________ Locate the **ambulacral ossicles** with the pores through which tube feet extend. Find the **radial canal,** a small tube under the **ambulacral groove,** and the **radial nerve** beneath the radial canal (Figure 17-4).

EXERCISE 17B

Class Ophiuroidea—Brittle Stars

Core Study

Brittle Stars

Phylum Echinodermata
  Subphylum Eleutherozoa
    Class Ophiuroidea
      Order Ophiurida
        Available genera

Where Found

Brittle stars, the most agile of echinoderms, are widely distributed in all oceans and at all depths. In many habitats, they are also the most abundant echinoderms yet are seldom seen by casual observers. They are secretive animals that hide under and between intertidal and subtidal rocks by day to escape predation by fish. Even at night, they may extend only their arms from hiding places to feed. If a diver should expose them, they will quickly retreat to safety, using rowing movements of their arms. Should a fish (or a human) catch a brittle star by one arm, the animal usually will simply cast off the arm (autotomize), leaving the predator with a wiggling arm while its erstwhile owner scurries to safety beneath a nearby rock. It is common to find brittle stars with missing or partly regenerated arms.

External Features

General Body Form. Note that the arms of brittle stars are sharply marked off from the central disc—a characteristic of ophiuroids (Figure 17-5). The arms are more flexible than those of sea stars. Does their appearance give a clue as to why? ____________ In both appearance and function, the arms resemble vertebral columns. In fact, internally they consist of a series of calcareous **vertebral ossicles,** each joined to the next by two pairs of muscles. Externally the arm is encased in a series of aboral, lateral, and oral **plates.**

Pull an arm off a preserved specimen and examine it in cross section. Locate the muscles and note the vertebral articulations.

Are there spines on the arms? ____________ How do the spines compare with those of asteroids? ____________ Do you find any pedicellariae or skin gills? ____________

Oral Surface. On the oral side, note five triangular **jaws** around the mouth. Find the five **oral shields** (also called **buccal shields**), which are oval plates located on the interradial area between the rays (Figure 17-6). One of these is slightly modified as a **madreporite plate,** and its tiny pores connect with a madreporite canal inside. Compare the location of the madreporite plate in asteroids and ophiuroids.

Table of Comparison

Complete a table of comparative characteristics of the external anatomy of the sea star and the brittle star on p. 280. After you have observed the sea urchin and the sea cucumber, add these to the table.

Distal to the oral shields and close to each arm is a pair of grooves, representing the openings of the **bursae.** (In *Ophioderma* a second pair is located distal to the first.) The bursae, composed of 10 saclike cavities within the disc, are peculiar to ophiuroids. Water is pumped in and out of them for respiratory purposes, and the gonads discharge their products into them. In some species, they also serve as brood pouches, but in *Ophioderma* development is external.

The tube feet (podia) are small, do not have suckers, and project laterally between the skeletal plates. They are largely sensory in function but may also assist in locomotion. Examine the rough spines used in gripping the substrate. How do they compare with those of asteroids? ____________

Behavior

Avoid rough handling of a brittle star because this may cause it to "freeze," becoming immobile, or even to cast off one or more of its arms.

Locomotion. Brittle stars "walk" by twisting, highly flexible arm movements. How are the arms used in locomotion? ____________ Watch their movements carefully.

©Cleveland P. Hickman, Jr.

A

©William C. Ober

B

Figure 17-5
A, Brittle star, *Ophiocoma aethiops,* from the Galápagos Islands. **B,** Basket star, *Astrophyton muricatum,* from the Caribbean. Ophiuroids are agile, have sharply defined arms, and are fragile.

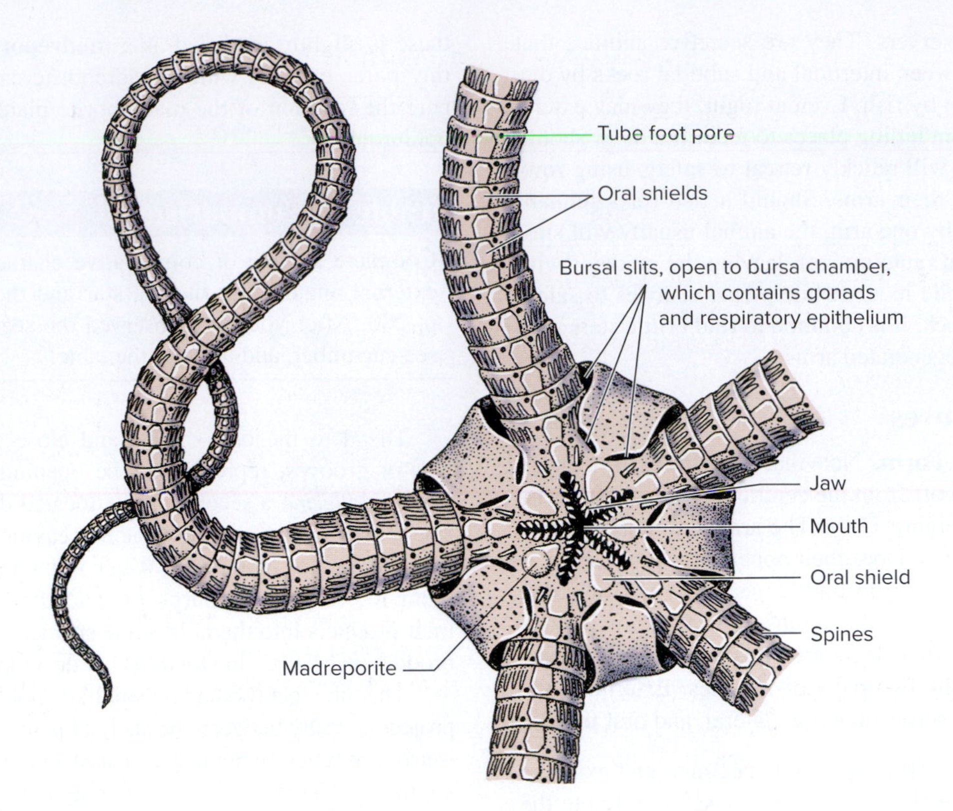

Figure 17-6
Oral view of central disc of a brittle star.

What provides the actual force for forward movement? ____________ Do the arms push or pull the brittle star? ____________ Do the arms work in pairs? ____________ How are the tube feet used—or are they used at all in locomotion? ____________ Does one arm always lead or follow? ____________ Note that locomotory patterns differ somewhat in different species of brittle stars.

Turn over the brittle star and watch its righting response. If placed on a sandy substrate, will the brittle star burrow? ____________

Feeding. Most ophiuroids are either active predators or selective deposit feeders. Deposit feeders capture organic material with the tube feet that secrete copious amounts of mucus. Mucus and food are rolled into a ball by a tube foot, then transferred to a small scale that lies adjacent to the tube foot. The next tube foot picks up the food and transfers it to the next scale, and so on until the food ball reaches the mouth. Active predators, such as species of the genus *Ophioderma,* capture benthic (bottom-dwelling) organisms by curling an arm around the prey and sweeping it into the mouth.

Drop very small bits of fish or shrimp into the aquarium near (but not touching) some brittle stars to see the method of feeding.

Do the animals appear to sense the presence of food before touching it? ____________ Is the species you are studying an active predator or a selective deposit feeder? ____________ The digestive system is much reduced, compared with that of sea stars. The mouth leads by a short esophagus to a stomach, the site of digestion and absorption. There is no intestine, anus, or hepatic cecum.

Reactions to Other Stimuli. Note how the animal reacts to mechanical stimulation. Touch the tip of an arm. Does it retreat from or advance toward the source of the stimulus? ____________ Touch a more proximal part of the arm. Is the reaction the same or different? ____________ Stimulate the base of an arm or the central disc. What happens? ____________

Can you determine whether ophiuroids—at least the species you are studying—respond positively or negatively to light? ____________

The water-vascular, hemal, and nervous systems are on a plan similar to that of asteroids.

Written Report

Record your behavioral observations on separate paper.

EXERCISE 17C
Class Echinoidea—Sea Urchins

Core Study

Sea urchins, like sea stars, are familiar denizens of seashores. Lacking arms, and with the body enclosed within a globose shell, or test, of interlocking plates which bear movable spines, sea urchins have evolved a body design quite unlike that of sea stars and brittle stars. Nevertheless, they bear the typical pentamerous plan of all echinoderms, with a water-vascular system and other characteristics that set the strange echinoderms apart from any other animal phylum. Sea urchins and their kin, heart urchins and sand dollars, all bear a spiny armament suggesting *echinos,* the Greek word for hedgehog, from which the echinoids take their scientific name. Unlike the carnivorous sea stars and brittle stars, sea urchins are herbivores that scrape incrusting algae from rock surfaces, nibble on plants, or trap and eat drifting food. Sea urchins are "regular" echinoids: radially symmetrical, globose in shape, and armed with long spines. Other echinoids, such as sand dollars and heart urchins, are "irregular" echinoids: bilaterally symmetrical with bodies variably shaped, such as flattened sand dollars or heart-shaped heart urchins.

Arbacia

Phylum Echinodermata
Subphylum Eleutherozoa
Class Echinoidea
Genus *Arbacia*
Species *Arbacia punctulata*

Where Found

Like sea stars and brittle stars, echinoids are strictly marine, benthic animals widely distributed in all seas, from the intertidal to deep sea. Some favor rocky, high-energy coastlines with pounding surf, but most echinoids are subtidal, grazing in turtle-grass buds or on coral reefs or (especially the irregular echinoids) burying themselves in sandy bottoms, where they feed on microscopic organic matter. *Arbacia punctulata* (Gr. Arbaces, first king of Media*), the purple sea urchin on the East Coast, is found from Cape Cod to Florida and Cuba. Other common species include *Strongylocentrotus drobachiensis* (Gr. *strongylos,* round, + *kentron,* spine), the green sea urchin of both East and West Coasts of North America (Figure 17-7); *Strongylocentrotus purpuratus,* purple urchin of the West

*The name of the genus *Arbacia,* bestowed by British zoologist John Edward Gray in 1835, is an example of a "nonsense" name that lacks any descriptive value. Gray apparently chose the name after reading Lord Byron's poem *Sardanapalus,* concerning Arbaces, who, according to legend, founded the Median empire (now part of northern Iran) about 830 B.C.

Figure 17-7
The green sea urchin *Strongylocentrotus drobachiensis.* Note the slender, suckered tube feet. Urchins often attach bits of shell, marine algae, and other debris to themselves for camouflage. Small, stalked, white-tipped pedicellariae can be seen surrounding the bases of spines near the center of the photograph. Larger, three-jawed pedicellariae are shown enlarged.

Coast; *Diadema antillarum,* long-spined urchin of the Caribbean and South Atlantic; and species of *Lytechinus* (Gr. *lytos,* broken, + *echinos,* urchin) on both coasts.

Behavior and External Structure

External features of a sea urchin are best observed in a living animal. However, if living forms are not available, submerge a preserved specimen in a bowl of water and use the following account, directed toward live urchins, to identify the external structures.

Place a living sea urchin on a glass plate, submerge in a bowl of seawater, and observe under a dissecting microscope.

Spines. Examine the long, movable **spines,** each attached at a ball-and-socket joint by two sets of ring muscles. Remove a spine (instructor's option) and note that its socket fits over a rounded **tubercle** on the test (Figure 17-8). An inner ring of **cog muscles** holds the spine erect. Hold the tip of a spine and try to move it. Do you feel the locking mechanism of the cog muscles? ____________ Of what advantage is such a locking mechanism? ____________ Now, with a probe, touch the epidermis near a spine. Does the spine move? ____________ In which direction? ____________ The outer ring of muscles is responsible for directional movement. Are all the spines the same length? ____________ Are they all pointed on the distal ends? ____________

Tube Feet. Notice that the tube feet all originate from rows of perforations in the five **ambulacral regions.** The tube feet can be extended beyond the ends of the spines. Do any possess suckers? Are any suckerless? ____________ Do they move? What happens when you touch one? ____________ When you jar the bowl? ____________

Mouth and Peristome. Examine the oral side of the urchin and find the **mouth,** with its five converging **teeth** and collarlike **lip** (Figure 17-8). The teeth are part of a complex chewing mechanism called **Aristotle's lantern,** which is operated internally by several sets of muscles.* The lip contains circular, or "purse-string," muscles.

The membranous **peristome** surrounding the mouth is perforated by five pairs of large oral tube feet. Do these podia have suckers? ____________ They are probably sensitive to chemical stimuli. The peristome also bears some small spines.

Pedicellariae. Notice on the peristome and elsewhere a number of three-jawed **pedicellariae** on the ends of long, slender stalks (see Figure 17-7). Smaller but more active pedicellariae are located among the spines. Stimulate some of them by touching gently with a camel's hair brush. You may want to pinch off some of the pedicellariae to examine them more closely on a slide, particularly if you are using a

*The curious name for the protrusible chewing mechanism of sea urchins derives from a passage in the writings of Aristotle (384–322 B.C), where he compared the chewing apparatus to the frame of a lantern.

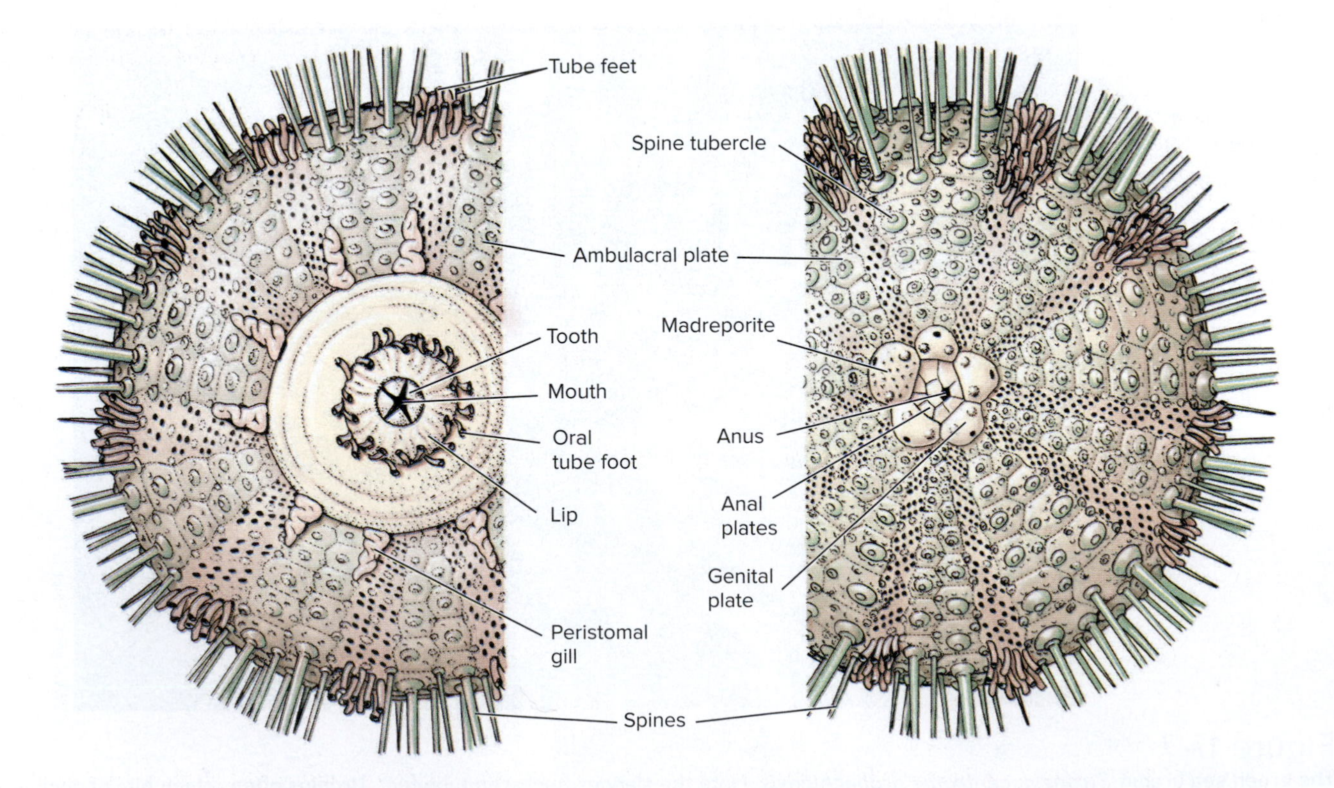

Figure 17-8
External structure of a sea urchin. Oral *(left)* and aboral *(right)* surfaces are shown with spines partly removed.

preserved specimen. Their functions are to discourage intruders and to help keep the skin clean. If you are working with a living sea urchin, drop some sand grains on the surface to see how the pedicellariae respond to clean the animal.

Locomotion. Note how the urchin uses its spines and tube feet in locomotion. Carefully, so as not to injure its tube feet, turn the urchin over (oral side up). Does it use its spines or its tube feet to right itself? ____________ Notice which ambulacra turn first and mark that row by removing some of its spines or by marking some of the spines with thread; then turn the animal over again and see if the same ambulacra turn first. Repeat once more.

Tilt the glass plate to determine if the urchin moves up or down. Tilt in the opposite direction and see what happens. Is it positively or negatively geotactic? ____________

Some sea urchins are adapted for burrowing into rock or other hard material by using both their spines and their chewing mechanisms. *Strongylocentrotus purpuratus,* common on the North American Pacific Coast, excavates cup-shaped depressions in stone.

Most echinoids have tiny modified spines called **sphaeridia** (Gr. *sphaira,* sphere, + *idion,* dim. suffix), believed to be organs of equilibrium. In *Arbacia* these are minute, glassy bodies located one in each ambulacrum close to the peristome. Try to find and remove the sphaeridia to see whether their removal affects the urchin's righting reaction or its geotactic responses.

Direct a beam of bright light toward an urchin. How does it react? ____________

Epidermis. The test, podia, pedicellariae, and spines are covered with **ciliated epidermis,** although the epidermis may have become worn off from the exposed spines. Drop a little carmine suspension on various parts of the sea urchin and note the direction of the ciliary currents. Of what advantage are such currents to the urchin? ____________

Gills. At the outer edge of the peristome, between the ambulacra, find five pairs of branching peristomial **gills,** which open into the coelomic cavity.

Written Report

Record your observations on sea urchin behavior on separate paper.

Further Study

Test (Endoskeleton)

Examine a dried sea urchin test from which the spines have been removed and a dried sand dollar and/or heart urchin.

The test, or endoskeleton, is composed of calcareous plates (ossicles) symmetrically arranged and interlocked or fused so as to be immovable. Note the tubercles to which the spines were attached. Note the arrangement of the plates into 10 meridional double columns—five double rows of **ambulacral plates** alternating with double rows of interambulacral plates. What is the function of the perforations in the ambulacral plates? ____________ In asteroids the tube feet were extended between the plates rather than through them.

Table of Comparison

Complete a table of the comparative characteristics of the external anatomy of the sea star, brittle star, and sea urchin on p. 280. After you have observed the sea cucumber, add that to the table.

Examine the test of a sand dollar, a heart urchin, or both. Can you find perforations in them similar to those of the urchin? ____________ Do they have ambulacra, and is their arrangement pentamerous? ____________ The ambulacra in echinoids are homologous to the ambulacra of asteroids and ophiuroids.

On the aboral surface, note the area that is free from spines, the **periproct.** The **anus** is centrally located, surrounded by four (sometimes five) valvelike **anal plates** (Figure 17-8). Around the anal plates are five **genital plates,** so called because each bears a **genital pore.** Note that one of the genital plates is larger than the others and has many minute pores. This is the **madreporite plate,** which has the same function in sea urchins that it has in sea stars, for the echinoids have a **water-vascular system** in common with all echinoderms.

The test grows both by the growth of plates and by the production of new plates in the ambulacral area near the periproct.

EXERCISE 17D

Class Holothuroidea—Sea Cucumbers

Core Study

Sea Cucumbers

Phylum Echinodermata
  Subphylum Eleutherozoa
    Class Holothuroidea
      Order Dendrochirotida
        Genus *Cucumaria* (or *Thyone*)
          or
      Order Aspidochirotida
        Genus *Parastichopus*

Where Found

Sea cucumbers, perhaps the oddest members of a phylum distinguished by strange animals, look remarkably like their vegetable namesake. They are characterized by an elongate

©Cleveland P. Hickman, Jr.

A

©Cleveland P. Hickman, Jr.

B

Figure 17-9
Two sea cucumbers from the Eastern Pacific. **A,** *Holothuria fuscocinerea* is a deposit feeder that grazes the bottom with its mouth and tentacles. **B,** *Holothuria arenicola,* with its tentacles extended, is a suspension feeder that captures organisms on its mucus-coated tentacles.

body, a leathery body wall with warty surface, an absence of arms, and a mouth and an anus located at opposite poles of the animal. They are benthic (bottom-dwelling), slow-moving animals found in all marine habitats. Two common genera on the East Coast of the United States are *Thyone* and *Cucumaria; Parastichopus* is a familiar genus on the West Coast (Figure 17-9).

Behavior and External Structure

If possible, study living specimens in an aquarium or in a bowl of seawater containing a generous layer of sand, and then examine a preserved specimen. Sea cucumbers are slow to react. They should be left undisturbed for some time before the laboratory period if you are to see them relaxed and feeding.

Note that a holothurian, unlike other echinoderms, is orally-aborally elongated and has a cylindrical body, with the **mouth** encircled by **tentacles** at one end and **anus** at the other.

A more detailed description and behavior analysis of the animal will depend somewhat on the species you are observing. Notice the tentacles. Are they branched and extensible (as in *Cucumaria* and *Thyone*), short and shield-shaped (as in *Parastichopus*), or of some other type? ____________ The tentacles, which are modified tube feet, are hollow and a part of the **water-vascular system;** they are connected internally with the radial canals. The type of tentacle structure is related to feeding habits. *Cucumaria* and *Thyone* are suspension feeders that stretch their mucus-covered tentacles into the water or over the substrate until they are covered with tiny food organisms; then they thrust the tentacles into the mouth, one by one, to lick off the food. Can you observe these actions? *Parastichopus* and some others are deposit feeders that simply shovel mud and sand into the mouth, digest organic particles, and void the remainder.

Have you noticed any rhythmic opening and closing of the anus? ____________ This is a respiratory movement coordinated with the pumping action of the cloaca, which pumps water into and out of the **respiratory trees** (internal respiratory organs).

Does the animal try to burrow into the sand? ____________ Does it cover itself completely or leave the ends exposed? ____________ *Thyone* may take 2 to 4 hours to bury its middle by alternate circular and longitudinal muscle contractions. It is likely to be more active in the late afternoon and night than in the morning. If you are watching *Parastichopus* move, what does its muscular action remind you of? ____________ Are there waves of contraction? ____________

Does the sea cucumber react to mechanical stimulus? ____________ Try touching a tentacle. Does one or more than one tentacle react? ____________ Touch several tentacles. What happens when you stroke the body or gently pick up the animal? Does it expel water when you pick it up? ____________ Having observed its movement and reactions, can you see the advantages of the hydraulic skeleton? What other phyla use the hydraulic skeleton to advantage? ____________

Note the **tube feet** (podia). Are they scattered all over the body or arranged in ambulacral rows? ____________ This pattern differs among different species. Are the tube feet all alike? That is, are ventral tube feet any different from dorsal tube feet? ____________ Are any of them suckered? ____________ If the pentamerous arrangement of ambulacra is evident in your specimen, how many rows make up the ventral **sole?** ____________ How many are on the dorsal surface? ____________ If you place the animal on a solid surface, do the tube feet attach themselves? ____________ Can the animal right itself if turned over? ____________ Are the tube feet involved in the righting action? Are muscles involved? ____________

Does the animal show any geotactic reaction if placed on a vertical or sloping surface? ____________ Some burrowing forms are positively geotactic and move downward; other species are negatively geotactic and climb upward. *Thyone* gives no geotactic response.

Do you find any pedicellariae or skin gills? ____________ Do you feel the presence of a test under the epidermis? ____________ That is because the skeleton of the holothurian is usually limited to microscopic ossicles embedded in the tough, leathery body wall.

Written Report

Record your observations on sea cucumber behavior on separate paper.

Table of Comparison

Complete the comparative table (p. 280) of the likenesses and differences between the external anatomy of the sea star, brittle star, sea urchin, and sea cucumber.

Further Study

Internal Structure

Locate the five longitudinal ambulacral areas of a preserved sea cucumber. The sole (the ventral side applied to the substrate) has three ambulacra with well-developed tube feet. The dorsal side has two ambulacra with smaller tube feet (in some species, tube feet are absent on the dorsal surface). With scissors, open the body by making a longitudinal incision on the ventral (sole) side of the animal, between the central and right ambulacra. Pin down the walls and cover with water.

Digestive System. Note the large **coelomic cavity.** Just behind the mouth is the **pharynx,** supported by a ring of calcareous plates (Figure 17-10). It is followed by a short, muscular **stomach** and a long, convoluted **intestine,** held in place by mesenteries and expanding somewhat at the end to form a **cloaca,** which empties at the **anus.**

Respiratory System. Two branched **respiratory trees** are attached to the cloaca; they serve as both respiratory and excretory organs. They are aerated by a rhythmic pumping of the cloaca. Several inspirations 1 minute or more apart are followed by a vigorous expiration that expels all the water.

Note the muscles between the cloaca and the body wall. Long **retractile muscles** (how many? ____________) run from the pharynx to join the longitudinal **muscle bands** in the body wall.

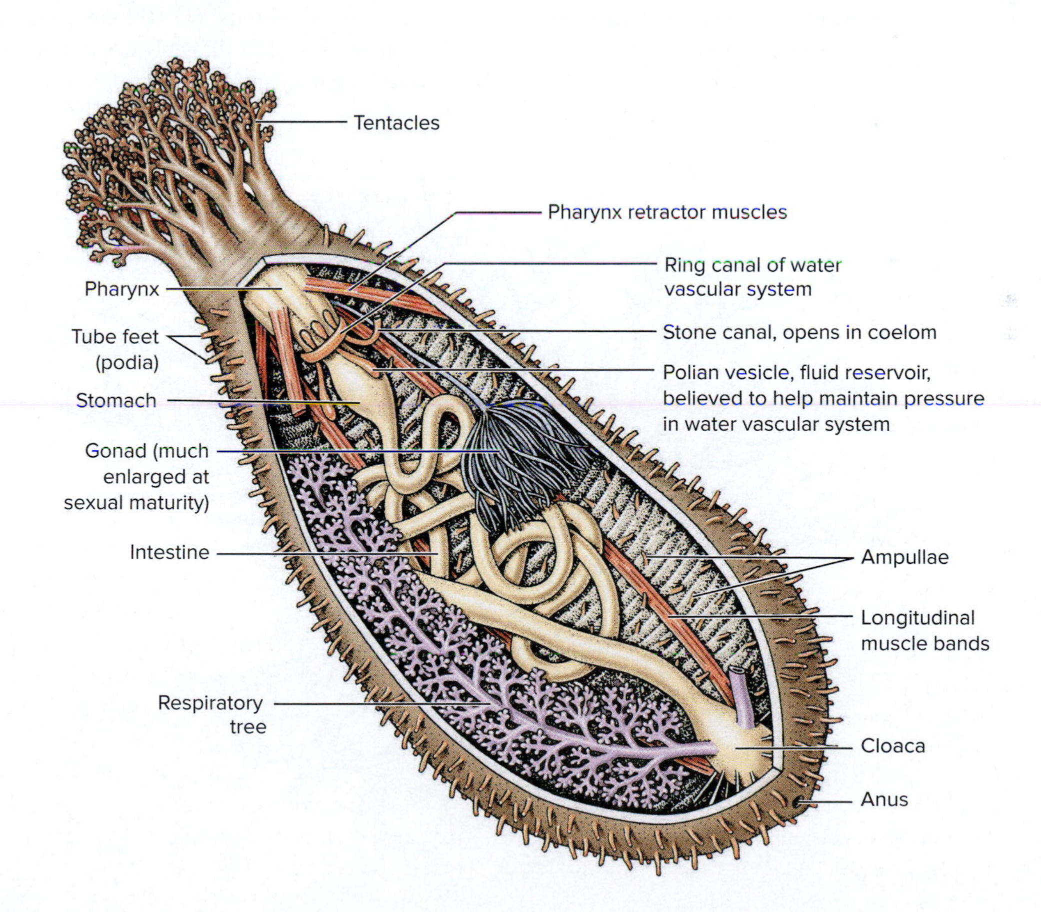

Figure 17-10
Internal anatomy of a sea cucumber.

Water-Vascular System. A **ring canal** surrounds the pharynx. One or more rounded or elongated sacs called **polian vesicles** hang from the ring canal into the coelom and open into the ring canal by a narrow neck. Polian vesicles are believed to function as expansion chambers in maintaining pressure within the water-vascular system. One or more **stone canals** also open into the ring canal from the body cavity. In adult sea cucumbers, the water-vascular system has usually lost contact with the seawater outside. Coelomic fluid, rather than seawater, enters and leaves the system.

Five **radial canals** extend from the ring canal forward along the walls of the pharynx to give off branches to the tentacles, which are actually modified tube feet. From there the radial canals run back along the inner surface of the ambulacra, where each gives off **lateral canals** to the **tube feet** and **ampullae.** Valves in the lateral canals prevent backflow. Note the ampullae along the ambulacra in the inner body wall.

Reproductive System. The **gonad** consists of numerous tubules united into one or two tufts on the side of the dorsal mesentery. These become quite large at sexual maturity. (Your specimen may have a much enlarged gonad.) A **gonoduct** passes anteriorly in the mesentery to the **genital pore.** The sexes are separate, and fertilization is external.

Endoskeleton. The **endoskeleton** consists largely of tiny, calcareous ossicles scattered in the dermis. These can be seen on a prepared slide.

EXERCISE 17E

Class Crinoidea—Feather Stars and Sea Lilies

Feather Stars

Phylum Echinodermata
Subphylum Pelmatozoa
Class Crinoidea
Order Articulata
Genus *Antedon*

Where Found

Crinoids are found in all seas except the Baltic and Black seas. Most are subtidal, although some are littoral, and some are found as deep as 5000 meters. The free-living feather stars (Figure 17-11) prefer rocky bottoms and are most abundant in shallow tropical lagoons. The sea lilies (about one-eighth of the species) are stalked and sessile and prefer muddy sea bottoms and deep waters.

General Structure

A feather star passes through a stalked stage but is free-moving as an adult (Figure 17-11). It is made up of the following three general regions: (1) 10 long, jointed **arms,** bearing jointed **pinnules;** (2) the **calyx,** containing the digestive and other organs; and (3) a circlet of short-jointed **cirri,**

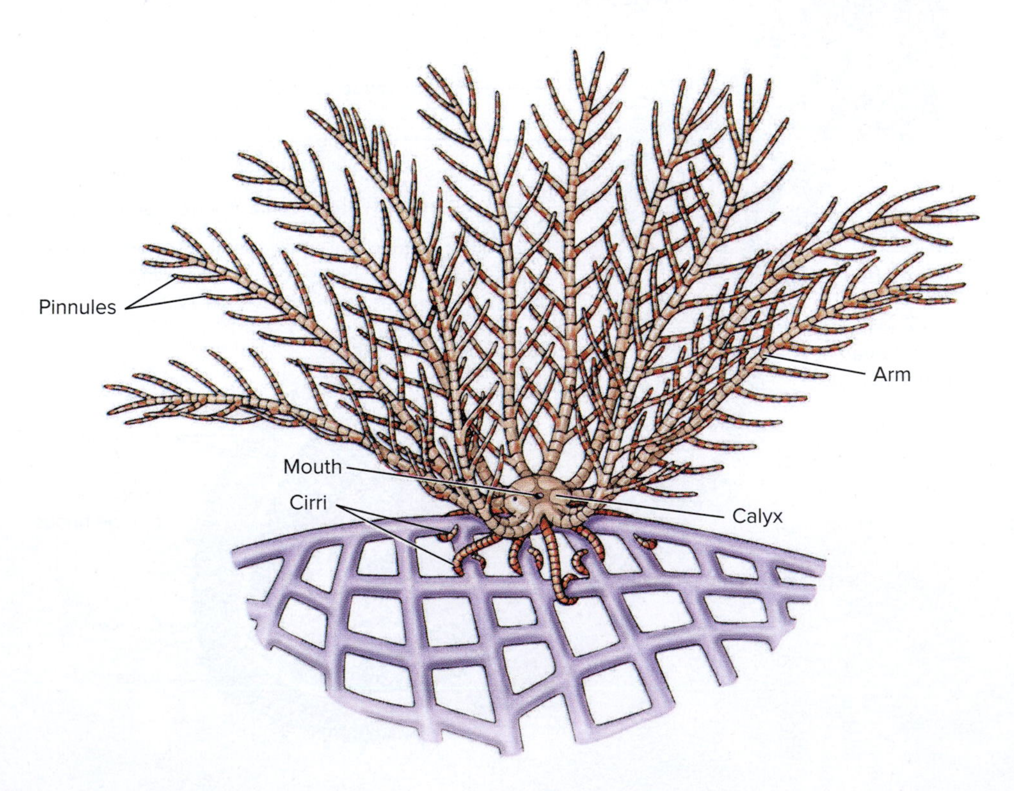

Figure 17-11
Feather star, *Antedon.*

on which the animal can rest or move about. The calyx and arms together are called the **crown.** The leathery skin covering the calyx is called the **tegmen** (L., covering).

Ambulacral grooves radiate out from the central mouth to the arms and pinnules, so that each arm and pinnule has a ciliated food groove. Tube feet are located along the food grooves in the pinnules.

Crinoids are suspension feeders, using a mucus-ciliary method of capture and food movement. Small organisms are caught in mucous nets and moved along the ambulacral grooves to the mouth. The pinnules can secrete a narcotizing toxin for quieting the prey.

Feather stars can crawl by holding onto objects with the adhesive ends of their pinnules and pulling themselves along by bending their arms. They can also swim by raising and lowering alternate sets of arms.

Classification

The following classification is abbreviated.

Phylum Echinodermata

Class Crinoidea (cry-noy′de-a) (Gr. *krinon,* lily, + *eidos,* form). Sea lilies and feather stars. Aboral attachment stalk of dermal ossicles. Anus on oral surface; five branching arms with pinnules; ciliated ambulacral groove on oral surface with tentacle-like tube feet for food collecting; spines, madreporite, and pedicellariae absent. Examples: *Antedon, Florometra.*

Class Asteroidea (as′ter-oy′de-a) (Gr. *aster,* star, + *eidos,* form). Sea stars. Star-shaped, with arms not sharply marked off from central disc; ambulacral grooves open, with tube feet on oral side; tube feet often with suckers; anus and madreporite aboral; pedicellariae present. Examples: *Asterias, Pisaster.*

Class Ophiuroidea (o′fe-u-roy′de-a) (Gr. *ophis,* snake, + *oura,* tail, + *eidos,* form). Brittle stars and basket stars. Star-shaped, with arms sharply marked off from central disc; ambulacral grooves closed, covered by ossicles; tube feet without suckers and not used for locomotion; pedicellariae absent. Examples: *Ophiura, Gorgonocephalus, Ophioderma.*

Class Echinoidea (ek′i-noy′de-a) (Gr. *echinos,* sea urchin, hedgehog, + *eidos,* form). Sea urchins, heart urchins, and sand dollars. More or less globular or disc-shaped, with no arms; compact skeleton, or test, with closely fitting plates; movable spines; ambulacral grooves closed and covered by ossicles; tube feet with suckers; pedicellariae present. Examples: *Arbacia, Strongylocentrotus, Lytechinus, Mellita.*

Class Holothuroidea (hol′o-thu-roy′de-a) (Gr. *holothourion,* sea cucumber, + *eidos,* form). Sea cucumbers. Cucumber-shaped; with no arms; spines absent; microscopic ossicles embedded in thick, muscular wall; anus present; ambulacral grooves closed; tube feet with suckers; circumoral tentacles (modified tube feet); pedicellariae absent; madreporite plate internal. Examples: *Thyone, Parastichopus, Cucumaria.*

LAB REPORT

Name ______________________

Date ______________________

Section ______________________

Comparative Table of Echinoderm Characteristics

Characteristic	Sea Star	Brittle Star	Sea Urchin	Sea Cucumber
Symmetry				
Body shape				
Tube feet (present, absent, with or without suckers)				
Endoskeleton (ossicles, spicules, location)				
Spines (present, absent)				
Respiration (papulae, tube feet, respiratory tree, bursae)				
Ambulacral groove (open, closed)				
Madreporite location (oral, aboral, internal)				
Feeding behavior (carnivore, herbivore, omnivore)				

Phylum Chordata A Deuterostome Group

18 EXERCISE

Protochordates
Subphylum Urochordata
Subphylum Cephalochordata

What Defines a Chordate?

Chordates show a remarkable diversity of form and function, ranging from **protochordates** to humans. Most chordates are vertebrates, but the phylum also includes a few invertebrate groups. All animals that belong to phylum Chordata must have at some time in their life cycle the following characteristics, often called chordate hallmarks.

1. **Notochord.** The notochord (Gr. *nōton,* back, + L. *chorda,* cord) (Figure 18-1) is a slender rod of cartilage-like connective tissue lying near the dorsal side and extending most of the length of the animal. It is regarded as an early endoskeleton and has the functions of such. In most vertebrates, it is found only in the embryo.
2. **Pharyngeal pouches and slits.** The pharyngeal pouches and slits (see Figure 18-1) are a series of paired slits in the pharynx, serving as passageways for water to the gills. In some vertebrates, they appear only in the embryonic stages.
3. **Dorsal tubular nerve cord.** A dorsal tubular nerve cord, with its modification, the brain, forms the central nervous system. It lies dorsal to the alimentary tract and has a fluid-filled cavity, in contrast to the invertebrate nerve cord, which is ventral and solid.
4. **Endostyle or thyroid gland.** The endostyle or its derivative, the thyroid gland, is found in all chordates, but in no other animals. The endostyle secretes mucus and traps small food particles for protochordates and lamprey larvae. Some cells in the endostyle are homologous with cells of the thyroid gland found in the remainder of vertebrates.
5. **Postanal tail.** A postanal tail projects beyond the anus at some stage and serves as a means of propulsion in water. It may or may not persist in the adult. Along with body muscles and stiffened notochord, it provides motility for a free-swimming existence.

These features vary in chordates. Some chordates have all of these structures throughout life. In many chordates; having more derived characteristics the gill slits never break through from the pharynx but merely form pouches that have no function, the notochord is replaced by the vertebral column and only the dorsal nerve cord actually persists in the adult as a diagnostic chordate character. **Protochordates** demonstrate each of the chief chordate characteristics at some point in their life cycle.

What Defines a Chordate?

CLASSIFICATION
Phylum Chordata

EXERCISE 18A
Subphylum Urochordata—*Ciona,* an Ascidian
Ciona

EXERCISE 18B
Subphylum Cephalochordata—Amphioxus
Amphioxus

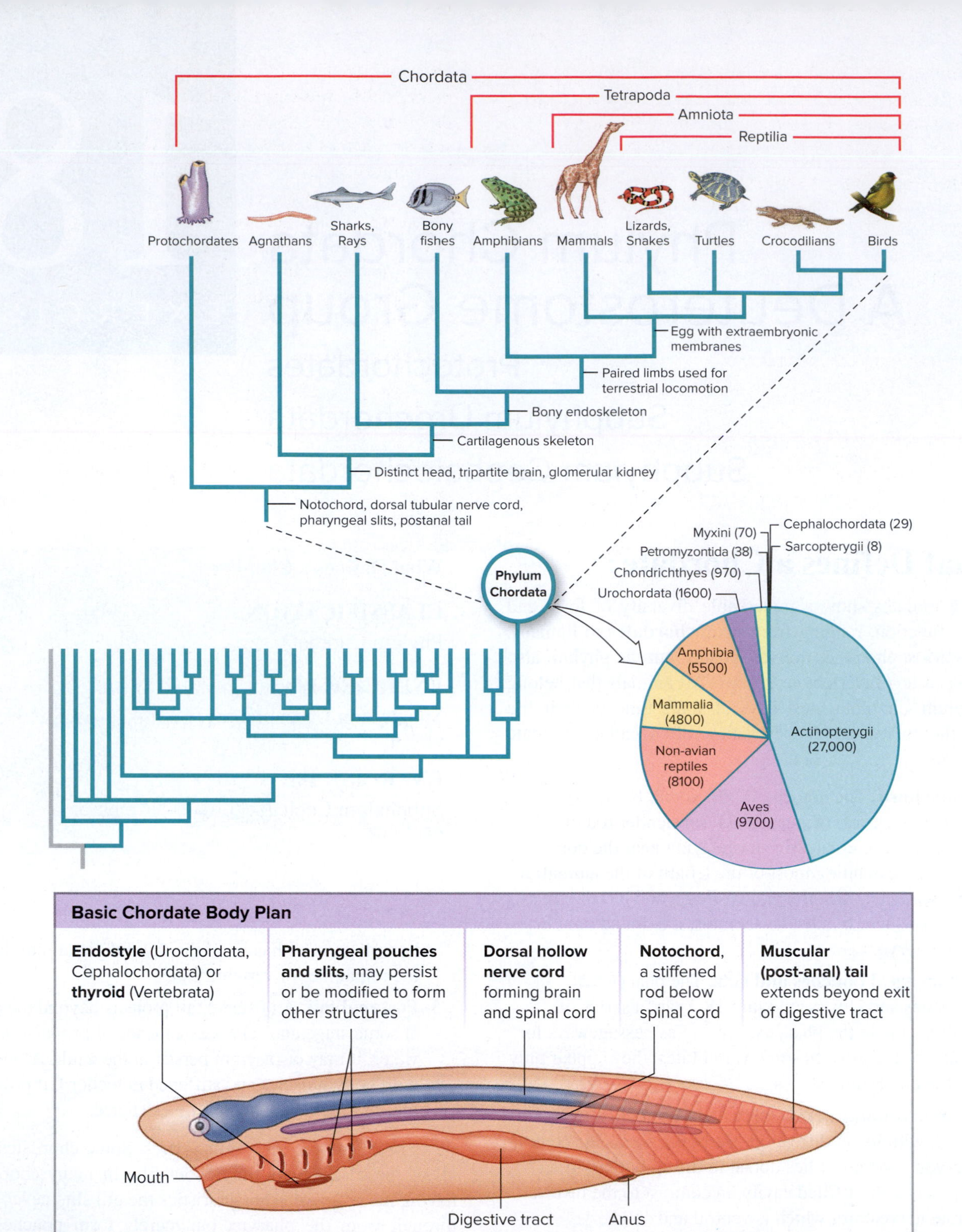

Figure 18-1

Basic chordate body plan, showing the five anatomical characters that distinguish a chordate and that appear at some time in the life cycle of all chordates.

Classification

Phylum Chordata

Subphylum Urochordata (u′ro-kor-da′ta) (Gr. *oura,* tail,+ L. *chorda,* cord). **(Tunicata.)** Tunicates. Only larval forms have all chordate characteristics; almost all adults sessile, without notochord and dorsal nerve cord; body enclosed in tunic. Example: *Molgula,* a sea squirt.

Subphylum Cephalochordata (sef′a-lo-kor-da′ta) (Gr. *kephalē,* head, + L. *chorda,* cord). Lancelet. Notochord and nerve cord persist throughout life; lance-shaped. Example: *Branchiostoma* (amphioxus).

Subphylum Vertebrata (ver′te-bra′ta) (L. *vertebratus,* backboned). **(Craniata.)** Vertebrates. Enlarged brain enclosed in cranium; nerve cord surrounded by bony or cartilaginous vertebrae; notochord in all embryonic stages and persists in adults of some fishes; typical structures include two pairs of appendages and body plan of head, trunk, and postanal tail.

Superclass Cyclostomata

Class Myxini (mik-sin′ee) (Gr. *myxa,* slime). Hagfishes.

Class Petromyzontida (pet′tro-my-zon′ti-da) (Gr. *petros,* stone, + *myzon,* sucking). Lampreys.

Superclass Gnathostomata (na′tho-sto′ma-ta) (Gr. *gnathos,* jaw, + *stoma,* mouth). Jawed fishes, all tetrapods. Jaws present; usually paired limbs; notochord persistent or replaced by vertebral centra.

Class Chondrichthyes (kon-drik′thee-eez) (Gr. *chondros,* cartilage, + *ichthys,* a fish). Sharks, skates, rays, and chimaeras.

Class Actinopterygii (ak′ti-nop-te-rij′ee-i) (Gr. *aktis*, ray, + *pteryx*, fin, wing). Ray-finned fishes.

Class Sarcopterygii (sar-cop-te-rij′ee-i) (Gr. *sarkos*, flesh, + *pteryx*, fin, wing). Lobe-finned fishes.

Class Amphibia (am-fib′e-a) (Gr. *amphi,* both or double,+ *bios,* life). Amphibians. Frogs, toads, and salamanders.

Class Reptilia (rep-til′e-a) (L. *repere,* to creep). Reptiles. Non avian reptiles: snakes, lizards, turtles, crocodiles, and others.

Class Aves (ay′veez) (L. pl. of *avis,* bird). Birds.

Class Mammalia (ma-may′lee-a) (L. *mamma,* breast). Mammals.

EXERCISE 18A

Subphylum Urochordata—Ciona, an Ascidian

Core Study

Ciona

Phylum Chordata
Subphylum Urochordata
Class Ascidiacea (sea squirts)
Order Enterogona
Family Cionidae
Genus *Ciona*
Species *Ciona intestinalis*

Urochordata (Gr. *oura,* tail, + L. *chorda,* cord) are commonly called **tunicates** because of their leathery covering, or tunic. They are divided into three classes: Ascidiacea, sea squirts; Thaliacea, salpians; and Appendicularia, larvaceans. The largest group is the ascidians (Gr. *askidion,* leather bag or bottle), which are also the most generalized. They are called **sea squirts** because of their habit, when handled, of squirting water from the excurrent siphon. Any of the small, translucent ascidians may be used for this exercise. Adult tunicates are sessile, whereas the larvae undergo a brief free-swimming existence.

Where Found

Tunicates are found in all seas and at all depths. Most of them are sessile as adults, although some are pelagic (found in the open ocean). *Ciona intestinalis* (Gr. *Chionē,* demigoddess of mythology) is a cosmopolitan species commonly found in shallow water. Like most shallow-water sea squirts, it attaches to almost any available rigid surface, especially wharf pilings, anchored and submerged objects (rocks, shells, and ship bottoms), and eelgrass. It grows to 15 cm in its largest dimension. Although sea squirts are common in marine environments, they are often overlooked. One of the best places to see sea squirts is on pilings, where they may cover the surface.

External Features and Behavior

Sea squirts are fairly hardy in a marine aquarium.

Examine, in a finger bowl of seawater, a living solitary tunicate, such as *Ciona* or *Molgula,* or a portion of a colony of *Perophora* or other ascidians as available.

Observe the use of the two openings, or **siphons.** When fully submerged and undisturbed, the siphons are open, and respiratory water, kept moving by ciliary action, enters the more terminal siphon (called the **incurrent,** or **oral, siphon**) at the mouth, circulates through a large pharynx, and leaves through the **excurrent,** or **atrial, siphon** on one side (the dorsal side) (Figures 18-2 and 18-3).

Release a little carmine suspension near the animal to verify this.

The outer covering of a tunicate is called the **tunic,** or **test** (Figure 18-3). It is secreted by the **mantle,** which lies just inside it, and it contains cellulose—an uncommon substance in animals. The mantle contains muscle fibers, by which the body can contract. If the tunic is translucent enough and the light is properly adjusted, you may be able to see some of the internal structure.

Neuromuscular System. The tunicate nervous system is reduced and not well understood. There is a cerebral ganglion (closely associated with a sub neural gland of uncertain function) located between the siphons. The tunic probably has no sensory nerves, but pressure on the tunic may be transmitted to nerves in the mantle. Both direct and crossed reflexes have been observed in some ascidians and

Figure 18-2

Ciona intestinalis, a solitary tunicate, showing its siphons in use. It remains anchored throughout life to one spot on the seafloor. Its free-swimming larva bears the chordate hallmarks: notochord, pharyngeal slits, dorsal tubular nerve cord, endostyle, and postanal tail.

can be tested in a living *Ciona* or other tunicate by touching selected areas with the tip of a glass rod or dissecting needle.

Direct reflexes result from mechanical stimulation of the *outer* surface of the siphons or tunic.

Gently stimulate various areas of the tunic and note the response. Gently touch the outer surface of one of the siphons, note the response, and then stimulate the other siphon.

What areas of the body are most sensitive? **Crossed reflexes** result from mechanical stimulation of the *inner* surface of the siphons. The normal response is to close the *other* siphon and then contract the body.

Gently touch the inner surface of the oral siphon. What happens? __________ Try a stronger stimulus of the same siphon. Do you get the same response? __________ How do these responses differ from those in which the outer surface was stimulated? __________ Repeat with the atrial siphon.

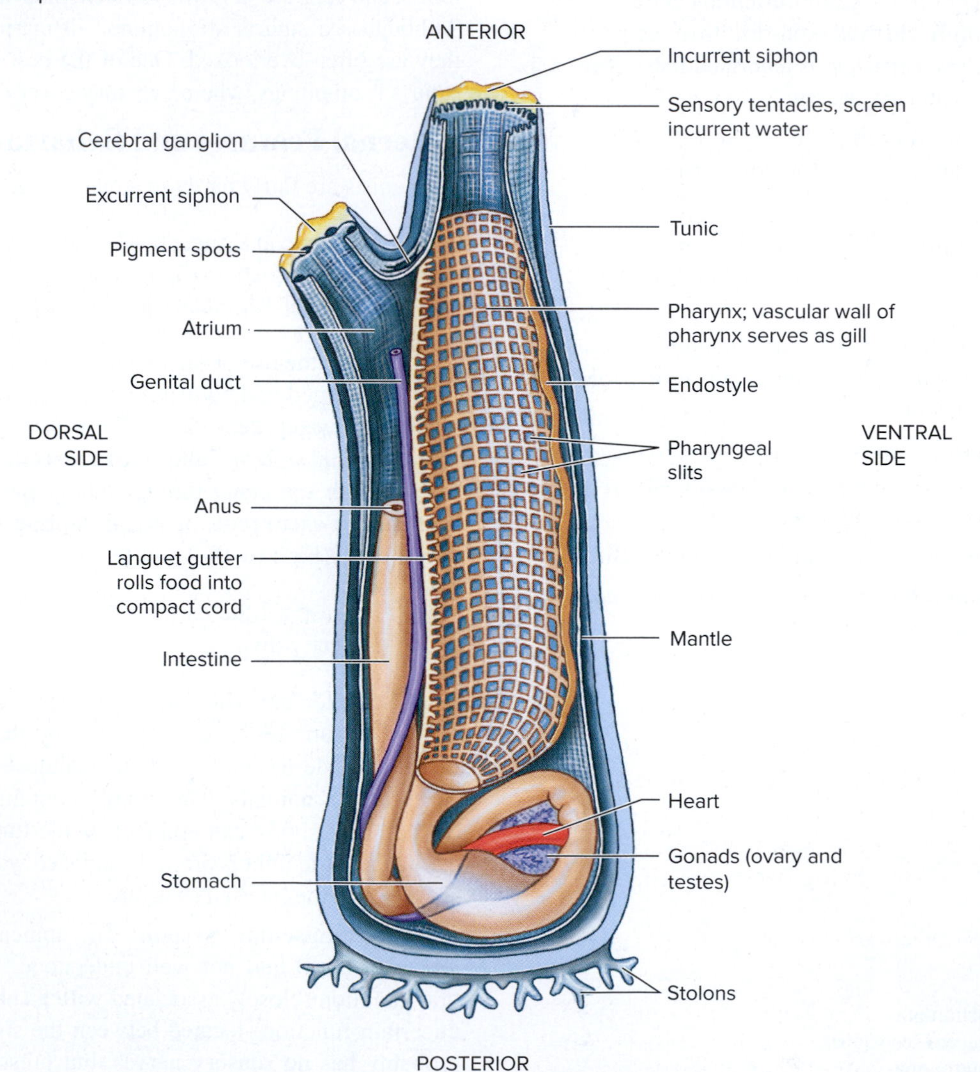

Figure 18-3

Structure of a solitary sea squirt, *Ciona*.

Of what protective value would these reflexes be to the animal? __________

Further Study

Internal Structure

Internal structure can be observed on either a living or a preserved specimen. Use fine scissors to slit the tunic longitudinally, beginning at the incurrent siphon and continuing the cut to the base of the pharynx. Be *very* careful to cut *only* the tunic and not the mantle beneath it. Slip the animal out of its tunic then return the animal to the bowl of seawater and study with a dissecting microscope.

Respiratory System. The **branchial sac,** or **pharynx,** is the largest internal structure, and the space between it and the mantle is the **atrium** (Figure 18-3). The pharyngeal wall is perforated with many pharyngeal slits, through which water passes into the atrium to be discharged through the excurrent siphon. The vascular wall of the pharynx serves as a gill for gas exchange.

Circulatory System. In a living specimen, with test removed and a light properly adjusted, you should be able to see the beating of the **heart,** located near the posterior end on the right side. The tubular heart empties into two vessels, one at each end. Its peristaltic waves are of particular interest, because they send the blood in one direction for a while and then reverse direction and pump the blood in the opposite direction. Apparently there are two pacemakers that initiate contractions, one at each end of the heart, and they alternate in dominance over each other. The tunicate vascular system is an open type of system. Blood cells are numerous and colorful. There are no respiratory pigments.

Digestive System. At the junction of the **mouth** and the **pharynx** is a circlet of **tentacles** that form a grid that screens the incurrent water. By dropping a grain of sand into the incurrent siphon of a living tunicate, you may be able to observe the ejection reflex.

Inside the pharynx along the midventral wall is the **endostyle,** a ciliated groove that secretes a great deal of mucus. Cilia on the walls of the pharynx distribute the mucus. Food particles become tangled in the mucus and are propelled by cilia to a dorsal gutter, where the mucus with its trapped food becomes rolled into a compact cord. In *Ciona* and a few other tunicates, the gutter is lined with a row of curved, ciliated, fingerlike processes called **languets** [F. *languette,* small tongue]. The cord is propelled posteriorly to the esophagus and stomach.

If living ascidians are available, examine one that has been submerged for some time in a suspension of carmine particles in seawater. Open the pharynx by cutting through the incurrent siphon and downward, a little to one side of the midventral line. Then cut around the base and lay the animal open in a pan of water. You should be able to see a concentration of carmine particles in the middorsal area. Cut out a small piece of the pharyngeal wall (free from the tunic and mantle) and mount on a slide to observe pharyngeal slits and beating cilia.

It may be difficult to differentiate among **esophagus, stomach,** and **intestine.** The **anus** empties into the atrium near the excurrent siphon.

Excretion. A ductless structure near the intestine is assumed to be a type of nephridium and to be excretory in function.

Reproduction. *Ciona,* like most tunicates, is hermaphroditic and has a single ovary and testis, each with a gonoduct that opens into the atrium.

Some tunicates are colonial. In some colonial tunicates zooids (distinct individuals) are separate and attached by a stolon, as in *Perophora.* In others, zooids are regularly arranged and partly united at the base by a common tunic; all incurrent siphons are at one side and excurrent siphons at the other. Another type of colonial tunicate has a system of zooids that share a common atrial (excurrent) chamber. Colonies are formed asexually by budding.

Solitary tunicates generally shed their eggs from the excurrent siphon, and development occurs in the sea. Colonial species usually brood their eggs in the atrium, and the microscopic larvae leave by the excurrent siphon. For a very brief period, the larvae are nonfeeding and live a planktonic, free-swimming existence; then they settle down, attach to the substrate, and metamorphose.

Ascidian Larvae (Study of Stained Slides)*

Ascidian larvae are free-swimming (Figure 18-4). Why are the larvae often called "tadpole" larvae? __________ They do not look like sessile adult sea squirts and are actually more characteristic of the chordates than are the adults. They possess not only pharyngeal slits but also a notochord, dorsal tubular nerve cord, and tail, structures that have been lost in the adult. You may be able to identify some of the following structures on a stained slide.

Adhesive papillae at the anterior end of the larva are used to attach to some object during metamorphosis, which occurs within a short time after hatching. The **notochord** can be identified in the long tail. This character, which is lost in the adult, gives subphylum Urochordata ("tail-cord") its name. A **nerve cord** dorsal to the notochord enlarges anteriorly into a neural vesicle. Can you identify a pigmented, photoreceptive **eyespot?** A smaller, pigmented area anterior to the eye is a **statocyst.** What is the function of the statocyst? __________ At metamorphosis these portions of the nervous system degenerate, and a ganglion serves as the nerve center.

Look for the anterior **oral (incurrent) aperture** and the more posterior **atriopore (excurrent aperture).** Perhaps you can identify the **branchial basket** (pharynx) with **pharyngeal slits, stomach, intestine, atrium,** and **endostyle.**

---

*Slides suitable for this exercise are available from Carolina Biological. See Appendix B, p. 420.

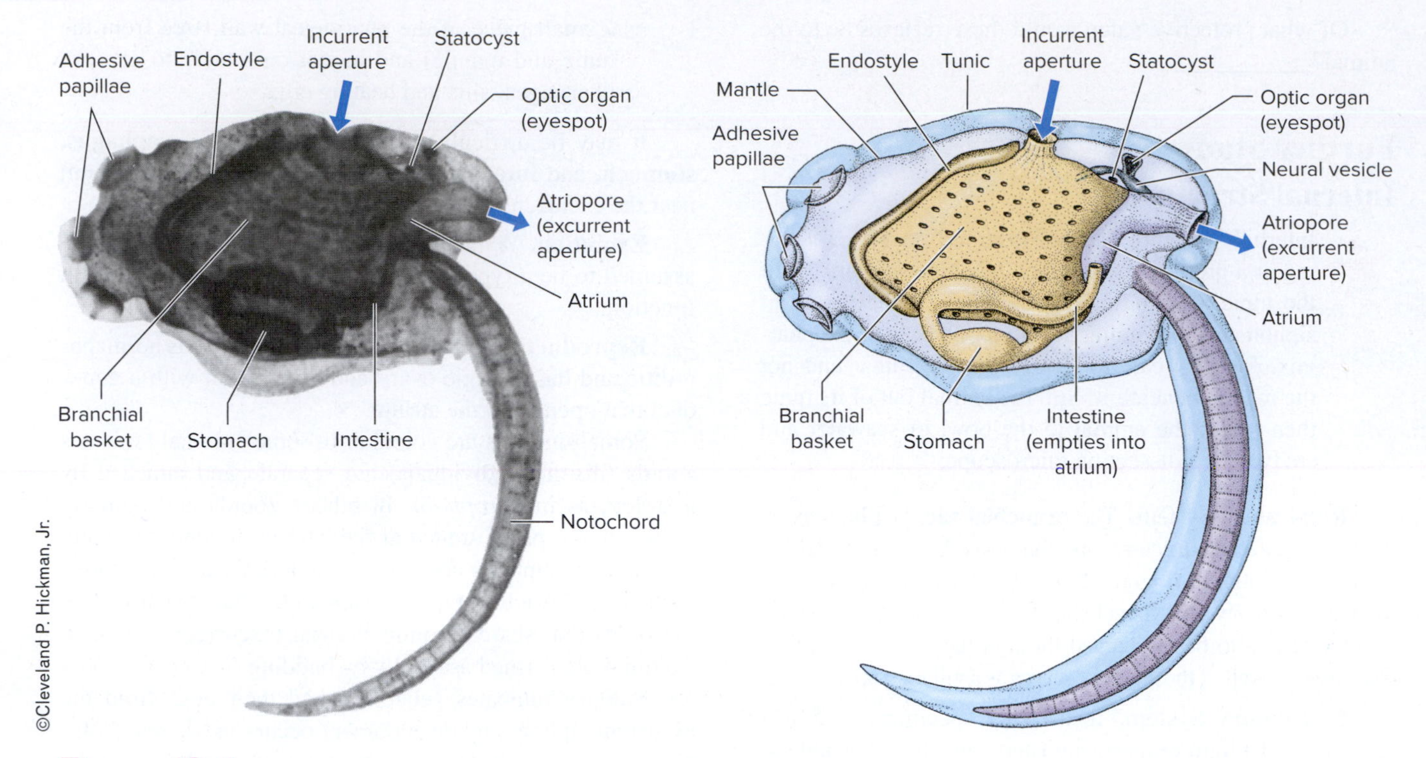

Figure 18-4
Tadpole larva of a tunicate, photographed from a stained slide specimen.

Review Your Knowledge

Test your understanding of the main features of the Urochordata covered in the exercise by filling out the summary table on pp. xi–xii. Try to complete the table without looking at your notes.

EXERCISE 18B

Subphylum Cephalochordata—Amphioxus

Core Study

Amphioxus

Phylum Chordata
    Subphylum Cephalochordata
        Genus *Branchiostoma* (= *Amphioxus*)
        Species *Branchiostoma lanceolatus*

The little lancelet, *Branchiostoma* (Gr. *branchia,* gills, + *stoma,* mouth), commonly called **amphioxus,** illustrates basic chordate structure and is considered similar to the ancestor of the vertebrates. Besides the basic characteristics—**notochord, pharyngeal slits, dorsal hollow nerve cord,** and **postanal tail**—it also possesses the beginning of a **ventral heart** and a **metameric arrangement** of muscles and nerves. There are only two genera of cephalochordates—*Asymmetron* (Gr. *asymmetros,* ill-proportioned) and *Branchiostoma.*

Where Found

Branchiostoma is common along the southern California and southern Atlantic coasts of the United States, as well as the coasts of China and the Mediterranean Sea. On sandy bottoms, it dives in headfirst, then twists upward so that the tail remains buried in sand and the anterior end is thrust upward into the water (Figure 18-5).

External Structure

Place a preserved mature specimen in a watch glass and cover with water. Do not dissect or mutilate the specimen.

How long is it? _________ Why is it called a lancelet? _________ Does it have a distinctive head? _________ Observe the **dorsal fin,** which broadens in the tail region **(caudal fin)** and continues around the end of the tail to become the **ventral fin.**

The anterior tip of the animal is the **rostrum** (Figure 18-6). With a hand lens, find the opening of the **oral hood,** which is fringed by a number of slender oral tentacles, also called **buccal cirri,** which strain out large particles of sand and are sensory in function.

On the flattened ventral surface are two **metapleural folds** of skin extending like sled runners to the ventral fin. Find the **atriopore,** which is anterior to the ventral fin. The atriopore is the opening of the atrium (L., entrance hall), a large cavity surrounding the pharynx.

Figure 18-5
Amphioxus in typical feeding position, with oral aperture facing upward.

The **anus** opens slightly to the left of the posterior end of the ventral fin.

In mature specimens, little, blocklike **gonads** (testes or ovaries) lie in the atrium anterior to the atriopore and just above the metapleural folds on each side. They can be seen through the thin body wall.

Study of the Whole Mount

Examine with low power a stained and cleared whole mount of an immature specimen.

Chevronlike **myotomes** (Gr. *mys,* muscle, + *tomos,* slice) along the sides of the animal are segmentally arranged muscles. Does the myotome of an amphioxus zigzag more or less than the myomere of a bony fish? __________ How might the differences in musculature help us predict whether amphioxus or the fish would be the better swimmer? __________ Identify the various parts of the **fin** and note its skeletal support, the transparent **fin rays.** You may have to reduce the light to see the fin rays.

Beneath the rostrum is a large chamber called the **buccal cavity,** which is bounded laterally by fleshy, curtainlike folds and is open ventrally. The rostrum and lateral folds

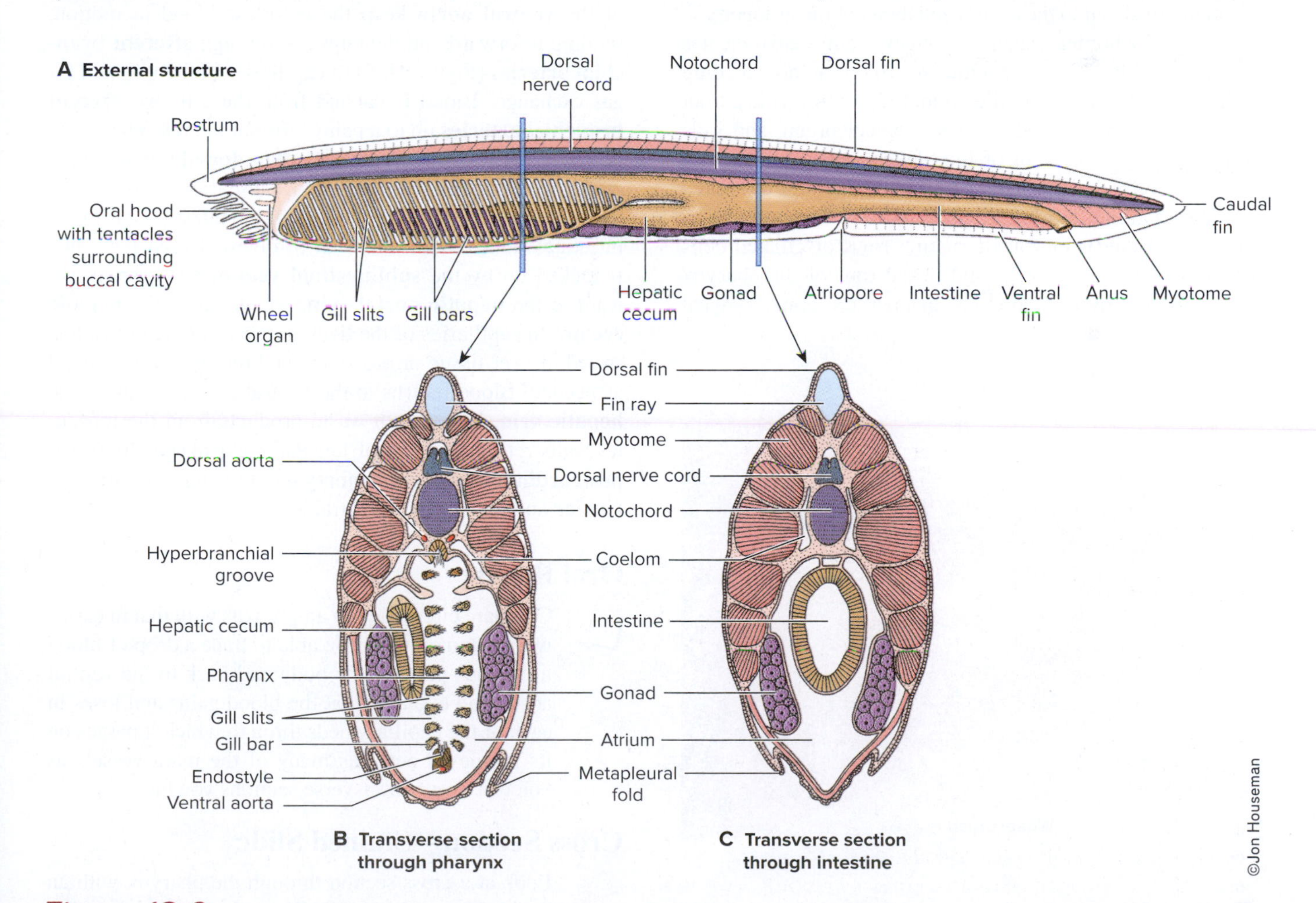

Figure 18-6
Structure of amphioxus, showing **A,** external structure and **B, C,** transverse sections through the pharynx and intestine.

together make up the **oral hood** (Figure 18-6). The roof of the oral hood bears the notochord, which may have a supporting function in spreading the hood open. Each of the oral tentacles is stiffened by a skeletal rod of fibrous connective tissue. Behind the buccal cavity is an almost perpendicular membrane, the **velum** (L., veil), pierced ventrally by a small opening, the true **mouth,** which is always open and leads into the **pharynx.** On the walls of the buccal cavity, projecting forward from the base of the velum, are several fingerlike, ciliated patches that compose the **wheel organ** (Figure 18-7). The rotating effect of its cilia helps maintain a current of water flowing into the mouth. Around the mouth, projecting posteriorly from the velum, are about a dozen delicate **velar tentacles,** also ciliated. Both oral tentacles and velar tentacles have chemoreceptor cells for monitoring incurrent water.

The large **pharynx** narrows into a straight **intestine** extending to the **anus** (Figure 18-6). The sidewalls of the pharynx are composed of a series of parallel, oblique **gill bars,** between which are **gill slits.** Just posterior to the pharynx is a diverticulum of the intestine called the **hepatic cecum,** or liver, which extends forward along one side of the pharynx. Surrounding the pharynx is the **atrium,** a large cavity that extends to the **atriopore.** Water entering the mouth filters through gill slits into the atrium and then out the atriopore.

Cephalochordates such as sponges, clams, and tunicates are filter feeders. They use a mucus-ciliary method, feeding on minute organisms. As the animal rests with its head out of the sand, the ciliated tentacles, wheel organ, and gills draw in a steady current of food-laden water, from which the cirri and velar tentacles strain out large and unwanted particles. On the floor of the pharynx is an **endostyle** (Figure 18-6B) consisting of alternating rows of ciliated cells and mucus-secreting cells, and in the roof of the pharynx is a ciliated **hyperbranchial groove** (= epipharyngeal groove). Particles of food entangled in the stream of mucus secreted by the endostyle are carried upward by cilia on the inner surface of the gill bars, then backward toward the **intestine** by cilia in the hyperbranchial groove. Digestion occurs in the intestine.

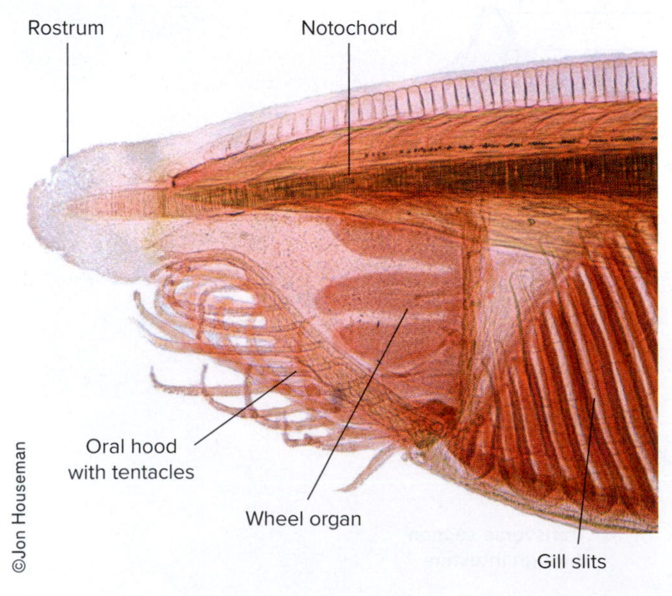

Figure 18-7
Photograph of head region of amphioxus.

Oxygen–carbon dioxide exchange occurs in the epithelium covering the gill bars. The **notochord** just dorsal to the digestive system is transversely striated and is best seen in the head and tail regions. It provides skeletal support and a point of attachment for muscles. Note that it extends almost to the tip of the rostrum. Above and parallel to the notochord is the **dorsal tubular nerve cord** (Figure 18-6). How far does it extend? __________ Is there any sign of a structure that you would call a brain? __________ The row of black spots in the nerve cord consists of pigmented **photoreceptor cells.** Chemoreceptors are scattered over the body but are particularly abundant on the oral and velar tentacles. Touch receptors are located over the entire body.

Further Study

Circulatory System

Amphioxus does not have a heart; peristaltic contractions of the **ventral aorta** keep the colorless blood in motion, sending it forward and then upward through **afferent branchial arteries** (Figure 18-8) to capillaries in the gill bars for gas exchange. Blood is carried from the gills by **efferent branchial arteries** up to a pair of **dorsal aortas.** These join posterior to the gills to form a **median dorsal aorta,** which gives off **segmented arteries** to capillaries of the myotomes and to capillaries in the wall of the intestine. From the intestinal wall, blood, now rich in digested food nutrients, is picked up by the **subintestinal vein** and is carried forward to the **hepatic portal vein,** which enters the **hepatic cecum.** In capillaries of the liver, nutrients are removed and stored in liver tissue or processed and returned to the blood as needed. Blood returns to the ventral aorta by way of the **hepatic vein.** Blood with waste products from the muscular walls returns by way of left and right **precardinal** and **postcardinal veins,** which empty into left and right **ducts of Cuvier** and then into the ventral aorta.

Oral Report

Compare circulation in amphioxus with that in earthworms and crayfish. Be able to trace a drop of blood to various parts of the body and back to the ventral aorta, and explain what the blood gains and loses in each of the capillary beds through which it passes on its journey. Locate as many of the main vessels as you can on any transverse sections you have.

Cross Section—Stained Slide

Look at a cross section through the pharynx with an unaided eye and understand how the section is cut with reference to the whole animal. Note the **dorsal fin,**

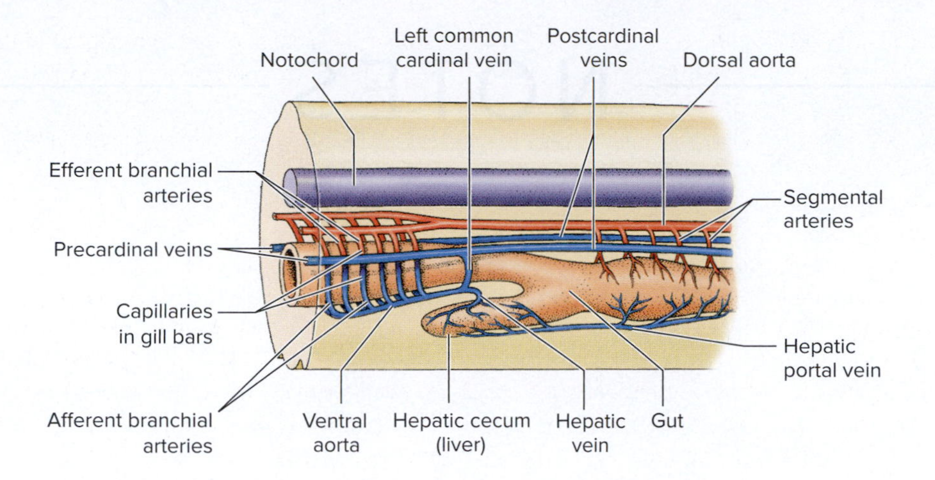

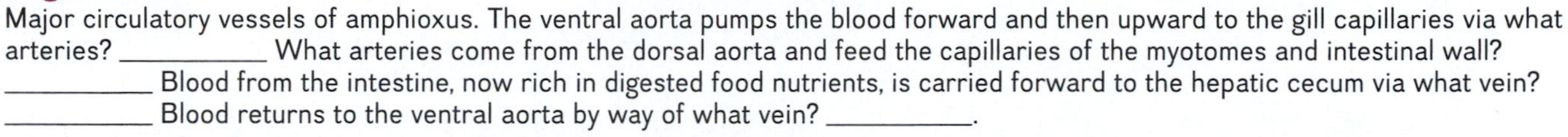

Figure 18-8

Major circulatory vessels of amphioxus. The ventral aorta pumps the blood forward and then upward to the gill capillaries via what arteries? __________ What arteries come from the dorsal aorta and feed the capillaries of the myotomes and intestinal wall? __________ Blood from the intestine, now rich in digested food nutrients, is carried forward to the hepatic cecum via what vein? __________ Blood returns to the ventral aorta by way of what vein? __________.

its supporting **fin ray,** and the ventral **metapleural folds** (see Figure 18-6). With a microscope, examine the **epidermis,** a single layer of columnar epithelial cells, and **dermis,** a gelatinous connective tissue layer. Large **myotomes,** or muscles, are paired, but members of a pair are not opposite each other. Myotomes are separated by connective tissue, **myosepta.**

The **nerve cord,** enclosed within the **neural canal,** has in its center a small **central canal,** which is prolonged dorsally into a slit. In some sections, dorsal **sensory nerves** or ventral **motor nerves** may be seen. These are given off alternately from the cord to the myotomes. The large, oval **notochord** with vacuolated cells is surrounded by a **notochordal sheath.**

The cavity of the **pharynx** is bounded by a ring of triangular **gill bars** separated by **pharyngeal slits** that open into the surrounding **atrium** (see Figure 18-6B). From your study of the whole mount, why do you think the cross section shows the gill bars as a succession of cut surfaces? __________ The somewhat rigid gill bars contain blood vessels and are covered by ciliated **respiratory epithelium,** where gaseous exchange is made. On the dorsal side of the pharynx, find the ciliated **hyperbranchial groove** on the ventral side, the **endostyle.** The latter secretes mucus, in which food particles are caught. What is the function of the cilia in the grooves? __________

Gonads (ovaries or testes) lie on each side of the atrial cavity. The reduced **coelom** consists of spaces, usually paired, on each side of the notochord and hyperbranchial groove, as well as on each side of the gonads and below the endostyle. In favorable specimens, little **nephridial** tubules may be found in the dorsal coelomic cavities (see Figure 18-6). What is their function? __________

Ventral to the notochord are paired **dorsal aortas.** The **ventral aorta** lies ventral to the endostyle.

Drawings

On separate paper, make drawings of such sections of amphioxus as your instructor requests.

Review Your Knowledge

Test your understanding of the main features of the Cephalochordata covered in the exercise by filling out the summary table on p. xii. Try to complete the table without looking at your notes.

NOTES

The Fishes—Lampreys, Sharks, and Bony Fishes

EXERCISE 19A

Class Petromyzontida—Lampreys (Ammocoete Larva and Adult)

EXERCISE 19A
Class Petromyzontida—Lampreys (Ammocoete Larva and Adult)
Lampreys

EXERCISE 19B
Class Chondrichthyes—Cartilaginous Fishes
Squalus, Dogfish Sharks
Demonstrations

EXERCISE 19C
Class Actinopterygii—Bony Fishes
Perca, Yellow Perch

EXPERIMENTING IN ZOOLOGY
Agonistic Behavior in Paradise Fish, *Macropodus opercularis*

EXPERIMENTING IN ZOOLOGY
Analysis of the Multiple Hemoglobin System in *Carassius auratus,* Common Goldfish

Core Study

Lampreys

Phylum Chordata
  Subphylum Vertebrata
    Superclass Agnatha
      Class Petromyzontida
        Genus *Petromyzon*
          Species *Petromyzon marinus*

The fishes comprise almost half of all living vertebrate species and include jawless fishes, cartilaginous fishes and the bony fishes. Hagfishes and lampreys are the only living descendants of the earliest known vertebrates, a group of Paleozoic jawless fishes collectively called **ostracoderms.** Hagfishes and lampreys are conventionally grouped together in superclass Agnatha ("without jaws") and share certain characteristics, including the absence of jaws, internal ossification, scales, and paired fins. In other respects, however, hagfishes and lampreys are radically different from each other.

Lampreys have a worldwide distribution and most are **anadromous** (an-ad′ruh-mus; Gr. *anadromos,* running upward), meaning that they ascend rivers and streams to spawn. The species *Petromyzon marinus* (Gr. *petros,* stone, + *myzon,* sucking, referring to its habit of holding its position in a current by grasping a stone with its mouth) is the sea lamprey that lives in the Atlantic drainages of Canada, the United States, Iceland, and Europe and is landlocked in the Great Lakes. It grows to be 1 m long and can live both in freshwater and in the sea. It is a marine species that migrates up freshwater streams to spawn. Young larvae, known as **ammocoetes** (sing. **ammocoete;** Gr. *ammos,* sand, + *koitē,* bed, Fr.? *keisthai,* to lie, referring to the preferred larval habitat), live in sand for 3 to 5 years as filter feeders. Then they metamorphose rapidly into adults and become parasites of fishes. Attaching themselves with their suckerlike mouth, they rasp away the fish's flesh with their horny teeth and suck out blood and body fluids. Adults grow rapidly for a year, spawn in winter or spring, and soon die. After invading the Great Lakes in the nineteenth century via man-made canals, sea lampreys devastated the important commercial fisheries there. Wounding rates of fishes are lower now, but sea lampreys remain a threat to the commercial fishing trade.

Freshwater lampreys, known as **brook** or **river lampreys,** belong to genera *Lampetra* (L. *lambo,* to lick or lap up) and *Ichthyomyzon* (Gr. *ichthyos,* fish, + *myzon,* sucking), of which there are about 33 species. They have larval habits similar to those of the marine form. Adults in about half the freshwater species are not parasitic. Nonparasitic forms do not eat as adults and live only a month or so after emerging from the sand to spawn.

Drawings

On separate paper, draw such transverse sections as required by your instructor. Label fully. How many features of the amphioxus adult do you find repeated in the ammocoete, the larval form of a vertebrate? __________ Do you think this might be interpreted as an example of homology? __________ Keep in mind the structure of these early chordate forms and be able to compare them with those of the fish, amphibian, and mammal forms that you will study later.

Adult Lampreys

External Structure

Examine a preserved specimen of an adult lamprey.

In addition to the obvious difference in size, it will be immediately evident that the adult lamprey differs in many anatomical details from the ammocoete. During the dramatic metamorphosis from larva to adult, the body becomes rounder and shorter, the pharynx becomes divided longitudinally, the larval hood is replaced by an oral disc with teeth, the eyes enlarge, and the nostril shifts to the top of the head. These changes are essential to the shift from a larval life of filter feeding to an adult existence as a parasite of fish.

Note the eel-like shape of the lamprey and its tough, scaleless skin. Among the epithelial cells are numerous gland cells that produce a protective slime. Identify the two **dorsal fins** and **caudal fin** (Figure 19-1A). There are no paired appendages. What does a lack of paired fins tell you about how a lamprey might swim, as compared with a bony fish? __________ Examine the hood-shaped **buccal funnel** supported by a cartilaginous ring that serves as a sucking disc for attachment to a host. The opening is fringed by numerous fingerlike sensory papillae, and the interior of the funnel bears horny "teeth"; these are actually epidermal thickenings and not homologous to true vertebrate teeth, which are derived from mesoderm. Locate the **mouth** at the back of the buccal funnel and dorsal to the **tongue.** The tongue also bears sharp, horny "teeth" used for rasping.

A single **nostril** located middorsally on top of the head opens into an olfactory sac (the latter visible in the sagittal section, below in Figure 19-1B). Just behind the nostril is a small, oval area marking the position of the so-called third eye, the **pineal organ.** It is not an eye in the true sense, but it does contain photoreceptors that detect changes in illumination and adjust internal activities of the lamprey. The pineal organ is present in most fishes but is better developed in lampreys than in any other living vertebrate except certain reptiles. Note the functional, lidless **eyes** and seven **external gill slits** just behind the eye on each side of the head.

The **lateral line system,** characteristic of nearly all fishes, consists of specialized receptors located in small patches on the head and trunk of the lamprey. They can be located as groups of pores extending below and caudally from the eye on each side of the head. These receptor cells

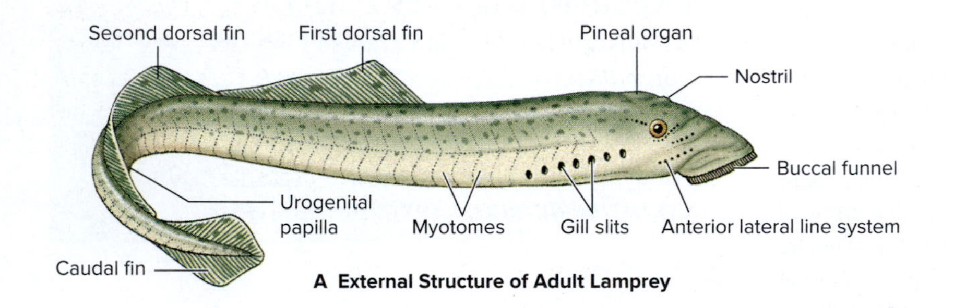

A External Structure of Adult Lamprey

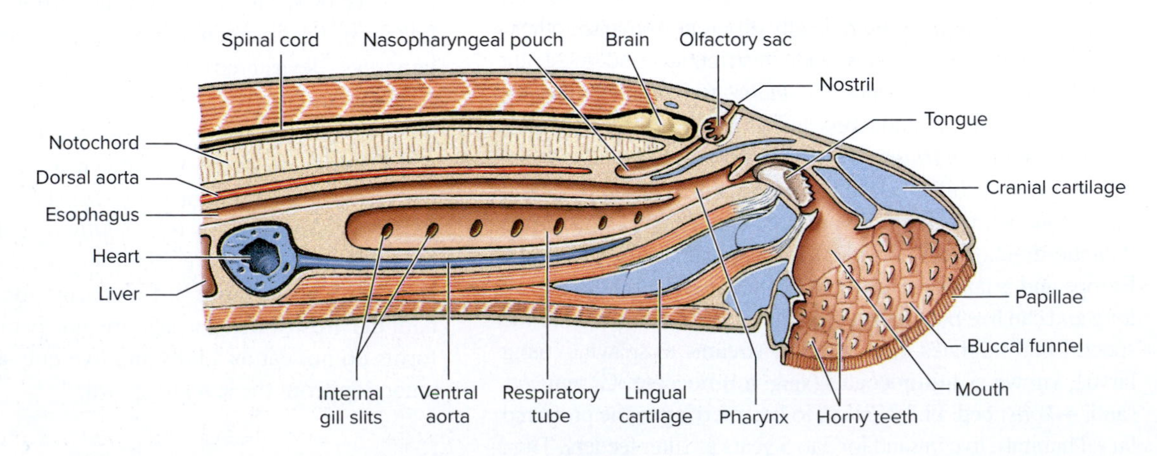

B Sagittal Section of Adult Lamprey

Figure 19-1
Adult lamprey.

are sensitive to currents and water movement. How would they be useful to the lamprey? __________ In the lamprey, the receptors are open to the exterior and not protected within canals as they are in bony fishes.

Find the **urogenital* sinus,** with its projecting urogenital papilla and the ventral juncture of the trunk and the tail. There is an **anal opening** in front of the urogenital sinus.

How does the structure of the myotomes compare with the structures of amphioxus? ______________________________

____

Internal Structure

Examine a sagittal section of the anterior portion of the lamprey body (Figure 19-1B).

Identify the **notochord,** a rodlike mass of vacuolated cells enclosed by a tough, fibrous sheath. As in amphioxus, the firm yet flexible notochord prevents the body from shortening when the muscles contract. The notochord remains well developed throughout the life of the lamprey. In addition to the notochord, the lamprey's skeleton consists of various elements of cartilage and fibrous connective tissue. Some skeletal elements can be seen in the sagittal section.

Follow the digestive tract through the oral hood, mouth, tongue, and **pharynx,** which leads into two tubes: the **esophagus** dorsally and a **respiratory tube** ventrally. The esophagus leads into the **intestine,** which continues as a long tube to the cloaca. There is no stomach, a primitive feature.

The wall of the respiratory tube is perforated by seven **internal gill slits.** Each gill slit leads into an enlarged branchial pouch lined with gill lamellae (these may not be visible on the sagittal section). Bony fish pass water through the mouth and then over the gills before it exits behind the gills. Why is this not possible for a feeding lamprey? __________ How might it solve the problem? __________________

The **brain,** a bilobed structure similar to the brain of sharks, and the **spinal cord** are seen lying above the notochord. The nostril leads first into the **olfactory sac** with a much-folded inner surface and then into the elongate **nasopharyngeal pouch.** Water is drawn in and squeezed out of the olfactory sac with each respiratory movement of the pharynx. The pineal organ may be visible just behind the nostril.

The **heart** lies within the **pericardial cavity,** a division of the coelom. Venous drainage from the body is received by the **sinus venosus,** which empties into the **atrium** on the left side of the pericardial cavity. Blood from the atrium passes into the **ventricle** on the right side of the pericardial cavity and is then pumped into the **ventral aorta.** The ventral aorta gives off eight pairs of afferent branchial arteries that lead to gill capillaries, where the blood is oxygenated, then collected by the **dorsal aorta** lying just ventral to the notochord. The dorsal aorta continues caudally, supplying blood to the viscera and body musculature. Usually these major blood vessels are difficult to see in a sagittal section.

Ammocoete Larvae

Although ammocoete larvae resemble amphioxus in, life habit, and many anatomical details, it bears several characteristics that anticipate the vertebrate body plan that are lacking in amphioxus. For example, how is blood propelled in amphioxus? ______________________________

____

The ammocoete has a two-chambered heart; two median eyes, each with lens and receptor cells; a three-part brain; thyroid and pituitary glands; and a pronephric kidney. Instead of the numerous pharyngeal slits found in the amphioxus, an ammocoete larva has only seven pairs.

Examine a preserved ammocoete larva as well as a stained whole mount of a small specimen.

Study of the Preserved Larva

Cover the preserved larva with water in a water glass. Use a hand lens or dissecting microscope.

List two physical differences between the preserved specimen and a mature amphioxus. __________ Note **myotomes** (segmental muscles) appearing faintly on the surface (Figure 19-2). Do myotomes have the same arrangement as those of amphioxus? __________ Note the **oral hood** with **oral papillae** attached to the roof and sides of the hood. They are used, as in amphioxus, for filter feeding. The **lateral groove** on each side contains seven small **gill slits.** The anus, or **cloacal opening,** is just anterior to the **caudal fin.** Is the caudal fin continuous with the **dorsal fin**? __________ Note **chromatophores** scattered over the body.

Study of the Stained Whole Mount

With low power, examine a stained whole mount of an ammocoete larva.

On a whole mount, find the darkly stained, dorsal, hollow **nerve cord,** which is enlarged anteriorly to form the **brain.** Immediately below it is the lighter **notochord** (Figure 19-2).

The oral hood encloses a **buccal cavity,** to the back and sides of which are attached oral papillae. Posterior to the buccal cavity is the **velum,** a large pair of flaps that create water currents. The large **pharynx** has **internal gill slits,** which open into **gill pouches.** Gill pouches open to the outside by small **external gill slits.** How many pairs of gill slits are there in the ammocoete? __________ Using low power, focus upward onto the outer surface of the animal to see a row of small external gill slits. Between the internal gill slits are cartilaginous rods, **gill bars,** which strengthen the pharynx walls. Note **gill lamellae** on the pharynx walls. They are rich in capillaries, through which the blood gives up its carbon dioxide and takes up its oxygen from the water.

---

*Urogenital = urinogenital. The stem **uro** as used in "urogenital" derives from the Greek *ouron,* meaning urine. Unfortunately, the stem is used in other terms to mean "tail" (Gr. *oura,* tail), as in *Urodela, urostyle,* and *uropod.*

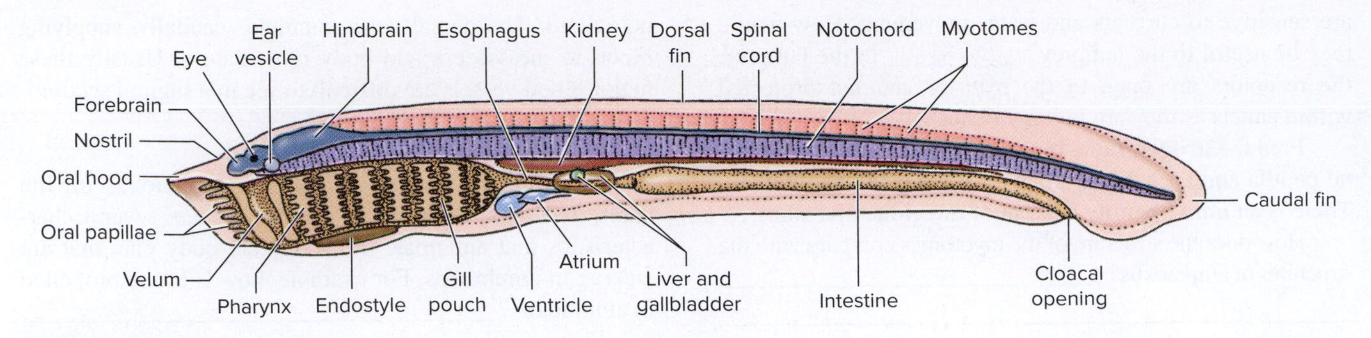

Figure 19-2
Ammocoete larva, sagittal section.

The ammocoete is a filter feeder. Was amphioxus also a filter feeder? __________ Water is kept moving through the pharynx by muscular action of both the velum and the whole branchial basket. This contrasts with amphioxus, in which water is moved by ciliary action.

The **endostyle** (subpharyngeal gland) in the floor of the pharynx is a closed tube the length of four gill slits. It secretes mucus by a duct into the pharynx. Food particles brought in by water currents are trapped in the mucus and carried by ciliary action to the **esophagus.** During metamorphosis of the larva, a portion of the endostyle becomes a part of the adult's thyroid gland.

The narrow esophagus widens to become the **intestine,** the posterior end of which, called the **cloaca,** also receives the kidney ducts. The **anus** opens to the outside a short distance in front of the postanal tail.

The **liver** lies under the posterior end of the esophagus, and embedded in it is the **gallbladder,** which appears as a clear, round vesicle. A two-chambered **heart** lies under the forepart of the esophagus.

Over the heart and around the esophagus is the **pronephric kidney,** consisting of a number of small tubules that empty into the cloaca by pronephric ducts (not easily distinguished on the whole mounts). Later a mesonephric kidney will develop above the intestine, using the same ducts, and the pronephros will degenerate.

The tubular dorsal **nerve cord** enlarges anteriorly into a three-lobed **brain,** visible in most slides. How does the nerve cord differ in location from that of the flatworms? __________ How does the nerve cord differ from those of annelids and arthropods? __________ The **forebrain** contains the **olfactory lobe.** In front of the forebrain is the **nasohypophyseal canal,** opening dorsally to the outside by a median **nostril.** The darkly pigmented **eyes** connect with each side of the **midbrain.** At this stage, the eyes are covered with skin and muscle and have little sensitivity to light; however, the ammocoete larva does have photoreceptors in the tail. Find the **hindbrain** and, with careful focusing, try to see one of the clear, oval **ear vesicles** that flank each side of it. Which lobe of the brain is chiefly concerned with the sense of smell? __________ sight? __________ Of hearing and equilibrium? __________ How is the nervous system of this agnathan advanced over that of the cephalochordate amphioxus? ____________________________________

Further Study

Circulation in Ammocoetes

The circulatory system is similar to that of amphioxus, but blood is driven by a **two-chambered heart.** The circulatory pattern is basically that of all fishes. It is a single-circuit system with two sets of capillaries: capillary beds of the gills and capillaries of the body. The **ventral aorta** carries deoxygenated blood forward from the heart, bifurcating at about the fourth gill pouch and giving off eight pairs of **afferent** (L. *ad,* toward, + *ferre,* to carry) **branchial arteries** to the gills' capillary beds, where the blood is oxygenated. Eight pairs of **efferent** (L. *ex,* out, + *ferre,* to carry) **branchial arteries** direct oxygenated blood to the **dorsal aorta,** which sends a pair of arteries forward to the brain and head. Most of the dorsal aorta blood, however, runs posteriorly, giving off **segmental arteries** to capillaries of the body walls and one large **intestinal artery** to capillary beds in the intestinal wall. The **hepatic portal vein** carries blood from the intestine to the liver capillaries; the **hepatic vein** carries blood from the liver capillaries to a thin-walled sac, the **sinus venosus,** at the back of the heart. Blood from the body wall and myotomes is picked up by cardinal veins. On each side, an **anterior cardinal vein,** lateral to the notochord, and a **posterior cardinal vein,** lateral to the dorsal aorta, unite to form a **duct of Cuvier,** which, with its mate from the other side, enters the sinus venosus. The sinus venosus is connected to the thin-walled **atrium** of the heart by valves that prevent backflow. Valves also guard the opening from the atrium to the thicker, more muscular **ventricle,** which pumps blood through valves into the ventral aorta. How does the ammocoete circulatory system differ from that of an earthworm? ____________________________________

____

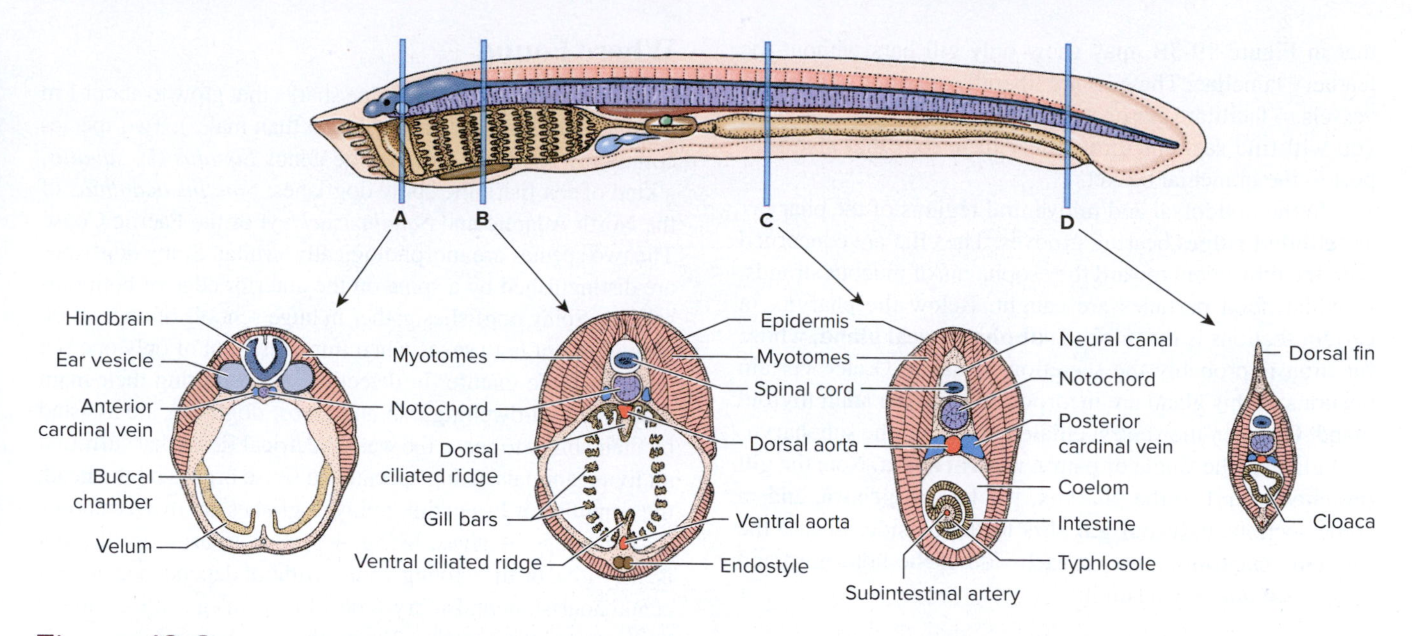

Figure 19-3

Transverse sections of ammocoete larva. A, Posterior part of buccal chamber. B, Pharynx. C, Intestine. D, Cloaca.

Transverse Sections of Ammocoetes

Your slide may contain four typical sections—one each through the brain, pharynx, intestine, and postanal tail. Or it may contain 15 to 20 sections through the body, arranged in sequence. As you study each section, refer to the whole mount again to interpret relationships. Use the low power of the microscope.

Sections Anterior to the Pharynx. Sections through the **forebrain** may include the **oral papilla** and **oral hood.** Sections through the **midbrain** may include **eyes, buccal chamber,** and portions of **velar flaps.** Sections through the **hindbrain** may include **ear (otic) vesicles** and buccal chamber (Figure 19-3) or the forepart of the pharynx. Compare the size of the brain with that of the spinal cord in more posterior sections. Do you find a **notochord** lying just below the brain in any of the sections? __________ Why is the notochord considered a primitive type of endoskeleton? ____

____

Sections Posterior to the Pharynx. Choose a section through the trunk posterior to the pharynx and identify the following (Figure 19-3): **epidermis; myotomes** (lateral masses of muscles); **nerve cord** surrounded by the **neural canal** and containing a cavity, the **neurocoel; notochord,** with large, vacuolated cells; **dorsal aorta** (probably contains blood cells); and **posterior cardinal veins,** one on each side of the aorta (blood cells are usually present). You may find the cardinals joining to form the duct of Cuvier. The **coelomic cavity,** lined with peritoneum, contains the visceral organs. What distinguishes a true coelom from other cavities in the body? ____

____

Visceral contents of the coelomic cavity will vary according to the location of the section.

1. Just behind the pharynx, you will find the **esophagus,** composed of columnar epithelium; the paired **pronephric kidneys,** appearing as sections of small tubules; and chambers of the **heart.**
2. Sections cut posterior to the heart will show the dark **liver** and possibly the hollow **gallbladder** ventral, or lateral, to the esophagus, with sections of pronephric kidneys or their ducts located under the posterior cardinals.
3. Sections through the intestine will reveal it as a large tube of columnar epithelium with a conspicuous infolding, the **typhlosole,** carrying the **subintestinal artery** (Figure 19-3C). Above the intestine, you may find the **mesonephric kidneys,** with their tubules, and a small **gonad** between the kidneys. The **cloaca** is located farther back (Figure 19-3D).
4. In sections posterior to the anus (postanal tail), identify the caudal fins. What other structures can you identify?

Sections through the Pharynx. The body wall, nerve cord, and notochord will be similar to these features in the preceding sections. The central part of the section is taken up by the large pharynx, on whose walls are the **gills,** with their platelike **gill lamellae,** extending into the pharynx or into the **gill pouch,** depending on how the section has been cut. Each lamella has lateral ridges. Some sections, such as that in Figure 19-3B, may show only gill bars without the feathery lamellae. The gills are liberally supplied with blood vessels to facilitate gas exchange. Outside the gill chambers, you will find sections through cartilage rods that give support to the branchial basket.

In the middorsal and midventral regions of the pharynx are **ciliated ridges** bearing grooves. The cilia are concerned with the movement toward the esophagus of mucous strands, in which food particles are caught. Below the pharynx in certain sections is the bilobed **subpharyngeal gland,** whose function is probably the secretion of mucus. Later, certain portions of this gland are incorporated into the adult thyroid gland. Between the ventral ciliated ridge and the subpharyngeal gland is the single or paired **ventral aorta.** Note the **gill pouches** lateral to the pharynx, the **lateral groove,** and in some sections **external gill slits** to the outside. Locate the **anterior cardinal veins** on each side of the notochord and the **dorsal aorta** beneath it.

EXERCISE 19B

Class Chondrichthyes—Cartilaginous Fishes

Core Study

Squalus, Dogfish Sharks

Subphylum Vertebrata
Superclass Gnathostomata
Class Chondrichthyes
Subclass Elasmobranchii
Order Squaliformes
Genus *Squalus*

Cartilaginous fishes are a compact, ancient assemblage of about 970 species characterized by cartilaginous skeletons, powerful jaws, and well-developed sense organs. They include sharks, skates, rays, and chimaeras. Subclass Elasmobranchii (e-laz′-mo-bran′kee-i; Gr. *elasmos,* metal plate, + *branchia,* gills) embraces sharks, skates, and rays—cartilaginous fishes with exposed gill slits opening separately to the outside. Most are carnivores and many are top predators. The dogfish shark is an excellent example of the generalized body plan of early jawed vertebrates.

Where Found

Dogfish sharks are small marine sharks that grow to about 1 m in length (females are slightly larger than males). Two species commonly studied belong to the genus *Squalus* (L. *squalus,* a kind of sea fish), the spiny dogfishes: *Squalus acanthias* of the North Atlantic and *Squalus suckleyi* of the Pacific Coast. The two species are morphologically similar. Spiny dogfishes are distinguished by a spine on the anterior edge of both dorsal fins. Spiny dogfishes gather in huge schools of up to 1000 individuals of both sexes when immature, but of only one sex when they are mature. In detecting and capturing their main foods, bottom-dwelling fish and crabs, dogfishes are assisted by their ability to sense the weak electrical fields that surround all living animals, using specialized sense organs on the head, the ampullae of Lorenzini. Spiny dogfishes are ovoviviparous (L. *ovum,* egg, + *vivus,* living, + *parere,* to bring forth); that is, they give birth to living young without dependence on placental nourishment. Embryos develop in an egg capsule in the oviduct until they hatch in the mother just before birth.

Heavy fishing pressures on dogfish sharks appear to have impacted population. These fish are consumed in some parts of the world, and body parts like their fins and tails are also marketed. They are currently considered "vulnerable" on the IUCN Red List (see also Lack, M. 2006. Conservation of Spiny Dogfish *Squalus acanthias*: a role for CITES? TRAFFIC International). Some will argue that we will soon reach a point where it will be difficult to justify continued use of these animals for anatomy study.

External Structure

Examine an intact, preserved dogfish to identify the following features.

The body is divided into the **head** (to the first gill slit), **trunk** (to the cloacal opening), and **tail.** The **fins** include a pair of **pectoral fins** (anterior), which control changes in direction during swimming; a pair of **pelvic fins,** which serve as stabilizers and which in males are modified to form claspers used in copulation; two medium **dorsal fins,** which also serve as stabilizers; and a **caudal fin,** which is **heterocercal** (asymmetric dorsoventrally) (Figure 19-4).

Identify the **mouth** with its rows of **teeth** (modified placoid scales), which are adapted for cutting and shearing; two ventral **nostrils,** which lead to olfactory sacs and are equipped with folds of skin that allow continual in-and-out movement

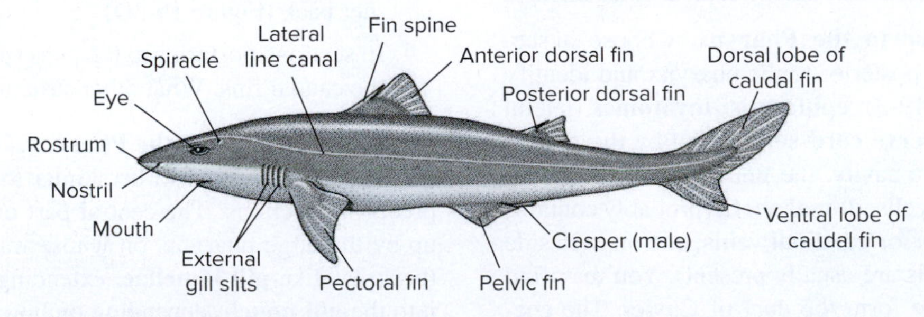

Figure 19-4
External anatomy of a dogfish shark, *Squalus* sp.

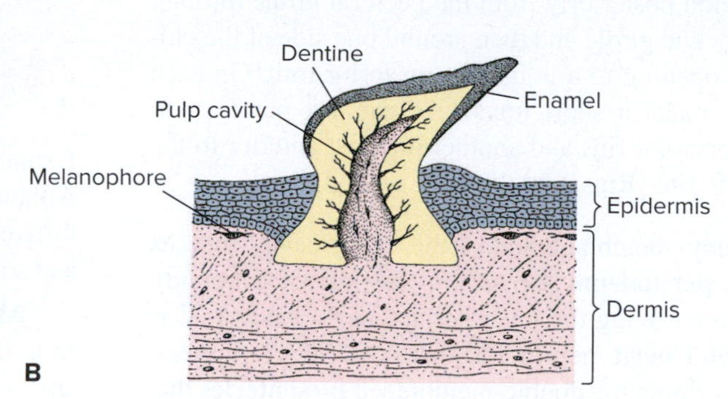

Figure 19-5

A, Surface view of shark skin, showing dermal denticles, or placoid scales, SEM, ×195. **B,** Lateral section of a single denticle.

of water; and lateral **eyes,** which lack movable eyelids but have folds of skin that can cover the eyeballs. The part of the head anterior to the eyes is called the **rostrum** (snout).

A pair of dorsal **spiracles** posterior to the eyes are modified gill slits that open into the pharynx. They can be closed by folds of skin during part of the respiratory cycle to prevent the escape of water. Spiracles serve for water intake when a shark is feeding. Five pairs of **gill slits** are the external openings of the gill chambers.

Insert a probe into one of the slits and notice the angle of the gill chamber.

The **pharynx** is the region in back of the mouth into which the gill slits and spiracles open. A **lateral line,** appearing as a white line on each side of the trunk, represents a row of minute, mucus-filled sensory pores used to detect differences in velocity of surrounding water currents and thus to detect the presence of other animals, even in the dark. Note the **cloacal opening** between the pelvic fins.

The leathery skin consists of an outer layer of epidermis covering a much thicker layer of dermis densely packed with fibrous connective tissue. Draw your finger lightly over the shark's skin to feel the spines of the **placoid scales** (Gr. *placos,* tablet, plate) (Figure 19-5). Each scale is anchored in the dermis and is built much like a tooth (the shark's teeth are, in fact, modified placoid scales). Each scale contains a pulp cavity and a thick layer of dentine, both derived from the dermis, and is covered with hard enamel, derived from the epidermis. Spiny scales help reduce friction-producing turbulence as a shark swims, thus lessening drag.

With its dark dorsal and light ventral surfaces **(countershading),** the animal's coloration is protective, making it difficult to see when viewed from above or from below.

Internal Structure

If you will not be doing a dissection, examine longitudinal and transverse sections of the shark. Note the cartilaginous skull and vertebral column. With the help of Figure 19-6, identify as many of the internal structures as possible.

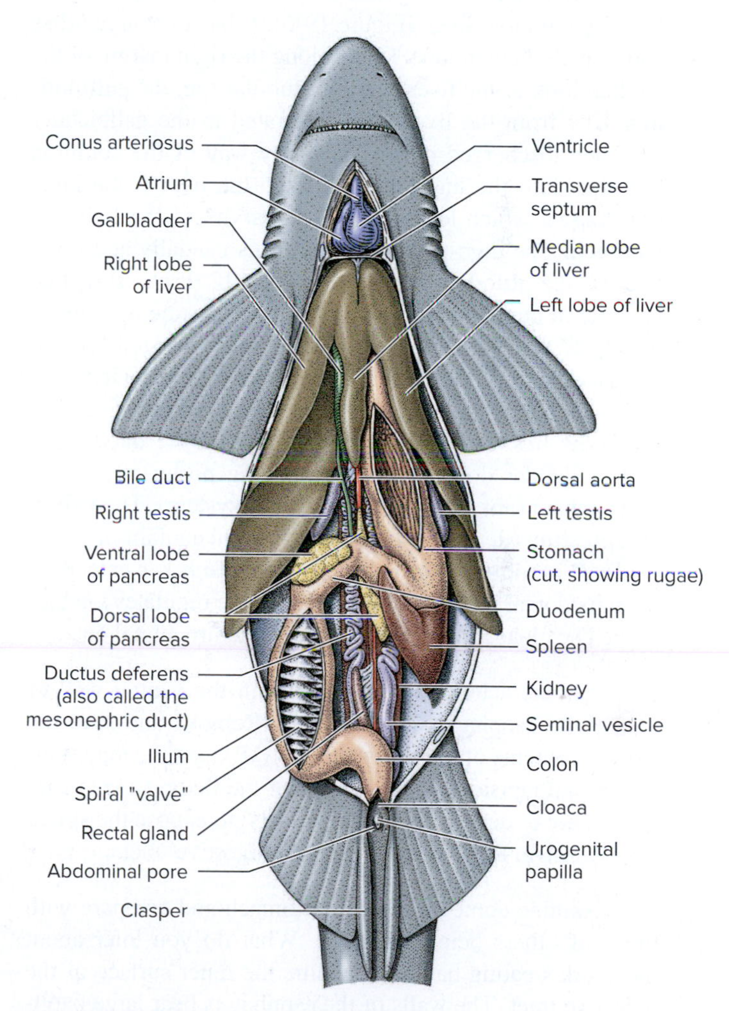

Figure 19-6

Internal anatomy of a dogfish shark, ventral view. What large organ likely assists in keeping the shark buoyant? __________ What structure slows the passage of food through the gut? __________ What organ is formed around the posterior end of the J-shaped stomach? __________ What gland is specific to the cartilaginous fishes and helps regulate the fishes' salt imbalance? __________ What organ functions as the primary source of red blood cell production? __________

Dissection of a Shark

Open the coelomic cavity by extending a midventral incision posteriorly from the pectoral girdle through the pelvic girdle and then around one side of the cloacal opening to a point just posterior to it. On each side, make a short transverse cut just posterior to the pectoral fins and another one just anterior to the pelvic fins. Rinse out the body cavity.

The shiny membrane lining the body cavity and its organs is the **peritoneum** (Gr. *peritonaios,* stretched around). The peritoneum lining the inner surface of the body wall is called **parietal peritoneum,** and that covering the visceral organs and forming the double-membraned **mesenteries** that suspend the digestive organs is called **visceral peritoneum.**

Digestive System

Identify the large **liver** (Figure 19-6). It has two large lobes and a small median lobe. Lying along the right margin of the median lobe of the liver is a thin, tubular sac, the **gallbladder.** Bile from the liver is concentrated in the gallbladder and then discharged during meals by way of the common bile duct into the intestine. Dorsal to the liver is the large **esophagus,** which leads from the pharynx to the J-shaped **stomach.** The digestive tract then turns caudally and gives rise to the **duodenum** (L. *duodeni,* 12 each, so-called because in humans this first part of the intestine is approximately 12 fingerwidths in length). Between the stomach and the duodenum is a muscular constriction, the **pyloric valve,** which regulates the entrance of food into the intestine. The **pancreas** lies close to the ventral side of the duodenum, with a slender dorsal portion extending posteriorly to the large **spleen** (not a part of the digestive system). The **valvular intestine** (ileum) is short and wide and contains a **spiral "valve,"** or **ridge.** The short, narrow **rectum** has extending from its dorsal wall a **rectal gland,** which regulates ion balance. The cloaca receives the rectum and urogenital ducts.

Make a longitudinal incision in the ventral wall of the esophagus and stomach. Remove and save the contents of the stomach, if any. Extend the longitudinal incision along the wall of the ileum (taking care not to destroy the blood vessels) to expose the **spiral valve.** Rinse out the exposed digestive tract.

Examine contents from the stomach and compare with those of others being dissected. What do you infer about the shark's eating habits? Examine the inner surface of the digestive tract. The walls of the esophagus bear large papillae, whereas the walls of the stomach are thrown into longitudinal folds called **rugae** (roog′ee; L. *ruga,* wrinkle). Note the structure of the pyloric valve. Observe the cone-shaped folds of the **spiral valve** and see if you can determine how materials pass through. The spiral valve, not really a "valve" but a fold of tissue that spirals down the ileum much like a spiral staircase, slows the passage of food through the gut. Why might slowing the passage of food be beneficial for the digestive process? ______________________________

____

If the intestine has been everted into the cloacal region, carefully pull it back into the body cavity.

Urogenital System

Although the excretory and reproductive systems have quite different functions, they are closely associated structurally and so are studied together (Figure 19-7).

Male. Soft, elongated **testes** lie along the dorsal body wall, one on each side of the esophagus (Figure 19-7). They are held in place by a mesentery called the **mesorchium** (Gr. *mesos,* middle, + *orchis,* testicle). A number of very fine tubules (vasa efferentia) in the mesentery run from each testis to a much convoluted **mesonephric duct** (also called Wolffian duct or ductus deferens). The **kidneys** (also called opisthonephroi) are long and narrow and lie behind the peritoneum on each side of the dorsal aorta. They extend from the pectoral girdle to the **cloaca.** The sperm ducts, which serve as both urinary ducts and sperm ducts, take a twisting course along the length of the kidneys to the cloaca and collect wastes from the kidneys by many fine tubules. The sperm ducts widen posteriorly into **seminal vesicles,** which dilate terminally into **sperm sacs** before entering the cloaca. The cloaca is a common vestibule into which both the rectum and the urogenital ducts empty. In the center of the cloaca, dorsal and posterior to the rectum, is a projection called the **urogenital papilla,** which is larger in males than in females. Seminal vesicles empty into the urogenital papilla, which empties into the cloaca.

Slit open the cloaca to see the urogenital papilla.

A groove along the inner edge of each **clasper** is used in transferring spermatozoa to a female at copulation.

Female. A pair of **ovaries** lie against the dorsal body wall, one on each side of the esophagus (Figure 19-7). Enlarged ova may form several rounded projections on the surface of the ovaries. A pair of **oviducts** (Müllerian ducts) run along the dorsal abdominal mesentery. The anterior ends join to form a common opening into the abdominal cavity, called the **ostium tubae.** The oviducts are anteroventral to the liver but may be difficult to find except in large females. Ripened ova leave the ruptured wall of the ovary and enter the abdominal cavity. They are drawn through the ostium into the oviducts, where fertilization may occur. An expanded area of each oviduct dorsal to the ovary is a **shell gland (oviducal gland),** which in *Squalus* secretes a thin membrane around several eggs at a time. The posterior end of each oviduct enlarges into a **uterus,** the caudal end of which opens into the **cloaca.** In immature dogfishes both the shell gland and the uterus may not be apparent.

One to seven eggs, depending on the species, may develop in each uterus. Vascularized villi on the wall of the uterus

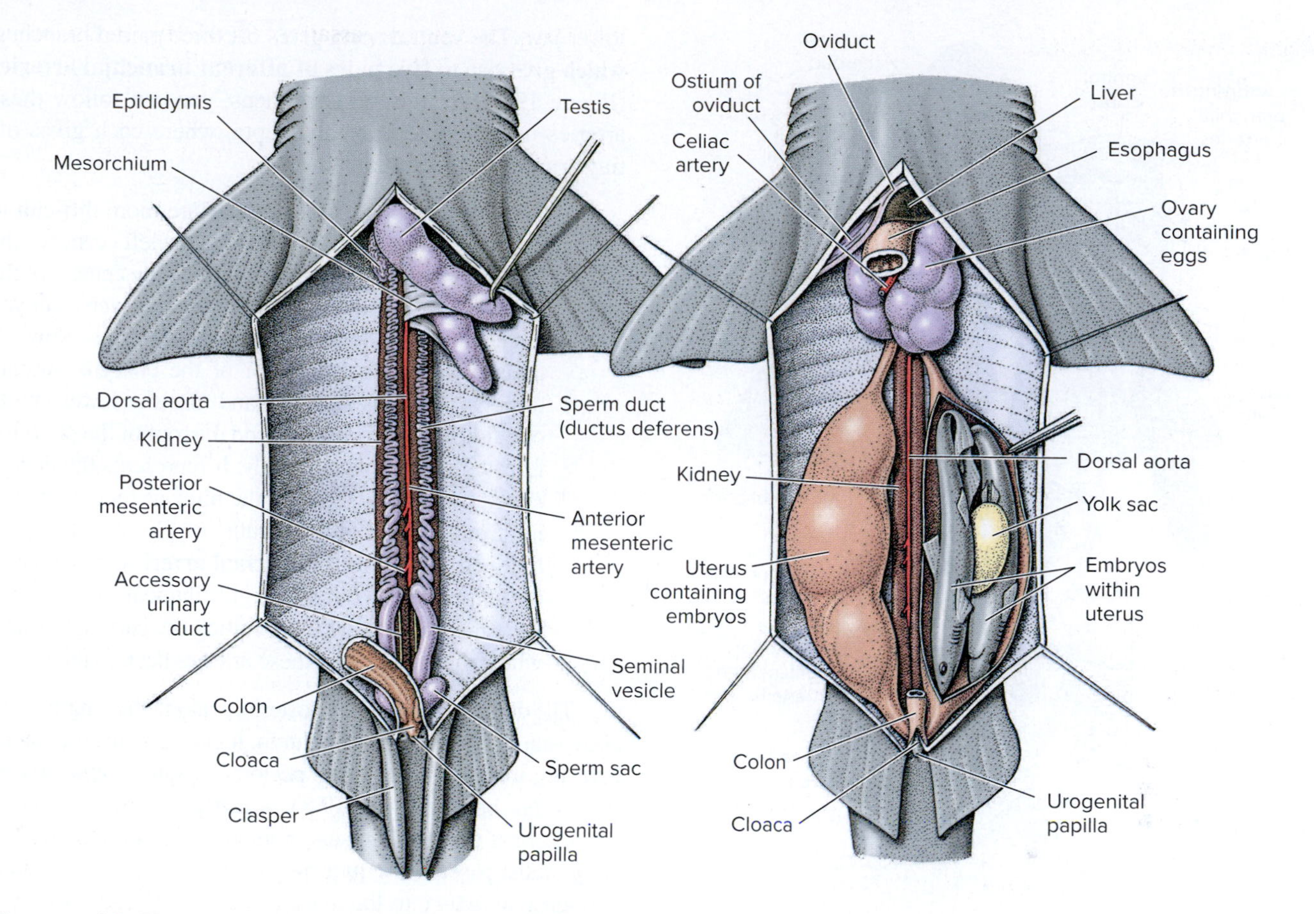

Figure 19-7

Urogenital system of a dogfish shark. **Left,** Male. **Right,** Female.

come in contact with the yolk sac of the embryo in a placenta-like manner. As mentioned earlier, spiny dogfishes are ovoviviparous because the embryo does not depend on placental nourishment. However, other sharks include some that are dependent on the mother for nourishment through the placental connection (**viviparous;** L. *vivus,* living, + *parere,* to bring forth) and still other sharks that lay shelled eggs containing a large amount of yolk (**oviparous;** L. *ovum,* egg, + *parere,* to bring forth). Gestation periods vary from 16 to 24 months, and the young at birth range from 12 to 30 cm in length.

Slender **kidneys** (opisthonephroi) extend the length of the dorsal abdominal wall, dorsal to the peritoneum. A very slender **Wolffian duct** embedded on the ventral surface of each kidney empties into the cloaca through a **urogenital papilla.**

Slit open the cloaca and identify the urogenital papilla, the entrance of the rectum, and, on the dorsal side, openings from the uteri.

Further Study

Circulatory System

Because sharks do not have bone marrow, the spleen is the location of red blood cell production. The spleen also filters blood much as does the lymph system of other animals. The basic plan of circulation in the shark is similar to that in the ammocoete larva.

Spread the ventral body wall to reveal the cavity containing the heart **(pericardial cavity).** Lift up the **heart** to see the thin-walled, triangular **sinus venosus.**

Blood passes from the sinus venosus to the **atrium,** which surrounds the dorsal side of the muscular **ventricle;** it then flows into the ventricle, which pumps it forward into the **conus arteriosus** (Figure 19-8). **Valves** prevent backflow between compartments.

Venous System

Slit open the sinus venosus transversely, extending the cut somewhat to the left; wash out its contents.

Look for openings into the sinus venosus of one of each of the following paired veins: (1) **common cardinal** (L. *cardinalis,* chief, principal) **veins (ducts of Cuvier)** (Figure 19-8), which extend laterally and into which empty large **anterior cardinal sinuses, posterior cardinal sinuses,** and **subclavian veins** (L. *sub,* under, + *clavus,* key); (2) **inferior jugular veins** (L. *jugulum,* collarbone) from the floor of the mouth and gill cavities (these veins are not shown in Figure 19-8); and

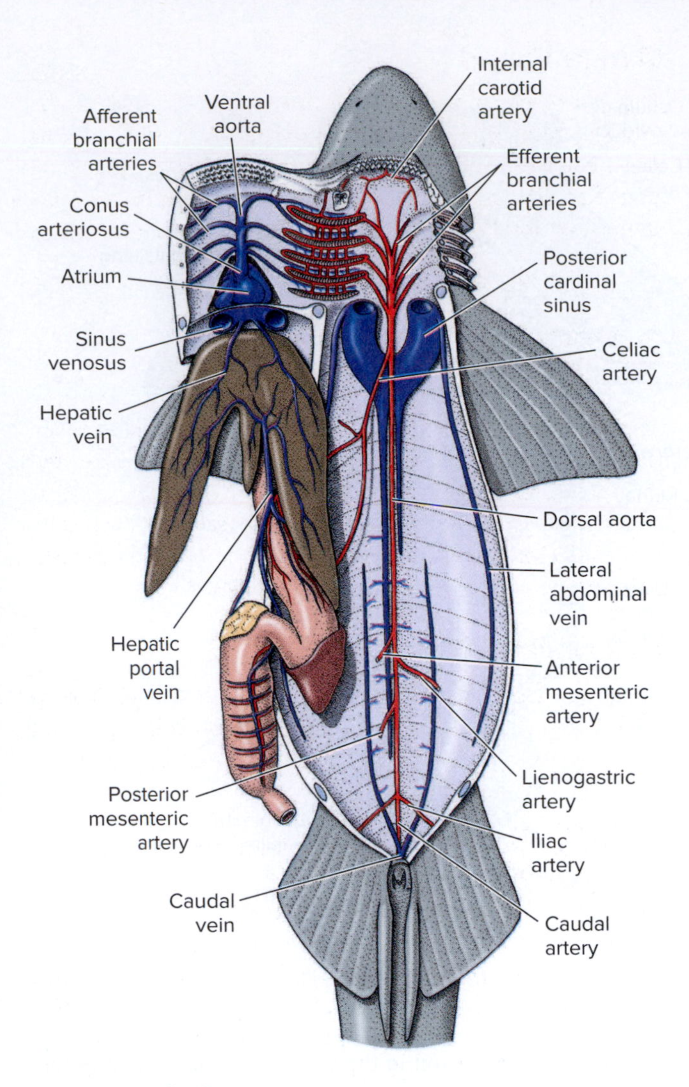

Figure 19-8
Circulatory system of a dogfish shark.

(3) **hepatic veins,** which empty near the middle of the posterior wall of the sinus venosus and bring blood from the liver.

The **hepatic portal vein** (Figure 19-8) gathers blood chiefly from the digestive system through a system of gastric, pancreatic, and intestinal veins. The hepatic portal vein enters the right lobe of the liver and divides into several small **portal veins** (trace some of these subdivisions); it then divides into a system of capillaries, from which some of the carbohydrates brought from the intestine may be stored in liver cells as glycogen (animal starch) until needed. Blood from capillaries flows into the **hepatic veins** and from there into the sinus venosus.

Renal portal veins arise from the **caudal vein** in the tail and carry blood to capillaries of the kidneys. Many small renal veins carry blood from the kidneys to the posterior cardinal sinuses and from there to the sinus venosus.

Arterial System. The arterial system includes (1) the afferent and efferent branchial arterial system and (2) the dorsal aorta and its branches.

From the conus arteriosus, trace the **ventral aorta** forward, removing most of the muscular tissue from the lower jaw. The ventral aorta gives off three paired branches, which give rise to **five pairs of afferent branchial arteries** (Figure 19-8). In injected specimens, you can follow these arteries into the interbranchial septa, where each gives off tiny arteries to the gill lamellae.

The **efferent branchial arteries** are more difficult to dissect. With scissors, cut through the left corner of the shark's mouth and backward through the centers of the left gill slits, continuing as far as the transverse cut you made earlier at the base of the pectoral fins. Now cut transversely across the floor of the pharynx straight through the sinus venosus, and turn the lower jaw to one side to expose gill slits and the roof of the pharynx. Locate the spiracle internally. It represents the degenerated first gill slit. Dissect the mucous membrane lining from the roof of the mouth and pharynx to expose four pairs of **efferent branchial arteries,** which carry oxygen-rich blood from the gill filaments and unite to form the **dorsal aorta.** By cutting the cartilages under which they pass, trace these arteries back to the gills.

The dorsal aorta extends posteriorly along the length of the body ventral to the vertebral column. It gives rise to **subclavian arteries,** which connect to the pectoral region; a **celiac artery** (Gr. *koilia,* belly) (Figure 19-8), which gives off branches to the intestinal tract and gonads; numerous **parietal arteries** to body walls; **mesenteric arteries** to the intestine and rectum; **lienogastric artery** to the intestine and the rectum; **lienogastric artery** to the intestine and the rectum; **renal arteries** to the kidneys; and **iliac arteries** (L. *ilia,* flanks) to the pelvic fins. The aorta continues to the tip of the tail as the **caudal artery.**

Respiratory System
In sharks, water taken in through both mouth and spiracles is forced laterally through five pairs of gills and leaves through five pairs of external gill slits (some elasmobranchs have a different number of gills).

On the shark's right (intact) side, separate the gill units by cutting dorsally and ventrally from the corners of each gill slit. Now you can examine the structure of the intact gills on this side and observe the gills in cross section on the other side.

The area between the **external gill slits** and **internal gill slits** comprises the **gill chambers** (**gill pouches** and **branchial chambers**). The incomplete rings of heavy cartilage supporting the gills and protecting the afferent and efferent branchial arteries are called **gill arches.** Short, spikelike projections extending medially from the gill arches are **gill rakers.** What might be the function of gill rakers? __________ Cartilaginous **gill rays** fan out laterally from the gill arches to support the gill tissues.

Remove an intact half of a gill arch, along with its gill tissue. Examine with a hand lens. Float a small piece of gill in water and examine with a dissecting microscope.

Primary lamellae (gill filaments) are small, platelike sheets of epithelial folds arranged in rows along the lateral face of each gill. Viewing with a microscope reveals that the primary lamellae are seen to be made up of rows of tiny plates, called **secondary lamellae,** which are the actual sites of gas exchange. Blood capillaries in secondary lamellae are arranged to carry blood inward, or in the opposite direction of seawater, which is flowing outward. This countercurrent flow encourages gas exchange between the blood and water. Gill lamellae are arranged in half-gills, or **demibranchs,** on each side of the branchial arch. The two demibranchs together form the gill unit, or **holobranch.** The spiracles are believed to be remnants of the first gill slit. They are usually larger in slow-moving, bottom-dwelling sharks than in fast-swimming sharks, in which, because of their motion, there is a more massive flow of water through the mouth.

Oral Report

Be able to identify both external and internal features of the shark and give the functions of each organ or structure.

Be able to trace the flow of blood from the heart to any part of the body (such as the pectoral region, the kidneys, and the tail) and back to the heart.

Review Your Knowledge

Test your understanding of the main features of the cartilaginous fishes covered in the exercise by filling out the summary table about vertebrata on p. xii. Try to complete the table without looking at your notes.

Demonstrations

1. *Dogfish uterus with developing pups or dogfish embryos with attached yolk sac*
2. *Various sharks, skates, and rays*
3. *Preparation of the skull and/or skeleton of a shark*
4. *Corrosion preparation of the arterial system*
5. *Shark teeth*
6. *Microslides of shark skin*

EXERCISE 19C

Class Actinopterygii—Bony Fishes

Core Study

Perca, Yellow Perch

Subphylum Vertebrata
Class Actinopterygii
Superorder Teleostei
Order Perciformes
Genus *Perca*
Species *Perca flavescens*

Where Found

The yellow perch is a common freshwater fish widely distributed through the lakes of the American Midwest, the American Northeast, and parts of Canada. They are a popular sport fish and are taken by both commercial and recreational anglers. A closely related species is found in Europe and Asia.

Characteristics

Actinopterygii (bony fishes) is the largest group of vertebrates, both in number of species (more than 28,000) and in number of individuals. By adaptive radiation, they have developed an amazing variety of forms and structures. They flourish in freshwater and seawater and in both deep and shallow water. Their chief characteristics usually are **dermal scales, operculum over the gill chamber** of each side, **bony skeleton, terminal mouth, swim bladder, homocercal tail,** and both **median** and **paired fins.**

External Structure

Obtain a preserved fish and, after you have studied its external anatomy, compare it with living fishes in the aquarium. What can their structure tell you of their living habits? __________

The body of the perch is fusiform, or torpedo-shaped. Is it compressed in any of its planes? __________ Identify the **head,** which extends to the posterior edge of the **operculum; trunk,** which extends to the anus; and **tail** (Figure 19-9). Identify the **pectoral, pelvic, anal, dorsal,** and **caudal fins.** How many of each are there? __________ Which of the fins are paired? __________ Note the **fin rays,** which support the thin membrane of each fin. Some of the rays are soft and some are spiny. How would you expect these spiny rays to look when the fish is threatened by a predator? __________ The caudal fin is **homocercal** (Gr. *homos,* same, + *kerkos,* tail), meaning that the upper and lower halves are equal.

The terminal **mouth** is adapted for overtaking prey while swimming. Fishes with superior mouths (those facing upward) are usually surface feeders, whereas those with inferior mouths (facing downward) are usually bottom feeders. Do the **eyes** have lids? __________ Could the perch have binocular vision? Why? __________ On each side in front of the eye, a pair of **nostrils** open into an olfactory sac. Water enters the sac through the anterior aperture, which is provided with a flaplike valve, and leaves through the posterior aperture. The **ears** are located behind the eyes, but they are not visible externally. A **lateral line** along the side of the body is a row of small pores or tubules connecting with a long, tubular canal bearing sensory organs. These are sensitive to pressure and temperature changes and are responsive to water currents. Many microscopic sense organs are found in the skin.

Lift a gill cover, or **operculum** (L., cover), and study its structure. Along the ventral margin of the operculum, find a membrane supported by bony rays. This membrane fits snugly against the body to close the branchial cavity during

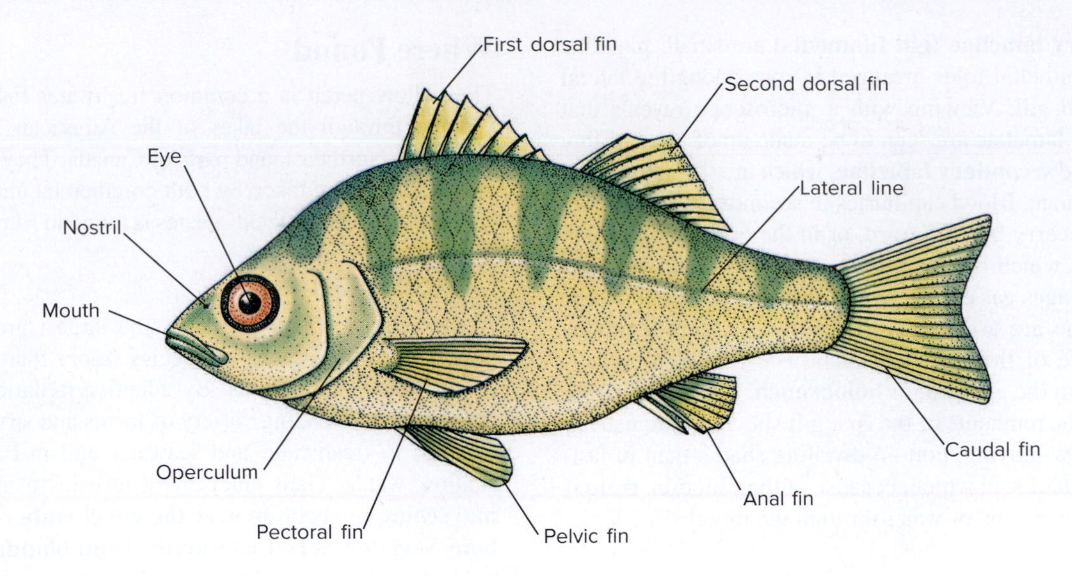

Figure 19-9
Yellow perch, external features.

certain respiratory movements. With your probe, examine the **gills** beneath the operculum.

Find the **anus** near the base of the anal fin and the small, slitlike **urogenital opening** just posterior to the anus.

Note the arrangement of the **scales.**

Remove a scale from the lateral line region, mount in water on a slide, and examine with low power.

The anterior, or embedded, side of the perch scale has radiating grooves. The posterior, or free, edge has very fine teeth. These are **ctenoid scales** (ten′oid; Gr. *kteis, ktenos,* comb). Note the fine, concentric lines of growth. The scales are covered with a very thin epidermis, which secretes mucus over the scales. This reduces friction in swimming and makes capture by a predator more difficult. Ctenoid scales are usually found on fishes with spiny rays in the fins, whereas soft-rayed fishes usually have cycloid scales, which lack marginal teeth.

Skeletal System

Examine the mounted specimen of a perch.

The bony skeleton of the perch consists of an **axial skeleton** (which includes the bones of the skull, vertebral column, ribs, and medial fins) and an **appendicular skeleton** (which includes the pectoral girdle and fins and pelvic girdle and fins; Figure 19-10). Examine the mounted perch skeletons on display, and compare them with skeletons of other fishes and of amphibians, birds, and mammals. Do you see any basic similarity?

Vertebral Column. The vertebral column typically has one vertebra per body segment. Like amphioxus, fish trunk muscles are arranged as zigzagging myotomes. The alternating side-to-side contractions of these muscles is what produces the undulating body movement that propels a fish through water. The vertebral column of bony fishes is well suited to withstand the forces placed on it during the contraction of these trunk muscles during swimming. Fishes have a pair of **ribs** for every vertebra; they serve as stiffening elements in the connective tissue septa that separate the muscle segments and thus improve the effectiveness of muscle contractions. Many fishes have both dorsal and ventral ribs, and some have numerous, riblike intermuscular bones as well.

Skull. The skull of bony fishes is highly variable. Bones that may be present in one group of bony fishes may be completely absent in another group. It is convenient to divide the skull into the bones related to the jaws, the **neurocranium,** and the **operculum.** Jaw bones have undergone significant changes in bony fishes to allow the diversity of feeding types. The upper jaw now consists of both the **maxilla** and the **premaxilla.** The lower teeth-bearing jaw bone is the **dentary.** The neurocranium is the most solid part of the skull and protects the brain. This portion of the skull is made up of many bones, and the exact number varies from one type of bony fish to another. The operculum bones are fairly consistent in form and function across bony fishes, and they protect the gills.

Appendicular Skeleton. This part of the bony fish skeleton consists primarily of support for the various fins. The **dorsal fins** consist of **fin rays** and **supports for the fin rays.** Depending on the species of bony fish, it may have both an anterior and a posterior dorsal fin or just one dorsal fin. The **caudal fin** and the **anal fin** have similar bony supports. Most prominent on the appendicular skeleton are the **pelvic girdle** and the **pectoral girdle.** The pectoral girdle consists of a series of small bones that support the pectoral

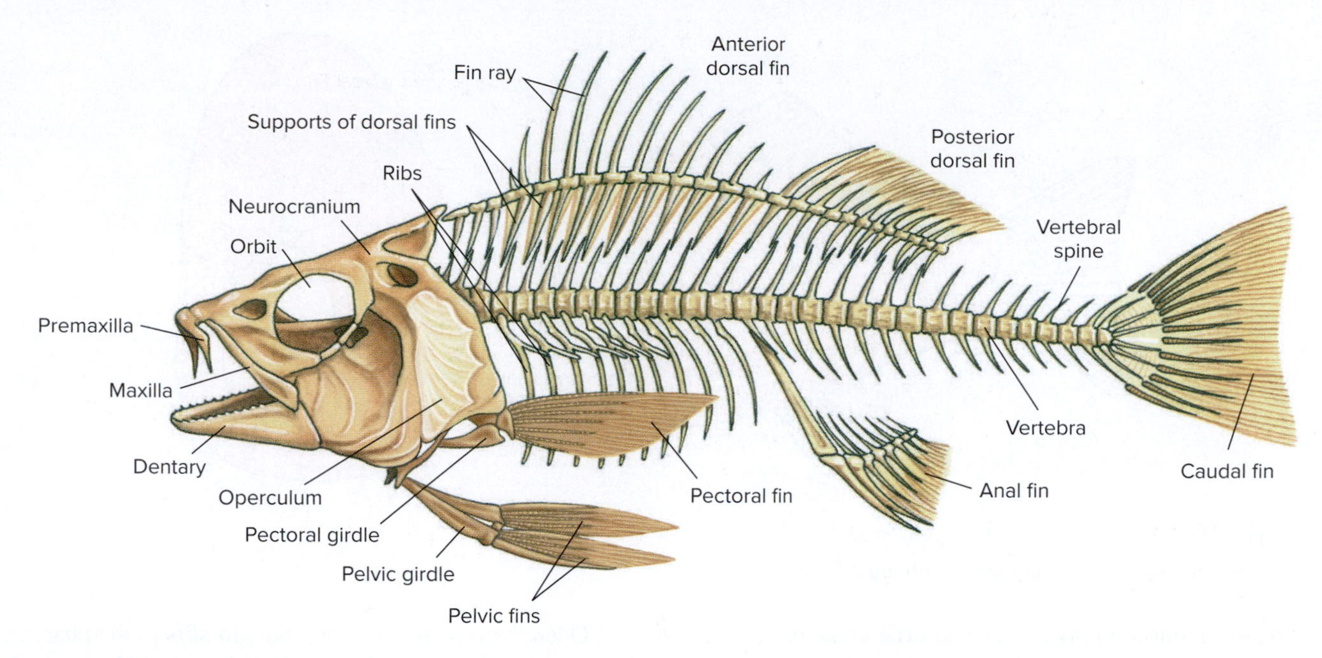

Figure 19-10
Skeleton of a perch.

fins and allow each fin to articulate independently. The pelvic girdle is simpler and usually consists of a bone on each side. In some bony fishes, the pelvic girdle is connected to the pectoral girdle. What do you think are the primary functions of each fin as they relate to fish swimming? Caudal fin? ____________ Dorsal fins? ____________ Anal fin? ____________ Pectoral fins? ____________ Pelvic fins? ____________

Muscular System

Although the muscles of a perch are less complex than those of land vertebrates, they make up a much larger mass in relation to body size. Tetrapod locomotion results largely from direct action of muscles on bones of the limbs, but fish locomotion results from indirect action of the segmental muscles—**myomeres**—on the vertebral column, a method by which a large muscle mass produces a relatively small amount of action. This type of movement is efficient in a water medium, but it would be less effective on land. Myomeres (derived from embryonic myotomes) consist of blocks of longitudinal muscle fibers placed on each side of a central axis, the vertebral column. Their contraction, therefore, bends the body, and the action passes in waves down the body, alternating on each side.

After cutting off the sharp dorsal and ventral spines, skin one side of the body and note the shape of the myomeres.

They resemble Ws that are turned on their sides and stacked together. A horizontal septum of connective tissue divides the muscles into dorsal **epaxial muscles** (Gr. *epi,* upon, + *axis,* axle, meaning above the axis, or vertebral column) and ventral **hypaxial muscles** (Gr. *hypo,* under, + *axis,* axle, meaning below the vertebral column) (Figure 19-11). Posteriorly both epaxial and hypaxial muscles are active in locomotion, but anteriorly hypaxial muscles serve more for support of body viscera than for locomotion. Try to separate the myomeres. Observe the direction of the muscle fibers. Do they run zigzag, as the myomeres seem to? Or are they all directed horizontally—or vertically?

Now watch the swimming motions of fishes in the aquarium and visualize the use of the body muscles in locomotion. What part do fins play in locomotion?

Dissection of individual muscles in the fish is difficult and will not be attempted here. Muscles operating the jaws, opercula, and fins are often named according to their function and, as in other vertebrates, include adductors, abductors, dilators, levators, and so on (see Exercise 23).

Mouth Cavity, Pharynx, and Respiratory System

Before starting the dissection, it is good, if you have not already done so, to cut off the sharp dorsal and ventral fins to protect the hands.

Cut away the operculum from the left side, exposing the gill-bearing bars, or arches. How many arches are there? ________________________________

Cut one gill arch, place in water, and examine with a hand lens or dissecting microscope.

If injected, branchial arteries will be colored. Note **gill filaments** borne on the posterior, or aboral, side of the arch (Figure 19-12A). Are the filaments arranged in a single or double row? __________ It is in these filaments, containing capillaries from the branchial arteries, that exchange of gases takes place. **Gill rakers** on the oral surface of each

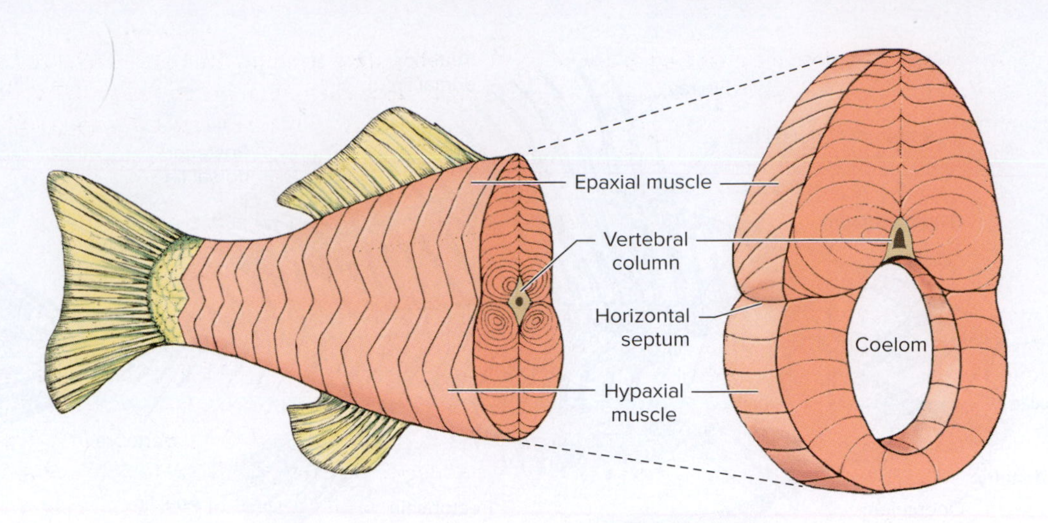

Figure 19-11
Diagram of the skeletal musculature of a teleost fish.

gill bar strain out food organisms and offer some protection to the gill filaments from food passing through the pharynx.

Cut through the angle of the left jaw, continuing the cut through the middle of the left gill arches to expose the mouth cavity and pharynx.

Open the mouth wide and note **gill slits** in the pharynx. In the mouth, locate the fine **teeth.** What would be the main function of these teeth? __________ Just behind the teeth, across the front of both the upper and the lower jaw, find the **oral valves.** These are transverse membranes that prevent the outflow of water during respiration. An inflexible **tongue** is

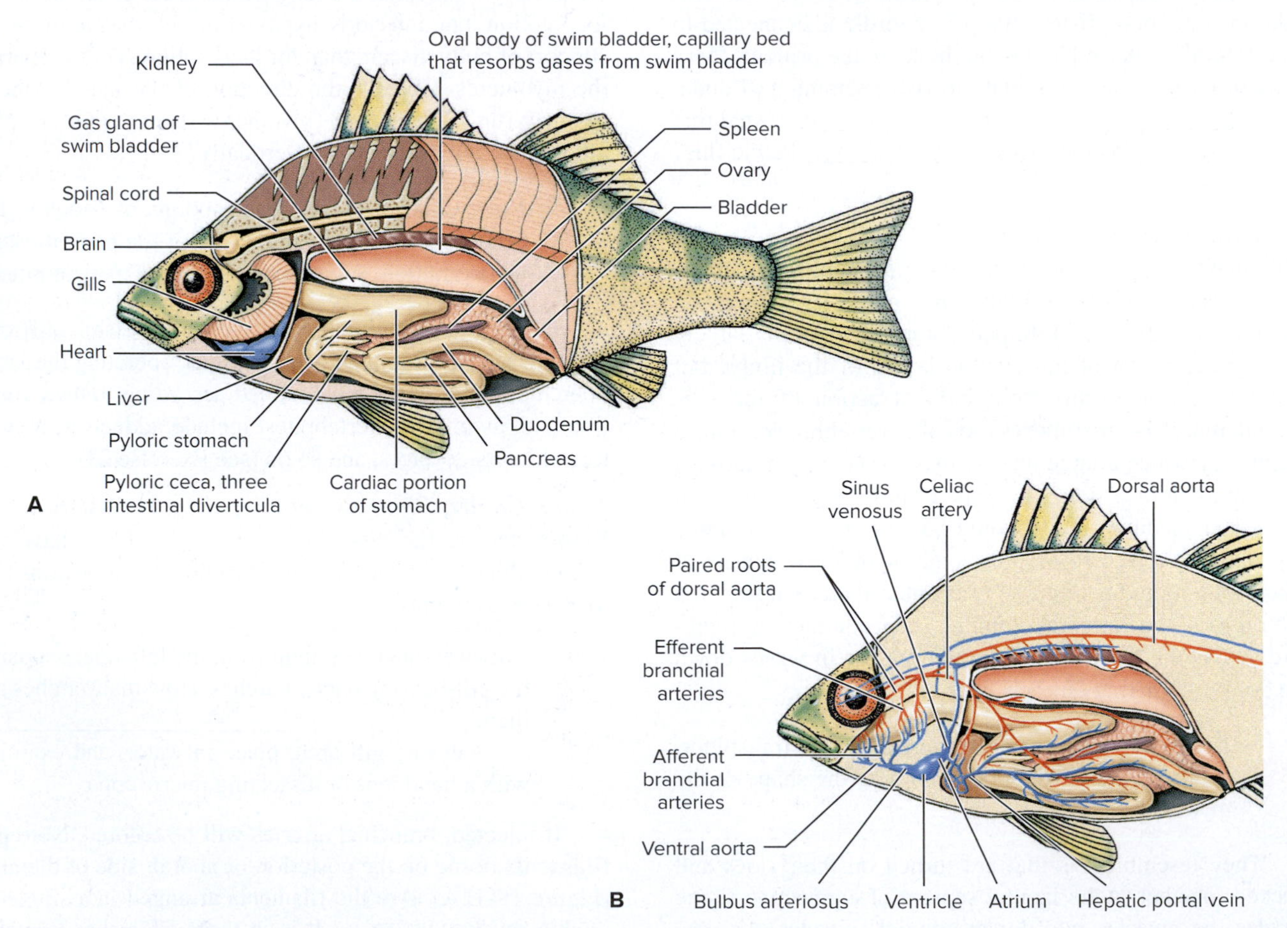

Figure 19-12
Yellow perch. **A,** Internal anatomy. **B,** Principal vessels of the circulatory system.

supported by the hyoid bone. Explore the **spacious** pharynx, noting the size and arrangement of gill bars and gill slits. Gills separate the **oral cavity** from the **opercular cavity.**

Water movement across the gills involves the combined pumping action of both the oral and opercular cavities—a "double-pump" system. The volume of the **oral pump** (mouth cavity) can be changed by raising and lowering the jaw and floor of the mouth. The volume of the **opercular pump** (opercular cavity) can be enlarged and decreased by muscles that swing the operculum in and out. Valves guard the opercular clefts, preventing the backflow of water. The action of the two pumps creates a pressure differential, which maintains a smooth flow of water across the gills throughout nearly the entire breathing cycle.

Now watch fishes in the aquarium, observing their respiratory movements until you understand the sequence.

Water movement across the gills is actually much smoother and less pulsatile than it appears from watching a fish respire. The reason is that the pressure in the opercular cavity is maintained *lower* than the pressure in the mouth cavity for about 90% of the respiratory cycle, and this provides the pressure that drives water across the gills.

Further Study

Abdominal Cavity

Starting near the anus and being careful not to injure the internal organs, cut anteriorly on the midventral line to a region anterior to the pelvic fins. Now, on the animal's left body wall, make a transverse cut, extending dorsally from the anal region; make another cut dorsally between the pectoral and pelvic fins; then remove the left body wall by cutting between these two incisions. On the right side, make similar transverse cuts, so that the right wall can be laid back, but do not remove it.

You have now exposed the abdominal cavity. This, together with the **pericardial cavity,** which contains the heart, makes up the **coelomic cavity.** What is the shiny lining of the coelom called? __________

Probably the first organ you will see will be the **intestine,** encased in yellow fat. Carefully remove enough fat to trace the digestive tract anteriorly. Find the **stomach,** lying dorsal and somewhat to the left of the intestine. Anterior to the stomach is the **liver,** dark red in life but bleached to a cream color by preservative. The **spleen** (Figure 19-12A) is a dark, slender organ lying between the stomach and the intestine. The **gonads** are in the dorsoposterior part of the cavity. The **swim bladder** lies dorsal to these organs and to the peritoneal cavity. It is long and thin-walled. Do not injure its walls or any of the blood vessels lying among the viscera. **Kidneys** are located dorsal to the swim bladder and will be seen later.

Digestive System. Run your probe through the mouth and into the opening of the esophagus at the end of the pharynx. Now lift up the liver and trace the **esophagus** from the pharynx to the large **cardiac portion** of the stomach. This ends posteriorly as a blind pouch. The short **pyloric portion** of the stomach, opening off the side of the cardiac pouch, empties by way of a **pyloric valve** into the **duodenum,** the s-shaped proximal part of the intestine. Three intestinal diverticula, the **pyloric ceca,** open off the proximal end of the duodenum near the pyloric valve (Figure 19-12A). Follow the intestine to the **anus.** Note the supply of blood vessels in the **mesentery.** The **pancreas,** a rather indistinct organ, lies in the fold of the duodenum. The **liver** is large and lobed, with the **gallbladder** located under the right lobe. Bile from the liver is drained by tubules into the gallbladder, which in turn opens by several ducts into the duodenum posterior to the pyloric ceca.

Cut open the stomach and place its contents in a glass dish with water. Compare with your neighbor's findings. Examine the stomach lining. How is its surface increased? ______________ Cut open and examine the pyloric valve. Open a piece of the intestine, wash it, and examine the lining with a dissecting microscope. What type of muscle would you expect to find in the intestine? __________________

____

Swim Bladder. The swim bladder is a long, shiny, thin, but tough-walled sac that fills most of the body cavity dorsal to the visceral organs. In some fishes (not perch), it connects with the alimentary canal. Cut a slit in it and observe its internal structure. In its anterior ventral wall, look for the **gas gland,** which contains a network of capillaries, and the **rete mirabile** (rē′tē muh-rab′uh-lē; L., wonderful net), which assists in secretion of gases, especially oxygen, into the bladder at high pressure. Another capillary bed, the **oval,** lies in the dorsal body wall. Gases are resorbed from the swim bladder in this area. The swim bladder is a "buoyancy tank," or hydrostatic organ, that adjusts the specific gravity of fish to varying depths of water, so that the fish is always neutrally buoyant. Can you think of disadvantages to this type of buoyancy system? __________ What will happen to the volume of the bladder and to a perch's specific gravity if the perch swims upward from some depth toward the surface? __________ When you dissected the dogfish shark, did you find a swim bladder? __________ What does a shark use for buoyancy? __________

Reproductive System. Sexes are separate, but it is difficult to distinguish them externally.

The **ovary** is single and lies in back of the stomach, just below the swim bladder and dorsal to the intestine. Size of the ovary varies seasonally, being largest during the winter months before spawning. A prolongation of the ovary posteriorly serves as a sort of **oviduct** for carrying eggs to the **urogenital pore** just posterior to the anus (Figure 19-13A).

In males, two elongated **testes** are attached to the swim bladder by mesenteries. They become greatly enlarged

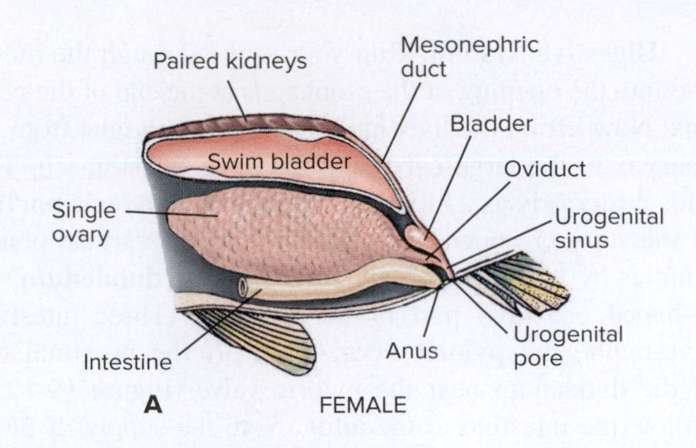

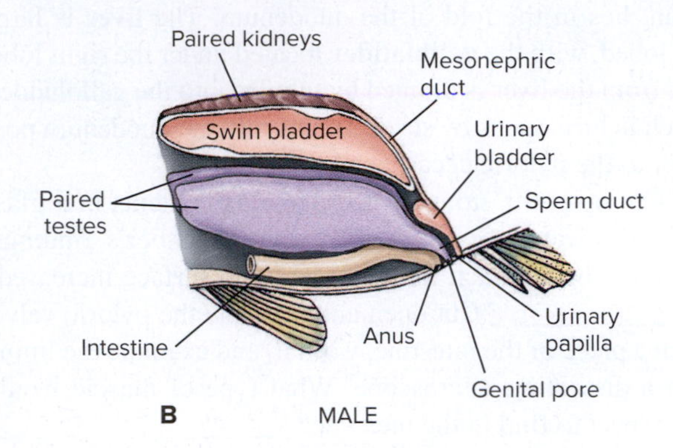

Figure 19-13
Urogenital system. **A**, Female. **B**, Male.

before spawning and are usually smallest during the summer months. A **sperm duct (vas deferens)** runs along a longitudinal fold in each testis adjacent to the spermatic artery. The two ducts join in the posterior midline and extend to the **genital pore** just posterior to the anus (Figure 19-13B).

Excretory System. The **kidneys (mesonephroi)** are paired masses that lie against the dorsal body wall and extend the whole length of the abdomen above the swim bladder. They are often fused posteriorly, but the anterior parts are usually separated by the dorsal aorta. The anterior ends consist largely of blood sinuses and have lost their renal function. In the posterior end, they follow the body wall ventrally. Here the posterior ends of the **mesonephric ducts** (Wolffian ducts) may be seen extending the short distance from the kidneys to a small **urinary bladder,** which lies posteriorly between the gonad and the swim bladder. In females the urinary bladder joins the oviduct to form a **urogenital sinus,** emptying through the **urogenital pore.** In males the bladder empties separately through a **urinary pore,** around which there may be a small, external projection of the bladder called the **urinary papilla.** The male urinary and genital pores lie close together posterior to the anus.

Circulatory System

Extend the midventral incision to the jaw to expose the heart. Enlarge the opening by removing a triangular piece of body wall on each side of the cut.

Heart. The **pericardial cavity** is separated from the abdominal cavity by a **transverse septum.** The septum is not homologous to the diaphragm of mammals.

The fish heart is often referred to as a two-chambered heart—a thin-walled **atrium** and a muscular **ventricle**—but it actually consists of four chambers in series (see Figure 19-12B). The first chamber is the **sinus venosus,** the thin-walled sac adjoining the atrium posteriorly. The sinus venosus serves as a receiving chamber for venous blood returning from the body, and it helps assure a smooth flow of blood into the second chamber in the series, the atrium. Blood is pumped from the atrium to the ventricle, which, with each contraction (called the **systolic** period of the heartbeat), ejects blood into the final chamber, the **bulbus arteriosus.** When the ventricle contracts, blood pressure rises and is transmitted to the bulbus arteriosus. Then, as the ventricle relaxes (the **diastolic** period of the heartbeat), the high pressure persists in the bulbus arteriosus and maintains an even flow of blood into the ventral aorta. Valves between the bulbus and the ventricle prevent backflow of the blood during the diastolic period.

Arterial System. From the bulbus arteriosus, blood flows into the short **ventral aorta** (see Figure 19-12B).

Remove the operculum and trace the aorta forward.

The ventral aorta gives off four pairs of **afferent branchial arteries** to the branchial arches. From capillaries in the gills, the oxygenated blood is collected by **efferent branchial arteries** and is emptied into two roots of the **dorsal aorta** above. These roots join immediately to form the dorsal aorta.

The dorsal aorta is the major distributing vessel of fish circulation. Branches of the dorsal aorta supply arterial blood to the head, body musculature, swim bladder, and internal organs, such as gut, liver, gonads, and kidney. Note that fish circulation is a **single-circuit** system. All blood leaving the heart passes through at least two sets of capillaries: capillaries of the gills and of the body organs. With the evolution of lungs in tetrapod vertebrates, blood circulation became drastically altered into a **double-circuit** system, with separate pulmonary and systemic circuits.

Examine fish blood on a slide; note oval **erythrocytes** and various types of **leukocytes.** Do the red cells have nuclei? __________

Oral Report

Be able to identify the organs and external structures of a perch and give their functions.

Make a comparison between various body systems of a perch and those of an ammocoete larva. Compare with those of a dogfish shark.

EXPERIMENTING IN ZOOLOGY

Agonistic Behavior in Paradise Fish, Macropodus opercularis

Many organisms compete for resources that are limited in the environment. These resources can be food, shelter, or mates. Males of many species of animals compete for females. Competition for females often leads to agonistic encounters between conspecific males, and these encounters can escalate into fights that involve physical contact. Thus, natural selection has favored agonistic behavior in males of some species, so much that their aggression is innate (instinctive). Males in these species interact aggressively even when females are not present. Scientists who have observed agonistic interactions between males have suggested that the intensity of encounters can depend on how closely matched the males are in size and strength. Small males can become injured when fighting larger males and may choose not to fight. However, when males are closely matched in size and strength, agonistic encounters tend to be intense as each male attempts to establish its dominance and ultimately its access to the resource (females, in this case).

Paradise fish, *Macropodus opercularis* (Figure 19-14), are tropical fish from Southeast Asia. They have organs in their mouth that allow them to breathe air as well as use their gills. Males are more colorful than females and have red and blue bars running vertically down their body. Several paradise fish can be kept together in a single tank; however, when two males are isolated from the other fish, they begin to display agonistic interactions.

Getting Ready

Work with a partner for this exercise. Try to work in a room that can be darkened. You should have access to a fish tank (approximately 10 to 40 liters) that has an overhead tank light. If an overhead light is not available, illuminate the tank from the side with a small desk lamp. Turning off the room light will minimize the ability of the fish to see you and will minimize their distractions. Have a stopwatch available for timing fish behavior during your experiment.

How to Proceed

From a communal tank, select two male paradise fish that appear similar in size and other physical traits. Make sure to add the male fish to the experimental tank at exactly the same time. For the next 15 minutes, quantify the interactions between the males. The easiest way to measure aggressive interactions is to record the amount of time the two males stay very close to each other. You might elect to record total time that the fish spend within one body length of one another. If the two males display aggression to each other, you will notice several different types of behaviors. For example, males might chase each other, raise their dorsal fins, swim parallel very closely together, or even come side to side and vibrate, thereby sending mechanical waves

Figure 19-14
Two male paradise fish interacting.

toward each other. If you decide to use categories of behaviors as well as total time spent near each other, you might want to consider observing two "practice" males before starting your experiment, so that you can become familiar enough with the various behaviors to be able to count them during the experiment.

After recording your observations and data from the two males that are closely matched in size, select two new males. This time, however, choose two males that differ conspicuously in size (but make sure you have two males and are not placing a male and female together). Repeat the experiment, keeping track of time the males spend together and the number of agonistic behaviors displayed.

Compile data from the entire class and calculate overall mean times males spend together in closely matched contests versus contests in which males differ in size.

Do males that are closely matched in size interact more than males that differ in size? __________ Do the closely matched males have encounters that escalate into more aggressive behaviors (e.g., biting, nipping) as compared to males that differ in size? __________

Questions for Independent Investigations

1. Keeping the fish matched in size, how might light levels affect agonistic interactions? __________
2. Scientists who study fish aggression have found that individuals that are tank residents are more aggressive than fish that have been recently added to a tank. How could you test this with paradise fish? __________
3. Many kinds of fish have chemical alarm pheromones that are released when a fish is injured. Gently remove a couple of scales from a paradise fish and rinse its body with filtered water. This rinse water may now contain alarm pheromones. Add these chemicals to a tank with two interacting males. How might these chemicals affect the aggression levels of the males? __________ Why? __________
4. What might happen to agonistic interactions if paradise fish are exposed to a potential predator? __________

References

Csanyi, V., J. Haller, and A. Miklosi. 1995. The influence of opponent-related and outcome-related memory on repeated aggressive encounters in paradise fish. Bio. Bull. **188:**83–86.

Francis, R. C. 1983. Experiential effects on agonistic behavior in the paradise fish, *Macropodus opercularis.* Behaviour **85:**292–313.

Gerlai, R. 1993. Can paradise fish recognize a natural predator? An ethological analysis. Ethology **94:**127–136.

Haung, W. B., and F. L. Cheng. 2006. Effects of temperature and floating materials on breeding by the paradise fish (*Macropodus opercularis*) in the non-reproductive season. Zool. Studies **45:**475–482.

Hotta, T., T. Takeyama, D. Heg, S. Awata, L.A. Jordan, and M. Kohda. 2015. The use of multiple sources of social information in contest behavior: testing the social cognitive abilities of a cichlid fish. Front. Ecol. Evol. **3**:85. doi:10.3389/fevo.2015.00085.

Jakobsson, S., O. Brick, and C. Kullberg. 1995. Escalated fighting behaviour incurs increased predation risk. Animal Behav. **49:**235–249.

EXPERIMENTING IN ZOOLOGY

Analysis of the Multiple Hemoglobin System in Carassius auratus, *Common Goldfish*

Carassius, Goldfish

Subphylum Vertebrata
Class Actinopterygii
Order Cypriniformes
Family Cyprinidae (carps and minnows)
Genus *Carassius*
Species *Carassius auratus*

Cells require oxygen to utilize aerobic pathways in the production of ATP. For large, multicellular animals, diffusion of oxygen from the environment into the animal is insufficient to supply their metabolic needs. This problem led to the development of complex oxygen exchange surfaces and organs in animals and to the development of oxygen transport molecules within circulatory systems. In vertebrates most oxygen is transported in the blood bound to hemoglobin within red blood cells. Oxygen is picked up by hemoglobin in the gills or lungs at the exchange surfaces and delivered to tissues by circulation of the blood. The hemoglobin molecule in vertebrates consists of four noncovalently linked subunits that form a tetrameric molecule. Typically this tetramer is assembled from two α (alpha) polypeptide chains and two β (beta) polypeptide chains. Each polypeptide subunit of hemoglobin binds a heme group containing a ferrous iron responsible for binding the oxygen molecule (Figure 19-15).

Oxygen availability is relatively constant in the terrestrial environment; however, in the aquatic environment, oxygen availability is highly variable because oxygen solubility in water is dependent upon temperature, ionic concentration, pH, and amount of biomatter consuming or producing oxygen. Cyprinid fishes, which include carp, minnows, and goldfish, often inhabit shallow ponds in which temperature may vary by as much as 10° or 20° C each day. Typically, as the temperature of water increases, oxygen solubility decreases. A change in water temperature from 0° C to 40° C drops dissolved oxygen to about one-half. An additional confounding factor is that animal tissue metabolism increases as temperature increases.

The decrease in oxygen availability, combined with increased oxygen demand as water temperature increases, poses a problem for maintaining aerobic respiration in fishes. One adaptation fish have evolved to deal with variable oxygen availability is to produce multiple forms of hemoglobin. These multiple forms of hemoglobin in fishes may be distinguished electrophoretically (di Prisco and Tamburrini, 1992; Riggs, 1970) and may differ in physiological properties. The differences in physiological properties may allow for enhanced oxygen transport when oxygen availability is low in the environment. Trout are known to have up to nine distinctly different hemoglobin forms (I–IX), whereas goldfish exhibit three different hemoglobin forms, called **G1, G2**, and **G3**. These three forms in goldfish vary in concentration as temperature varies, presumably brought about by rapid changes in aggregation of different subunits (Houston and Cyr, 1974; Houston and Rupert, 1976; Houston et al., 1976).

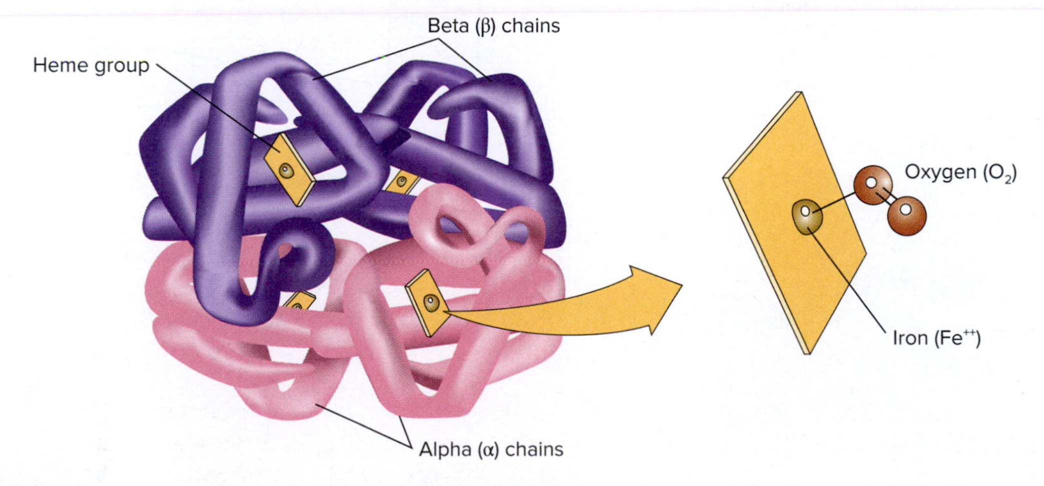

Figure 19-15
Human hemoglobin molecule, showing the four polypeptide chains (two α and two β), each associated with a heme group to which an oxygen molecule will bind.

How to Proceed*

We can visualize the multiple hemoglobin systems found in common goldfish using simple native polyacrylamide gel electrophoresis. In this procedure, we will remove red blood cells from goldfish that have been maintained in the laboratory at about 20° C to 25° C using a heparinized syringe. (Heparin is a compound that prevents coagulation of blood removed from the circulatory system.) The cells will be washed in isotonic saline (0.9% NaCl) to remove serum proteins and lysed in distilled water to release the hemoglobins. (Distilled water is hypotonic to the red blood cells; therefore, the cells will swell and burst, releasing hemoglobin and other

*Additional suggestions for implementation of this exercise are found in Appendix A, pp. 387–419.

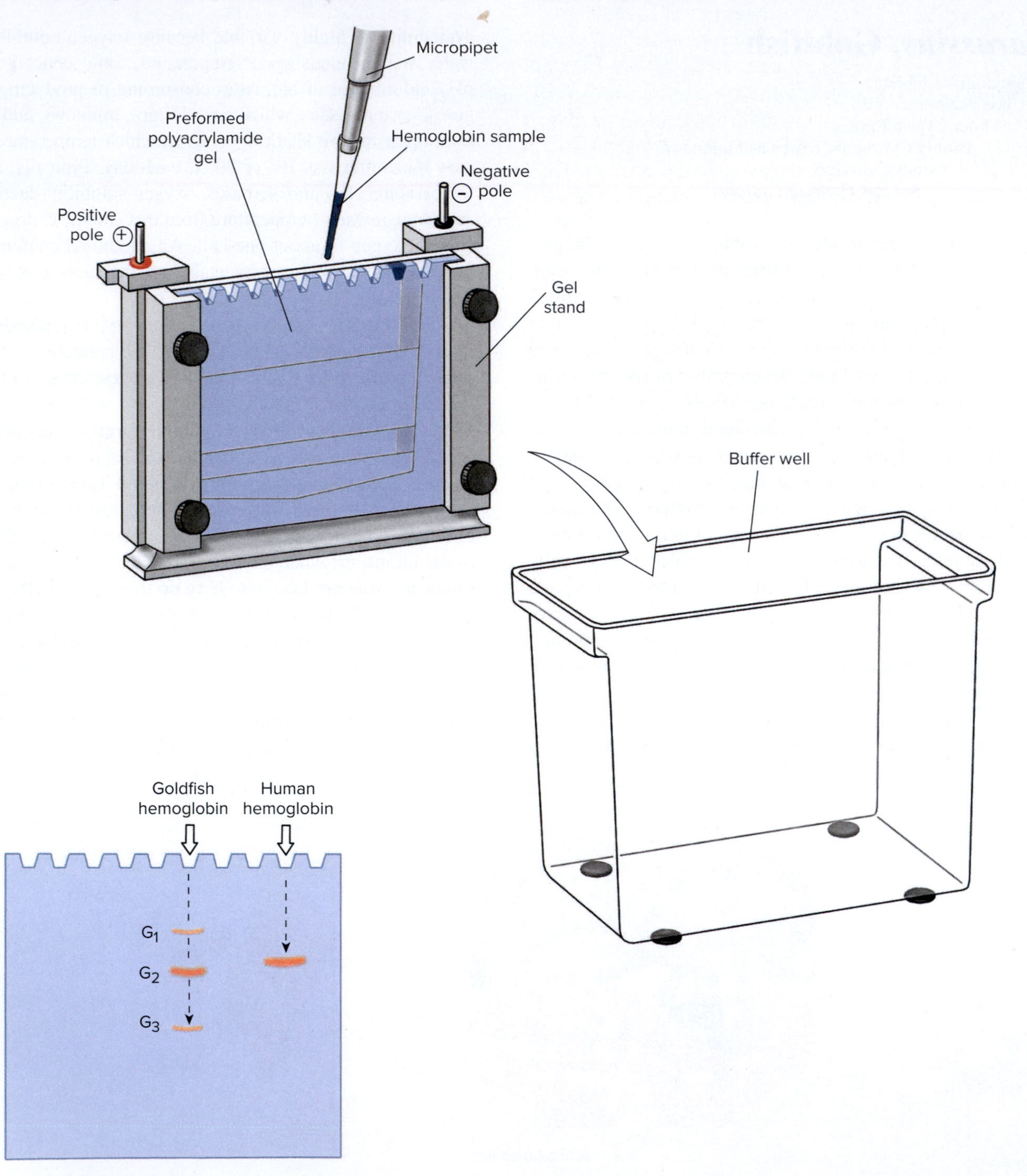

Figure 19-16
Diagram outlining the preparation, loading, and electrophoresis of a vertical polyacrylamide gel to separate the isoforms (G_1, G_2, and G_3) of hemoglobin from goldfish.

cellular components.) The hemoglobin lysate will be mixed 1:1 with a loading buffer and a 20 µl sample applied to a well on an 8% native polyacrylamide gel (Figure 19-16). A sample of human hemoglobin prepared in the same way as the goldfish hemoglobin will serve as a marker on the gel. Electrophoretic separation will proceed for 30 to 45 minutes at 80 to 90 volts. Following electrophoresis, you should be able to see three faint reddish-orange bands in the goldfish lane indicating the presence of the three isoforms of hemoglobin. Typically the G_2 band (middle band) predominates, whereas the G_1 (upper) and G_3 (lower) bands are fainter. Only one band will appear in the human hemoglobin lane. To assist in visualization of the hemoglobin bands, the gel will be removed from its case, stained with Coomassie Blue dye for a few minutes, and then destained to remove dye not bound to protein. A picture of the gel may be taken if appropriate equipment is available.

Questions for Thought

1. How might possession of multiple hemoglobin isoforms allow for adaptation to different environments? ______________________________
2. Why do you think trout have so many different forms of hemoglobin? ______________________________
3. Can you suggest any other species of fish that might show multiple forms of hemoglobin? ______________

References

de Souza, P. C., and G. O. Bonilla-Rodriguez. 2007. Fish hemoglobins. Brazilian J. Med. Biol. Research **40:**769–778.

di Prisco, G., and M. Tamburrini. 1992. The hemoglobins of marine and freshwater fish: the search for correlations with physiological adaptation. Comp. Biochem. Physiol. **102B:**661–671.

Houston, A. H., and D. Cyr. 1974. Thermoacclimatory variation in the haemoglobin systems of goldfish *(Carassius auratus),* and rainbow trout *(Salmo gairdneri).* J. Exp. Biol. **61:**455–461.

Houston, A. H., K. M. Mearow, and J. S. Smeda. 1976. Further observations upon the hemoglobin systems of thermally acclimated freshwater teleosts: pumpkinseed *(Lepomis gibbosus),* white sucker *(Catostomus commersoni),* carp *(Cyprinus carpio),* goldfish *(Carassius auratus),* and carp-goldfish hybrids. Comp. Biochem. Physiol. **54A:**267–273.

Houston, A. H., and R. Rupert. 1976. Immediate response of the hemoglobin system of the goldfish, *Carassius auratus,* to temperature change. Can. J. Zool. **54:**1737–1741.

Riggs, A. 1970. Properties of fish hemoglobins. In Hoar, W. S., and D. J. Randall, (eds.). Fish physiology. Vol. 4. New York, Academic Press.

NOTES

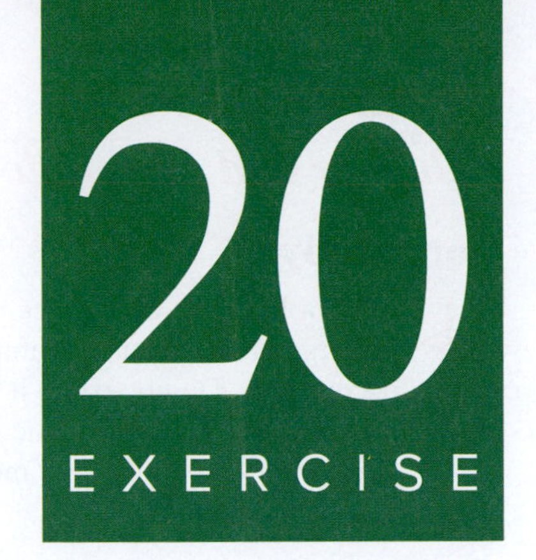

The Amphibians

Frogs

Amphibians are a transition group between aquatic and strictly land vertebrates. They have the soft, moist epidermis of aquatic forms and therefore cannot stray far from water or moist surroundings. Amphibian eggs lack the tough protective shell and specialized extraembryonic membranes characteristic of eggs of terrestrial vertebrates and so remain adapted to an aquatic habitat. Many amphibians have developed lungs for air breathing but still have aquatic larvae with external gills. A moist skin serves also as a respiratory organ. Evolution of lungs has brought a change in circulation that includes a pulmonary circuit as well as a systemic circuit. Amphibians have a three-chambered heart (two atria and one ventricle). They are usually four-limbed for walking or jumping but many have webbed feet for swimming.

Frogs and toads belong to order Anura (Gr. *an,* without, + *oura,* tail), the largest and most diverse group of living amphibians. Anurans differ from salamanders (order Caudata) and tropical caecilians (order Gymnophiona) in several distinctive ways that are associated with a specialized jumping mode of locomotion. A frog's body is extremely shortened and virtually fused with the head, an adaptation that provides rigidity to the skeletal framework. It lacks true ribs. Caudal vertebrae (which would normally form the tail) are fused into a single, pillarlike bone, the urostyle. The hindlegs are much larger and more powerful than the forelegs. Anurans are the most diverse of the amphibians and are commonly used in the general zoology laboratory because of their availability. The large bullfrog, *Lithobates*[1] *catesbeiana,* and the small leopard frog, *Lithobates pipiens,* are commonly used for dissection.

Frogs

EXERCISE 20A
Behavior and Adaptations

EXERCISE 20B
Skeleton
Shaping the Body for Life on Land
Axial Skeleton
Appendicular Skeleton

EXERCISE 20C
Skeletal Muscles
Directions for Study of Frog Muscles

EXERCISE 20D
Digestive, Respiratory, and Urogenital Systems
Mouthparts
Dissection of a Frog

EXERCISE 20E
Circulatory System
Venous System
Arterial System
Heart

EXERCISE 20F
Nervous System
Brain and Spinal Cord
Spinal Nerves

Frogs

Phylum Chordata
Subphylum Vertebrata
Class Amphibia
Order Anura (= Salientia)
Family Ranidae
Genus *Lithobates*

Where Found

Ranid frogs (family Ranidae) are almost worldwide in distribution. Favorite habitats are swamps, low meadows, brooks, and ponds, where they feed on flies and other insects. Their young, tadpoles, develop in water and are herbivorous. It should be noted that ranids, like many groups of frogs, appear to be declining in number around the world. It appears that there are several possible causes for the decline, and many scientists are now investigating the conservation biology of amphibians. Companies that supply frogs for study report that animals are farm-raised and not wild-caught.

[1]Until recently these frogs were considered to be in the genus *Rana.*

EXERCISE 20A

Behavior and Adaptations

Core Study

Place a live frog on a piece of wet paper towel in a jar large enough not to cramp it. Do not let its skin become dry. Do not excite it unnecessarily. Observe its adaptations for its aquatic lifestyle. Make notes of your observations. Have a mounted frog skeleton at your table for comparison.

Is the skin smooth or rough? Moist or dry? Examine the frog's feet. How are they adapted for jumping? For swimming? For landing on a slippery rock or log? Note the sitting position and compare with the mounted skeleton. In what ways is the body form of a frog adapted as a lever system that can catapult the animal into the air? ________________

Compare the color of the dorsal and ventral sides. Imagine a predator approaching a swimming frog from above or from below. What is the protective advantage in this dorsal-ventral difference in coloration? ________________

Knowing that a frog captures prey by striking at it with its protrusible tongue, how is the position of the eyes advantageous to the animal? ____________ How do the eyes close? ____________ Examine the skeleton and see how this is possible. Note the transparent nictitating membrane. How is it used? ________________________________

____

Feeding Reactions. Place live fruit flies or mealworms in the jar with the frog. If the frog is hungry and is not excited, it may feed. Describe how it seizes prey.

Breathing. Observe movement of the throat and nostrils. Air is drawn into the mouth; then the nostrils close and throat muscles contract to force the air into the lungs. This form of breathing is called **positive pressure breathing.** The nostrils (nares) can be opened or closed at will. Record the number of movements per minute of the throat and then of the nostrils. ____________ Do their rates coincide? ____________ Do the sides of the body move in breathing? ____________ Excite the animal by prodding; then, as soon as it becomes quiescent, count the rate of breathing again.

Righting Reaction. Place the frog on its back, release it, and note how it rights itself. Repeat this experiment, but hold the frog down gently with your hand until it ceases to struggle. When you release it, the frog may remain in this so-called hypnotic state for some time.

Locomotion. Observe a frog in an aquarium. What is the floating position? How does it use its limbs in swimming? How does it dive from a floating position? Note how it jumps.

Written Report

Record your observations on p. 317.

External Structure

Study a preserved frog and compare it with a live frog and a mounted skeleton.

Note the **head** and **trunk.** Note the **sacral hump** produced by the protrusion of the pelvic girdle. Find this hump on the mounted skeleton. The **cloacal opening** is at the posterior end of the body.

On the **forelimbs,** identify the arm, forearm, wrist, hand, and digits. On the **hindlimbs,** find the thigh, shank, ankle, foot, and digits. How does the number of digits compare with your own? ________________ Can you find the rudimentary thumb (prepollux) and the rudimentary sixth toe (prehallux)? During breeding season, the inner (thumb) digit of males is enlarged into a nuptial pad for clasping a female. Observe the **webbed** toes of the hindfoot. How is a long ankle advantageous to a frog? ________________

____

The **eyes** are protected by **eyelids** and by a transparent **nictitating membrane** (L. *nictare,* to wink). Look in the corner of your neighbor's eye for a vestige of this membrane, the semilunar fold.

The **tympanic membrane** (eardrum) is a circular region of tightly drawn skin located just back of the eye. The frog has no external ear—only the middle and internal ears.

Now examine a salamander. Salamanders come in many body forms (Figure 20-1). Some have very tiny legs and external gills. In what type of habitat do you think these animals live? ________________

Do you see an external eardrum? How do you think most salamanders communicate? ________________

Some terrestrial salamanders do not have lungs. Many of these animals have very large tails. What purpose would a large tail serve for these animals? ________________

Identification

Label the external parts of the frog in the drawing on p. 316.

Figure 20-1

(A–D) Various body forms of salamanders. **A**, *Eurycea longicauda*. **B**, *Aneides flavipunctatus*. **C**, *Ambystoma mavortium*. **D**, *Notophthalmus viridescens*

LAB REPORT

Name ____________________ Class ____________________

Date ____________________ Order ____________________

Section ____________________ Genus ____________________

External Anatomy of Leopard Frog

Observations on Behavior of a Frog

Feeding reaction ______________________________

____

____

____

Breathing ______________________________

____

____

____

Reaction to touch ______________________________

____

____

____

Righting reaction ______________________________

____

____

____

Locomotion ______________________________

____

____

____

Other observations ______________________________

____

____

____

____

____

EXERCISE 20B

Skeleton

Core Study

Shaping the Body for Life on Land

In amphibians, as in fishes, a well-developed skeleton provides a framework for the muscles in movement and protection for the viscera and nervous systems. Movement onto land and the development of tetrapod legs capable of supporting the body's weight introduced a new set of stress and leverage problems. These changes are most noticeable in frogs and toads, where the entire musculoskeletal system is specialized for jumping and swimming by simultaneous extensor thrusts of the hindlimbs. Adult frogs no longer move in the typical sinuous, or fishlike, motion of their aquatic ancestors. Consequently, the vertebral column has lost its flexibility and, together with the enlarged pelvic girdle, has become a rigid frame for transmitting force from the hindlimbs to the body.

Prepared skeletons are brittle and delicate. Handle them with care and do not deface the bones in any way. Use a probe or dissecting needle, not a pencil, in pointing.

The **axial** (L., axis) **skeleton** includes the skull, vertebral column, and sternum. The **appendicular** (L. *ad,* to, + *pendare,* to hang) **skeleton** includes the pectoral and pelvic girdles, forelimbs, and hindlimbs.

For a discussion of the structure and growth of bones and their articulations, see Exercise 23A.

Axial Skeleton

Skull. The skull and jaws of a frog serve as protection for the brain and special sense organs. As with the rest of the frog skeleton, it is vastly altered as compared with early amphibian ancestors. It is much lighter in weight, is more flattened in profile, and contains fewer bones and less ossification. The front part of the skull, with locations for the eyes, nose, and brain, is better developed, whereas the back of the skull, which in fishes contains gill apparatus, is much reduced. Lightening of the skull was essential to mobility on land, and the other changes fitted the frog for its improved senses and means of feeding and breathing.

The skull (Figure 20-2) includes the **cranium,** or braincase, and the **visceral skeleton,** made up of bones and cartilage of the jaws; hyoid apparatus; and little bones of the ears. All these elements of the visceral skeleton* are derived from

*The term "visceral" (from the Latin *viscera,* bowels) refers to elements and structures of the body associated with the gut tube. The visceral skeleton represents specialized parts of the primitive gut of jawed vertebrates.

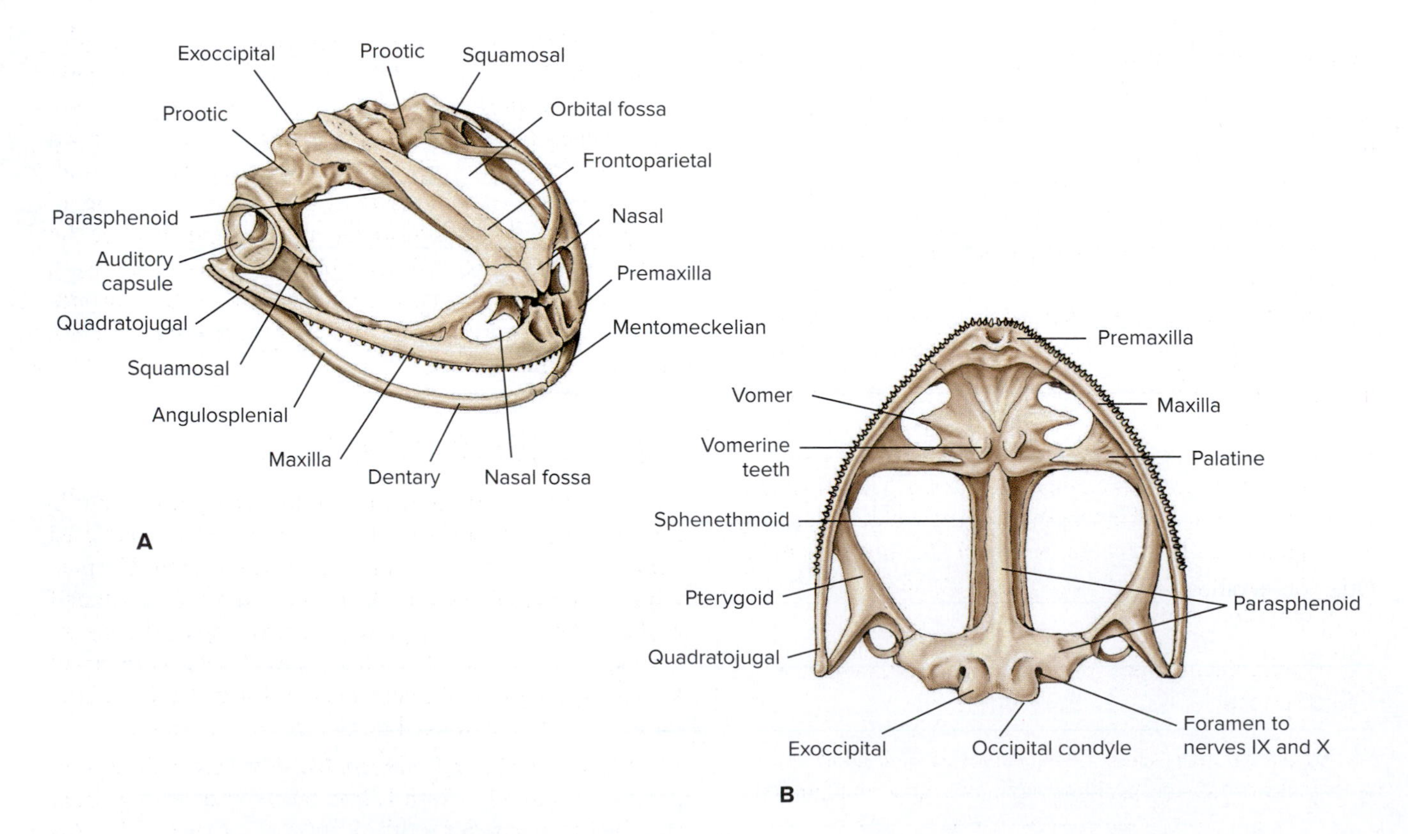

Figure 20-2
The skull of a frog. **A,** Dorsolateral view. **B,** Ventral view.

the jaws and gill apparatus of fish ancestors. The **orbital fossae** and **nasal fossae** are dorsal openings where the eyes and external nares are located.

Locate on the dorsal side of the skull the **nasal** bones; the single **sphenethmoid** (Gr. *Sphēn,* wedge, + *ēthmos,* sieve, + *edios,* from); the long **frontoparietals,** which cover much of the brain; the **prootics** (pro'ah-ticks; Gr. *pro,* before, + *ous,* ear), which enclose the inner ears; the **exoccipitals** (L. *ex,* without, + *occiput,* back of the head), which surround the hindpart of the brain; and the **foramen magnum,** the opening for the spinal cord.

The jaws of a frog combine lightness and strength with large mouth size for eating insects and other prey. (Some tropical frogs resemble walking mouths!); for example, use the Internet to look up a picture of a frog in the genus *Ceratophrys.* The upper jaw is formed by the **premaxillae, maxillae** (L., jaw), and **quadratojugals** (L. *quadratus,* squared, + *jugam,* yolk). Which bones bear teeth? ____________
The **squamosal** (L. *squama,* scale) supports the cartilaginous **auditory capsule.** The three-pronged **pterygoid** (ter'uh-goid; Gr. *pteryx,* wing, + *eidos,* form) articulates with the maxillary, prootic, and quadratojugal (Figure 20-2).

On the ventral surface of the skull, find the wing-shaped **vomers** (L., ploughshare) (**vomerine teeth** are projections of these bones); slender **palatines** (L. *palatum,* palate); and dagger-shaped **parasphenoid** (Gr. *para,* beside, + *sphēn,* wedge, + *eidos,* form), which forms the floor of the braincase. Cartilages form the sides of the braincase.

The lower jaw **(mandible)** consists of small **mentomeckelians** (L. *mentum,* chin, + J. F. Meckel, German anatomist), long **dentary** bones, and **angulosplenials** (L. *angulus,* corner, + *splenium,* patch).

The **hyoid** (Gr. *hyoeidēs,* Y-shaped) **apparatus,** the much reduced and transformed remnants of gill arches, lies in the floor of the mouth. It may be missing in prepared skeletons, or it may be mounted separately. It is cartilaginous and supports the tongue and larynx.

In some preparations, the **columella,** a small bone used in the transmission of sound, is found in the auditory capsule.

Vertebral Column. The backbone of a frog consists of only nine vertebrae and a **urostyle** (Gr. *oura,* tail, + *stylos,* pillar), the latter representing fusion of several caudal vertebrae. The first vertebra, the **atlas** (Gr. *Atlas,* a Titan of Greek mythology, who bore the heavens on his shoulders), articulates with the skull. The ninth, or **sacral*** (L. *sacer,* sacred), vertebra has transverse processes for articulation with the ilia of the pelvic girdle.

With the help of Figure 20-3, identify the parts of a vertebra.

No frogs of the large family Ranidae have ribs either as larvae or as adults; ribs do appear, however, in two other frog families.

*The origin of the term "sacrum," meaning holy bone, is unknown. It is speculated that the curved appearance of the human sacrum suggested to Renaissance anatomists a resemblance to an obsidian knife used in ancient sacrifice.

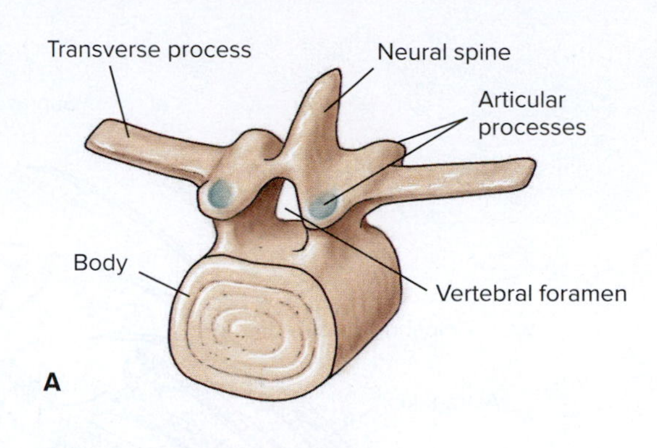

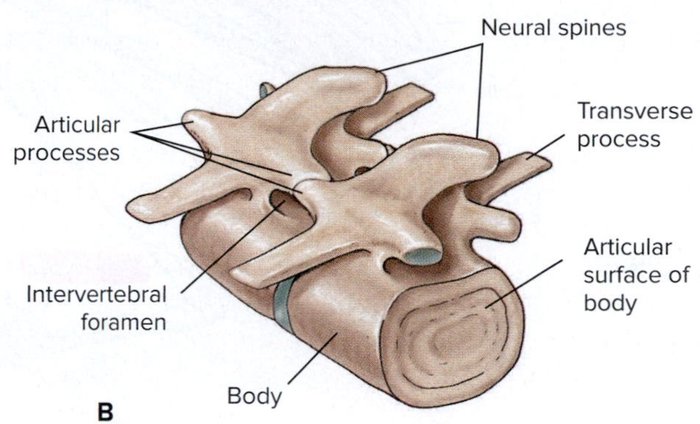

Figure 20-3
Frog vertebrae. **A,** Anterior view. **B,** Two vertebrae, posterior lateral view showing articulations.

The sternum (L., breastbone) provides ventral protection for the heart and lungs and a center for muscular attachment. Its four parts, beginning at the anterior end, are the **episternum** (Gr. *epi,* upon), a cartilaginous, rounded end often not seen in prepared skeletons; **omosternum** (Gr. *omos,* shoulder); **mesosternum** (Gr. *mesos,* middle), the section located posterior of the coracoid bone of the pectoral girdle; and **xiphisternum** (Gr. *xiphos,* sword), a cartilaginous, heart-shaped end often not present on prepared skeletons. The episternum and omosternum are not visible in Figure 20-4.

Appendicular Skeleton

Pectoral Girdle and Forelimbs. The pectoral girdle serves as support for the forelimbs, which, in frogs, are used mainly to absorb the shock of landing after a jump. The pectoral girdle articulates with the sternum ventrally. Each half of the girdle includes a suprascapula and scapula (L., shoulder blade) and a clavicle (collar bone) (L. *clavicula,* small key) lying anterior to the coracoid (Gr. *korax,* crow, + *eidos,* form). Consult Figure 20-4 for the bones of the forelimb.

Pelvic Girdle and Hindlimbs. The pelvic girdle supports the hindlimbs. Each half is made up of a long **ilium** (L., flank), an anterior **pubis** (L. *pubes,* mature), and a posterior **ischium** (Gr. *ischion,* hip). Consult Figure 20-4 for the bones of the hindlimb.

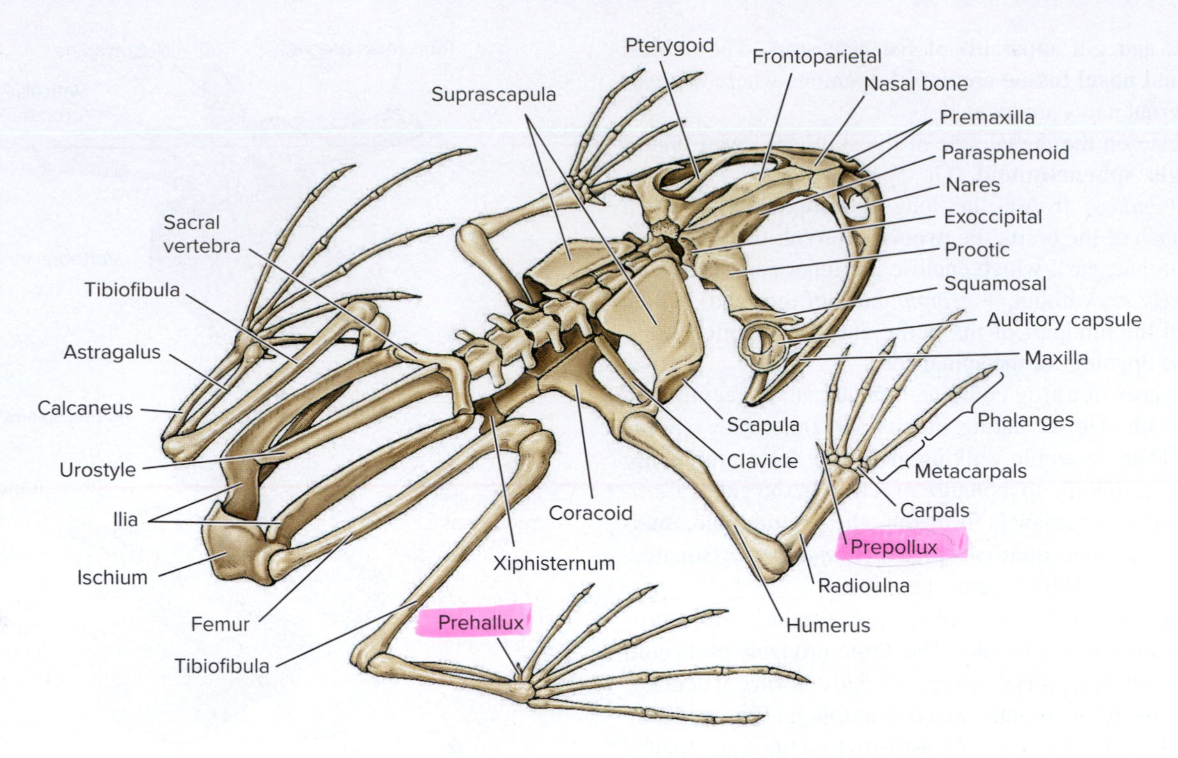

Figure 20-4
Skeleton of a frog.

Oral Report

Be familiar with the parts of the skeleton and the purpose served by each part.

EXERCISE 20C

Skeletal Muscles

Core Study

A frog has hundreds of muscles, but this exercise describes only the most important ones (more correctly, the most conspicuous ones, since they are all important to a frog).

If you learn a few of the Greek and Latin roots of some of the muscle names, they will give you clues to their orientation or action. For example, **rectus** means straight, so the fibers of rectus muscles generally run along the long axis of the body. **Gracilis** means slender. **Triceps** means three heads, and this muscle has three tendons of origin; the **biceps** has two heads of origin. Long muscles are called **longus** and short muscles are called **brevis;** large muscles are termed **magnus** or **major. Anticus** means anterior. Other muscles are named for specific movements. The **sartorius** is derived from the Latin word for tailor and is homologous to a human muscle of the same name that is active in crossing the legs (tailors, before the days of sewing machines, sat on the floor with crossed legs). The **gastrocnemius** (Gr. *gaster,* stomach, + *kneme,* tibia) is named for its fat "belly."

NOTE: You will also need some familiarity with the terminology relating to muscle connections and their actions. Refer to the general discussion of skeletal muscles in Exercise 23B.

Directions for Study of Frog Muscles

To skin the frog, slit the skin midventrally from the cloacal region to chin, keeping the scissors point up to prevent injuring the underlying muscles. Make a transverse cut completely around the body just above the hindlegs and another one anterior to the forelegs. A middorsal cut the length of the back will divide the skin into portions that can be peeled off easily. Loosen the skin with a blunt instrument and carefully pull off over a leg. Be careful not to tear thin muscle attached to the skin. The skin can be pulled over the head and eyes in the same way.

The large spaces between skin and muscle where the skin is not attached are **subcutaneous lymph sacs.**

In separating muscles from each other, first observe the direction of the muscle fibers and the extent of the muscle; then use your fingers, a blunt probe, or the *handle* of a scalpel to loosen the tissues. *Never use scissors, a scalpel blade, or a needle for dissecting muscles.* Never cut a muscle unless instructed to do so. If it is necessary to cut superficial muscles to find deep muscles, cut squarely across the belly (middle fleshy portion) of the muscle, leaving the origin and insertion in place.

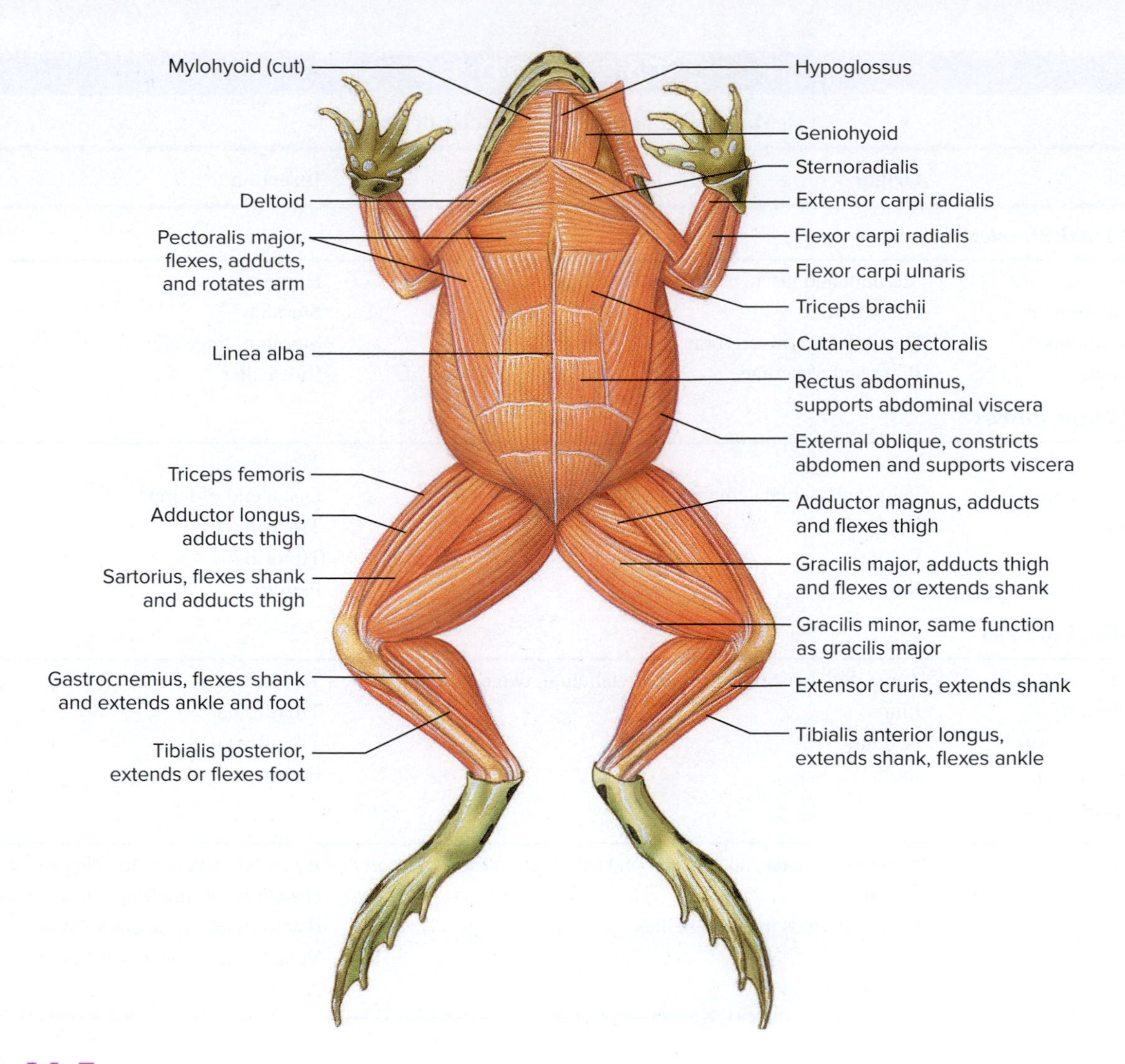

Figure 20-5
Muscles of a frog, ventral view.

Trunk Muscles

The trunk muscles of a frog, no longer required to produce the lateral flexion movements of the amphibians' swimming ancestors, have been modified to brace the back and to support the viscera in air. The ventral trunk muscles (Table 20.1) are arranged in layers that run in different directions. The **rectus abdominis** runs longitudinally, forming a sling from pubis to sternum (Figure 20-5). In the midventral line is a thin but tough band of connective tissue, the **linea alba** (literally, "white line"). Inserting on the linea alba are two oblique muscle bands, the **external oblique** and the **transversus,** the latter lying beneath the external oblique and rectus abdominis. (The transversus is not shown in Figures 20-5 or 20-6.) These assist the rectus abdominis in supporting the viscera.

Thigh Muscles

A frog's limb muscles are derived from muscles that raised and lowered the fins of fishes, now much modified to brace and move the limbs for walking and thrust-swimming. Not surprisingly, limb muscles are complex because they perform several actions. Nevertheless, we can recognize two major groups of muscles on any limb: an anterior and ventral group that pulls the limb forward (protraction) and toward the midline (adduction) and a second set of posterior and dorsal muscles that draws the limb back (retraction) and away from the body (abduction). Only the more conspicuous thigh muscles are included in this study (Table 20.1).

In the first group (protractors and adductors) are the **sartorius, adductor magnus, gracilis major, adductor longus** (visible on the ventral surface, Figure 20-5), **biceps femoris,** and **gracilis minor** (visible on dorsal surface, Figure 20-6). All of these draw the limb forward and toward the midline or flex more distal parts of the limb. In the second group (retractors and abductors) is the **triceps femoris,** a large muscle of three divisions, all of which abduct the thigh and extend the shank (Figure 20-6).

Shank Muscles

The most conspicuous shank muscle is the **gastrocnemius,** which extends from the femur to the foot and inserts by the Achilles tendon; it extends the ankle during jumping and swimming. Because the gastrocnemius is easily dissected

TABLE 20.1

Major Trunk and Leg Muscles of Frogs

Muscle	Origin	Insertion
Ventral Trunk Muscles		
Pectoralis	Sternum and fascia of body wall	Humerus
Rectus abdominis	Pubic border	Sternum
External oblique	Dorsal fascia of vertebrae, also ilium	Sternum, linea alba
Transversus	Ilium and vertebrae	Linea alba
Ventral Thigh Muscles		
Sartorius	Pubis	Tibiofibula
Adductor magnus	Pubic and ischial symphysis	Distal end of femur
Gracilis major	Ischium	Tibiofibula
Gracilis minor	Ischium	Tibiofibula
Adductor longus	Ilium	Femur
Dorsal Thigh Muscles		
Triceps femoris	Three divisions: one head on acetabulum, two on ilium	Tibiofibula
Biceps femoris	Ilium	Tibiofibula
Semimembranosus	Ischium	Tibiofibula
Gluteus	Ilium	Femur
Shank Muscles		
Gastrocnemius	Two heads: distal end of femur and tendon from triceps femoris	By Achilles tendon to sole of foot
Peroneus	Femur	Distal end of tibiofibula; head of calcaneus
Tibialis anterior longus	Femur; divides into two bellies	By two tendons on ankle bones
Extensor cruris	Femur	Ventral surface of tibiofibula
Tibialis posterior	Side of tibiofibula	Ankle

out together with the sciatic nerve that innervates it, it is commonly used in physiological studies of skeletal muscle. The ankle is flexed by the **tibialis anticus longus** and other muscles not shown in Figure 20-6. The major action of the **peroneus** is to flex the ankle joint and extend the shank.

If the frog you are dissecting is in the *Rana* group, it likely has well-developed leg muscles for swimming and jumping. Toads, another large group of frogs, are toxic and have less developed hindlimb muscles. Why might a toxic animal have less developed leg muscles? ________________

____

Other Muscles

With the help of Figures 20-5 and 20-6, you can identify many of the muscles of the back, shoulder, head, and arm.

How the Muscles Act

A muscle has only one function—to contract. For effective action, muscles must be arranged in antagonistic pairs. The gastrocnemius and tibialis anticus longus represent such an antagonistic pair. Loosen the body of each of these muscles, pull on the gastrocnemius, and see what happens. Now pull on the tibialis anticus longus. Which of these muscles would be used in jumping or diving? ____________ Which in sitting? _____________ See whether you can locate other antagonistic pairs.

For most movements, groups of muscles rather than single muscles are required. By varying the combination of these groups, many complicated movements are possible.

Oral Report

Be able to demonstrate a careful dissection of the muscles and to name the muscles and their actions.

Written Report

On separate paper, tell (1) what principal muscles of the hindleg are involved when a frog leaps and (2) when the frog resumes a sitting position, what principal muscles contract.

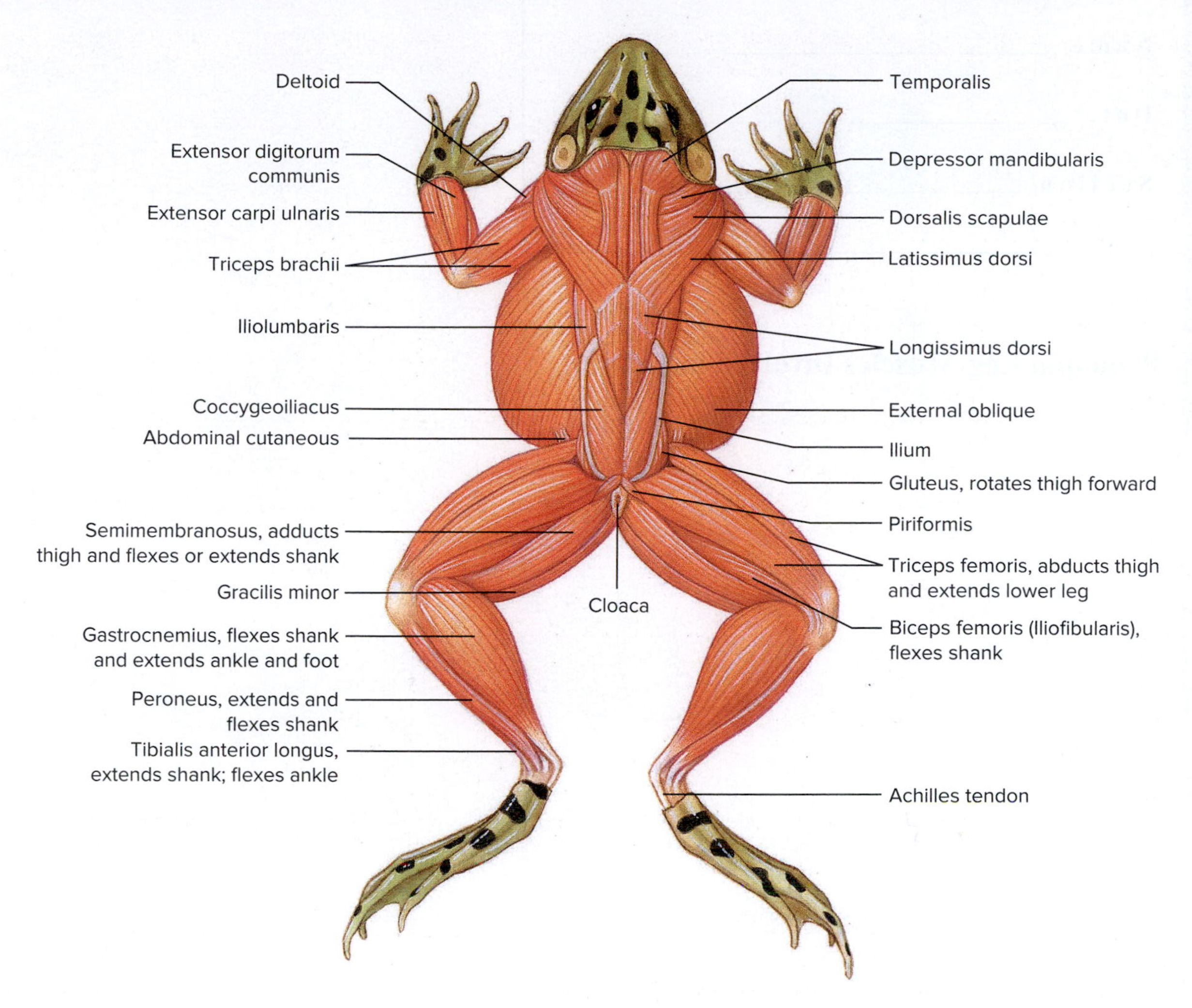
Deltoid
Extensor digitorum communis
Extensor carpi ulnaris
Triceps brachii
Iliolumbaris
Coccygeoiliacus
Abdominal cutaneous
Semimembranosus, adducts thigh and flexes or extends shank
Gracilis minor
Gastrocnemius, flexes shank and extends ankle and foot
Peroneus, extends and flexes shank
Tibialis anterior longus, extends shank; flexes ankle
Temporalis
Depressor mandibularis
Dorsalis scapulae
Latissimus dorsi
Longissimus dorsi
External oblique
Ilium
Gluteus, rotates thigh forward
Piriformis
Triceps femoris, abducts thigh and extends lower leg
Biceps femoris (Iliofibularis), flexes shank
Cloaca
Achilles tendon

Figure 20-6

Muscles of a frog, dorsal view.

LAB REPORT

Name ______________________________

Date ______________________________

Section ______________________________

Principal Leg Muscles Involved in Leaping

Principal Leg Muscles Involved in Sitting

EXERCISE 20D

Digestive, Respiratory, and Urogenital Systems

Core Study

Mouthparts

Pry open the mouth, cutting the angle of the jaw if necessary, and wash in running water.

The posterior portion of the mouth cavity is the **pharynx,** which connects with the **esophagus.** Feel the **maxillary teeth** along the upper jaw and the **vomerine teeth** in the roof of the mouth (Figure 20-7). Are these better adapted for biting and chewing or for holding prey to prevent escape? ______________ Find the **internal nares** (sing. **naris**) in the roof of the mouth and note how they connect with the external nares. Note how the ridge on the lower jaw fits into a groove in the upper jaw to make the mouth closure airtight. This is important for a frog's respiratory movements.

Eustachian tubes, which connect with and equalize air pressure in the middle ear, open near the angle of the jaws. In male frogs, openings on the floor of the mouth slightly anterior to the eustachian tubes lead to **vocal sacs,** which, when inflated, serve as resonators to intensify the mating call. Examine the **tongue** and note where it is attached. Which end of the tongue is flipped out to catch insects? ________________ Feel the **sensory papillae** on the tongue surface. The free end of the tongue is highly glandular and produces a sticky secretion that adheres to the prey. Behind the tongue is a slight elevation in the floor of the mouth, containing the **glottis,** a slitlike opening into the **larynx.**

Dissection of a Frog

Make an incision through the abdominal wall from the junction of the hindlegs to the lower jaw, cutting through the bones of the pectoral girdle as you go. Make transverse cuts anterior to the hindlegs and posterior to the forelegs, and pin back the flaps of muscular tissue.

Note the three layers of the body wall: **skin, muscles** (with enclosed skeleton in some places), and **peritoneum,** which lines the large **coelom.**

In mature females, the ovaries, with their dark masses of eggs, may fill much of the coelomic cavity. In this case, remove the left ovary and its white, convoluted oviduct.

Note the **heart** enclosed in its **pericardial sac** and surrounded by lobes of the **liver.** Lift up the heart to find the **lungs.**

Digestive System

The digestive tract is relatively short in adult amphibians, a characteristic of most carnivores. However, the larval (tadpole) stages of the frog are herbivorous, feeding on algae and other vegetation. They have a relatively long digestive tract because their bulky food must be submitted to time-consuming fermentation before useful products can be absorbed.

You have seen the **mouth** and **pharynx.** Lift the heart, liver, and lungs to see where the **esophagus** empties into the stomach (Figure 20-8). A **pyloric valve** controls movements of food into the **small intestine.** Note the blood vessels in the mesentery, which holds the stomach and small intestine in place. Why must the digestive tract be so well supplied with blood? ______________________________

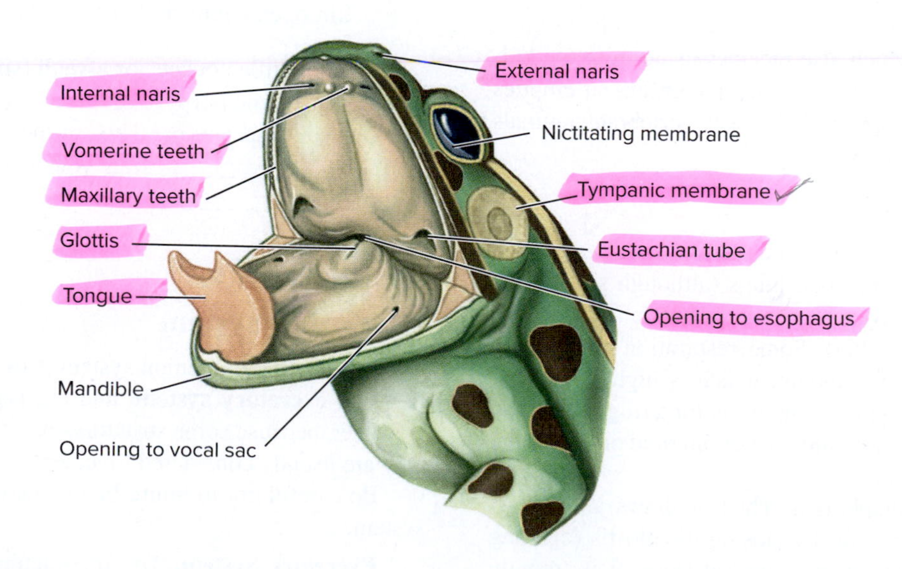

Figure 20-7
Mouthparts of a frog.

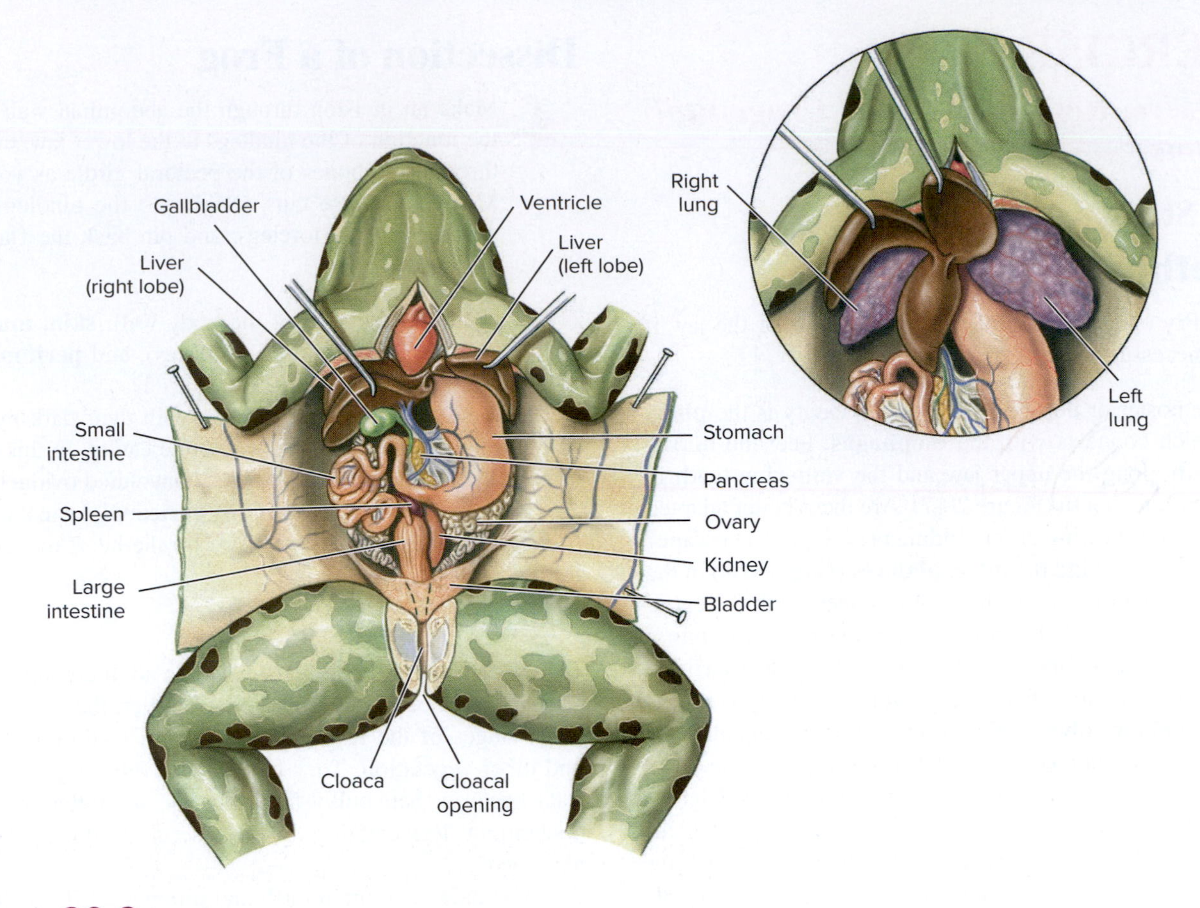

Figure 20-8
Abdominal cavity of a frog, ventral view. **Right,** the liver has been lifted up and turned back to expose the lungs.

The **liver,** the largest gland in the body, secretes bile, which is carried by a small duct to the **gallbladder** for storage. Find the gallbladder between the right and median lobes of the liver. The **pancreas** is thin and inconspicuous, lying in the mesentery between the stomach and the duodenum.

The **large intestine** narrows down in the pelvic region to form the **cloaca,** which also receives urine from the kidneys and products from the reproductive organs. It empties through the **cloacal opening.** Not all vertebrate animals have cloacas; mammals do not.

Respiratory System

Amphibians breathe through their skin **(cutaneous respiration)** as well as with their lungs (although some adult amphibians do not have lungs and rely primarily on gas exchange through the skin). Some respiration also occurs through the lining of the mouth, which is highly vascular. Cutaneous respiration is very important for a frog, especially in winter, when it burrows into the bottom mud of ponds and ceases all lung breathing.

A frog has no diaphragm. Thus, it draws air into its mouth cavity through nares by closing the **glottis** (opening into the windpipe) and depressing the floor of its mouth. Then, by closing the nares and raising the floor of its mouth cavity, it forces air from the mouth through the glottis into the lungs (positive pressure breathing). Air is expelled from the lungs by contraction of the muscles of the body wall and elastic recoil of the stretched lung.

Probe through the **glottis** into the **larynx.** Find the short **bronchus,** connecting each **lung** to the larynx. Slit open a lung and observe its internal structure.

Note the little pockets, or **alveoli** (sing. **alveolus**), in the lining. What is the purpose of this arrangement?

Did you find any parasites in the lungs of your specimen? _______________ Can you identify them (refer to p. 156)? _________________.

Further Study

Urogenital System

Functionally the urogenital system is two systems—the **urinary,** or **excretory system,** and the **reproductive system.** However, because some structures function in both systems, they are usually considered together.

Be careful not to injure blood vessels as you study this system.

Excretory System. The **mesonephric kidneys** separate urine from blood. They rid the body of metabolic wastes (aided by the lungs and skin), and they maintain a proper water balance in the body and a general constancy of content in the blood.

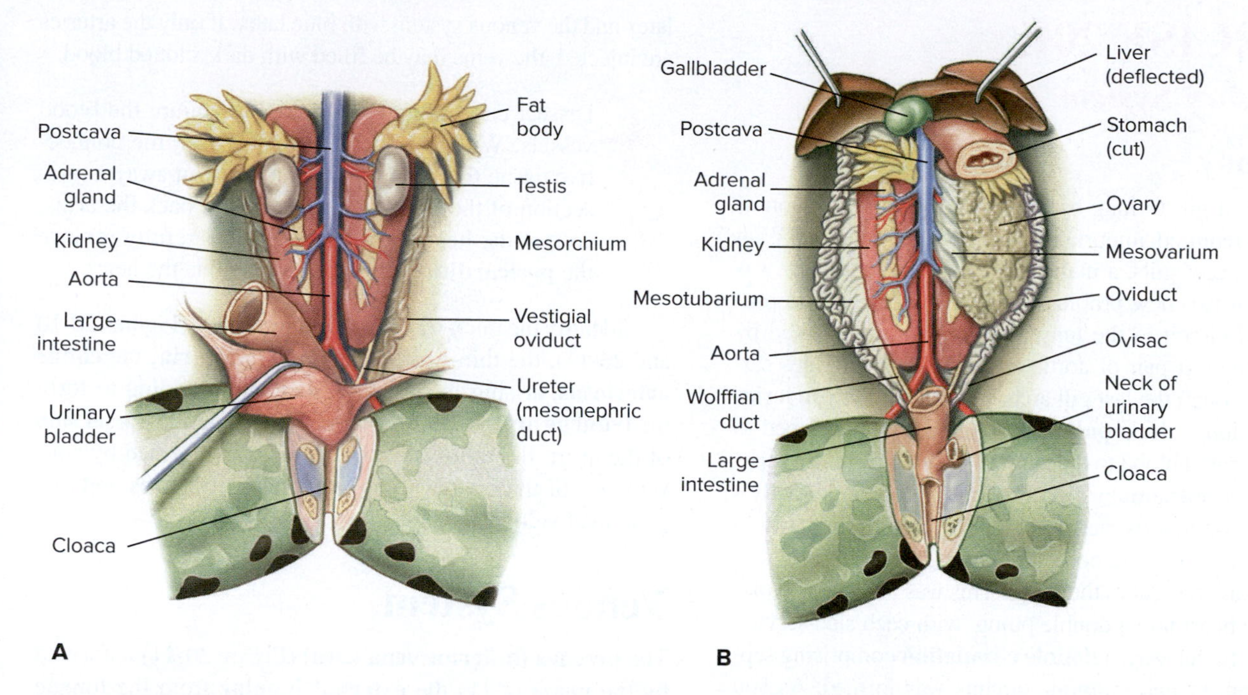

Figure 20-9

Urogenital system of a frog, ventral views. **A,** Male. **B,** Female.

The kidneys (Figure 20-9) lie close to the dorsal body wall, separated from the coelom by a thin peritoneum. The **urinary bladder,** when collapsed, appears as a soft mass of thin tissue just ventral to the large intestine. It is bilobed and empties into the cloaca. The **ureters** connect the kidneys with the cloaca.

To expose the cloaca, cut through the ischiopubic symphysis with a scalpel and push the pelvic bones aside. If you wish, you may open the cloaca just left of the midventral line, separate the cut edges, and find the bladder opening on the ventral wall of the cloaca and the openings of the ureters on the dorsal wall.

Adrenal glands, a light stripe on the ventral surface of each kidney, are endocrine glands, not urogenital organs. **Fat bodies** attached to the kidneys, but lying in the coelom, are for fat storage. They may be large in fall and small or absent in spring. Why? ______________________

____

Male Reproductive System. A small, pale **testis** lies on the ventral side of each kidney (Figure 20-9). Sperm pass from the testis into some of the kidney tubules and then are carried by the ureter to the cloaca and hence to the outside. Thus, male ureters serve also as genital ducts. In leopard frogs, a small **vestigial oviduct** runs parallel to the ureter (this is absent in some species of *Rana*).

Female Reproductive System. Ovaries are attached by mesenteries to the dorsal wall of the coelom. In winter and early spring, the ovaries are distended with eggs. If the specimen was euthanized in summer or early fall, the ovaries will be small, pale, and fan-shaped. Convoluted **oviducts** widen anteriorly (dorsal to the lungs) into funnel-like **ostia** and posteriorly into **uteri,** which empty into the cloaca. Eggs are released from the ovary into the coelom, carried in coelomic fluid to the ostia, and then down the oviducts by ciliary action to the outside. At **amplexus** (the courtship embrace) a male clasps a female and fertilizes the eggs externally as they are laid in the water.

During breeding season, the thumbs of the male frogs of some species enlarge, presumably to assist in clasping a female during amplexus. Why would a male need to clasp a female so firmly? ______________ What might be trying to displace the male? ______________________

____

Recently, research has found that a top-selling weed killer (atrazine) in the United States dramatically affects the sexual development of frogs in both the laboratory and the field. Male frogs exposed to the herbicide show indications of being hermaphrodites, with the development of both testes and ovaries. Does your frog show any of these signs? ______________________

Oral Report

Be able to demonstrate your dissection and state the functions of the structures you have studied.

Trace the route of food through the digestive tract and tell what happens to it at each stage.

Trace the route of eggs and spermatozoa and be able to explain the anatomy and physiology of excretion and reproduction.

Understand the mechanics and physiology of respiration and tell how the mechanics of respiration in frogs differs from that in humans.

EXERCISE 20E

Circulatory System

Core Study

The shift from gill to lung breathing during the evolution of amphibians required important changes in circulation. With the elimination of gills, a major resistance to blood flow was removed. But two new problems arose. The first was to provide a blood circuit to the lungs. This was accomplished by converting the last pair of aortic arches (which in fishes carried blood through the last gill arch) into **pulmonary arteries** to serve the lungs. New **pulmonary veins** then developed to return oxygenated blood to the heart. The second change separated this new **pulmonary circuit** from the rest of the body's circulation such that oxygenated blood from the lungs would be selectively sent to the body and deoxygenated blood from the body would be sent to the lungs. This was achieved by partitioning the heart into a double pump, with each side serving each circuit. In this way, a **double circulation** comprising separate **pulmonary** and **systemic** circuits was formed. As you will see in your dissection of a frog, this modification was not completed: the atrium, which was single in fish, is completely divided into two atria, but the ventricle remains undivided. Complete separation is seen in birds and mammals, which have completely divided hearts of two atria and two ventricles.

To assist your study of circulation, you will dissect a frog that has had the arterial system injected with red or yellow latex and the venous system with blue latex. If only the arteries are injected, the veins may be filled with dark, clotted blood.

Dissect carefully and do not cut or injure the blood vessels. With a probe, you may loosen the connective tissue that holds them in place. Cut away a midsection of the pectoral girdle and pin back the arms, so that the heart is fully exposed. Carefully remove the **pericardium,** the sac that contains the heart.

Identify the thick-walled, conical **ventricle** (Figures 20-10 and 20-11); the thin-walled **left and right atria;** the **conus arteriosus,** arising from the ventricle and dividing to form the **truncus arteriosus** on each side; and, on the dorsal side of the heart, the thin-walled **sinus venosus,** formed by convergence of three large veins—two **precaval veins** and one **postcaval vein.***

Venous System

The **precava (anterior vena cava)** (Figure 20-11) is formed by the union of (1) the **external jugular** from the tongue and floor of the mouth; (2) the **innominate vein** (L. *in,* not,

*"Caval" derives from the Latin *cavus,* meaning hollow, and refers to the sinus venosus, into which both the precaval veins (also called the anterior vena cava) and the postcaval vein (= posterior vena cava) drain. The caval veins of tetrapod vertebrates replace the cardinal veins of fishes (see pp. 294–295).

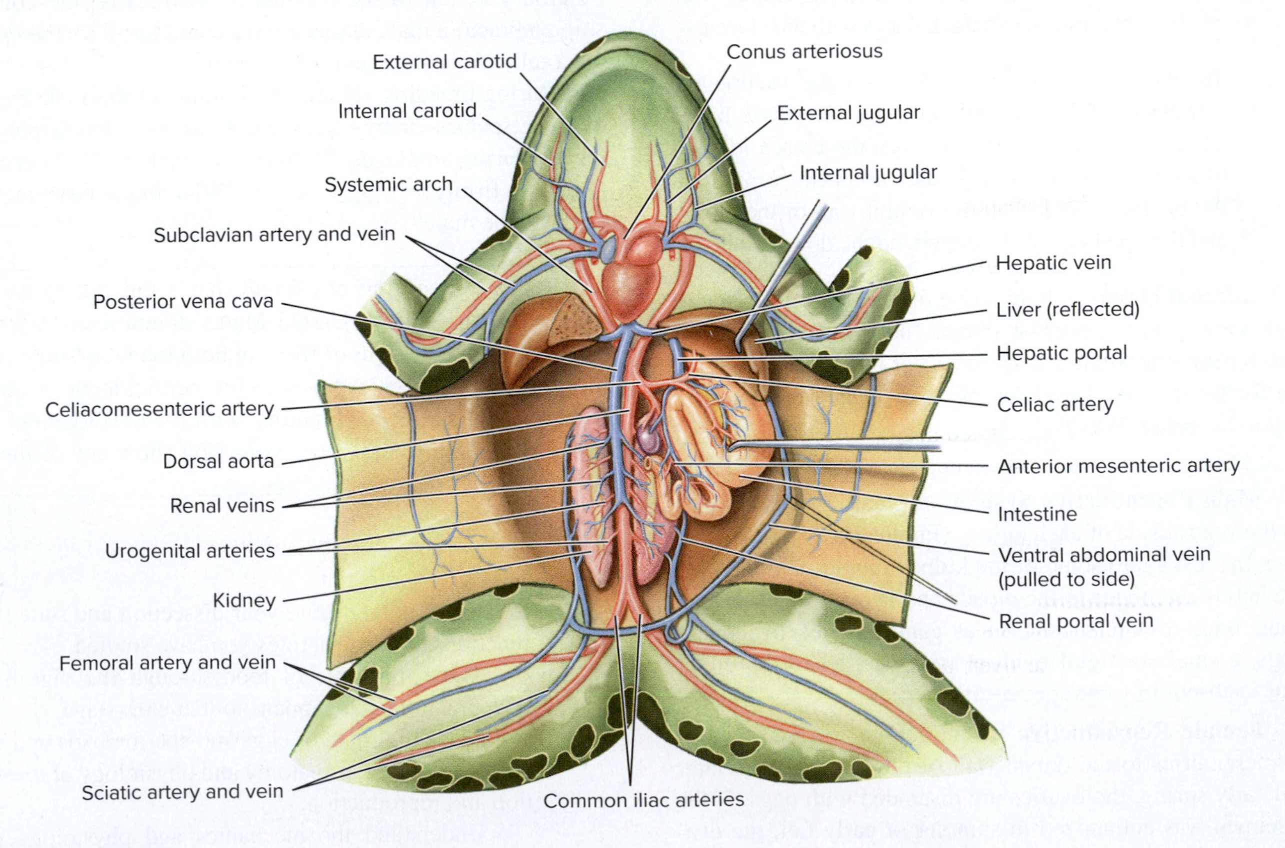

Figure 20-10
Circulatory system of a frog. Arterial, red; venous, blue.

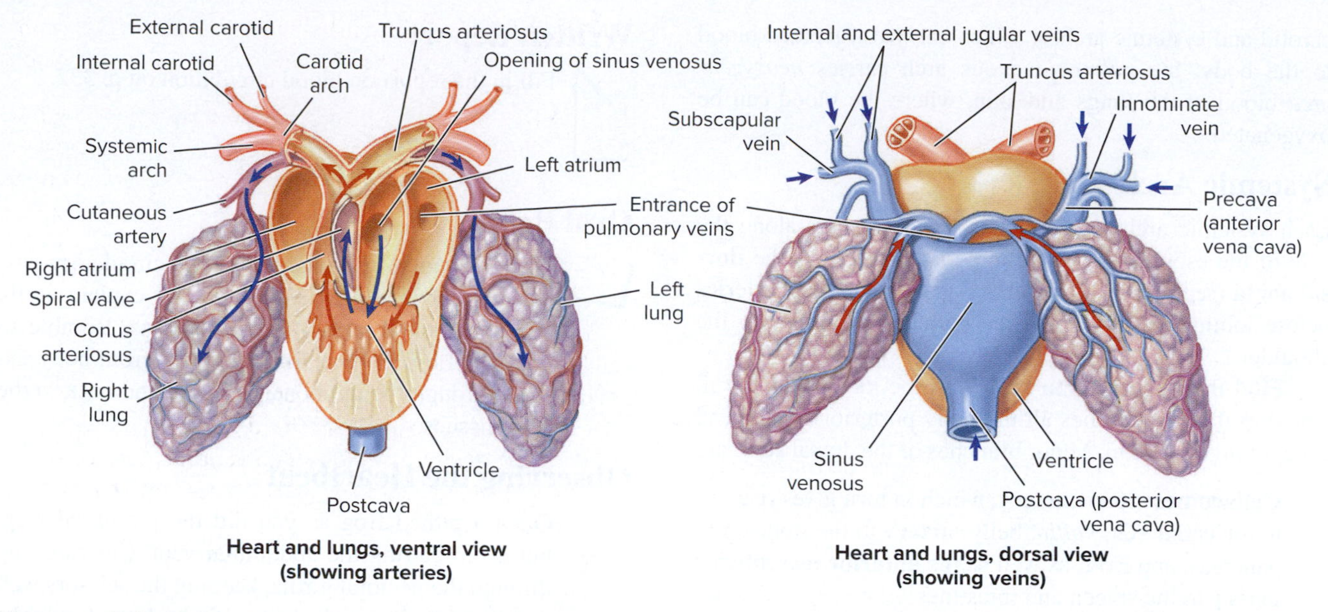

Figure 20-11

Structure of a frog heart. **Left,** Ventral view of frontal section. **Right,** Dorsal view. What structure collects deoxygenated blood returning to the heart and is formed by the convergence of three large veins? ____________ What structure may help prevent mixing of oxygenated and deoxygenated blood as it leaves the heart (use your textbook to help you if necessary)? What artery carries deoxygenated blood to the skin, where the blood can be oxygenated? ____________

+ *nomen,* named), made up from the **subscapular vein** from the shoulder and the **internal jugular** (L. *jugulum,* collarbone) from the brain; and (3) the **subclavian** (L. *sub,* under, + *clavus,* key, below the clavicle), which receives blood from the arm and dorsal body wall.

The **postcava (posterior vena cava)** extends from the sinus venosus through the liver to the region between the kidneys. It receives **hepatic veins** from the liver, **renal veins** from each kidney, and **ovarian** or **spermatic veins** from the gonads.

Pulmonary veins transport blood from the lungs to the left atrium. How does oxygen content of the blood in these veins differ from that in any other vein? ____________

____

Portal Systems

Ordinarily, veins carry blood directly from a capillary bed to the heart. This plan is interrupted in amphibians by capillary beds in two portal systems—the hepatic portal and renal portal systems.

Hepatic Portal System. In the hepatic portal system, blood is carried to capillaries of the liver by two veins: (1) the **ventral abdominal vein** in the ventral body wall collects from the pelvic veins, which are branches of the femoral veins; it empties into the liver; and (2) the **hepatic portal vein** receives the splenic, pancreatic, intestinal, and gastric veins. From the capillary bed in the liver, the blood is picked up by **hepatic veins** and is carried to the postcava and then the sinus venosus. The hepatic portal system, present in all vertebrates, is of great importance because it delivers nutrients absorbed from the gut directly to the liver. This guarantees that the liver will have first opportunity to store and process food materials before they are released into general circulation. In this way, blood leaving the liver remains relatively uniform, regardless of digestive activities underway.

Renal Portal System. Amphibians also inherited a *renal* portal system from their fish ancestors. Most of the blood returning from the hindlegs is interrupted in its journey toward the heart to be diverted into a network of capillaries in the kidneys. The **renal portal vein,** found along the margin of each kidney, is formed by union of the **sciatic** and **femoral veins** (see Figure 20-10). Then, from the kidney, blood is collected by the **renal veins** and carried to the **postcava.** All vertebrates except mammals have a renal portal system. You will see, then, that both hepatic and renal portal systems fit the definition of a portal system as one that begins and ends in capillaries.

Arterial System

The **carotid, pulmocutaneous,** and **systemic arches** (known collectively as the **aortic arches**) arise from the **truncus arteriosus** (Figure 20-11).

Carotid Arch

The **common carotid artery** divides into the (1) **internal carotid,** which leads to the roof of the mouth, eye, brain, and spinal cord and (2) **external carotid (lingual)** to the floor of the mouth, tongue, and thyroid gland (see Figure 20-10).

Pulmocutaneous Arch

The third arch divides into a short **pulmonary artery** to the lungs and a longer **cutaneous artery** to the skin. Unlike the carotid and systemic arches, which carry oxygenated blood to the body, the pulmocutaneous arch carries *deoxygenated* blood to the lungs and skin, where the blood can be oxygenated.

Systemic Arch

Each systemic arch (one from each side) passes along the side of the esophagus to join middorsally to form the **dorsal aorta** (see Figure 20-10). Each gives off several arteries before joining, among these the **subclavian artery** to the shoulder.

Find the **dorsal aorta** by lifting the kidneys to see it. The dorsal aorta supplies all the body posterior of the head except lungs and skin. Major branches of the dorsal aorta are

1. **Celiacomesenteric artery,** which in turn gives rise to the **celiac** (Gr. *koilia,* belly) **artery** to the stomach, pancreas, and liver, as well as the **anterior mesenteric artery** to the spleen and intestines
2. Six pairs of **urogenital arteries** to the kidneys, fat bodies, and gonads
3. **Lumbar arteries** (not shown in Figure 20-10) to the muscles of the back
4. **Common iliac** (L. *ilia,* flanks) **arteries,** formed by the division of the dorsal aorta. Each iliac gives off a **femoral artery** to the thigh, as well as branches to the urinary bladder, abdominal wall, and rectum. The iliac then continues into the hindleg as the **sciatic artery**

Heart

Make a frontal section of the heart (Figure 20-11), dividing it into dorsal and ventral valves.

Find the opening from the sinus venosus into the **right atrium** and the opening from the pulmonary veins into the **left atrium.** Why is the ventricle more muscular than the atria? __________ The **conus arteriosus,** which receives blood from the **ventricle,** divides to form a left and right truncus arteriosus. Valves to prevent backflow of blood guard entrances to the atria and conus.

Even though an amphibian heart is three-chambered, with two atria and one undivided ventricle, oxygenated and deoxygenated blood are effectively separated in the heart. Oxygenated blood from the lungs is sent preferentially to the body, whereas deoxygenated blood from the body is directed toward the pulmocutaneous arch. This partitioning is aided by a spiral fold inside the conus arteriosus and by pressure changes within the heart with each heart contraction. Technically the frog heart can be thought of as having five chambers, since both the sinus venosus and the conus arteriosus are contractile and assist in pumping blood.

Written Report

Fill in the report on blood circulation on p. 332.

Oral Report

Demonstrate your dissection of the circulatory system to your instructor and be able to explain orally any phase of its anatomy or functions. Be able to trace the flow of blood into the heart from body tissues, through the pulmonary circuit, and back to the body tissues.

Observing the Heartbeat

Open a pithed frog as you did the preserved frog, but avoid cutting the abdominal vein. Cut carefully through the pectoral girdle, keeping the scissors well up to avoid injuring the heart. Pin back the forelimbs and keep the heart well moistened with frog Ringers solution.

Identify the ventricle, left and right atria, truncus arteriosus, and aortic arches. Watch the heartbeat and note the series of alternating contractions—first the two atria, then the ventricle, and finally the arterial trunk. Raise the ventricle carefully to view the contraction of the sinus venosus immediately before contraction of the atria.

A frog is an **ectothermic** animal; that is, its body temperature is governed by the environmental temperature, rather than by internal metabolic means (endothermic). You can examine the effect of temperature changes on heart rate by performing a simple experiment.

Count and record the number of beats per minute at room temperature. Flood the abdominal cavity with ice-cold frog saline solution. Immediately count the beats again; then wait a little while until the maximum effect of the cold is achieved, and then count and record again.

Replace the cold saline solution with frog saline solution warmed to about 40° C. Count the beats immediately and record. Allow the warmth to take effect; then count and record again. Replace the warmed saline with saline at room temperature, and count the heart rate again. Record your results on p. 333.

What effect does temperature change have on the heart rate? ____________ What advantage would this reaction to cold have for the frog? ____________ Why should the solution not be warmed to more than 40° C? ____________ What is the effect of fever on human heart rate? ____________

NOTES

LAB REPORT

Name ____________________

Date ____________________

Section ____________________

Frog Circulation

1. Trace the shortest route a corpuscle could take on each of the following trips, underscoring each place where it would go through a **capillary bed.**

   a. Ventricle to lung and return ____________________

   b. Ventricle to brain and return ____________________

   c. Systemic arch to intestine to right arm ____________________

   d. Hindleg to left atrium by way of renal portal ____________________

2. What are the chief gains and losses that take place in the blood in the following organs?

   a. Lung ____________________

   b. Kidney ____________________

   c. Intestinal wall ____________________

d. Liver ____

____

____

e. Muscles ____

____

____

3. Describe the shortest route from the **left atrium** to an **arm** and back to the **left atrium.**

____

____

____

____

____

4. Effect of temperature on heart rate

Rate of contraction

at room temperature	____________	/min (first count)
of cooled heart	____________	/min (first count)
of cooled heart	____________	/min (later count)
of warmed heart	____________	/min (first count)
of warmed heart	____________	/min (later count)
at room temperature	____________	/min (later count)

LAB REPORT

EXERCISE 20F

Further Study

Nervous System

The nervous and endocrine systems are the coordinating systems of the body, integrating activities of the various organ systems.

The nervous system is composed of (1) the **central (cerebrospinal)** nervous system, consisting of the brain and spinal cord, which are housed in the skull and spinal column and are concerned with integrative activity, and (2) the **peripheral** nervous system, made up of the paired cranial and spinal nerves and the **autonomic** nervous system. These make up a system for the conduction of sensory and motor information throughout the body. The autonomic system consists of a pair of autonomic nerve cords together with their ganglia and nerve fibers, which innervate the viscera.

Brain and Spinal Cord

Place the frog *dorsal surface up.* Remove the skin from the head and back. Cut through the skull just back of the nares. Beginning here, use the tip of a scalpel or forceps to chip away small pieces from the top of the cranium, being careful not to injure the delicate tissue beneath. Expose the entire brain from the olfactory nerves to the vertebral column. Now, beginning with the first vertebra, snip through each side of each vertebra between the neural spine and the articular processes (see Figure 20-3A) and remove the dorsal piece to expose the spinal cord. Continue until the whole cord is exposed (Figure 20-12).

The central system is enclosed in two membranes called the **meninges.** The tough **dura mater** usually clings to the cranial wall and neural canal; the thinner **pia mater** adheres to the brain and cord.

Remove the dura mater from the brain and cord and identify the following parts.

Dorsal View of the Brain and Spinal Cord

1. The **forebrain** consists of cerebral hemispheres and diencephalon (Figure 20-13). The **cerebral hemispheres** constrict anteriorly to form **olfactory lobes,** from which the **olfactory nerves** (cranial nerves **1**) extend to the nares. The **diencephalon** is a depressed region behind the cerebrum. In the center of the diencephalon is a small stalk that was attached to the **parietal complex;** this structure, which lies just beneath the roof of the skull, is usually torn off when

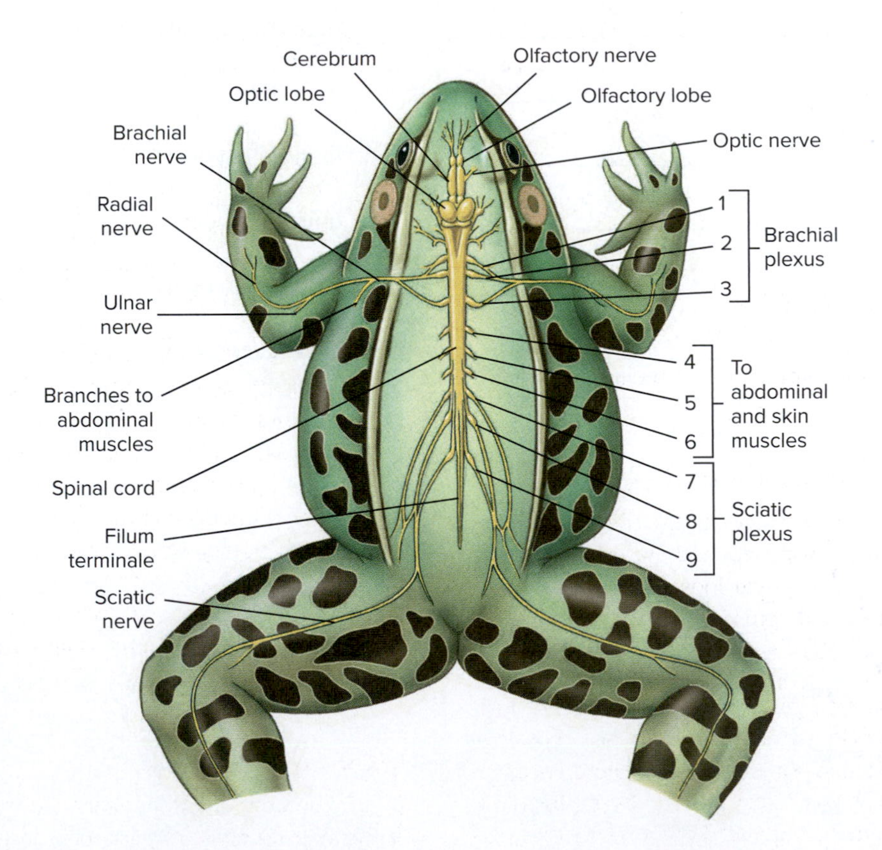

Figure 20-12
Nervous system of a frog, dorsal view.

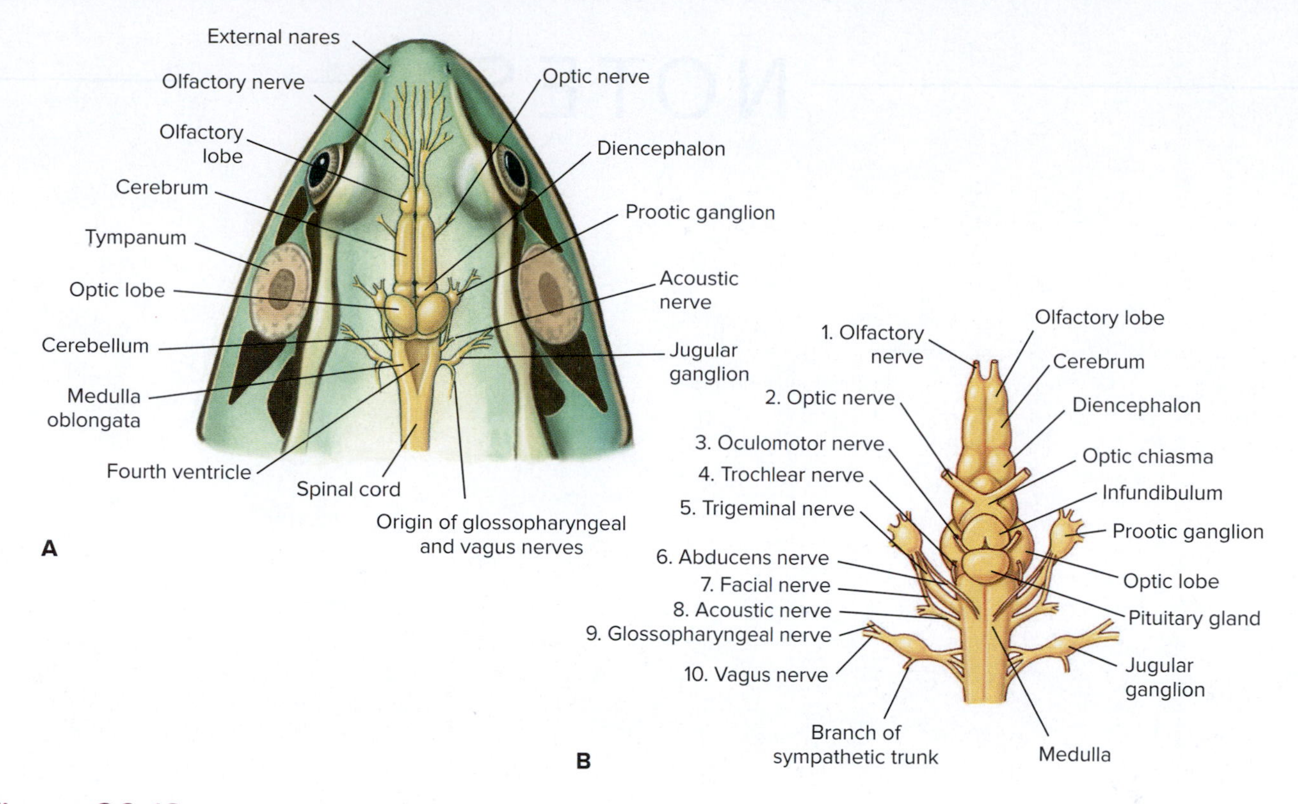

Figure 20-13
Brain of a frog. **A,** Dorsal view. **B,** Ventral view.

the cranium is removed. The parietal complex consists of an **epiphysis,** or **pineal gland,** that produces the hormone melatonin, and the **parietal organ,** which is a rudimentary third eye. The optic nerves **(2)** arise on the ventral side of the diencephalon.

2. The **midbrain** bears on the dorsal side two prominent **optic lobes.** Cranial nerves **3** and **4** arise in the midbrain.
3. The **hindbrain** consists of the **cerebellum** and **medulla.** The last six cranial nerves arise from the medulla.
4. The **spinal cord** is a continuation of the medulla and ends in the urostyle.

Ventral View of the Brain and Spinal Cord

With the brain under water, cut the olfactory nerves and carefully lift the anterior end of the brain, gently working it loose. Continue loosening and lifting to remove the brain and spinal cord in one piece; then place them in a dish of water.

On the ventral side (Figure 20-13B), locate the **optic chiasma,** where the **optic nerves** meet and cross. Posterior to the optic chiasma is a slight extension of the diencephalon, to which is attached posteriorly the **hypophysis,** or **anterior lobe** of the pituitary gland. This is sometimes broken off in dissection, remaining in the floor to the cranium. There are 10 pairs of cranial nerves.

Consult your textbook for the functions of the various parts of the brain.

Spinal Nerves

Lift the tip of the urostyle and cut around it carefully, preventing injury to the spinal nerves below. Continue your dissection, removing bone and muscle above the spinal nerves until they are completely exposed.

The 10 pairs of spinal nerves (see Figure 20-12) emerge through small openings between the vertebrae and appear as white threads in the dorsal body wall. The first three nerves on each side form a **brachial plexus** (a network of nerves that interchange fibers) to the arm, neck, and shoulder. The next three are in the body wall. The seventh to tenth (the tenth is a very small nerve) form the **sciatic plexus** to the leg.

The sciatic plexus on each side gives off to the hindleg a **femoral nerve** and a large **sciatic nerve,** the largest nerve in the body.

Oral Report

Be prepared to demonstrate from your dissection all the component parts of a frog's nervous system so far studied. From your textbook, find out their various functions.

NOTES

The Reptiles

Turtles

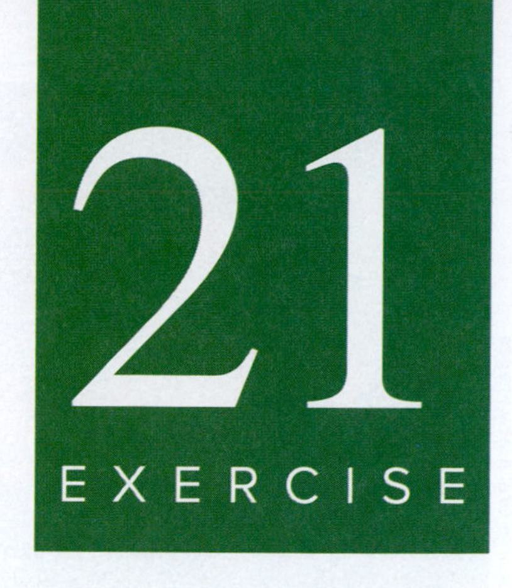

EXERCISE 21

Painted Turtle

Core Study

Amphibians and nonavian reptiles are similar in that both are ectothermic tetrapods. However, they differ in several important characteristics. One of the most important differences is that nonavian reptiles have an amniotic egg, an egg that allows reproduction and development outside of water. Nonavian reptiles also have a tough protective skin, better-developed lungs, a more efficient circulatory system, and a more complex nervous system. Still, nonbiologists often confuse the terms salamander (an amphibian) and lizard (a reptile). For example, although many salamanders and lizards share a basic body structure—four legs and a tail (Figure 21-1)—they are still very different. Salamanders will always have soft, permeable skin. Lizards, on the other hand, will always be covered with scales, a characteristic that protects the skin and helps prevent water loss.

Modern nonavian reptiles—lizards, snakes, crocodilians, turtles—belong to the **diapsid** lineage, a group characterized by having a skull with two pairs of windowlike openings in the cheek (temporal) region. This is a derived condition that lightened the skull, furnished edges for jaw muscle attachment, and provided space that allowed jaw muscles to bulge when the jaw was closed. Interestingly, turtles lack these openings and are often considered to be living representatives of an early **anapsid** (no opening) group. Research published in the last 15 years using morphological and genetic evidence suggests that the skull openings were lost in early turtle evolution and that the current condition evolved secondarily from ancestors having temporal openings.

Turtles have changed little over the past 200 million years, and, although they are highly specialized and much modified from the earliest known amniote fossils, they reveal several features that distinguish the reptilian lineages. A turtle's shell, the anatomical feature that makes it instantly recognizable, is undoubtedly one secret of the group's success, providing protection to an otherwise ungainly animal.

EXERCISE 21
Painted Turtle

Painted Turtle

Phylum Chordata
Subphylum Vertebrata
Class Reptilia
Order Testudines (Chelonia)
Genus *Chrysemys*
Species *Chrysemys picta*

Where Found

Painted turtles, *Chrysemys picta,* are familiar aquatic turtles of ponds, marshes, lake edges, and slow streams. They are widely distributed in the central and northern United States, as far south as Georgia and Louisiana and north along the southern edge of the Canadian provinces. Painted turtles feed on aquatic vegetation, crayfish, snails, and insects. They are relatively common throughout most of their range; however, their numbers decrease wherever aquatic habitats (rivers, ponds, lakes) are modified or destroyed. Their numbers have declined in Canada to the point at which they are considered a species of concern.

External Structure

If living turtles are available, study locomotion and external features for adaptations that distinguish this group. If available turtles are not painted turtles, based on their adaptations would you say that these turtles are aquatic turtles or terrestrial turtles? _______. Why? ________________________________

The anatomical form of a turtle is unusually broad and flattened, compared with that of other reptiles, with limbs extending laterally from the sides and bent downward at elbow and knee. As a result, a turtle's gait is slow and seemingly awkward, although not as inefficient as it might appear—it has, after all, served this successful group for some 200 million years while numerous, more agile reptiles

©George Grall/Getty Images
A

©Lee Kats
B

Figure 21-1
Comparison of salamander and lizard. **A,** Salamander, and **B,** lizard.

have disappeared. The protective shell of a turtle, its exoskeleton, includes the dorsal **carapace** (Sp. *carapacho,* shell) and the ventral **plastron** (Fr., breastplate). Some turtles have hinged plastrons and can close their plastrons tightly. Other species rely more on their snapping jaws for defense and have greatly reduced plastrons. The bones of the carapace are covered with horny **scutes** (homologous to the epidermal scales of other reptiles).

Although the turtle's shell is confining, it is not responsible for the awkward gait—a sprawling posture with the body dragging on the ground was characteristic of primitive land-dwellers in general. However, the subsequent evolution of the shell did prevent any further refinement in locomotion.

The head is stoutly protected with bone. A peculiarity of turtles is their lack of teeth. Instead, the edges of the jaws are formed into sharp ridges covered with strong, horny beaks. Locate the **external nares** (nostrils). Note that they are set close together at the tip of the snout, enabling the turtle to breathe while remaining almost completely submerged. The eyes are well developed. (Turtles, like lizards, have an abundance of cones in the retina, providing them a colorful view of the world.) The eyes are provided with upper and lower eyelids and an additional eyelid at the anterior corners, the transparent **nictitating** (L. *nictare,* to wink) **membrane,** which can be pulled across the eye to moisten and cleanse the cornea.

In most turtles, the neck is long and flexible and folds dorsoventrally to allow retraction of the head within the protective confines of the shell. In one suborder of turtles, distributed in the Southern Hemisphere, the neck folds sideways when the head is retracted.

Skeleton

Examine a mounted skeleton of a painted turtle. Handle with care and use a probe or dissecting needle, not a pencil, in pointing.

Examine carapace and plastron of the shell (Figure 21-2). The plastron, normally united to the carapace by bony bridges, will have been removed in the mounted skeleton. Note that bones of the carapace are covered by horny scutes that do not coincide in number or position with the underlying bone. On the outside, there are five central scutes of epidermal origin, but on the inside there are eight central bony plates (of dermal origin) that are fused to the thoracic vertebrae. Notice the growth lines in the scutes, formed as new keratin is laid down at the edges of the expanding scutes. These cannot be used to determine a turtle's age because new growth is influenced by many environmental factors, including food availability and temperature.

As in other vertebrates, the **endoskeleton** comprises two divisions, an **axial skeleton** consisting of skull, vertebrae, and ribs, and an **appendicular skeleton** consisting of the pectoral and pelvic girdles and limb bones.

A turtle skull is completely roofed over with the braincase (Figure 21-2B). Locate openings for the eyes, nostrils, and ears. Note wide excavations at the rear where the skull has been scalloped out to receive the neck muscles. There are also tunnels on each side at the rear of the upper jaw that provide space for powerful jaw muscles. These excavations for muscles in the anapsid skull of turtles are not true skull openings ("fenestrations") of the kind found in the diapsid skull of other reptiles. Beneath the lower jaw is the **hyoid** (Gr. *hyoeides,* Y-shaped) **apparatus,** a complex of bones, mostly fused, that support the tongue (Figure 21-2A). The hyoid apparatus is derived from remnants of gill arches of fish ancestors.

The vertebral column consists of 8 **cervical vertebrae** that form the bony support of a turtle's flexible neck; 10 **thoracic vertebrae** fused to the carapace; 2 **sacral vertebrae** with expanded ends fused to the pelvic girdle (frogs and other amphibians never have more than 1 sacral vertebra); and 25 to 30 **caudal vertebrae.** Note the strong articulating processes (called **zygopophyses**) of the elongate cervical vertebrae. The articulations are arranged to allow a turtle to withdraw its neck into an s-shaped bend. The thoracic (trunk) vertebrae have elongate centra, and each gives rise to a rib on each side. Note that the central 8 pairs of ribs

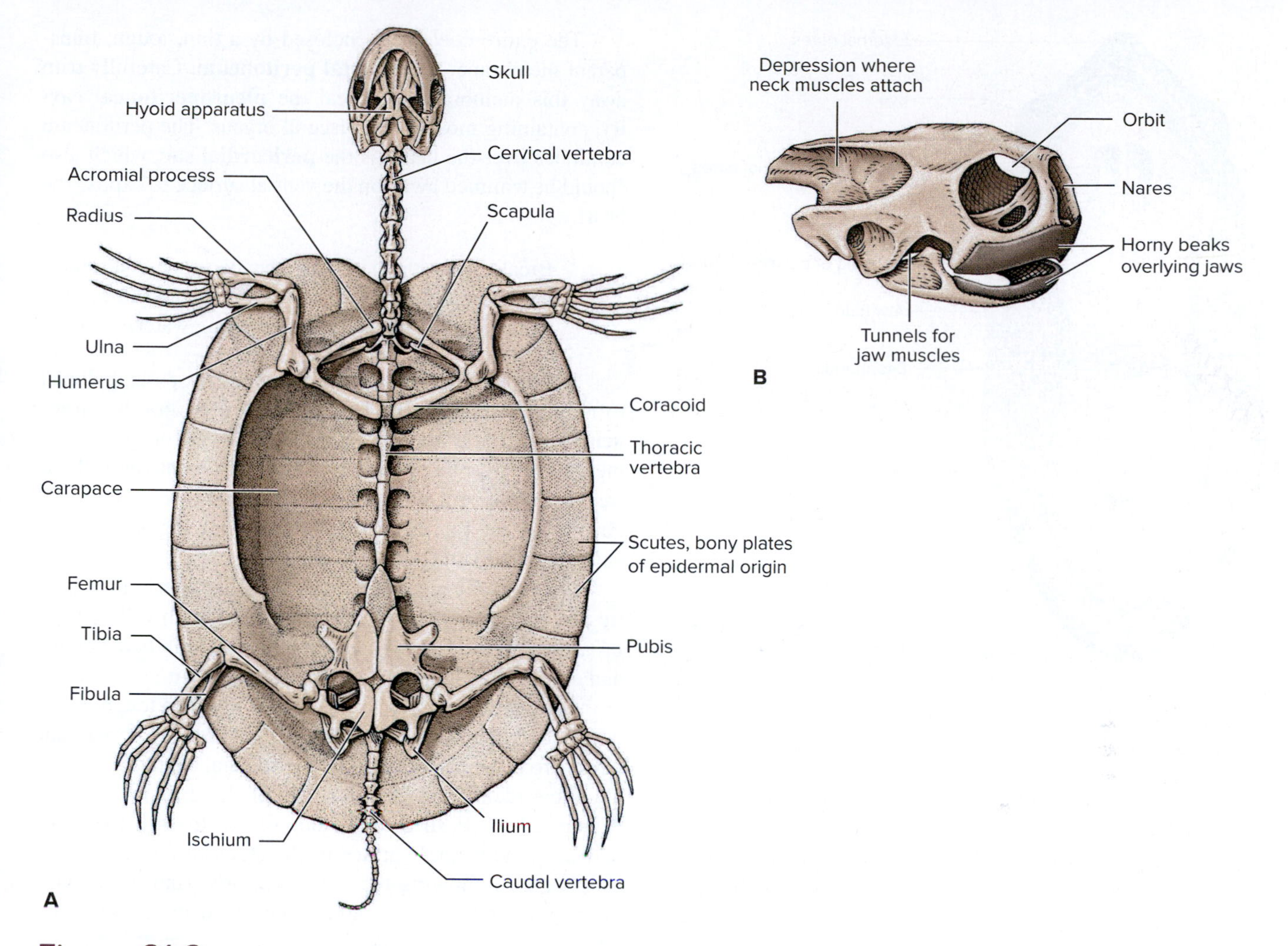

Figure 21-2

Skeleton of a painted turtle. **A,** Ventral view of entire skeleton. **B,** Lateral view of skull.

extend laterally from the vertebrae like flying buttresses to fuse with, and lend strength to, the shell.

Turtles have stout pectoral and pelvic girdles to support the limbs and bear the animal's weight. Both girdles are much modified, compared with those of other reptiles. The pectoral girdle lacks a sternum (because the plastron serves a sternum's function), and the strutlike bones of the pectoral girdle form an odd tripartite scaffold between the carapace and plastron. An elongate **scapula** meets and articulates with the carapace dorsally. Fused to the scapula is the **acromial** (Gr. *akros,* summit, + *omos,* shoulder) **process** projecting anteriorly and at right angles to the scapula. Also lying at right angles to the two scapulas are the paddlelike **coracoid** (Gr. *korax,* crow, + *eidos,* form) **processes,** which are joined by cartilage at their medial margins. The pelvic girdle consists of the **ilia** (sing. **ilium**) attached to the ribs of two sacral vertebrae and, on either side, a broad **ischium** and **pubis.** The limbs of a turtle are stout but otherwise typical of reptiles and contain the same elements found in other vertebrate tetrapod pentadactyl limbs.

Internal Structure

Internal structure will be studied with preserved turtles.

Oral Cavity and Pharynx

To open the mouth, first trim away some of the neck skin if necessary to expose the entire head. Cut through the temporal muscle on either side of the jaw; then force a strong knife blade between the jaws. Force the jaws apart enough to allow cutting the angle of the jaws with heavy scissors or bone shears.

Open the mouth widely. The oral (buccal) cavity (Figure 21-3), the space enclosed by the jaws, leads into the **pharynx,** a cavity extending from the posterior border of the tongue to the esophagus. Note the horny beaks covering the toothless jaws and the triangular **tongue** that is firmly attached throughout its length. The **glottis** is a narrow, slitlike opening posterior to the tongue that leads into the larynx. The external nares lead through the nasal passages to the **internal nares,** which open on the roof of the mouth. There

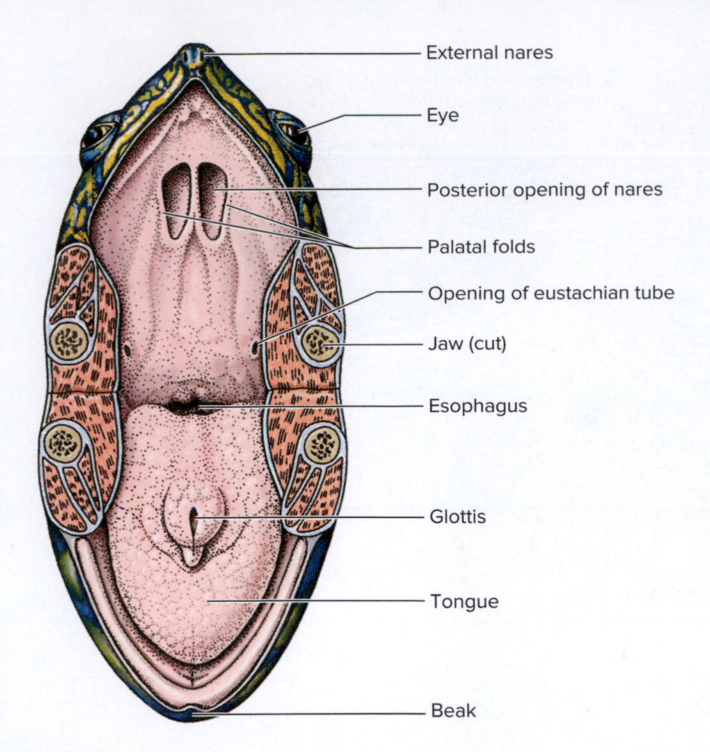

Figure 21-3
Oral cavity and pharynx of a painted turtle.

is no secondary palate in turtles, but the roof of the mouth is vaulted and bears a pair of **palatal folds** that help direct the air toward the glottis. (The secondary palate, present in crocodilians and mammals, is a bony plate that completely separates respiratory and food passages, thus providing a channel for free passage of air from nose to pharynx; the secondary palate allows an animal to manipulate food in its mouth while breathing through its nose.) Locate the **eustachian (auditory) tubes** in the roof of the pharynx; these lead to the middle ear cavity on each side of the head. At the posterior margin of the pharynx is the **esophagus.**

Digestive System

If not already done, cut through the bridges uniting the carapace and plastron with a bone saw or hacksaw. Strong bone shears are usually required to break through the anterior and posterior margins of these bony bridges to complete the separation. Remove the plastron by slicing carefully through adhering tissue and muscle, holding the scalpel *close* to the undersurface of the plastron.

With the plastron removed, you will see much of the viscera covered by several muscles of the pectoral and pelvic girdles, with their origins on the plastron.

Peel back and cut off the large, fan-shaped pectoralis major muscle near its insertion on the humerus. Cut through and remove the coracoid bone of the pectoral girdle and its muscles on both sides to expose the anterior viscera. Trim away the superficial muscles of the pelvic girdle.

The entire coelom is enclosed by a thin, tough, transparent membrane, the **parietal peritoneum.** Carefully trim away this membrane to reveal the **pleuroperitoneal cavity,** containing most of the visceral organs. The peritoneum continues over the heart as the **pericardial sac,** which also should be trimmed away on the ventral surface to expose the heart.

Break away the sides of the carapace with bone shears to expose more of the viscera. Wash out the pleuroperitoneal cavity with running water.

The brown **liver** extends across the body, passing dorsal to the heart (Figure 21-4). The left lobe covers the **stomach,** to which it is bound by a mesentery (gastrohepatic ligament). Use a probe or another blunt instrument carefully to free the stomach from the liver. The right lobe of the liver covers part of the **duodenum** (the initial segment of the small intestine) and part of the colon.

If the specimen is a female, the pleuroperitoneal cavity just posterior to the liver may be filled with yellow eggs of various sizes. The eggs are contained within two **ovaries** and confined by a mesentery, the **mesovarium.**

Look beneath (dorsal to) the heart to find a tough membrane, the **transverse septum,** separating the pericardial cavity from the rest of the pleuroperitoneal cavity.

Now identify the components of the digestive system (Figure 21-4). Push the left lobe of the liver to the right and trace the ventral surface of the stomach forward to the **esophagus.** Following the gut posteriorly, find the duodenum; the **pancreas,** a slender, pale gland lying along the anterior border of the duodenum; and the **gallbladder,** a dark green sac embedded in the dorsal side of the right liver lobe. It may be necessary to break away the posterior margin of the right liver to expose the gallbladder. The gallbladder stores and concentrates bile from the liver; the green color is contributed by a variety of pigments present in the bile. The duodenum gives rise to the **small intestine** proper, which coils several times before emptying into the **colon** (large intestine). The colon terminates at the **cloaca,** which also receives ducts from the ovaries (oviduct) and kidneys (ureter, or metanephric duct).

Respiratory System

With bone shears, cut through the acromial processes on both sides of the thoracic region and trim away the tissue in the neck region. Make a light midventral incision in the neck and retract tissue to expose the trachea.

Follow the **trachea** anteriorly, carefully scraping away muscle and connective tissue from the trachea and hyoid bone, beneath which the trachea disappears. Cut through the two horns of the hyoid bone on both sides, lateral to the trachea, and carefully free the hyoid from underlying trachea and larynx (Figure 21-4).

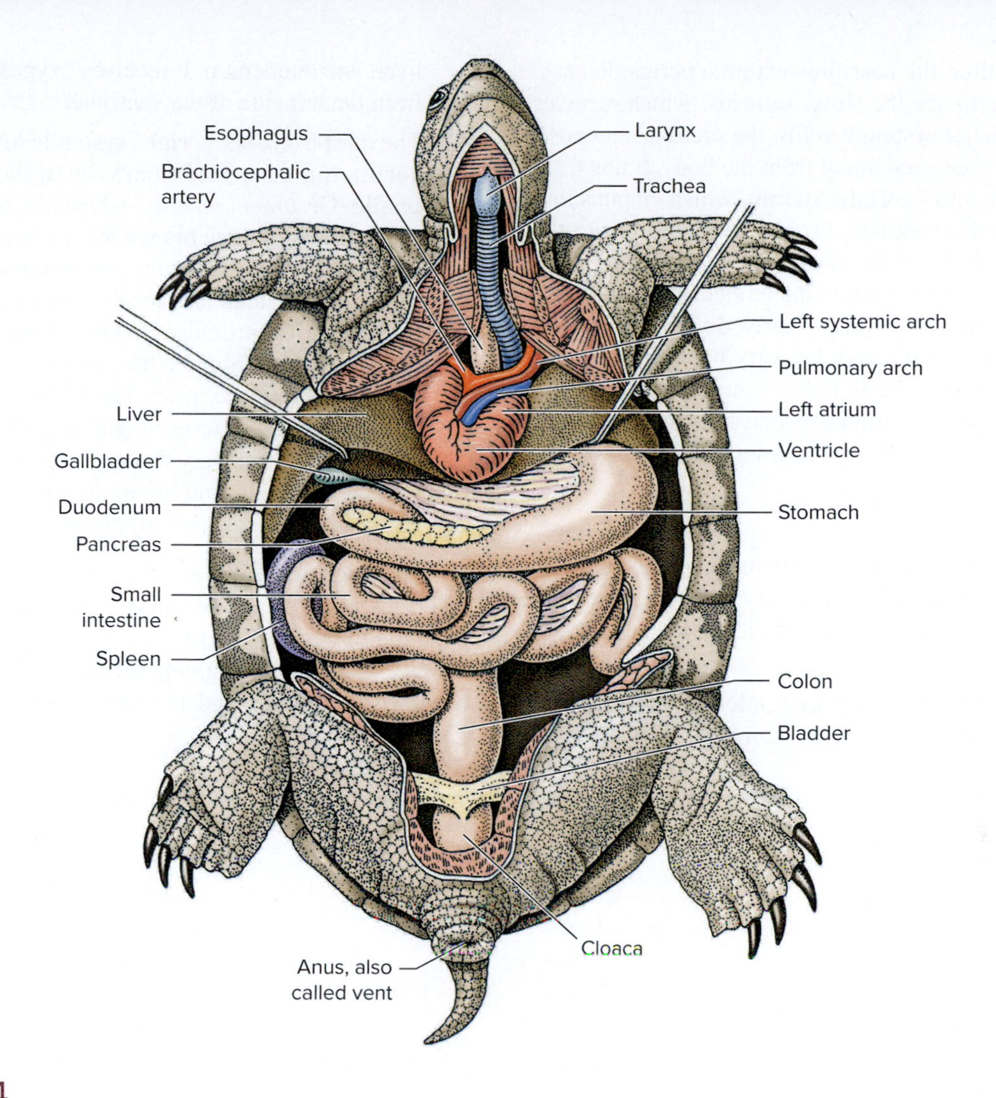

Figure 21-4

Internal anatomy of a painted turtle. In desert species, the bladder can be large and used as a water reservoir. Notice how tightly the organs sit in the body cavity. Why might turtles have physiological adaptations for dealing with extended periods of breath-holding? (Hint: how is the physiology impacted when its legs and head are withdrawn into the shell?) ____________________

____

The glottis, seen earlier, opens into the **larynx,** an enlarged vestibule that leads into the trachea. Note the complex of cartilages and associated muscles (the latter partly removed with the hyoid bone) that open and close the glottis. In many animals, the larynx contains vocal cords, but most turtles are voiceless, able to do little more than make a hissing sound by rapidly expelling air from their lungs. If turtles cannot vocalize, how might they communicate with each other? ____________________

____

Follow the trachea posteriorly to the point where, with the esophagus, it drops middorsally between the large blood vessels emerging from the heart. At this point the trachea divides into two **bronchi** (sing. **bronchus**). Trace the left bronchus to the lung. Reflect the left lobe of the liver and stomach to the right and cut through the peritoneum to expose the **left lung.** The left bronchus will be seen lying between the left pulmonary artery and vein.

Slice open the lung with a lateral incision. The lung is basically saclike, although modestly subdivided into **alveolar pockets,** giving the inner wall a honeycomb appearance. The lung is mostly confined to a separate pleural cavity by an extension of the transverse septum, the **pleuropericardial membrane.** In mammals these membranes become incorporated into the muscular diaphragm. But turtles have no diaphragm and must use other muscles to expand and compress the lungs. The volume of the body cavity is relatively fixed by the shell. What would happen to the internal volume if the turtle extended its head and legs? ____________

____

Would air go into the lungs or be pushed out? ____________

____

Circulatory System

We will confine our study of the circulatory system to the heart and major vessels entering and leaving it (Figure 21-4).

As noted earlier, the heart lies within a pericardial sac. Lift the ventricle to see the **sinus venosus,** which receives blood from the major **systemic veins,** the precaval and postcaval veins. Deoxygenated blood from the body drains from the sinus venosus into the **right atrium,** which empties into the right side of the ventricle. Oxygenated blood from the lungs enters the left atrium via **pulmonary veins** and is pumped from the left atrium into the left side of the heart.

The turtle ventricle is only partly divided by a septum (complete in crocodilians), but very little admixture of oxygenated and deoxygenated blood occurs. From the ventricle, three large arterial trunks extend forward; these represent the subdivision of the conus arteriosus of fishes and amphibians.

1. The trunk farthest to the turtle's left is the **pulmonary arch,** which divides immediately into right and left pulmonary arteries. The pulmonary arch takes deoxygenated blood from the right ventricular chamber and distributes it to the lungs.
2. The middle trunk is the **left systemic arch (left aorta),** which gives off branches to the stomach, pancreas, liver, and duodenum. It receives oxygenated blood from the left side of the ventricle.
3. The third trunk is the **right systemic arch (right aorta).** It is concealed from view by the **brachiocephalic** (Gr. *brachion,* arm, + *kephale,* head) **(innominate) artery,** a large branch that the right systemic arch gives off immediately after leaving the heart. The right systemic arch also receives oxygenated blood from the left ventricular chamber. Small **coronary arteries** spring from the base of the brachiocephalic and branch over the heart. The brachiocephalic divides at once into four arteries: **right** and **left subclavian arteries,** which supply the neck region, pectoral girdle, shoulder muscles, and forelimbs; and **right** and **left carotid arteries,** which supply the head.

Although you will not trace the arteries farther, the left and right systemic arteries unite posteriorly in a V shape to form the **dorsal aorta.** This vessel continues posteriorly in the median dorsal line, giving off branches that supply most of the visceral organs and all of the muscles of the trunk and hindlimbs.

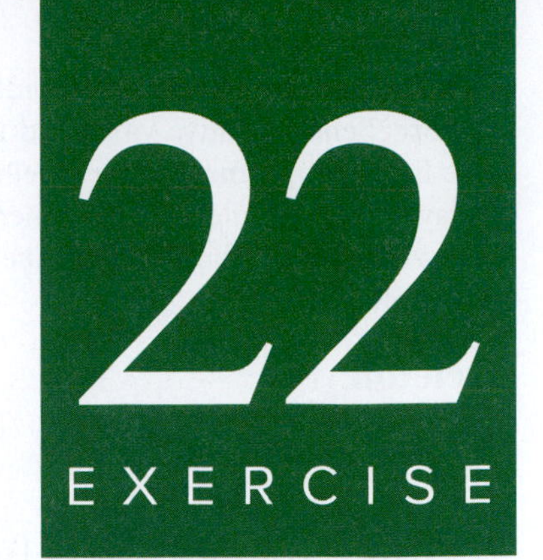

The Birds

EXERCISE 22
Pigeon

Core Study

Birds, described by English zoologist Thomas Huxley as "glorified reptiles," do indeed bear the stamp of their reptilian heritage in many subtle ways. But birds have become so highly specialized into flying machines with all the constraints that design for flight requires that the truth of Huxley's comment is not instantly evident to one gazing casually at a bird.

The demands of flight, far more than anything else, have shaped the form and function of birds. Virtually every adaptation found in flying birds focuses on two features: more power and less weight. This will be the central theme of this exercise.

Pigeon (Rock Dove)

Phylum Chordata
Subphylum Vertebrata
Class Aves
Neognathae
Order Columbiformes
Family Columbidae
Genus *Columba*
Species *Columba livia,* rock dove

Where Found

The common pigeon (rock dove), so familiar to city dwellers, is of Old World origin but has been introduced throughout the world. It is not migratory, yet it has excellent navigational proficiency. In fact, it is the homing pigeon that was used to carry messages in wars from the time of Caesar's conquest of Gaul through World War II. The pigeon is also one of the swiftest birds in flight. Many domesticated varieties have been developed over centuries of breeding; Charles Darwin was himself a pigeon fancier. The "city pigeon" is highly variable in coloration. The rock dove is known to be susceptible to infection from the West Nile virus, although it commonly survives such infections.

EXERCISE 22
Pigeon
Pigeon (Rock Dove)

Feathers

Examine one of the larger contour feathers that assist in flight.

Identify the central **shaft** (also called the **rachis** [Gr. *rhachis,* spinel]), which is a continuation of the **quill** that is thrust into the feather follicle of a living bird. The shaft bears numerous **barbs,** which spread laterally to form the feather's expansive webbed surface, the **vane.**

Flex the feather in your hands, noting its resilience and toughness, despite its remarkable light weight. If you run your fingers down the vane toward the quill, you will separate some of the barbs that are normally linked together by tiny **barbules.** Note that considerable force is needed to separate the barbs. Because detachment happens in the course of a bird's daily activities, the bird spends time each day preening: zipping the barbs back together by drawing the feather through its bill. You can do the same with your fingers. Why would preening be so important for a bird? ______________

Examine a prepared slide of a contour feather, using a dissecting microscope or low power of a compound microscope.

The microscope reveals barbules that extend out at roughly 45° angles in both directions from each barb and are equipped with numerous tiny hooks. Note how adjacent rows of barbules overlap and fit into each other. The hooks of barbules on one barb become fixed to the grooves in barbules on the neighboring barb.

There are several types of feathers that may have been placed on display. The **contour feathers** (such as the one you just examined) give the bird its outward form; contour feathers used in flight are called **flight feathers. Down feathers** are soft tufts, lacking hooks, that are found on young birds or beneath the contour feathers of adult birds. Down feathers have excellent insulative value and function mainly to conserve heat. **Filoplume feathers** are hairlike feathers thought to play a sensory role. These feathers, consisting of a weak shaft with a tuft of short barbs at the tip, are the "hairs" visible on plucked fowl.

Skeleton

Study the mounted pigeon skeleton and unmounted bird bones on display.

A pigeon skeleton (Figure 22-1), like that of other birds, is a marvel of lightness combined with strength. It is even lighter than it looks because many of the limb and girdle bones are hollow. If a long bone, such as the humerus, has been broken or cut open for examination, note its tubular form and the internal struts that have developed where stresses must be borne. Many of the bird's bones are "pneumatized": that is, they are penetrated by extensions of the air sac system and thus contain buoyant warm air rather than the bone marrow typical of mammalian bones. (Some bird bones do contain marrow, but much of a bird's red and white blood cell production occurs in the spleen and liver.)

A striking feature of a bird skeleton is its rigidity. Of the axial skeleton, only the neck remains flexible. The remaining vertebrae are fused together with the pelvic girdle to form a stiff, boxlike framework to support the legs and provide rigidity for flight. The double-headed **ribs** are also mostly fused with the **thoracic vertebrae** and with the **sternum.** Unlike in mammals, nearly all of which have 7 cervical vertebrae, the number of **cervical vertebrae** in birds varies from 8 to 24, depending on the species (long-necked birds have the most). Note the complex articulations of the cervical vertebrae of this most flexible part of the pigeon's body. How many cervical vertebrae does the pigeon have? __________

____

The last rib-bearing thoracic vertebra is fused with five **lumbar,** two **sacral,** and five **caudal** vertebrae to form a thin, platelike structure, the **synsacrum.** The **ilium** of the pelvic girdle is also fused with the synsacrum. This very

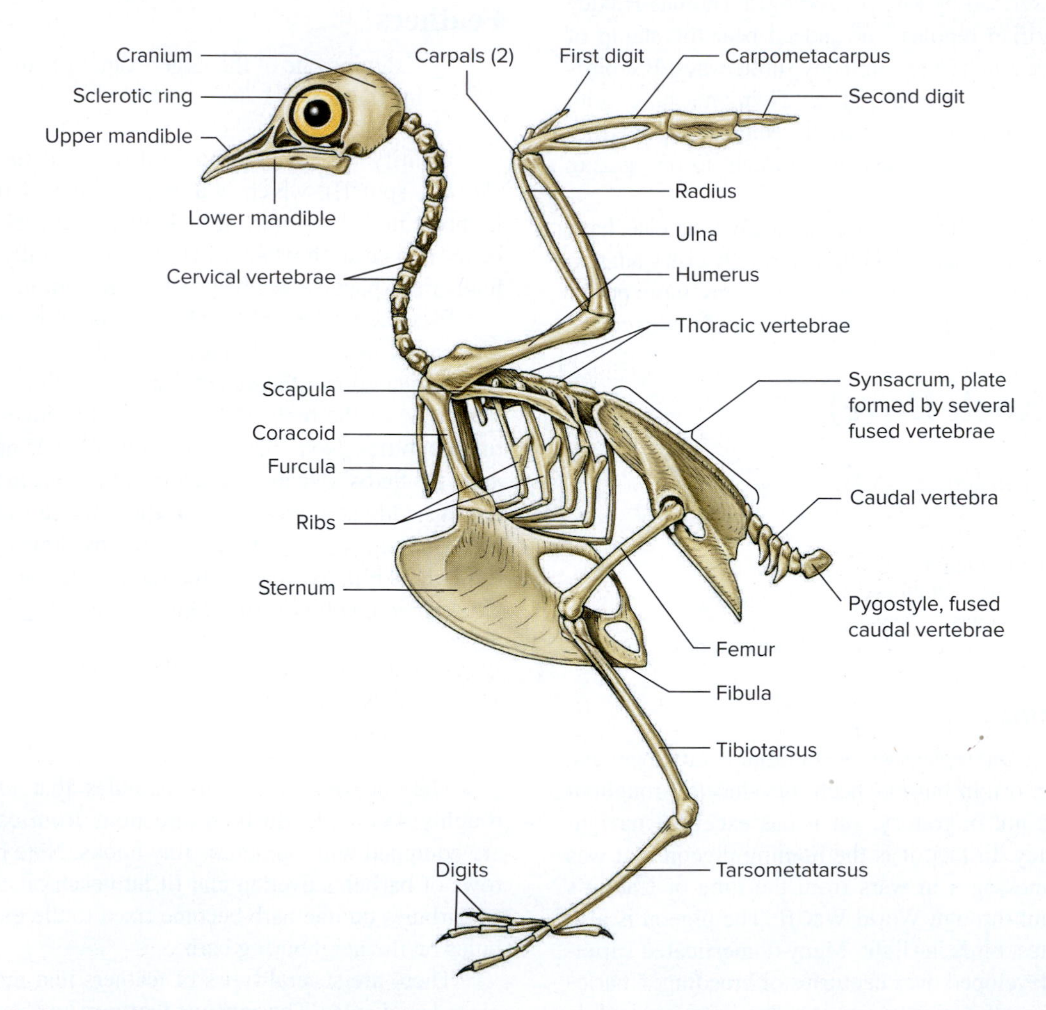

Figure 22-1

Skeleton of a pigeon. Why is the keel of the sternum so large? __________ What bone is formed by the fusion of 13 vertebrae? __________ What bone is sometimes called the "wishbone"? __________ How would you describe the fibula of the bird, as compared with that of other vertebrates (about the same, reduced, enlarged)? __________

light but stout arrangement provides further rigidity to the body frame. Finally, note the short bony tail, consisting of five free **caudal vertebrae** that carry four caudal vertebrae fused into a **pygostyle,** which supports the tail feathers.

A pigeon skull is composed of individual bones that are completely united in adults to form a single thin-walled, lightweight structure. Birds are descended from archosaurian reptiles, which belong to the diapsid lineage of amniotes; this lineage is characterized by skulls having two openings, or fenestra, in the temporal region. Birds are so highly specialized, however, that it is difficult to see any trace of diapsid origin in their skulls. The large, bulging **cranium** encloses the brain, which is much larger, relative to body size, than the brain of a turtle because the complex movements of bird flight require far more central nervous coordination than does the locomotion of a turtle. The pigeon's large eyes are housed in sockets and encircled in front by a protective ring of shingle-like bony plates, the **sclerotic ring.** The **beak** consists of a **lower mandible** hinged to the skull in a way that provides wide-gaping action. The **upper mandible** (also called the **maxilla**) is fused to the skull in pigeons, but some birds (parrots, for example) have kinetic skulls with movable bony elements that allow the upper mandible to tilt upward when the bird opens its mouth. This allows parrots to handle and manipulate awkwardly shaped fruits and nuts.

We will turn our attention now to the appendicular skeleton. Examine the pectoral girdle, composed of a tripod of paired bones: **scapula, coracoid,** and **furcula** ("wishbone"). The scapula is a thin, bladelike bone tied to the ribs by ligaments. The stout coracoid bone unites the scapula and **sternum.** Describe the sternum. ____________ Why is it so large? ____________ Both muscles that depress the wing (pectoralis) and those that raise the wing (supracoracoideus) are attached to the sternum. Where scapula and coracoid unite, there is a hollow depression, into which the ball of the chief bone of the wing, the **humerus,** fits. The supracoracoideus is attached by a tendon to the upper side of the humerus so that it pulls from below by an ingenious "rope-and-pulley" kind of arrangement (Figure 22-2). In this way, muscle weight is kept below the center of gravity, providing greater flight stability.

The pelvic girdle, as we have seen, is a fused structure that is almost paper thin but is strengthened by bony ridges that can be seen by looking at the underside of the girdle. The **femur** (thighbone) is directed forward and is virtually buried in the flesh of a living bird. The **tibia** is the main bone of the shank ("drumstick"); the **fibula** is reduced to a thin splint. The ankle is greatly modified. Some of the proximal pebblelike tarsal bones of the tetrapod limb are united with the tibia to form the tibiotarsus, and the distal tarsal bones are fused with the metatarsals to form a single elongate **tarsometatarsus.** The pigeon, like most other birds, has four digits, three directed forward and the fourth directed backward.

The bones of the forelimbs are highly modified for flight. Note how the wing folds into a compact Z shape when the bird is at rest. Identify the **humerus,** and locate the expanded dorsal surface for the attachment of the pectoral muscles. The **radius** and **ulna** are longer than the humerus, and the ulna, the larger of the two, carries the secondary flight feathers. Most modified are the wrist and digits, which carry the primary flight feathers. Identify the two **carpals** (wrist bones) and two elongate **carpometacarpals** (palm bones), so-called because they are formed by the fusion of three carpal and three metacarpal bones. There are only three **digits** (fingers). The first digit, or "thumb," carries the feathers of the **alula.** The second finger is by far the largest; this, together with the palm bones, carries the primary flight feathers. The third digit, like the first, is reduced to a small bone (not shown in Figure 22-1) that carries a single, outermost flight feather.

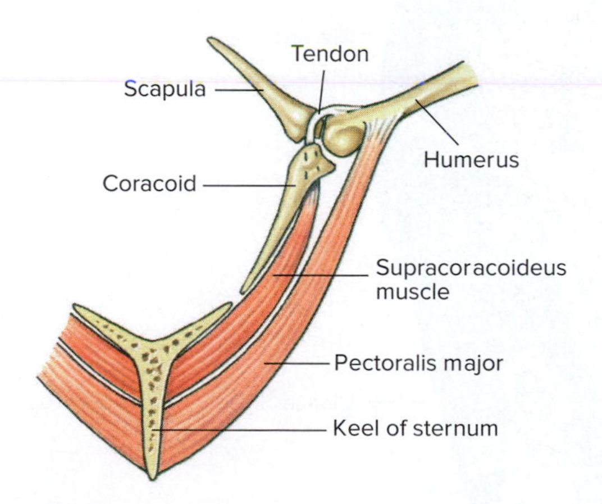

Figure 22-2

The major flight muscles of a bird are arranged to keep the center of gravity low in the body. Both the supracoracoideus and the pectoralis are anchored on the sternum keel. Contraction of the pectoralis muscle pulls the wing downward. As the pectoralis relaxes, the supracoracoideus muscle contracts and, acting as a pulley system, pulls the wing upward.

Internal Structure and Function of the Digestive System

We will limit our study of a pigeon's internal anatomy to the digestive system, which is readily accessible by dissection. Begin by removing the skin and feathers from the ventral side of a pigeon.

With a scalpel, make a midventral incision from the upper neck to the cloaca. Using the blunt end of the scalpel, separate the skin and feathers from the underlying musculature of the neck, breast, and abdomen. Be careful in the neck region not to damage the crop, which lies just beneath the skin at the base of the neck. Cut away the loosened skin from the breast musculature down to the wing insertion and from the abdomen. It is not necessary to remove the skin and feathers from the head, back, wings, legs, and tail.

To examine the oral cavity, cut through the angle of the jaws with heavy scissors.

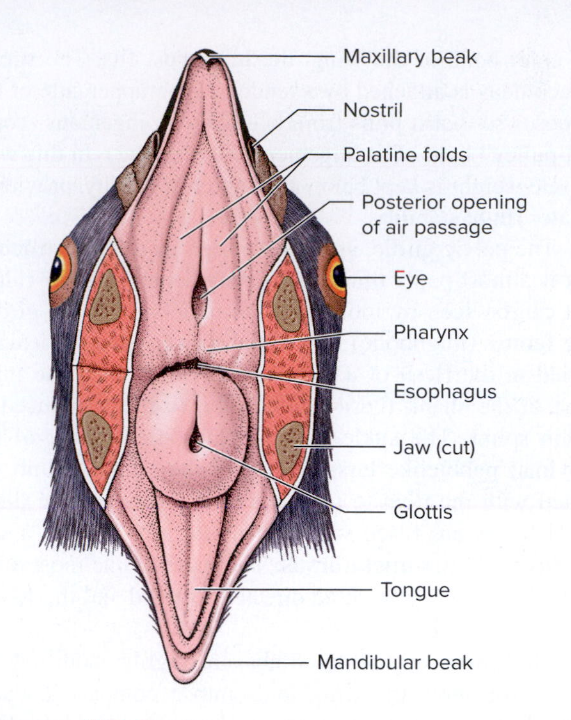

Figure 22-3
Oral cavity of a pigeon.

The muscular **tongue,** used for manipulating food, is narrow and sharply pointed in front (Figure 22-3). The tongue of most birds, pigeons included, is poorly supplied with taste buds, but all birds can probably taste food to some extent. Seed eaters especially use vision and touch rather than taste as the principal means of evaluating their food; the inside of the mouth is liberally supplied with touch endings.

On the roof of the mouth, find the **palatine folds** (Figure 22-3). These are fleshy folds that extend from the lateral borders of the upper jaw to the midline. Probe between the midline slit of the palatine folds to expose a dorsal passage through which air passes from the nostrils to the glottis. The **pharynx** is the common chamber of the mouth and nasal cavity, beginning at the caudal end of the palatine folds. The **glottis,** the opening to the **larynx,** is a slitlike opening on a raised area just posterior to the tongue.

Find the opening to the **esophagus,** which extends from the pharynx to the stomach. In seed-eating birds, such as pigeons, the esophagus swells into a storage chamber, the **crop,** located at the base of the neck (Figure 22-4).

Locate the bilobed crop just anterior to the keel of the sternum and make a midventral incision with a scalpel to expose the interior.

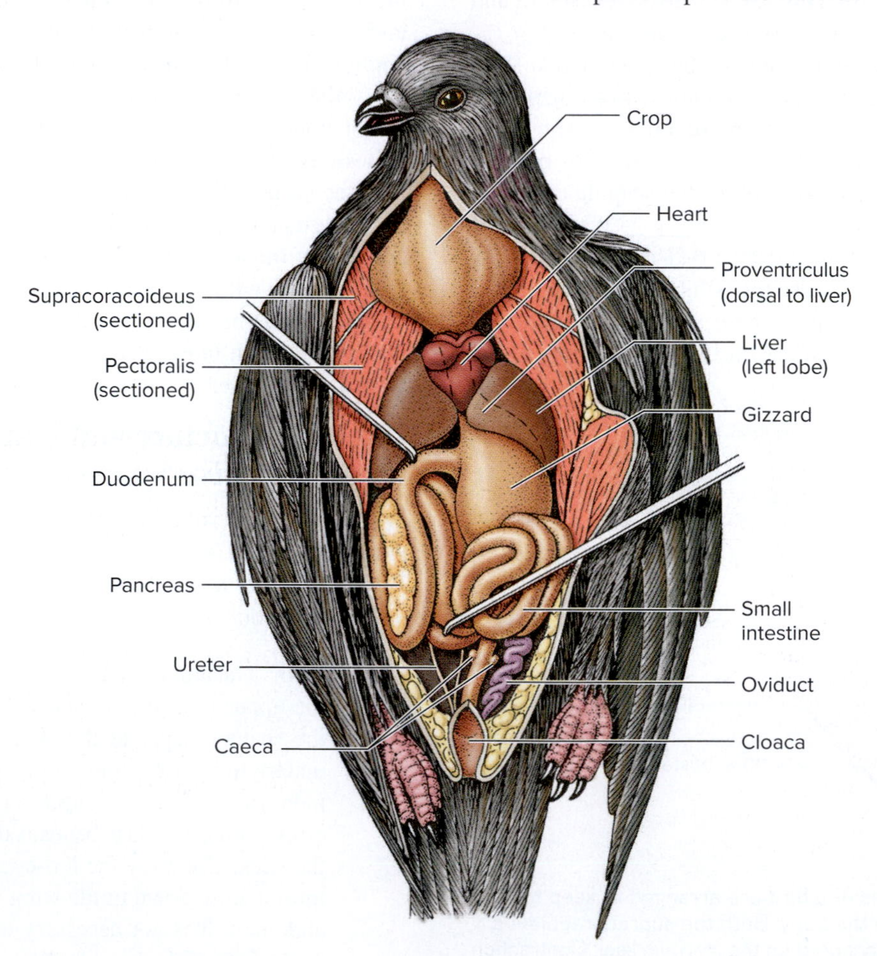

Figure 22-4
Internal anatomy of a pigeon.

An expansible crop allows seed eaters to swallow food quickly and store it for a period while seeds are softened with mucus before entering the stomach. In pigeons the crop not only stores food but also secretes a mixture of fluid-filled cells called "pigeon milk," which the parents regurgitate to feed their young. Flamingos and some penguins also produce a type of "milk." As in mammals, in addition to providing nutrition for the young, the "milk" assists in the development of the immune system.

To expose more of the digestive system, the sternum must be removed.

Pass a sharp scalpel down both sides of the breastbone keel to cut free the breast musculature.

The large breast muscles are the **pectoralis** and the **supracoracoideus,** the strong flight muscles mentioned earlier in your study of the skeleton. Use your fingers to force these muscles away from the keel. Note the silvery fascia that separates these two muscles.

Continue to dissect away the breast muscles from the keel, base of the sternum, and coracoid. With heavy scissors, and using care to avoid damaging underlying organs, cut through the ribs, anterior end of the sternum, and coracoid on both sides and remove to expose the viscera. Remove any remaining skin from the ventral side of the abdomen as far posteriorly as the cloaca.

From Figure 22-4, identify the exposed visceral organs: heart, liver (two lobes), gizzard, and intestine. Note that the body organs are rather firmly supported by mesenteries. The body wall is lined with **parietal peritoneum,** and the visceral organs (except the kidneys) are covered with **visceral peritoneum.**

Remove the heart after severing the pulmonary and aortic arches and the sinus venosus.

All birds have two stomachs: an anterior glandular stomach, the **proventriculus,** and a posterior muscular stomach, the **gizzard** (Figure 22-4). Pass a probe through the opening in the crop and into the spindle-shaped proventriculus. The proventriculus produces an acid-rich secretion containing a peptide enzyme that attacks protein foods. The muscular stomach, or gizzard, is especially thick in seed eaters, such as pigeons, and lined with a horny, keratinous material that serves as grinding plates for mechanically reducing food. Seed and grain eaters also swallow grit (rough granules of sand or stone). Why would birds intentionally swallow grit? ________________

A bird's intestine is the principal organ of digestion. Its length and dimensions differ with the bird's diet: long, coiled, and relatively thick-walled in seed eaters but shorter and thinner-walled in meat-and fruit-eating birds. Find the origin of the looped **duodenum,** the initial segment of the small intestine, where it emerges from the gizzard beside the junction of the proventriculus with the gizzard. The **pancreas** lies in a ligament that connects the halves of the duodenal loop. Trace the small intestine to the straight large intestine. Locate a pair of budlike **caeca** at the junction of small and large intestines. These small sacs are of uncertain function in pigeons, but, in some birds (grouse, for example), they are much longer and appear to function in the absorption of water and amino acids.

The large intestine empties into the **cloaca,** which also receives the ureters and genital ducts.

The bilobed **liver,** noted earlier, is as large as or larger than the liver of mammals of equal size. Its function in birds is similar to its function in mammals: it is the central organ of metabolic regulation, which monitors and adjusts the circulating levels of metabolites. The liver also has numerous other functions, including the synthesis of bile, storage of glycogen, inactivation of toxins, synthesis of plasma proteins, removal of damaged blood cells, and storage of iron- and fat-soluble vitamins.

Although the respiratory system will not be explored in this exercise, the **lungs** can be seen flattened against the body wall by retracting the digestive organs to one side. Although the lungs are the center of the breathing system, birds have a complex system of air sacs (usually collapsed and difficult to locate in a preserved pigeon) which serve as reservoirs for fresh air. These are interconnected in such a way that perhaps 75% of inspired air bypasses the lungs and flows directly into air sacs. On expiration some of this fully oxygenated air is shunted through the lung, while used air passes directly out. As a consequence of this remarkable system, the lungs receive fresh air during both inspiration and expiration. Birds are negative pressure breathers like mammals, but unlike mammals they lack a muscular diaphragm. Air is pulled in (inspiration) by thoracic and abdominal muscles that expand the thoracic cavity. During expiration abdominal muscles contract to push air out of the air sacs.

Oral Report

Familiarize yourself with a pigeon skeleton and digestive system. Be prepared to name the major elements of the skeleton and explain how a pigeon skeleton is adapted for the demands of flight. Be prepared to trace the route of food through the digestive system and explain its fate en route. How is a pigeon's digestive system adapted for a herbivorous diet?

NOTES

The Mammals

Fetal Pig

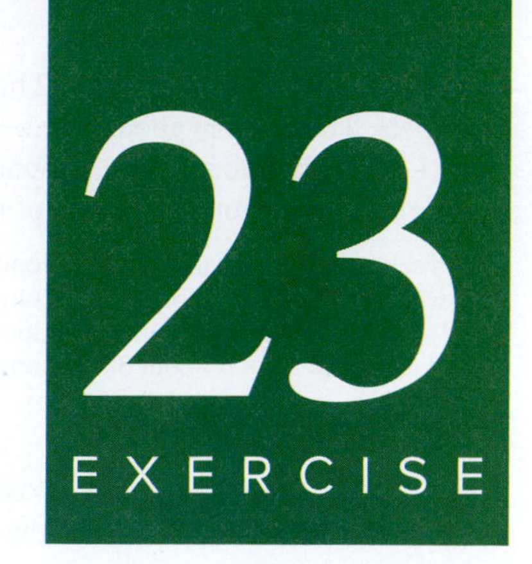

Fetal Pig

Class Mammalia
Subclass Theria
Infraclass Eutheria
Order Artiodactyla
Genus *Sus*
Species *Sus domesticus*

Mammalia are those animals whose young are nourished by milk from the breasts of the mother. Mammals have a muscular diaphragm, a structure found in no other class, and a four-chambered heart. Most are covered with hair. Their nervous system is especially well developed. Their eggs develop in a uterus, with placental attachment for nourishment (except the monotremes, which lay eggs, and in the marsupials, which have only a weakly and briefly developed placental attachment).

Order Artiodactyla includes the even-toed, hoofed mammals, such as deer, sheep, cattle, and camels. These usually have two toes, but some, such as hippopotamuses and pigs, have four toes.

Fetal pigs are an especially desirable laboratory example of a mammal. They are easy to obtain, relatively inexpensive, and easily stored in individual plastic bags. Fetal pigs are obtained from the uteri of sows slaughtered for market. Because they are unborn, their bones are still largely cartilaginous, which makes the specimens pliable and easy to handle. They have an umbilical cord, by which they were attached to the placenta in the uterus.

The embryo depends on maternal blood to bring it nutrients and oxygen and to carry off the waste products of metabolism because its own organs cannot serve these functions until birth. This exchange of materials between fetal blood and maternal blood takes place within the placenta of the mother's uterus. The difference between fetus and adult is largely physiological, but there are also a few morphological differences, especially in the circulatory system, which you will observe in your specimen.

The period of gestation in a pig is 16 to 17 weeks, as compared with 20 days in a rat, 8 weeks in a cat, 9 months in a human being, 11 months in a horse, and 22 months in an

Fetal Pig

EXERCISE 23A
Skeleton
Axial Skeleton
Appendicular Skeleton
Structure of a Long Bone
Growth of a Bone
Articulations

EXERCISE 23B
Muscular System
Organization of Skeletal Muscles
Dissection of Fetal Pig Muscles
Muscles of the Forequarter
Muscles of the Hindquarter

EXERCISE 23C
Digestive System
Head and Throat
Abdominal Cavity
Digestive Tract
Thoracic Cavity and Neck Region

EXERCISE 23D
Urogenital System
Urinary System

EXERCISE 23E
Circulatory System
Heart
General Plan of Circulation
Blood Vessels

EXERCISE 23F
Nervous System
Spinal Nerves
Autonomic Nervous System
Brain

EXERCISE 23G
Respiratory System
Demonstration

elephant. Pig litters average 7 to 12 but may have as many as 18 piglets. The pigs are about 30 cm long at birth and weigh from 1 to 1.5 kg (about 2 to 3 pounds). The age of a fetus may be estimated from the length of its body:

At 3 weeks, the fetus is about 1.3 cm long.
At 7 weeks, the fetus is about 3.8 cm long.
At 14 weeks, the fetus is about 23 cm long.
At full term, the fetus is about 30 cm long.

External Structure

Before proceeding with the regular exercises, look at the external structure of a pig.

On the head, locate the **mouth,** with fleshy lips, and **nostrils** at the tip of the snout. The snout has a tough rim for rooting and bears **vibrissae,** stiff sensory hairs (whiskers). Each eye has two lids and a small membrane in the medial corner, which represents the nictitating membrane. Fleshy pinnae, or ear flaps, contain the external auditory opening.

On the trunk, locate the **thorax,** supported by ribs, sternum, and shoulder girdle with forelimbs attached; **abdomen,** supported by a vertebral column and muscular walls; **sacral region,** comprising the pelvic girdle with hindlimbs attached; **umbilical cord;** five to eight pairs of **mammae,** or nipples, on the abdomen; and **anus** at the base of the tail.

What is the sex of your specimen? __________ In the **male,** the **urogenital opening** is just posterior to the umbilical cord; the **scrotal sacs** form two swellings at the posterior end of the body (the **penis** can sometimes be felt under the skin as a long thin cord passing from the urogenital opening back between the hindlegs). In the **female,** the urogenital opening is just ventral to the anus and has a fleshy tubercle projecting from it.

Examine the cut end of the umbilical cord. Note the ends of four tubes in the cord. These represent an umbilical vein, two umbilical arteries, and an allantoic duct, all of which during fetal life are concerned with transporting food, oxygen, and waste products to or from the placenta of the mother's uterus.

Notice that the entire body of a fetal pig is covered with a thin cuticle called the **periderm.**

EXERCISE 23A

Skeleton

Core Study

The skeleton of an adult pig, although different in size and proportion, is nonetheless quite similar to the skeletons of other mammals, such as a dog, a cat, or a human. The bones are homologous, and the origins and insertions of muscles are usually comparable.

Because of its immature condition, the skeleton of a fetal pig is unsuitable for classroom study. However, if you are planning to dissect the muscles of a fetal pig, it is essential to be familiar with the bones. Skeletons of a cat or dog can be used quite satisfactorily for this purpose. Figures 23-1 and 23-2, showing a cat skeleton, will help you in your identification. For the purpose of comparison, a human skeleton illustration (Figure 23-4) is also provided.

As you study the mounted skeleton of a dog or cat, compare the parts with those of the fetal pig skeleton (Figure 23-3). Then try to locate these parts and visualize their relationships within the flesh of a preserved pig. As you do, notice the similarities with a human skeleton (Figure 23-4).

As in a frog, a pig skeleton consists of an **axial skeleton** (skull, vertebral column, ribs, and sternum) and an **appendicular skeleton** (pectoral and pelvic girdles and their appendages).

Axial Skeleton

Skull. The skull can be divided into a **facial region,** containing the bones of the eyes, nose, and jaws, and a **cranial region,** which houses the brain and ears. The smooth, rounded occipital condyles (Gr. *kondylos,* knuckle) of the skull articulate with the ring-shaped first cervical vertebra (called the **atlas,** named for Atlas of Greek mythology, condemned to hold the heavens on his shoulders for all eternity). The foramen magnum ("great opening") is the opening at the posterior end of the braincase for the emergence of the spinal cord. Many important bones of the skull can be identified with the help of Figure 23-1.

Vertebral Column. Note the five types of vertebrae: **cervical,** in the neck; **thoracic,** bearing the ribs; **lumbar,** without ribs but with large transverse processes; **sacral,** fused together to form a point of attachment for the pelvic girdle; and **caudal,** the tail vertebrae. In humans three to five vestigial caudal vertebrae are fused to form the coccyx of the tail bone. All vertebrae are built on the same general plan but, with recognizable individual differences (see Figures 20-2, 23-2, and 23-3).

Ribs. Observe the structure of a rib and its articulation with a vertebra. Each rib articulates with both the body of the vertebra and a transverse process. The **shaft** of the rib ends in a **costal** (L. *costa,* rib) **cartilage,** which attaches to the sternum or to another costal cartilage and so indirectly to the sternum. A cat has 1 pair of free, or floating, ribs. A pig has 14 or 15 pairs of ribs, of which 7 pairs attach directly to the sternum and 7 or 8 attach indirectly. How many does a human have? ___________

Sternum. The sternum is composed of a number of ossified segments, the first of which is called the **manubrium,** and the last of which is called the **xiphisternum.** Those in between are collectively called the **body** of the sternum.

Appendicular Skeleton

Pectoral Girdle and Its Appendages. The pectoral girdle comprises a pair of triangular **scapulae,** each with

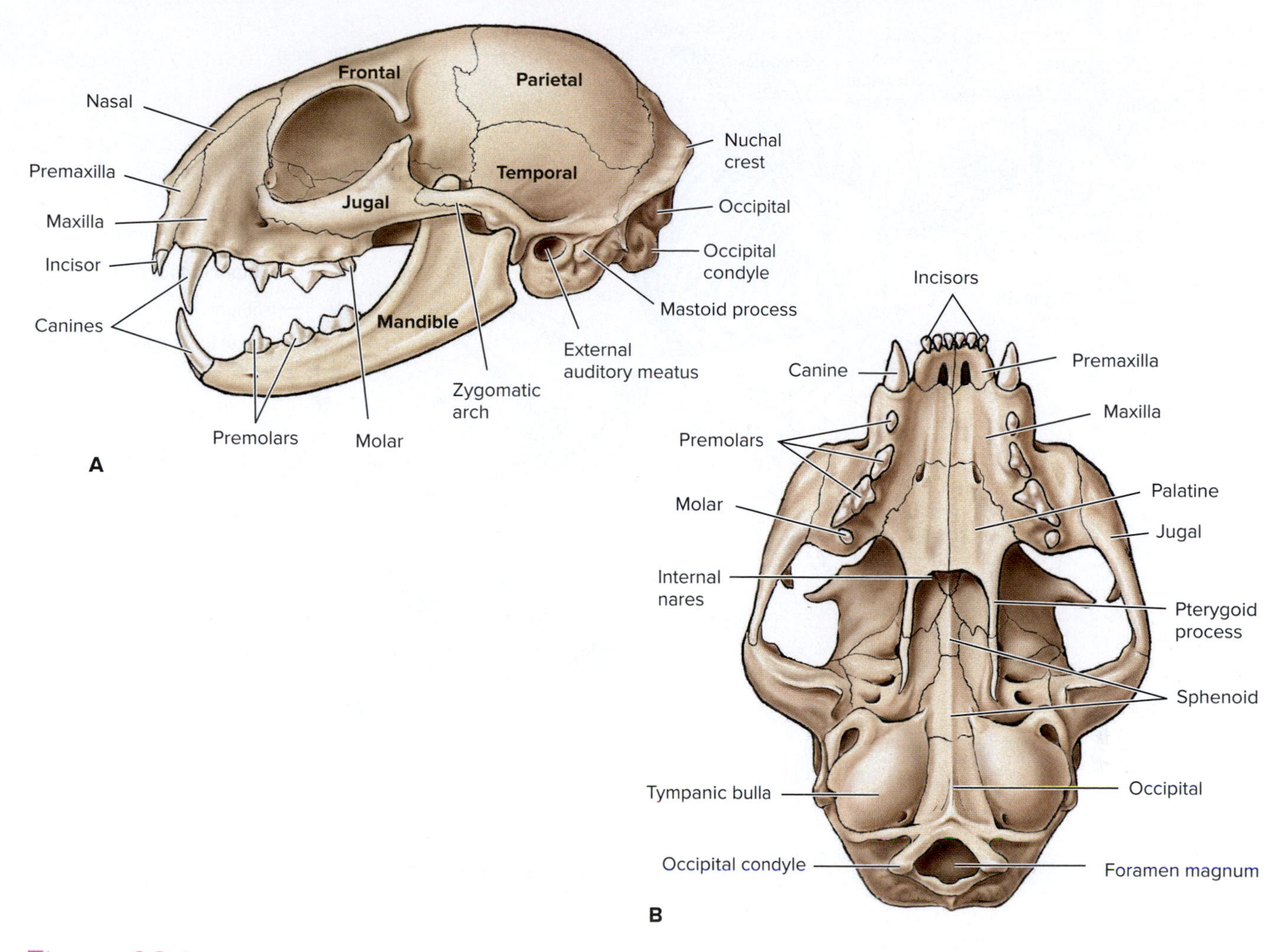

Figure 23-1
Skull of a cat. **A**, Lateral view. **B**, Ventral view.

a lateral **spine** and a **glenoid fossa** at the ventral point for attachment with the head of the humerus. The **forelimb** (see Figures 23-2 to 23-4) includes (1) the **humerus;** (2) two forearm bones—a shorter, more medial **radius** and a longer, more lateral **ulna,** with an **olecranon** (ol-ek′re-non) **process** at the proximal, or elbow, end; (3) the **carpus,** consisting of two rows of small bones; (4) the **metacarpals** (five in a cat and human and four in a pig); and (5) the **digits,** or toes, made up of **phalanges.**

Pelvic Girdle and Its Appendages. The pelvic girdle in adult mammals consists of a pair of **innominate** (L., without name) **bones,** each formed by the fusion of the **ilium** (L., flank), **ischium** (Gr. *ischion,* hip), and **pubis** (L., mature). A lateral cavity, the **acetabulum** (L., vinegar cup), accepts the head of the femur. The pubic bones and the ischial bones of opposite sides unite at **symphyses,** and the ilia articulate with the sacrum so that the innominates and the sacrum together form a complete ring, or **pelvic canal.** Each pelvic appendage includes (1) the **femur,** or thigh bone; (2) the larger **tibia** and more slender **fibula** of the shank; (3) the ankle, or **tarsus,** comprising seven bones, and fibular tarsal (**calcaneus;** cal-ka′nee-us; L. *calx,* heel), forming the projecting heel bone; (4) the **metatarsals** (five in a cat, of which the first is very small, and four in a pig); and (5) the **digits,** or toes (four in both a cat and a pig), composed of **phalanges.**

Structure of a Long Bone

Longitudinal and transverse sections through a long bone (Figure 23-5) illustrate a shell of **compact bone,** within which is a type of **spongy bone (cancellous bone),** the spaces of which are filled with **marrow.** The **shaft** (also called the **diaphysis;** di-af′uh-sis; Gr. *dia,* through, + *phyein,* to grow) is usually hollowed to form a **marrow cavity.** In a young mammal, there is only red marrow, a blood-forming substance, but this is gradually replaced in adults with yellow marrow, which is much like adipose tissue. The ends of the bone, or **epiphyses** (e-pif′uh-sees; Gr. *epi,* upon, + *phyein,* to grow) usually bear a layer of **articular cartilage.** The rest of

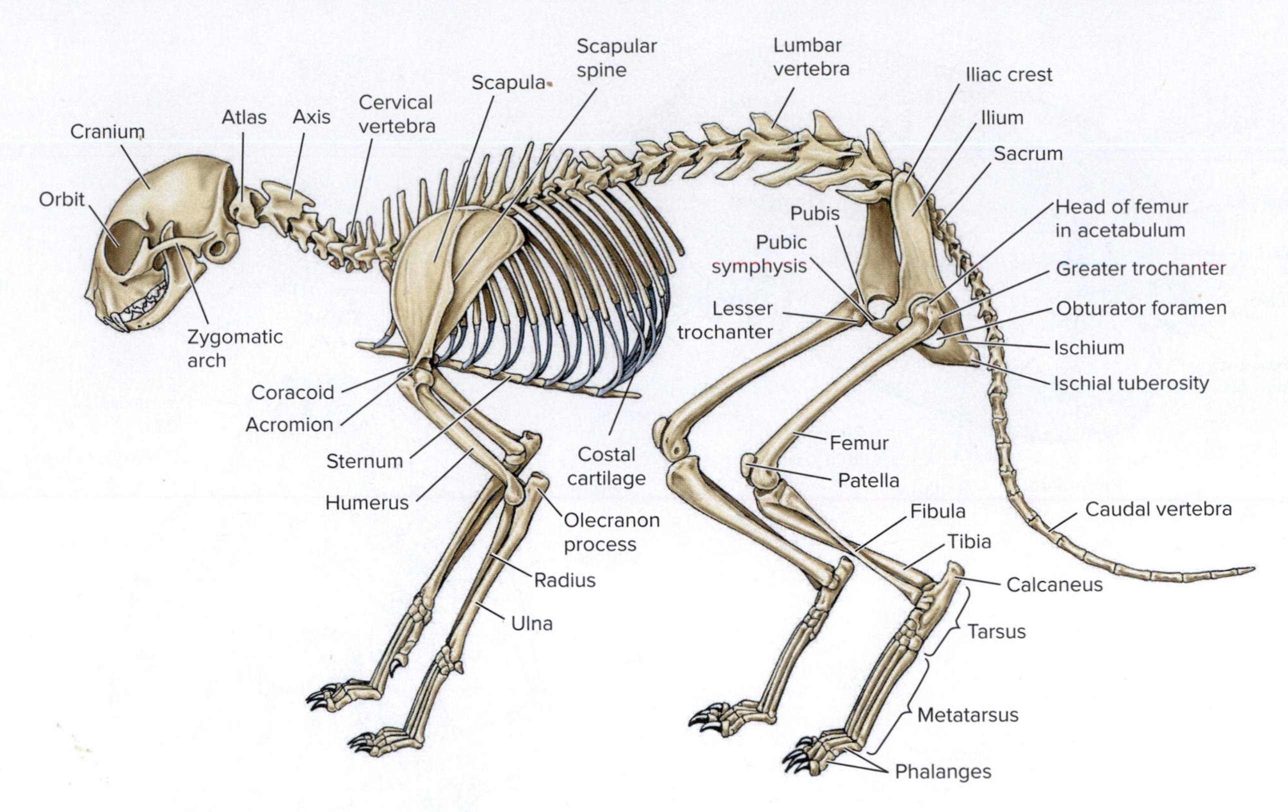

Figure 23-2
Skeleton of a cat.

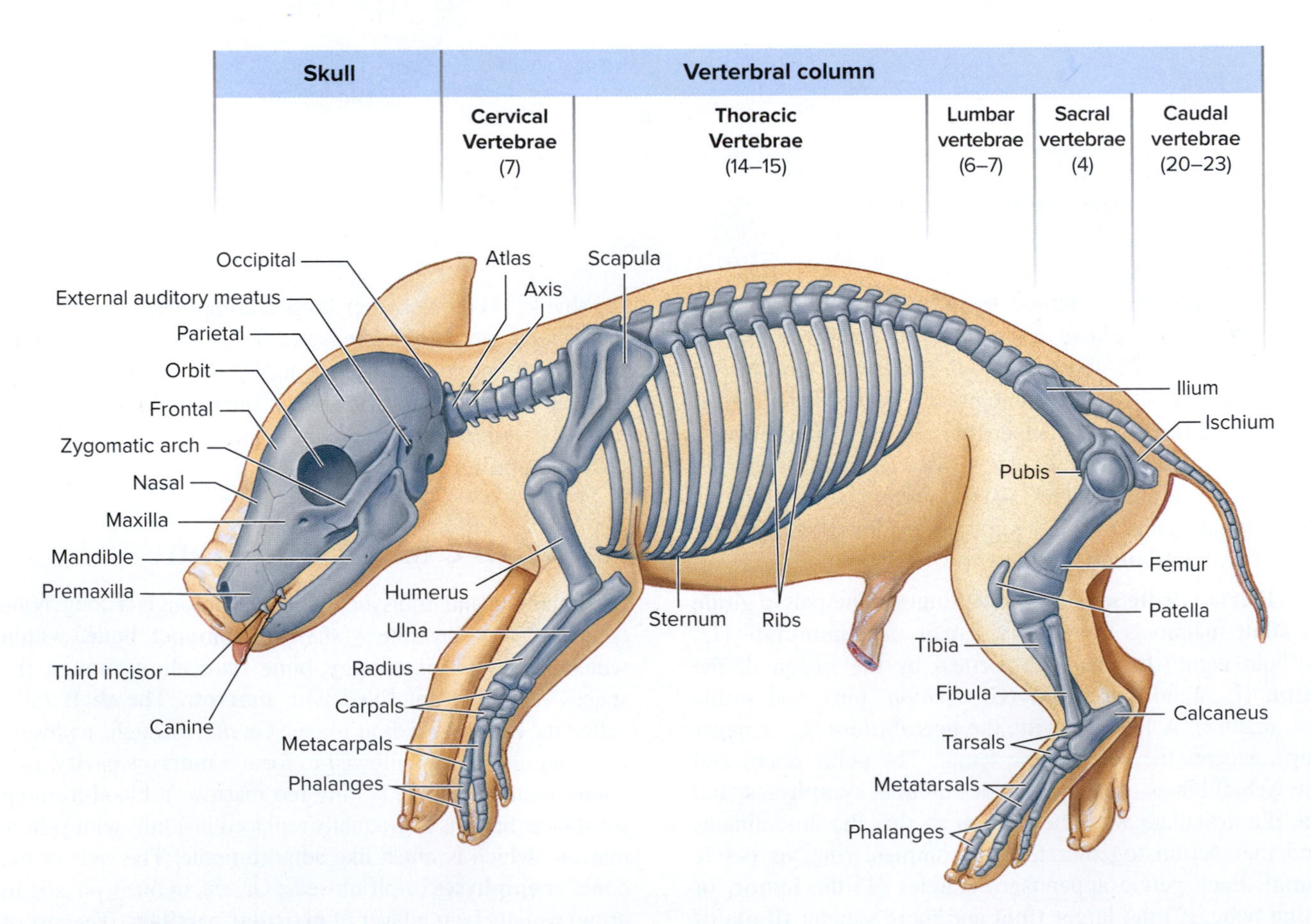

Figure 23-3
Skeleton of a fetal pig.

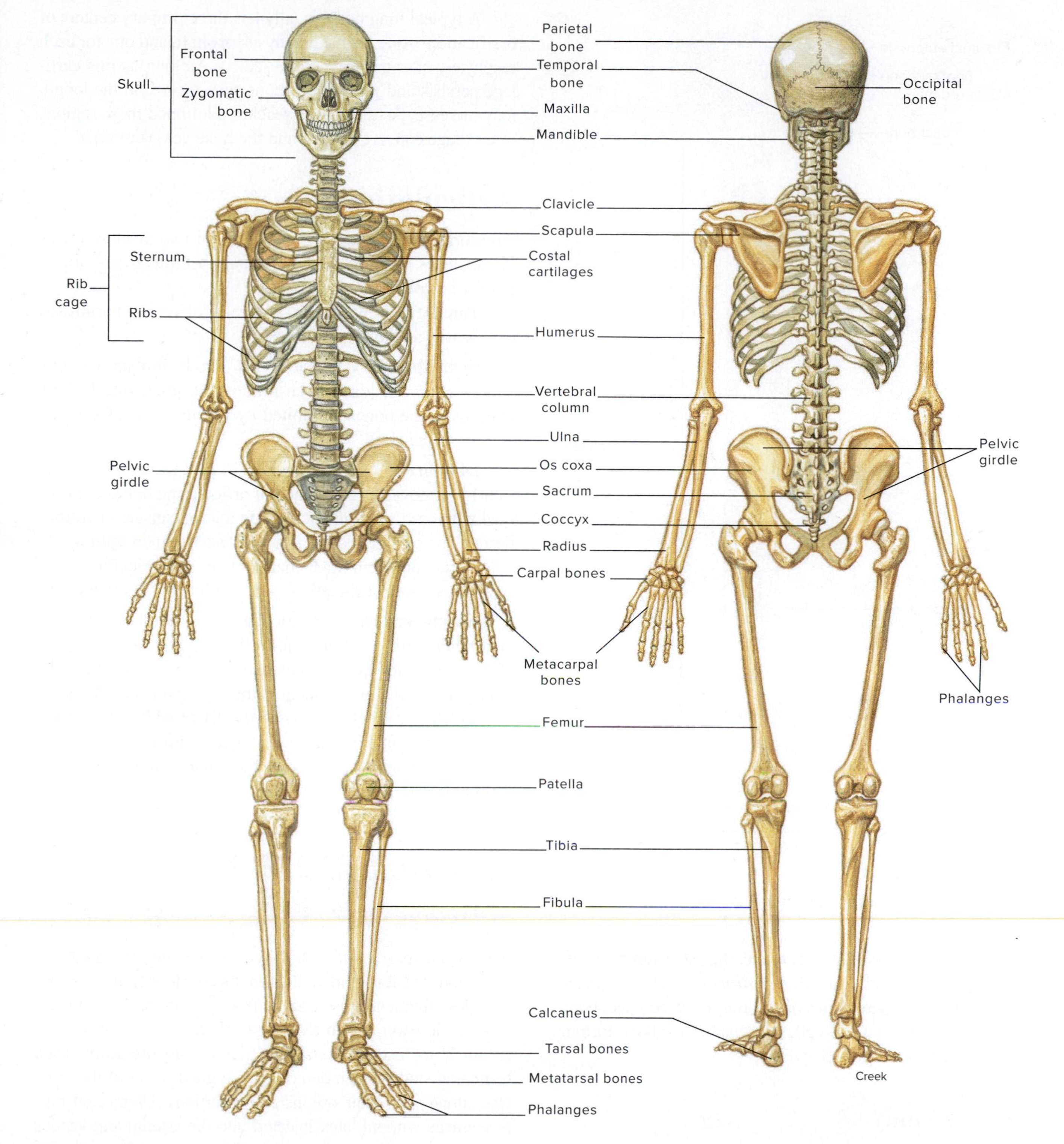

Skull
Frontal bone
Zygomatic bone
Parietal bone
Temporal bone
Maxilla
Mandible
Occipital bone
Clavicle
Scapula
Sternum
Costal cartilages
Rib cage
Ribs
Humerus
Vertebral column
Ulna
Pelvic girdle
Os coxa
Sacrum
Coccyx
Radius
Carpal bones
Metacarpal bones
Phalanges
Femur
Patella
Tibia
Fibula
Calcaneus
Tarsal bones
Metatarsal bones
Phalanges
Creek

Figure 23-4
Human skeleton.

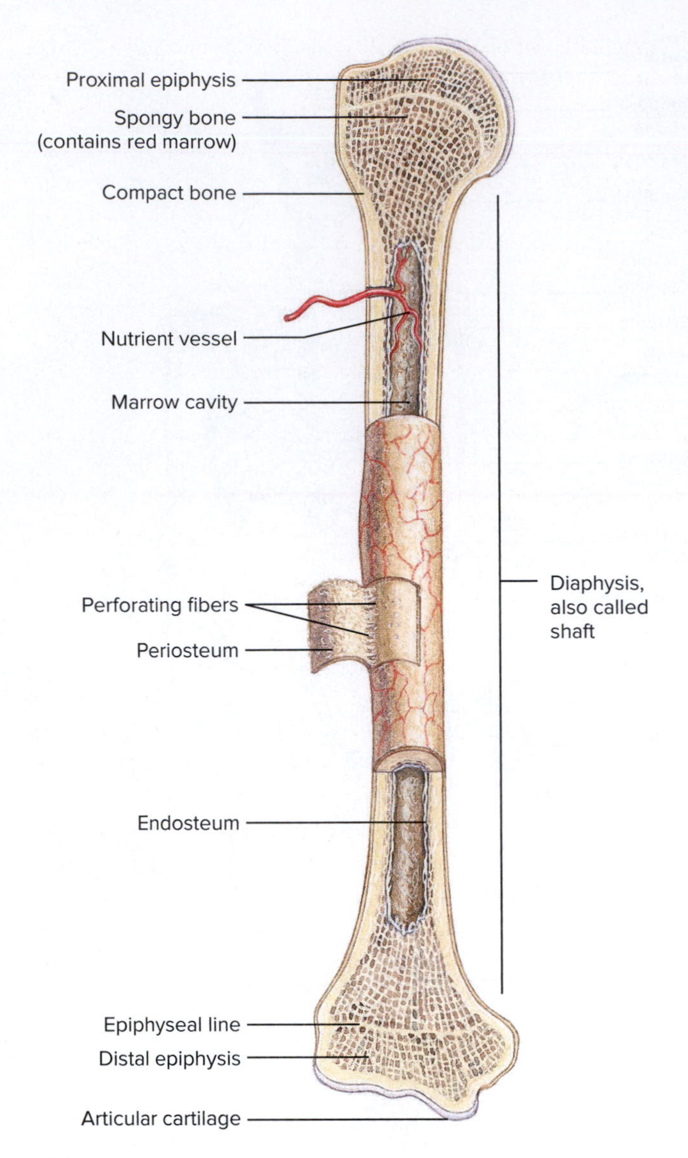

Figure 23-5
Diagram of longitudinal section of a long bone.

the bone is covered with a membrane, the **periosteum** (pear-ee-os′te-um; Gr. *peri,* around, + *ostrakon,* shell). Arteries, veins, nerves, and lymphatics pass through the compact bone to supply the marrow and cells responsible for bone formation and maintenance.

Further Study

Growth of a Bone

The primitive skeleton of an embryo consists of cartilage and fibrous tissue, in which bones develop by a process of ossification. Bones that develop in fibrous tissue—namely, some bones of the cranium and face—are called **membranous bones.** Most bones of the body develop from cartilage and are designated as endochondral ("within cartilage") bones.

A typical long bone usually has three primary centers of ossification—one for the diaphysis, or shaft, and one for each epiphysis, or extremity (Figure 23-5). As long as this cartilage persists and grows, new bone may form, and the length may increase. As the animal reaches adulthood these regions of cartilage convert to bone and the bone ceases to grow.

Articulations

An articulation, or joint, is the union of two or more bones or cartilages by another tissue, usually fibrous tissue or cartilage or a combination of the two.

Three types of joints are recognized—synarthrosis, diarthrosis, and amphiarthrosis.

Synarthrosis. A synarthrosis (sin-ar-thro′sis; Gr. *syn,* with, + *orthron,* joint) is an immovable joint. Interlocking margins of the bones are united by fibrous tissue. Example: sutures of the skull.

Diarthrosis. A diarthrosis (Gr. *dis,* twice, + *arthron,* joint) is a movable joint. Ends of articulating bones are covered with cartilage and enclosed in a joint capsule of fibrous tissue. The capsule contains a joint cavity lined with a vascular synovial membrane that secretes a lubricating fluid. Examples: most of the joints—knee, elbow, and others.

Amphiarthrosis. An amphiarthrosis (am-fee-ar-thro′sis; Gr. *amphi,* both, + *arthron,* joint) is a slightly movable joint. The bones are joined by a flattened disc of fibrocartilage. The bones of the joint are bound together by ligaments. These are tough bands or sheets composed mostly of white, fibrous tissue and are pliable but not elastic (except for the nuchal ligament at the back of the neck). Examples: pubic symphysis and joints between the vertebrae.

EXERCISE 23B

Muscular System

Because the muscles of a fetal pig are softer and the separations of the muscles less evident than in a lean cat or frog, a pig has been less frequently used for muscle dissection in beginning classes. However, if full-term pigs (30 cm or more in length) are used, and careful attention is given to the dissection, even beginning students can demonstrate a great many of the muscles, along with their origins and insertions. Uninjected pigs (specimens without latex injected into the arterial and venous systems) are quite satisfactory for this work; in fact, in some ways they are easier to work with than injected pigs.

A human being has approximately 700 identified and named skeletal muscles. A pig probably has fewer (because it lacks our five-fingered manual dexterity) but still possesses hundreds of muscles. Here only about 40 of the largest and most conspicuous will be studied. Even so, for the sake of time, your instructor may be selective, or some of the dissection may be done on an extra-credit basis.

Organization of Skeletal Muscles

Skeletal muscle is under voluntary control—that is, it is innervated by motor fibers. Skeletal muscle can contract effectively to about 30% of its resting length. However, the force that a muscle develops is not the same throughout its shortening length. If a muscle is stretched out fully, little force can be developed. Similarly, when a muscle is fully contracted, its force again diminishes. You know from your own experience that it is difficult to lift a heavy object with your forearm fully extended but that it becomes easier as your arm approaches a 90° angle. But, when the muscle is maximally shortened, the force you can exert again declines.

Skeletal Muscle Is Organized into Functional Bundles. Muscle fibers are bound together by a fibrous connective tissue called **fascia** (fa′shē-uh; pl. **fasciae,** fa′shē-ē; L., bundle) into bundles called **fasciculi** (fa-sick′yu-li; L. *fasciculus,* small bundle). Fasciculi are in turn organized in various ways into an entire muscle. The most common type of muscle is a **parallel muscle,** in which the fasciculi are arranged side by side parallel to the long axis of the muscle. Other muscles are arranged in sheets with broad attachments; still others are arranged in a circle, such as those that form sphincters around orifices.

How Muscles Are Connected. Skeletal muscles are connected to cartilage, bone, ligaments, and skin either directly by their investing fascia or indirectly by means of tendons or aponeuroses, never by actual muscle fibers themselves. If the muscle fibers come very close to the bone, we say the muscle has a "fleshy attachment." A **tendon** is a narrow band of tough, fibrous connective tissue, and an **aponeurosis** (Gr. *apo,* from, + *neuron,* sinew) is a broad, thin sheet of tough connective tissue that connects the muscle to its place of attachment.

Origin, Insertion, and Action. A muscle begins at its **origin,** which in general is the stationary end. It ends at its **insertion,** which is the end that moves. The movement produced by a muscle is its **action.** For example, the biceps brachii of a human arm originates on the scapula and inserts on the radius of the forearm. Its action is to flex the forearm. Sometimes these rules for origin and insertion do not easily apply. If, for example, a muscle extends between a broad aponeurosis and a narrow tendon, the aponeurosis is considered the origin and the tendon the insertion. Some muscles have several tendons at one end and only one at the other end; in this case, the muscle has multiple origins but only one insertion. Multiple origins in such instances are called **heads.** Even these rules do not apply to every muscle, and in the end it is better to know what a muscle does (its action) than to try to know every origin and insertion.

Naming Actions. Anatomists have long used descriptive terms for dynamic motion; such terms prevent having to use complicated notations such as "bends the foreleg toward the body" or "raises the arm toward the shoulder." **Flexion** (flek′shun) moves a distal part of a limb toward the next proximal part—for example, bending the fingers or the elbow. "Flexion" also refers to the bending of the head or trunk toward the ventral surface. **Extension** is the opposite movement; it *increases* the angle between articulating elements. Straightening the arm or the fingers is an extension movement. Muscles causing these movements are called **flexors** and **extensors.** Figure 23-6 illustrates muscle actions.

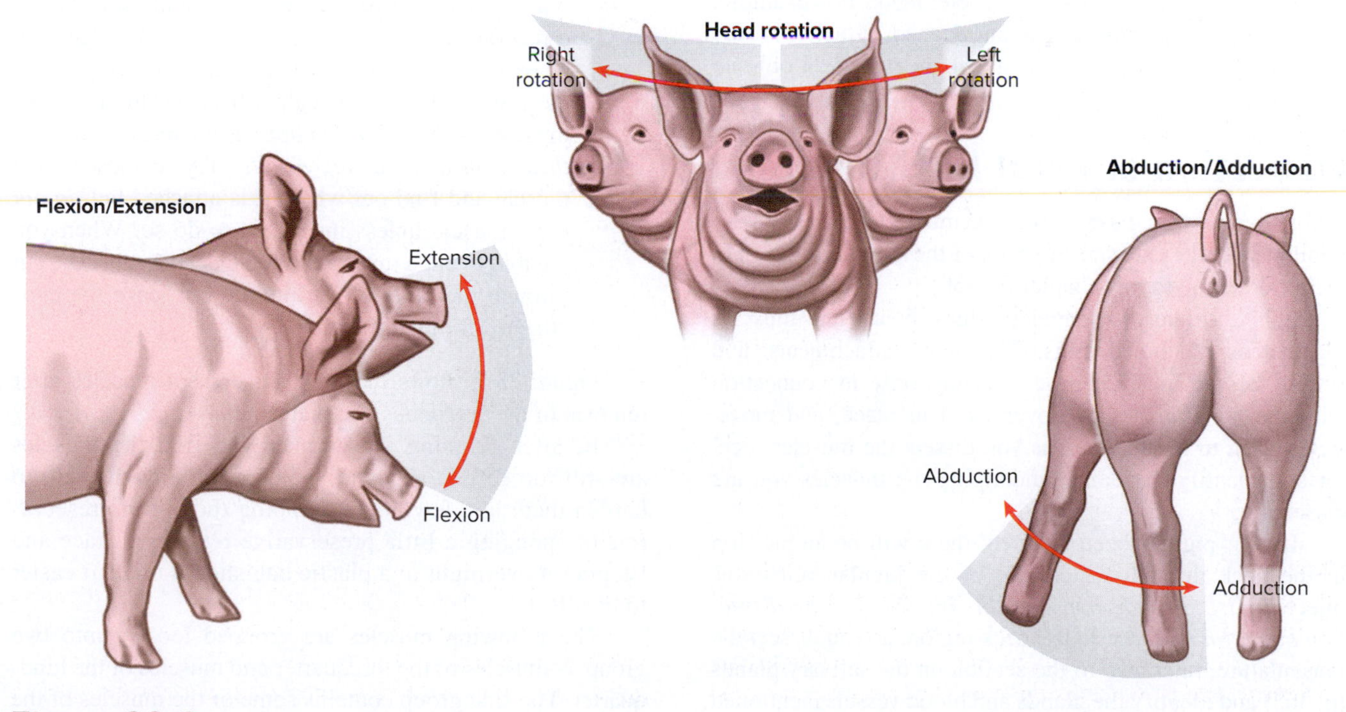

Figure 23-6
Illustrates muscle actions.

Adduction (L. *ad,* to, + *ducere,* to lead) refers to a movement of the distal end of a bone—for example, the humerus or femur—that brings it closer to the ventral median line of the body. In **abduction** (L. *ab,* from) such a bone is moved farther away from the ventral median line, as in raising the arm out to one side. Muscles that cause these movements are **adductors** and **abductors,** respectively. (It will be helpful if you can remember the meanings of the prefixes *ad* [meaning *to*] and *ab* [meaning *from*]).

Protractors in a tetrapod (an animal with four legs) may be defined as muscles that move the distal ends of bones—for example, the femur or humerus—forward longitudinally; **retractors** move them backward. A **depressor** may cause a part, such as the mandible or an eyelid, to be lowered; a **levator** raises such a part. **Rotation** is the turning of a part: for example, the rotation of the radius on the ulna or the first vertebra (atlas) on the second (axis). Human anatomists use several other muscle action terms as well (for example, "inversion," "eversion," "pronation," and "supination"), but learning the actions defined in this section will suffice for our study of a fetal pig.

Naming Muscles. The names of muscles are descriptive and may be derived from one or more of the following characteristics: (1) **position** (the brachialis is an arm muscle, the pectoralis is a chest muscle, and the cutaneous lies just under the skin); (2) **action** (adductor longus, depressor rostri); (3) **shape** (deltoid); (4) **direction** (transversus abdominus); (5) **number of divisions** (the biceps is two-headed, the digastric is two-bellied); and (6) **attachments** (the sternomastoid is attached to the sternum and to the mastoid process of the skull). Many names combine two or more of these descriptive elements. For example, the extensor carpi obliguus is an extensor attached to the carpus, which it extends, and its fibers run in an oblique direction.

Dissection of Fetal Pig Muscles

Before beginning a dissection of the muscles be sure you are familiar with the external structure of the animal, as given in the introduction to this chapter (p. 350).

It is absolutely essential to study bones and muscles together as functional units. The shape, attachments, and actions of the muscles have meaning only in connection with the bones that they cover, hold in place, and move. Refer often to the skeleton as you dissect the muscles. Feel for and identify the bones underlying the muscles you are dissecting.

If your pig has been injected, there will be an incision in the neck through which one of the jugular veins was injected. *Use the side opposite this incision for the dissection of lateral muscles.* In the neck region, as you dissect the musculature, refer also to the section on the salivary glands (p. 363) and identify the glands and blood vessels mentioned there.

The animal first must be skinned. Make a longitudinal middorsal incision through the skin from head to tail (making sure the cut is through the skin only). Make a midventral incision from chin to groin and circular incisions through the skin around the neck, chest, and groin. Then, starting with a corner of skin in the dorsal neck region, lift the skin with forceps and use a probe or the *blunt edge* of a scalpel to push back the muscle underneath. Continue removing pieces of skin until you have uncovered the entire body and legs (unless the instructor asks you to do only one side or only the forequarter).

The outer layer of muscle is a thin, superficial layer of **cutaneous muscle,** many fibers of which are attached directly to the skin. *Try not to remove this muscle layer with the skin.*

After the pig is skinned, there is usually a great deal of fat and connective tissue still covering the muscles. Much of this can be removed by gentle rubbing with a paper towel or by careful scraping with a scalpel.

Identify the outer layer of cutaneous muscle. In the neck and face region, this cutaneous layer is called the **platysma** (pla-tiz′ma; Gr., flat piece); in the trunk region, it is called the **cutaneous maximus.** These muscles are used in twitching the skin to shake off insects, dirt, or other irritants. This thin layer of muscle is *not shown in the illustrations.* After identifying this layer, *remove it carefully* to identify the superficial muscles underneath.

To locate the borders of muscles, scrape off the overlying connective tissue and fascia and look for the direction of the muscle fibers. A muscle edge may be seen where the fibers change direction. Try to slip the flat handle of the scalpel between the layers of muscle at this point. *Do not cut the muscles or tear them with a dissecting needle.* Try to loosen each muscle and find out where it is attached but *do not cut* a muscle unless instructed to do so. When you are told to cut a muscle to locate deeper muscles, cut through the belly of the muscle but leave the ends attached for identification.

Figure 23-7 shows the more superficial muscles after removal of the cutaneous and platysma layer.

If, after skinning the animal, you find the muscles are still too soft to separate, exposure to air will help to harden them in a few hours. Dipping the pig in preservative or sponging a little preservative over its surface and keeping it overnight in a plastic bag should make it easier to handle.

The following muscles are grouped loosely into two groups—muscles of the forequarter and muscles of the hindquarter. The first group contains some of the muscles of the face and neck as well as those of the shoulder, chest, and

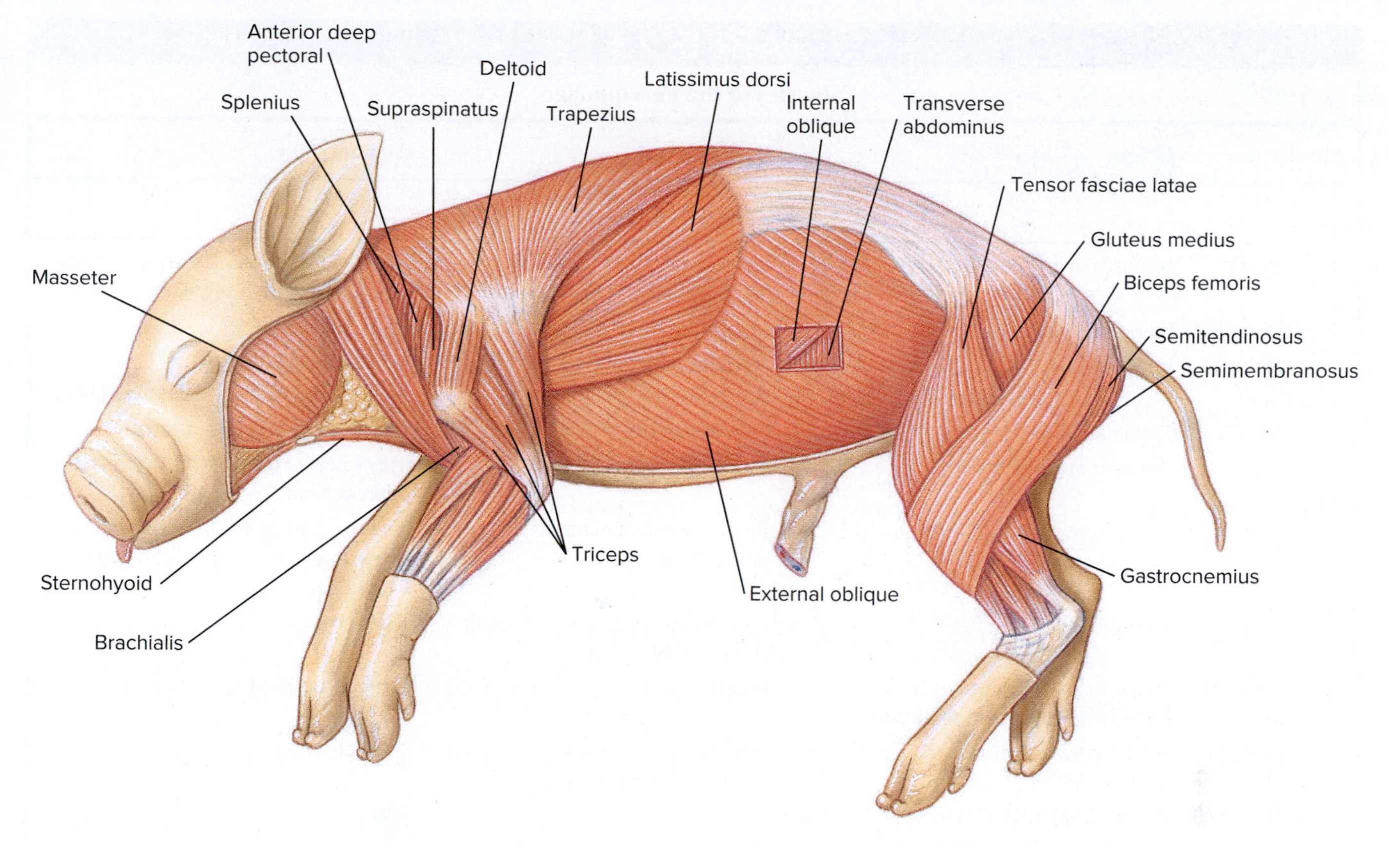

Figure 23-7

Superficial muscles of a fetal pig, lateral view. What muscle is most responsible for chewing? ____________ Clench your teeth and feel your own muscle that closes your jaw. What muscle elevates the shoulder? ____________ What anterior superficial thigh muscle flexes the hip joint and extends the knee? ____________

forelimb. The second group includes muscles of the back, abdomen, and hindlimb. This is not an exhaustive list, but it includes the chief superficial muscles and many of the muscles of the second layer.

Muscles of the Forequarter

Muscles of the Face, Neck, Chest, and Shoulder (Table 23.1)

Beginning with the throat muscles, locate the most ventral pair, the **sternohyoids,** marking the ventral midline of the neck and covering the larynx (Figure 23-8). These muscles retract and depress the hyoid and the base of the tongue, as in swallowing. Immediately internal to the sternohyoid on each side is the long **sternothyroid,** which retracts the larynx. Note that each has a lateral and a medial branch; separate these muscles to locate the small, dark, compact **thyroid gland** lying on the ventral side of the trachea.

The **digastric** muscle is the major depressor of the mandible—that is, it opens the jaw. It originates by a strong tendon from the base of the skull. Stretching ventrally between the mandibles as a thin transverse sheet is the **mylohyoid,** which compresses the floor of the mouth and assists in swallowing. The large muscle of the cheek is the **masseter.** This muscle and another (the temporal, not shown in Figure 23-9) are the major muscles that elevate the jaw and close the mouth.

Locate the **sternocephalic,** a flat muscle band that passes diagonally across the throat posterior to the submaxillary gland and beneath the parotid gland; it turns the head. When both sternocephalics contract together, the head is depressed. Posterior to the sternocephalic is the **brachiocephalic,** a large, band-shaped muscle that originates on the skull and inserts on the shoulder. This muscle raises the head, or if the head is fixed in position by other muscles, it draws the forelimb forward. It originates on two different processes (nuchal crest and mastoid process) of the skull.

Examine the muscles that position the shoulder girdle and move the forelimb (Figure 23-8). The **superficial pectoral** (equivalent to the pectoralis major of humans) adducts the humerus. The **posterior deep pectoral** and **anterior deep pectoral** both retract and adduct the

TABLE 23.1

Muscles of the Forequarter

Muscle	Origin	Insertion	Action
	Muscles of the Face, Neck, Chest, and Shoulder		
Sternohyoid	Anterior end of sternum	Hyoid bone	Retracts and depresses hyoid and base of tongue, as in swallowing
Sternothyroid	Sternum	Larynx	Retracts larynx
Digastric	By tendon from mastoid process of skull	Medial surface of mandible	Depresses mandible
Mylohyoid	Medial surface of mandibles	Hyoid bone	Raises floor of mouth and hyoid bone
Masseter	Zygomatic arch	Lateral surface of mandible	Elevates jaw and closes mouth
Sternoce-phalic	Anterior end of sternum	Mastoid process of skull	Turns head; two muscles together depress head
Brachioce-phalic	Two origins: nuchal crest and mastoid process of skull	Proximal end of humerus and fascia of shoulder	Singly, inclines head; when head is fixed, draws limb forward; together extend head
Superficial pectoral	Sternum	By a broad aponeurosis on medial surface of humerus	Adducts humerus
Posterior deep pectoral	Posterior half of sternum and cartilages of fourth to ninth ribs	Proximal end of humerus	Retracts and adducts forelimb
Anterior deep pectoral	Anterior part of sternum	Scapular fascia and aponeurosis that covers dorsal end of supraspinatus	Adducts and retracts limb
Trapezius	Nuchal crest of skull and neural spines of first 10 thoracic vertebrae	Spine of scapula	Elevates shoulder
Latissimus dorsi	Some of the thoracic and lumbar vertebrae and four ribs preceding last rib	Medial surface of humerus	Draws humerus upward and backward and flexes shoulder
Deltoid	Scapular aponeurosis	By an aponeurosis on proximal end of humerus	Flexes shoulder and abducts arm
Rhomboideus	Second cervical to ninth or tenth thoracic vertebrae	Medial surface of dorsal border of scapula	Draws scapula mediodorsally or rotates it
Rhomboideus capitis	Occipital bone	Dorsal border of scapula	Draws scapula forward and rotates shoulder
Splenius	First four or five thoracic neural spines	Occipital and temporal bones and first few cervical vertebrae	Singly, inclines head and neck to one side; together, elevate head and neck
Ventral serratus	Cervical part on transverse processes of last four or five cervical vertebrae; thoracic part on lateral surfaces of last eight or nine ribs	Medial surface of scapula	Singly, cervical part draws shoulder forward and thoracic part backward; together, shift weight to limb of contracting side; both sides together form elastic support that suspends trunk between scapulas; raise thorax
Supraspinatus	Anterior and dorsal portion of scapula and scapular spine	Proximal end of humerus	Extends humerus
Infraspinatus	Lateral surface and spine of scapula	Lateral surface of proximal end of humerus	Abducts and rotates forelimb
Triceps brachii	Long head: posterior border of scapula	All three heads insert on medial and lateral surfaces of olecranon process of ulna	Extends forearm
	Lateral head: lateral side of proximal end of humerus		
	Medial head: medial surface of proximal end of humerus, covering insertion of teres major		
Brachialis	Proximal third of humerus, ventral to lateral head of triceps	Medial surface of distal end of radius and ulna	Flexes elbow
Biceps brachii	Ventral surface of scapula near glenoid fossa	Proximal ends of radius and ulna	Flexes elbow

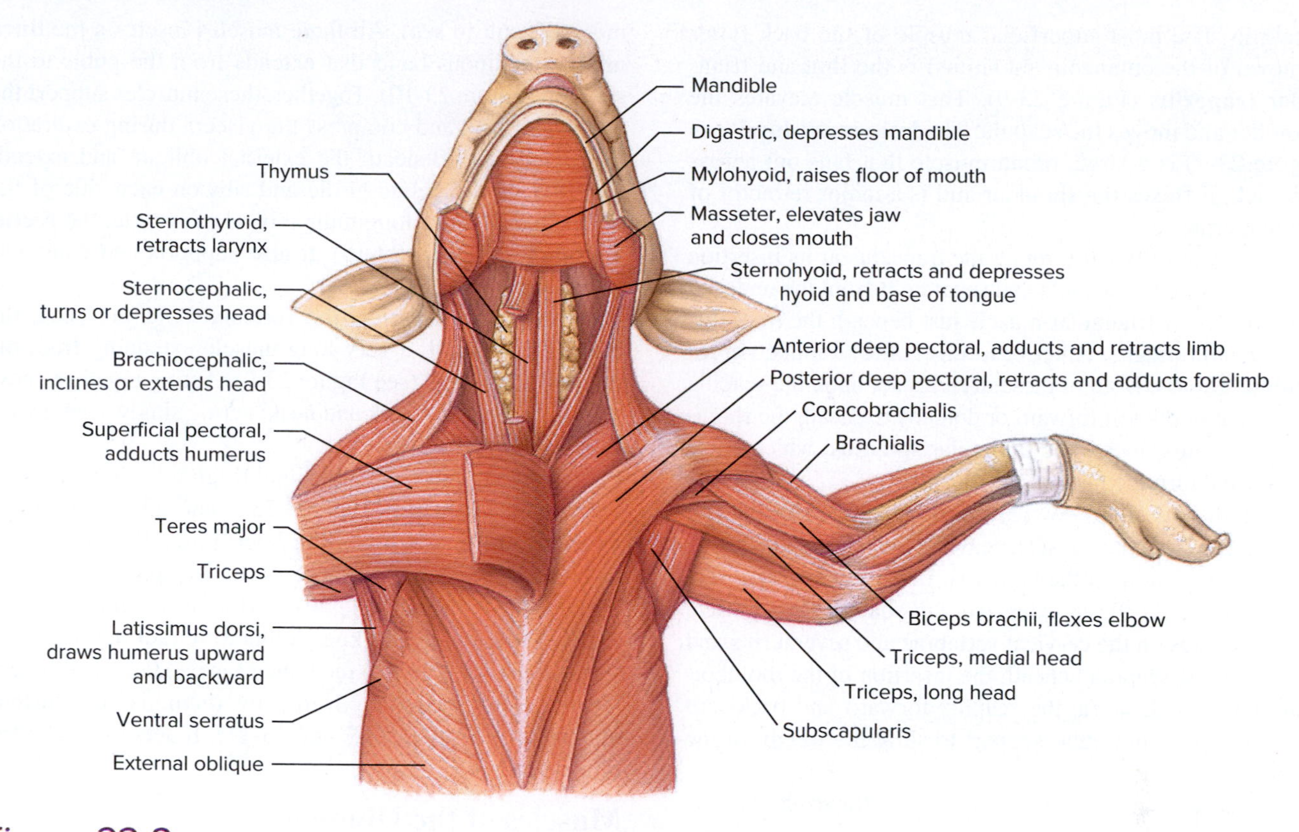

Figure 23-8

Muscles of the ventral thoracic region of a fetal pig. The pectoral muscle has been removed from the pig's left side to reveal the underlying musculature.

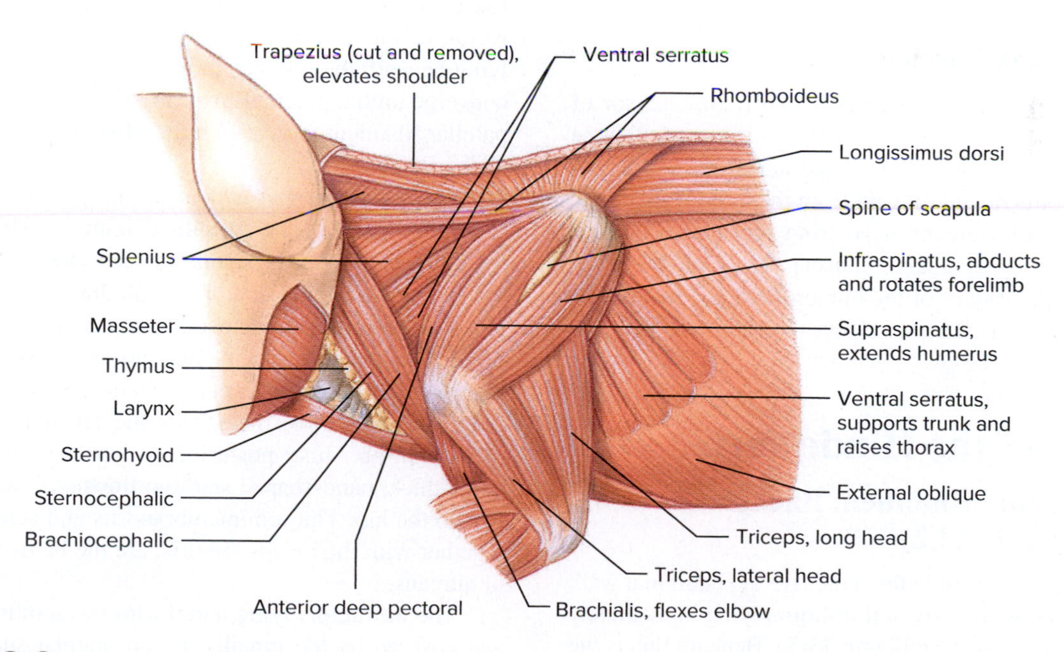

Figure 23-9

Muscles of the neck and shoulder of a fetal pig, lateral view. The trapezius muscle has been removed to reveal underlying musculature.

forelimb. The most superficial muscle of the back (after removal of the cutaneous maximus) is the thin and triangular **trapezius** (Figure 23-9). This muscle elevates the shoulder and moves the scapula. The **latissimus dorsi** (see Figure 23-7) is a large, broad muscle that fans out across the back; it flexes the shoulder and is a major retractor of the humerus.

Now carefully cut through the trapezius at its insertion on the spine of the scapula and separate it from the underlying muscles. A triangular muscle just beneath the trapezius is the **rhomboideus.** Anterior to this is the straplike **rhomboideus capitis.** Both of these muscles act upon the scapula to rotate it or draw it forward or dorsally. Beneath the rhomboideus capitis, locate the triangular **splenius,** which helps elevate and turn the head.

Cut the latissimus dorsi at its origin along the spine and remove most of the muscle, leaving only a centimeter or two at its insertion on the humerus. Locate the **ventral serratus** (Figure 23-9), an extensive, fan-shaped chest muscle that originates on the cervical vertebrae and several ribs and inserts on the scapula beneath the insertion of the rhomboideus muscles. It shifts the scapula forward and backward and serves as a muscular support to sling the weight of the trunk.

Arising near the insertion of the trapezius on the scapula is the **deltoid.** It inserts on the humerus and acts to protract (move forward) the upper foreleg. Carefully trim away the deltoid muscle at both origin and insertion. Beneath it lies the **infraspinatus,** which abducts and rotates the forelimb. Anterior to this is the **supraspinatus,** a fleshy muscle on the anterior surface of the scapula; it extends the humerus.

Muscles of the Foreleg

The largest muscle is the **triceps brachii,** an extensor of the forearm (Figures 23-8 and 23-9). It arises from three heads: a triangular-shaped long head at the posterior border of the scapula, a lateral head from the lateral surface of the humerus, and a medial head from the medial surface of the humerus. Two smaller muscles lying on the anterior and ventral surfaces of the humerus are the **brachialis** and **biceps brachii;** both act upon the elbow to flex the forearm.

Muscles of the Hindquarter

Muscles of the Abdomen, Back, and Hip (Table 23.2)

Three thin sheets of muscle lie in the lateral abdominal wall. Most superficial is the **external oblique,** lying immediately beneath the cutaneous (see Figure 23-7). Beneath this is the **internal oblique,** this is revealed by cutting a window high up on the external oblique. Then, by separating fibers of the internal oblique, you will see the **transverse abdominal,** the deepest layer and thinnest of the three muscles (and most difficult to see). All three muscles insert on the **linea alba,** a tendinous band that extends from the pubis to the sternum (Figure 23-10). Together, these muscles support the abdominal wall and compress the viscera during expiration and defecation. Beneath the external oblique and extending between the pelvic girdle and ribs on each side of the midventral line is a longitudinal band of muscle, the **rectus abdominus** (Figure 23-10). It also supports and constricts the abdomen.

Dorsal and lateral to the vertebral column, locate the **longissimus dorsi,** a very long muscle extending from the sacrum to the neck (see Figure 23-9). Acting together, these muscles extend the back and neck; acting singly, each flexes the spine laterally.

The most anterior superficial thigh muscle is the **tensor fasciae latae** (see Figures 23-7 and 23-11); it flexes the hip joint and extends the knee joint. Posteriorly the most superficial thigh muscle is the **biceps femoris** (see Figures 23-7 and 23-11). Its action is complex, acting across both the hip and knee joints to retract the thigh and flex the shank. Between the tensor fasciae latae and biceps femoris, and partially covered by them, is the **gluteus medius** (see Figures 23-7 and 23-11). It acts to abduct the thigh.

Muscles of the Hindleg

The **quadriceps femoris** is a large muscle group covering the anterior and lateral sides of the femur. It comprises four muscles: **rectus femoris,** a thick muscle on the anterior side of the femur; **vastus lateralis** (Figure 23-11), lateral to the rectus femoris and partly covering it; **vastus medialis** (Figure 23-10), on the medial surface of the rectus femoris; and **vastus intermedialis,** a deep muscle lying beneath the rectus femoris. All four of these muscles converge on the patella (kneecap) and then continue as the patellar ligament to insert on the tibia. These are extensors of the shank.

The posteromedial half of the thigh is covered with a thin, wide muscle, the **gracilis** (Figure 23-10). It adducts the thigh and flexes the shank. Just anterior to the gracilis is the **sartorius,** a thin band of muscle that covers the femoral blood vessels; it is delicate and easily destroyed if not identified. Cut through the gracilis and sartorius to reveal the large **semimembranosus** muscle in the medial portion of the thigh (Figure 23-10). It extends the hip joint and adducts the hindlimb. Just posterior to the semimembranosus is the thick, band-shaped **semitendinosus;** it acts mainly to extend the hip. The semimembranosus and semitendinosus, together with the biceps femoris, are the hamstring muscles of humans.

The **adductor,** lying anterior to the semimembranosus and covered by the gracilis, is a triangular-shaped muscle that, as its name suggests, adducts the femur—that is, draws it toward the midline.

Also on the medial side of the thigh are three smaller muscles: the triangular-shaped **pectineus,** an adductor of the

TABLE 23.2

Muscles of the Hindquarter

Muscle	Origin	Insertion	Action
Muscles of the Abdomen, Back, and Hip			
External oblique	Lateral surface of last 9 or 10 ribs and lumbodorsal fascia	Linea alba, ilium, and femoral fascia	Compresses abdomen, arches back; singly, it flexes trunk laterally
Internal oblique	Similar to external oblique	Similar to external oblique	Similar to external oblique
Transverse abdominal	Similar to external oblique	Similar to external oblique	Similar to external oblique
Rectus abdominus	Pubic symphysis	Sternum	Constricts abdomen
Longissimus dorsi	Sacrum, ilium, and neural processes of lumbar and thoracic vertebrae	Transverse processes of most vertebrae and lateral surfaces of ribs except the first	Singly, flexes spine laterally; together, extend back and neck; rib attachments may aid in expiration
Tensor fasciae latae	Crest of ilium	Fascia over knee, patella, and crest of tibia	Flexes hip joint and extends knee joint
Biceps femoris	Lateral part of ischium and sacrum	By a wide aponeurosis to patella and fascia of thigh and leg	Abducts and extends limb; may also flex knee joint
Gluteus medius	Fascia of longissimus dorsi, ilium, and sacroiliac and sacrosciatic ligaments	Proximal end of femur	Abducts thigh
Muscles of the Hindleg			
Quadriceps femoris, a large muscle group consisting of			
1. Rectus femoris	Ilium	Patella and its ligament	Extends shank
2. Vastus lateralis	Proximal end of femur	Patella and its ligament	Extends shank
3. Vastus medialis	Proximal end of femur	Patella and its ligament	Extends shank
4. Vastus intermedialis	Proximal end of femur	Patella and its ligament	Extends shank
Gracilis	Pubic symphysis and ventral surface of pubis	Patellar ligament and proximal end of tibia	Adducts hindlimb
Sartorius	Iliac fascia and tendon of psoas minor (external iliac vessels lie between two heads)	Patellar ligament and proximal end of tibia	Adducts hindlimb and flexes hip joint
Semimembranosus	Ischium	Distal end of femur and proximal end of tibia, both on medial side	Extends hip joint and adducts hindlimb
Semitendinosus	First and second caudal vertebrae and ischium	Proximal end of tibia and calcaneus	Extends hip and tarsal joint and flexes knee joint
Adductor	Ventral surface of pubis and ischium and tendon of origin of gracilis	Proximal end of femur	Adducts hindlimb and extends and rotates femur inward
Pectineus	Anterior border of pubis	Medial side of shaft of femur	Adducts hindlimb and flexes hip
Iliacus	Ventral surface of ilium and wing of sacrum	Proximal end of femur together with psoas major	Flexes hip and rotates thigh outward
Psoas major	Ventral sides of transverse processes of lumbar vertebrae and last two ribs	With iliacus on proximal end of femur	Flexes hip and rotates thigh outward

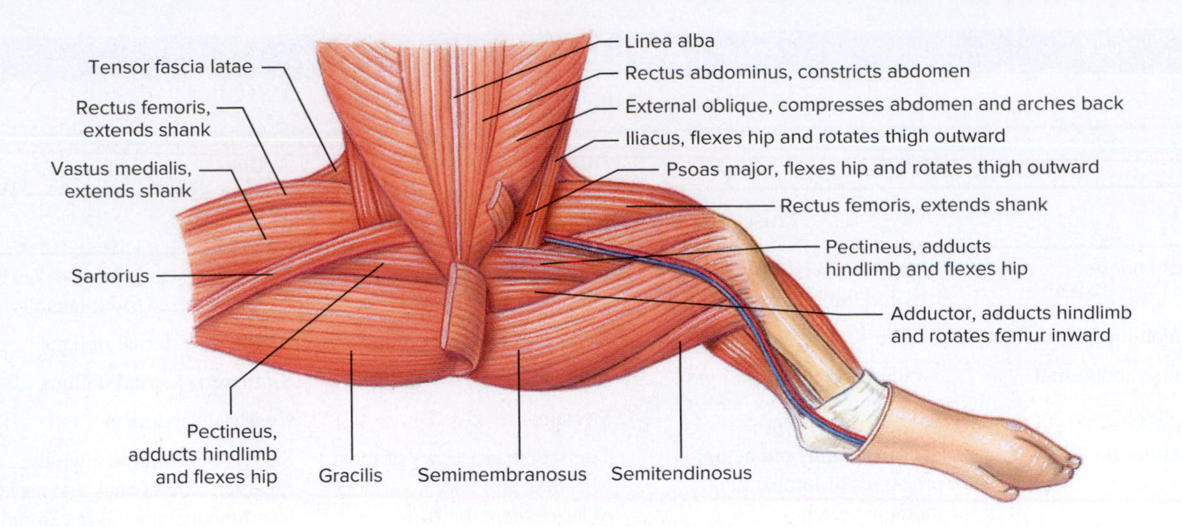

Figure 23-10

Muscles of the hindlimb of a fetal pig, ventral view. The gracilis and sartorius have been cut and removed from the pig's left leg.

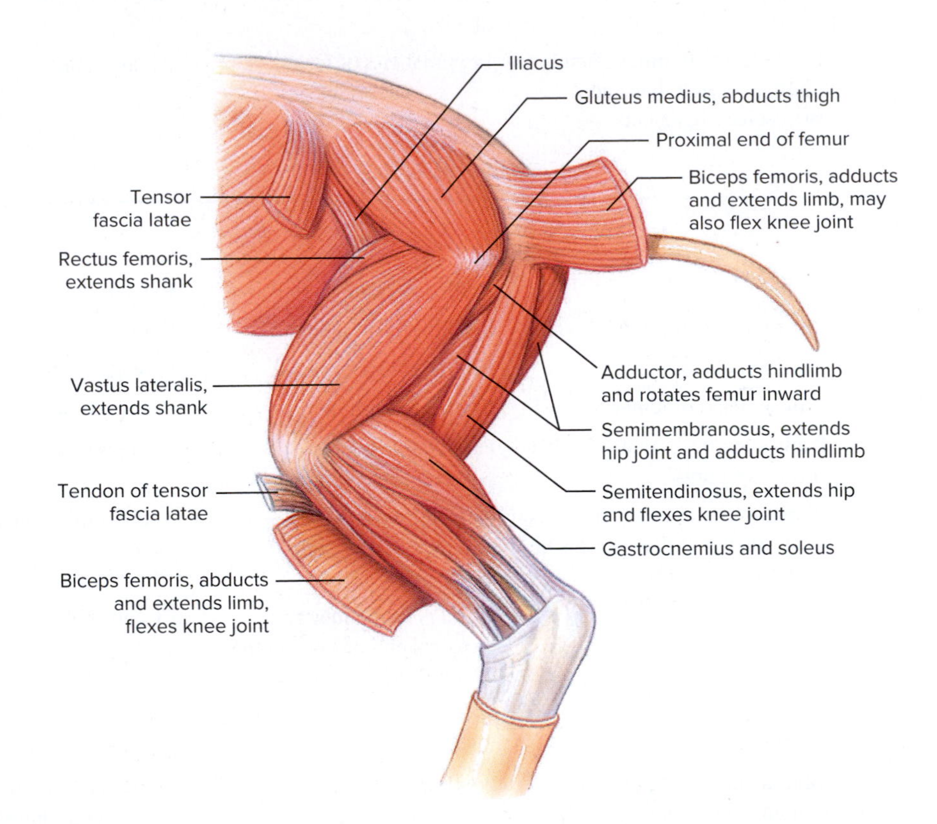

Figure 23-11

Muscles of the hindlimb of a fetal pig, lateral view.

thigh; the **iliacus,** which flexes the hip and rotates the thigh outward; and the **psoas major,** which acts the same as the iliacus.

Some of the shank muscles are shown in Figure 23-11 but are not described in this exercise.

Identification

Be able to identify and explain the origin, insertion, and action of as many of the aforementioned muscles as your instructor has assigned.

EXERCISE 23C

Digestive System

Core Study

Head and Throat

Salivary Glands

Three pairs of salivary glands produce a continual background level of fluid secretions containing lysozymes and immunoglobulins that flush the teeth and mouth cavity and help keep bacterial growth under control. During meals, much larger quantities of saliva are produced. These secretions contain lubricating glycoproteins called **mucins** and a large amount of **salivary amylase** (α-amylase), which begins the breakdown of complex carbohydrates such as starch.

On the right side of the face, neck, and chin, carefully remove the skin if you have not already done so. A muscle layer will tend to adhere, but push this layer back into place gently so as not to destroy the glands beneath. Now carefully remove the thin muscles behind the angle of the jaw and beneath the ear to uncover the **parotid gland.** Do not destroy any large blood vessels.

The triangular parotid (pa-rot′id; Gr. *para,* beside, + *ous,* ear) gland is broad, thin, and rather diffused, extending from almost the midline of the throat to the base of the ear (Figure 23-12). Do not confuse the salivary glands, which are choppy and lobed in appearance, with the lymph nodes, which are more smooth and shiny. The **parotid duct** comes from the deep surface of the gland and follows the ventral border of the masseter (cheek) muscle along the external maxillary vein to the corner of the mouth (Figure 23-12).

The **submaxillary (mandibular) gland** lies under the parotid gland and just posterior to the angle of the jaw. It is darker, compact, and oval. Its duct comes from the anterior surface of the gland and passes anteriorly, medial to the mandible, and through the sublingual gland to empty into the floor of the mouth. This duct is very difficult to trace.

To find the **sublingual glands,** remove the mylohyoid muscle and the slender pair of geniohyoid muscles immediately beneath it.

On each side of the head, a whitish, elongated sublingual gland is located between the diagastric muscle, which lies inside the mandible, and the genioglossus, which is one of the muscles of the base of the tongue. A sublingual artery and vein will be seen along the ventral side of each gland. The sublingual glands empty by way of several short ducts to the floor of the mouth.

Mouth Cavity and Pharynx

Cut through the angle of the mouth on both sides with scissors. Cut posteriorly, pulling open the mouth as you proceed. Follow the angle of the tongue and do not cut into the roof of the mouth. Continue the cuts to the esophagus to expose the oral cavity and pharynx fully (Figure 23-13).

Teeth may not be erupted yet, although the canines and third pair of incisors may be seen in older fetuses. (The third incisors and the canines are the first to erupt; the second incisors are the last.) A young pig will have three incisors, one canine, and four premolars on each side of each jaw.

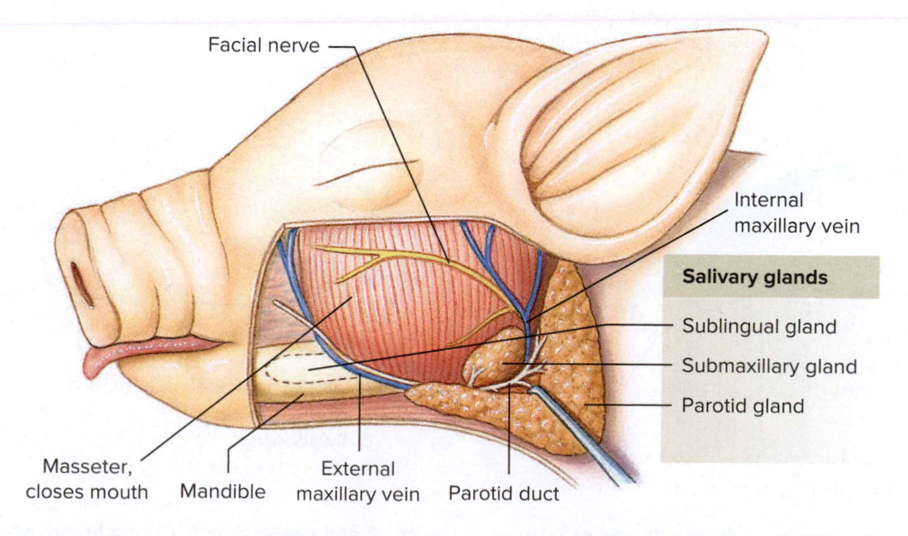

Figure 23-12
Dissection of the head and neck of a fetal pig to show some superficial veins, nerves, and salivary glands.

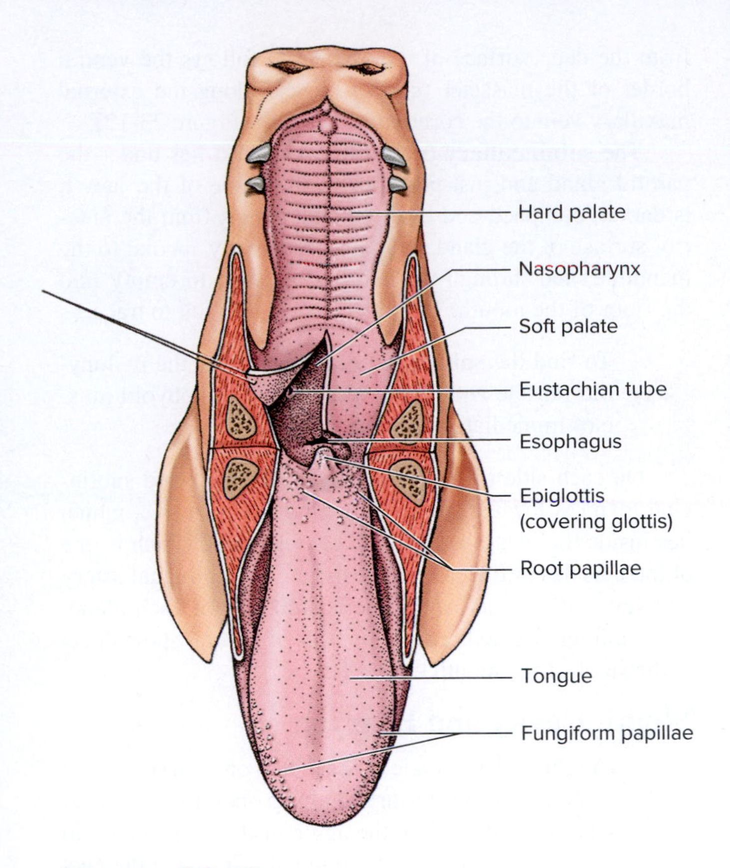

Figure 23-13
Oral cavity of a fetal pig.

Remove the flesh along the right jaws and carefully cut away enough of the jawbone to expose buds of the embryonic teeth.

The mouth cavity is roofed by a narrow, bony **hard palate,** sheathed ventrally with mucous membranes ridged into transverse folds. Extending posteriorly from the hard palate is the **soft palate** composed of thick membrane. The hard and soft palates of mammals completely separate the oral cavity from the air passages above, an innovation that allows a mammal to chew a mouthful of food at leisure while breathing freely through its nose. Among other vertebrates, only crocodilians have a hard palate; the soft palate is unique to mammals.

Open the mouth wide, drawing down the tongue, to locate at the posterior end of the soft palate the opening into the **nasopharynx** (fair′inks; Gr. *pharyngx,* gullet), the space above the soft palate. It connects with the nasal passages from the nostrils.

To expose the nasopharynx, make a midline incision of the soft palate. Locate the small openings on each side of the roof of the nasopharynx.

From these openings, the **eustachian tubes** (named after B. Eustachio, an Italian physician) lead to the middle ear.

Posterior to the nasopharynx is the **laryngeal pharynx,** which connects the oral cavity with the **esophagus.** Both nasal and laryngeal pharynx are derived from the pharynx of ancestral chordates, which evolved as a filter-feeding apparatus. Pharyngeal (gill) pouches develop in this region in all vertebrate embryos. In fishes, these pouches break through to develop into gill chambers, but in tetrapods they become transformed into other structures: middle ear cavity and glandular tissue (thyroid, parathyroid, and thymus).

The **larynx** lies in the floor of the laryngeal pharynx. Locate the flaplike **epiglottis,** which folds up over the **glottis** (open end of the larynx) to close it when food is being swallowed. Note that in the mouth air passages are *dorsal* to the food passage. In the throat, however, air is carried through the larynx and trachea, which are *ventral* to the food passage (esophagus). These passageways cross in the pharyngeal cavity (Figure 23-14). When the animal is respiring, the epiglottis fits up against the opening into the nasopharynx, allowing air

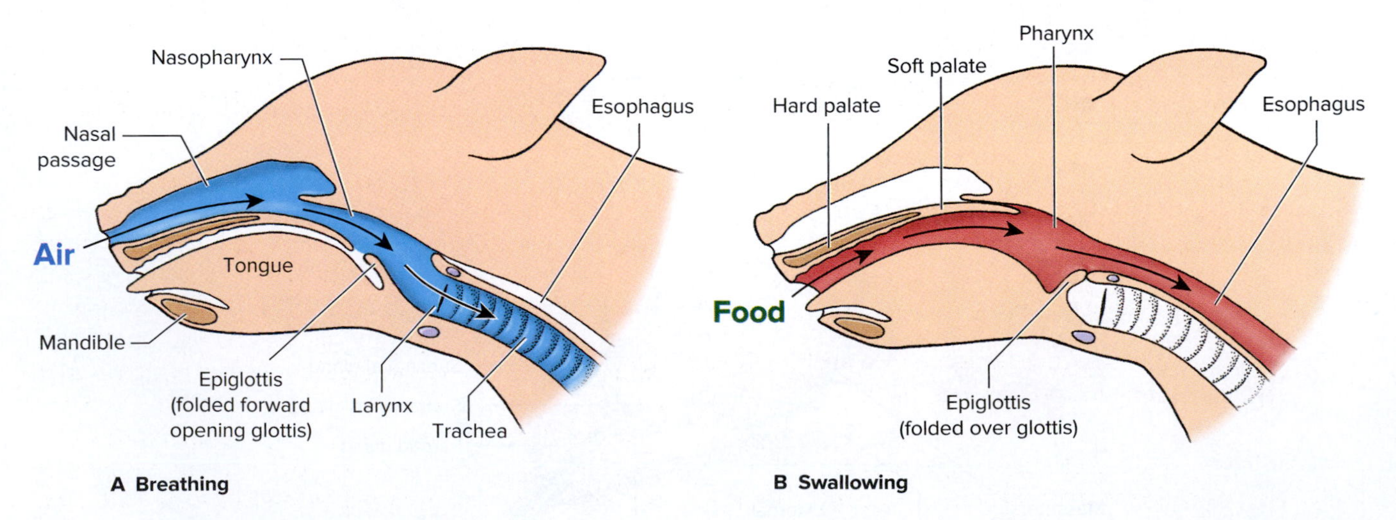

Figure 23-14
Relationship of respiratory passage to the mouth and esophagus, in breathing and swallowing. **A,** During breathing, the glottis is open to receive air from nostrils and is protected from food and saliva by the epiglottis. **B,** For swallowing, the larynx is pushed anteriorly, causing the epiglottis to fold over the glottis, thus closing air passage to the lungs. Feel your Adam's apple (larynx) as it moves up when you swallow.

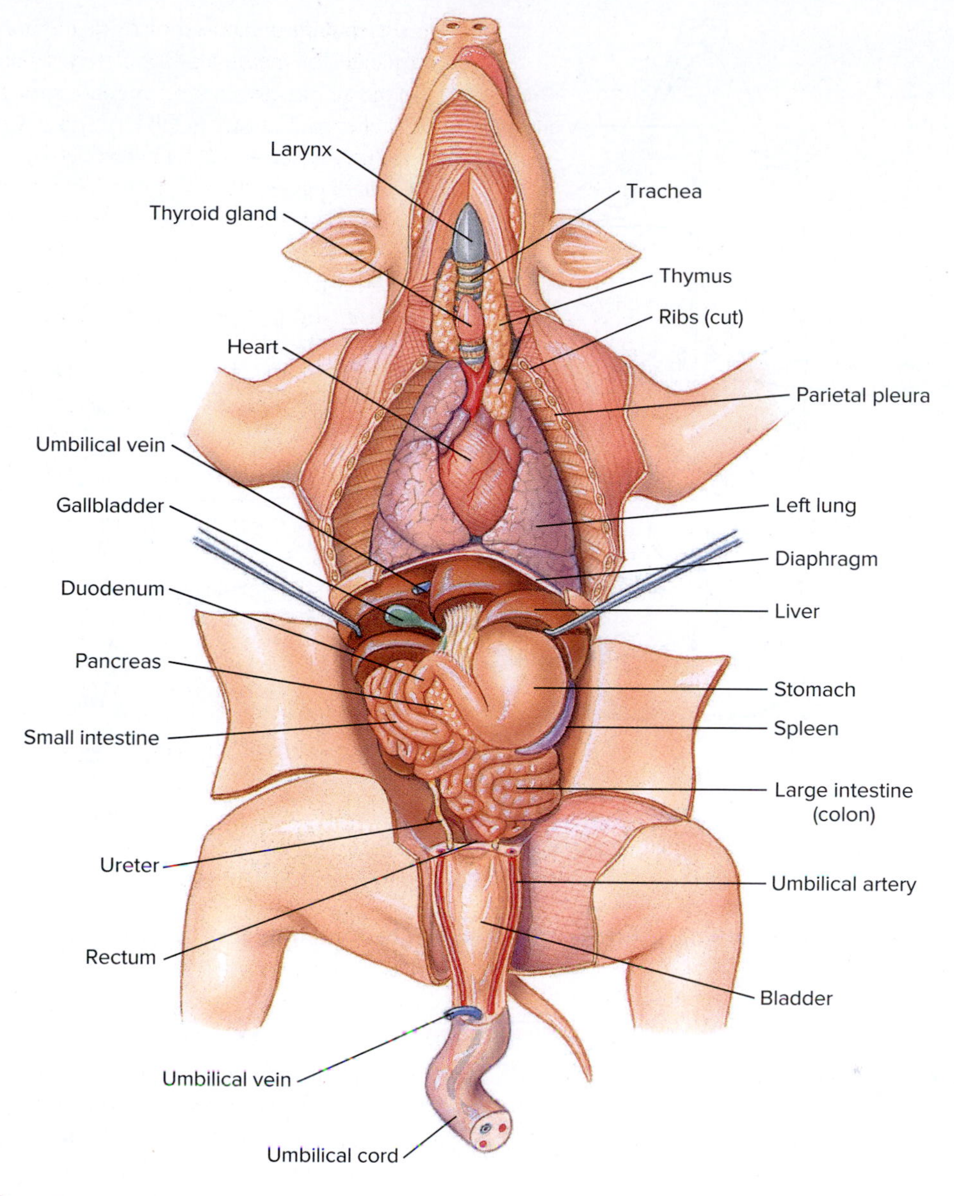

Figure 23-15

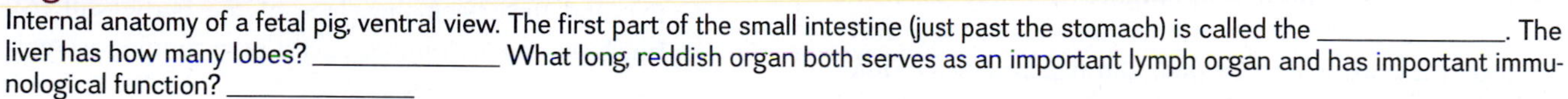

Internal anatomy of a fetal pig, ventral view. The first part of the small intestine (just past the stomach) is called the ______________. The liver has how many lobes? ______________ What long, reddish organ both serves as an important lymph organ and has important immunological function? ______________

into the larynx but preventing saliva or food from entering the mouth. During swallowing the larynx is pushed forward, causing the epiglottis to fold over the glottis, thus opening the food passage while closing off the air passage.

Continue your dissection of the neck region by making a midventral incision down the neck.

After parting the skin and clearing away some tissue around the larynx, you will expose the **thymus,** a large, soft, irregular mass of glandular tissue lying lateral to the sternohyoid muscles (Figure 23-15). It is an extensive gland in the fetus and young animal. After puberty it decreases in size, but it continues to function throughout life. The thymus extends caudally under the sternum, with its posterior portion overlying the heart. The thymus is part of the body's lymphatic system and is filled with lymphocytes of all sizes, especially T cells, which are important in immunological responses.

As you clear tissue from the larynx, find the **trachea** (windpipe) extending caudally from it. The trachea is stiffened by a series of C-shaped cartilage rings, which are incomplete dorsally where the trachea lies against the esophagus.

Tracing the trachea posteriorly, you will see the **thyroid gland,** a small, dark red, oval gland lying on the trachea beneath the sternothyroid muscles (Figure 23-15). The thyroid is an endocrine gland that produces thyroxin and triiodothyronine, two hormones that promote growth and development and regulate the metabolic rate.

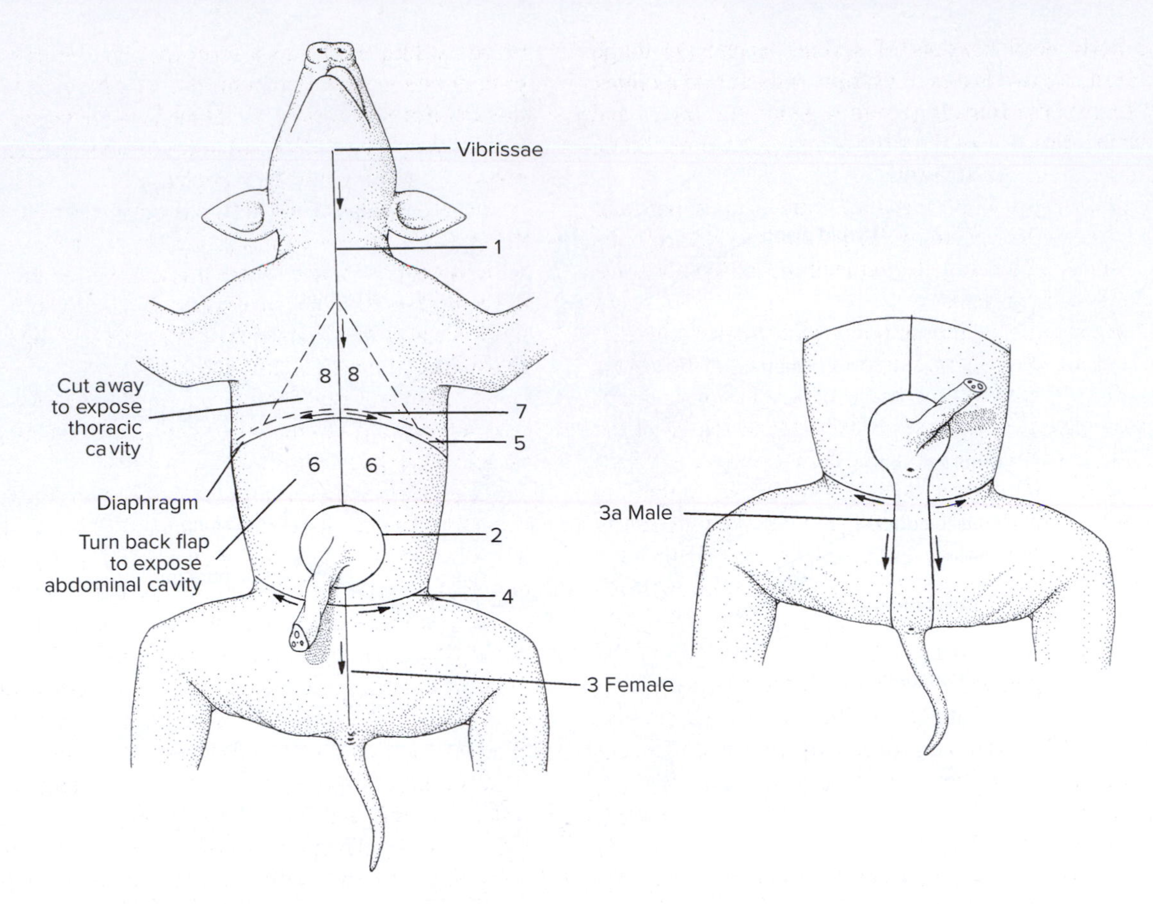

Figure 23-16
Cutting diagram. The numbers indicate the order in which each incision is to be made. Incisions 1 through 6 expose the abdominal cavity; 7 and 8 expose the thoracic cavity.

Abdominal Cavity

Directions for Dissection

To proceed further, it is necessary to expose the organs of the abdominal cavity. Place the pig ventral side up in the dissecting pan. Tie a cord or rubber band around one forelimb, loop the cord under the pan, and fasten it to the other forelimb. Do the same to the hindlegs.

It is important to remember that instructions referring to the "right side" refer to the animal's right side, which will be on your left as the animal lies ventral side up in the dissection pan.

With a scalpel, make a midventral incision through the skin and muscles, but not into the body cavity, continuing the incision already made in the neck posteriorly to within 1 cm of the umbilical cord (incision 1, Figure 23-16). Cut around each side of the cord (2). If your specimen is female, continue down the midline from the cord to the anal region (3). If it is male, make two incisions, one on each side of the midline, to avoid cutting the penis, which lies underneath (3a). Now, in either sex, deepen the incisions you have made in the abdominal region through the muscle layer to reach the body cavity, taking care not to injure the underlying organs. With scissors, make two lateral cuts on each side, one just anterior to the hindlegs (4) and the other posterior to the ribs (5), and turn back the flaps of the body wall (6). Flush out the abdominal cavity with running water.

All visceral organs are invested in mesentery and held in place with connective tissue. Loosen this tissue carefully to separate organs, tubes, and vessels, being careful not to cut or tear them. *Do not remove any organs unless you are specifically directed to do so.* Be careful in all your preliminary dissection not to destroy blood vessels or nerves, to keep them intact for later dissection of the circulatory and nervous systems.

Notice that the umbilical cord is attached anteriorly by a tube, the **umbilical vein.**

Tie a string around the umbilical vein in two places and sever the vein between the two strings.

The strings will identify this vein later. Lay the umbilical cord between the hindlegs and identify the following parts.

The **body wall** consists of several layers: (1) tough external **skin,** (2) two layers of **oblique muscle** and an inner layer of **transverse muscle** (try to separate the layers and determine the direction of the fibers), and (3) an inner lining of thin, transparent **peritoneum.**

The **diaphragm** is a muscular, dome-shaped partition separating the peritoneal cavity (abdominal cavity) from the thoracic cavity, which together constitute the coelom. *Do not remove the diaphragm.*

The peritoneum is the smooth, shiny membrane that lines the abdominal cavity and supports and covers the organs within it. That which lines the body walls is called the **parietal peritoneum.** It is reflected off the dorsal region of the body wall in a double layer to form the **mesenteries,** which suspend the internal organs, and then continues around the organs as a cover, where it is called the **visceral peritoneum.**

The **liver** is a large, reddish gland with four main lobes lying just posterior to the diaphragm. The greenish, saclike **gallbladder** may be seen under one of the central lobes (Figure 23-15).

The **stomach** is nearly covered by the left lobe of the liver. The **small intestine** is loosely coiled and held by mesenteries. Note the blood vessels in the mesentery that supports the digestive tract. The **large intestine** is compactly coiled on the left side posterior to the stomach.

The **spleen** is a long, reddish organ attached by a mesentery to the greater curvature of the stomach. The spleen contains one of the largest concentrations of lymphatic tissue in the body. It phagocytizes spent blood components and salvages the iron from hemoglobin for reuse. It is also immunologically important in initiating immune responses by B cells and T cells.

Umbilical arteries are two large arteries extending from the dorsal wall of the coelom to and through the umbilical cord (Figure 23-15).

The **allantoic bladder,** the fetal urinary bladder, is a large sac lying between the umbilical arteries. It connects with the allantoic duct in the umbilical cord.

Kidneys are two large, bean-shaped organs attached to the dorsal wall and lie dorsal to the intestines. They are outside the peritoneal cavity in the **cisterna magna** and are separated from the other abdominal organs by the peritoneum.

With scissors, begin just anterior to the diaphragm and cut along the midventral line through the sternum to a point midway between the forelegs. Keep the lower blade of the scissors up to prevent injuring the heart underneath. Now make a lateral cut on each side just anterior to the diaphragm (see incision 7, Figure 23-16). This exposes the thoracic cavity but leaves the diaphragm in place.

Digestive Tract

The digestive system consists of the alimentary canal, extending from mouth to anus, and glands (such as salivary glands, liver, and pancreas) that assist in its function of converting food into a form that can be assimilated for growth and energy requirements. You have already studied the anterior portions of the alimentary canal: mouth cavity and salivary glands. We will now consider the digestive tract proper, beginning with the esophagus.

The **esophagus** is a soft, muscular tube that leads from the pharynx to the stomach. Locate it in the neck region posterior to the larynx, where it is attached to the dorsal side of the trachea by connective tissue. Find the esophagus in the thoracic cavity posterior to the lungs; in the abdominal cavity, find where it emerges through the diaphragm at the cardiac end of the stomach. The muscles at the anterior end of the tube are striated (voluntary), gradually changing to smooth muscle. How does this affect swallowing? ______________

The **stomach** (see Figure 23-15) is a large, muscular organ that breaks up food and thoroughly mixes it with gastric juice. Identify its **cardiac end** near the heart, its **pyloric end** that joins the duodenum, its **greater curvature** where the spleen is attached, and its **lesser curvature.** The **fundus** (L., bottom) is the anterior blind pouch. The contents of the fetal digestive tract, made green by pigments in the bile salts, are called **meconium** and contain epithelium sloughed from the mucosa lining, sebaceous secretions, and amniotic fluid swallowed by the fetus. Open the stomach longitudinally, rinse out, and find (1) the **rugae,** or folds, in its walls; (2) the opening from the esophagus; and (3) the **pyloric** (Gr. *pylōros,* gatekeeper) **valve,** which regulates the passage of food into the duodenum.

The **small intestine** includes the **duodenum** (doo-uh-de′num or doo-od′in-um; L., "twelve-each," referring to its length in humans, which equals about 12 fingerwidths), or first portion, which lies next to the pancreas and receives the common bile duct and pancreatic duct, and the **jejunum** and **ileum,** indistinguishable in the fetal pig, which make up the remainder of the small intestine.

Remove a piece of the intestine, open it, and examine it *under water* with a hand lens or dissecting microscope. Observe the minute, fingerlike **villi** (sing. **villus;** L., shaggy hair), which greatly increase the absorptive surface of the intestine.

Most digestion and absorption take place in the small intestine.

The **large intestine** includes a long, tightly coiled **colon** and a straight, posterior **rectum,** which extends through the pelvic girdle to an **anus.** Its primary function is absorption of water and minerals from the liquified chyme that enters it.

Find the **cecum,** a blind pouch of the colon at its junction with the ileum. In humans and anthropoid apes, the cecum has a narrow diverticulum (a tube, blind at distal end) called the **vermiform** (L., worm-shaped) **appendix.** Open the cecum (on its convex side opposite the ileum) and note how the entrance of the ileum forms a ring-shaped **ileocecal valve.** The posterior end of the rectum will be exposed in a later dissection.

Digestive Glands. The **liver,** a large, brownish gland posterior to the diaphragm, has four main lobes: left and right lateral lobes and left and right central lobes. One of the many important functions of the liver is the production of **bile,** a fluid containing bile salts, which are steroid derivatives responsible for the emulsification of fats. Bile is stored and concentrated in the **gallbladder,** a small, greenish, oval sac embedded in the dorsal surface of the right central lobe of the liver. The liver is connected to the upper border of the stomach by a tough, transparent membrane, the **gastrohepatic ligament,** in which are embedded blood vessels (in the left side) and ducts (in the right side). Carefully loosen the gallbladder and note its tiny **cystic duct.** This unites with **hepatic ducts** from the liver to form the **common bile duct,** which carries bile to the duodenum. Probe the gastrohepatic ligament and adjoining liver tissue carefully to find these ducts. Do not injure the blood vessels lying beside them.

The **pancreas** is a mass of soft glandular tissue in the mesentery between the duodenum and the end of the stomach. Push the small intestine, except the duodenum, to the left to explore the gland. Its pancreatic juice is carried by a **pancreatic duct** to the pyloric end of the duodenum. The pancreas is a double gland having both endocrine and exocrine portions. Its endocrine portion produces two hormones, insulin and glucagon, which are of great importance in carbohydrate and fat metabolism. The exocrine portion secretes pancreatic juice—a mixture of water, electrolytes, and enzymes. The enzymes include carbohydrases, which digest sugars and starches; lipases, which split lipids; and proteases, which break down proteins. The pancreas is the only source of lipases in the digestive system.

Other digestive juices are secreted by the **mucosa** lining of the stomach and of the small intestine.

Histological Study of the Intestine. Examine a cross section of human or other mammalian small intestine and compare it with the histological structure of amphibian intestine. Examine slides of the mammalian liver and pancreas.

Identification

Be able to locate and give the functions of the parts of the digestive system.

Further Study

Thoracic Cavity and Neck Region

The **mediastinal septum,** which separates the right and left lung cavities, is a thin, transparent tissue attached to the sternal region of the thoracic wall (Figure 23-17).

Carefully separate the mediastinal septum from the body wall. Now lift up one side of the thoracic wall and look for the small **internal thoracic artery** and **vein** (also called sternal or mammary) embedded in the musculature of the body wall. Carefully separate these vessels on each side and lay them down over the heart and lungs for future use. Now you may cut away some of the ventral thoracic wall (see incision 8, Figure 23-16) to allow a better view of the thoracic cavity containing the left and right **lungs** and the **heart.**

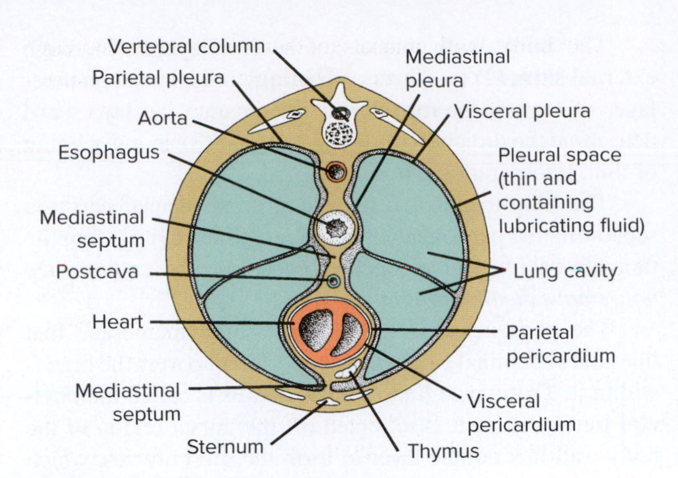

Figure 23-17

Diagrammatic transverse section through the thorax in the region of the ventricles to show the relations of the pleural and pericardial membranes, mediastinum, and lung cavities.

Pleura (Gr., side) is the name given to the peritoneum that lines each half of the thoracic cavity and covers the lungs (Figure 23-17). The peritoneum lining the thoracic cavity is the **parietal** (L. *paries,* wall) **pleura;** the part applied to the lungs is the **visceral** (L., bowels) **pleura.** The small space between is the **pleural cavity,** in which lubricating **pleural fluid** prevents friction. The portions of the parietal pleurae on the medial side next to the heart are called the **mediastinal pleurae.** The **mediastinum** is the region between the mediastinal pleurae. It contains the pericardium and heart and roots of the big arteries and veins, as well as the trachea, the esophagus, part of the thymus, and other parts.

The double-walled **pericardium** enclosing the heart is made up of an outer **parietal pericardium** and a **visceral pericardium** applied to the heart with pericardial fluid in the space between.

EXERCISE 23D

Urogenital System

Urinary System

Read the directions carefully and dissect cautiously. Do not tear or remove any organs, blood vessels, or ducts. Instead, separate them carefully from the surrounding tissues. You will dissect the urogenital system of only one sex, then exchange your dissected specimen with that of another student who has dissected the opposite sex. Therefore, prepare your specimen with the care you would give to a demonstration dissection.

The urinary system consists of a pair of kidneys, pair of ureters, urinary bladder, and urethra (shared with the reproductive system in males).

The fetal **urinary bladder** is the **allantoic** (Gr. *allas,* sausage) **bladder,** a long sac located between the umbilical arteries (Figure 23-18). It narrows ventrally to form the **allantoic duct,** which continues through the umbilical cord and is the fetal excretory canal. The bladder narrows dorsally to empty into the **urethra,** the adult excretory canal. The urethra will be dissected later. After birth the allantoic end of the bladder closes to form the urinary bladder.

Kidneys are dark and bean-shaped. They lie *outside* the peritoneum on the lumbar region of the dorsal body wall. Uncover the right kidney carefully. The depression on the median side of each kidney is called the **hilus.** Through it pass the renal blood vessels and the **ureter,** or excretory duct. Follow the left ureter posteriorly to its entrance into the bladder. Be careful of small ducts and vessels that cross the ureter. (Note the small **adrenal gland,** an endocrine gland, lying close to the medial side of the anterior end of the kidney and embedded in fat and peritoneum.)

Slit open the right kidney longitudinally, cutting in from the outer border. Remove the ventral half of the kidney and lay it in a dish of water.

Study the kidney section with a hand lens or dissecting scope. Identify the **cortex** (L., bark), or outer layer, containing microscopic renal corpuscles; **medulla** (L., marrow, pith), or deeper layer, containing the radially arranged blood vessels and collecting tubules; and **renal pyramids,** which contain groups of collecting tubules coming together to empty through **papillae** into the pelvis. The **pelvis** is a thin-walled chamber that connects with the ureter. Divisions of the pelvis into which the papillae empty are referred to as **calyces** (ka'luh-sez; sing. **calyx**) (Figure 23-18).

For a description of the anatomy and physiology of excretion, read your textbook. Examine the demonstration specimens of sheep kidneys on display. You will be expected to understand the structure and functioning of the kidney and its functional unit, the nephron.

Male Reproductive System

The location of the testes in the fetus depends on the stage of development of the fetus. Each testis originates in the abdominal cavity near the kidney. During the development

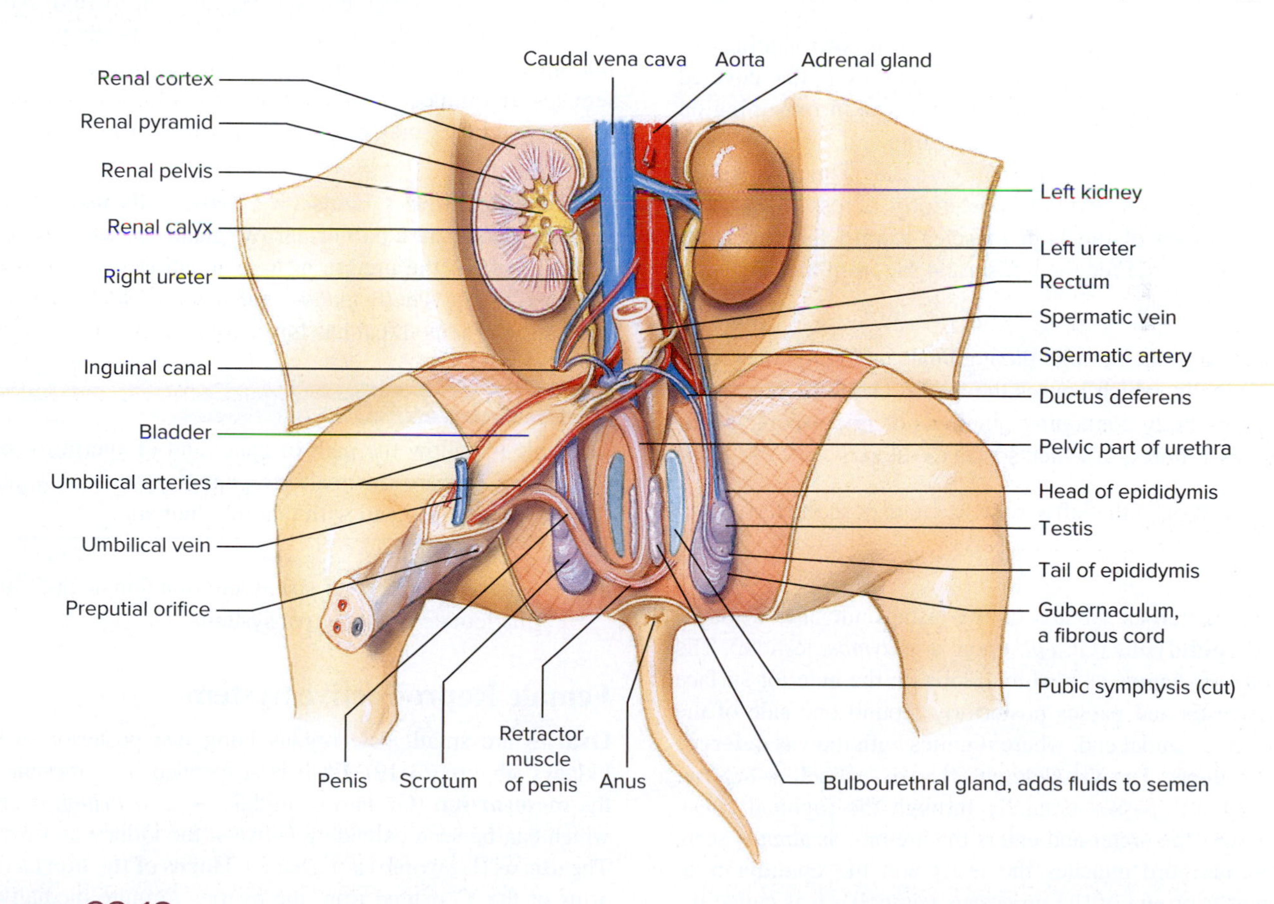

Figure 23-18
Male urogenital system of a fetal pig. The right kidney has been longitudinally sectioned to show internal structure.

of the fetus, a prolongation of the peritoneum, the **processus vaginalis,** grows down into each half of an external pouch, the scrotum. Later the testis "descends" into the **scrotum** through the **inguinal canal,** where it lies within the sac, or sheath, formed by the processus vaginalis. The inguinal canal is the tubular passage connecting the abdominal cavity with the scrotal sac (Figure 23-18). The descent of both testes into the scrotum is usually completed shortly before birth. Part of the cutaneous scrotum can be seen ventral to the anus.

Lay the umbilical cord and allantoic bladder between the legs and locate the **urethra** (Gr. *ouron,* urine) at the dorsal end of the bladder. The urethra bends dorsally and posteriorly to disappear into the pelvic region. Find the sperm ducts (pl. **vasa deferentia;** sing. **vas deferens**), two white tubes that emerge from the openings of the **inguinal** (L., groin) **canals,** cross over the umbilical arteries and ureters, and come together medially to enter the urethra. Also emerging from each inguinal ring are the spermatic artery, vein, and nerve. Together with the vas deferens, these make up the **spermatic cord,** which leads to the testis.

Now lay the bladder up over the abdominal cavity and locate the thin, hard, cordlike **penis** under the strip of skin left posterior to the urogenital opening. The penis lies in a sheath in the ventral abdominal wall, ending at the **urogenital opening.**

Carefully separate the penis from surrounding tissue; then cut away the skin and muscle that covered it. Complete the removal of any skin remaining on the inside of the thigh and rump and carefully separate away the underlying fascia. The thin-walled scrotal sac extends posteriorly across the ventral surface of the high muscles toward the cutaneous scrotum. Free the left scrotal sac from the surrounding tissues.

Pass a probe through the inguinal canal into the processus vaginalis, which houses the **testis.** The testis is a small, hard, oval body containing hundreds of microscopic **seminiferous tubules,** in which sperm develop.

Cut open the left scrotal sac to expose the testis.

The seminiferous tubules of the testis unite into a much-coiled **epididymis** (Gr. *epi,* upon, + *didymos,* testicle). The epididymis begins as a whitish lobe on the anterior surface of the testis and passes posteriorly around one side of the testis to its caudal end, where it unites with the **vas deferens** (sperm duct). The vas deferens (L. *vas,* vessel, + *deferre,* to carry off) passes cranially through the inguinal canal, loops over the ureter and enters the urethra, as already seen. A fibrous cord attaches the testis and the epididymis to the posterior end of the processus vaginalis. It is called the **gubernaculum** (L., rudder). A narrow band of muscle (the cremaster) runs along the lateral and posterior part of the processus vaginalis parallel with the vas deferens.

Separate tissues on each side of the penis in the pelvic region; then, being careful not to injure the penis or cut too deeply, use a scalpel to cut through the cartilage of the pelvic girdle.

Spread the legs apart to expose the **urethra** and its connection with the penis.

The urethral canal extends throughout the length of the penis and serves as a common duct for both sperm and urine.

Now, beginning at its juncture with the bladder, follow the urethra posteriorly and locate the following **male glands.**

Seminal Vesicles. Seminal vesicles are a pair of small glands on the dorsal side of the urethra. These glands contribute a secretion to the semen that is rich in fructose, a six-carbon sugar that stimulates previously inactive but mature spermatozoa to become highly motile. In humans, secretion of the seminal vesicles makes up more than 60% of the volume of the semen.

Prostate Gland. The prostate gland is poorly developed in the fetus; it lies between and often partly covered by the seminal vesicles but may be difficult to find. Alkaline secretions of the prostate assist in neutralizing acids normally present in the urethra, as well as in the vagina of females. In humans, prostate secretion is known to contain a compound that may help prevent urinary tract infections in males.

Bulbourethral Glands (= Cowper Glands). Bulbourethral glands are a pair of narrow glands about 1 cm long on each side of the urethra near its junction with the penis (Figure 23-18). These glands add a thick, sticky alkaline mucus to the semen that has lubricating properties.

Identification

Be able to follow the path of urine and of sperm to the outside. Compare the male urogenital system of a pig with that of a frog and with that of a human.

Now exchange your specimen for a female and study the female reproductive system.

Female Reproductive System

Ovaries are small, pale organs lying just posterior to the kidneys (Figure 23-19). Each is suspended by a mesentery, the **mesovarium** (Gr. *mesos,* middle, + L. *ovarium,* ovary), which can be seen extending between the kidney and ovary. The **uterus** (L., womb) is Y-shaped. **Horns of the uterus** (the arms of the Y) extend from the ovaries to unite medially at the **body of the uterus,** which leads to a **vagina** (L., sheath).

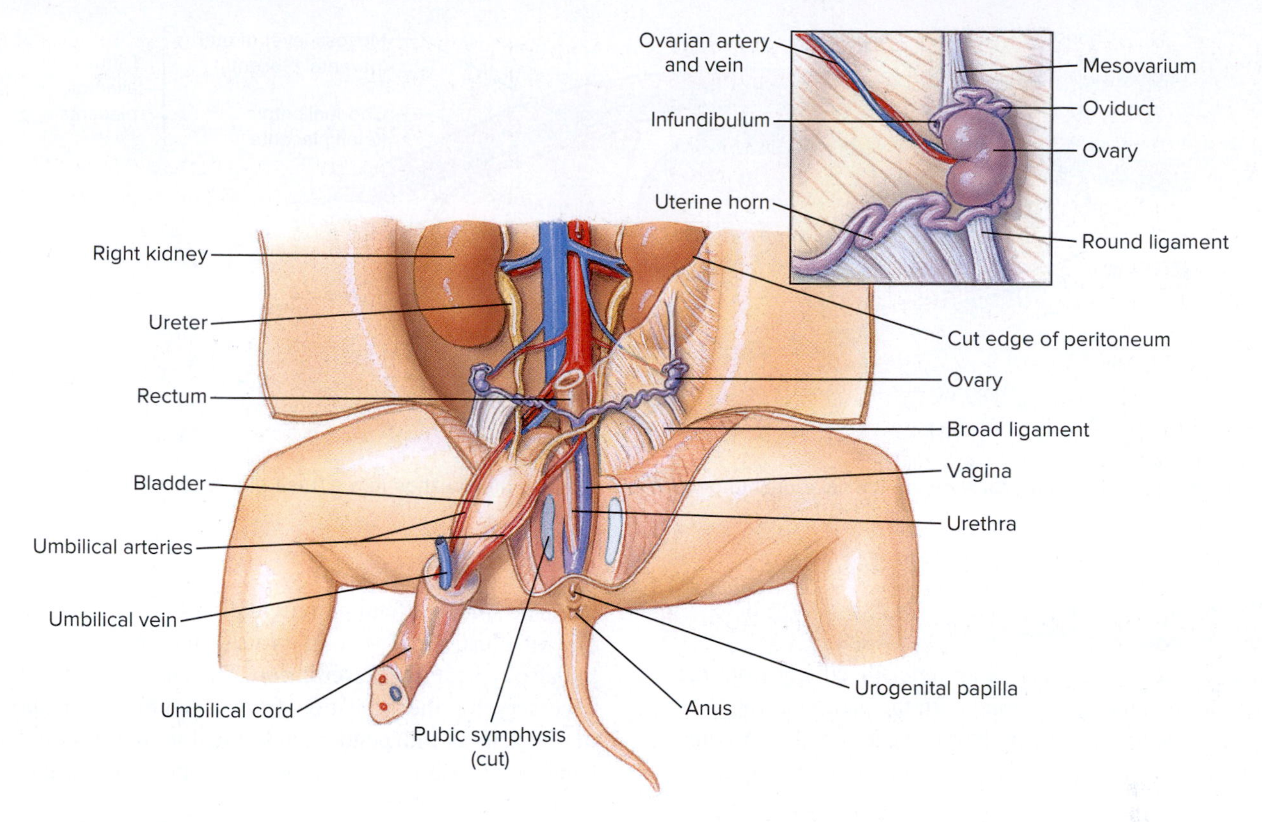

Figure 23-19

Female urogenital system of a fetal pig. What structures serve as excretory ducts for the kidneys? ____________ What structure drains the bladder? ____________ Embryonic pigs develop in what paired structures? ____________

It is in the horns of the uterus, not the body of the uterus, that fetal pigs develop. (Most mammals have a similar Y-shaped uterus, called **bicornuate** ["double-horned"], but in higher primates, including humans, the two uterine horns are completely fused, a condition called **simplex.**) Each uterine horn is suspended by a mesentery, the **broad ligament.** Follow the uterine horn anterior to the ovary, where it gives rise to a highly convoluted **oviduct** (also called **fallopian tube,** after G. Fallopio, Italian anatomist). The oviduct coils around the ovary and terminates at a wide, ciliated funnel, the **infundibulum** (L., funnel). In adult pigs, eggs released from the ovary at ovulation are swept into the opening **(ostium)** of the infundibulum by ciliary currents.

To expose the rest of the reproductive system, the pelvis must be cut open.

Lay the allantoic bladder anteriorly over the abdominal viscera. Cut through the muscle medially between the legs and through the pelvic girdle. Be careful not to cut through the urethra or vagina. Spread the legs apart and separate the urethra from the surrounding tissue. (Note where the ureters join the dorsal end of the bladder.)

Lay the bladder and urethra to one side. Follow the body of the uterus posteriorly to a slight constriction called the **cervix.** From here the tube widens and is called the **vagina.** The vagina and urethra soon join to form a **urogenital sinus,** which is a short common passageway for the two systems.

The **vulva** is the external opening of the urogenital sinus, ventral to the anus. The ventral side of the vulva extends out to form a pointed **genital papilla.** A small, rounded **clitoris** may be seen extending from the ventral floor of the urogenital sinus but is not always evident.

In adults, at copulation, the male penis places the spermatozoa, contained in a seminal fluid, into the vagina. The sperm must pass through the uterus to the oviduct to fertilize the egg. After fertilization the zygote passes down to the horn of the uterus to develop. Note how the horns are adapted for carrying a litter. How does this compare with the human uterus? ____________ How is the developing fetus nourished? ____________ How are waste products from the fetus disposed of? ____________

The placenta of a pig is known as a **chorioallantoic,** or **diffuse, placenta** (Figure 23-20). In the uterus of a dog and some other carnivores, the placenta is called a **zonary placenta** because the chorionic villi are located in a girdle-like zone, or band, around the middle of each pup rather than over the whole surface of the chorion as in the pig. The human placenta is disc-shaped.

If you have not studied the male system, trade your specimen for a male and make a thorough study of the male system.

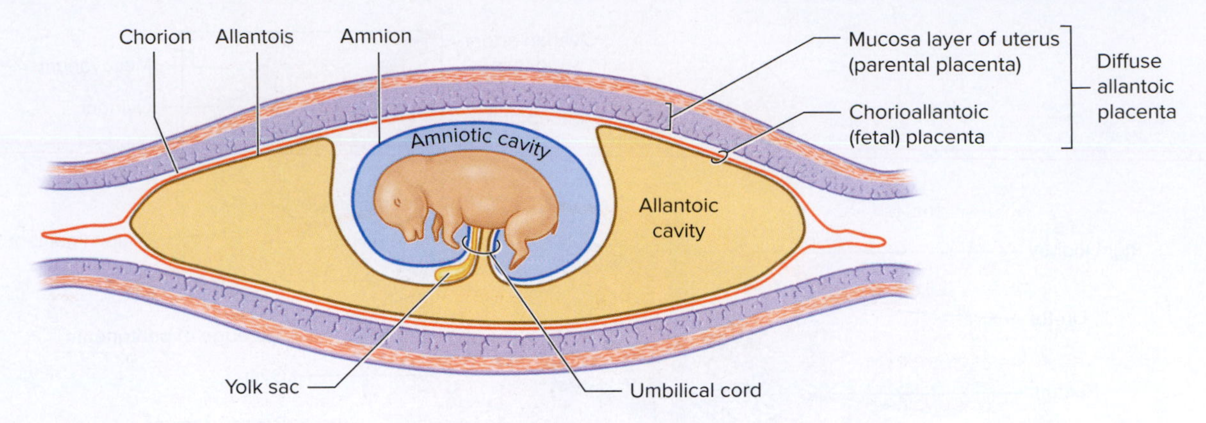

Figure 23-20
Diagram of a pig fetus in utero, showing the relationship of fetal and parental membranes in a diffuse placenta.

Identification

Be able to trace the path of an unfertilized egg from the ovary to the uterus. Compare with the path in a frog.

Be able to trace the path of urine in females. Are any parts shared by both urinary and reproductive systems?

___

___

___

Oral Report

Be able to identify and give the functions of the reproductive organs of both male and female pigs.

Written Report

On separate paper, compare the mammalian and amphibian reproductive systems as illustrated by your study of a pig and of a frog. How is each adapted to its own type of reproduction? (For instance, is their fertilization internal or external?)

EXERCISE 23E

Circulatory System

Core Study

The circulatory system of a pig is quite similar to that of other mammals, including humans. In contrast to the single-circuit system of a fish, with its two-chambered heart, and the incomplete double circuit of a three-chambered amphibian heart, a mammal has an effective four-chambered heart, which allows the blood two complete circuits: a **systemic** circuit through the body, followed by a **pulmonary** circuit to the lungs for oxygenation. The right side of the heart receives oxygen-poor blood returning from the body tissues and pumps it to the lungs; the left side receives oxygen-rich blood returning from the lungs and pumps it to the body tissues.

The study of fetal pig circulation has the added advantage of illustrating not only typical mammalian circulation but also typical **fetal circulation.** The changes in circulation necessary for the transition from a nonbreathing, noneating fetus to an independent individual with a fully independent circulation are both crucially important and elegantly simple.

Uninjected vessels will contain only dried blood (or they may be empty) and therefore will be fragile, flattened, and either brown or colorless. If injected, the arteries will have been filled with latex or a starchy injection mass through one of the arteries in the cut umbilical cord and will be firm and pink or yellow. The veins will have been injected through an external jugular vein in the neck and will be blue. Sometimes, however, the injection medium does not fully penetrate the vasculature. The lymphatic system will not be studied.

Uncover and separate the vessels with a blunt probe and trace them as far into the body as possible, but be careful not to break or remove them. Nerves often follow an artery and a vein and will appear as tough, shiny, white cords. Do not remove them. You may, in fact, find it efficient to identify the major nerves at this time as you find them. As you identify a vessel, separate it and carefully remove investing muscle and connective tissue, taking care not to break the vessel or destroy other vessels that you have not yet identified. Do not remove any body organs unless specifically directed to do so.

As it is often difficult to trace the arterial system without damaging the venous system, which lies above it, you will study the veins first. However, because corresponding arteries and veins usually lie side by side, it is often convenient to study both systems at the same time.

Keep in mind that there is considerable variation among individual pigs in the points of vessel bifurcation, especially in the venous system of the neck and shoulder region. *The venous arrangement in your pig almost certainly will not look exactly like the manual illustrations.* Make notes or

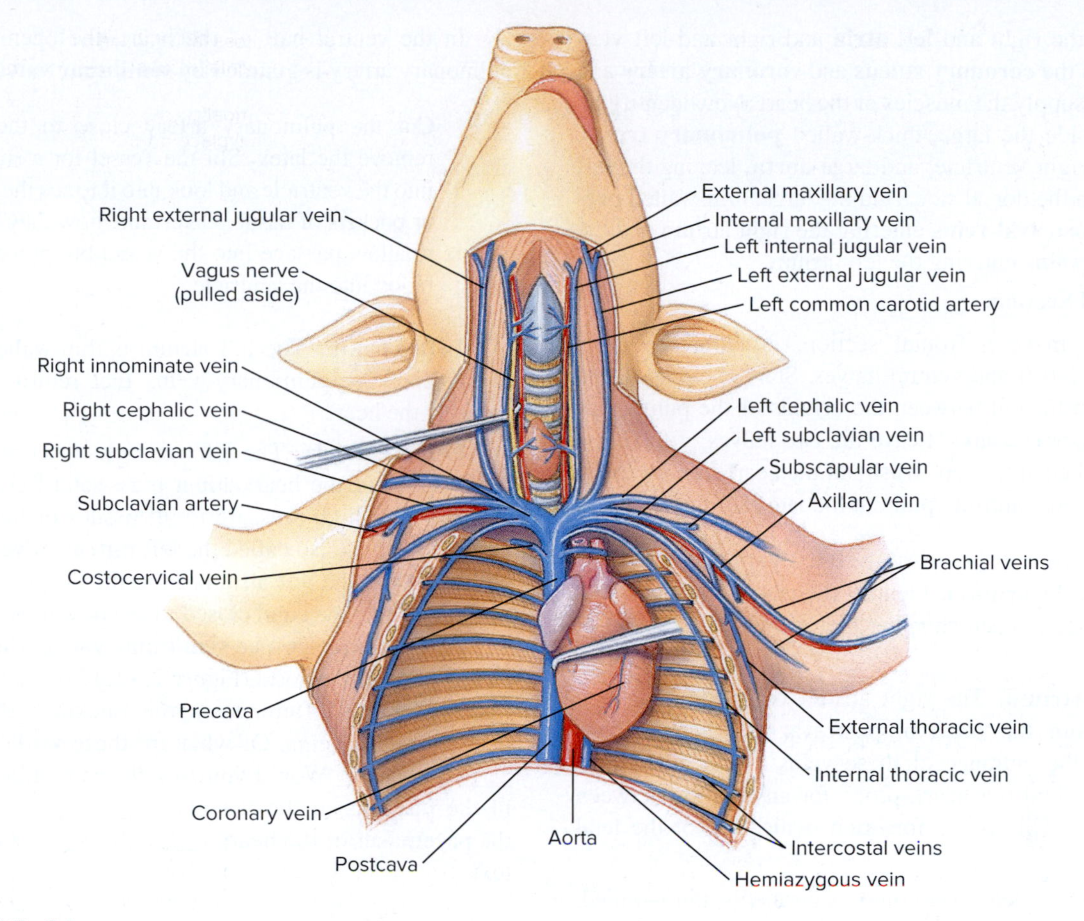

Figure 23-21
Veins of head, shoulders, and forelimbs of a fetal pig. The internal thoracic veins, shown laterally on the opened thoracic cavity, actually lie ventral to the heart.

sketches of any variations that you find in your specimen. With careful dissection, both veins and arteries can be left intact.

Heart

Note carefully the shape and slope of the diaphragm and how it forms the posterior boundary of the thoracic cavity. Then cut the diaphragm away from the body wall to make entrance into the thoracic cavity more convenient.

Open the pericardial sac and examine the heart. It has two small, thin-walled atria and two larger, muscular-walled ventricles.

Right Atrium (Anterior and Ventral). Lift the heart (Figure 23-21) and see the precaval and postcaval veins that empty into the right atrium. The **postcava** (posterior vena cava) from the abdominal region emerges through the diaphragm; the **precava** (anterior vena cava) comes through the space between the first ribs (Figure 23-21).

Left Atrium (Anterior and Dorsal). The left atrium receives the **pulmonary veins** on the dorsal side. Note the conspicuous, earlike right and left auricles (L. *auricula,* ear) lying on each side of the heart. The term "auricle" is often incorrectly used as a synonym for "atrium"; however, the term should be reserved for the earlike flap that protrudes from each atrium.

Right Ventricle. The right ventricle is large and thick-walled. The **pulmonary artery** leaves the right ventricle and passes over the anterior end of the heart to the left, where it divides back of the heart to go to the lungs.

Left Ventricle. The left ventricle is larger and covers the apex of the heart (posterior). It gives off the large **aorta,** which rises anteriorly just behind (dorsal to) the pulmonary artery. The **coronary sulcus** is the groove on the surface of the heart between the right and left ventricles. It contains the **coronary artery** and **vein,** which supply the tissues of the heart itself.

Structure of the Heart

External View.

Use fresh or preserved pig or sheep hearts and compare with the heart of a fetal pig.

Locate the right and left **atria** and right and left **ventricles.** Find the **coronary sulcus** and **coronary artery** and **vein,** which supply the muscles of the heart. Now identify on the ventral side the large, thick-walled **pulmonary trunk,** leaving the right ventricle, and large **aorta,** leaving the left ventricle. On the dorsal side, find the large, thin-walled **precaval** and **postcaval veins** entering the right atrium and the **pulmonary veins** entering the left atrium.

Frontal Section.

Now make a frontal section, dividing the heart into dorsal and ventral halves. Start at the apex and direct the cut between the origins of the pulmonary and aortic trunks. Leave the two halves of the heart attached at the top. Wash out the heart cavities (or, if injected, carefully pull out the latex filling).

The cavities of the heart are lined with a shiny membrane, the **endocardium.** Identify the chambers and valves at the entrance to each chamber that prevent a backflow of blood.

Right Atrium. The right atrium is thin-walled, with openings from the precava, postcava, and hemiazygous veins. Find the entrance of these veins. Now, in the dorsal half of a fetal pig heart, probe for an opening between the two atria. This is the **foramen ovale,** one of the fetal shortcuts.

Right Ventricle. The right ventricle is thick-walled. The tricuspid valve (also called the right atrioventricular valve) prevents backflow of blood to the atrium (Figure 23-22). The valve is composed of three **cusps,** or flaps of tissue, that extend from the floor of the atrium and are connected by fibrous cords, the **chordae tendineae** (ten-din′ee-ee), to **papillary muscles** projecting from the walls of the ventricle.

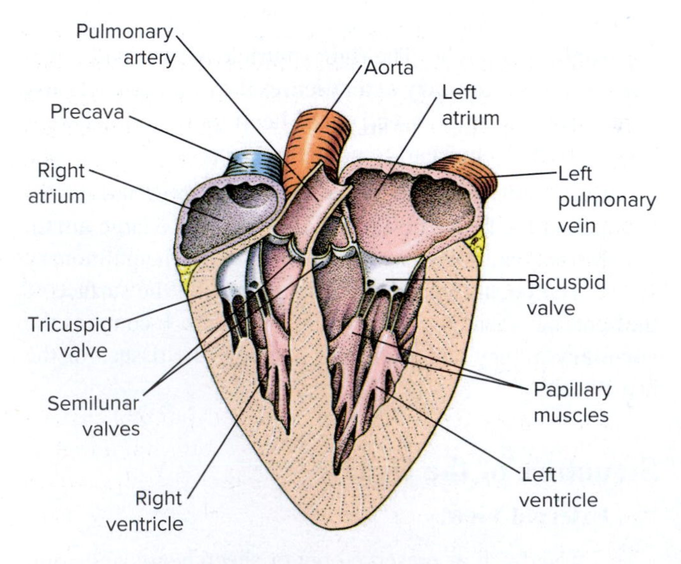

Figure 23-22
Frontal section of a sheep heart.

In the ventral half of the heart, the opening into the pulmonary artery is guarded by **semilunar valves.**

Cut the pulmonary artery close to the heart and remove the latex. Slit the vessel for a short distance into the ventricle and look into it to see the three cusps, or pockets of tissue. Determine how they would work to allow passage into the vessel but prevent return of blood into the ventricle.

Left Atrium. The left atrium is thin-walled. Find the entrance of the pulmonary veins that return oxygen-rich blood to the heart.

Left Ventricle. The left ventricle is the most muscular chamber of the heart, for it must send freshly oxygenated blood at high pressure to all tissues of the body. The **bicuspid valve** (also called the left **mitral valve**) guards the entrance from the left atrium. Find the cusps of this valve. Push the valve open and closed to see how it works.

Find the three-cusped **semilunar valve,** which prevents backflow from the aorta (Figure 23-22).

Notice the difference in the thickness of the walls of arteries and veins. Of what are these walls composed? ____________ Would you find the same kind of muscle in the heart as in the vessels? ____________ What is the pacemaker of the heart? ____________ Consult your text.

Further Study

General Plan of Circulation

In mammalian circulation (after birth), the systemic and pulmonary systems of circulation are separate. Blood from all parts of the body except the lungs returns by way of the large **precava** (anterior vena cava) and **postcava** (posterior vena cava) to the **right atrium.** From there it goes to the **right ventricle** to be pumped out through the **pulmonary arteries** to the lungs to be oxygenated. The blood returns through the **pulmonary veins** to the **left atrium** and then to the **left ventricle,** which sends it through the **aorta** to branches that carry blood finally to capillaries of all parts of the body. The venous system returns it again to the right side of the heart to begin another circuit. Mammals have a **hepatic portal system** but no renal portal system as in amphibians.

Before birth, when the lungs are not yet functioning, a fetus depends on the placenta for nutrients and oxygen, which are brought to it through the **umbilical vein.** Therefore, the general plan of circulation is modified in a fetus, so that most of the pulmonary circulation is short-circuited directly into the systemic bloodstream to be carried to the placenta instead of the lungs. Only enough blood goes to the lungs to nourish the lung tissue until birth. These modifications will be mentioned as they arise in later descriptions.

Blood Vessels

Veins of the Head, Shoulders, and Forelimbs

Carefully remove the remainder of the sternum and the first rib without damaging the veins beneath.

You will find considerable variation in the veins of individual specimens.

The precaval division of the venous circuit carries blood from the head, neck, thorax, and forelimbs to the heart.

Begin by locating the **internal jugular vein,** found lying along the trachea and adjacent to the common carotid artery and the vagus nerve (Figure 23-21). It drains the brain, larynx, and thyroid.

Trace the internal jugular vein posteriorly to its confluence with the **external jugular vein** and **subclavian vein;** these join to form the **innominate vein.** Follow the two innominate veins (right and left) through the opening at the level of the first rib into the chest cavity. Here they join to form the **precava,** which enters the right atrium. The precava also receives at this level a pair of **internal thoracic veins** lying on each side of the sternum. This is the vein that, along with the accompanying artery, you detached from the ventral muscle wall of the chest cavity (Exercise 23C). You will also find its continuation in the abdominal muscle wall.

Another vessel entering the precava dorsolaterally close to the heart is the **costocervical trunk.** It receives several veins from the neck and dorsal thorax.

Lift the heart and left lung and push them to the right. The unpaired **hemiazygous (= azygous) vein** lies in the chest cavity along the left side of the aorta and receives from the left and right intercostal veins. Follow the hemiazygous anteriorly to the point at which it crosses the aorta and goes under the heart. Lift up the heart to see the vein cross the pulmonary veins and enter the right atrium along with the postcava.

Return now to the neck region and locate the **external jugular vein.** This vein was probably used to inject the venous system and will be tied and severed on one side of the neck. The external jugular is formed by the union of the **internal maxillary** (dorsal) and the **external maxillary** (ventral) near the angle of the jaw. These vessels drain superficial parts of the head.

Joining the base of the external jugular (or sometimes the subclavian) is the **cephalic vein,** a large superficial vein from the arm and shoulder. It lies just beneath the skin.

The **subclavian vein** extends from the shoulder and forelimb. It is a deeper vein, which follows the subclavian artery. (There may be from one to three subclavian or brachial veins. If there are more than one, there may be considerable anastomosing [networklike branching and rejoining of vessels] among them.) The subclavian, as it continues into the arm, is called the **axillary** (L. *axilla,* armpit) in the armpit and the **brachial** (Gr. *brachion,* upper arm) in the upper arm. Large nerves of the brachial nerve plexus overlie these veins.

Arteries of the Head, Shoulders, and Forelimbs

Arteries carry blood away from the heart, and, with the exception of the pulmonaries, all branch from the main artery, the aorta.

The **aorta** begins at the left ventricle (ascending aorta), curves dorsally (aortic arch) behind the left lung, and extends posteriorly along the middorsal line (descending, or dorsal, aorta). The first two branches are the **brachiocephalic trunk** and, just to the left of it, the **left subclavian artery** (Figure 23-23).

Brachiocephalic Trunk. The brachiocephalic is a large, single artery that extends anteriorly a short distance and branches into the carotid trunk and the right subclavian artery.

The carotid trunk may extend anteriorly for as much as a centimeter, or it may divide at once into the **left** and **right common carotid** arteries, which form a **Y** over the trachea and extend up each side of it. Each common carotid artery follows an internal jugular vein and the vagus nerve toward the head, giving off branches to the esophagus, thyroid, and larynx. Near the head, each common carotid gives off an **internal carotid,** a deep artery passing dorsally to the skull and brain and an **external carotid,** which is a continuation of the common carotid that supplies the tongue and face.

The **right subclavian artery** arises from the brachiocephalic and continues into the right arm as the **axillary** in the armpit, **brachial** in the upper arm, and **radial** and **ulnar** in the lower arm. The **subscapular** extends to the deep muscles of the shoulder and branches off the axillary part of the artery.

Left Subclavian Artery. The left subclavian arises directly from the aorta. Otherwise, it is similar to the right subclavian artery.

Each of the subclavian arteries gives off several branches, which include the **internal thoracic artery** and **thyrocervical artery.** The internal thoracic artery (also called the **mammary** or **sternal artery** and, in humans, an artery used in coronary bypass operations) is the artery you detached earlier. It supplies the ventral muscular wall of the thorax and abdomen.

The thyrocervical artery arises from the subclavian at the same level as the internal thoracic. It supplies the thyroid and parotid glands and some of the pectoral muscles.

Pulmonary Circulation

Pulmonary Trunk. The pulmonary trunk arises from the right ventricle, ventral to the aorta. Follow it as it branches into **right** and **left pulmonary arteries,** which carry oxygen-poor blood to the lungs. Scrape away some lung tissue to find branches of these vessels.

Pulmonary Veins. Pulmonary veins empty oxygen-rich blood into the left atrium.

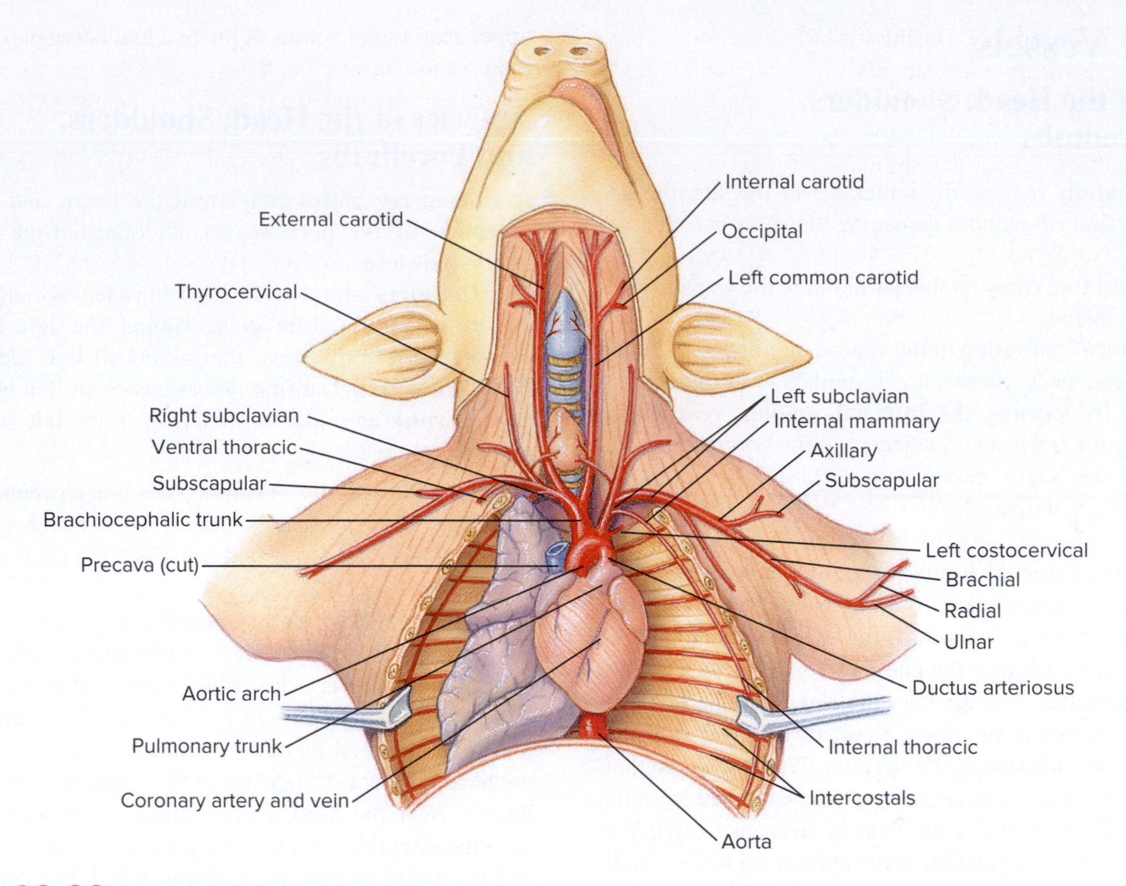

Figure 23-23

Arteries of head, shoulders, and forelimbs of a fetal pig. The internal thoracic arteries, shown laterally, actually lie ventral to the heart. What pair of arteries supplies blood to the tongue and face? ____________ What artery supplies blood to the thyroid and parotid glands? ____________ Near the head, the common carotid artery splits into what two arteries? ____________

Probe gently under the left atrium to expose the veins.

You should be able to find the large trunk entering the heart just under the hemiazygous vein and a pair of vessels servicing each lobe of the lungs.

Fetal Shortcuts

Foramen Ovale. This is a fetal opening in the wall between the right and left atria. Part of the blood from the right atrium can pass through the foramen ovale directly to the left atrium, where it can go to the left ventricle, to the aorta, and back into the systemic circulation without going to the lungs at all (Figure 23-24). The remainder of the blood enters the pulmonary trunk.

Ductus Arteriosus. This is the short connection between the pulmonary trunk and the aorta. Trace this connection, which begins where the smaller pulmonary arteries branch off toward the lungs. Part of the blood from the right ventricle goes to the lungs, and part is shunted through the ductus arteriosus to the aorta (Figure 23-24).

Ductus Venosus. This is a third fetal shortcut that connects the umbilical vein with the postcava, passing through a

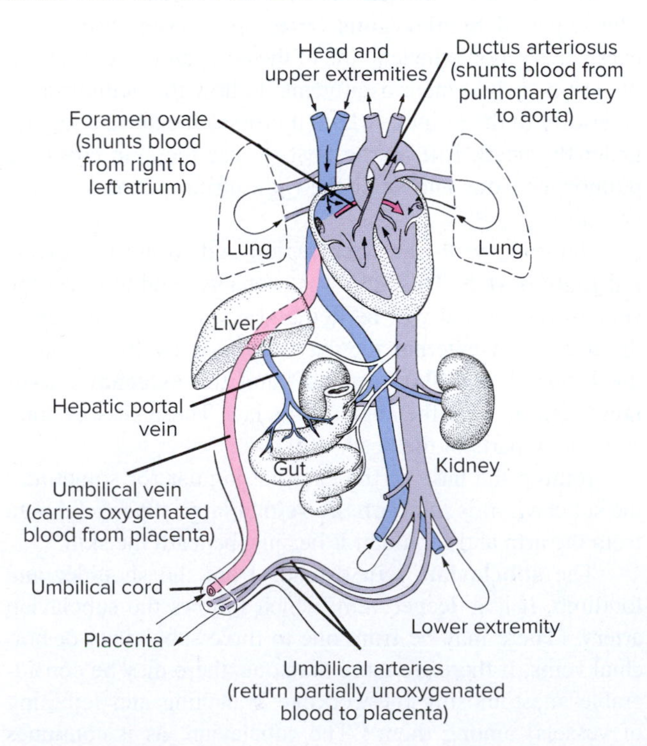

Figure 23-24

Scheme of fetal circulation.

channel in the liver tissue. During fetal life, the anterior part of the postcava carries a mixture of oxygen-poor blood from body tissues and oxygen-rich blood (and nutrients) from the placenta by way of the umbilical vein (the vein tied and cut earlier). After birth both the umbilical vein and the ductus venosus degenerate.

You will be able to see the arrangement of the vessels in the liver tissue by using a probe to scrape or "comb out" the liver tissue to separate it from the vessels. Wash out the loose tissue.

Some of the vessels you see will probably be hepatic ducts.

Changes in Circulation at Birth

Two crucial events happen at the moment of birth: (1) the placental bloodstream upon which the fetus has depended is abruptly cut off, and (2) pulmonary circulation immediately assumes the task of oxygenating the blood. One of the most extraordinary aspects of mammalian development is the perfect preparedness of the circulatory architecture for this event.

When a newborn piglet (or human) takes its first breath, vascular resistance through the lungs is suddenly lowered as the lungs expand. The ductus arteriosus functionally constricts almost immediately, and the lungs receive full blood circulation. Blood then returns from the pulmonary veins to the left atrium, raising pressure in this chamber enough to close the flaplike valve over the foramen ovale. With blood no longer passing between the right and left atria, the foramen ovale gradually grows permanently closed, leaving a scar called the **fossa ovalis.** The ductus arteriosus also closes permanently, becoming a ligament between the pulmonary artery and aortic arch.

With severing of the umbilical cord, flow through the umbilical vein and arteries ceases immediately, and these vessels are eventually reduced to fibrous cords.

Postcaval Venous Circulation

The **postcava** carries blood to the heart from the hindlimbs and trunk. Follow it from the right atrium through the intermediate lobe of the lung and through the diaphragm to the liver (Figure 23-25). Push the intestines to the left side, clean away most of the dorsal portion of the liver, and note where the postcava emerges and continues posteriorly. Clean away the peritoneum that covers the postcava, and find the following veins that enter it on each side.

One or two **renal veins** extend from each kidney. Do both right and left veins enter at the same level? __________ The

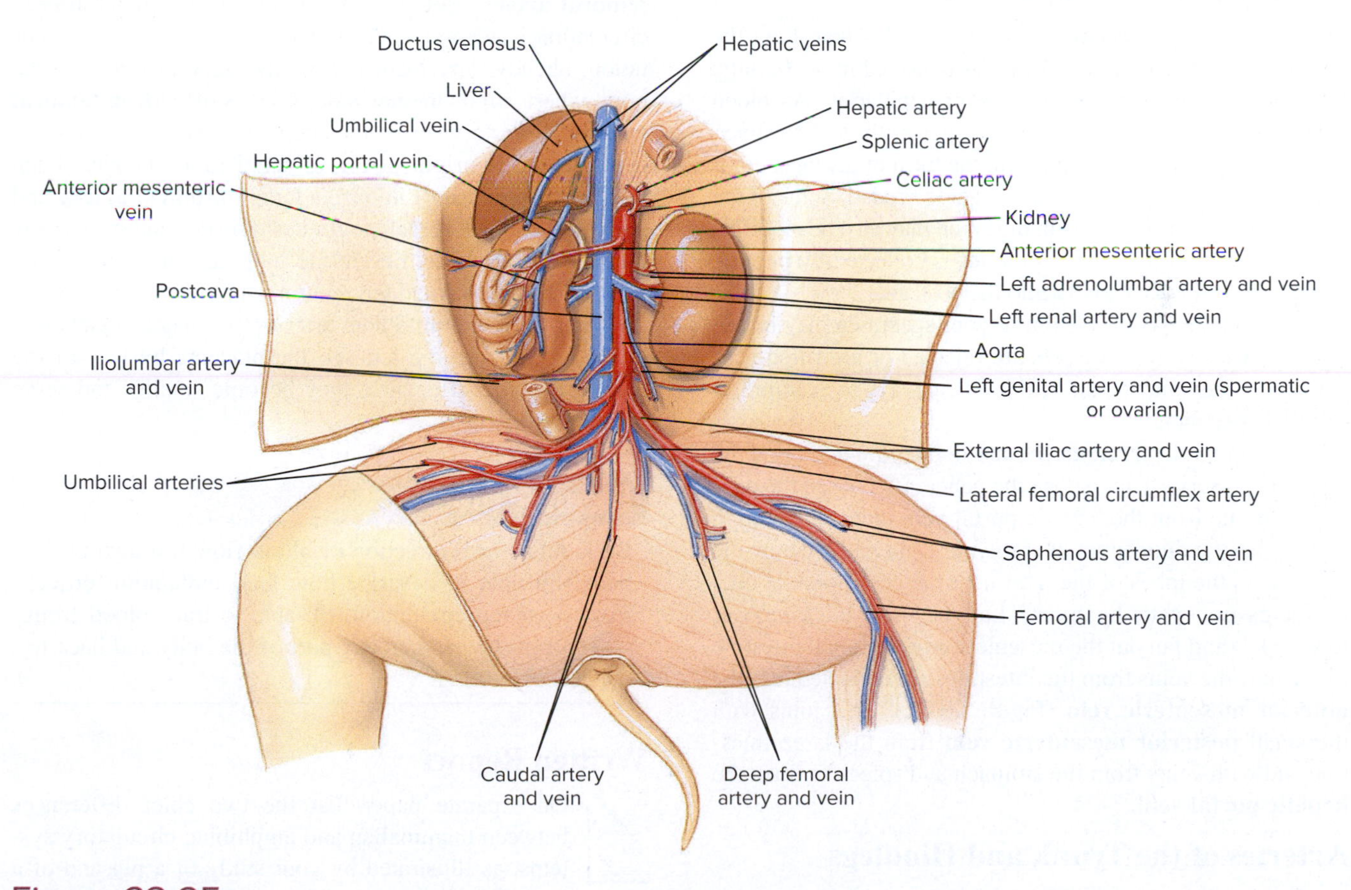

Figure 23-25
Circulation of trunk and hindlegs of a fetal pig.

narrow, band-shaped adrenal gland (= suprarenal gland) lying against the anteromedial border of each kidney is drained by a small **adrenal vein** that may enter the renal vein or may empty directly into the postcava. It is accompanied by the adrenal artery. Just posterior to the adrenal vein and at about the same level as the renal veins, find the **parietal vein** extending to the body wall, accompanied by the parietal artery. The **genital vein** and **artery** serve the testes and ovaries. In males the vein is called the **spermatic vein** and is incorporated together with the spermatic artery into the spermatic cord to the testis. In females it is called the **ovarian vein;** it lies suspended in the mesovarium along with the ovarian artery.

Follow the postcava posteriorly to its bifurcation into the **common iliac veins.** These in turn divide into **internal** and **external iliac veins.** The internal iliac vein (hypogastric) is the medial branch extending dorsally and posteriorly into the deep tissue and draining blood from the rectum, bladder, and gluteal muscles. The external iliac vein is the lateral branch and the largest vein entering the common iliac. It drains the foot and leg and is formed by the union of the large **femoral vein,** which extends ventrally toward the knee, and **deep femoral vein,** which extends dorsally and posteriorly into the deep muscles of the thigh. It also receives a lateral branch, the **circumflex iliac vein,** which drains the muscles of the abdomen and upper thigh.

Hepatic Portal System

The hepatic portal system is a series of veins that drain the digestive system and spleen. Blood is collected in the **hepatic portal vein,** which carries it to liver capillaries. As blood laden with nutrients from the intestine passes through the liver, the liver may store excess sugars in the form of glycogen, or it may give up sugar to the blood if sugar is needed. Some of the amino acids formed by protein digestion may also be modified here. Blood from the liver capillaries is collected by **hepatic veins,** short veins that empty into the postcava.

The hepatic portal system in your specimen may not be injected. Unless these vessels are injected or are filled with dry blood, they may be difficult to locate. Try to examine an injected specimen.

Lift up the liver, stomach, and duodenum and draw the intestine posteriorly to expose the pancreas. Loosen the pancreatic tissue from the hepatic portal vein that runs through it. Now lay the duodenum over to your right and see how the vein enters the lobes of the liver near the common bile duct (in the gastrohepatic ligament). Lay the small intestine over to your left and fan out the mesenteries of the small intestine to see how the veins from the intestines are collected into the **anterior mesenteric vein** (Figure 23-25). This joins with the small **posterior mesenteric vein** from the large intestine and with veins from the stomach and spleen to form the **hepatic portal vein.**

Arteries of the Trunk and Hindlegs

Descending Aorta. The descending aorta follows the vertebral column posteriorly, first lying dorsal and then ventral to the postcava (Figure 23-25). After passing through the diaphragm, the aorta gives rise to a single large artery, the **celiac artery.** To find it, clip away the diaphragm and remove the tissue around the aorta at the anterior end of the abdominal cavity. Be very careful not to destroy the celiac ganglion and sympathetic nerves (described in Exercise 23F). The celiac artery divides almost immediately into arteries serving the stomach, spleen, pancreas, liver, and duodenum.

Just posterior to the celiac, find another unpaired artery, the **anterior mesenteric artery.** This vessel must be dissected from the surrounding tissue. It sends branches to the pancreas, small intestine, and large intestine.

Locate next the paired **parietal arteries, renal arteries,** and **spermatic** or **ovarian arteries.** Posterior to these is the single **posterior mesenteric artery,** which divides and sends branches to the colon and rectum. It will probably have been broken off during earlier dissection of the abdominal cavity.

At its posterior end, the aorta now divides into two large lateral **external iliac arteries** to the legs, two medial **internal iliac arteries** to the sacral region, and a small continuation of the aorta into the tail, the **caudal artery.** Trace one of the external iliac arteries. It first gives off a lateral **circumflex artery** that supplies some of the pelvic muscles, then penetrates the peritoneal body wall to enter the leg as the **femoral artery.** Note that nearly all of the caudal arterial circulation is accompanied by corresponding venous circulation, already described. Follow the femoral artery to the point where, on its median side, it gives off a **deep femoral artery.** This artery serves the deep muscles of the thigh.

Locate the origin of the **internal iliac arteries.** Each internal iliac gives off at once a large **umbilical artery** and then, as a smaller vessel, continues dorsally and posteriorly into the sacral region beside the internal iliac vein, giving off branches to the bladder, rectum, and gluteal muscles.

In a fetus, the umbilical arteries are major vessels that return blood from the fetus to the placenta by way of the umbilical cord. After birth they become smaller and serve the urinary bladder.

Identification

Be familiar with direction of blood flow throughout the body. Be able to describe how fetal circulation differs from postnatal circulation. Be able to trace blood from any part of the heart to any part of the body and back to the heart.

Written Report

On separate paper, list the two chief differences between mammalian and amphibian circulatory systems, as illustrated by your study of a pig and of a frog. How are these differences adaptive to an all-terrestrial or half-aquatic life?

EXERCISE 23F

Nervous System

Core Study

The nervous system can be divided into the **central nervous system (CNS),** containing the spinal cord and brain, and the **peripheral nervous system (PNS),** containing all the neural tissue outside the central nervous system. The PNS is the link between an animal's environment and its CNS, and it contains both **sensory** components that bring sensory information to the CNS and **motor** components that carry neural commands from the CNS to peripheral effectors (usually muscles). Such neural information, in the form of nerve impulses, is conducted over axons that are bundled together into **peripheral nerves** (also called nerve trunks). Peripheral nerves comprise both **spinal nerves** that communicate with the spinal cord and **cranial nerves** that are connected to the brain.

The **autonomic nervous system** is a subdivision of the PNS that provides automatic control over visceral activities, such as coordination of the cardiovascular, respiratory, digestive, and reproductive functions.

Consult your textbook for additional information on the nervous system and be able to relate this material to your dissection. In your dissection of the nervous system of a fetal pig, you will first look for the spinal nerves and plexuses and then look for the sympathetic nerve trunks and some of their ganglia. These you will find in the neck region and body cavities. Then you will study a preserved mammalian brain.

Spinal Nerves

Each spinal nerve arises from the spinal cord as two roots: a **dorsal root** containing the axons of sensory nerves and a **ventral root** containing motor neurons that control peripheral effectors (Figure 23-26). Each dorsal root bears a prominent **ganglion** that contains the cell bodies of the sensory neurons. The ventral roots bear no ganglia because the cell bodies of the motor neurons are contained within the spinal cord. Spinal nerves emerge from the vertebral column through openings between the individual vertebrae called **intervertebral foramina** (sing. **foramen**). After exiting through a foramen, each spinal nerve divides into three branches: a small **dorsal branch** that supplies the muscles and skin of the dorsal part of the body, a large **ventral branch** that supplies the skin and muscles of the ventral and lateral part of the body and limbs, and a small **communicating branch** that joins ganglia of the autonomic nervous system. The major named spinal nerves to be studied in this exercise are all ventral branches.

There are 33 pairs of spinal nerves in a fetal pig (31 pairs in humans) and each is identified by its association with the vertebra through which it exits: 8 cervical, 14 thoracic, 7 lumbar, and 4 sacral (Figure 23-27).

Brachial Plexus

In your search for arteries and veins in the foreleg, you saw a group of tough, white nerves extending into the arm. These are the sixth, seventh, and eighth cervical and first thoracic nerves, which, with their interconnecting branches, form the

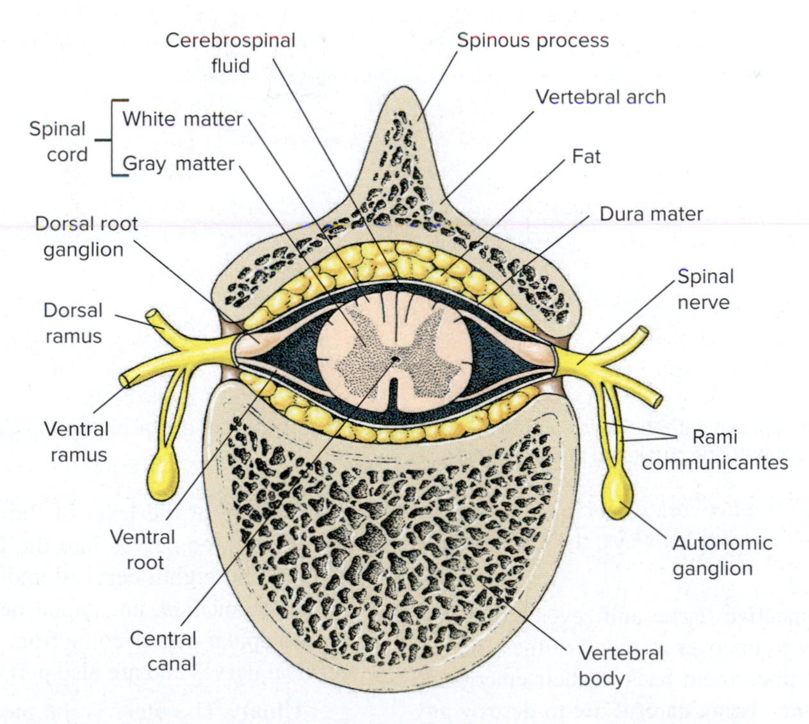

Figure 23-26
Section of spinal column of a mammal, showing spinal roots and branches of spinal nerve.

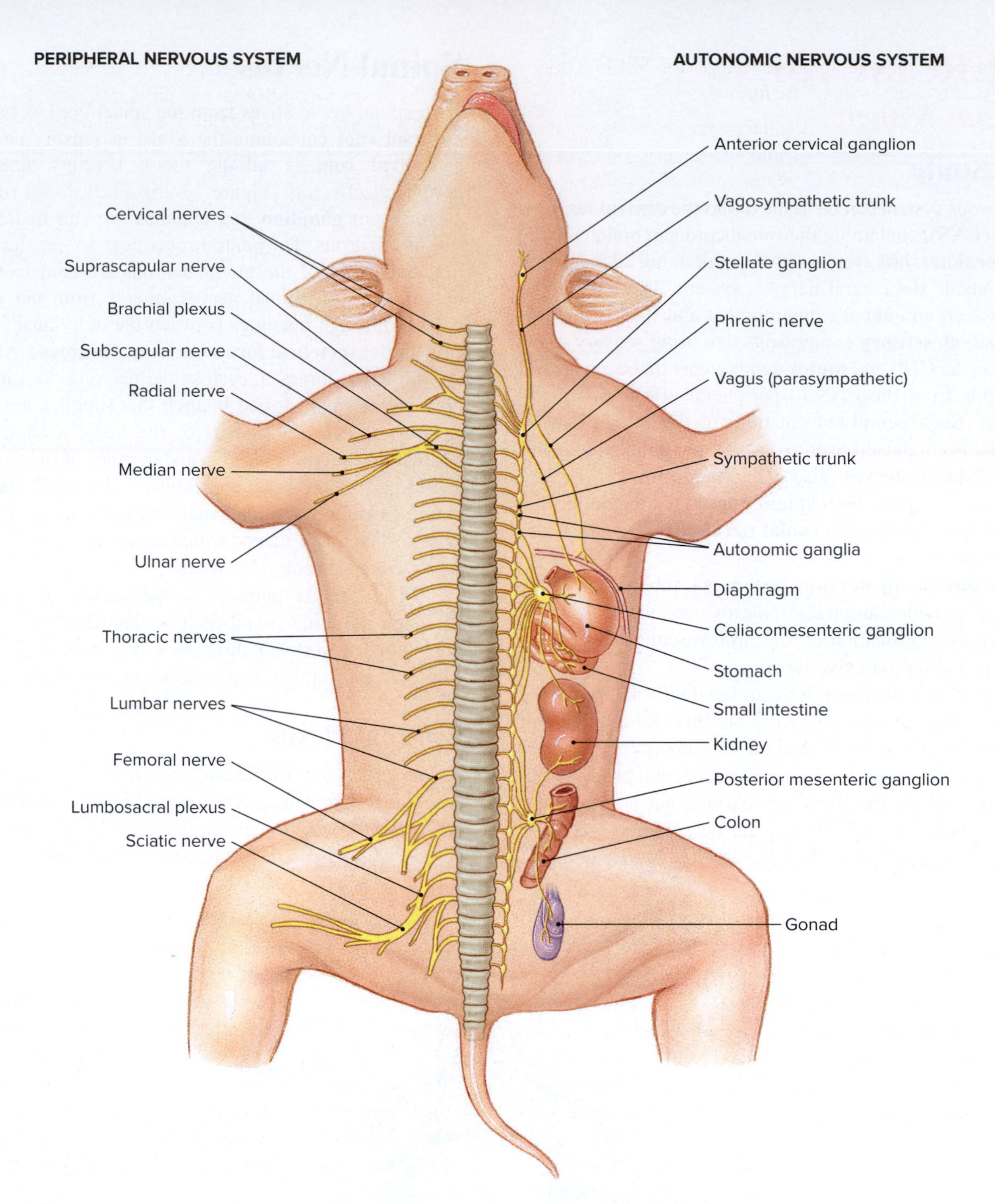

Figure 23-27

Peripheral nervous system and sympathetic division of the autonomic nervous system of a fetal pig. **Left (pig's right),** peripheral nervous system. **Right (pig's left),** sympathetic trunk and major ganglia.

brachial plexus (a plexus is a network of nerves produced by the convergence of fibers in complex ways; the term literally means interwoven).

Clear away connective tissue and sever blood vessels if necessary to uncover as many of these nerves as possible and trace them back to their emergence from the vertebrae, being careful not to destroy any of the branches. Each nerve sends branches to join neighboring nerves.

Starting at the level of the first rib and moving anteriorly are three nerves into the foreleg that come from the seventh and eighth cervical and the first thoracic nerves—the ulnar, median, and radial nerves. The subscapular and suprascapular nerves come from the fifth, sixth, and seventh cervical nerves and are also part of the brachial plexus.

Ulnar. The ulnar is the most posterior of the brachial nerves. It follows the bend of the elbow to the underside of the foreleg and foot.

Median. The median nerve follows the brachial artery and vein along the inside of the foreleg.

Radial. The radial is the larger nerve. It lies underneath the median and passes into the deep muscles of the foreleg. By skinning the lateral side of the arm and shoulder and dissecting under the muscles (triceps), you may find the radial nerve where it crosses over to the lateral side of the foreleg and passes down the dorsal (radial) side of the foreleg and foot.

Subscapular Nerve. The subscapular nerve is from parts of the sixth and seventh cervical nerves. It branches into the muscles under the scapula.

Suprascapular Nerve. This large nerve is from the fifth and sixth cervical nerves to the muscles above the scapula.

You can probably also locate the fourth, third, and second cervical nerves, but the first is difficult to find.

Thoracic and Lumbar Nerves

Thoracic nerves lie in the intercostal spaces, running parallel with the intercostal arteries and veins. Trace one or two back to the vertebral column.

Posterior to the ribs in the dorsal body wall, the first four lumbar nerves lie just under the peritoneum. They angle out posterolaterally from the vertebral column. Trace some of these.

Lumbosacral Plexus

The lumbosacral plexus is formed by the ventral branches of the fifth, sixth, and seventh lumbar and first sacral nerves; it sends large nerves into the hindlegs. To identify the nerves making up the plexus, you may have to clip the iliac artery and vein from the postcava and remove some of the muscle along the backbone. The **femoral nerve** and **saphenous nerve** (which branches from the femoral) follow arteries of the same names. The large **sciatic nerve** follows the internal iliac artery and vein into the deep sacral region. The nerve emerges with the vessels close to the backbone and travels around the end of the ischium to follow the femur (Figure 23-27).

If you wish to trace the sciatic into the leg, skin the dorsal side of the pelvis and thigh and lay aside the layer of muscle along the backbone for 3 to 4 cm above the tail.

Further Study

Autonomic Nervous System

The autonomic nervous system (ANS) is concerned with involuntary body functions that are not controlled by the cerebrum and do not ordinarily affect consciousness. It governs (1) heart muscle; (2) smooth involuntary muscle, such as that of the digestive tract, and blood vessels; and (3) secretions of various glands. The ANS is composed of two divisions, **sympathetic** and **parasympathetic,** and each of these divisions has a characteristic anatomical and functional organization. One important characteristic of *both* divisions is that there is always a synapse imposed between the CNS where the autonomic fibers originate and the peripheral effectors where they end. The first neuron with its body in the spinal cord is called a **preganglionic neuron.** Its axon extends out some distance from the spinal cord to a ganglion, where it synapses with the second neuron in the series, the **postganglionic neuron.** This neuron extends the remaining distance to the effector organ, a muscle or gland.

In the sympathetic division of the ANS, preganglionic fibers all emerge from the thoracic and lumbar segments of the spinal cord and pass to **sympathetic ganglia** located close to the spinal cord. Many of these ganglia are interconnected into an elongate **sympathetic trunk** (Figure 23-27).

In the parasympathetic division of the ANS, preganglionic fibers emerge from the cranial (brain stem) and sacral segments of the spinal cord. Preganglionic fibers are quite long, extending all the way from the CNS to the peripheral effector organ. Here they synapse with short postganglionic fibers. Although the parasympathetic outflow travels over several different cranial and sacral nerves, nearly 75% of its outflow is carried by a single nerve, the **vagus nerve** (tenth cranial nerve).

Begin by finding the **vagosympathetic trunk.** Clean away some of the peritoneum if necessary to uncover one of these trunks lying on the dorsomedial side of the carotid artery.

Each vagosympathetic trunk is made up of the **vagus (tenth cranial) nerve** and **sympathetic trunk,** which are united in a common sheath and enter the thoracic cavity at the level of the first rib. The vagus nerve is a mixed nerve carrying both sensory and motor (parasympathetic) fibers. It follows the esophagus to the abdominal cavity, sending branches to the heart, bronchi, stomach, and intestines. The sympathetic trunk continues along the vertebral column under cover of the peritoneum and is connected by communicating branches (rami communicantes) to the spinal nerves.

Sympathetic Ganglia

There are several large ganglia in the sympathetic system that you may be able to find with some care (Figure 23-27). These are usually paired and appear as irregular, whitish swellings along the trunks.

Anterior Cervical Ganglion. This is a small ganglion at the base of the skull near the division of the external and internal carotids.

Stellate Ganglion. This is made up of the posterior cervical and first thoracic ganglia and located under cover of the first rib on the side of the trachea near the origin of the subclavian artery. It is at this ganglion that the vagus and sympathetic trunks separate. From this region, try to separate the two nerves and trace them both back to the skull. Now follow the vagus to the stomach and the sympathetic trunk into the dorsal thoracic wall.

Thoracic and Lumbar Ganglia. These are segmentally arranged swellings on the main sympathetic trunks. Lift the trunk slightly to see the rami communicantes dipping into the muscle to connect with spinal nerves where they emerge from the vertebrae.

Celiacomesenteric Ganglia. These are large, elongated, irregular ganglia, one located on each side of the aorta at the origin of the celiac and anterior mesenteric arteries. They send branches to the stomach, intestine, kidney, and genital organs.

Posterior Mesenteric Ganglion (Unpaired). This is a small ganglion on the posterior mesenteric artery. It sends branches to the intestine and to the ovaries or testes.

Do you see any pattern of similarity between the nervous systems of a pig and a frog? ____________ By observing charts or diagrams, compare the nervous system of a pig with that of a human being. What conclusions can you draw about the nervous system of vertebrates in general?

____

____

____

What conclusions can you draw about the nervous systems of the vertebrates versus that of the invertebrates? __________

____

____

Brain

Directions will be given for dissection and study of a sheep brain (Figure 23-28).

The sheep brain provided to you may be in the skull case or may already be dissected out of the skull case. If it is in the skull case, the skull case should be cracked to make removal of the brain easier. Gently use forceps or the tip of a scalpel to remove pieces of the cracked skull case, being careful not to puncture the membrane covering the brain underneath. Gently loosen the brain from the remaining pieces of the braincase. To remove the brain fully from the braincase, you will have to cut the cranial nerves, but leave nerve stumps as long as possible.

Both the brain and the spinal cord are covered with protective membranes that carry blood vessels to nourish them. These are called **meninges** and include the following:

Dura mater—tough outer membrane; tends to cling to the skull in places

Arachnoid—clings to the pia mater; too thin to identify

Pia mater—fine vascular membrane; adheres closely to the brain and spinal cord and contains many blood vessels that nourish the brain

Cerebrospinal fluid flows in the space between membranes and in the cavities, or ventricles, of the brain and the central canal of the spinal cord.

Carefully remove **dura mater** from the brain. Now the brain will be very soft and quite easily damaged, but it can be removed from the braincase. Lift it gently and look for the cranial nerves on the ventral side where they pass into the floor of the braincase. Clip the nerves, leaving the stubs on the brain cut as long as possible. Cut the spinal cord 2.5 to 5 cm below the brain, leaving one or two stubs of spinal nerves in place.

Dorsal Surface of the Brain

Cerebrum (Cerebral Hemispheres). The cerebrum is divided into left and right oval lobes separated by the **longitudinal fissure.** The cortex (outer layer) is made up of raised convolutions called **gyri** (ji′rē), separated by fissures called **sulci** (sul′kē). The cortex contains most of the gray matter of the brain. A groove, the **transverse fissure,** marks off the cerebrum posteriorly from the cerebellum.

Cerebellum. The cerebellum lies behind the cerebrum and is somewhat triangular. Its folds and fissures are much finer than those of the cerebrum.

Medulla Oblongata. The medulla oblongata is small and lies behind and ventral to the cerebellum. It narrows posteriorly to continue as the spinal cord.

Corpus Callosum. Spread apart the lobes of the cerebrum to see, at the bottom of the fissure, a broad white band of nerve fibers passing between the hemispheres. This band is the corpus callosum.

Corpora Quadrigemina (Midbrain). Gently spread apart the cerebrum and cerebellum to see four small, rounded knobs, the corpora quadrigemina.

Ventral Surface of the Brain

Olfactory Lobes. Olfactory lobes (= bulbs) extend anteriorly from the cerebral hemispheres (Figure 23-28). They fit into depressions in the floor of the skull and send nerves to the nostrils.

Optic Chiasma. Optic nerves posterior to the olfactory lobes cross here to form the optic chiasma.

Pituitary Gland (Hypophysis). Posterior to the optic chiasma, a swelling called the **tuber cinereum** bears a stalk, the **infundibulum,** on which the **pituitary** is attached. The pituitary gland fits into a depression in the floor of the cranium.

Pons. The pons is a broad, raised area of transverse fibers posterior to the pituitary and joining the medulla

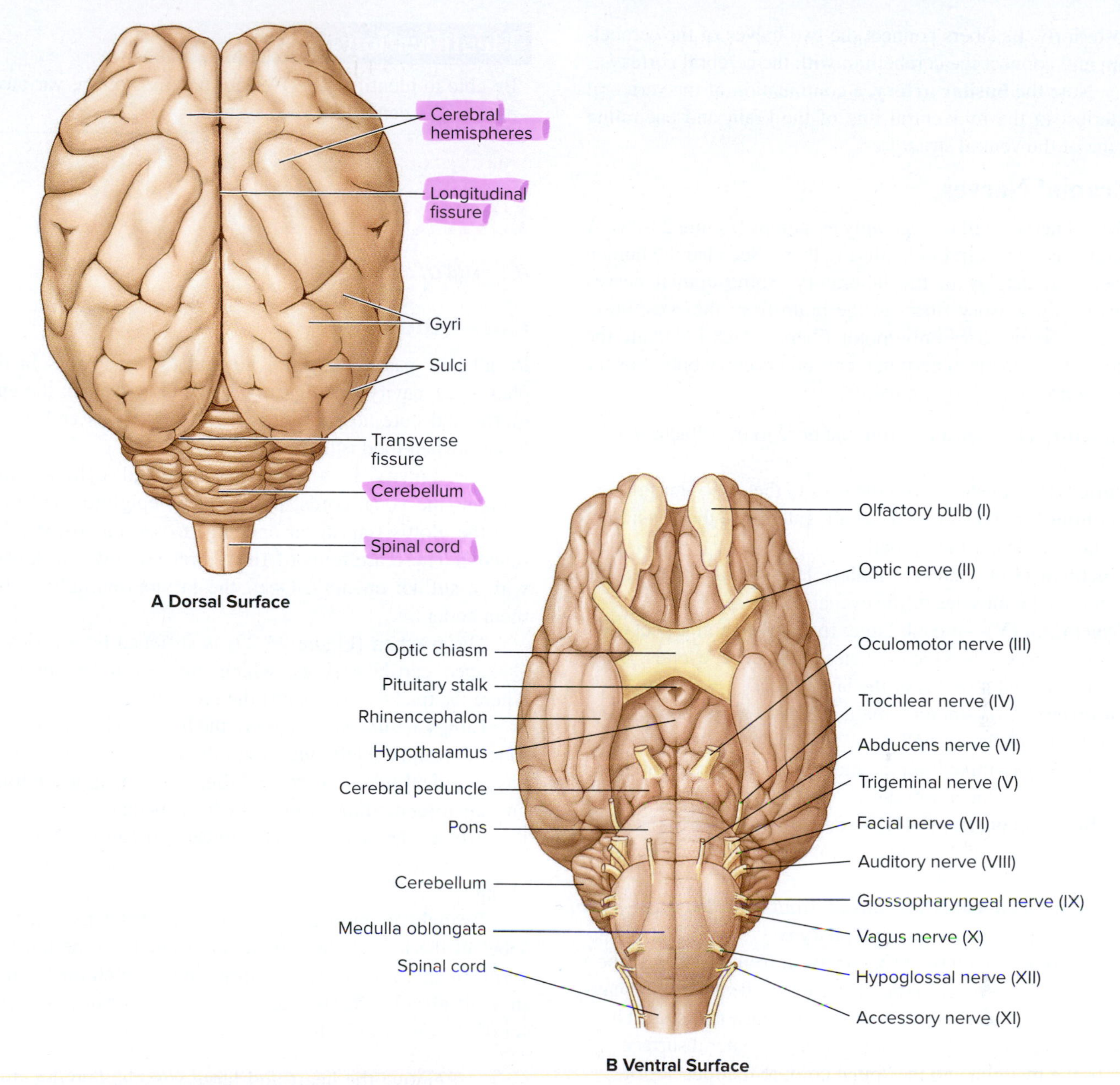

Figure 23-28

Sheep brain, ventral view. What part of the brain lies ventral to the cerebellum and narrows to form the nerve cord? _____________
What part of the brain connects the two halves of the cerebellum and connects the cerebellum to the cerebral cortex? _____________
What large cranial nerve sends branches to the nose, eyelids, face, tongue, and jaw muscles? _____________

posteriorly. Its fibers connect the two halves of the cerebellum and connect the cerebellum with the cerebral cortex.

Note the **basilar artery,** a continuation of the vertebral arteries, in the midventral line of the brain and encircling some of the ventral structures.

Cranial Nerves

Cranial nerves will be seen only as stumps (Figure 23-28). A hand lens may help in identifying them. See also the human brains on display in the laboratory. Some cranial nerves carry only sensory fibers to the brain from the eyes, nose, or ears. Some carry only motor fibers, which innervate the muscles or glands of certain areas; others carry both sensory and motor fibers. All are paired.

Olfactory (I)—sensory; from the nose to the olfactory lobes
Optic (II)—sensory; from the eyes to the optic chiasma
Oculomotor (III)—motor; small; anterior to the pons; to the muscles of the eyeball
Trochlear (IV)—motor; smallest; lateral to the oculomotor; to the muscles of the eyeball
Trigeminal (V)—mixed; large; from the lateral aspect of the pons; sends three branches to the nose, eyelids, face, tongue, and muscles of the jaw
Abducens (VI)—motor; small; posterior to the pons; to the muscles of the eyeball
Facial (VII)—mixed; on the medulla; lateral to the abducens; to the facial muscles
Auditory (acoustic) (VIII)—sensory; posterior to the facial nerve on the anterolateral aspect of the medulla; from the inner ear
Glossopharyngeal (IX)—mixed; from the side of the medulla; to the tongue and pharynx
Vagus (pneumogastric) (X)—mixed; from the side of the medulla; passes into the thoracic and the abdominal cavities to the pharynx, larynx, heart, lungs, and stomach
Spinal accessory (XI)—motor; from the lateral surface of the medulla and the spinal cord; to the muscles of the neck
Hypoglossal (XII)—motor; from the ventral surface of the medulla; to the muscles of the tongue and neck

Medical students for generations have used the following mnemonic aid (with variations) to help them remember the order of the cranial nerves: "*O*n *O*ld *O*lympus' *T*owering *T*ops *A* *F*inn *A*nd *G*erman *V*iewed *S*ome *H*ops." The first letter of each word stands for the initial letter of a nerve.

Sagittal Section of the Brain

On half-sections of preserved brain, find the (1) gray and white matter; (2) corpus callosum—cut section, curved; (3) corpora quadrigemina—between the cerebrum and cerebellum; (4) arbor vitae—white fibers, tree-shaped, in the cerebellum; (5) third ventricle—cavity above the optic chiasma; and (6) fourth ventricle—small cavity with a very thin roof, dorsal in the medulla, under the cerebellum.

Identification

Be able to identify and give the functions of the various parts of the central nervous system.

EXERCISE 23G
Respiratory System

Core Study

Recall how passageways for air and for food cross in the pharyngeal cavity (see Figure 23-14). Locate again the **epiglottis** and note how it closes the **glottis** (open end of the larynx) when food is being swallowed.

The **larynx** is a cartilage-reinforced cylinder that contains the vocal cords. Displace the epiglottis and look into the glottis (laryngeal opening) to see the **vocal folds (cords).** These are folds of flesh directed downward, each with a slitlike opening. Open the larynx ventrally to see them better.

The **trachea** (Figure 23-29) is stiffened by a series of C-shaped cartilage rings, which are incomplete dorsally where the trachea lies against the esophagus.

Lungs are unequally paired, the larger right lung having four lobes and the left lung two or three lobes. The right lung has an **apical lobe,** a **cardiac lobe,** a **diaphragmatic lobe,** and an **intermediate lobe,** which is median and notched to surround the postcava. The apical and cardiac lobes are usually fused on the left lung. Lungs are spongy and highly elastic.

Bronchi are branches of the trachea that extend into the lobes of the lungs. These branch repeatedly in the lungs to form, finally, microscopic **respiratory bronchioles,** which give off **alveolar ducts,** each of which terminates in a cluster of alveoli, or air cells.

Remove the heart and large vessels, leaving stubs of the pulmonary vessels attached to the lungs. Remove the larynx, trachea, and lungs. Slit the larynx and trachea and follow the bronchi as far into the lungs as possible. Trace a blood vessel into the lung tissue. Cut a thin section from a lung and examine with a hand lens or dissecting microscope.

Demonstration

The *butcher's pull.*[*] Examine the fresh lungs, trachea, and heart of a small sheep. Use a blowpipe to expand a lung. Trace one of the bronchi into the lung and follow some of its

[*]The butcher's pull is referred to as the sheep pluck in the Carolina catalog (228830).

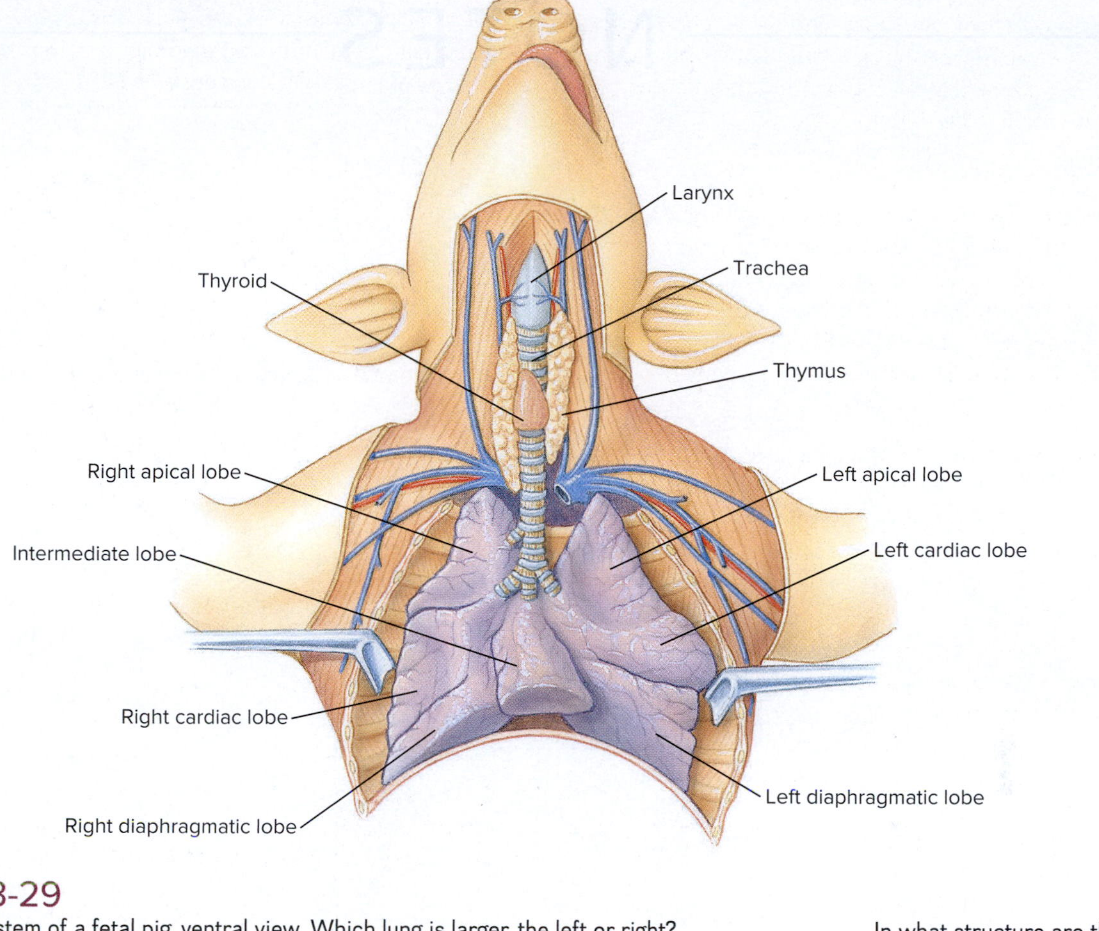

Figure 23-29

Respiratory system of a fetal pig, ventral view. Which lung is larger, the left or right? ______________ In what structure are the vocal cords found? ______________ What gland is important for regulating growth, development, and metabolic rate? ______________

branches as far as possible into the lung tissue. Identify the pulmonary vessels. Trace a blood vessel as far as possible into the lung tissue. Why is the tissue so highly vascular? ______________ Cut a cross section from a large artery and from a large vein and contrast their walls. Contrast the trachea with the esophagus. Where do you find the cartilage? ______________ Of what advantage is the cartilage? ____
____
____

NOTES

APPENDIX A

Instructor's Resources for Implementing Exercises

Exercise 1: The Microscope

Exercise 1A: Compound Light Microscope

Materials

Compound microscopes	Salt or sugar crystals
Microscope lamps	Distilled water
Blank slides and coverslips	Materials suitable for wet mounts
Prepared slides	Pipette
Typewritten letters (e, a, h, or k)	Gum arabic solution
Colored threads	Pond water

Notes

1. **Koehler illumination.** Students find it easier to learn to focus properly than to adjust the illumination properly. To avoid problems later, we spend a few minutes talking the students through the procedure. Correct adjustment of the illumination system is commonly referred to as Koehler (or Köhler) illumination. Here in more detail are the steps involved:
   a. Focus on a specimen (preferably a stained section) with low power.
   b. It is best to first center the condenser. Most condensers have little adjustment screws used for centering, and students sometimes fiddle with these, not knowing what they are for. To center the condenser, move the condenser with its focus knob up toward the top of its range. Close down the iris diaphragm; then move the condenser up and down until a sharp image of the diaphragm is seen. Center the circle of light using the condenser centering screws.
   c. Once the condenser is centered, open the iris diaphragm until the dark edges are just outside the field of view; that is, the entire field of view should be filled with light.
   d. Remove the ocular (or one of the oculars, if using a binocular microscope) and look into the tube. Adjust the iris diaphragm so that the circle of light covers ⅔ to ¾ of the illuminated area on the objective lens.
   e. Replace the eyepiece. Once the students have done this, they are able to correctly adjust the illumination in just a few seconds. This adjustment procedure should be repeated when students switch to high power, but, with a bit of practice, students learn how to use the iris diaphragm effectively without having to remove the ocular each time they switch objectives. Students should be advised *not* to change the condenser position to get more or less contrast. Lowering the condenser tends to reduce resolution. They *should* be encouraged instead to use the diaphragm to readjust illumination.

2. **Techniques for wet mounts of living specimens.** There are a number of very useful techniques for maintaining live microscopic forms on slides that you might wish to demonstrate to the class. Some of these can be used to keep unicellular eukaryotes or other microscopic forms alive on a slide for hours or even for days if kept in a cool place.
   a. **Temporary wet mount.** Prepare a temporary wet mount (Figure A-1). Select a clean slide and coverslip. Mount a piece of insect wing or appendage, a bit of feather, or a similar object on the slide. Add a drop of distilled water with a pipette. Hold the coverslip at an angle with one edge on the slide and the other held up with forceps or dissecting needle. Slowly lower the raised edge onto the drop of water. This helps prevent air bubbles from forming. Examine with low power and sketch what you see in the field of vision. Switch to high power without moving the slide and draw the portion now visible.
   b. **Petroleum jelly ring.** Using a hypodermic syringe filled with melted or softened petroleum jelly, make a ring on a slide, add the desired culture material, and cover with a coverslip. The depth of the ring may vary to accommodate different-sized forms or amounts of fluid.

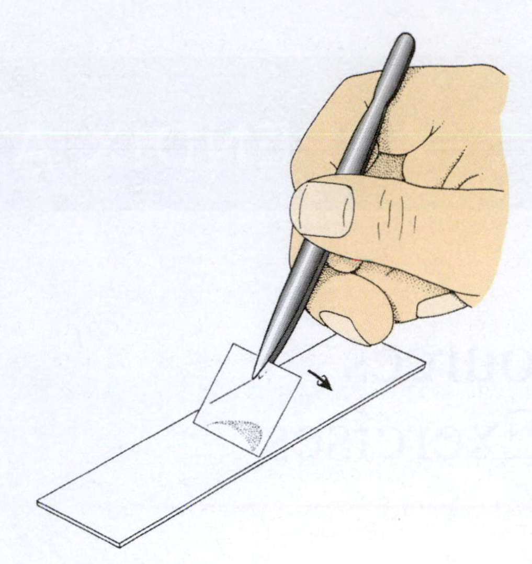

Figure A-1
Preparation of a wet mount.

c. **Sealed coverslip.** Place a drop of the culture on a microscope slide in the usual way, apply a coverslip, and then use fingernail polish, melted petroleum jelly, or ordinary 3-in-1 machine oil to seal the edges of the coverslip. If algae are present in the culture, they will provide oxygen for the animal life.

d. **Hanging drop.** Place a drop of the culture on a coverslip; then invert the slip over a deep-well depression slide, forming a hanging drop. By sealing the coverslip with nail polish, melted petroleum jelly, or oil, you can retard evaporation. Air in the depression cavity supplies oxygen for the animals in the hanging drop.

e. **Holding small invertebrates for observation.** Use a small piece of plastic wrap as a coverslip over a drop of liquid on a slide. This holds the organisms quiet, retards evaporation, and allows the organisms to be returned unharmed. Larger forms may be mounted between two pieces of the wrap and then placed on a slide. The preparation can then be turned over and viewed on both sides.

f. **Demoslides.** Connecticut Valley Biological offers "demoslides," which can be used to culture unicellular eukaryotes and watch the progress of the culture in a thinly compressed observation chamber at the end of a larger growth chamber without transferring the organisms to a separate slide.

3. **Some points on use and limitations of light microscopy.** An article in *American Scientist* explains the many new ways the light microscope is being used in biological research (Taylor, D. L., et al., 1992).

All references to papers and publications mentioned in Appendix A are available as part of the Instructor Resources section in Connect for the main text (available only to instructors who are logged into Connect). Visit www.mheducation.com/highered/connect for more information.

Exercise 1B: Stereoscopic Dissecting Microscope

Materials

Binocular dissecting microscopes
Microscope lamps
Prepared slides of fluke, tapeworm, or other whole mount
Crayfish gills, pieces of preserved sea star test, and so on
Finger bowls, watch glasses, or both

Note

1. **How to hold an unmounted specimen for stereomicroscopic study in any desired position.** Place the specimen in a small culture dish or Stender dish on clean, washed sand (preferably rounded rather than sharp grains) and add just enough alcohol to cover it. Push the specimen into the sand just enough to hold it in the desired position for study. Plasticene (modeling clay) can be substituted for sand in a pinch, but it is not as good. If the specimen is to be secured more or less permanently in one position, heat wax in a dissecting dish and press the animal's appendages gently into it, so that the animal is held firmly when the wax cools.

Exercise 2: Cell Structure and Division

Exercise 2A: The Cell—Unit of Protoplasmic Organization

Materials

Prepared stained slides
  Immature sea star eggs
  Amphiuma liver
  Mitochondria from *Amphiuma* or other amphibian
  Golgi complex
Slides
Cover glasses
Toothpicks
Stain (methylene blue, methyl green, gentian violet, or other)
Microscopes

Notes

1. **Prepared slides of mitochondria and Golgi.** Carolina and Ward's offer mitochondria slides prepared from amphibian or reptile liver. The best have been stained with Janus green, which is thought to react with cytochrome oxidase in the mitochondria to keep the dye oxidized. In the surrounding cytoplasm, the dye is reduced to a colorless form. Although students will see the mitochondria only as granules or short, rodlike structures with no internal structure, the light microscope does reveal cytoplasmic distribution and relative size of the mitochondria and shows that they are abundant in the cell.

Carolina offers Golgi complex prepared from *Amphiuma* or *Necturus,* stained with silver-hematoxylin-eosin. In the slides we have seen, staining was nearly perfect and the Golgi obvious. Ward's offers Golgi prepared from nerve ganglia. Other suppliers of prepared microslides are listed in Appendix B.

Students usually need help in locating the Golgi apparatus, but in good slide preparations it is relatively prominent. In stomach and intestinal tissue, it is always found in a supranuclear position. Its appearance is variable but it is usually seen as a more or less dense network of parallel anastomosing strands, deeply stained. As with the mitochondria slides, students must relate what they see on the slide to the electron micrograph of Golgi (Figure 2-5).

2. **Demonstrating cyclosis.** Cyclosis, the streaming movement of cytoplasm, can be demonstrated by mounting an *Anacharis* (formerly *Elodea*) leaf in 10% sugar solution.

Exercise 2B: Cell Division—Mitosis and Cytokinesis

Materials

Stained slides showing mitotic stages:
- Whitefish blastula or
- Fertilized *Ascaris* eggs

Compound microscopes

Notes

1. **Slide study of mitosis** is preferably done with the whitefish blastula, with mitosis in the fertilized *Ascaris* egg offered as an alternative. It is not intended that both should be studied.
2. **Separating mitosis and meiosis.** The study of mitosis is deliberately separated from the study of meiosis (Exercise 3A). Students should firmly understand mitosis and, in both lecture and laboratory, be given some assimilation time before being introduced to meiosis. Presenting the two concepts together may appear to be an effective and practical teaching approach, but, as we discovered in our own classes, for many students the two concepts become seriously confused.

Exercise 3: Gametogenesis and Embryology

Exercise 3A: Meiosis—Maturation Division of Germ Cells

Materials

Prepared slides
- Grasshopper testis, showing stages of spermatogenesis
- *Ascaris* uterus, showing sperm entrance and maturation stages

Microscopes

Note

1. **Success with the grasshopper testis exercise** depends principally on securing good slide material for the students. It may be necessary to sample prepared longitudinal sections from several suppliers of microslides and be resolute about returning substandard material. Suppliers of microslides are listed in Appendix B.

Exercise 3B: Cleavage Patterns—Spiral and Radial Cleavage

Materials

Prepared slides
- *Cerebratulus* early development (from Carolina; see "Notes")
- *Cerebratulus* late development (from Carolina; see "Notes")

Sea star development
Compound microscopes

Notes

1. **Prepared slides.** Whole mounts of *Cerebratulus* early and late development are available only from Carolina Biological Supply Co., as far as we can determine. All stages are on two slides, cat. # 309556 and 309562. Different stages are mixed together in mounting medium and pipetted onto individual slides, so it is a matter of chance how many of each stage will appear on the slide. Those we have seen are of excellent quality.

   Whole mounts of sea star development are available from most suppliers of prepared microslides (see Appendix B for listing of suppliers). Slides are available of all stages through late gastrula or bipinnaria combined on single slides.
2. **Choice of material to illustrate spiral cleavage.** Spiral cleavage is typical of virtually all annelids, nearly all molluscs except the cephalopods, turbellarian flatworms, all nemerteans, acanthocephalans, some rotifers, some gastrotrichs, sipunculans, echiuroids, and some brachiopods. A modified form intermediate between spiral and radial is found in the nematodes. The nemertean *Cerebratulus* was chosen for this exercise because it cleaves spirally with almost textbook perfection. The embryos of other readily available materials, such as *Crepidula* and *Mytilus,* develop a specialized cytoplasmic component, the polar lobe, in which important morphogenetic factors are segregated before the first cleavage. Because of the large polar lobe's presence, the 2-cell stage looks like 3 cells and the 4-cell stage looks like 5 cells. This is confusing to students and complicates the comparison of spiral and radial cleavage.

3. **Radial and bilateral cleavage.** Our treatment of the cleavage types would be considered simplified by many embryologists who recognize bilateral cleavage as a major modification of the radial pattern. According to this distinction, bilateral cleavage is found among the nematodes, some hemichordates (Enteropneusta), protochordates (Urochordata and Cephalochordata), amphibians, and higher mammals. The remaining vertebrates show discoidal cleavage. Thus, of the Deuterostomia, only the chaetognaths and the echinoderms show "true" radial cleavage. However, bilateral cleavage is basically radial in pattern, except that only one plane passing through the major axis will yield any symmetry. In other words, distinguishable right and left halves of the future embryo are formed by the first two cleavages. Because radial and bilateral cleavage patterns closely resemble each other, we have combined them for this exercise.

   It is important that students understand that, whereas all of the spiralia are protostomes, not all protostomes have spiral cleavage. Superficial cleavage is typical of insects, and radial or bilateral cleavage, as well as other patterns not easily categorized, are scattered among various protostome groups. A summarization of these varieties is found in Blackwelder and Garoian (1986, pp. 124–125).
4. **Project study of the early development of mud snail *Ilyanassa obsoleta*.** *Ilyanassa* (= *Nassarius*), one of the most common intertidal gastropod molluscs on the Atlantic Coast, was the subject of some of the earliest experimental embryology studies. Like most other molluscs, the egg develops with a polar lobe in which important cytoplasmic determinates are segregated. The adults are easily collected in the lower intertidal zone in almost any quiet estuary with a sandy-mud bottom, or they can be purchased from the Marine Biological Laboratory at Woods Hole, Massachusetts (see Appendix B for address). Living snails are easily maintained in the laboratory and can be brought to spawning readiness by feeding them marine bivalves. The egg capsules are laid on the sides of the aquarium: these may be opened and the eggs removed for study. All these methods, as well as a description of development, are found in the excellent paper by J. R. Collier, 1981, "Methods of obtaining and handling eggs and embryos of the marine mud snail *Ilyanassa obsoleta*," in *Laboratory Animal Management: Marine Invertebrates,* Committee on Marine Invertebrates, Washington, D.C., National Academy Press.
5. **Fertilization and cleavage in sea urchin eggs.** When in breeding condition, sea urchins can be induced to spawn by injecting with dilute KCl or acetylcholine or by stimulating with a weak electric current. Eggs and sperm are mixed to begin development. The methods are well known to embryologists and can be found in almost any experimental embryology manual (e.g., Johnson and Volpe, 1973).

All references to papers and publications mentioned in Appendix A are available as part of the Instructor Resources section in Connect for the main text (available only to instructors who are logged into Connect). Visit www.mheducation.com/highered/connect for more information.

Exercise 3C: Frog Development

Materials

Preserved frog eggs in development stages
Prepared stained slides
  Frog ovary, cross section
  Frog development stages, cross sections
Models of frog development (optional)
Microscopes
Hand lenses or binocular dissecting microscopes

Exercise 4: Tissue Structure and Function

Materials

Compound microscopes
Prepared slides (the selection will depend on time and materials available)
  Frog skin, cross section
  Amphibian small intestine, cross section (*Amphiuma* is best)
  Artery, vein, and nerve, cross section
  Trachea, cross section
  Ascaris, cross section through intestine
  Ground bone, cross section
  Cartilage (can be seen on trachea slide)
  Skeletal muscle
  Smooth muscle (can be seen on intestine or artery)
  Cardiac muscle
  Human blood
  Amphibian or reptile blood
  Areolar connective tissue
  Others as desired

Notes

1. **Selecting material from this exercise.** Instructors should not feel constrained to cover all the material in this exercise in a single laboratory period. The exercise contains more material than can be covered in a 2-hour laboratory period and perhaps even in a 3-hour period. For a 2-hour period, we suggest restricting the exercise to epithelial tissue, some examples of connective tissue (e.g., areolar and adipose), and vascular tissue. Cartilage and bone tissue could be examined during study of the skeletal system of any of the vertebrate types (e.g., frog or fetal pig). Dense fibrous and areolar connective tissue could be introduced before approaching muscle dissection of the fetal pig. Examination of vascular, muscle, and nervous tissues is effectively combined with study of these systems in fish, frog, or fetal pig. Similarly, examination of tissues combined into organs (frog skin,

amphibian intestine, and section through artery, vein, and nerve) is probably more meaningful to the student if done when these systems are studied in a frog or fetal pig. Suppliers of microslides are listed in Appendix B.

2. **Demonstrations.** Following are suggestions for demonstrations for this exercise and additions or substitutions for the slide material selections. See Appendix B for sources of microslide material.
   a. **Prepared slide of mesentery.** A spread preparation of epithelium stripped from the underlying connective tissue provides a surface view of simple squamous epithelium. Slides specially prepared with silver stains reveal the interdigitating boundaries of the epithelial cells. A bit of sloughed-off epidermis of a frog placed in a drop of water on a slide will also show this type of epithelium.
   b. **Prepared slide of kidney.** Tubules of kidney are lined with simple cuboidal epithelium. Glandular tissue of thyroid or submaxillary gland may also be used to demonstrate cuboidal epithelium.
   c. **Prepared slide showing motor nerve endings on skeletal muscle.** Slides prepared by gold impregnation clearly reveal the synapses between motor nerve fibers and skeletal muscle fibers. Point out to students that each terminal branch ends on a muscle fiber. Here it loses its myelin sheath and ends in a little crowfoot pattern that may resemble a tiny disc on the muscle fiber. The terminal part of the axon divides into several branches, each terminating in a muscle fiber; thus, a single nerve fiber innervates numerous muscle fibers.
   d. **Prepared slide showing decalcified bone.** Decalcified preparations reveal the morphology of the cells (osteocytes) and the organic components of compact bone. Slides stained with hematoxylin and eosin are deeply colored because of the high collagen content of the organic matrix. Although osteocytes are present in the lacunae (they are absent in ground sections of bone) and stain densely, they tend to shrink during preparation and pull away from the lacunae borders. Canaliculi are better seen on ground sections of bone.
   e. **Long bone cut in longitudinal section.** With such preparations, one can distinguish among compact bone, spongy bone, and marrow cavity.

Exercise 5: Ecological Relationships of Animals

Exercise 5A: A Study of Population Growth, with Application of the Scientific Method

Materials

For each group of four students:

5 half-pint, wide-mouth mason jars
5 screw-on rings
5 pieces of aluminum window screen, cut to fit snugly into the rings
5 pieces of paper towel, same size as the screen
Resource, consisting of 5 pounds stoneground whole-wheat flour mixed with ¼ pound powdered brewer's yeast (both available at health food stores); this will supply enough resource for 30 groups of 4 students each
Inert filler; vermiculite works well (available at garden stores); sawdust seems to inhibit growth; vermiculite should be screened with window screen or finer mesh to remove large particles
250 adult beetles; these can be purchased from scientific supply houses or secured from a culture kept in the laboratory; fill several quart jars ¾ full with flour/yeast mixture (95:5), add some beetles, cover with screen and paper towel top; as the medium is used up, add more; every year or so, discard half the medium (and included beetles) and replace with fresh medium

Notes

1. **This is a project exercise requiring most of a term to complete.** Because it is an excellent introduction to the scientific method, which students are characteristically introduced to at the beginning of a zoology or biology course, we suggest setting aside part of the first laboratory period to get it under way. Only about an hour is required for each student group to separate the beetles into jars, but students will need more reading time to understand the background to the exercise and to be properly acquainted with the methodology. Consequently it is best to require the students to have read the exercise before coming to the laboratory. A promise of a quick quiz at the beginning of the laboratory period over the scientific method and the procedures is usually sufficient to ensure compliance.
2. **This lab is a good place to discuss basic population biology** and/or show a film on population biology.
3. **One way to help separate flour from beetles** is to fashion a top by putting three pieces of screen (and no paper towel) in one screw-on ring, screw it onto the jar, then gently shake out the flour. The screens must be closely opposed to prevent beetles' slipping through. The adults, pupae, and larger larvae will be caught on the three layers of screen and remain in the jar (a few may escape through the screening, so students should watch for this during the separation). At the end of the experiment, you may want to combine and save all beetles and flour to add to your culture for next term.
4. **The report.** This project is a good subject for a scientific paper of approximately three to five pages in length. Assign it to be written using the standard format of scientific articles: introduction, materials and methods, results, discussion, and literature cited.

You may want to provide a handout that fully describes what is expected. Encourage students to look at their data in original ways. Also encourage students to explore the literature. You may wish to place the article by Peters and Barbosa (1977) on reserve at the library.

5. **Some tips.** There are several ways to simplify this laboratory. One is to count only adults. Another is to use only three jars per group, containing either 3, 10, and 50 g of resource with filler, or the same without filler.

Exercise 5B: Ecology of a Freshwater Habitat

Materials

Essential collecting equipment is given in the exercise; however, consult the appendix in E. B. Klots' field book for collecting methods and suggestions for useful collecting and sorting equipment, much of it homemade (Klots, E. B. 1966. The new field book of freshwater life. New York, G. P. Putnam & Sons).

Notes

1. **This exercise and the dichotomous key** were designed to be used anywhere in North America, although obviously local available pond and stream conditions will influence the biotic composition. Species diversity will be restricted in temporary ponds or in ponds subject to pollution, agricultural runoff, stream flow-through, or use by farm animals. Even a small pond can support rich fauna and flora if it is permanent and protected from these disturbances. Indeed, to show students how persistent and intrusive life is, you might consider sampling diversity in a drainage ditch, marshy meadow, swamp, or small prairie slough.
2. **Complete the identifications immediately after collection** if at all possible. If time constraints forbid this, place the material in a *small* amount of water in a refrigerator for work-up within the next day or two or, if the exercise cannot be completed until the following week, preserve everything in 70% alcohol.

Exercise 6: Introduction to Animal Taxonomy

Materials

Animal specimens representing several phyla and classes, suitable reference texts or field guides

Notes

1. **Cladogram exercise.** For Exercise 6A on how to read and compare cladograms, no additional materials are required. The instructor may choose to require students to submit a brief written report showing the character matrix and resulting cladogram. Students sometimes struggle with the concept that the interpretation of a branching pattern is critical to interpretation of a tree. The idea that branches can rotate within a tree without changing the tree's meaning is difficult. However, constructing simple model trees using ball and socket toys, sticks and Styrofoam balls, or similar items is often extremely helpful to students. Manipulating a physical model of a cladogram reaches students with a wide range of learning styles.
2. **Selection of specimens.** For Exercise 6B on the use of the taxonomic key, select specimens to represent a wide range of animal taxa. Although students will recognize by common name many of the specimens easily available to you, they usually will not be able to classify them correctly without aid of the key. Specimens that will provide some challenge to students are hydroids (slide mounted), sea pansy or sea whip, ctenophore, freshwater turbellarians (slide), tapeworm, liver fluke, *Ascaris,* spiny-headed worm (acanthocephalan), brachiopod, sea cucumber, chiton, marine segmented worm such as *Arenicola* or *Aphrodite,* chaetognath, horseshoe crab, spider, tunicate, and amphioxus. Less challenging, perhaps, but still providing keying experience, are a variety of vertebrates, insects, sponge, octopus, bivalve, snail, sea star or brittle star, earthworm, and marine crab or freshwater crayfish. All of these are available as preserved material from biological supply houses.

   If insects are used with the key in Exercise 16, select relatively robust preserved specimens that can withstand handling. Suggestions are silverfish, cockroach, cricket, grasshopper, dobsonfly, mayfly, dragonfly, stonefly, giant water bug, leafhopper, tiger beetle, caddisfly, moth or butterfly, horse fly, bumble bee or honeybee, and termite. Preserved specimens of all these may be obtained from biological supply houses. These suggestions represent 14 different orders, quite enough for a class. The key is more comprehensive than this, permitting certain other orders to be included if desired. Slides are also available of whole mounts of fleas, body lice, and representatives of other insect orders not included in these suggestions.
3. **Organization.** A workable system is to place several specimens of each species in a separate tray or container, marked with a letter of the alphabet, on a service table. Students return specimens to the tray or container after identification.
4. **The report.** The written report requires students to verify the identification of each insect order by consulting a reference, such as an insect field guide or other reference. This exercise acquaints students with the reference, introduces them to the variety of forms within a taxon, and emphasizes the importance of confirming a tentative identification.

Exercise 7: Unicellular Eukaryotes

Exercise 7A: Phylum Amoebozoa—*Amoeba* and Others

Materials

Amoeba culture
Other ameba cultures as available
Slides, ringed or plain
Coverslips
Petrolatum
Thread
10% nigrosin (for demonstration #3)
Radiolarian shells, preserved or on permanent slides
Prepared slides
  Amoeba
  Entamoeba
  Other amebas as available
Microscopes

Notes

1. **Finding amebas in the culture.** Amebas will be found on the bottom of the shipping container or petri dish into which they are poured for the laboratory. Advise students not to agitate the container by squirting unused contents of a pipette back into the culture. It helps to pour off most culture water before the lab, leaving only a few millimeters in the container. Many instructors find it best to pipette amebas for the students.
2. **Subculturing amebas.** Amebas are rather easily subcultured. Several methods are described in Needham (1959), all using some variation of the classical hay or rice infusion. To see what the Society of Protozoologists is recommending for culture establishment, see pp. 2–3 in Lee et al. (1985).
3. **Nigrosin vital stain,** 10%, is a good relief stain against a background of which ameboid movement and contractile vacuole activity show up well.
4. **Entamoeba histolytica** when viewed at high power (430) with student microscopes is too small to reveal much detail, but the exercise is useful to give students an impression of the small size of *E. histolytica,* as compared with free-living amebas, and shows them what one is looking for in diagnosing amebic dysentery. We always provide a demonstration scope showing *E. histolytica* under oil immersion; at this magnification, most of the organelles shown in Figure 7-3 are discernible.
5. **Demonstration of osmoregulation.** How would the osmoregulatory requirements of marine unicellular eukaryotes compare with those of freshwater forms? This might be demonstrated by placing a drop of rich *Amoeba* culture in each of four deep-well slides. Fill three of the wells with culture water that has been made up to 2%, 4%, and 6% saline (by adding 2, 4, or 6 g of salt/100 ml of water) and the fourth with plain culture water. After a few minutes, place a drop from the bottom of each well on a slide and examine. Time rate of discharge of contractile vacuoles in specimens from each osmotic concentration. If there is a difference, can you explain it?

All references to papers and publications mentioned in Appendix A are available as part of the Instructor Resources section in Connect for the main text (available only to instructors who are logged into Connect). Visit www.mheducation.com/highered/connect for more information.

Exercise 7B: Phylum Euglenozoa and Viridiplantae—*Euglena, Volvox,* and *Trypanosoma*

Materials

Euglena culture
Volvox culture
Stained slides of *Euglena, Volvox,* and *Trypanosoma*
Other flagellate cultures or slides as available
Methylcellulose, polyvinyl alcohol, Protoslo, or Detain
Methyl violet or Noland's stain
Microscopes
Slides and coverslips

Notes

1. ***Euglena gracilis* and *Euglena viridis,*** the two most commonly supplied of more than 100 species of genus *Euglena* distributed in organically rich freshwater ponds and ditches, differ in several ways. *Euglena gracilis,* which is 35 to 65 μm in length, has 6 to 12 shield-shaped chloroplasts, each bearing a prominent central pyrenoid region (containing enzymes that catalyze conversion of photosynthetically produced glucose into starch and paramylon). The pyrenoid of each chloroplast is covered with a watch-glass-shaped complex of starch grains and paramylon bodies (paramylon, also spelled paramylum, is an unbranched nutrient polysaccharide related to starch, but the glucose molecules are β-linked rather than α-linked as they are in starch). In the slightly larger *E. viridis* (40 to 80 μm in length), the chloroplasts form ribbons that radiate from a single large, central pyrenoid body. *E. gracilis* is a more rapid, smoother swimmer with pronounced euglenoid movement when swimming stops; the swimming and euglenoid movements of *E. viridis* are more jerky than those of *E. gracilis.*
2. **Preparing methylcellulose.** Ten percent methylcellulose for slowing unicellular eukaryotes is prepared by adding 10 g methylcellulose to 50 ml of water. Bring to a boil, cool, and let stand for 30 minutes. Add cold distilled water to make 100 ml.

   Sodium carboxymethylcellulose (2%) is an alternative medium for slowing unicells. Bring 100 ml distilled water to a boil and then slowly add 2 g sodium carboxymethylcellulose.

Suitable methylcellulose products are available from Carolina Biological Supply Co. ("Protoslo"), Ward's ("Detain"), and Connecticut Valley (10% methylcellulose). With any of these, make a small ring the size of a dime on a slide, add a drop of culture to the center of the ring, and cover.

3. **Preparing Noland's stain for flagella and cilia.** Moisten 20 mg of gentian violet with 1 ml distilled water; add 80 ml of a saturated solution of phenol in water; then add 20 ml of formalin (40% formaldehyde); finally, add 4 ml glycerin. Mix these constituents together and add a drop of the solution to the drop of culture to be examined.
4. **How to concentrate *Euglena*.** *Euglena* may be concentrated in a culture by shining a bright light through a narrow slit in a piece of cardboard placed against the shaded culture container. After a few minutes, the positively phototropic *Euglena* will be concentrated in a narrow band at the container's edge, where they can be removed easily with a pipette. (Do not use *too* bright a light; *Euglena* become negatively phototactic in intense light.)
5. ***Volvox globator,*** the species usually offered by supply houses, is found in nitrogen-rich freshwater habitats. Most colonies range from 350 to 500 µm in diameter and contain from 5000 to 15,000 cells. This species is monoecious, so both microgametes and macrogametes are formed by the same colony, as described in the manual. *V. aureus,* sometimes supplied commercially for laboratory use, has both dioecious and monoecious varieties. The somewhat smaller *V. tertius,* a dioecious species used in experimental work, is not ordinarily supplied commercially for use in teaching laboratories.

   Remind students to look for *Volvox* colonies in the bottom of the container—they are easily seen—rather than aimlessly pipetting from the culture container.
6. **Avoiding cross-contamination of cultures.** If several different unicellular eukaryote species are used in the same laboratory over a period of several days, be careful to use separate pipettes for each to avoid cross-contamination of cultures. *Euglena* especially will quickly build up populations in other cultures if accidentally introduced.

Exercise 7C: Phylum Apicomplexa—*Plasmodium* and *Gregarina*

Materials

Mealworms (*Tenebrio* larvae) or
Cockroaches *(Periplaneta, Blatta)*
Stained slides of gregarines
Stained human blood smears containing *Plasmodium* stages
Watch glasses
Invertebrate Ringer's solution, or 0.65% saline solution
Slides and coverslips
Teasing needles
Pipettes
Microscopes

Notes

1. **Preparing insect saline.** Insect physiological saline, especially balanced for cockroaches, is prepared by dissolving 10.9 g NaCl, 1.6 g KCl, 0.8 g $CaCl_2$, and 0.17 g $MgCl_2$ in 1 liter of distilled water.
2. **Prepared slides of gregarines and *Plasmodium*.** If living material for the study of gregarines is not available, prepared slides may be obtained from some biological supply houses (see Appendix B for suppliers of prepared microslides).

   Prepared slides of *Plasmodium* are also available from biological supply houses. Students may be provided with a blood smear containing various red blood cells' stages (trophozoites and merozoites) for study; other stages, such as sporozoites, gametocytes, and a cross section of a mosquito gut showing oocysts, can be placed on display with accompanying explanatory legends. Mounted slides of male and female *Anopheles* mosquitoes are also available as a demonstration.

   Figure 7-13 in the manual illustrates *Plasmodium falciparum,* considered the most virulent of the *Plasmodium* species in humans. The ring stage is smaller than in either *P. malariae* or *P. vivax,* and the crescent shape of the gametocytes is distinctive for *P. falciparum.* If species other than *P. falciparum* or *P. vivax* are placed on display, you may wish to provide color plates of the red cell stages (found in most parasitology texts), so that the students can see the species differences. Excellent color plates of these and other *Plasmodium* species are also found in Coatney, et al. (1971).
3. **Demonstration of living *Monocystis*.** *Monocystis lumbrici* is a gregarine that can be demonstrated quite easily by removing seminal vesicles from live earthworms and examining portions teased in 0.65% saline solution. If trophozoites are present, note their feeble movements. Stained slides of sections or smears of earthworm seminal vesicles showing developmental stages of *Monocystis* are available from dealers of microslides (see Appendix B).
4. **Demonstration of coccidia slides.** *Eimeria* causes coccidiosis, an important cause of death in domestic rabbits and chickens. A demonstration of the parasite can be arranged using stained slides of *Eimeria stiedae* from infected rabbit bile ducts or *E. tenella* from infected chicken intestine, available from biological supply houses.

   On a stained slide of infected rabbit bile duct, developing trophozoites appear as spherical bodies, large and small, embedded in the outer ends of the columnar cells that line the ducts. You may find some

All references to papers and publications mentioned in Appendix A are available as part of the Instructor Resources section in Connect for the main text (available only to instructors who are logged into Connect). Visit www.mheducation.com/highered/connect for more information.

of them undergoing multiple fission. The largest stages are male gametocytes, in which large numbers of microgametes develop. Female gametocytes are somewhat smaller and have a nucleus and darkly staining granules around the periphery. A female gametocyte encysts (oocyst) and becomes fertilized. The encysted zygote escapes into the lumen of the bile duct and is shed with the feces. You may find a number of oocysts lying free in the lumen of the duct.

Exercise 7D: Phylum Ciliophora—*Paramecium* and Other Ciliates

Materials

Stained slides of paramecia—normal, undergoing binary fission, and in conjugation
Paramecia cultures
Vorticella, Stentor, Spirostomum, or other ciliate cultures
Protoslo or 10% methylcellulose
Congo red–yeast mixture or Congo red–milk mixture
Acidified methyl green
0.25% NaCl solution
Dilute picric acid
Weak acetic acid
Salt crystals
Cotton
Toothpicks or pins
Slides and coverslips
Microscopes

Notes

1. **Preparing Congo red–stained yeast.** Boil some dry yeast in a small amount of water in a test tube; then cool. Add a 1% solution of Congo red and allow it to diffuse into the yeast for a few minutes. Then decant most of the dye solution to concentrate the yeast cells.
2. **Preparing acidified methyl green.** Combine 1 g methyl green, 100 ml distilled water, and 1 ml glacial acetic acid.
3. **Slowing paramecia for study.** In our experience, students have better luck using cotton fibers than methylcellulose (Protoslo) to slow movement. Some of the paramecia will become trapped beneath or between the fibers, where they can be observed. This is especially important for the study of contractile vacuoles and feeding. However, some instructors may elect to have the students make their initial observations on swimming behavior using Protoslo.
4. **Watching feeding in paramecia.** Paramecia when offered yeast will eat rapidly and soon become sated. To watch actual feeding, encourage students to begin their observations as soon as possible (preferably within a few seconds) after adding Congo red–stained yeast.

   Congo red turns blue in an acid environment. Food vacuoles become acidic by fusion with acidosomes as digestion progresses, but seeing this happen by watching for a color change (which is subtle at best) may require more time and patience than most students are willing to invest.

   Be certain to emphasize to students the importance of using a *very small* amount of Congo red–stained yeast when examining feeding in *Paramecium.*
5. **Mating reaction and conjugation.** Many ciliates have been found to have mating types within each species or variety. Members of one mating type will mate with members of another mating type, but not with their own type. Pure lines of mating types of paramecia, together with instructions for mating them, can be obtained from biological supply houses (e.g., Carolina 131540 *Paramecium multimicronucleatum,* 131546 *P. aurelia,* 131554 *P. caudatum;* Ward's 87-V-1310 *P. caudatum,* 87-V-1305 *P. bursana,* 87-V-1300 *P. aurelia*). These cultures should be used within 24 hours of delivery.
6. **How to concentrate unicellular eukaryotes.** Select a large-mouthed jar provided with a two-hole stopper or with a lid in which holes have been cut. Into one hole of the stopper, insert the stem of a small funnel in an inverted position, with its broad end covered by fine-meshed monofilament nylon netting (available as "laboratory sifters" or bolting cloth from biological supply houses). Into the other hole of the stopper, insert a larger funnel in an upright position. Put the stopper snugly in place in the jar and pour the culture water into the upright large funnel. As the jar fills, the water will filter out through the small funnel, and organisms will be retained by the bolting cloth. An indefinite quantity of water may be run through this mechanism. The device may be used in the field to transport concentrated unicellular eukaryotes to the lab. If the water being filtered contains algae or scum, it may be necessary to strain this out by tying a piece of cheesecloth over the entrance funnel.
7. **How to obtain abundant dividing stages of *Paramecium.*** During fission most paramecia tend to settle toward the bottom of the culture. Put a concentrated culture of *Paramecium* in a large funnel or other vessel provided with a stopcock at its lower end (e.g., separatory funnel). After being provided with suitable food (see #8), three or more generations are produced each day. By opening the stopcock at intervals and drawing off a few milliliters of culture, you can obtain large numbers of fission stages.

   Early stages appear to be spindle-shaped. The whole process of dividing usually requires about 25 minutes.
8. **Culturing paramecia and mixed unicellular eukaryotes.** Fill a finger bowl two-thirds full of distilled water and add 4 kernels of *boiled* wheat or rice and 15 or 20 pieces of boiled timothy hay, 1 or 2 cm long. Inoculate immediately. If bacterial film forms over water surface, break it up. Allow plenty of diffused light.

Several approaches to culturing paramecia and other ciliates are detailed in the compendium edited by J. G. Needham (1937).

The "demoslide" tubes supplied by Connecticut Valley Biological (cat. # LW2262) are useful for holding and viewing paramecium cultures (or cultures of any other unicellular eukaryote).

9. **Project experiments with *Stentor* and *Dileptus*.** Numerous ideas for research projects, mostly simple in nature, are detailed in the book by Goldstein and Metzner (1971). They are, as the authors proclaim in the foreword, "intended for the science-minded student in need of a project, for the amateur biologist with a home laboratory, [or] for someone who is just getting the feel of biological research. . . ." Some of these suggestions could easily be assigned as extra-credit projects.
10. **Identifying species complexes of *Paramecium*.** *Paramecium caudatum* and *P. multimicronucleatum,* both commonly supplied by biological supply houses, are difficult to distinguish and often confused by suppliers. A reliable diagnostic micronuclear staining procedure is described by Cole, T. A., R. Sehra, and W. H. Johnson, 1992, "Species identity of commercial stocks of *Paramecium* in the U.S." *Amer. Biol. Teacher* **54**(5):299–302.

Experimenting in Zoology: Effect of Temperature on the Locomotor Activity of *Stentor*

Materials

Cultures of *Stentor* (each student pair will require 3 or 4 *Stentor*)
Shallow depression slides and coverslips
Finely divided graph paper (millimeter grid) and cellophane tape
Culture dish (4½ inch or 8 inch) or plastic dish for water bath
Petroleum jelly (Vaseline) and cotton swabs (Q-tips)
Pasteur pipette (glass)
Supply of crushed ice for each student pair
Thermometer
Optional: hand tally counter for each student pair

Notes

1. **Preparations.** This experiment is not difficult to conduct and can be completed in about 1½ hours if students are provided depression slides with the *Stentor* already selected. Allow another hour if students are to do all preparations (prepare slides, transfer *Stentor,* and set up viewing assembly). Results are predictable and usually turn out with gratifying uniformity among different student pairs in a class. Refrigerator dishes also serve nicely for the water bath, as do standard 4½-inch culture dishes. Each student pair will require a dish of crushed ice, a beaker of warm water, and a thermometer.

   Tape a 20 mm by 20 mm grid (cut from graph paper) under the depression slide, using clear packaging tape that is waterproof. Trim excess tape from the bottom of the slide with a razor blade.

   Be certain that the depression slides are perfectly clean before the *Stentor* are added.
2. **Selecting the *Stentor*.** Each student pair needs about three *Stentor* so that one can be chosen that is actively swimming. Sometimes a *Stentor* becomes trapped between the coverslip and edge of the depression slide concavity; others are just not inclined to swim. The *Stentor* must be removed from the culture and transferred to the depression slide individually with a glass Pasteur pipette (the 9-inch disposable variety are a little easier to use than the 5¾-inch), but, with a little practice, this is accomplished quickly. If time is short, this should be done in advance by the instructor or laboratory assistant; however, we find that students do it themselves nearly as well with little practice. *Stentor* are rather easily seen with the naked eye (over a white background) and can be drawn up individually with Pasteur pipette and bulb. The petroleum jelly must be added carefully in an unbroken circle just outside the depression, and the coverslip must be pressed firmly down over the slide to form a perfect seal.
3. **Potential problems.** If too much fluid is placed in the depression, the *Stentor* may be washed out when the coverslip is added. If too little water, a bubble will be left in the field of view. Sometimes the *Stentor* get trapped in the petroleum jelly. At the highest temperatures, the *Stentor* often slow down or stop altogether; this happens when the temperature approaches the incipient lethal temperature for the organism. Students will often note that the *Stentor* change shape at the higher temperature, becoming elongate (trumpet-shaped); they may continue to swim or they may attach themselves to the glass surface. The thermal history of the culture will determine the maximum temperature that can be used. This is usually either 25° or 30° C.
4. **Swimming rate** can be counted easily without hand tally counters, but, if you have them, students enjoy using them, perhaps because it provides an aura of precision to the procedure.
5. **Temperature equilibrium.** Students should be advised to stabilize the water temperature for 2 to 3 minutes at each measurement point. However, the *Stentor* come into temperature equilibrium quickly, probably in less than a minute at a new temperature.
6. **Converting proportional swimming speeds to true swimming speeds.** It is a simple matter to convert all the measured rates (lines crossed per minute) to true

All references to papers and publications mentioned in Appendix A are available as part of the Instructor Resources section in Connect for the main text (available only to instructors who are logged into Connect). Visit www.mheducation.com/highered/connect for more information.

swimming speeds by setting up proportionalities between the two—for example,

$$\frac{\text{TSS at } 30°}{\text{LCM at } 30°} = \frac{\text{TSS at } 20°}{\text{LCM at } 20°}$$

and

$$\text{TSS at } 20° = \frac{\text{TSS at } 30° \times \text{LCM at } 20°}{\text{LCM at } 30°}$$

where TSS = true swimming speed in mm/sec and LCM = lines crossed per minute. If true swimming speed is measured at the lowest temperature at the start (where the slow movement makes accurate measurements easier), students can convert all the lines-crossed-per-minute values to true swimming speeds as they progress through the experiment.

Experimenting in Zoology: Genetic Polymorphism in *Tetrahymena*

Materials

Two or three strains of *Tetrahymena thermophila* or genomic DNA from two or three strains
Proteose peptone medium
Culture tubes
Autoclave
Ice buckets
High-speed centrifuge
Microcentrifuge
Spectrophotometer and quartz cuvettes
1.5 ml microcentrifuge tubes
20, 200, and 1000 µl micropipettes and disposable tips
Chemicals for DNA extraction and purification (see "Notes")
Chemicals for DNA amplification (see "Notes")
RAPD primers
DNA thermocycler
Agarose
6× gel loading buffer
Ethidium bromide
50× stock TAE buffer
DNA size markers
Mini-horizontal gel electrophoresis apparatus
Low-voltage power supplies
UV transilluminator

Notes

1. **Culturing *Tetrahymena thermophila*.** Cultures of *Tetrahymena thermophila* strains I–VII may be purchased through American Type Culture Collection (ATCC) (www.atcc.org). The axenic cultures are grown in sterile proteose peptone medium at room temperature.

Proteose Peptone Medium
5.0 g proteose peptone (Difco 0120)
5.0 g tryptone
0.2 g K_2HPO_4
dH_2O to 1.0 L, adjust to pH 7.2 before autoclaving

Use 5.0 ml per 15-ml tube. Inoculate each culture with about 100 µl of stock culture. Allow animals to grow for 2 to 3 days before harvesting for DNA.
T. thermophila mating type I (ATCC #30007)
T. thermophila mating type II (ATCC #30008)
T. thermophila mating type III (ATCC #205042)
T. thermophila mating type IV (ATCC #205043)
T. thermophila mating type V (ATCC #30305)
T. thermophila mating type VI (ATCC #30306)
T. thermophila mating type VII (ATCC #30307)

2. **DNA extraction and purification.** After cultures have grown for 2 to 3 days, place them on ice for 5 minutes; then warm them to about 50° C for 2 minutes to inactive swimming; then chill on ice for 2 minutes. Harvest cells by centrifugation at 10,000 g for 10 minutes at 4° C. Immediately decant supernatant from tubes and resuspend the cells in 200 µl of $T_{10}E_1$ buffer (10 mM Tris, 1 mM EDTA, pH 8.0) per original 5 ml tube. Transfer the samples to 1.5 ml polypropylene microfuge tubes. Freeze the cells in liquid nitrogen or on dry ice and then thaw on ice. Add 2 µl of Proteinase K (10 mg/ml stock) and 2 µl RNase A (10 mg/ml stock) to each tube; vortex and incubate at 37° C for 15 to 30 minutes. Add 1 volume of phenol/chloroform/isoamyl alcohol (25:24:1 v/v/v, equilibrated with $T_{10}E_1$, pH 8.0), vortex and incubate for 5 minutes at 37° C. Centrifuge the samples at 12,000 g for 5 minutes (4° C or room temperature). Transfer the upper aqueous layer to a clean centrifuge tube and add 1 volume of chloroform/isoamyl alcohol (24:1); vortex and centrifuge at 12,000 g for 5 minutes. Transfer the upper aqueous layer to a clean centrifuge tube add 1/10 volume of 3 M NaAcetate, pH 5.2, and 2 volumes of 100% ethanol. Vortex and incubate the samples at −20° C for 1 hour or −80° C for 10 minutes. Centrifuge the sample at 12,000 g for 20 minutes at 4° C. Decant supernatant and wash the pelleted DNA twice with 70% ethanol. Air dry (or vacuum dry) the DNA and then resuspend in 50 µl $T_{10}E_1$, pH 8.0. Remove a 5 µl sample, dilute to 0.5 µl in dH_2O, and measure the A_{260} in a quartz cuvette using a spectrophotometer. Calculate the stock DNA concentration as follows: (A_{260} × 50 µg DNA/ml × 100-fold dilution = stock concentration in µg DNA/ml).

Note: Instructors may test this protocol with samples of *Tetrahymena thermophila* DNA obtained from the authors. DNA samples of two strains will be provided (~1 µg each), along with samples of two different RAPD primers sufficient for several amplification reactions. To obtain a set of samples, please e-mail a request to Dr. Lee Kats at lee.kats@pepperdine.edu. Users must provide all other equipment, solutions, and supplies.

3. **Amplification of RAPD Markers.** Amplifications are carried out in 100 μl reaction volumes, with each reaction consisting of the following components. (50 μl reactions may be used.) All solutions and supplies must be sterile and DNase-free.
   10 μl 10× amplification buffer (100 mM Tris-HCl, 500 mM KCl, pH 8.3)
   8 μl 25 mM Mg2Cl stock
   X μl template DNA from *Tetrahymena* = to 100 ng
   1 μl 10 mM dATP
   1 μl 10 mM dGTP (final concentration of dNTP's = 100 μM)
   1 μl 10 mM dCTP
   1 μl 10 mM dTTP
   1 μl 0.1% gelatin
   5 μl RAPD primer (= 25 pmol) (only one primer in each reaction)
   X μl distilled H_2O (to achieve a final volume of 100 μl)
   1 μl Taq polymerase DNA _____ (= 5 units)
   100 μl total volume

   Each strain of *Tetrahymena* DNA should be amplified with at least two or three different RAPD primers to ensure a variety of RAPD marker products for visualization. Amplifications are performed in an automated thermocycler set for 45 cycles each consisting of 1 minute at 94° C (denaturation), 1 minute at 35° C (primer annealing), and 2 minutes at 72° C (primer extension). A final 5 minute extension at 72° C is optional.

   Note: If the thermocycler used for amplification does not possess a heated cover, it is necessary to layer 100 μl of mineral oil on the surface of the reaction mixtures to prevent evaporation.

4. **Horizontal submarine gel electrophoresis of RAPD marker products.** Amplification products may be visualized by electrophoresis in a 1.4% agarose gel and detected by ethidium bromide* staining of the DNA. To prepare the gel, use 1.4 g of electrophoresis-grade agarose for each 100 ml of a 1× Tris/acetate/EDTA buffer (TAE). (50 × stock TAE = 2 M Tris acetate, 100 mM EDTA, ~pH 8.5. To make, add 242 g Tris base, 57.1 ml glacial acetic acid, and 37.2 g $Na_2EDTA \cdot 2H_2O$; bring to 1.0 liter with distilled water.) Bring buffer and agarose to a boil twice to completely dissolve agarose. Cool to approximately 55° C and add ethidium bromide to a final concentration of 0.5 μg/ml. Pour gel into mold with well comb in place and let cool to room temperature. Remove comb from solidified gel and place into a submarine gel electrophoresis apparatus with the wells toward the negative pole. Cover the gel with 1× TAE buffer prior to loading samples.

   Each RAPD sample consists of 10 μl to which 2 μl of a 6× gel loading buffer has been added (6× gel loading buffer = 0.25% xylene cyanol in 30% glycerol). A suitable DNA marker (0.5 to 1.0 μg) should also be available to load in a separate lane. (A 100-base-pair DNA ladder is available from Promega [www.Promega.com], catalog #G2101.) Electrophorese samples toward the positive pole (red) at 80 volts for about 1 hour. (*Note:* voltage and run time vary with gel size and thickness.) Examine gel on UV transilluminator** to visualize the RAPD marker band products.

*Caution: ethidium bromide is a known carcinogen. Wear gloves when loading and analyzing the gel.

**Caution: ultraviolet radiation from the transilluminator may cause eye or skin damage. Wear protective face and eye shields while viewing gel.

Exercise 8: The Sponges

Exercise 8: Class Calcispongiae—*Sycon*

Materials

Preserved or living *Sycon* (= *Scypha, Grantia*)
Examples of asconoid and leuconoid sponges
Prepared slides
Chlorine bleach (sodium hypochlorite)
Microslides
Coverslips
   Sycon, transverse sections
   Leucosolenia, transverse sections
Spicule stew
Single-edged razor blade
Watch glasses
Hand lenses or dissecting microscopes
Compound microscopes

Notes

1. **Preparation of spicule samples.** Calcareous and siliceous sponges are identified principally from their spicules. To prepare a spicule sample, place a small piece (2- to 3-mm-square block) of the sponge in a tube and add about 2 ml of Clorox (sodium hypochlorite solution). Allow an hour or so for the organic matter to dissolve and the spicules to settle. Pipette off the Clorox, add water to wash, allow the spicules to settle, and then remove the water with a Pasteur pipette. Repeat the washing with water, transfer the spicules by pipette to a microscope slide, and add a drop of water and coverslip. Examine at about 100× or more.

   Permanent spicule mounts are easily prepared by following the water washes described with one or two washings with 95% alcohol. After the final wash, transfer the spicules with a pipette to a microscope slide and allow the alcohol to evaporate. Apply a drop of mounting medium, add a coverslip, and label.

2. **Study of gemmules.** Gemmules, the overwintering stage of freshwater sponges of family Spongillidae, are composed of masses of living amebocytes enclosed within a tough, highly resistant shell. The shell is perforated by a pore, through which living cells can exit in spring.

   Freshwater sponges are abundant in ponds and streams; one zoologist commented that, if one cannot find them, one is not looking hard enough! Look for shriveled remains of a sponge colony in early winter. The gemmules, resembling fig seeds clustered among the spicules, may be scraped into a jar of pond water. Place a few clean glass slides in Petri dishes and cover with the pond water. Put the dishes in a safe place out of direct sunlight and drop three or four gemmules on each glass slide. Leave undisturbed for a few days at room temperature, adding water to replace evaporation. The gemmules should stick to the glass within a day or so, indicating that the amebocytes have moved out of the gemmule and are beginning to divide actively. Soon an elevation will appear around the gemmule, becoming chimneylike as the sponge takes form. Spicules can be seen within 5 or 6 days, and, by adding a bit of carmine dye to the water, you can see water moving into the ostia and out through the osculum.

3. **Making permanent mounts of gemmules.** Place gemmules in a test tube containing cold concentrated nitric acid and let stand 1 to 6 hours, or until they turn a translucent orange or yellow. Then wash several times with water, dehydrate with alcohol, and mount as indicated in project 1 for sponge spicules.

4. **Drying whole freshwater sponges.** Remove the sponges from water and place in a warm, shady place to dry out completely. Dried sponges are very fragile and must be handled carefully. Pack in soft, crushed tissue paper or toilet tissue (never in cotton) for shipping or storing. **WARNING:** wash your hands and do not rub your eyes after handling a dry sponge.

5. **Demonstrations.**

   a. Different types of sponge skeletons, including some of the glass sponges

   b. Prepared slides of spicules, spongin, gemmules, and sections of whole mounts of various sponges

   c. Commercial (bath) sponges. A common genus of this leuconoid type is *Spongia.* Commercial sponges have a complicated organization (Figure 8-4). When brought up alive from the sea bottom and sliced open, the inner surface looks raw and slimy, resembling liver. Dry, macerated sponges of commerce (though seldom seen now) are composed only of the spongin skeleton.

Exercise 9: The Radiate Animals

Exercise 9A: Class Hydrozoa—*Hydra, Obelia,* and *Gonionemus*

Materials

Living material
- Hydras
- *Artemia* larvae, *Daphnia,* or enchytreid worms
- Marine hydroids, if available

Preserved material
- *Obelia*
- *Gonionemus*

Prepared slides
- Stained hydras
- Budding hydras
- Male and female hydras
- Cross sections of hydras
- *Obelia* colonies
- *Obelia* medusae

Clean microscope slides
Coarse thread
Watch glasses and depression slides
10^{-4} reduced glutathione
Bouin's fluid
Compound microscopes
Dissecting microscopes or hand lenses
Coverslips

Notes

1. **Geographic distribution of hydra.** Pennak (1989) lists 16 species of hydras in North America, all but 5 of which are of local or scattered distribution (so far as is known). In addition to the 3 species mentioned in the manual, Pennak lists 2 other broadly distributed species: *Hydra oligactis,* widely distributed in the northern states from Montana east to the Atlantic, and *Hydra carnea,* distributed from the East Coast as far west as Nebraska. Both of these species doubtless have similar east-west distributions in southern Canada.

   Hydra littoralis differs ecologically from *H. carnea* in preferring swift waters or wave-swept shores; *H. carnea* is found in standing waters.

2. **Green vs. brown hydras for class use.** The green hydras, *Chlorohydra viridissima,* appear green because their cells are packed with the symbiotic green algae of genus *Chlorella.* A single hydra may harbor 150,000 algal cells contained within vacuoles of the hydra's cells. The algae provide the hydra with photosynthetic products, such as maltose, which the hydra rapidly converts to glucose for its own metabolic needs. The algae also supply oxygen, a by-product of their photosynthesis. In return, the hydra provides the algae with amino acids and nucleotides, which the algae use for protein

and nucleic acid synthesis. Green and brown hydras share the same habitat and normally multiply at similar rates. If food is scarce, however, green hydras have the advantage.

Green hydras make an interesting demonstration for the class. Brown hydras, however, are larger and more suitable for class study of movement and feeding behavior.

3. **Reduced glutathione** is available from suppliers of biochemicals. 10^{-4} glutathione is approximately 0.03 g/liter.
4. **Symbionts of hydras.** *Kerona* and *Trichodina* are two of the more common symbionts of hydras. *Kerona* looks much like paramecia but is much shorter and has frontal cirri curving along the anterior edge of the oral groove. *Trichodina* looks entirely different: a squat ciliate with an aboral holdfast disc bearing hooks.
5. **Living *Obelia* colonies** on seaweed fronds may be purchased from suppliers of marine material (e.g., Marine Biological Laboratory, Woods Hole, Mass.; see Appendix B for address).

   In our experience, the polyps of living *Obelia* are rapidly cropped by several predators on the seaweed, especially small nudibranchs. If not examined within 2 or 3 days after collection, one may find little left but the hydrocaulus. Numerous other marine forms are present on the colonies (polychaetes, microcrustaceans, ectoprocts, etc.) which will be of great interest to students. Often the *Obelia* colonies are covered with diatoms.

   Other colonial hydroids (e.g., *Eudendrium, Pennaria, Tubularia, Bougainvilla)* are available as living material from marine suppliers. Of these we have found *Tubularia* especially suitable for classroom work. *Tubularia crocea,* known as the pink-hearted hydroid, is common on wharves, pilings, and bridges on both the Atlantic and Pacific Coasts. This species is mentioned briefly in the delightful book by Richard Headstrom (1984), along with comments on many other tubularian species. It is described in detail by T. H. Waterman (1950).
6. **Raising brine shrimp.** Brine shrimp *(Artemia)* are excellent for demonstrating feeding behavior in hydras. Brine shrimp eggs can be obtained from any biological supply company and hatched in a day or two. They are widely used as food for aquarium animals. Use a shallow, rectangular pan or tray of glass, plastic, or enamel (not metal). Cut a divider of glass, Plexiglas, or wood that will extend across the width of the pan, fitting snugly against the sides of the pan but lacking ½ to 1 inch of reaching the bottom. Fill the pan three-quarters full of salt water (natural or artificial seawater, or simply a tablespoon of sodium chloride into a quart of tap water). The divider should extend above the surface of the water.

   Add about ¼ to ½ teaspoon of dry eggs to the water in one side of the pan; they will float on top of the water and be confined to that side of the pan. Place an opaque cover over the pan except for 1 inch or so at the end farthest from the eggs. The hatched nauplii will be attracted to the light, swim under the divider, and cluster at the open end, where they are easily siphoned or drawn off into a brine shrimp net or finger bowl. At 23.8° to 26° C (75° to 80° F), they will hatch in 24 to 48 hours. Two or three trays started daily on a rotating basis will keep a constant supply available for daily feeding.

   If larvae are to be fed to freshwater forms, they should be rinsed well first using a small-mesh aquarium net.

   If larvae are to be held alive for a period, they should be removed to another container of salt water where they are not crowded and are fed small amounts of yeast suspension or green one-celled algae daily. The water should be changed often to prevent fouling. Although aeration is not essential, it is helpful in keeping successful cultures.

   Additional information on rearing *Artemia* may be found in the classic compendium edited by J. G. Needham (1937).
7. **Preparing Bouin's fixative.** Combine 75 ml saturated aqueous picric acid (about 1 g will dissolve), 25 ml of formalin (40% formaldehyde), and 5 ml of glacial acetic acid. Add the acetic acid just before use. After fixation of tissues, wash in ethyl alcohol, 45% or stronger, until the yellow color is removed.
8. **Demonstrating freshwater jellies.** Freshwater jellies *(Craspedacusta sowerbyi)* are found occasionally in various parts of the United States, especially in the East. They are small but make an interesting demonstration. The life cycle includes polyps and medusae; the polyp is a small (2 mm), nontentacled simple tube present in colonies. The medusa (15 to 20 mm in diameter) is provided with more than 200 marginal tentacles lacking adhesive discs. Life cycle and stages of this species are described by Pennak (1989), who includes a drawing of the medusa.
9. **Experiments with hydras.** How to raise hydras and carry out simple experiments with them is explained in Section 4 of the book by Goldstein and Metzner (1971).

Exercise 9B: Class Scyphozoa—*Aurelia,* a "True" Jelly

Materials

Preserved *Aurelia*
Living jellies if available
Hand lenses or dissecting microscopes
Finger bowls
Ladle

Notes

1. **Obtaining living scyphozoans.** These are most readily obtained from suppliers of marine materials during spring and summer months; few are available during fall and winter, when most zoology courses are offered.

   The **scyphistoma** of *Aurelia,* however, can be obtained from the Marine Biological Laboratory (see Appendix B for address) October through April. These are relatively hardy and can be maintained in cooled seawater aquaria and even in aerated culture dishes. They are fed ground shrimp or particles of meat. The Gulf Specimen Company offers *Chrysaora* (stinging nettle jelly), and other species may be available upon inquiry. Additional references are provided in Brown, J. G. (ed.), 1937, *Culture Methods for Invertebrate Animals,* Comstock Publishing Company.
2. **The "upside-down jelly."** *Cassiopeia* (order Rhizostomeae) is common in the shallow waters around Florida. They usually can be obtained from Florida marine supply houses and can be kept for several days or weeks in a marine aquarium. This interesting jelly has eight thick, gelatinous oral lobes that are fused in such a manner as to obliterate the central mouth and form numerous canals with small openings in the oral lobes. The oral lobes also bear many small tentacles with nematocysts. There are 16 rhopalia, but no tentacles around the scalloped margin.

   Although *Cassiopeia* can swim by rhythmic pulsations of the bell, it spends much of its time lying oral side up on the bottom of lagoons and tidal pools, anchored by a suckerlike action of its aboral surface (hence the name "upside-down jelly"). Here, as it pulsates, it draws water over its oral lobes, bringing in food and oxygen. Small organisms are paralyzed by nematocysts, entangled by mucus, and swept into the canals by flagellar action.

   Cassiopeia can be cultured in marine tanks of artificial seawater. The scyphistomae reproduce readily by two asexual methods—one by budding off young medusae, the other by budding off small planuloid larvae that detach and swim about and finally settle down and develop into scyphistomae.

Exercise 9C: Class Anthozoa—*Metridium* and *Astrangia*

Materials

Metridium, preserved
Living sea anemones and corals, if available
Finger bowls and/or dissecting pans
Dried corals
Astrangia, preserved

All references to papers and publications mentioned in Appendix A are available as part of the Instructor Resources section in Connect for the main text (available only to instructors who are logged into Connect). Visit www.mheducation.com/highered/connect for more information.

Note

1. **Sea anemone behavior.** We find that 4 or 5 sea anemones will serve a class of 20 students. Allow 45 minutes for the behavioral study. We place the sea anemones at least a day before the laboratory in clear refrigerator boxes, if the anemones are large, or in finger bowls if they are small. These are kept submerged in the marine aquarium until the laboratory period, when they are lifted out and gently placed on the lab benches at the beginning of the period. Students are warned not to touch the anemones until they begin the behavioral study, since once an anemone strongly contracts from clumsy handling it may not relax for a long time.

   Touching an anemone with saliva on a coverslip may cause it to contract, so that should be done only near the end of the exercise.

   Nematocysts discharged on a coverslip can be observed more easily if a drop of methylene blue or acid fuchsin is added.

   The demonstration with acontia and nematocyst discharge is dramatic and is worth the exercise of itself. It may be done as a demonstration with a single anemone if living material is in short supply. Prodding the anemone strongly to stimulate acontia discharge will not injure the animal, but it will remain strongly contracted for some time thereafter. The acontia are equipped with cilia, which cause the acontia to move slowly in snakelike fashion on the slide.

   Nematocysts of *Metridium senile* acontia are grainlike capsules, measuring 6×10^{-2} mm in length, and about 5×10^{-3} mm in width. They are visible at low power but must be viewed at high power to see the discharged threads.

Experimenting in Zoology: Predator Functional Response: Feeding Rate in *Hydra*

Materials

Brine shrimp eggs and shrimp hatchery
Deep depression slides
Pond water
Brown hydra (enough for about 10 per student) that have not been fed for at least 24 hours
55 mm disposable Petri dishes
Disposable pipettes
Dissecting microscope (low power)
Clock or stopwatch

Note

1. The most convenient prey to use for this experiment is *Artemia,* the brine shrimp. Brown hydra readily feed on the first stage nauplii of this arthropod but may have trouble handling them as they increase in size during

later development. Therefore, brine shrimp should be hatched no earlier than 1–2 days before they are needed. Functional response rate will be recorded as the number of prey eaten during a 10-minute period as prey density increases.

Exercise 10: The Flatworms

Exercise 10A: Class Turbellaria—Planarians

Materials

Live planarians
Prepared slides
  Planaria, stained whole mounts
  Bdelloura, stained whole mounts
  Planaria, cross sections
Powdered carmine or talc
Black paper
  Snap-cap vials
  Camel hair brushes
  Small mirrors
Small flashlights (pocket-size) or narrow-beam electric lamp
Modeling clay
Petri dishes
Spring, pond, well water, or dechlorinated tap water (not distilled water or demineralized water)
Raw liver (or cut-up mealworms)
Ice cubes
Watch glasses or depression slides
Dissecting microscopes or hand lenses

Notes

1. **Sources of planaria for class study.** Biological supply houses such as Carolina, Ward's, and Connecticut Valley offer "brown planaria" *(Dugesia trigrina),* probably the best species for routine class study as well as the regeneration experiment. Most supply houses additionally offer "black planaria" (usually *Phagocata gracilis*) and semitransparent "white planaria" (usually *Procotyla fluviatilis*), which have highly visible digestive tracts.

   Planarians may be kept for some days in the plastic shipping containers *if* the water is changed daily and the inside surfaces of the containers are wiped clean of accumulated slime.
2. **Collecting and keeping planarians.** Planarians also may be collected on pieces of raw beef tied to stones or plants near the water's edge and can be kept in the laboratory if fed once a week on small amounts of fresh beef liver or hard-boiled egg yolk, pieces of earthworm, crushed snails, etc. Remove any uneaten food after an hour or so. The water should be changed frequently to prevent pollution.
3. **Reactions to stimuli.** When observing response to food by smearing a bit of liver on a coverslip inverted over a depression slide, caution students that the amount of liver must be *very* small. If more than just a spot of meat is placed on the coverslip, the water will quickly become clouded and permeated with liver particles; with the odor of meat everywhere, the planarians then seem unable to locate the meat.
4. **Suggestions for the planaria regeneration experiment.** This may be done as an extra-credit exercise. We recommend using a tissue-covered ice cube over other methods for quieting and extending planarians; cuts can be made with precision under the dissecting microscope and the animals are not harmed by chilling; 30-ml snap-cap plastic vials make excellent containers and are reusable. Punch a single small hole in the lid to allow air exchange. Alternatively use the screw-cap specimen jars in which unicellular eukaryotes and other invertebrates are shipped from biological supply houses. Partial longitudinal cuts will almost always tend to grow back together and must be recut two or even three times during the first 24 hours for success (using a tissue-covered ice cube). Even so, some will rejoin even after repeated resectioning.

   Keep planarians in separate containers after cutting to prevent whole specimens from cannibalizing cut specimens.

   Brown planarians *(Dugesia tigrina)* are best for the regeneration exercise. Fortunately, since students are often forgetful, planarians will live for weeks without much attention.

   There are excellent discussions of planaria regeneration in Pearse et al. (1987, pp. 214–221) and in Buchsbaum et al. (1987, pp. 171–179). This material should be made available to students to assist them in their written reports.

   Ward's offers a "planaria regeneration study kit" with all materials required for the experiment. We have not tried it.
5. **Response to light and dark backgrounds in nondirectional illumination.** This exercise may be used to supplement the three experiments on reactions of planarians to stimuli (p. 150).

   Prepare the lower half of a Petri dish by painting its sides and one-half of its bottom on the outside with black paint (or cover with black tape or black paper). Set the dish on a white surface so that half of the bottom is black and the other half is white. Cut a circle of black paper a little larger in diameter than the top of the Petri dish. Cut out the center of the circle, leaving a ring of paper about 3 cm wide, or wide enough to extend inward from the edge of the dish about 1.5 cm, to shade the sides of the dish from overhead illumination. Clean and rinse the dish thoroughly; then place it

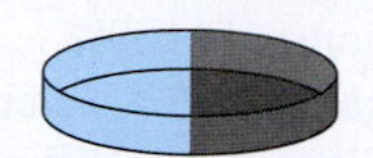

Paint one-half of Petri dish on outside with flat black paint.

After introducing planaria, cover with black paper with a hole cut in the center.

Preparing painted dishes for the nondirectional illumination study.

in a dark room or box, with a light source several feet above the dish. Place a few planarians in the center of the dish and leave for a while. When they have ceased moving about, count the animals on the dark surface and those on the white surface and compare the numbers. Now remove the animals from the dish, add a suspension of carmine to the water, and rotate the dish gently. Drain off the carmine suspension and rinse the dish *gently* with water. The movements of the planarians can be seen where the carmine particles adhere to the mucous trails left on the bottom of the dish. Do you note any difference in the length of the trails on the black and white surfaces? ____________ Is there a directional response? ____________

6. **Anterior-posterior activity gradients in planaria.** A variation on the planarian regeneration theme (Experimenting in Zoology, p. 167) is a project study of gradients in planarians. Section each of several live planarians into three parts—cephalic, body, and caudal pieces—and place the parts in separate dishes or vials containing pond water and appropriately labeled. In 3 to 5 days, colorless blastemas will protrude from the cut edges. Submerge the pieces in a 1% methylene blue solution for 5 to 8 hours or until the specimens are stained. Place stained pieces on a depression slide under anaerobic conditions by sealing the cover glass with petroleum jelly (Vaseline). When the oxygen is used up, the color disappears. Note which end of the pieces shows the color response first. Upon reexposure to air, blastemas at the more active (anterior) ends should regain color more rapidly. This experiment demonstrates polarity, the gradients of activity along the anterior-posterior axis.

Exercise 10B: Class Trematoda—Digenetic Flukes

Materials

For study of *Clonorchis*:
- Stained slides
- *Clonorchis*
- Miracidia
- Cercariae

All references to papers and publications mentioned in Appendix A are available as part of the Instructor Resources section in Connect for the main text (available only to instructors who are logged into Connect). Visit www.mheducation.com/highered/connect for more information.

Compound microscopes (dissecting microscopes useful for viewing entire whole mount)

For study of *Schistosoma*:
- Prepared slides
- *Schistosoma mansoni* in copula
- *Schistosoma* eggs
- Compound microscopes

For observations of living flukes:
- Pithed leopard frogs *(Rana pipiens)*
- Live snails (*Physa, Certhidea,* others)
- Dissecting tools
- Normal saline solution (0.65%)
- Syracuse watch glasses
- Microscope slides
- Compound and dissecting microscopes

Note

1. **Prepared microslides.** See Appendix B for a listing of suppliers. Most suppliers offer whole mounts of *Clonorchis* and *Schistosoma mansoni.* Miracidiae, sporocysts (in section of snail tissue), and cercariae are also available and make useful demonstrations to supplement the schistosome exercise.

Exercise 10C: Class Cestoda—Tapeworms

Materials

Preserved tapeworms

Prepared slides of tapeworms showing scolex and immature, mature, and gravid proglottids

Microscope

Notes

1. **Choice of tapeworm.** This exercise has been written to serve as a guide for either *Taenia pisiformis* or *Dipylidium caninum;* (instructor's preference). If new slides are being purchased (see Appendix B for listing of microslide suppliers), it is prudent to check slide quality carefully before accepting an order; many commercially available tapeworm slides are of poor quality, suffering from a variety of problems: overstaining, understaining, or selection of mature segments that do not clearly reveal reproductive structures.
2. **Measly meat demonstration.** In earlier editions of this manual, we suggested studying a piece of "measly" meat infected with tapeworm cysts or bladder worms. However, we have been unable to locate commercial suppliers of measly meat. Nevertheless, measly meat makes a dramatic demonstration, and it may be worth the effort to try to obtain a sample of formalin-preserved infected meat through a veterinarian.

Exercise 11: Nematodes and Four Small Protostome Phyla

Exercise 11A: Phylum Nematoda—*Ascaris* and Others

Materials

Preserved *Ascaris* specimens
Prepared slides
- Stained *Ascaris* cross sections
- *Trichinella* cysts in pork muscle
- *Trichinella,* male and female
- *Necator americanus*
- *Enterobius vermicularis*
- *Turbatrix*

Living *Turbatrix*
Soil samples, if desired
Slides and coverslips
Razor blades
Dissecting pans and pins
Microscopes

Notes

1. **A caution.** The eggs of female *Ascaris* may remain viable for many weeks in formalin-preserved females. While chances of infection are remote, students should be cautioned to wash their hands after dissecting worms.
2. **How to quiet *Turbatrix*.** We find that *Turbatrix* can be quieted for study most effectively by *gently* warming the slide over a lamp. Others prefer to add a drop of 1 N HCl to the culture. With this alternative, the worms will remain active for several minutes before becoming quiescent. An advantage to this alternative is that students sometimes see the birth of living juvenile worms as the mother slowly succumbs to the acid.

   Turbatrix may be weakly stained for study with Nile blue sulfate in 70% alcohol.

   Some simple experiments with *Turbatrix* are suggested by Goldstein and Metzner (1971, pp. 161–165).
3. **Population growth in vinegar eels.** Material needed for this exercise: Living *Turbatrix aceti* stock culture; apple cider vinegar; dechlorinated tap water; petri dishes; disposable bulb pipets; gridded 1-ml Sedgewick-Rafter counting chamber; dissecting microscope (or compound microscope with scanning power). Gridded Sedgewick-Rafter counting cells are available from several suppliers, including Cole-Palmer, Forestry Supplies, and eBay. One or two chambers is usually sufficient for the class.

   Mix three vinegar/water dilutions: (1) 100% apple cider vinegar; (2) 75% vinegar and 25% water; (3) 50% vinegar and 50% water. Fill one petri dish with 100% vinegar to 3/4 full. Fill another petri dish with 75% vinegar, and third with 50% vinegar. Inoculate each with 5 drops of the stock culture and mix well.

   To count the worms per ml, stir the dish and then immediately put 1 ml of sample into the counting chamber and cover with coverslip. Count all worms in the 1-ml sample. If the numbers are too high to get an accurate count, you can count a subsample of grids and multiply to get an estimate of total worms in the 1-ml sample.

   Students should count samples immediately after preparing the dilutions, and then twice each week. After one month students turn in their counts, and the data for entire class can be distributed to each student. Students then count the average number of worms per ml for 100%, 75%, and 50% vinegar concentrations. Population growth in each concentration can be plotted on a graph with error bars.

   Students may write a lab report, or the instructor may provide students with questions to answer.
4. **Methods for collecting and demonstrating soil nematodes.**
   a. *Boiled potato.* Leave some pieces of boiled potato for several days in various places—under a rock, plank, or bit of soil or in a garden, a meadow, a marsh or an empty lot—and then place them in sterile test tubes, plug with cotton, and set in a warm room for several days before examining.

   b. *Baermann apparatus.* Wrap a sample of soil in several layers of cheesecloth and suspend from the arm of a ring stand. Fit a short rubber tube with petcock clamp to the spout of a funnel, and attach the funnel to an arm of the ring stand below the bag of soil. Pour warm water (20° to 26°C) into the funnel, so that 4 to 6 cm of the soil in the bag is immersed. After 1 to several hours, drain off a little of the water from the tube into a small test tube. Let the nematodes settle, and then pipette off the supernatant fluid. Examine the nematodes under the microscope. More information on the use of the Baermann apparatus for collecting soil nematodes is provided by Goldstein, P., and J. Metzner. 1971. *Experiments with microscopic animals.* Garden City, NY, Doubleday.
5. **Methods for demonstrating *Rhabditis maupasi.*** The larval stages of *Rhabditis maupasi* are a common nematode parasite living in the nephridia of earthworms (other congenerics occur in freshwater and in soil). The larvae never develop to adults in living earthworms, but after the earthworm's death the nematodes feed on bacteria in the decaying tissue and rapidly mature to adults. Fill some Petri dishes half-full of agar jelly (made by boiling 2 g of agar per 100 ml of water) and allow to cool. Into each dish, cut 6-mm lengths from the posterior third of several fresh earthworms. Cover and leave for 2 to 5 days or until nematodes are seen on the putrefying pieces of

All references to papers and publications mentioned in Appendix A are available as part of the Instructor Resources section in Connect for the main text (available only to instructors who are logged into Connect). Visit www.mheducation.com/highered/connect for more information.

earthworm. Adding small bits of raw meat from time to time and subculturing will maintain cultures indefinitely.

Another method is to slit freshly killed *Lumbricus* down the middorsal line posterior to the fifteenth segment. Pin back the walls, cover with water, and examine the nephridia. Larval nematodes may be found in or on some nephridia. Encysted stages also occur in coelomic "brown bodies." If parasites are found, place the earthworm in a jar on wet paper towels and cover. Keep the preparation wet for 2 to 4 days, depending on the temperature, while the earthworm decomposes and the nematodes reproduce. Examine some of the culture in a few drops of water. There should be much of the life cycle present in the culture: fresh eggs, eggs containing larvae, immature worms, and mature worms. To preserve the worms for storage, put the culture into a test tube and let settle. Pour off most of the water and add an equal amount of boiling 10% formalin to kill and straighten the worms.

Exercise 11B: A Brief Look at Some Other Protostomes

Materials

Living materials
- *Philodina,* or mixed rotifers
- *Chaetonotus,* or mixed gastrotrichs
- *Gordius,* or other nematomorphs, if available

Preserved or plastic-mounted material
- "Horsehair worms" (such as *Gordius*)
- Spiny-headed worms (such as *Macracanthorhynchus*)

Prepared slides of any of the previously listed types
Slides, depression slides, and coverslips
Microscopes

Exercise 12: The Molluscs

Exercise 12A: Class Bivalvia (= Pelecypoda)—Freshwater Clam

Materials

Living bivalves in an aquarium
Living or preserved clams for dissection
Clean, empty bivalve shells
Dissecting pans
Pasteur pipettes
Glass rod
Carmine suspension

Notes

1. **Living clams are recommended for this exercise.** They are odorless, the tissues retain their natural color, and students can make observations on ciliary action and the beating heart. It is best for the instructor to open the clams, which, with a bit of practice, can be done rapidly and with minimal damage to internal structures. Heating the clams in warm (not hot!) water causes them to relax and facilitates insertion of the knife between the shells. Use a stout, strong-bladed knife that can be forced with safety between the valves, while holding the clam against a firm surface. Scalpels should never be used to open clams.

   After opening a living clam, it should be covered with pond water or dechlorinated tap water.
2. **Fine-tipped dissection scissors** should be used to open the pericardium and visceral mass. The student-quality scissors commonly provided in dissection kits are not satisfactory.

Exercise 12B: Class Gastropoda—Pulmonate Land Snail

Materials

Living snails (*Helix,* others)
Preserved or freshly killed snails
Assortment of snail shells
Squares of glass plate
Finger bowls
Dissecting microscopes

Note

1. **Narcotizing snails.** To narcotize or kill pulmonate snails or slugs fully relaxed for study or preservation, seal the specimens for 24 hours in a jar of water, capped so as to exclude all air. Boiling the water beforehand to drive out air will shorten the asphyxiation time. Animals thus treated will be fully relaxed with antennae and foot extended. This procedure, recommended by Knudsen (1966), works much better than others we have tried.

Exercise 12D: Class Cephalopoda—*Loligo,* the Squid

Materials

Preserved or freshly killed squids
Dissecting instruments
Dissecting pans

Notes

1. Many instructors will restrict the study of the squid to external structure (pp. 195–198).
2. **Demonstrations** appropriate for this exercise include the following:
   a. Microslides showing spermatophores of *Loligo*
   b. Preserved octopus and cuttlefish *(Sepia)*
   c. Shells of Nautilus
   d. Dried cuttlebone of *Sepia*

e. Dissection of an injected cephalopod to show the circulatory system
f. Dissection of a cephalopod brain
g. Living cephalopod, if available

Exercise 13: The Annelids

Exercise 13A: Errantia—Clamworm

Materials

Preserved clamworms *(Nereis)*
Living nereids and/or other polychaetes as available
Dissecting tools
Dissecting pans
Clean slides
Hand lenses or dissecting microscopes
Compound microscopes

Notes

1. **Preparing specimens of *Nereis* with the proboscis extended.** *Nereis* will die with the proboscis fully extended if allowed to asphyxiate in a bottle of seawater, capped to exclude all air.
2. **Cross sections of *Nereis*.** Some instructors include in this exercise the study of the cross section of *Nereis,* using prepared microslides available from biological supply houses. However, with the exception of the presence of parapodia in *Nereis,* the earthworm cross section (Exercise 13B) reveals the same principal annelid features. The cross section of *Nereis* is pictured and described on p. 274 in Brown (1950). Pierce and Maugel (1987) provide a photograph and accompanying interpretive drawing on p. 148.

Exercise 13B: Sedentaria—Earthworm

Materials

Living earthworms
7% ethanol for anesthetizing living worms for dissection (immerse worms in anesthetic 30 minutes before class use)
Stained cross section of earthworms
Paper towel, or large-sheet kimwipes
Glass plates
Dissecting microscopes (or hand lenses)

Notes

1. **Earthworm behavior.** Worms respond better to a strong light (negative phototaxis) if they are kept in subdued light first.

   If a worm is not responding as expected to moisture and light, try covering it with a second paper towel (or kimwipes) also moistened in the center, so that the worm is gently "sandwiched" between the two towels. It will better sense the difference between moist and dry areas of the towels.
2. **Earthworm dissection.** This exercise applies to living earthworms, anesthetized or freshly killed. It is easily adapted to preserved worms, but because living worms are actually cheaper than preserved worms, it makes little sense to use the preserved. Living worms offer several advantages: organs have their natural color and are more easily identified; the circulation is prominent and easily traced; and the gut, especially the gizzard, reveals its natural peristaltic movements.

   Take care that worms are fully anesthetized in 7% ethanol before students begin dissections. Ethanol *cannot* be added to the saline once the worms are opened because it kills them.

   For those who may object to using living animals for dissection, but who want students to see the natural internal pigments of the living animal, living earthworms may be killed by immersion for 30 minutes in 10% to 15% alcohol just before the laboratory.
3. **Saline for dissection.** Living earthworms should be covered with near-isotonic saline after they are opened. Frog saline (0.6% NaCl, or 6 g NaCl per liter of solution) almost perfectly matches the osmolarity of earthworm body fluid. Several liters will be needed for a laboratory. We find it convenient to place the saline in a 5-gallon jug provided with a spigot at its base. Elevate the jug and provide a pan beneath the spigot to catch spilled saline.

   We use dissecting microscopes for this laboratory. The dissecting pan with the pinned-out and saline-covered earthworm is supported on plywood "bridges" that span the microscope stage.
4. **Freshwater oligochaetes.** Several species are available from biological supply houses, or they may be collected from the mud and debris of streams, lakes, and ponds. *Chaetogaster* lives on small crustaceans, insect larvae, and the like. *Aeolosoma* uses cilia around the mouth to sweep in food particles. In *Stylaria* the prostomium is drawn out into a long proboscis. *Tubifex* is a reddish oligochaete 2 to 3 cm long that builds burrows in the mud, where it lives head down and waves its extended tail back and forth to stir up water currents. It is especially common in sluggish and polluted streams and lakes.

   Gaseous exchange in most oligochaetes occurs through the thin body wall, but *Dero* and *Aulophorus* have ciliated anal gills. Gilled species lie quietly hidden in the substratum or in a tube with the posterior end projecting into the water.
5. **Slides of earthworm nephridia.** Prepared slides of nephridia are available from biological supply houses, but many are poorly prepared and so badly twisted that they are virtually useless for study. Have students examine a stained nephridium preparation, noting the narrow tubules, peritubular circulation (usually more deeply stained than the tubule and often fragmented during preparation), bladder, and nephrostome.

6. **How to isolate annelid setae.** Boil pieces of the worm in 5% potassium hydroxide solution until the tissue has dissolved. Allow the setae to settle and decant the fluid. Wash by adding water, allow to settle, and then decant; repeat several times. There are several types of setae—capilliform, straight, and curved and those having the ends bifurcated, hooked, pectinate, and the like. The type, number, and location of the setae are frequently of taxonomic significance in the classification of annelids.

Exercise 13C: Family Hirudinida—Leech

Materials

Living leeches
Preserved *Hirudo*

Notes

1. **How to see the nephridiopores.** These can be demonstrated by gently squeezing an alcohol-narcotized (7% ethanol) specimen, causing a little fluid to be exuded from the nephridial bladders.
2. **Medicinal leeches.** Medicinal leeches are available from Carolina Biological, from Ward's, and from Leeches U.S.A. (see Appendix B). These are large specimens, easily maintained in pond or dechlorinated water in large stacking dishes; the dishes must be kept covered, since these leeches show a strong propensity to wander. Students find them fascinating. However, one must use caution in allowing these leeches to fasten themselves for any length of time to students' hands or arms. If hungry (and all that we have received are), they will waste little time in slicing into the skin with their triradiate, sawlike chitinous jaws to initiate blood flow. The salivary glands of medicinal leeches contain a complex mixture of anticoagulant (hirudin, an antithrombokinase), anesthetic (leeching is nearly painless), vasodilators that cause tissues surrounding the wound to swell, and other compounds that liquefy coagulated blood (hence the effectiveness of medicinal leeches in plastic surgery for reestablishing circulation in replanted appendages or skin-flaps that are failing due to blood engorgement).

   We have had students willingly "leech" themselves in our laboratory to the utter captivation of all. But this should be approached with caution for two reasons: (1) the Y-shaped wound will continue to bleed for several hours, and (2) there is a marginal danger of infection from the bacterium *Aeromonas hydrophilia,* which is present in the saliva of leeches (Whitlock et al., 1983). Wound seepage can be stopped easily with a compression bandage, which must be kept in position for at least 24 hours. The possibility of wound infection, however, will doubtless dissuade most instructors in our increasingly litigious American society from subjecting students to "leeching." In fact, the chances of infection appear to be remote (see, for example, Rao et al., 1985).
3. **How to kill relaxed leeches.** Leeches may be killed in a relaxed condition for study by leaving them in a 7% ethanol solution.

All references to papers and publications mentioned in Appendix A are available as part of the Instructor Resources section in Connect for the main text (available only to instructors who are logged into Connect). Visit www.mheducation.com/highered/connect for more information.

Experimenting in Zoology: Behavior of Medicinal Leeches, *Hirudo medicinalis*

Materials

Medicinal leeches
Stopwatch
Tongue depressor or short wooden dowel
500 ml glass beaker
500 ml plastic beaker or container
Ring stand with clamp
Thermometer
Plastic tub
Pond water, bottled water, or dechlorinated water

Notes

1. **Sources of medicinal leeches.** Carolina Biological offers these animals. They also can be obtained from Biopharm Leeches (www.biopharm-leeches.com).
2. **Drawing conclusions from leech behavior.** Students should be encouraged to run as many behavioral trials with their leech as time allows. Occasionally some leeches will not swim the length of the tub and will attach to the side of the tub in the area of their release. Students should not draw conclusions about leech behavior until they have examined data from the entire class. After compiling the data, students could use a chi-square test to see if leeches chose one beaker significantly more than the other.
3. **Keeping medicinal leeches.** Medicinal leeches do well in captivity and can live for a long time if they are fed fresh beef liver and supplied with clean water.

Exercise 14: The Chelicerate Arthropods

Exercise 14: Chelicerate Arthropods—Horseshoe Crab and Garden Spider

Materials

Living materials, if available
  Limulus, in marine aquarium
  Spiders, in terrarium
Preserved materials
  Limulus
  Golden garden spiders
Dissecting microscopes or hand lenses

Notes

1. **Time required for exercise.** Study of external anatomy of both horseshoe crab and garden spider requires little more than 1 hour. We combine this exercise with Exercise 15 for a 4-hour laboratory period.
2. **Large golden garden spiders,** *Argiope aurantia,* are available from Connecticut Valley Biological. This is the best source we have found. Those offered as *Argiope* by Carolina are a different species, much less suitable for this exercise.
3. **Suggestions for spider dissection.** Some instructors prefer to have students keep the spider covered with water during dissection, but they may be examined in air without damage if glycerol is added to the alcohol preservative (about 5 cc per 100 ml of 70% ethanol) to reduce drying. Deep Petri dishes (e.g., 100 × 20 mm) containing a thin layer (about 0.5 cm thick) of dissecting pan wax make excellent containers. The spider can be secured with insect micropins. This exercise is much more successful if the students have good dissecting microscopes (rather than hand lenses) and good illumination from a bright focusing spotlight.

   Another excellent way to hold spiders in any position for study is to place them in a Syracuse dish containing some washed sand. Cover with alcohol. The spider is held in the desired position by pushing it gently into the sand.

Demonstrations

1. **Living or preserved specimens of tarantulas, trapdoor spiders, black widows, or other spiders as available.** (Living tarantulas are easily maintained in the laboratory and a source of fascination to all.) Captive garden spiders will spin webs in almost any container, which need be no more sophisticated than a large glass jar covered with cheesecloth or other netting.
2. **Various mites and ticks**
3. **Scorpions, alive or preserved**
4. **Fossil trilobites and/or eurypterids**
5. **Mounted, stained preparations of "trilobite larva"**

Exercise 15: The Crustacean Arthropods

Exercise 15: Subphylum Crustacea—Crayfish, Lobsters, and Other Crustaceans

Materials

Living materials
- Crayfish or lobsters
- Cultures of developing *Artemia* in different developmental stages
- Other crustaceans, such as *Eubranchipus, Daphnia, Cyclops,* ostracods, barnacles, and crabs, as available

Preserved materials
- Crayfish or lobsters (may be injected)
- Barnacles, crabs, and the like, as available

Bowls or small aquaria
Dissecting pans
Slides and coverslips
Compound microscopes

Notes

1. **Time required for exercise.** Allow 2 hours to complete the core study with additional time as needed for individual oral demonstrations. In 3- and 4-hour laboratory periods, students will be able to further study internal structures.
2. **Mounting appendages.** We use 4 × 10 inch pieces cut from white Bristol board. Students glue the appendages to the board in the same order they appear in Figure 15-3, using Elmer's glue.
3. **Latex-injected crayfish.** Usually the heart is destroyed in such preparations and perfusion of the arteries with latex is frequently poor. Uninjected specimens should be used to demonstrate the anatomy of the undamaged crayfish heart.
4. **Raising brine shrimp:** see notes for Exercise 8A.

Experimenting in Zoology: The Phototactic Behavior of *Daphnia*

Materials

Daphnia
Stopwatch
Open-ended glass column (2 to 3 cm diameter, 25 to 30 cm long)
Ring stand and two clamps
Dissecting scope illuminator or other small light
Pipette
Wax pencil

Notes

1. **Sources of *Daphnia.*** Living *Daphnia* may be obtained from any biological supply house (see Appendix B for listings). They also can be collected by using a plankton net in shallow ephemeral pools that do not contain fish.
2. **Light source.** Dissecting scope illuminators work well because they can be mounted on the ring stand and have a narrow beam. If the illuminator has a switch for variable beam strength, make sure that all students are using the same intensity and that they do not vary it from trial to trial. Other lamps can be used as long as they have a narrow beam that can be directed through the glass toward the rubber stopper.
3. **Drawing conclusions about the data.** After the class data have been compiled, a chi-square test can be used on the control and lighted experiments to see if *Daphnia* are demonstrating phototaxis.

Exercise 16: The Arthropods

Myriapods and Hexapods

Exercise 16A: Myriapods—Centipedes and Millipedes

Materials

Living materials, if available
  Centipedes
  Millipedes
Preserved materials
  Centipedes, such as *Scolopendra* or *Lithobius*
  Millipedes, such as *Spirobolus* or *Julus*
Small terrariums or jars for living materials
Dissecting pans and instruments
Dissecting microscopes

Note

1. **Time required for this exercise.** An examination of centipede external anatomy requires 25 to 30 minutes. Allow another 15 minutes for both centipede and millipede. This exercise can be combined with either Exercise 16B or 16C for a 2-hour laboratory period. For a 3- or 4-hour period, combine this exercise with both Exercises 16B and 16C.

Exercise 16B: Insects—Grasshopper and Honeybee

Materials

Living grasshoppers, if available
Preserved materials
  Grasshoppers, such as *Romalea*
  Apis, workers
  Apis, queen, drones, larvae, pupae (for demonstration only)
Dissecting pans or dishes (see note 4 for Exercise 16C)
Dissecting microscopes

Notes

1. **Demonstration mounts.** Demonstration mounts of insect parts, such as legs, wings, antennae, and mouthparts, can be prepared very simply. Clean a microslide with acetone, spray on a very thin coat of clear lacquer or varnish from a pressurized can from enough distance to prevent formation of bubbles, orient the parts on the slide while the lacquer is still wet, and label. A coverslip is not necessary. The mount is dry and usable within 30 minutes and can be stored flat in a dust-protected tray for future use.

   Slides that show the various types of insect antennae can be purchased or may be prepared by this method.
2. **Demonstrations of adult and immature insects.** These may be selected to illustrate the types and growth stages—direct development and gradual and complete metamorphosis.

Exercise 16C: Insects—House Cricket

Materials

Crickets, alcohol-preserved
Crickets, living
Dissecting dish with wax base
Insect pins
Fine-tipped scissors
Fine-tipped forceps
Carbon dioxide cylinder and small jars for anesthetizing crickets
Ground dog food mixed with carmine and moistened (potato slices sprinkled with carmine make an acceptable substitute)
Plastic squeeze bottles
Methylene blue stain, 0.5% in insect or amphibian saline
Insect saline (7.5 g NaCl/liter) or amphibian saline (6.0 g NaCl/liter)
Dissecting microscopes and focused lighting

Notes

1. **Time required for this exercise.** About 2 hours is needed, with additional time for oral demonstrations. We combine this laboratory with Exercise 16A and most of 16B in one 4-hour laboratory period.
2. **The cricket is an ideal insect for this exercise** because of the ease with which it can be cultured and contained, its lack of distastefulness (compared with cockroaches), its relatively large size, and the ease of determining age and sex.
3. **Rearing crickets.** House crickets are omnivorous and are easily reared on chicken starter mash. Crickets can be kept in any container, such as a steel or fiberglass rat cage or an aquarium. No top is necessary if stainless steel or glass containers are used because the crickets cannot climb the walls, or unless the container is less than 20 cm high. If plywood is used, glue a 2- to 3-cm-wide strip of heavy duty aluminum foil around the inner edge to form an escape-proof barrier. Do not cover the bottom of the container with sand, which tends to get wet and moldy. Egg cartons make good hiding places. Water is provided using plastic vials (55 to 75 ml capacity) with shallow slits cut in the top edge. These are filled with water and inverted over plastic Petri dish bottoms, which are then filled with small gravel to keep the crickets from drowning. Temperature should be 27°

All references to papers and publications mentioned in Appendix A are available as part of the Instructor Resources section in Connect for the main text (available only to instructors who are logged into Connect). Visit www.mheducation.com/highered/connect for more information.

to 30° C for best growth. More information on rearing methods is found in the paper by Clifford et al. (1977).

4. **Dissecting dishes for this exercise** are 20 × 100 mm glass Petri dishes containing about 5 mm of dissecting pan wax.
5. **Reproductive behavior.** Two male crickets placed together in a jar usually display and sing competitively. Squeezing a male usually causes extrusion of a spermatophore, which, when placed in water, extrudes a stream of sperm that is visible with a microscope. If several pairs are separated in glass jars, a little patience will be rewarded with the female mounting the male and with the transfer of a spermatophore.

Exercise 16D: Metamorphosis of *Drosophila*

Materials

Culture of *Drosophila*
Vials, microscope slides, culture medium, yeast solution, cotton stoppers
Ether for anesthetizing fruit flies
Microscopes

Notes

1. **Vial for culture of fruit fly metamorphosis.** Place warm culture medium on one side of the microscope slide. A pair of fruit flies is then added to each vial.
2. **Metamorphosis of the blow fly may be substituted here if desired.** Blow flies may be purchased from biological supply companies, or you can collect eggs by exposing in a dish in a suitable location two pieces of fresh beef, so that one strip of meat slightly overlaps the other. The meat should be kept moist, and some sugar should be added. Flies will usually lay their eggs in the crevices between the two pieces of meat within 3 to 6 days. Place the meat with the eggs in culture test tubes that contain a small amount of raw lean meat and stopper the tubes with cotton. The eggs will hatch within 24 hours. When the maggots are fully grown, remove the cotton stoppers and place the test tubes in a quart glass jar that contains sand or paper to a depth of about 2 cm. Cover the jars with muslin. The maggots will leave the test tube and pupate in the sand or paper (4 or 5 days). Then place the pupae in small screen cages, in which they will emerge as adults.

Exercise 16E: Collection and Classification of Insects

Materials

For collecting:
- Insect nets (aerial, sweep, and water nets)
- Cheesecloth
- Collecting bottles
- Cellucotton or envelopes

For killing:
- Small screw-top bottles
- Ethyl acetate or carbon tetrachloride
- Cotton

Cardboard or blotting paper
For preserving:
- Mounting boxes
- Cotton
- Transparent cover (glass or acetate)
- Insect pins
- Cork pinning boards or insect spreading boards
- Labels
- 70% alcohol
- KAAD (optional)

Note

1. **KAAD.** KAAD mixture is used to kill insect larvae.

Kerosene	10 ml
Glacial acetic acid	20 ml
95% ethyl alcohol	70 to 100 ml
Dioxane	10 ml

Larvae should be ready to transfer to alcohol for storage in ½ to 4 hours. For soft-bodied larvae, such as maggots, the amount of kerosene should be reduced.

Exercise 17: The Echinoderms

Exercise 17A: Class Asteroidea—Sea Stars

Materials

Living sea stars (*Asterias, Pisaster,* and others) in seawater
Preserved (or anesthetized) sea stars for dissection
Dishes for live material
Fresh or frozen seafood for feeding sea stars
Carmine suspension
Pieces of dried sea star tests
Camel hair brushes
Dissecting pans and tools
Dissecting microscopes or hand lenses

Notes

1. **Sea star demonstrations.** Several different types of dried sea stars placed on demonstration will emphasize diversity and adaptive radiation within the group, all variations on the basic pentamerous body plan. Some species have numerous rays, basically multiples of 5, although above 10 rays a sea star's arithmetic is not perfect and one finds stars with 17 or 32 rays, for example. Point out that many sea stars are large and quite colorful, especially those of the West Coast, in contrast

to most Atlantic Coast species, which tend toward more muted colors. If available, a sea star with one or more stumpy rays illustrates the capacity to regenerate lost limbs.

2. **Microslides of asteroid larval forms.** Microslides of bipinnaria and brachiolaria larvae and of pedicellariae that may be used for microscope demonstrations are available from biological supply companies. See Appendix B for listing of suppliers of prepared microslides.
3. **Using living sea stars for external structure.** Observations of the aboral and oral surfaces can be accomplished to advantage on living sea stars. Place a sea star in a large culture dish or refrigerator dish, cover with seawater, and examine with a dissecting microscope. We use a wooden support bridge to support the dish on the stage of standard dissecting microscopes.
4. **Demonstrating ampullar action with living sea stars.** If living sea stars are available, cut off the arm of a partly anesthetized animal (see note 5) and remove the dorsal surface. The disembodied arm will continue to move around in a dish of seawater for a long time, and students can see the action of the exposed ampulla. This provides a captivating demonstration that is well worth the mutilation of one animal. Further, if the sea star is large enough to have mature gonads, a smear of testis or ovary will reveal mature or developing sperm or eggs. Sea star spermatozoa have relatively large heads and long tails.
5. **Anesthetizing living sea stars for dissection.** Few schools are blessed with an easy supply of living sea stars, but, for those that are, living material is obviously to be preferred to preserved. Sea stars *must* be anesthetized before dissection, using $MgCl_2$ or $MgSO_4$ dissolved in fresh water at a concentration of about 7%. We find it requires 40 to 60 minutes for full anesthetization. Sea stars will recover if returned to seawater. The exercise requires only slight modification for use with living material.
6. **Demonstrating action of tube feet.** Another effective demonstration of tube feet is provided by fastening a living sea star, oral side up, on a piece of plate glass by means of rubber bands. Lay a small piece of celluloid or thick polyethylene film on top of the tube feet and have students note action of the tube feet and direction in which the tube feet move.
7. **Demonstrating the action of coelomic cells.** Inject into the coelom of a live star 5 ml of carmine suspension in seawater. Set the animal aside for about 8 hours; then examine under a dissecting microscope for the appearance of circulating particles in the skin gills. Pinch off some of the gills and examine under a compound microscope. Some of the carmine particles will have been picked up by phagocytic cells of the coelom, and these may be seen migrating through the thin walls of the gills. Such cells appear to have an excretory function. Examine drops of coelomic fluid for presence of other coelomic cells.
8. **Preparing echinoderm ossicles and pedicellariae for demonstration.** Put leftover skeletal parts (even whole specimens) into 5% to 10% potassium hydroxide solution or into full-strength commercial bleach. Warm over a Bunsen burner or boiling water to dissolve away the flesh. Let stand several hours; then decant carefully. Add fresh water, let stand again, and decant. The very tiny skeletal fragments settle slowly. Repeat until free of potassium hydroxide and debris. Cover with alcohol and again wash and decant, using two or three changes of alcohol. Now add 90% alcohol, shake, and put a drop on a clean slide. Quickly ignite the alcohol with a match. Add a drop or two of mounting medium and cover.

Exercise 17B: Class Ophiuroidea—Brittle Stars

Materials

Preserved brittle stars
Dissecting microscopes
Living brittle stars, if available
Fresh seafood for feeding, if living forms are used

Notes

1. **Observations on locomotion of living brittle stars.** This is best done in an aquarium or a tray large enough to allow free movement and containing enough seawater to enable the animal to right itself when turned over. Feeding is unpredictable. Some brittle stars will burrow if placed on a sandy substrate.
2. **Demonstrating brittle stars.** Dried brittle stars seldom survive student handling; plastic-embedded specimens are available. A dried basket star makes an interesting comparison with "conventional" brittle stars for demonstration. Basket stars (e.g., *Gorgonocephalus* sp. of the North Pacific), with their highly branched arms and tendril-like tips, belong to a separate order (Phrynophiurida) from that containing most other brittle stars (Ophiurida).
3. **How to visualize the vertebrae and vertebral articulations.** Place a brittle star arm in commercial bleach for several hours; then examine with a dissecting microscope.

Exercise 17C: Class Echinoidea—Sea Urchins

Materials

Living sea urchins in seawater
Preserved sea urchins
Dried or preserved sand dollars and/or sea biscuits
Dried tests of sea urchins
Glass plates
Large finger bowls or deep, clear plastic refrigerator dishes.
Carmine suspension
Dissecting pans and tools

Notes

1. **Demonstrations.** Provide dried or plastic-embedded echinoides, such as sand dollars (or key-hole urchins), sea biscuits, heart urchins, and various examples of sea urchins; preserved and/or dried Aristotle's lantern; microslides of sea urchin pedicellariae and pluteus larvae.
2. **Preparing the internal organs of a sea urchin for study.** Submerge only the aboral half of the animal in 2% nitric acid for 24 to 48 hours. Transfer to and submerge in water; cut a circle outside the periproct and extend the cuts through the ambulacral areas to the equator. The untreated portion of the test will hold the animal together for study.
3. **Preparing dried tests.** Submerge the animals in fresh commercial bleach (sodium hypochlorite). Complete removal of organic matter with a toothbrush.
4. **A study of living sand dollars.** If live sand dollars, such as the East Coast *Mellita,* Caribbean *Leodia,* or West Coast *Dendraster,* are available, they should be kept in sandy-bottomed containers (e.g., plastic refrigerator dishes). Place one in a bowl of seawater and examine with a hand lens or long-arm dissecting microscope. Have students examine spines on the aboral surface. Are they movable? Note the petal-shaped ambulacra; they are called petaloids. The tube feet are adapted for gas exchange rather than locomotion. Look at the flattened oral surface. Are the spines movable? How do they differ from those of *Arbacia?* Note the central mouth and the five-toothed chewing apparatus. Return the animal to the sandy bottom and observe how it burrows under the sand, using its oral spines to move the sand. Sand dollars feed on minute organic particles from the sand. They are passed back by tiny, club-shaped aboral spines, caught in mucus, and carried by ciliary currents to food grooves on the oral side that lead to the mouth.

Exercise 17D: Class Holothuroidea—Sea Cucumbers

Materials

Living sea cucumbers in aquarium or in bowls of seawater
Preserved or relaxed sea cucumbers for dissection
Prepared slides of holothurian ossicles

Notes

1. **How to relax sea cucumbers for dissection.** Inject living animals with 5 to 10 ml of 10% magnesium chloride an hour or so before use. Tentacles then may be gently forced out the anterior end.
2. **How to see sea cucumbers eviscerate themselves.** While sea cucumbers are famous for their radical evisceration, this is not readily accomplished merely by handling the animal roughly—at least not with *Thyone.* W. M. Reid (in Brown, op. cit.) suggests placing the animal for about a minute in 0.1% ammonium hydroxide made up in seawater. "Violent circular muscular contractions usually occur. If these movements do not begin spontaneously, hold the animal up by the posterior end for a few seconds. The entire anterior end of the animal with tentacles, calcareous ring, and the digestive system is violently expelled." This works. These structures will regenerate in 2 to 3 weeks, but the regenerated animal is much smaller than before. According to Pierce and Maugel (1987), *Thyone* can be made to eviscerate by injecting potassium chloride into the coelomic fluid.
3. **Demonstrations.** Appropriate demonstrations to accompany this exercise include microslides of sea cucumber ossicles and holothurian larval forms (e.g., auricularia and doliolaria larvae).
4. **How to prepare holothurian ossicles for study.** Holothuroid taxonomy is based in large part on the structure of ossicles in the body wall. These consist of microscopic discs, rods, buttons, and "tables," many of intricate form. They make a fascinating demonstration for students when prepared as permanent slides. Ossicles are destroyed by formalin, so they must be prepared from alcohol-preserved sea cucumbers. To prepare ossicles for examination, cut a small piece from the body wall, place it in a small test tube, and add a few ml of Clorox. Let stand until all the flesh has dissolved (an hour or two), leaving a sediment of whitish particles on the bottom. Pour off the supernatant fluid carefully and add clean water. Allow to settle, decant the water carefully (or use a Pasteur pipette), and resuspend in a few ml of alcohol. Again allow to settle; then pipette some of the sediment onto a slide. Allow to dry. Add a drop of mounting medium to the dried sediment and add a coverslip. The same method can be used for the ossicles and pedicellariae of other echinoderms.

Exercise 17E: Class Crinoidea—Feather Stars and Sea Lilies

Materials

Antedon—preserved or plastic-mounted

Exercise 18: Phylum Chordata: A Deuterostome Group

Exercise 18A: Subphylum Urochordata—*Ciona*, an Ascidian

Materials

Living (or preserved) *Ciona, Molgula, Corella,* or other small, translucent ascidian
Whole mounts of ascidian larvae

Carmine suspension in seawater
Finger bowls
Compound and dissecting microscopes

Notes

1. ***Ciona intestinalis*** has been the subject of several excellent morphological studies (see especially Miller, 1953, and Roule, 1884), so its anatomy is well understood. Goodbody (1974) should be consulted for ascidian physiology. The cosmopolitan *Ciona intestinalis* is probably the only species of the genus. The description in our exercise is largely "generic" in that it will describe genera other than *Ciona.*
2. **Whole mounts of ascidian tadpole larvae** are available from Carolina Biological, cat. 308260. The specimen used for the preparation of Figure 18-4 came from the series in Carolina's stock. They are far superior in quality to tadpole larvae that we sampled from other suppliers of microslides.
3. **Demonstrations.** Place on demonstration a variety of tunicates, both single and colonial (available from biological supply companies), and microslides of various stages of metamorphosis of the ascidian tadpole. Good slides are difficult to find; you may have to order samples from several suppliers to get satisfaction.
4. **Living tunicates.** Both solitary and colonial tunicates are available from marine suppliers. With transparent forms such as *Ciona intestinalis* or different species of *Clavelina,* the branchial sac can be viewed directly through the test.
5. **Behavioral observations.** Those suggested in the exercise are described in more detail by L. H. Kleinholz (in Brown, 1950, p. 554). The so-called crossed reflex is involved in coughing. If the inside of the siphon is tickled, the normal response is to close the *other* siphon and then contract the body. This would force out the source of an irritation (see Goodbody, 1974, p. 99). Sensory cells in *Ciona* have not been precisely identified but appear to be concentrated in the siphons.
6. **Methods for collecting and fertilizing eggs for observing the development of tadpole larvae.** These are described by Costello et al., 1957. It involves slitting open the test of living *Molgula* or *Ciona* and extending the animal by cutting the superficial muscles. With a Pasteur pipette, remove eggs from the oviduct (eggs are peach-colored when mature) to a watch glass of seawater. Put through several changes of seawater. From another individual, remove sperm from the sperm duct and make a suspension of the sperm in seawater. Add a drop or two of sperm suspension to the eggs to impart a milky appearance and let stand for 15 minutes. Now wash away the sperm. The tadpole larvae develop in about 24 hours. See the Costello paper for more details.
7. **Collecting eggs from colonial tunicates.** Brooding colonial tunicates usually release their larvae at dawn. In the laboratory, healthy specimens of *Amaroucium* or *Botryllus* may be kept overnight in a dark room or dark container and exposed to the light 15 to 20 minutes before needed, at which time swarms of larvae should appear. Locate with a dissecting microscope.

   Or squeeze a portion of a colony over a container of seawater in order to force out the eggs and larvae. Locate live larvae with a dissecting microscope and transfer to clean seawater.

   Ciona is oviparous, and eggs and sperm are discharged into the sea where fertilization occurs. According to early studies, eggs and sperm are discharged about 90 minutes *before* sunrise. The tadpole larva hatches about 25 hours after fertilization and settles to attach after 6 to 36 hours of free life.

All references to papers and publications mentioned in Appendix A are available as part of the Instructor Resources section in Connect for the main text (available only to instructors who are logged into Connect). Visit www.mheducation.com/highered/connect for more information.

Exercise 18B: Subphylum Cephalochordata—*Amphioxus*

Materials

Preserved mature amphioxus
Live animals can be obtained from Gulf Specimen Marine Lab (www.gulfspecimen.org)
Slides
Stained and cleared whole mounts of immature amphioxus
Stained cross sections of amphioxus
Watch glasses
Microscopes

Note

1. ***Branchiostoma virginiae*** is the common species of amphioxus along the southeastern coast of the United States. This is the species usually supplied as preserved and mounted material. *Branchiostoma californiense* occurs along the Pacific Coast from San Diego southward. Most detailed anatomical and physiological studies have been made on *Branchiostoma lanceolatum,* the European species. *Asymmetron,* the only other genus of cephalochordate, is so named because the gonads exist on the right side only.

   Note that adult preserved specimens of amphioxus usually have blocks of gonads visible through the body wall; these are shown in Figure 18-6. However, the juveniles that are usually chosen for whole mounts because of their smaller size lack gonads.

Exercise 19: The Fishes—Lampreys, Sharks, and Bony Fishes

Exercise 19A: Class Petromyzontida Lampreys (Ammocoete Larva and Adult)

Materials

Preserved material
- Ammocoetes (lamprey) larvae
- Adult lamprey specimens
- Longitudinal and transverse sections of lampreys

Watch glasses

Slides
- Stained whole mounts of ammocoetes
- Cross sections of ammocoetes

Compound and dissecting microscopes

Notes

1. **Life history set of a lamprey.** Ward's Natural Science offers a plastic-embedded life-history sequence of *Petromyzon* (eggs, several developmental stages, and section through head of the adult). Carolina offers embedded ammocoetes.
2. **Preparation of transverse sections of adult lampreys (or other large animals).** Freeze the preserved and injected animal; then cut into 5 cm transverse sections with a sharp hacksaw. Clean the sections carefully and secure the viscera in place with insect pins. Place each section in a container of slightly larger diameter than the section and cover with a melted 2.5% solution of agar, being careful while the solution is still warm to place the organs properly and to expel trapped air bubbles. When they have cooled, cut away excess agar from the surface. These preparations can be stored in a formaldehyde solution for long periods. Longitudinal sections can be made by the same method.

Exercise 19B: Class Chondrichthyes—Cartilaginous Fishes

Materials

Preserved dogfish sharks or

Longitudinal and transverse sections of a shark

Exercise 19C: Class Actinopterygii—Bony Fishes

Materials

Preserved, injected perch (other species of teleosts may be substituted)

Living fishes, any kind, in aquarium

Stained slides of fish blood

Mounted fish skeletons and skeletons of other vertebrates

Longitudinal and cross sections of preserved perch

Notes

1. **Watching living fishes.** An aquarium containing living fish should be made available for observation of swimming and respiratory movements.
2. **Demonstrating diversity within bony fishes.** Diversity, especially within the teleosts, can be illustrated by putting on display a variety of preserved specimens.
3. **Demonstration of fish scales and chromatophores.** To demonstrate fish scales and the chromatophores overlying them, remove with fine forceps a few scales from a living fish (e.g., goldfish, minnow, or perch). The fish does not have to be anesthetized, although this can be done easily by immersing the fish in tricaine methanesulfonate (MS-222), 0.2 to 0.05 g/liter (there is a wide latitude of permissible dosage). Mount the scales in a drop of water on a slide under a coverslip and observe under low power. Note the different types of chromatophores and the amount of pigment dispersal.

    If desired, the effect of drugs or hormones on the chromatophores can be studied by adding a drop of the drug or hormone to the water on the slide. Any of the following can be used: epinephrine (1 mg/ml), acetylcholine (100 mg/ml), melatonin (0.5 mg/ml), or pituitary extract (1 g beef pituitary powder [whole gland] shaken up in 10 ml of frog Ringers solution and filtered after letting stand 30 minutes).

Experimenting in Zoology: Analysis of the Multiple Hemoglobin System in *Carassius auratus,* Common Goldfish

Materials

Goldfish

0.9% saline

1.5 ml microcentrifuge tubes

1 ml syringes with 26-gauge needles

Powdered heparin

Microcentrifuge

Ice buckets

10 cm × 10 cm 8% native polyacrylamide gels (Novex)

2 × polyacrylamide gel loading buffer

Tris-glycine running buffer

Low-voltage power supplies

20, 200, and 1000 μl micropipets and disposable tips

Notes

1. **Animals: *Carassius auratus.*** The common goldfish may be purchased at any local aquarium supplier for modest cost. Animals in the 10 to 20 g size are

recommended; however, very little blood is needed, so small animals will suffice. It is recommended that animals be treated in accordance with institutional and governmental guidelines for the humane care and treatment of laboratory animals for teaching and research.

Anesthetized animals should be dispatched by medullary transection and 1 ml of blood removed from the caudal vein or heart into a syringe that contains a small amount (about 1 mg) of heparin. Transfer the blood sample to a 1.5 ml microcentrifuge tube and centrifuge at 10,000 g for 5 minutes to pellet the red blood cells. Remove the serum and discard. Add 1.0 ml of 0.9% saline, resuspend the cells, and repeat the centrifugation. Remove the saline wash and discard. Add 200 μl (about 3 to 4 volumes) of distilled water to the red blood cell pellet and resuspend. The cells should lyse in 1 to 2 minutes. Centrifuge the sample at 12,000 g for 5 minutes to pellet the cell ghosts. Transfer the clear red supernatant to a clean tube and place on ice.

Human hemoglobin may be collected from a finger prick using a sterile lancet after cleansing the fingertip with a 70% alcohol swab. Collect about 1 ml of blood into a syringe containing a small amount (about 1 mg) of heparin. (A needle is not necessary for this collection.) Prepare the hemoglobin lysate as described for the goldfish hemoglobin.

2. **Sample preparation.** For each sample, mix equal volumes of the hemoglobin lysate with polyacrylamide 2× gel loading buffer (100 mM Tris, pH 8.3, 10% glycerol, 0.0025% bromophenol blue) (available from Novex, see note 3). Load 20-μl samples onto the gel using a micropipet.
3. **Vertical polyacrylamide gel electrophoresis of hemoglobin.** Necessary equipment includes a mini–vertical gel electrophoresis apparatus appropriate for running 10 cm × 10 cm polyacrylamide gels and a low-voltage power supply. Hemoglobin isoforms may be separated on an 8% native Tris glycine polyacrylamide gel. These gels may be purchased from Invitrogen (www.catalog.invitrogen.com). The 1-mm-thick, 10-well comb format is appropriate for this application (Novex cat. # EC6015). The gel is electrophoresed in 1× Tris glycine running buffer (25 mM Tris, 192 mM glycine pH 8.3) for approximately 30 to 60 minutes. Running buffer stocks (10×) and sample buffer (2×) may also be purchased from Novex (cat. # LC2672 and LC2673).

Exercise 20: The Amphibians

Exercise 20A: Behavior and Adaptations

Materials

Living frogs	Pond or dechlorinated water
Jars or bowls	Paper towel
Aquarium	Fruit flies (for feeding frogs)

Notes

1. **The animal rights question.** It may happen at this point in the zoology laboratory that one or more students will object to dissecting preserved frogs, perhaps asking, do we have the right to use vertebrate animals in research? How can it be justified? The issue is addressed in a statement following the General Instructions of this manual (p. x), but the objection may still arise. We can respond by asking whether a snake has the right to eat the frog or whether the frog has the right to eat an insect. Just as the snake eats the frog to survive, humans not only use animals as food for survival but also apply our abilities through biomedical research with animals to learn ways to minimize human pain, disease, and suffering—and to benefit other animals. It does not hurt to emphasize how we have all benefited from biomedical research on animals.

    The matter is a touchy one because there have been several cases in which high school students have sued their schools over animal dissection requirements, and there has been at least one university case (see *Science,* 18 May 1990, p. 811), yet the direct dissection of vertebrate animals is the only way that one can understand vertebrate anatomy, and most biologists believe that such an understanding is essential for anyone oriented toward a career in animal biology or medicine. Direct examination of the organ systems of animals cannot be matched by any other approach. Computer simulation programs are *not* appropriate substitutes for direct dissection, despite claims to the contrary. To learn morphology from computer simulations would be rather like attempting to learn to play a musical instrument by watching videotapes of others playing the instrument. For more background on the animal rights issue, we suggest the books listed on the website. We especially recommend Pringle's book and the report of the Commission on Life Sciences, Committee on the Use of Laboratory Animals in Biomedical and Behavioral Research. As one of the committee members states, this report is the nearest thing we have to a national consensus and statement of policy on the animal use issue.

    How does one deal with the occasional student who refuses to dissect? Try to find some alternative means for the student to learn the material, with as little fuss as possible. Perhaps the student will accept the option of observing while a partner does the dissection. If not, assign a library project that requires researching some aspect of vertebrate adaptations. Stress that such a project is an alternative to dissection and not a penalty for refusing to dissect.

All references to papers and publications mentioned in Appendix A are available as part of the Instructor Resources section in Connect for the main text (available only to instructors who are logged into Connect). Visit www.mheducation.com/highered/connect for more information.

2. **This exercise may be supplemented** by providing demonstrations of living specimens of frogs, toads, and salamanders common to your area.

Exercise 20B: Skeleton

Materials

Frog and other vertebrate skeletons
Individual vertebrae
Frog skulls

Notes

1. **The exercise may be supplemented** by placing on display the skeletons of other vertebrates for comparison with the frog skeleton.
2. **Skeletons of embryos and *small* vertebrates may be stained *in situ*** by the following method. First fix small specimens (skinned small frogs are excellent) in 95% alcohol for 2 to 4 days. Then place them in 2% potassium hydroxide solution until the bones are visible through the tissue. Check frequently to make sure that the specimens are not macerated. Now transfer them to the following solution for 24 hours: 1 part of alizarin to 10,000 parts of 2% potassium hydroxide. Allow the stain to act until the desired intensity is obtained. It may take longer than 24 hours. Finally, clear the specimens in increasing concentrations of glycerin (10% to 50%). Excessively stained bones may be destained with 1% sulfuric acid made up in 95% alcohol.

Exercise 20C: Skeletal Muscles

Materials

Preserved frogs
Dissecting pans
Dissecting tools

Exercise 20D: Digestive, Respiratory, and Urogenital Systems

Materials

Preserved frogs
Prepared slides
Frog kidney — Frog ovary
Frog testis — Sperm smears

Notes

The following projects and demonstrations are appropriate supplements to this exercise.

1. **Cross sections of a frog body.** Cross sections should be made after the frog has been thoroughly frozen at a low temperature. Sections are easily cut with a hacksaw and completed according to directions given in the notes to Exercise 19A. Revealing relations of organs may be seen from sections made at the level of (a) a region a short distance anterior to the hind legs and (b) a region just posterior to the forelegs.
2. **Study of cross sections of intestine.** Compare slides of the cross section of the human intestine with that of the frog.
3. **Peristalsis in a frog.** Pith the brain of a frog that has been fed an hour or two previously. Open the abdominal cavity and flood with warm 0.6% saline solution. Peristalsis should be observed. The students should understand what type of muscle is involved (smooth muscle) and what directions the fibers run.

   Tie a thread tightly around the pyloric end of the stomach. Open the stomach near the cardiac end and with a pipette introduce physiological salt solution into the lumen. Close the opening with a second ligature; then cut out the stomach. Suspend the stomach by the pyloric end in Ringer's solution. Observe the wavelike peristaltic movement passing over the stomach. Record the rate per minute.

Exercise 20E: Circulatory System

Materials

Injected preserved frogs
Living frogs
Frog Ringer's solution
Frog holders
Ice
Pins
Paper towel
Masking tape

Notes

1. **A demonstration of a frog's beating heart,** arranged to record on a chart recorder or smoked kymograph drum, can also be used to demonstrate the effects of temperature and of adrenaline and acetylcholine. The procedures are well understood by most instructors and will not be repeated here. An even simpler procedure is to demonstrate the durability of the isolated frog heart submerged in a dish of Ringer's physiological saline. The saline must be well aerated. Adrenaline and acetylcholine may be added in concentrations of about 1/10,000. Flush with Ringer's between each test.
2. **Capillary bed.** If a latex-injected frog is available, remove a piece of the skin for examination under a dissection microscope or low power of a compound microscope. The capillary bed should show up well.

Exercise 20F: Nervous System

Materials

Preserved frogs

Note

1. **Maceration preparation of frog nervous system (for demonstration or special student project).** Skin a frog and immerse it in a bath of 30% nitric acid solution, with the exception of the head, which is kept out of the bath by a glass hook. Leave in this solution for about 24 hours and then submerge the head, leaving the whole frog in the bath for about 8 hours. Check frequently to determine the degree of maceration. Then tease as much of the muscle as possible from the nervous system. Transfer the specimen to an empty dish and allow water from a faucet to fall gently on the tissue from a short distance. This will carry away the remaining part of the muscle from the nervous system. The critical part of this preparation is the length of time in the immersion bath. Too long a time will dissociate the nervous system along with the other tissues. Preserve the completed preparation in a flat dish containing glycerin.

Demonstrations

1. **Cross section of spinal cord (prepared slide).** Note the surrounding meninges (dura and pia mater), ventral and dorsal fissures, dorsal septum from the dorsal fissure to near the center of the cord, inner fibers, and central canal.
2. **Cross section of frog eye (slide).** Find and identify the cornea, anterior and posterior cavities, crystalline lens, and retina.

Exercise 21: The Reptiles

Exercise 21: Painted Turtle

Materials

Turtle skeleton
Living painted turtles or other species
Preserved turtles

Notes

1. **Why turtles?** Turtles have long been chosen over lizards and snakes for comparative anatomy for several good reasons: they are large and the organs easily observed (once the plastron is removed); the skeleton is stoutly built and resistant to the abuses of student handling; they provide an interesting transition to birds, covered in Exercise 22 (turtles are "birds in shells" in that the two groups share, among other things, a highly flexible neck on a fused body frame, a toothless beak, and good vision); they are inexpensive; they represent a highly successful body plan, clumsy as they may appear; and students find them interesting and do not harbor any unreasonable fear of the creatures (as many do of snakes).
2. **How much time?** This exercise will require 2 to 3 hours to complete in its entirety, but instructors may choose to limit the exercise to a 1-hour study of external structure and skeleton.
3. **Singly injected turtles** are not required for this exercise, although having one or more available for the class will be helpful for visualizing the circulation.
4. **Exposing the viscera for dissection** is a cumbersome procedure using a bone saw and bone shears, and turtles should be prepared for the students before class. The bridge is partly sawed through with a bone saw and the break completed with bone cutting forceps (both instruments available from biological supply companies). Miniature saw blades that can be used with an electric drill are available (but difficult to find) and will vastly ease cutting through the bony bridge between the plastron and carapace. Singly injected turtles have had the plastron removed.

Exercise 22: The Birds

Exercise 22: Pigeon

Materials

Flight feathers
Other feather types as available for display
Pigeon skeleton
Preserved pigeons, plain
Preserved pigeons, air-sac-injected (optional, for demonstration)

Notes

1. **Allow about 2½ hours** for this exercise if internal structure is reviewed as well as feathers and skeleton. A study of skeleton alone will reveal many of the adaptations for flight.
2. **Pigeon internal anatomy** can be covered by using prepared specimens or models available from biological supply houses, rather than having students dissect preserved specimens. Ward's offers a "Bio-Mount" of triple-injected and dissected pigeons (mounted on a plate and sealed in a museum jar). Models of chicken internal anatomy are also available from biological suppliers (although expensive).

Exercise 23: The Mammals

Exercise 23A: Skeleton

Materials

Skeletons, cat or dog and human
Sections of bones (if possible, a fresh joint)
Other vertebrate skeletons

Note

1. **Demonstrations appropriate to this exercise** include (a) a fresh joint cut in longitudinal section with a bandsaw; (b) a long bone cut in longitudinal section; (c) a piece of skull or face bone available for examination of membrane bone; (d) skeletons of other vertebrates.

Exercise 23B: Muscular System

Materials

Fetal pig, embalmed

Although it is stated in the exercise, remind concerned students that fetal pigs are taken from the uteri of sows that have been slaughtered for market. Fetal pigs are commonly used in zoology laboratories because they are one of the by-products of the meat–packing industry. Consequently, they are inexpensive and serve as substitutes for other animals that would have to be killed to supply preserved material for dissection.

Notes

1. **Use the largest available fetal pigs** for this exercise; this facilitates muscle separation, never easy on a fetal pig. The muscle boundaries become more evident once the cutaneous muscle is removed. Sometimes it helps to use dry paper towel to rub off the cutaneous muscle—but caution the energetic student not to rub through the underlying musculature as well.
2. **Muscle names** make more sense to the students if you take a few minutes to explain what some of them mean. Refer also to Exercise 20C, where several of the Greek and Latin roots to common muscle names are explained.
3. **The text has been written to lead a student** through the dissection; distal musculature of the limbs has been omitted, and mention of origin, insertion, and action has been placed in tabular form (Tables 23.1 and 23.2), so that the descriptive text is not burdened with information that many instructors will not require their students to learn.

Exercise 23C: Digestive System

Materials

Fetal pigs, embalmed
Prepared slides
  Human intestine, cross section
Frog intestine, cross section
  Mammalian liver
  Mammalian pancreas

Notes

1. **Demonstration of peristaltic movement.** About 30 to 45 minutes after a rat has finished feeding, anesthetize it in an ether jar, open the abdominal cavity, and submerge the contents in warm physiological saline solution. Note the peristaltic movement of the intestinal tract.
2. **Demonstration of the ruminant stomach.** The ruminants are cud-chewing animals (cattle, deer, camels, and their kin). All are members of order Artiodactyla, although not all Artiodactyla are ruminants (the pig, for example). If a ruminant stomach can be obtained, preferably fresh, it makes a fascinating demonstration. Food swallowed after brief mastication and generous addition of saliva passes to the **rumen** for preliminary fermentation by a specialized microflora. Formed into small balls of cud, it is returned to the mouth for further mastication. Further fermentation follows in the rumen. When broken down to a pulp, it is passed to the **reticulum,** the second chamber with its honeycombed epithelium, where fermentation continues. The pulp next passes to the **omasum,** where water, soluble food, and microbial products are absorbed. Finally the smallest products pass to the **abomasum,** where proteolytic enzymes are added and normal digestion occurs in an acid environment.
3. **Demonstration slides.** Cross sections of mammalian esophagus, stomach, intestine, salivary glands, pancreas, and liver.

Exercise 23D: Urogenital System

Materials

Fetal pigs
Pregnant pig (or dog or cat) uteri for dissection or
  demonstration
Preserved sheep kidneys (optional)

Exercise 23E: Circulatory System

Materials

Fetal pigs
Pig or sheep hearts, fresh or preserved

Notes

1. **Microscope demonstrations.** These might include prepared slides of mammalian blood and of cross sections of artery and vein.
2. **The beef heart as a demonstration.** One reviewer of this exercise tells us that a fresh beef heart makes an excellent demonstration of heart valves. It is also possible to see the fossa ovalis and the ligamentum arteriosum (of the ductus arteriosus) with the beef heart.
3. **Demonstration of the action of heart valves.** If you are in the fortunate position of being able to procure a fresh sheep's heart with the roots of the great vessels uncut, the following makes a fine demonstration of valve action. Tie glass tubes in the aorta, the pulmonary artery, and the veins of the atria. Suspend the heart by clamping the glass tube tied in the aorta to a stand. Water poured into the atria will rise through the pulmonary artery and aorta. By means of a long pipette or tubing inserted through the veins of the atria, remove some of the water from the ventricles. The water will still remain at the same height in the aorta and pulmonary arteries because of the action of the semilunar valves. Empty the heart of water. With a long pipette shoved down the aorta, place some water in the left ventricle. When the left ventricle is filled, the mitral valve prevents the water from entering the left atrium, and it will now rise up through the aorta.

Exercise 23F: Nervous System

Materials

Preserved sheep brains

Notes

1. **Sheep brains should be purchased** with the dura mater, hypophysis, and cranial nerves intact (e.g., Carolina 228710) or with the brain still in its cranial case (e.g., Carolina 228720). We no longer include directions for dissection of a fetal pig brain because its softness makes dissection difficult.
2. **Demonstrations** might include models of a mammalian eye and ear and fresh or preserved eyes for dissection.

APPENDIX B

Sources of Living Material and Prepared Microslides

Sources of Living Material

Carolina Biological Supply Co.
2700 York Road
Burlington, NC 27215-3398
800-334-5551
www.carolina.com/

Connecticut Valley Biological Supply Co., Inc.
P.O. Box 326
82 Valley Road
Southampton, MA 01073
800-628-7748
www.ctvalleybio.com/

Fisher Scientific Co.
300 Industry Drive
Pittsburgh, PA 15275
800-766-7000 (orders)
www.fishersci.com/

Gulf Specimen Marine Laboratories., Inc.
222 Clark Drive
P.O. Box 237
Panacea, FL 32346
850-984-5297
www.gulfspecimen.org/

Insect Lore Products
132 S Beech Avenue
P.O. Box 1535
Shafter, CA 93263-1535
800-548-3284 (orders)
Fax: 661-746-0334
www.insectlore.com/

Leeches USA
300 Shames Drive
Westbury, NY 11590
800-645-3569
www.leechesusa.com/

NASCO (California)
P.O. Box 101
Salida, CA 95368
800-558-9595
www.enasco.com/

NASCO (Wisconsin)
901 Janesville Avenue
P.O. Box 901
Fort Atkinson, WI 53538-0901
800-558-9595
www.enasco.com/

Nebraska Scientific
3823 Leavenworth St.
Omaha, NE 68105-1180
800-228-7117
www.nebraskascientific.com/

Parco Scientific Co.
P.O. Box 851559
Westland, MI 48185
877-592-5837
www.parcoscientific.com/

Ward's Natural Science
5100 West Henrietta Road
P.O. Box 92912
Rochester, NY 14692-9102
800-962-2660 (orders)
www.wardsci.com/

Sources of Living Marine Material

Carolina Biological Supply Co.—see "Sources of Living Material"
Connecticut Valley Biological Supply Co.—see "Sources of Living Material"
Gulf Specimen Marine Labs—see "Sources of Living Material"

Marine Biological Laboratory
Supply Department
7 MBL Street
Woods Hole, MA 02543
508-548-3705
www.mbl.edu/

Sources of Prepared Microslides

Carolina Biological Supply Co.
2700 York Road
Burlington, NC 27215-3398
800-334-5551
www.carolina.com/

Connecticut Valley Biological Supply Co., Inc.
82 Valley Road
P.O. Box 326
Southampton, MA 01073
800-628-7748
www.ctvalleybio.com/

Insect Lore Products
132 S Beech Ave.
P.O. Box 1535
Shafter, CA 93263-1535
800-548-3284 (orders)
www.insectlore.com/

NASCO
901 Janesville Avenue
P.O. Box 901
Fort Atkinson, WI 53538-0901
800-558-9595
www.enasco.com/

INDEX

Note: All entries in **boldface** type indicate figures or tables.

A

Abdomen
  chelicerate arthropods, 220
  crickets, 251
  crustaceans, 228, **229**
  fetal pig, 350
  grasshoppers, 245
  honeybee, 247
  horseshoe crabs, 220, **222**
  spider, 224, **224,** 225
Abdominal cavity, 366–367
Abdominal cutaneous muscle, **323**
Abdominal extensor muscle, **233**
Abdominal flexor muscle, **233**
Abdominal pore, **297**
Abducens nerve, **335, 383,** 384
Abduction, 355, **355,** 356
Abductor muscles, 356
Acanthocephala, 85
Accessory gland, 254
Accessory urinary duct, **299**
Acetabulum, 156, 351
Acheta domesticus, 251–254
Achilles tendon, **323**
Acicula, **202**
Acoelomate animals, 150, 152
Acontia, 141
Acorn barnacle, 235
Acorn worms, 86
Acoustic nerve, **335**
Acromial process, 339, **352**
Actin, 91
Actinophrys, 93, **93**
Actinopterygii, 86, 283, 301–306
Actions of muscles, 355
Adduction, 355, **355,** 356
Adductor longus muscle, 321, **321, 322**
Adductor magnus muscle, 321, **321, 322**
Adductor muscles
  actions, 356
  of bivalves, 185, **187**
  fetal pig, 360, **361, 362**
  frogs, 321, **321, 322**
Adhesive pads, 136, **137**
Adhesive papillae
  ascidian larvae, 285, **286**
Adhesive threads, 191
Adipose connective tissue, 50, 57
Adjustable iris diaphragm, 4, 5
Adrenal glands, 327, **327,** 369, **369**
Adrenal vein, 378
Adult fruit flies, 254
Aedeagus, 254
Afferent branchial arteries
  amphioxus, 288, **289**
  bony fishes, **304,** 306
  lampreys, 294
  shark, 300
African sleeping sickness, 100
Agarose gel electrophoresis, 117, 118, **119**
Aggression in fishes, 307–308
Agnatha, 86, 291
Agonistic behavio, 307–308
Agranulocytes, 52
Air sacs, 347
Alcohol, avoiding use on microscopes, 5
Alcyonaria, 144
Alderflies, 73
Alimentary canal, 209, 367
Allantoic bladder, 367, 369
Alula, 345
Alveolar ducts, 384
Alveolar pockets, 341
Alveoli, 326
Ambulacral groove, 267, 270
Ambulacral ossicles, 268, **269,** 270
Ambulacral plates, **274,** 275
Ambulacral pores, 268
Ambulacral regions, 274
Ambulacral ridges, 268, **268**
Ambulacral spines, 267
Ambystoma mavortium, **315**
Amebocytes, 124–125
Ameboid cells, **125**
Ametabolous development, 254
Ammocoetes, 291, 293–296, **294**
Amniote fossils, 337
Amniotic eggs, 337
Amoeba
  feeding, 92
  general features, 91–92
  habitats, 89–90
  identifying, **70,** 84, **91,** 113–114
  locomotion, 91
  overview, 89–90, **90**
  shelled, **93,** 93–94
Amoeba proteus, 89–94
Amoebozoa, **70,** 89–94, 114. *See also* *Amoeba*
Amphiarthrosis, 354
Amphibians. *See also* Frogs
  identifying, 86, 283
  overview, 313
  respiration, 326
  skeleton, 318–319, **320**
Amphiblastula larvae, 123
Amphioxus, 86, 286–289
Amphipoda, **73**
Amphitrite, 204
Amphiuma, 14–16
Amplexus, 327
Ampullae
  sea cucumbers, **277,** 278
  sea stars, 266, 268, **268, 269,** 270
Ampullae of Lorenzini, 296
Anadromous species, 291
Anal canal tissues, **49**
Anal fins, 301, 302, **302, 303**
Anal opening of lamprey, 293
Anal papillae, **224**
Anal plates, **274,** 275
Anal pore, 111
Analysis of variance (ANOVA), 66
Anaphase
  meiosis, **26, 27**
  mitosis, 17, **18, 19**
Anapsids, 337
Ancestral state, 79
Ancylostoma, 177
Aneides flavipunctatus, **315**
Angulosplenials, **318,** 319
Animal classification, 77–86
Animal pole of frog eggs, 44
Animals, 84
  protozoans *vs.,* 77
Annelids
  clamworm, 202–204
  classification, 216
  earthworm, 204–210
  identifying, 71, 85
  leeches, 205, 215–218
  overview, 201–202
Annuli, 216
Anopheles mosquito, 106, **107**
Anoplura, 260
Anostraca, 234, **235**
Ant lions, 259
Antedon, 278
Antenna cleaner, **246,** 247
Antenna comb, 247
Antennae
  as basic features of insects, 243
  crickets, **252**
  crustaceans, 227, 228, **229,** 230, **231, 233,** 234
  grasshoppers, 244, **245**
  honeybee, **246**
  myriapods, **242,** 243, **243**
Antennal arteries, 232, **233**
Antennal glands, 234, 236
Antennules, **229,** 230, **233,** 234
Anterior adductor muscles, 185, **187**
Anterior aorta, **187,** 189
Anterior cardinal sinuses, 299
Anterior cardinal vein, 294, **295**
Anterior cervical ganglion, **380,** 381
Anterior deep pectoral muscle, 357, **357, 358, 359,** 360
Anterior lateral line systems, **292**
Anterior lobe, 335
Anterior mantle vein, 198
Anterior mesenteric artery
  fetal pig, **377,** 378
  frogs, **328,** 330
  shark, **299**
Anterior mesenteric vein, **377,** 378
Anterior trunk, planarian, 151
Anterior vena cava, 198, 328
Anthozoans, 84, 140–143, 144
Anticus muscles, 320
Ants, 260
Anurans, 313. *See also* Frogs
Anus
  amphioxus, 287, **287,** 288
  Ascaris, 172, **173,** 174
  bivalves, **187,** 190
  bony fishes, 302, 305, **306**
  chitons, 195
  clamworm, 202, **203**
  crustaceans, 228, **229, 233,** 234
  earthworm, 205, 206
  embryonic development, 46
  fetal pig, 350, 367, **369, 371**
  frogs, **323**
  honeybee, **247**
  horseshoe crabs, 221
  lampreys, 294
  larval formation, 40
  leeches, 216
  myriapods, 241
  sea cucumbers, 276, **277**
  sea stars, 267, **268**
  sea urchins, **274,** 275
  snails, **192, 193**
  spider, **224,** 225
  squids, 198
  tadpole, **45**
  tunicates, **284,** 285
  turtles, **341**
  vinegar eels, 175, **176**
Aorta
  bivalves, **187,** 189
  bony fishes, **304,** 305
  fetal pig, 368, **369,** 373, **376, 377,** 378
  frogs, **327**
  lampreys, **292,** 293, 294, **295**
  shark, **297, 299,** 300
  sheep, 374, **374**
  squids, 198
  turtle, 342
Aortic arches
  earthworm, 206, **207**
  fetal pig, **376**
  frogs, **329,** 329–330
Apertures of snail shells, 192, **193**
Apex of snail shells, 192, **193**
Aphids, 260, 261
Apical lobe, 384, **385**
Apicomplexans, 106–108, **108,** 114
Apis, 245–248, **246**
Aplysina, **124**
Aponeurosis, 355
Apopyles, **122,** 123, **125**
Appendicular skeleton
  birds, 345
  bony fishes, 302
  frogs, 318, 319, **320**
  mammals, 350–351
  turtles, 338
Arachnida, **72,** 86, 223–225
Araneae, **72**
Arbacia, 273, 275
Arcella, **70,** 93, **93**
Archaea, 77
Archaeocytes, **123, 125**
Archenteron, **37,** 38
Arctic ground squirrel, **1**
Areolar connective tissue, **50,** 54–55
Argiope, 223–225, **224**
Aristotle's lantern, 274
Arms
  brittle stars, 271
  feather stars, **278**
  sea stars, 266, 267, **268**
  squids, 196
Arolium, 245
Arrow worms, 85
Artemia, 147, 236
Arteries
  amphioxus, 288, **289**
  bony fishes, **304,** 306
  crustaceans, 232, **233,** 234
  fetal pig, **376**
  frogs, 328, 329–330
  lampreys, 294
  shark, **299,** 299–301
  squids, **197,** 198
  tissues, 54, 57, **57**
  turtles, **341**
Arthropods
  chelicerate arthropods, 220–224
  classification, 225, 258–262
  crickets, 251–254
  crustaceans, 227–237
  fruit fly, 254–255
  general features, 219–220
  grasshoppers and honeybees, 243–248
  identifying, **72,** 85

Arthropods (*Continued*)
myriapods, 241–243
subphylum Crustacea, 227–237
Articular cartilage, 351, **354**
Articular processes, **319**
Articulations, 354
Artiodactyla, 349
Ascaris
cell cycle studies using, 17, 19–20, **20**
general features, 172–174, **173, 175**
oogenesis and fertilization, 28–30, **30**
Ascidian larvae, 285, **286**
Asconoid sponges, 123, **124**
Asexual reproduction
Amoeba, 92
by cnidarians, 134
in sponges, 123
in *Volvox,* 99
Aster, 17, **18, 19**
Asterias, 38, **39**
Asteroidea, 84, 266–270, 279
Astragalus, **320**
Astrangia, 142–143, **143**
Astrophyton muricatum, 271
Asymmetron, 286
Atlas, 319, 350, **352**
Atrazine, 327
Atrial siphons, 283, **284**
Atriopore (excurrent aperture), 285, 286, **286,** 288
Atrium
amphioxus, 288, 289
ascidian larvae, 285, **286**
bony fishes, **304,** 306
frogs, 328
lampreys, 293, 294, **294**
mammalian heart, 373, 374
shark, **297**
tunicates, **284,** 285
turtles, **341**
Auchenorrhyncha, 260
Auditory capsule, **318,** 319, **320**
Auditory nerve, **383,** 384
Aurelia, **138**–140, **139**
Auricles
bivalves, **186, 187,** 189
fetal pig, 373
planarian, 150, **151**
Autonomic ganglia, **379, 380**
Autonomic nervous system, 334, **379, 380,** 381–382
Aves, 86, 283
Avian Reptiles, 86
Axial skeleton
birds, 344, **344**
bony fishes, 302
frogs, **318,** 318–319, **320**
mammals, 350
turtles, 338
Axillary arteries, 375, **376**
Axillary veins, **373,** 375
Axis, **352**
Axons, **55**
Axopodia, 94

B

Bacteria, 77
Balanus, 235
Bar lice, 260, 261
Barbs (feather), 343
Barbules, 343
Barnacles, 235–236
Basal body, **97**
Basal disc, 132, 141
Base (microscope), 3
Basement membrane, **48, 49**
Basket stars, 279
Basophils, **52**
Bdelloura, 152
Beaks
birds, **344,** 345, **346**
turtles, 338, 339, **339, 340**
Bees, 258
Beetles, **71, 74, 256,** 262
Biceps brachii, **358, 359,** 360
Biceps femoris
fetal pig, **357,** 360, **361, 362**
frogs, **322, 323**
Biceps muscles, 320
Bicornuate uterus, 371
Bicuspid valve, 374, **374**
Bilateral cleavage, 44
Bilateral symmetry, 149, 220
Bile, 368
Bile ducts, **297**
Bilharzia, 156
Binary fission, 92, 111
Binocular microscopes, **4,** 6, 8, **8**
Binomial nomenclature, 77
Bipinnaria larva, **39,** 40
Birds, 86, 343–347
Bispira brunnea, 204
Bivalvia
freshwater clams, 184–191
identifying, **72,** 85, 199
predation by sea stars, 269
Bivium, 267
Blaberus, 248
Bladder. *See also* Urinary bladder
bony fishes, **304,** 306
earthworm, **207**
trematodes, **154**
Bladder flukes, 156–157
Blastocoel, 38, **39,** 40, 44
Blastomeres, 36
Blastopores, **39,** 40, 46
Blastulas, 19, **37,** 38, **39,** 40, 44
Blattodea, 262
Blood cells
basic features, **52**
crustaceans, 235, 237
parasitic infections, 106, **107**
Blood flukes, 155–156
Blood vessels. *See also* Arteries; Circulation and circulatory systems; Veins
in bone, **51,** 351, **354**
earthworm, 209
parasitic infections, 155–156
tissues, **48,** 54–55, 57, **57**
Body cavity tissues, **48**
Body covering
flagellates, 97
Body of uterus, 370
Body tube (microscope), 3, 4
Body wall
fetal pig, 367
Metridium, 142
Bones, **51,** 344, 350. *See also* Skeletal systems
Bony fishes, 301–306
skull, 301, 302
Book gills, 221, **222**
Book lice, 260, 261
Book lungs, 225
Bottom habitats, 67
Brachial arteries, 375, **376**
Brachial nerve, **334**
Brachial plexus, **334,** 335, 379–381, **380**
Brachial veins, 375
Brachialis, **357, 358, 359,** 360
Brachiocephalic artery, **341,** 342
Brachiocephalic muscle, 357, **358, 359**
Brachiocephalic trunk, 375
Brachiolaria larva, **39,** 40
Brachiopoda, 85
Brain
bony fishes, **304**
frogs, 334–335
lampreys, **292,** 293
sheep, 382–384, **383**
Branchial arteries
bony fishes, **304,** 305
lampreys, 294
shark, 300
squids, **197,** 198
Branchial basket, 285, **286**
Branchial heart, **197,** 198
Branchial sac, 285
Branchial systems, crustaceans, 230
Branchial vein, **197,** 198
Branchinecta, 234, **235**
Branching, 78, **79, 80, 83**
Branchiopoda, 234–235, **235**
Branchiostegites, 228, **229**
Branchiostoma, 283, 286–289
Breathing, Respiration and respiratory systems
Brevis muscles, 320
Brine shrimp, 73, 147, 236
Bristletails, 260
Brittle stars, 84, 270–272, **272,** 279
Broad ligament, 371, **371**
Bronchi, **49,** 341, 384
Brood chamber of bivalves, 187
Bryozoa, **71**
Buccal cavity, 287, 294
Buccal chamber, 295, **295**
Buccal cirri, 286
Buccal funnel, 292, **295**
Buccal membrane, 197
Buccal podia, 274
Buccal shields, 271
Budding, 112, 134, 158
Bulbourethral glands, **369,** 370
Bulbus arteriosus, **304,** 306
Bursae, **253,** 254, 271, **272**
Butcher's pull, 384
Butterflies, **256,** 258

C

Caddisflies, 73, 74, 258, 259
Caeca, **346,** 347
Caecilians, 86
Calcaneus, **320**
Calcaneus bone, 351, **352, 353**
Calciferous glands, 206, **207**
Calcispongiae, 121–127
Calibrating ocular micrometers, 6–7, **8**
Calyces, **278**
Canal systems, 121
Canaliculi, **51**
Cancellous bone, 351
Cancer, 16
Canines, **351, 352**
Capillaries, 54, 248
Capillary circulation, 248
Carapace
crustaceans, 228, **229,** 232, **233,** 234
horseshoe crabs, 220, **221**
turtles, 338, **339,** 340
Carassius auratus, 309–311
Cardiac lobe, 384, **385**
Cardiac muscle, **54**
Cardiac stomach
bony fishes, **304,** 305
crustaceans, 232, **233,** 234
fetal pig, 367
sea stars, **268,** 269
Carotid arch, **328,** 329
Carotid artery
fetal pig, 375, **376**
turtle, 342
Carotid trunk, 375
Carpals
birds, 344, **344,** 345
frogs, **320**
mammals, **352, 353**
Carpometacarpals, **344,** 345
Carpometacarpus, 344
Carpus, 351
Carrying capacity, 64
Cartilage, **50,** 57, **58,** 199
Castes, honeybee, 247–248
Cat skeletal system, **351, 352**
Cattails, 67
Caudal artery, 300, **377,** 378
Caudal fins
amphioxus, 286, **287**
bony fishes, 302, **302, 303**
lampreys, 292, 293, **294, 295**
shark, 296, **296**
Caudal vein, 299, 300, **377**
Caudal vena cava, **369**
Caudal vertebrae
birds, **344,** 344–345
mammals, 350, **352**
turtles, 338, **339**
Cecropia moths, 248
Cecum, **197,** 198, 253, 367
Celiac artery
bony fishes, **304**
fetal pig, **377,** 378
frogs, **328,** 330
shark, **299,** 300
Celiacomesenteric artery, **328,** 330
Celiacomesenteric ganglia, 382
Celiacomesenteric ganglion, **380**
Cell cycle, 16–20
Cell differentiation in *Volvox,* 99
Cell division, 16–20
Cell membranes, 44
Cell organelles, 17, **18, 19**
Cell structure, 13–16
Cellular level of organization, 121
Centipedes, 86, 241, **242**
Central axis, 230
Central canal, 289, **379**
Central disc, 267, **272**
Central nervous system, 334, 379
Centrohelida, 114
Centrohelids, 94
Centromere, 17
Cephalic aorta, 198
Cephalic vein, 375
Cephalochordata, 86, 283, 286–289
Cephalopoda, 85, 195–199
Cephalothorax, 220–221, **222, 223, 224, 228**
Ceratium, 101, **101**
Cercariae, 154, 155–156
Cercozoa, 114
Cercus, 245, **245, 252, 253,** 258
Cerebellum, 335, **335,** 382, **383**
Cerebral ganglion
earthworm, **207,** 208
planarian, 152
Cerebral hemispheres, 334, 382
Cerebratulus, 37–38
Cerebropleural ganglia, 190–191
Cerebrospinal fluid, **379,** 382
Cerebrospinal nervous system, 334
Cerebrum, **334, 335,** 382
Ceriantipatharia, 144
Cerithidea californica, 157
Cervical groove, 228, **229**
Cervical nerves, **380**
Cervical vertebrae
birds, 344, **344**
mammals, 350, **352**
turtles, 338, **339**
Cervix, 371
Cestoda, 85, 158–161
Chaetognatha, 85
Chaetonotus, 180
Chaetopterus variopedatus, 203–204
Chagas' disease, 100
Character states, 79
Characters, 78
Cheek cells, 13
Chelae, 221
Chelicerae, 220, **222, 223**
Chelicerata, **72,** 220, 223, 225
Chelicerate arthropods, 220–224
Chelipeds, 228, **229,** 230, **231**
Chemotactic responses, 111
Chemotaxis, 150
Chewing lice, 260
Chiasmata, 27
Chilaria, 221, **222**
Chilopoda, 86, 225, 241
Chironex fleckeri, 138
Chitin, 220
Chitons, 85, 194–195
Chloragogue cells/tissues, 206, 209, **209**
Chlorophyll, 97

Chloroplasts, 97, **97,** 98, 99
Choanocytes, 121, 124, **124, 125**
Chondrichthyes, 86, 283, 296–301
Chordae tendineae, 374
Chordates, 281–289
Chorioallantoic placenta, 371
Christmas tree worms, 204
Chromatids, 17, **19,** 27, **28**
Chromatin, 14
Chromatin granules, 44
Chromatophores, 54, 195, 293
Chromoplasts, 70
Chromosomes
 crossing over, 25, 27, **28**
 daugter, 17
 division, 17–19
 maternal, 30
 number linked to species, 16
 paired, 25, **26**
 paternal, 30
Chrysemys picta, 337
Cicadas, 260
Cilia
 in bivalves, 185, **189,** 191
 in *Paramecium,* 109, 110
 planarian, 150
 in snail intestine, 194
 of trachea, **58**
 in *Vorticella,* 112
Ciliary muscle, 199
Ciliated epidermis, 275
Ciliated ridges, **295**
Ciliates
 basic features, 109–113
 effects of temperature on, 115–116
 genetic polymorphism in, 117–118
Ciliophora, 70, 84, 109–113, 114
Ciona, 283–286
Ciona intestinalis, 283, **284**
Circular muscles, 152, **153,** 209, **209**
Circulation and circulatory systems
 amphioxus, 288, **289**
 bivalves, **187,** 188–189, **190**
 bony fishes, **304,** 306
 crustaceans, 232, **233**
 earthworm, 206
 fetal pig, 372–378
 frogs, **328,** 328–330
 lampreys, 294
 shark, 299–301
 squids, **197,** 198
 tunicates, **284,** 285
 turtles, 341–342
Circumflex artery, 378
Circumflex iliac vein, 378
Cirri, 202, **203, 278**
Cirripedia, 235–236
Cisterna magna, 367
Cisternae, **15,** 16
Cladocera, **73,** 235, **235**
Cladogenesis, 78
Cladograms, **78,** 78–83
Clams, **72,** 184–191. *See also* Bivalvia
Clamworms, 202–204
Claspers, **296, 297,** 298, **299**
Classes, 77
Classification
 annelids, 216
 arthropods, 225, 258–262
 basic principles, 77–86
 chordates, 282–283
 cnidarians, 143–144
 echinoderms, 279
 mollusca, 199
 platyhelminthes, 161
 unicellular eukaryotes, 113–114
Clathrulina, 93, **93**
Clavicle, **320, 353**
Claws, **224,** 225, 245
Cleaning microscopes, 5, 6
Cleavage furrow, 18
Cleavage patterns, 36–40, **44,** 45
Cliona, 126
Clitellata, 85, 216
Clitellum, 205, 206, **208,** 216
Clitoris, 371
Cloaca
 Ascaris, **173,** 174
 birds, **346**
 frogs, 326, **326,** 327, **327**
 lampreys, 293, **294, 295**
 shark, 297, **297,** 298, **299**
 turtles, 340, **341**
Cloacal opening, 314, 326
Clonorchis, 153–154
Closed circulatory systems, 206
Clotting agent, 220
Clypeus, **246, 252**
Cnidarians
 anthozoans, 140–143
 classification, 143–144
 Hydrozoa, 132–137
 identifying, **71, 84**
 overview, 131–132
 scyphozoans, **138,** 138–140, **139**
Cnidocils, 132–133
Cnidocytes, 132, **133,** 136, **136**
Coarse-adjustment knob, 4, 5
Coccidians, 106
Coccygeoiliacus, **323**
Coccyx, **353**
Cockroaches, 107, 248, 262
Cocoons, 207, **208,** 248
Coelom
 amphioxus, 289
 bivalves, 189
 earthworm, 209
 frogs, 325
 lampreys, **295**
 squids, 198
Coelomic cavity
 bony fishes, 305
 lacking in flatworms, 150
 lampreys, 295
 molluscs, 189
 sea stars, 269, 270
 types, 172
Coelomic fluid, 269
Coelomic pouches, 40
Coelomic vesicles, **39,** 40
Coenobita clypeatus, 237
Coenosarc, 135, **135,** 136
Cog muscles, 274
Coleoptera, **71,** 74, 262
Collagen, 173
Collagenous fibers, 54
Collar, 194, 196
Collembola, 254, 261
Colon
 crickets, **253**
 fetal pig, 367, **380**
 shark, **297, 299**
 turtles, 340, **341**
Columella, 193, 319
Columnar cells, 54
Columnar epithelium, 54
Comb jellies, 84
Common bile duct, 368
Common carotid artery, 329, 375
Common iliac arteries, **328,** 330
Common iliac veins, 378
Communicating branch, 379
Community, 67
Compact bone, **51,** 351, **354**
Complete cleavage, 36
Complete metamorphosis, 254
Complete septa, 142
Compound eyes
 crickets, **252**
 crustaceans, **233,** 234
 grasshoppers, 244, **245**
 honeybee, **246**
 horseshoe crabs, 220, **222**
Compound light microscopes, 3–8, **4**
Concave mirrors, 4
Condenser (microscope), **4, 5**
Confirmation, 63
Conjecture, 63
Conjugation, 111
Connective tissues, 47, **50–52,** 54
Contour feathers, 343
Contractile vacuoles, **91,** 92, **97,** 112, **113**
Controls, 63
Conus arteriosis
 frogs, 328, **328, 329,** 330
 shark, **297**
Convergent evolution, 196
Copepoda, 73, 235, **235**
Copulation, earthworm, 206–207, **208**
Copulatory bursa, 177
Copulatory spicules, 176
Copulatory swimmerets, 229, **229,** 230
Coracobrachialis, **359**
Coracoid bone
 birds, **344,** 345, **345**
 frogs, 319, **320**
 turtles, 339, **339**
Coracoid process, 339, **352**
Corallites, 142, **143**
Corals, 84, 132, 142–143, **143**
Coregonus, 19
Cornea, 199
Corona, 179
Coronary arteries, 342, 373, **376**
Coronary sulcus, 373
Coronary veins, 373, **376**
Corpora quadrigemina, 382, **383**
Corpus callosum, 382
Costal cartilage, 350, **352, 353**
Costocervical arteries, **376**
Costocervical trunk, 375
Countershading, 297
Cowper glands, **369,** 370
Coxa
 grasshoppers, 245, **245**
 honeybee, **246**
 in spider leg, **224,** 225
Coxosternite, **243**
Crabs, **235**. *See also* Horseshoe crabs
Cranial cartilage, **292**
Cranial nerves, 378, **383,** 384
Cranial region, 350
Craniata, 283
Cranium. *See also* Skull
 birds, **344,** 345
 cat, **352**
 frogs, 318
Crayfish, **73,** 228–234, 236
Crickets, 251–254, **252–253,** 262
Crinoids, 84, 278–279
Cristae, 15, **15,** 16
Crocodilians, 86
Crop
 birds, 346, **346**
 crickets, **253**
 earthworm, 206, **207**
 snails, **193**
Crossed reflexes, tunicates, 283–284
Crossing over, 25, 27, **28**
Crown, feather stars, 279
Crustacea, 85, 227–237
Crustaceans
 brine shrimp, 236
 crayfish, **73,** 228–234, 236
 development pattern, 236
 experiments with, 239–240
 general features, 227–228
 identifying, 73, 225
 selected types, 234–236
Crystalline style, 190
Ctenidia, 195
Ctenoid scales, 302
Ctenophores, 84, 132
Cuboidal epithelium, 57
Cubozoa, 144
Cucumaria, 276
Cusps, mammalian heart, 374
Cutaneous arteries, **329,** 329–330
Cutaneous maximus, 356
Cutaneous muscle, 356
Cutaneous pectoralis, **321**
Cutaneous respiration, 326
Cuticle
 arthropod feature, 220
 Ascaris, 173, 174, **176**
 clamworm, 202, **203**
 earthworm, 209, **209**
 rotifers, 179
 vinegar eels, **176**
Cutting plates, 177
Cuttlefishes, 85
Cyanea capillata, 138
Cyclops, 235, **235**
Cyclostomata, 86, 283
Cystic ducts, 368
Cysts (in testes), 31
Cysts *(Trichinella),* 177, **178**
Cytokinesis, 16, 18
Cytopharynx, 109, 111, 112
Cytoplasm, 13, 14, 19–20
Cytoprocts, 111
Cytostomes, 109, 111, 112, **113**

D

Dactylozooids, 137
Damselflies, **74,** 259
Daphnia, 235, **235,** 237, 239–240
Darwin, Charles, 100
Daughter chromosomes, 17, **18**
Daughter colonies, 99
Decapoda, 73
Deep femoral artery, **377,** 378
Deep femoral vein, 378
Deep femoray vein, **377**
Deltoids
 fetal pig, **357, 358,** 360
 frogs, **321, 323**
Demospongiae, **121,** 123, 125, 126, **126**
Dendrites, **55**
Dendrocoelopsis vaginata, 150
Dense connective tissue, 47, **50**
Dentary bones, **303, 318,** 319
Denticles, 297, **297**
Dentine, 297, **297**
Deoxyribonucleic acid (DNA), 15, 17, 117
Deposit feeders, 272
Depressor mandibularis, **323**
Depressor muscles, 356
Derived state, 79
Dermal branchiae, 265–266, 267, **269,** 270
Dermal endoskeleton, 265
Dermal ossicles, **268, 269,** 270
Dermal ostia, 123
Dermal scales, 301
Dermaptera, 261
Dermis
 amphioxus, 289
 frog, 54, **56**
 sea stars, 268, **269,** 270
 shark, **297**
Descending aorta, 378
Deuterostomia, 36, 38
Deutomerite, 107
Dextral shells, 192–193, **193**
Diadema antillarum, 274
Diameter of microscope field of view, 7, **8**
Diaphragm, **365, 366,** 367, **380**
Diaphragmatic lobe, 384, **385**
Diaphragms (microscope), 4, 5
Diaphysis, 351, **354**
Diapsids, 337
Diarthrosis, 354
Diastolic period, 306
Dichotomous keys, 68–69, **70–74**
Diencephalon, 334, 335, **335**
Difflugia, 93, **93**
Diffuse placenta, 371
Digastric muscle, 357, **358, 359**
Digenetic flukes, 85, 153–158
Digestive glands, 232, **233,** 234
Digestive systems
 Ascaris, **173,** 174
 birds, 345–346
 bivalves, **187,** 189–190
 bony fishes, 305
 crickets, **253**

Digestive systems (*Continued*)
crustaceans, 232, **233**, 234
earthworm, 206, **207**
fetal pig, 363–368
frogs, 325–326, **326**
planarian, 151–152
sea stars, 269
shark, 298
snails, **193**
squids, 198
tissues, 54–58
tunicates, **284**, 285
turtles, 340
Digits, **344**, 345, 351
Dimorphism, 132
Dinoflagellates, 101, **101**, 114
Dioecious species, 134
Diopatra, 204
Diphyllobothrium latum, 158
Diploblastic phyla, 131
Diplodiscus, **157**
Diploid chromosomes, 25
Diplomonada, 114
Diplopoda, 86, 225, 241–243
Diplostraca, 235, **235**
Diptera, **71**, 258
Dipylidium, 158–161
Direct reflexes, tunicates, 283–284
Discoidal cleavage, 36
Diseases involving cell cycle, 16
Displaying insects, 255–257, **256**
Dissecting microscopes, **8**, 8–9
Distal epiphysis, **354**
Dobson flies, 73
Dobsonflies, 259
Dog heartworm, 178
Dog hookworm, **177**
Dogfish sharks, **296**, 296–301, **297, 299**
Domains, 77
Dorsal abdominal artery, 232, **233**
Dorsal aortas
amphioxus, 288, 289, **289**
bony fishes, **304**, 306
frogs, **328**, 330
lampreys, **292**, 293, 295, **295**
shark, **297, 299**, 300
turtles, 342
Dorsal blood vessel, 206, **207**, 209, **209**
Dorsal branch, 379
Dorsal ciliated ridge, **295**
Dorsal fins
amphioxus, 286, **287**, 288
bony fishes, 302, **302, 303**
lampreys, 292, 293, **294, 295**
shark, 296, **296**
Dorsal hollow nerve cord, 286
Dorsal median lines, 173
Dorsal pores, 206
Dorsal ramus, **379**
Dorsal root ganglion, 379, **379**
Dorsal surface of brain, 382
Dorsal tubular nerve cord, 281, 288
Dorsalis scapulae, **323**
Dorsoventral muscle fibers, 152
Double-circuit circulatory systems, 306, 328
Down feathers, 343–344
Dragonflies, 74, 259
Drones, 27–248
Drosophila melanogaster, **11**, 254
Ducts of Cuvier, 288, 294, 299
Ductus arteriosus, 376, **376**
Ductus deferens, **297, 369**
Ductus venosus, 376–377, **377**, 378
Dugesia dorotocephala, 150
Duodenum
birds, **346**, 347
bony fishes, **304**, 305
fetal pig, **365**, 367
shark, **297**, 298
turtles, 340, **341**
Dura mater, 334, **379**, 382
Dyads, 27, 29
Dysentery, 92, **93**

E

Ear vesicles, **294**, 295, **295**
Ears, 301, 314, 350, 364
Earthworm, 71, 85, 204–210. *See also* Annelids
Earwigs, 261
Ecdysis, 220, 251
Ecdysozoa, 172
Echinoderms
basic features, 265–266
brittle stars, 270–272
classification, 279
crinoids, 278–279
embryonic development, 38
sea cucumbers, 275–278
sea stars, 266–270
sea urchins, 273–275
Echinoidea, 84, 273–275, 279
Ecosystems, 67
Ectoderm, **37**, 38, **39**
Ectoplasm, 91, **91**, 97, 109
Ectotherms, 115, 330
Eels, 174–176
Efferent branchial arteries
amphioxus, 288, **289**
bony fishes, **304**, 306
lampreys, 294
shark, 300
Eggs
amphibians, 313
Ascaris, 19–20, 174
crickets, **253**
earthworm, 207, **208**
frogs, 44
fruit fly, 254
hydras, 134
insect, 248
oogenesis, 26, 28–30, **30**
schistosomes, **155**, 156
sea star, 13–14, **14**
sea stars, 269–270
shells, 19, 28
tapeworms, 158, **159**
trematodes, **154**
tunicates, 285
Eight-cell stage, 44, **45**
Eimeria, 106
Ejaculatory duct, *Ascaris*, **173**, 174
Elasmobranchii, 296
Elastic fibers, 54, 57
Electron microscopes, 9–11
Electrophoresis, 310–311
Embryonic development, **30**, 36–40, 44–46
Emergent plants, 67
Enamel, 296, **297**
Encysted phase of *E. histolytica*, 92, **93**
End sacs, 234
Endocardium, 374
Endochondral bones, 354
Endoderm, **37**, 38, **39**, 40, **45**, 46
Endoplasm, 91, **91**, 97, 109
Endoplasmic reticulum, **14**, 15, **15**
Endopods, 230, **231**
Endoskeleton
echinoderms, 265
sea cucumbers, 278
sea stars, 267–268
sea urchins, 275
turtles, 338
Endosteum, **354**
Endostyle
amphioxus, 288, 289
ascidian larvae, 285, **286**
chordates, 281
lampreys, 294, **294, 295**
tunicates, 285
Endothelium, **48**, 57
Endotherms, 115
Entamoeba gingivalis, 92
Entamoeba histolytica, 92, **93**
Entamoeba species, 92, **93**
Enterobius vermicularis, 178, **178**
Eosinophils, **52**
Epaxial muscles, 303, **304**
Ephemeroptera, 74, 258
Ephyrae, 138, **139**
Epiboly, 40, 44–46
Epidermal nerve plexus, 270
Epidermis
amphioxus, 289
Ascaris, 174
clamworm, **203**
cnidarians, **133**, 134, 136
crustaceans, 232
earthworm, 209, **209**
frogs, 54, **56**
lampreys, **295**
Metridium, 142
planarian, 150, 152, **153**
sea stars, 266, 268, **269**, 270
sea urchins, 275
shark, **297**
Epididymis, **299, 369**, 370
Epiglottis, 364, **364**, 384
Epigynum, **224**, 225
Epipharynx, **246**
Epiphyseal line, **354**
Epiphyses, 351, **354**
Epiphysis, 335
Episternum, 319
Epithelial tissues
cell structure, 13
development, 38
in frog skin, 54, **56**
functions, 47
lampreys, 292
types, **48–49**
Epitheliomuscular cells, **133**, 134
Equatorial plane, 38
Errantia, 85, 216
Esophagus
birds, 346, **346**
bivalves, **187**, 190
crickets, **253**
crustaceans, 234
earthworm, 206, **207**
fetal pig, 364, **364**, 367, **368**
frogs, 325
lampreys, **292**, 293, **294**
larval formation, 40
shark, 298, **299**
squids, **197**, 198
tissues, **49**
trematodes, 154
tunicates, 285
turtles, 340, **340, 341**
Ethidium bromide, 117, 118
Ethyl alcohol, 255
Eubranchipus, 234, **235**, 236
Eudorina, **101**
Euglena, **70, 97**, 97–98
Euglena gracilis, 97–98
Euglena viridis, 97–98
Euglenoid movements, 97, **98**
Euglenozoa, 100–101, 114
general features, 97
identifying, **70**, 84, 114
parasitic, 100
Eukarya, 77
Eumetazoans, 131
Euplectella, 126
Euplotes, **112**
Eurycea longicauda, **315**
Eustachian tubes
fetal pig, 364, **364**
frogs, 325, **325**
turtles, 340, **340**
Evolutionary trees, 78–83
Excretory canals, 158, 174
Excretory pore, 174
Excretory systems
Ascaris, **173**, 174, **175**
bivalves, 189
bony fishes, 306
chitons, 195
crustaceans, **233**, 234
earthworm, 206, **207**, 207–208
fetal pig, 368
frogs, 326–327
planarian, 149, 152
shark, 298
squids, 198
tapeworms, 159, **160**
trematodes, 154, **154**
tunicates, 285
Excurrent aperture, 185, **186, 187**, 285, **286**
Excurrent siphons, 283, **284**
Exoccipitals, **318**, 319, **320**
Exopods, 230, **231**
Exoskeleton
arthropod feature, 220
bivalves, 184
crustaceans, 228
grasshoppers, 244
horseshoe crabs, 220
spider, 223
Experimental procedures, 65–66
Exponential growth, 64
Extant taxa, 78
Extension, 355, **355**
Extensor carpi radialis, **321**
Extensor carpi ulnaris, **323**
Extensor cruris, **321, 322**
Extensor digitorum communis, **323**
Extensor muscles, 232, 355
External auditory meatus, **351, 352**
External carotid artery, **328**, 329, **329**, 375, **376**
External gill slits, 292, 294, **296**, 300
External gills, 46
External iliac artery, 378
External iliac vein, 378
External jugular vein
fetal pig, 375
frogs, 328, **328, 329**
External maxillary vein, 363, 375
External nares
frogs, 325, **325, 335**
turtles, 338, **340**
External oblique
fetal pig, **357, 359**, 360, **361, 362**
frogs, 321, **321, 322, 323**
Extracellular digestion, **133**, 133–134, 151
Exumbrella, 136, **137, 139**
Eyelids, 314
Eyepieces (microscope), 3, **4, 5**
Eyes
birds, **346**
bony fishes, 301, **302**
clamworm, 202, **203**
crickets, **252**
crustaceans, 228, **229, 233**
embryonic development, 46
fetal pig, 350
frogs, 314
grasshoppers, 244, **245**
honeybee, **246**
horseshoe crabs, 220, **222**
lampreys, 292, 294, **294**
myriapods, 241, **243**
shark, **296**, 297
snails, 192, **192**
spider, 224, **224**
squids, 196, 199
turtles, 338, **340**
Eyespots
ascidian larvae, 285, **286**
flagellates, 97
planarian, **151**, 152
sea stars, **267, 268**, 270

F

Facial nerve, **335**, 363, **383**, 384
Facial region, 350
Fairy shrimp, **73**, 234, **235**
Fallopian tubes, 371
False cornea, 199
Families, 77
Fan worms, 85
Fangs, 224, **224**, 241, **243**

Fanworms, 204, **204**
Fasciae, 355
Fasciculi, 355
Fat bodies, 251, 327, **327**
Fat storage cells, 50
Feather stars, 84, **278**–279
Feathers, 343–344
Feeding behavior
  Amoeba, 92
  amphioxus, **287**
  apicomplexans, 106
  bivalves, 188
  brittle star, 272
  ciliates, 110–111
  cnidarians, 133–134
  crinoids, 279
  euglenoids, 97
  frogs, 314
  planarian, 150
  predator functional response, 147
  schyphozoans, 140
  sea cucumbers, 276
  sea stars, 266, 269
Feeding polyps, 136, 137
Female pores, 206, **215,** 216
Female pronucleus, 29, 30
Femoral artery, **328,** 330, 378
Femoral nerve, 335, **380,** 381
Femoral veins, **328,** 329, 378
Femur
  birds, **344,** 345
  cat, **352**
  crickets, **252**
  fetal pig, **352**
  frogs, **320**
  grasshoppers, 245, **245**
  honeybee, **246**
  human, 353
  mammals, 351
  in spider leg, **224,** 225
  turtles, **339**
Fertilization, 28–30, **30**
Fertilization membrane, 38, **39**
Fetal circulation, 372
Fetal pigs
  circulatory system, 372–378
  digestive systems, 363–368
  external structures, 350
  muscular systems, 354–362
  nervous system, 379–384
  overview, 349–350
  respiratory system, 384–385
  skeleton, 350–351, **352**
  urogenital system, 368–372
Fibers, muscle, **53**
Fibrous connective tissue, 57
Fibula
  birds, **344,** 345
  mammals, 351, **352, 353**
  turtles, **339**
Fields of view (microscope), 7
Fieldwork, 67–69
Filament, 230, **231**
Filoplume feathers, 344
Filopodium, 91
Filter feeders, 188
Filum terminale, **334**
Fin rays, 287, 289, 301, **303**
Fin spine, **296**
Fine-adjustment knob, 4, 5
Fins
  amphioxus, 286, 287, **287**
  bony fishes, **302, 303**
  lampreys, 292, 293, **294**
  shark, 296
Fireflies, 248
First maxillae, 241
First polar body, 28–29, **30**
Fishes
  agonistic behavio, 307–308
  bony fishes, 301–306
  lampreys, 291–296
  multiple hemoglobin systems, 309–311
  sharks, 296–301
Fishflies, 73, 259
Fishing polyps, 137
Fission, 92, 98, 106, 111
Fixing, **17**
Flagella, 77, **97,** 97–101, 98, **125**
Flagellates, 97–101
Flame cells, 152
Flatworms
  flukes, 153–158
  general features, 149–150
  identifying, **71,** 85, 161
  regeneration experiment, 167
  tapeworms, 158–161
  turbellarians, 150–152, **151, 153**
Fleas, 260
Flexion, 355, **355**
Flexor carpi radialis, **321**
Flexor carpi ulnaris, **321**
Flexor muscles, 232, 355
Flies, displaying, **256**
Flight feathers, 343
Flight muscles, **253**
Floaters, 6
Floating life and emergent plant habitat, 67
Floscularia, 179
Flour beetle, 65–66
Flukes, 153–158
Fly larva, **71**
Focusing microscopes, 5, 6
Follicular cell layers, 44
Food vacuoles
  in *Amoeba,* **91,** 92, **93**
  in ciliates, 111, 112
  in *Vorticella,* 112
Foot
  bivalves, 185, **186**
  common to molluscs, 184
  rotifers, 179
  snails, 193, **193**
Foot gills, 230
Foot protractor muscles, 185, **186**
Foot retractor muscles, 185, **186, 187**
Foramen magnum, 319, 350, **351**
Foramen ovale, 374, 376, **376**
Foraminiferans, 93, **93,** 114
Forebrain, 294, **294,** 334
Foregut, **253**
Forelimbs
  fetal pig, 357–360, **358, 359**
  frogs, 314, 319
  honeybee, **246,** 246–247
  mammals, 351
Forewings, 244, **245**
Fork-tailed worm, **72**
Fossa ovalis, **377**
Four-cell stage, 44, **45**
Fourth ventricle, **383**
Freeze fracturing, 10
Freshwater habitats, 67–69
Freshwater invertebrates, **70–74**
Frogs
  behavior and adaptations, 314
  circulatory system, **328,** 328–330
  classification, 78, 86
  digestive system, 325–326, **326**
  embryonic development, 44–46
  nervous system, **334,** 334–335, **335**
  overview, 313
  skeletal muscles, 320–322, **321, 323**
  skeletal system, 318–319, **320**
  skin cross section, 54, **56**
  trematode parasites, 156–157
Frons, **245**
Frontal bone, **351, 352, 353**
Frontoparietals, **318,** 319, **320**
Fruit flies, **11,** 254
Fumigants, 255
Functional response, 147
Fundus, 367
Fungiform papillae, **364**
Funnel, 195, **197**
Funnel retractor muscles, 198
Furcula bone, **344,** 345

G

G1 phase, 17
G2 phase, 17
Gallbladder
  bony fishes, 305
  fetal pig, **365,** 367, 368
  frogs, 326, **326, 327**
  lampreys, 294, **294**
  shark, **297,** 298
  turtles, 340, **341**
Gamete formation, 25–32. *See also* Eggs; Ovaries; Testes
Gametogenesis, 25–32
Gammarus, 236
Ganglia, 190–191
Garden spider, 223–225, **224**
Gas glands, **304,** 305
Gastric filaments, **139,** 140
Gastric ligaments, 269
Gastric mill, 234
Gastric muscles, 232
Gastric pouches, **139,** 140
Gastric teeth, **233**
Gastrocnemius
  fetal pig, **357, 362**
  frogs, 320, **321,** 321–322, **322, 323**
Gastrodermis, **133,** 134, 136
Gastrohepatic ligament, 368
Gastroliths, 232
Gastroncnemius, **362**
Gastropoda, **72,** 85, 191–194, 199
Gastrotricha, **72**
Gastrotrichs, 180
Gastrovascular cavities
  Astrangia, 142
  common to radiates, 132
  Gonionemus, 136, **137**
  hydras, 132, **133**
  Metridium, **141,** 142
  Obelia, 136
  planarian, 151–152
Gastrovascular systems, 140
Gastrozooids, 137
Gastrula stage, 38–40, **39,** 44–46
Gemmules, 123, **123**
Gena, **245**
Genera, 77
Genetic polymorphism, 117–118
Genetic recombination, 27
Geniohyoid, **321**
Genital aperture, 193
Genital artery, **377,** 378
Genital bulb, 254
Genital chamber, **159, 160**
Genital ducts, 327
Genital opening, 229, **229**
Genital operculum, 221, **222**
Genital papillae, 371
Genital plates, **274,** 275
Genital pores, 152
  Ascaris, **173**
  bony fishes, 306
  horseshoe crabs, 221, **222**
  sea cucumbers, 278
  sea stars, 269
  sea urchins, 275
  tapeworms, 158, **159**
  trematodes, **154**
Genital vein, **377,** 378
Geotaxis, 111
Germ layers
  of flatworms, 149
  frog embryo example, **45,** 46
  functions, 38–40
  of radiates, 132
Giant fibers, 209
Giardia intestinalis, 101
Giardia lamblia, **100,** 101
Gill arches, 300
Gill bailer, 230, **231**
Gill bars
  amphioxus, 288, 289
  lampreys, 294, **295**
Gill chamber, 228, 300
Gill filaments, 188, **189,** 230, 303
Gill lamellae, **189,** 191, 295
Gill opercula, **222**
Gill pouches, **294,** 295, 300, 364
Gill rakers, 300, 303–304
Gill rays, 300
Gill slits
  amphioxus, 288, **288**
  bony fishes, 304
  frogs, 46
  lampreys, 292, 293, 295, **295**
  shark, **296,** 297, 300
Gills
  arthropod feature, 220
  bivalves, **186,** 187, **187,** 188, **189**
  bony fishes, **304**
  chitons, 195
  common to molluscs, 184
  crustaceans, 228, 230, **231, 233**
  embryonic development in frogs, **45,** 46
  horseshoe crabs, **222**
  sea urchins, 275
  squids, **197,** 198
Girdle, 194
Gizzard, 206, **207, 346,** 347
Gland cells, **133,** 134
Gland tissues, **48,** 54, **56,** 57
Glass sponges, 125, **127**
Glenoid fossae, 351
Glial cells, 51
Globigerina, 93, **93**
Glochidia, 190, **191**
Glomeruli, **10**
Glossa, **246**
Glossopharyngeal nerve, **335, 383,** 384
Glottis
  birds, 346, **346**
  fetal pig, 364, 384
  frogs, 325, **325,** 326
  turtles, 339, **340,** 341
Glutathione, 133
Gluteus medius, **357,** 360, **361, 362**
Glycogen, **9, 14**
Gnathobases, 221, **222**
Gnathostomata, 86, 283
Goblet cells, 54, **56,** 57, **58**
Goldfish, 309–311
Golgi complex, 9
Gonads
  amphioxus, 287, **287,** 289
  bivalves, **187,** 190
  bony fishes, 305, **306**
  crustaceans, 232
  fetal pig, **380**
  Gonionemus, 136
  Metridium, **141,** 142
  schyphozoans, **139,** 140
  sea stars, **268,** 269, **269**
  tunicates, **284**
Gonangia, 135, **135,** 136
Gonionemus, 132, 136–137
Gonium, **101**
Gonoducts, 278
Gonopores, **135,** 136, 241, 243
Gonyaulax, 101
Gooseneck barnacle, **235**
Gordius, 180
Gorgodera, 156, **157**
Gorgoderina, 156, **157**
Gracilis major, 321, **321, 322**
Gracilis minor, 321, **321, 322, 323**
Gracilis muscles, 320, 360, **361, 362**
Gradual metamorphosis, 251, 254
Granulocytes, **52**
Grasshoppers, 107
  classification, 262
  displaying, **256**
  general features, 243–245, **244, 245,** 260, 262
  spermatogenesis in, 31–32
Gravid proglottids, 158, 159, **160**
Gray crescent, 44, **45**
Greater curvature of stomach, 367
Greater trochanter, **352**

Green glands, 230, 234
Gregarina, 106–108, **108**
Gregarines, 106–108, **108**
Ground substance, **50**
Gubernaculum, **369,** 370
Gullet, 109, 111, 112, 136, 140
Gymnodinium, **101**
Gynecophoric canal, 156
Gyri, 382, **383**

H

Habitats, freshwater, 67–69
Haemal canals, **268**
Haematoloechus, 156–157, **157**
Hagfishes, 86, 291
Haploid cells, 25, 27
Hard palate, 364, **364**
Head
  bony fishes, 301
  chitons, 194
  clamworm, 202, **203**
  crickets, 251
  fetal pig, 363
  frogs, 314
  grasshoppers, 244, **244, 245**
  honeybee, 246, **246**
  planarian, 150
  shark, 296
  snails, 192, **193**
  squids, 196
  turtles, 338
Head retractor muscles, 198
Heads of muscles, 355
Heart
  birds, **346**
  bivalves, **187,** 188–189, 191
  bony fishes, **304,** 306
  crustaceans, 232, **233,** 236
  fetal pig, **365, 368,** 373–374
  frogs, 325, **329,** 330
  lampreys, 292, 293, 295
  snails, **193**
  squids, **197**
  tunicates, **284,** 285
  turtles, 340, **341,** 342
Heart urchins, 279
Heartbeat, 236–237, 306
Hectocotyly, 196–197
Helicodiscus, **193**
Heliozoan, **70**
Helix, 192
Hemal canals, **269**
Hemiazygous vein, 375
Hemichordata, 86
Hemimetabolous development, 254
Hemiptera, 74, 260, 261, 262
Hemocoel, 220
Hemocyanin, **52,** 221
Hemoglobin, 309–311
Hemolymph, 220
Hepatic arteries, 232, 377
Hepatic cecum, 288
Hepatic ducts, 368
Hepatic portal system, 329, 374
Hepatic portal vein
  amphioxus, 288, **289**
  bony fishes, **304**
  fetal pig, **376, 377,** 378
  frogs, **328,** 329
  lampreys, 294
  shark, 299–300
Hepatic veins
  amphioxus, 288, **289**
  fetal pig, **377,** 378
  frogs, **328,** 329
  lampreys, 294
  shark, 299–300
Hepatopancreas, 232, 234
Hermaphroditic duct, **193**
Heterocercal fins, 296
Heteroptera, 261
Heterotrophic flagellates, 97, 100
Hexacorallia, 144
Hexactinellida, 126
Hexapoda, 85, 225
High-power objectives, 3, 5, 7
Hilus, 369
Hindbrain, 294, **294, 295,** 335
Hindgut, **253**
Hindlimbs
  fetal pig, 360–362, **361, 362**
  frogs, 314, 319
  honeybee, **246,** 247
Hindwings, 244, 246
Hinge ligament, 185, **187**
Hirudinea, **71,** 85
Hirudinidae, **215,** 215–216
Hirudo, **215,** 215–218
Histology, 47
Holobranchs, 300
Holometabolous development, 254
Holophytic nutrition, 97–98
Holothuria arenicola, **276**
Holothuria fuscocinerea, **276**
Holothuroidea, 84, 275–278, **276,** 279
Holozoic nutrition, 98
Homarus americanus, 228
Homeotherms, 115
Homocercal tail, 301
Homologous chromosomes, 25, **26,** 27, **28**
Homologs, 25, **26,** 27, **28**
Homoscleromorpha, 126
Honeybees, 245–248, **246**
Hooks, **157,** 158, **161**
Hookworms, **177**
Hoppers, 260, 262
Horizontal submarine gel electrophoresis, **119**
Horns of uterus, 370–371
Horsehair worms, 85, 180
Horseshoe crabs, **52,** 86, 220–222
House cricket, 251–254
Human blood flukes, 155–156
Humerus
  birds, **344,** 345, **345**
  cat, **352**
  fetal pig, **352**
  frogs, **320**
  human, **353**
  mammals, 351
  turtles, **339**
Hyaline cap, 91, **91**
Hyaline cartilage, **50**
Hyaline cortex, 91
Hyalonema, 126
Hydra, identifying, **71,** 143. *See also* Hydrozoa
Hydra littoralis, 147
Hydracarina, **72**
Hydranths, 135, **135**
Hydroids, 84, 132–137
Hydromedusae, 84, 136
Hydrostatic skeletons, 132
Hydrozoa
  Gonionemus, 132, 136–137
  Hydras, 132–135
  identifying, **71,** 84, 143
  Obelia, 132, 135–136
  predator functional response, 147
Hymenoptera, 258, 260
Hyoid apparatus, 319, 338, **339,** 340
Hypaxial muscles, 303, **304**
Hyperbranchial groove, 288, 289
Hypoglossal nerve, **383,** 384
Hypoglossus, **321**
Hypopharynx, **243,** 244, **245,** 251, **252**
Hypophysis, 335, 382
Hypostomes, 132, **134, 135,** 136
Hypothalamus, **383**
Hypotheses, 63–65, 78, 79

I

Ichneumons, 258
Ichthyomyzon, 291
Ileocecal valve, 367
Ileum, 367
Iliac arteries, 300
Iliac crest, **352**
Iliacus, **361,** 362, **362**
Iliolumbar arteries, **377**
Iliolumbaris, **323**
Ilium
  birds, 344
  cat, **352**
  fetal pig, **352**
  frogs, 319, **320, 323**
  mammals, 351
  shark, **297**
  turtles, 339, **339**
Illuminators (microscope), 4
Image-forming optics, 3
Immersion oil, 5
Incisors, **351, 352**
Inclusions in cells, 13
Incomplete cleavage, 36, 38
Incomplete septa, 142
Incurrent aperture, 185, **186, 187,** 285, **286**
Incurrent canals, 122, **122,** 123
Incurrent siphons, 283, **284**
Inferior jugular veins, 299
Infraspinatus, **358, 359,** 360
Infundibulum
  fetal pig, 371, **371**
  frogs, **335**
  sheep, 382
Inguinal canal, 369, **369,** 370
Ink sacs, **197,** 198
Innominate artery, 342
Innominate bones, 351
Innominate vein, 328–329, **329,** 375
Insects
  collecting, 255–257, **256**
  displaying, 255–257, **256**
  fruit flies, 254–255
  grasshoppers, 243–245, **244, 245**
  honeybees, 245–248
  house cricket, 251–254
  identifying, **71, 72,** 258–262
  killing, 255
  relaxing for mounting, 255
Insertions of muscles, 355
Instars, 254
  insect, 254
Intercalated discs, **54**
Intercostal arteries, **376**
Interlamellar junctions, 188, **189**
Intermediate lobe, 384, **385**
Internal carotid artery
  fetal pig, 375, **376**
  frogs, **328,** 329, **329**
Internal gill slits, **292,** 293, 294, 300
Internal gills, 46
Internal iliac artery, 378
Internal iliac vein, 378
Internal jugular vein
  fetal pig, 375
  frogs, **328,** 329, **329**
Internal mammary arteries, **376**
Internal maxillary vein, 363, 375
Internal nares, 325, **325,** 339, **351**
Internal oblique muscles, **357,** 360, **361**
Internal thoracic artery, 368, 375, **376**
Internal thoracic veins, 368, 375
Interphase, 17, **18**
Interstitial cells, **133,** 134
Intervertebral foramina, **319,** 379
Intestinal arteries, 294
Intestines
  amphioxus, 288
  Ascaris, **173,** 174, **175**
  ascidian larvae, 285, **286**
  birds, **346,** 347
  bivalves, **187,** 190
  crustaceans, 232, **233**
  earthworm, 206, **207, 209**
  embryonic development, 46
  fetal pig, **365,** 367, **380**
  frogs, 325, 326, **326, 327, 328**
  lampreys, 293, **294,** 295, **295**
  larval formation, 40
  parasitic organisms, 92, 101, 172–174, 177–178, 179
  planarian, **151, 153**
  sea stars, **268,** 269
  shark, 298
  squids, 198
  trematodes, **154, 157**
  tunicates, **284,** 285
  turtles, 340, **341**
  vinegar eels, 175, **176**
Intracellular digestion, **133,** 134
Invagination, 38
Involuntary muscle, **53, 54**
Iris diaphragm, 4, 199
Irregular echinoids, 273
Ischial tuberosity, **352**
Ischium
  cat, **352**
  fetal pig, **352**
  frogs, 319, **320**
  mammals, 351
  turtles, 339, **339**
Isolecithal eggs, 36, 38
Isopods, **73**
Isoptera, 259, 261

J

Jaws. *See also* Maxillae
  birds, **346**
  brittle stars, 271, **272**
  clamworm, 202, **203**
  squids, 197
  turtles, **339, 340**
Jejunum, 367
Jellies, 84, 138–140
Jericho, 156
Joint gills, 230
Jointed appendages, 219
Joints, types, 354
Jugal bone, **351**
Jugular ganglion, **335**
Julus, 241, **242**

K

Katharina tunicata, 194
Katydids, 262
Keratin, **49**
Kidneys
  bivalves, **187,** 189
  bony fishes, **304,** 305, **306**
  fetal pig, 367, 369, **369, 371, 377, 380**
  frogs, **326,** 326–327, **327, 328**
  lampreys, **294**
  microscopic views, **10**
  shark, **297,** 298, 299, **299**
  snails, **193**
  squids, **197,** 198
  tissues, **48**
Killing insects, 255
Kinetochore, 17
Kinetoplasts, **100**
Kinetosomes, **100**

L

Labellum, **246, 252**
Labial palps
  bivalves, **186,** 188, 189–190
  crickets, **252**
  grasshoppers, **245**
  honeybee, **246**
Labium
  grasshoppers, 244, **245**
  honeybee, 246
  myriapods, 243
Labrum
  crickets, **252**
  grasshoppers, 244, **244, 245**
  honeybee, **246**
  myriapods, 241, **243**

Lacewings, 259
Lacunae, **50, 51**
Ladder nervous systems, 152
Lamellae, **51,** 221
Lamp shells, 85
Lampetra, 291
Lampreys, 86, 283, 291–296, **292**
Lancelets, 286–289
Land snails, 191–194
Languets, 285
Lappets, 139
Large intestine
fetal pig, **365,** 367
frogs, 326, **326, 327**
Larvae
ammocoetes, 291, 293–296, **294**
ascidian, 285, **286**
cleavage patterns, 36, 38–40
crustaceans, 234
frogs, 325
lampreys, 291, 293–296, **294**
molluscs, 190, **191**
radiates, **135,** 136, 138
sea cucumber, **266**
sponges, 123
tapeworms, 158
Tribolium confusum, **65**
trilobite larvae, 222, **223**
Laryngeal pharynx, 364
Larynx
birds, 346
fetal pig, **359,** 364, **364, 365,** 384, **385**
frogs, 325
turtles, 341, **341**
Lateral canals, **268, 269,** 270, 278
Lateral femoral circumflex artery, 378
Lateral fins, 196
Lateral groove, 293
Lateral line, 174, 297
Lateral line systems
Ascaris, 173
bony fishes, 301, **302**
lampreys, **292,** 292–293
sharks, **296**
Lateral mantle arteries, 198
Lateral nerves, 152
Lateral neurals, 209
Lateroneural vessels, 206
Latissimus dorsi
fetal pig, **357, 358, 359,** 360
frogs, **323**
Laurer's canal, **154**
Leeches, **71,** 85, 205, 215–218
Left adrenolumbar artery, **377**
Left adrenolumbar vein, **377**
Left atrium
fetal pig, 373, 374
frogs, 328, **329,** 330
sheep, **374**
turtles, **341**
Left carotid artery, 342
Left common carotid artery, 375, **376**
Left lung, turtle, 341
Left pulmonary artery, 375
Left pulmonary vein, **374**
Left subclavian arteries, 342, 375, **376**
Left systemic arch, **341,** 342
Left ventricle, 373, 374, **374**
Legs. *See also* Forelimbs; Hindlimbs; Locomotion; Walking legs
arthropod, **246,** 246–247, **252**
horseshoe crab, **222**
Lens, in squids, 199
Lens paper, 5, 6
Lentic habitats, 67
Lepas, **235,** 236
Lepidoptera, 258
Lesser curvature of stomach, 367
Lesser trochanter, **352**
Leuconoid sponges, 123, **124, 125**
Leucosolenia, 123, **124,** 126
Leukocytes, **52**
Levator muscles, 356
Lice, 260
Lienogastric artery, 300
Light responses, 111, 150, 239–240
Lighting for microscopic studies, 5–6
Limax, 192
Limpets, **72**
Limulus, 220–222
Linea alba, 321, **321,** 360, **362**
Lineage splitting, 78
Lines of growth, 185, 192
Lingual cartilage, **292**
"Lion's mane jelly," 138
Lipid storage cells, **50**
Lips, 172, 274, **274**
Lithobius, 241
Liver
birds, **346,** 347
bony fishes, **304,** 305
crustaceans, 232
fetal pig, **365,** 367, 368
frogs, 325, 326, **326, 327, 328**
lampreys, **292,** 294, **294**
parasitic infections, 106, **107**
shark, **297,** 298, **299**
squids, **197,** 198
turtles, 340, **341**
Liver cells, **9, 14,** 14–16
Liver fluke, 153–158
Liver glands, 234
"Living fossils" (horseshoe crabs), 220–222
Lizards, 86, **338**
Lobes of liver, **297**
Lobes of lungs, 384, **385**
Lobsters, 228–234, 236
Locomotion
Amoeba, 91
annelids, 202
anurans, 313
bony fishes, 303
brittle stars, 271–272
ciliates, 109, 110, 116
clamworm, 202
earthworm, 205
euglena, 97
feather stars, 279
flagellates, 97, **98,** 99
frogs, 314
myriapods, 243
nematodes, 175
planarians, 150
sea stars, 267, 270
sea urchins, 275
squids, 195, 197
turtles, 337
Logistic growth, 64
Loligo, 195–199
Long bones, 351, 354, **354**
Long monaxons, 125
Longissimus dorsi, **323, 359,** 360, **361**
Longitudinal fission, 98
Longitudinal fissure, 382, **383**
Longitudinal lines, 173, 174
Longitudinal muscles
Ascaris, 174, **175**
earthworm, 209, **209**
planarian, 152, **153**
Longitudinal nerve, **160**
Longus muscles, 320
Loose connective tissue, 47, **50**
Lophotrochozoa, 172
Lorica, 179
Lotic habitats, 67
Lower mandible, **344,** 345
Low-power objectives, 3, 5, 6, 7
Lubber grasshopper, 31–32, 243–245, **244, 245**
Lumbar arteries, 330
Lumbar ganglion, 382
Lumbar nerves, **380,** 381
Lumbar vertebrae, 344, 350, **352**
Lumbosacral plexus, **380,** 381
Lumbricidae, 216
Lumbricus, 204–210
Lumen (small intestine), 54
Lumen (trachea), **58**
Luminescence, 248
Lung flukes, 156, **157**
Lung slits, **224**
Lungs
birds, 347
fetal pig, **365, 368,** 384
frogs, 325, 326, **326,** 328–330, **329**
snails, 194
tissues, **48**
turtles, 341
Lymnaea stagnalis, 194
Lymphocytes, **52**
Lysosomes, 92
Lytechinus, 274, 279

M

Macracanthorhynchus hirudinaceus, 179–180
Macrogametes, 99
Macrogametocytes, 106, **107, 108**
Macromeres, 44
Macronucleus, 112, 113, **113,** 117
Macrophages, **10,** 10–11
Macropodus opercularis, 307–308
Madreporite plate
brittle stars, 271, **272**
sea stars, 267, **268,** 270
sea urchins, **274,** 275
Maggots, 254
Magnification in microscopes, 3, 6
Magnus muscles, 320
Major muscles, 320
Malacostraca, **235,** 236
Malaria, 106, **107, 108**
Male accessory gland, **253**
Male genital bulb, **253**
Male pores, 205, **215,** 216
Male pronucleus, 30
Malpighian tubule, 251, **253**
Malthus, Thomas, 64
Mammae, 350
Mammals. *See also* Fetal pigs
circulatory system, 372–378
digestive systems, 363–368
identifying, 86, 283
muscular systems, 354–362
nervous system, 379–384
overview, 350
respiratory system, 384–385
skeletal systems, 350–354
urogenital system, 368–372
Mammary artery, 375
Mandibles, 319, **325**
birds, **344,** 345
cat, **351**
crickets, **252**
crustaceans, 230, **231**
fetal pig, **352, 359, 363, 364**
grasshoppers, 244, **244, 245**
honeybee, 246, **246**
human, **353**
myriapods, 241, **243**
Mandibular beak, **346**
Mandibular gland, 363
Mandibular muscles, 232
Mandodea, 262
Mantids, 248, 262
Mantle
bivalves, 184, 185–188, **188**
common to molluscs, 184
snails, **192,** 194
squids, 196, **197**
tunicate, 283
Mantle cavity
bivalves, 185–188
chitons, 195
common to molluscs, 184
snails, 194
squids, 197–198
Manubrium, 136, 350
Marrow, **51,** 351
Marrow cavities, 351, **354**
Marsupials, 349
Masseter, 357, **357, 358, 359, 363**
Mastax, 179
Mastoid process, **351**
Maternal chromosomes, 30
Mating in earthworms, 206–207, **208**
Maturation in gametogenesis, 25, 28–30
Maxillae
birds, 345
bony fishes, 302, **303**
cat, **351**
crustaceans, 230, **231**
fetal pig, **352**
frogs, **318,** 319, **320**
grasshoppers, **245**
honeybee, **246**
human, **353**
myriapods, 241, **243**
Maxillary beak, **346**
Maxillary palp, **245, 252**
Maxillary teeth, 325, **325**
Maxillipeds
crustaceans, 228, **229,** 230, **231**
myriapods, 241, **243**
Mayflies, 256, 258
Mayfly, **74**
Mealworms, 107
Meconium, 367
Mecoptera, 259
Median dorsal aorta, 288
Median mantle artery, 198
Median nerve, **380,** 381
Mediastinal pleurae, 368, **368**
Mediastinal septum, 368, **368**
Mediastinum, 368
Medicinal use, 220
horseshoe crabs, 220, **221**
leeches, **215,** 215–218
Medulla, 335, **335**
Medulla oblongata, **335,** 382, **383**
Medusa buds, **135,** 136
Medusae, 132, 135, **135,** 136, 138
Megaloptera, 73
Meiosis, mitosis *vs.,* 25
Meiosis I, 25, **26, 29**
Meiosis II, 25–26, **27, 29**
Membranous bones, 354
Meninges, 334, 382
Mentomeckelians, **318,** 319
Merostomata, 86, 220, 225
Merozoites, 106, **107, 108**
Mesenchyme, **39,** 40
Mesenteric arteries, 300
Mesenteries
bony fishes, 305
fetal pig, 366, 367
sea stars, **269,** 270
shark, 298
Mesoderm
of flatworms, 149
frog embryo, **45,** 46
mother cell, 40
sea stars, 38–40
Mesodon, **193**
Mesoglea, **133,** 134, 136
Mesohyl, 123, 124
Mesolecithal eggs, 44
Mesonephric ducts, **297,** 298, 306, **327**
Mesonephric kidneys, 326, **327**
Mesorchium, 298, **299, 327**
Mesosternum, 319
Mesothoracic legs, **252**
Mesothoracic wings, **252**
Mesothorax, 244, **245,** 246, 251
Mesotubarium, **327**
Mesovarium, **327,** 340, 370, **371**
Metacarpals, **320,** 351, **352, 353**
Metameric arrangement, 286
Metamerism, 201, 220
Metamorphosis, 46, 251, 292
complete, 254
defined, 254
gradual, 251, 254
Metanauplius, 236
Metaphase
meiosis, **26,** 27, **27, 30**
mitosis, 17, **18, 19, 20**

Metaphase plate, 17, 18, 19
Metapleural folds, 286, 289
Metatarsals, 351, **352, 353**
Metatarsus
honeybee, 246, **246**
mammals, **352**
spiders, **224,** 225
Metathoracic leg, **252**
Metathoracic wing, **252**
Metathorax, 244, **245,** 246, 251
Metazoans, 13, 77
Metridium, 140–142, **141**
Microaquariums, 113
Microgametes, 99
Microgametocytes, 106, **107,** 108
Micromeres, 44
Micrometers, ocular, 7, **8**
Micronucleus, 112, 117
Micropyle, 123
Microsatellite, 117
Microscopes, 3–11
Microtubules, 97, **98**
Midbrain, 294, 335
Middle leg, **246,** 247
Midgut, 253
Millipedes, 86, 241, **242,** 243
Minisatellite, 117
Miracidia, 156
Mitochondria, **9, 14, 15,** 15–16
Mitosis, 16–20, 25
Molars, **351**
Molgula, 282
Mollusca
bivalves, 184–191
chitons, 194–195
classification, 199
identifying, **72,** 85
land snails, 192–194
overview, 183–184
squids, 85, 195–199
Molting, 228, 236, 251, 254
Monaxons, 125
Monkey intestine, **56**
Monocytes, **52**
Monoecious species, **133,** 134, 152, 205
Monogenea, 161
Monostyla, 179
Monotremes, 349
Moon jellies, **138,** 138–140
Morula stage, 38, **39,** 44
Mosaic cleavage, 36
Mosquitos, 106, **107**
Moss animals, **71**
Moths, 248, 256, **256,** 258
Motor nerves, 289
Motor neurons, **55**
Mouth
amphioxus, 288
Ascaris, 172
birds, 345, **346**
bivalves, **187,** 189–190
bony fishes, 301, **302,** 303–304
brittle star, **272**
chitons, 194–195
crustaceans, 230, **233**
earthworm, 205, 206
embryonic development, 46
feather stars, 278
fetal pig, 350, 363–364
frogs, 325, **325**
grasshoppers, 244, **244**
honeybee, **246**
horseshoe crabs, 220, **222**
lampreys, 292, **295**
larval formation, 40
leeches, 216
Metridium, **141,** 142
myriapods, **243**
planarians, 150, **151**
schyphozoans, **139,** 140
sea cucumbers, 276
sea stars, 266, **268**
sea urchins, 274, **274**
sharks, **296,** 297
snails, 193, **193**
squids, 197
tissues, **49**
trematodes, 154, **154, 157**
tunicates, 285
turtles, 339–340, **340**
vinegar eels, 175, **176**
Movement. *See* Locomotion
Mucins, 363
Mucous glands, **48,** 54, **56,** 193–194
Mucous membrane, 54–55, **56**
Mud snail, 157
Multicellular organisms, 77
Multiple hemoglobin systems, 309–311
Muscle tissue, 47, **53–54,** 185, **187**
Muscular hydrostats, 188
Muscular systems
bony fishes, 303, **304**
crustaceans, 232, **233**
fetal pigs, 354–362
frogs, 320–322, **321, 323**
Mutualism, 101
Myelin, 51, **55**
Mylohyoid, **321**
Mylohyoid muscle, 357, **358, 359**
Myofibrils, **53**
Myomeres, 303
Myosepta, 289
Myosin, 91
Myotomes
amphioxus, 287, **287,** 289
lampreys, **292,** 293, **294, 295**
Myriapoda, 86, 225, 241–243
Mysis, 236
Myxini, 86, 283

N

Nacre, 193
Nacreous layer, 185, **188**
Naming muscles, 356
Nares, **320, 339, 340.** *See also* Nostrils
external (*See* External nares)
internal (*See* Internal nares)
Nasal bones
cat, **351**
fetal pig, **352**
frogs, **318,** 319, **320**
Nasal fossa, **318,** 319
Nasal passages, 364
Nasohypophyseal canal, 294
Nasopharyngeal pouch, **292,** 293
Nasopharynx, 364, **364**
Nauplius, 236
Nautiluses, 85
Necator americanus, 177
Neck
crickets, **252**
mammals, 368
tapeworms, 158
Nematocysts, 132, **133,** 134, 136, 138, 140
Nematoda, **71,** 85, 172–178
Nematomorpha, 85
Nemertea, 37, **71,** 85
Nephridia, 206, 207, **207, 209**
Nephridial tubules, 289
Nephridiopores
earthworm, 206, **207,** 208
leeches, 216
Nephrostomes, 207, **207,** 208
Nereis, 202–204
Nerve cords
amphioxus, 289
Ascaris, 174, **175**
ascidian larvae, 285, **286**
crickets, **253,** 254
crustaceans, **233**
earthworm, **207, 209**
lampreys, 294
planarian, 152, **153**
tapeworms, 159, **159, 160**
Nerve fibers, 57
Nerve nets, 134
Nerve rings, 270
Nervous systems
bivalves, 190–191
crustaceans, 234
earthworm, 208–209
fetal pig, 379–384
flatworms, 152
frogs, **334,** 334–335, **335**
sea stars, 270
squids, 198–199
Nervous tissue, 51–54, **55**
Neural canal, 289
Neural fold, 46
Neural groove stage, 46
Neural plate, **45,** 46
Neural spines, **319, 352**
Neural tube stage, **45,** 46
Neurilemma, 51, **55**
Neurocranium, 302, **303**
Neuroglia, 51
Neuromuscular system, tunicates, 283–284
Neurons, 51, **55**
Neuropodium, **203**
Neuroptera, 73, 259
Neutrophils, **52**
Nictitating membrane, 314, **325**
Nictitating membranes, 338
Nidamental glands, **197,** 198, 199
Noctiluca, 101, **101**
Nonavian Reptiles, 86
Nosepiece, 3, **4**
Nostrils
birds, **346**
bony fishes, 301, **302**
fetal pig, 350, **364**
frogs, 314, **320,** 325, **325**
lampreys, 292, **294, 295**
sharks, **296,** 297
turtles, 338, **340**
Notochord
amphioxus, 286, **288,** 289
ascidian larvae, 285, **286**
as basic feature of chordates, 281
frog embryo, **45,** 46
lampreys, **292,** 293, **294, 295**
Notochordal sheath, 289
Notophthalmus viridescens, **315**
Notopodium, **203**
Notostraca, 73
Nuchal crest, **351**
Nuclear envelope, 14
Nuclei (cell)
Amoeba, **91,** 92, **93**
basic features, 13, **14**
ciliates, **109,** 110, 112
euglenoids, **97, 100**
formation, 20
larval formation, **39**
structure of, 14–15
Vorticella, 112
Nucleoli, 14, **14,** 15
Nucleoplasm, 14
Nudibranchs, 85
Nutrient vessels, **354**
Nutritive-muscular cells, **133,** 134
Nymphs, 254

O

Obelia, 132, **135,** 135–136
Objectives (microscope), 3, 4, 5, 7
Obligate heterotrophs, 100
Oblique muscle, 367
Observation, 63, 64
Obturator foramen, **352**
Occipital artery, **376**
Occipital bone, **351, 352, 353**
Occipital condyle, **318, 351**
Ocelli
bivalves, 191
crickets, **252**
grasshoppers, 244, **245**
honeybee, **246**
myriapods, 241, **242,** 243
planarians, 150
spiders, 224
Octocorallia, 144
Octopuses, 85
Ocular micrometers, 6, **8**
Oculars (microscope), 3, **4**
Oculomotor nerve, **335, 383**
Odonata, 74, 259
Odontophores, **192**
Oil-immersion objectives, 3, 5
Olecranon process, 351, **352**
Olfactory bulb, **383**
Olfactory lobes, 334, **334, 335**
lampreys, 294
Olfactory nerves, 334, **334, 335,** 384
Olfactory sac, **292,** 293
Oligochaeta
general features, 204–210, 216
identifying, **71,** 85
Ommochromes, 195
Omosternum, 319
Onchospheres, 159
One-cell stage, 44, **45**
Oocysts, 106, **107**
Oogenesis, 26, 28–30, **30**
Oogonia, 28, **29**
Oothecae, 248
Ootype, **154**
Open circulatory systems, 189, **190**
Opercula, 46, 301, 302, **302, 303**
Opercular cavity, 305
Opercular pumps, 305
Ophiocoma aethiops, 271
Ophiuroidea, 84, 270–272, 279
Opisthaptor, 157, **157**
Opisthobranchs, 194
Opisthokonta, 114
Opisthokonts, 77
Opisthosoma, 221, **222**
Opthalmic artery, 232, **233**
Optic chiasma, 335, **335,** 382, **383**
Optic lobe, **334,** 335, **335**
Optic nerves, **334,** 335, **335,** 382, **383**
Oral (incurrent) aperture, 285, **286**
Oral arms, **139,** 140
Oral cavity. *See* Mouth
Oral disc, 112
Oral groove, 109, 111
Oral hood
amphioxus, 286, **287,** 288, **288**
lampreys, 293, **294,** 295
Oral lobes, 136, **137**
Oral nervous systems, 270
Oral papillae, 293, **294,** 295
Oral plate, 46
Oral pumps, 305
Oral shields, 271, **272**
Oral siphons, 283, **284**
Oral suckers, 153, **157**
Oral valves, 304
Oral-aboral flattening, 266
Orbital fossa, **318,** 319
Orbits, **303, 339, 352**
Orconectes rusticus, 228
Orders, 77
Organelles, 13
Organs, defined, 47
Origins of muscles, 355
Orthoptera, 262
Os coxa, **353**
Osculum, 121, 122, **122, 124, 125**
Osmoregulation, 92, 97, 109, 152
Osmotrophic nutrition, 100, 106
Ossicles, 267–268
Ossification centers, 354
Osteon, **51**
Ostia
bivalves, 188, **189**
crustaceans, 232, **233**
fetal pig, 371
frogs, 327
sponges, **124**
squids, 199
Ostium tubae, 298
Ostracoderms, 291

Ostracods, **73,** 235, **235**
Otic vesicles, 295, **295**
Outgroup method, 79
Ova. *See* Eggs
Ovaducal gland, **197**
Oval capillary bed, 305
Ovarian arteries, **371,** 378
Ovarian veins, 329, **371,** 378
Ovaries
Ascaris, **173,** 174, **175**
bivalves, 190
bony fishes, **304,** 305, **306**
crustaceans, 232
earthworm, 206, **207, 208**
fetal pig, 370, **371**
frogs, **326,** 327, **327**
hydras, **133,** 134
insect, 254
Obelia, **135**
planarian, **151**
sea stars, 269
shark, 298, **299**
squids, **197,** 199
tapeworms, **159, 160**
trematodes, **154, 157**
tunicates, **284**
turtles, 340
vinegar eels, 176, **176**
Oviducal gland, 199, 299
Oviducts
Ascaris, **173,** 174, **175**
birds, **346**
bony fishes, 305, **306**
crickets, **253,** 254
crustaceans, **229,** 232
earthworm, 206, **207, 208**
fetal pig, 371
frogs, 327, **327**
grasshoppers, 245
planarian, **151**
shark, 298, **299**
snails, **193**
tissues, **48**
vinegar eels, **176**
Oviparous species, 299
Ovipositor, 245, **245, 252,** 254
Ovisacs, **327**
Ovoviviparous species, 175–176, 296
Oxygen availability in water, 309

P

Painted turtle, 337–342
Palatal folds, 340, **340**
Palatine bones, **318,** 319, **351**
Palatine folds, 346, **346**
Pallial cartilages, 197, **197**
Pallial line, 185, **187, 188**
Pallial muscle, 185, **187, 188**
Palps, 202, **203, 231, 245**
Pancreas
birds, **346,** 347
bony fishes, **304,** 305
fetal pig, **365,** 368
frogs, 326, **326**
shark, **297,** 298
squids, 198
turtles, 340, **341**
Pancreatic duct, 368
Panulirus, 228
Papillae
bony fishes, 305, **306**
fetal pig, **364,** 369
frogs, 325
lampreys, **292,** 293, 294, **294**
shark, **297,** 298, 299, **299**
Papillary muscles, 374, **374**
Papulae, 266
Parabasala, 114
Paradise fish, 307–308
Paragordius, 180
Parallel muscle, 355
Paramecium, **109,** 109–111, **110**
Paramylon granule, **97,** 98
Parapodia, 202, 203, **203**
Paraproct, **252**
Parasites
flukes, 153–158
lampreys, 291–296
nematodes, 172–174, 177–178, 179
protozoan, 16, 100, 106–108, **108,** 113–114
tapeworms, 158–161
Parasitic *Amoeba,* 92–93
Parasphenoids, **318,** 319, **320**
Parastichopus, 276
Parasympathetic division, 381
Parenchyma, 150, 152, **153,** 154
Parfocal lenses, 5
Parietal arteries, 300, 378
Parietal bone, **351, 352, 353**
Parietal complex, 334–335
Parietal organ, 335
Parietal pericardium, 368, **368**
Parietal peritoneum
birds, 347
earthworm, 209
fetal pig, 367
shark, 298
turtles, 340
Parietal pleura, **365,** 368, **368**
Parietal vein, 378
Parotid duct, 363, **363**
Parotid gland, **363**
Parsimony principle, 79, 81
Patella
mammals, **352, 353**
in spider leg, **224,** 225
Paternal chromosomes, 30
Pauropoda, 225
PCR (polymerase chain reaction) technique, 117–118
Pearl formation, 185
Pecten, **246**
Pectineus, 360, **361,** 362, **362**
Pectoral fins, 296, **296, 302, 303**
Pectoral girdle
birds, 345, **345**
bony fishes, 302–303, **303**
frogs, 319
mammals, 350–351
turtles, 338, 339, 340
Pectoral muscles
birds, 345, **345, 346,** 347
frogs, **321, 322**
mammals, 356, **357, 358, 359**
Pedal disc, 141
Pedal ganglia, 191, **193**
Pedicellariae, 265, **273,** 274–275
Pedicels, 224, **224**
Pedipalps, 220, **222, 223, 224**
Pellicles, 97, **97,** 109, 111, 112
Pelomyxa, 90
Pelvic canal, 351
Pelvic fins, 296, **296, 302, 303**
Pelvic girdle
bony fishes, 302, **303**
frogs, 319
human, **353**
mammals, 351
turtles, 338, 339, 340
Pen, squids, **197,** 199
Penis
fetal pig, 350, **369,** 370
planarian, **151**
snails, **193**
squids, **197,** 199
Pentaradial symmetry, 266
Peranema, 97
Perch, 301–306
Pereiopods, 230
Perforating fibers, **354**
Pericardial cavity, 293, 303–304, 305
Pericardial membrane of bivalves, 185
Pericardial sac, 325, 340
Pericardial sinus, 232
Pericardium, 232, 368, **368**
Periderm, 350
Periosteum, **51,** 354, **354**
Periostracum, 185, **188,** 193, **193**
Peripheral nervous system, 334, 379, **380**
Periplaneta, 248
Periproct, 275
Perisarc, 135, **135**
Peristalsis, 251
Peristomal gills, **274**
Peristomes, 112, 141, 202, **203**
Peristomial membrane, 197, 267
Peristomium, 202, **203,** 205, 206
Peritoneum
earthworm, 209, **209**
fetal pig, 367
sea stars, 268, **269,** 270
shark, 298
Perivitelline space, 19, 28, **30**
Peroneus, 322, **322, 323**
Perophora, 285
Petromyzon marinus, 291
Petromyzontida, 86, 283, 291–296
Phagocytes, 92
Phagocytosis, 92
Phalanges
frogs, **320**
mammals, 351, **352, 353**
Pharyngeal bulb, 175, **176**
Pharyngeal cavity, **153,** 364, **364**
Pharyngeal chamber, 152
Pharyngeal pouches, 281, **282**
Pharyngeal sheath, 151, **151, 153**
Pharyngeal slits, 281, **282,** 285, 286, 289
Pharynx
amphioxus, 288, 289
Ascaris, **173,** 174
ascidian larvae, 285, **286**
birds, 346, **346**
bony fishes, 303
clamworm, 202, **203**
earthworm, 206, **207**
fetal pig, 363–364
frogs, 325
lampreys, **292,** 293, **294**
Metridium, **141,** 142
Paramecium, 109
planarian, 150, 151, **151, 153**
rotifers, 179
sharks, 296
tissues, **49**
trematodes, 154
tunicates, **284,** 285
turtles, 339
vinegar eels, 175, **176**
Phasmatodea, 262
Philodina, 179
Phosphorescence, 101
Photoreceptors, 202, 209, 288, 292. *See also* Eyes; Ocelli
Photosynthesis, 97–98
Phototactic responses, 239–240
Phototaxis, 111, 150
Phrenic nerve, **380**
Phthiraptera, 260
Phyla, 77
Phylogeny reconstruction, 78–83
Physa, **193**
Physalia pelagica, 137
Pia mater, 334, 382
Pigeon, 343–347
Pigeon milk, 346
Pilidium, 40
Pinacocytes, 124
Pineal gland, 335
Pineal organ, 292, **295**
Pinning insects, 255–257, **256**
Pinnules, 278
Pinocytosis, 92
Pinworms, 178, **178**
Piriformis, **323**
Pituitary gland, 335, **335,** 382, **383**
Pituitary stalk, **383**
Placoid scales, 297, **297**
Plakina, 126
Planarians, 71, 85, 150–152, **151, 153,** 167
Plane surfaces, 4
Plankton, 98, 101
Planospiral shells, **193**
Planula larvae, **131,** 135, 139, **139**
Plasma membrane, **91,** 109
Plasma membranes, 13, 14, 20
Plasmodium, 106, **107, 108**
Plastrons, 338
Platelets, **52**
Plates, brittle stars, 271
Platyhelminthes
flukes, 153–158
general features, 149–150
identifying, **71, 85,** 161
regeneration experiment, 167
tapeworms, 158–161
turbellarians, 150–152, **151**
Platyias, 179
Platysma, 356
Plecoptera, 74, 259
Pleistoannelida, 216
Pleopods, 228, **229,** 230
Pleura, 368
Pleural cavity, 368, **368**
Pleural fluid, 368
Pleuron, 251
Pleuropericardial membrane, 341
Pleuroperitoneal cavity, 340
Plumatella, **71**
Pneumatophores, 137
Pneumostome, **192,** 194
Podia, 276, 278
Podial pores, **272**
Poikilotherms, 115
Pointers (microscope), 3
Poison fangs, 241, **243**
Poison glands, 54
Poison sac, 247
Polar bodies, 26, 28–30, **29**
Polarity of frog eggs, 44
Pollen basket, **246,** 247
Pollen brush, 246, **246**
Pollen comb, **246,** 246–247
Pollen packer, **246,** 247
Polyacrylamide gel electrophoresis, 310–311
Polyaxons, 125
Polychaetes, 216
Polygyra, 192
Polymerase chain reaction (PCR) technique, 117–118
Polymorphism, 117–118, 137
Polyphemus moths, 248
Polyplacophora, 85, 194–195, 199
Polyps, 132, 134, 135, **135, 139,** 142, **143**
Polyribosomes, **15**
Polystoma, 157, **157**
Pond stewardship, 68
Pons, 382–384, **383**
Pore (sea star), 270
Pores (nuclear), 14, **15**
Pores (sponges), 121
Porifera, 84, 121–127
Porocytes, **124**
Portal veins. *See* Hepatic portal vein; Renal portal vein
Portuguese man-of-war, 137
Positive pressure breathin, 314
Postanal tail, 281
Postcardinal veins, 288, **289**
Postcava
fetal pig, **368,** 373, 374, **377,** 377–378
frogs, **327,** 329, **329**
Postcaval vein, 328, 373
Posterior adductor muscles, 185, **187**
Posterior aorta, 189, **190,** 198
Posterior cardinal vein, 295, **295**
Posterior deep pectoral muscle, 357, **358, 359,** 360
Posterior mesenteric artery, **299, 378**
Posterior mesenteric ganglion, **380,** 382
Posterior mesenteric vein, 378
Posterior trunks, 151, **151**
Posterior vena cava, **197,** 198, **328,** 329
Postganglionic neurons, 381
Precardinal veins, 288, **289**

Precava, 373, **374, 376**
Precaval veins, 328, **329,** 373, 374
Predator functional response, 147
Prediction in scientific method, 63, 65
Preganglionic neurons, 381
Prehallux, **320**
Premaxillae
bony fishes, 302, **303**
cat, **351**
fetal pig, **352**
frogs, **318,** 319, **320**
Premolars, **351**
Prepollux, **320**
Preputial orifice, **369**
Primary lamellae, 300
Primary oocytes, 28, **29**
Primary septa, 142
Primary spermatocytes, 27, **29, 31,** 31–32
Primordial germ cells, 25, 26, 28
Principle of Parsimony, 79, 81
Prismatic layer, 185, **188,** 193
Proboscis worms, **71**
Procambarus clarkii, 228
Processus vaginalis, 370
Proglottids, 85, 158, 159, **160**
Promarginal teeth, **224**
Pronephric kidneys, 294
Pronotum, **252**
Pronuclei, 20
Prootic ganglion, **335**
Prootics, **318,** 319, **320**
Prophase
meiosis, **26,** 26–27, **27**
mitosis, 17, **18, 20**
Prosobranch snails, 194
Prosoma, 220–221, **223**
Prosopyles, **125,** 122, **122,** 123
Prostate gland, 370
Prostomium, 202, **203,** 205
Prothoracic leg, **252**
Prothorax, 244, **245,** 246, 251
Protochordates, 281
Protomerite, 107
Proton pumps, 92
Protonephridia, 152
Protoplasmic strands, 99
Protopods, 230, **231**
Protostomia, 36, 172–180
Protozoans. *See also* Amoeba; *Amoeba*
amebas, 89–94
animals *vs.,* 77
apicomplexans, 106–108, **108**
cell structure, 13
ciliates, 109–113
classification, 113–114
flagellates, 97–101
genetic polymorphism in, 117–118
parasitic, 16, 100, 106–108, **108,** 113–114
temperature effects on, 115–116
Protozoea, 236
Protractor muscles, **209, 247,** 356
Proventriculus, **253, 346,** 347
Proximal epiphysis, **354**
Pseudocardinal tooth, **187**
Pseudocoel, 172, **173,** 174, **175, 176**
Pseudopodia, 91, **91, 92,** 93
Pseudostratified ciliated epithelium, **58**
Pseudostratified columnar epithelium, **49**
Pseudostratified epithelium, 57
Psoas major, **361,** 362, **362**
Psocoptera, 260, 261
Pterygoid process, **351**
Pterygoids, **318,** 319, **320**
Ptychodiscus, 101
Pubic symphysis, **352, 369, 371**
Pubis
cat, **352**
fetal pig, **352**
frogs, 319
mammals, 351
turtles, 339, **339**
Pulmocutaneous arch, 329–330
Pulmonary arch, **341,** 342
Pulmonary arteries
amphibians, 328
fetal pig, 373, 374, 375
frogs, 329–330
sheep, **374**
turtles, 342
Pulmonary circuit, 328, 372
Pulmonary trunk, 373, 375, **376**
Pulmonary veins
amphibians, 328, 329
fetal pig, 373, 374, 375
turtles, 342
Pulp cavity, shark denticles, 296, **297**
Pupae, **65,** 254
Pycnogonida, 225
Pygostyle, **344,** 345
Pyloric ceca
bony fishes, **304,** 305
sea stars, **268,** 269, **269,** 270
Pyloric ducts, 269
Pyloric stomach
bony fishes, **304,** 305
crustaceans, 232, **233,** 234
fetal pig, 367
sea stars, **268,** 269
Pyloric valve, 298, 305, 325, 367

Q

Quadratojugals, **318,** 319
Quadriceps femoris, 360, **361**
Queen bees, 247
Quills, 343

R

Rachis, 343
Radial arteries, 375, **376**
Radial canals
Gonionemus, 136
sea cucumbers, 278
sea stars, **268, 269,** 270
sponges, 122, **122,** 123
Radial cleavage, 36, **36,** 38–40
Radial nerves, 270, **334, 380,** 381
Radiate phyla, 131. *See also* Cnidarians
Radiating canals, 109
Radiolarians, 94, 114
Radioulna, **320**
Radius
birds, **344,** 345
mammals, 351, **352, 353**
turtles, **339**
Radula, 192, **192,** 194
Ragworms, 202–204
Rami communicantes, **379**
Randomly amplified polymorphic DNA markers (RAPDs), 117–118
Ranid frogs, 313
RAPDs (randomly amplified polymorphic DNA markers), 117–118
Ray-finned fishes, 86
Rays, 86
Rays (arm) of sea stars, 266, 267
Recombination of genes, 27
Rectal ceca, **268,** 269
Rectal gland, **297,** 298
Rectum
bivalves, **186, 187,** 190
crickets, **253**
fetal pig, **365,** 367, **369, 371**
honeybee, **247**
shark, 298
squids, **197,** 198
vinegar eels, **176**
Rectus abdominus, 321, **321, 322,** 360, **361, 362**
Rectus femoris, 360, **361, 362**
Rectus muscles, 320
Red blood cells, **10, 52,** 106
Red marrow, **51**
Red swamp crayfish, 228
Red tides, 101
Reflected light in microscopes, **8,** 8–9
Reflecting mirrors, 4
Reflexes, tunicates, 283–284
Regeneration, 167
Regular cleavage, 36
Regular echinoids, 273
Relaxing insects for mounting, 255
Renal arteries, 300, **377,** 378
Renal calyces, 369, **369**
Renal cortex, 369, **369**. *See also* Kidneys
Renal medulla, 369
Renal opening, 230
Renal papillae, 198
Renal pelvis, 369, **369**
Renal portal system, 329
Renal portal veins, 299, 300, **328,** 329
Renal pyramids, 369, **369**
Renal veins, 329, 377
Replicates, 66
Reproduction and reproductive systems
Amoeba, 92
apicomplexans, 106, **107**
Ascaris, **173,** 174, **175**
bivalves, 190
bony fishes, 305–306
ciliates, 110, 111, 112
cnidarians, 134–135
crickets, **253**
crustaceans, 232
earthworm, 206–207, **207, 208**
euglenoids, 98
fetal pig anatomy, 369–372
frogs, 327, **327**
gametogenesis, 25–32
horseshoe crab, 221–222, **222**
planarians, **151,** 152
schistosomes, 155–156
scyphozoans, 138, **139**
sea cucumbers, 278
sea stars, 269–270
shark, 298–299
snails, **193**
sponges, 123
squids, 199
tapeworms, 158, 159, **160**
trematodes, 154
tunicates, **284,** 285
Volvox, 99
Reproductive cells, 99
Reproductive ducts, 269
Reptiles, 86, 283, 337–342
Reservoir, 97, **97**
Respiration and respiratory systems
birds, 347
bivalves, 188
crustaceans, 230
frogs, 313, 314, 326
mammals, 384–385
multiple hemoglobin systems, 309–311
sea cucumbers, 276, 277
shark, 300–301
spiders, 224
squids, 198
tunicates, 285
turtles, 340–341
Respiratory bronchioles, 384
Respiratory epithelium, 289
Respiratory trees, 276, **277**
Respiratory tubes, **292,** 293
Response to stimuli, 111
Rete mirabile, 305
Retina, 199
Retractor muscles, 356
Revolving nosepiece, 3, **4**
Rhabdites, 152
Rhabditis maupasi, 179
Rhabdodermella, 122
Rhinencephalon, **383**
Rhomboideus capitis, **358,** 360
Rhomboideus muscles, **358, 359,** 360
Rhopalium, 139
Rib cage, **353**
Ribbon worm, 37–38, 85
Ribonucleic acid (RNA), 15
Ribosomes, 15
Ribs
birds, 344, **344**
bony fishes, 302, **303**
fetal pig, **352, 365**
human, **353**
mammals, 350
Right atrium
fetal pig, 373, 374
frogs, 328, **329,** 330
sheep, **374**
turtles, 342
Right carotid artery, 342
Right common carotid artery, 375
Right pulmonary artery, 375
Right subclavian arteries, 342, 375, **376**
Right systemic arch, 342
Right ventricle, 373, 374, **374**
Righting reactions, 314
Ring canals
Gonionemus, 136
schyphozoans, **139,** 140
sea cucumbers, **277**
sea stars, **268,** 270
Rock dove, 343–347
Rock lobster, 228
Romalea, 31–32, 243–245, **244, 245**
Root papillae, **364**
Rostellum, 158, **160, 161**
Rostrum
amphioxus, 286, **287, 288**
crustaceans, 228, **229, 233**
shark, **296,** 297
Rotation, **355,** 356
Rotifers, **72, 179**
Rough endoplasmic reticulum, **9, 14,** 15
Round ligament, **371**
Roundworms
cell cycle studies using, 17, 19–20, **20**
identifying, **71,** 85
oogenesis and fertilization, 28–30, **30**
Rouns worms
general features, 172–174, **173, 175**
Rugae, 298, 367
Rusty crayfish, 228

S

S phase, 17
Sacral hump, 314
Sacral region, 350
Sacral vertebrae
birds, 344
frogs, 319, **320**
mammals, 350
turtles, 338, 339
Sacrum, **352, 353**
Sagittal section of brain, 384
Salamanders, 86, **315,** 338
Salivary amylase, 363
Salivary glands
crickets, 251
fetal pig, 363, **363**
snails, **193**
tissues, **48**
Saltatory conduction, 54
Sand dollars, 84, 279
Sandworms, 202–204
Saphenous artery, 378
Saphenous nerve, 381
Saphenous vein, 378
Saprophagous nematodes, 176
Saprozoic nutrition, 98
Sarcomeres, **53**
Sarcopterygii, 283
Sartorius
fetal pig, 360, **361, 362**
frogs, 320, 321, **321, 322**
Sawflies, 258
Scales, 302
Scanning electron microscopes (SEM), 10–11
Scanning objectives, 3, 6
Scaphopoda, 85
Scapula
birds, 344, **344,** 345, **345**
cat, **352**

fetal pig, **352**
frogs, **320**
human, **353**
mammals, 350–351
turtles, 339, **339**
Scapular spine, **352, 359**
Scarabeidae, 180
Schistosomes, 155–156
Schistosomiasis, 155
Schizogony, 106
Schizonts, 106, **108**
Schwann cells, 51, **55**
Sciatic artery, **328,** 330
Sciatic nerve, **334,** 335, **380,** 381
Sciatic plexus, **334,** 335
Sciatic veins, **328,** 329
Scientific method, 63–66
Scientific names, 77
Sclerites, 244
Sclerosepta, 142
Sclerotic ring, **344,** 345
Scolex, 158, **161**
Scolopendra, **242, 243**
Scoring, 79
Scorpionflies, 259
Scorpions, 86
Scrotal sacs, 350
Scrotum, **369**
Scuds, 73
Scutes, 338, **339**
Scyphistomae, 138, **139**
Scyphomedusae, 136, 138
Scyphozoans, 84, **138,** 138–140, **139,** 143–144
Sea anemone, 84, 140–142, **141**
Sea biscuits, 84
Sea cucumbers, 84, **266,** 275–278, **277,** 279
Sea lampreys, 291
Sea lilies, 84, 278–279
Sea squirts, 282, 283–286
Sea stars
anatomy, **268**
basic features, 84, 266–270
egg cells, 13–14, **14**
embryonic development, 38, 40
identifying, 84, 279
Sea urchins, 84, 273–275, 279
Sea walnuts, 84
Sea wasp, 138
Second maxillae, 241
Second polar body, **29,** 29–30
Secondary lamellae, 300
Secondary oocytes, 28, **29, 30**
Secondary spermatocytes, 27, **29, 31,** 32
Sedentaria, 85, 216
Segmental arteries, 294
Segmental ganglia, 254
Segmentation
basic to annelids, 201–202
basic to arthropods, 220
crickets, 251
earthworm, 205
grasshoppers, 244
myriapods, 241, **242**
Segmentation cavity, 44
Segmented arteries, 288
Semen, 370
Semicircular notch, 247
Semilunar valves, 374, **374**
Semimembranosus
frogs, **322, 323**
Semimembranosus muscle
fetal pig, **357,** 360, **361, 362**
Seminal grooves, 205
Seminal receptacle
crickets, **253**
crustaceans, 229, **229**
earthworm, 206, **207, 208**
planarian, **151**
trematodes, **154, 157**
vinegar eels, 176, **176**
Seminal vesicle
Ascaris, **173,** 174
earthworm, 206, **207, 208**
fetal pig, 370
shark, **297,** 298, **299**
trematodes, **154**
Seminiferous tubules, 370
Semitendinosus muscle, **357,** 360, **361, 362**
Sensillae, 216
Sensory hairs, 223
Sensory nerves, 289
Sensory organs. *See also* Eyes
crustaceans, 234
sea stars, 270
squids, 199
Sensory papillae, 325
Sensory tentacles, **267, 268, 284**. *See also* Tentacles
Septal filaments, **141,** 142
Septum, **207**
Serial homology principle, 229–230
Setae, 203, **203,** 205, 209, **209**
Sex cells, in *Volvox,* 99
Sexual reproduction. *See also* Reproduction and reproductive systems
cnidarians, 134–135
gametogenesis, 25–32
schistosomes, 155–156
sponges, 123
Volvox, 99
Shaft, of long bone, 351, 354
Shaft, of rib, 350
Shaft of feather, 343
Shallow-water habitats, 67
Shank muscles, frogs, 321–322
Sharks, 86, 296–301
Sheep
brain, 382–384
heart, **374**
Shell gland, 298
Shelled amebas, **93,** 93–94
Shellfish, 72
Shells
egg, 19, 28
molluscs, 184, 185, **187, 188, 192**–193
turtle, 337–338
Shipworm, 191
Short monaxons, 125
Signet-ring stage, 106, **107, 108**
Silverfish, 254, 260
Simple columnar epithelium, **48**
Simple cuboidal epithelium, **48**
Simple eyes, 220, **222**
Simple squamous epithelium, **48**
Simplex uterus, **371**
Single-circuit circulatory systems, 306
Sinistral shells, 193, **193**
Sinus venosus
bony fishes, **304,** 306
frogs, 328, 329, **329**
lampreys, 293, 294
turtles, 342
Sinusoids, 14
Siphonaptera, 260
Siphonoglyph, 141, **141**
Siphons, 184, 197, 283
Sister chromatids, 17, 25–26, **26,** 27, **27, 28**
Sister taxa, 78
Skates, 86
Skeletal muscle, **53,** 355–356. *See also* Muscular systems
Skeletal systems
birds, 344–345
bony fishes, 302–303, **303**
frogs, 318–319, **320**
mammals, 350–354, **352**
sponges, 125–126, **127**
squids, 199
turtles, 338–339
Skin. *See* Dermis; Epidermis
Skin gills, 266, **267**
Skin tissues, **49,** 54, **56**
Skull
birds, **344,** 345
cat, **351, 352**
frogs, **318,** 318–319
human, **353**
mammals, 350
turtles, 338, **339**
Sleeping sickness, 100
Slides, caring for, 6
Slugs, 85
Small intestine
birds, **346**
fetal pig, **365,** 367, **380**
frogs, 325, **326**
tissues, **48,** 54–58
turtles, 340, **341**
Smooth muscle, 47, **53,** 55
Snail fever, 156
Snails
anatomy, 191–194
classification, 72, 85
as fluke hosts, 155–156
Snakes, 86
Social featherduster worm, **204**
Soft palate, 364, **364**
Soil nematodes, 176, 178–179
Sole, 276
Soleus, 362
Somatic cells, in *Volvox,* 99
Specialization in arthropods, 220
Species
chromosome numbers, 16
classification, 77
identifying, 68–69, **70–74,** 84–86
Sperm, sea stars, 269–270
Sperm ducts
bony fishes, 306, **306**
crustaceans, 232
earthworm, 206, **207, 208**
shark, **299**
Sperm funnel, **207**
Sperm receptacle, 199
Sperm sacs, 298, **299**
Spermathecal duct, **193**
Spermatic arteries, **369,** 378
Spermatic cord, 370
Spermatic veins, 329, **369,** 378
Spermatids, 27, **29, 31,** 32
Spermatogenesis, 26-27, **29,** 31-32, **31**
Spermatogonia, 26, **29, 31,** 32
Spermatophoric gland, **197,** 199
Spermatophoric sac, 199
Spermatozoa, hydras, 134
Spermiogenesis, 27
Sphaeridia, 275
Sphenethmoid bone, **318,** 319
Sphenoid, **351**
Spicules
Ascaris, 172, **173,** 174
sponges, 121, 122, **123,** 125–126, **126**
Spiders, 86, 223–225, **224**
Spinal accessory nerve, 384
Spinal column. *See* Vertebral column
Spinal cord
bony fishes, **304**
fetal pig, **379**
frogs, **334,** 335, **335**
lampreys, **292,** 293, **294, 295**
sheep, **383**
Spinal nerves, 335, **379,** 379–381
Spindle fibers, 17, **18, 19**
Spine (vertebral). *See* Vertebral column
Spines, 351
brittle stars, **272**
of scapula, **359**
sea stars, 266, **267, 268**
sea urchins, **273, 274**
Spinnerets, **224,** 225
Spinous process, **379**
Spiny dogfishes, 296
Spiny lobster, 228
Spiny-headed worms, 85, 179–180
Spiracles, 46
grasshoppers, 244, **245**
myriapods, 241, 243, **243**
shark, 296, **296**
Spiral cleavage, **36,** 36–40
Spiral valve, 198, **297,** 298, **329**
Spirobranchus giganteus, 204, **204**
Spirogyra, 93
Spirostomum, 113, **113**
Spirotrichonympha, **100**
Spleen
bony fishes, **304,** 305
fetal pig, **365,** 367
frogs, **326**
shark, **297,** 298
turtle, **341**
Splenic artery, **377**
Splenius, **357, 358, 359,** 360
Sponges, 84, 121–126, **127**
Spongia, 126
Spongilla, 126
Spongin, 121, **126**
Spongocoel, 121, 122, **122,** 123, **124**
Spongy bone, **51,** 351, **354**
Sporocysts, 155
Sporogony, 106
Sporozoites, 106, 107, **107**
Spreading boards, 256–257
Springtails, 254, 261
Spur, **246,** 247
Squalus, 296–301
Squamosals, **318,** 319, **320**
Squamous epithelial cells, 13
Squamous epithelium, **49,** 54, **56,** 57
Squids, 85, 195–199
Stage (microscope), 3, 4, **4**
Stage micrometers, 6
Stalks, 112
Star coral, 142–143
Starfish. *See* Sea stars
Statocyst
ascidian larvae, 285, **286**
Statocysts
crustaceans, 234
Gonionemus, 136
squids, 199
Staurozoa, 144
Stellate ganglia, 198, **380,** 381–382
Stem cell research, 150
Stentor, **70, 112,** 112–113, **113,** 115–116
Stereoscopic dissecting microscopes, **8,** 8–9
Sternal artery, 232, **233,** 234, 375
Sternocephalic muscle, 357, **358, 359**
Sternohyoids, 357, **357, 358, 359**
Sternoradialis, **321**
Sternorrhyncha, 261
Sternothyroids, 357, **358, 359**
Sternum
birds, 344, **344,** 345
cat, **352**
crickets, 251, **252**
crustaceans, **229**
fetal pig, **352, 368**
grasshoppers, **245**
human, **353**
mammals, 350
spider, **224**
Stewardship, 68
Stigma, 97, **97,** 98, 99
Stimuli response, 111, 150
Sting of honeybee, 247, **247**
Stinging organelles, 132, 140
Stomach
ascidian larvae, 285, **286**
birds, 347
bivalves, **187,** 190
bony fishes, **304,** 305
crustaceans, 232, **233,** 234
fetal pig, **365,** 367, **380**
frogs, **326**
larval formation, 40
schyphozoans, 140
sea stars, 267, **268,** 269
shark, **297,** 298
snails, **193**
squids, **197,** 198
tissues, **48**
tunicates, **284,** 285
turtles, 340, **341**

Stomodeum, 40
Stone canals, **268,** 270, **277,** 278
Stone crab, **235**
Stoneflies, 259
Stonefly, **74**
Stony corals, 142–143, **143**
Stramenopiles, 14, 114
Stratified squamous epithelium, **49,** 54, **56**
Striated muscle, 47, **53, 54,** 220
Stridulating, 251
Strobila, 138, **139,** 158
Strobilation, 138
Strongylocentrotus drobachiensis, 273, **273**
Strongylocentrotus purpuratus, 273
Subclavian arteries
  fetal pig, 375, **376**
  frogs, **328,** 330
  shark, 300
  turtles, 342
Subclavian veins
  fetal pig, 375
  frogs, **328,** 329
  shark, 299
Subcutaneous lymph sacs, 320
Subesophageal ganglia, 234
Subintestinal artery, **295**
Subintestinal vein, 288
Sublingual glands, 363, **363**
Submaxillary gland, 363, **363**
Submerged plant habitat, 67
Submucosa, 55, **56, 58**
Subneural vessel, 206, 209
Subpharyngeal ganglion, **207**
Subscapular arteries, 375, **376**
Subscapular nerves, **380,** 381
Subscapular vein, 329, **329**
Subscapularis, **359**
Subspecies, 77
Substage condenser, 3, **4, 5**
Substage illuminator, 4, **4**
Subumbrella, 136, **137, 139**
Suckers
  leeches, 215
  tadpoles, 46
  tapeworms, 158, **160, 161**
  trematodes, **154, 157**
Sucking lice, 260
Sulci, 382, **383**
Superficial cleavage, 36
Superficial muscles of fetal pig, **357**
Superficial pectoral muscle, 357, **358, 359**
Suprabranchial chamber, **186, 187,** 188, 190
Supracoracoideus, 345, **345, 346,** 347
Supraesophageal ganglia, 234
Suprascapula, **320**
Suprascapular nerves, **380,** 381
Supraspinatus, **357, 358, 359,** 360
Suspension feeders, 279
Swim bladders, **304,** 305, **306**
Swimmerets, 228, **229,** 230, **231**
Sycon, 121–123, **122,** 126
Syconoid sponges, **122,** 122–123
Symbiotic mutualism, 101
Sympathetic division, 381
Sympathetic ganglia, 381–382
Sympathetic trunk, **380,** 381
Symphyses, 351
Synapsis, 27, 28, **28**
Synarthrosis, 354
Synsacrum, 344, **344**
Systemic arch, **328,** 329, **329,** 330
Systemic circuit, 372
Systemic heart, **197,** 198
Systemic veins, 342
Systems, defined, 47
Systolic period, 306

T

Tactile hairs, 234
Tadpole larvae, 285, **286**
Tadpole shrimp, **73**
Tadpole stage, **45,** 46
Taenia, 158–161
Tagmata
  arachnids, 223–224
  crickets, 251
  grasshoppers, 244
  myriapods, 241
  origins, 220
Tail fan, 230
Tails, 296, 301. *See also* Caudal fins; Caudal vertebrae
Tapeworms, 85, 158–161
Tarsals, **352, 353**
Tarsometatarsus, **344,** 345
Tarsus
  crickets, **252**
  grasshoppers, 245, **245**
  honeybee, **246**
  mammals, 351, **352**
  in spider leg, **224,** 225
Taxa, 77, 78–83, 84–86
Taxis, in ciliates, 111
Taxonomic keys, 68–74, 84–86
Taxonomy, 78–83
Tectibranchs, 85
Teeth
  bony fishes, 304
  cat, **351**
  fetal pig, 363
  frogs, **318,** 319, 325, **325**
  lampreys, 292, **292**
  sea urchins, 274, **274**
  shark, 297
Tegmen, 279
Tegument, 154, 158
Telolecithal eggs, 36
Telophase
  meiosis, **26, 27**
  mitosis, 18, **18, 19, 20**
Telson, 221, **222, 223**
Temperature coefficient, 116
Temperature effects on oxygen availability, 309
Temperature effects on protozoans, 115–116
Temporal bone, **351, 353**
Temporalis, **323**
Tendons, 345, **345,** 355
Tensor fasciae latae, **357,** 360, **361, 362**
Tentacles
  amphioxus, 288
  clamworm, 202, **203**
  Gonionemus, 136
  hydras, 132, **133**
  Metridium, 141, **141**
  Obelia, 136
  schyphozoans, **139,** 140
  sea cucumbers, 276, **277**
  sea stars, **267, 268**
  snails, 192, 193
  squids, 196, **196**
  tunicates, 285
Tentacular bulb, 136, **137**
Teredo, 191
Teres major, **359**
Tergum, **245,** 251
Termites, 100, 101, 259, 261
Testes
  Ascaris, **173,** 174, **175**
  bivalves, 190
  bony fishes, 305, **306**
  crickets, **253,** 254
  crustaceans, 232, **233**
  earthworm, 206, **207, 208**
  fetal pig, **369,** 370
  frogs, 327, **327**
  hydras, 134, **134**
  Obelia, **135**
  Romalea, 31
  sea stars, 269
  shark, **297,** 298, **299**
  squids, **197,** 199
  tapeworms, **159, 160**
  trematodes, **154, 157**
  tunicates, **284**
  vinegar eels, 176
Testing predictions, 63–64, 65
Tests, 275, 283
Tetrahymena, **112,** 117–118
Theca, 142
Thecostraca, 235–236
Theileria, 16
Thigh muscles, frogs, 321, **321, 322**
Thigmotaxis, 111, 150
Third ventricle, **383**
Thoracic cavity, 368
Thoracic ganglion, 382
Thoracic nerves, **380,** 381
Thoracic vertebrae
  birds, 344, **344**
  mammals, 350
  turtles, 338, 339, **339**
Thorax
  crickets, 251
  fetal pig, 350
  grasshoppers, 244–245
Thrips, 258
Thrombocytes, **52**
Thymus, **359,** 365, **365, 368, 385**
Thyone, 275, 276, 279
Thyrocervical artery, 375, **376**
Thyroid gland
  chordates, 281
  fetal pig, 357, 365, **365, 385**
Thyroxin, 364
Thysanoptera, 258
Thysanur, 260
Thysanura, 254, 260
Tibia
  birds, 345
  crickets, **252**
  grasshoppers, 245, **245**
  honeybee, **246,** 247
  mammals, 351, **352, 353**
  spider, **224,** 225
  turtles, **339**
Tibialis anterior longus, **321, 322**
Tibialis anticus longus, 322
Tibialis posterior, **321, 322**
Tibiofibula, **320**
Tibiotarsus, **344,** 345
Ticks, 86
Tissues
  defined, 47
  major types, 47–53
  of radiates, 131
  of various organs, 54–58, **56–58**
Toads, 86, 313. *See also* Frogs
Tongue
  birds, 346, **346**
  bony fishes, 304–305
  fetal pig, **364**
  frogs, 325, **325**
  honeybee, **246**
  lampreys, 292, **295**
  turtles, 339, **340**
Tooth shells, 85
Touch stimuli, 150
Trache
  crickets, **253**
Trachea
  fetal pig, **364,** 365, **365,** 384, **385**
  tissues, **49,** 57, **58**
  turtles, 340, **341**
Tracheal spiracle, **224,** 225
Tracheal systems in arthropods, 220
Tracheal tubes, 243, 251
Transitional epithelium, **49**
Transmission electron microscopes (TEM), 9–10
Transmitted light in microscopes, **8,** 8–9
Transparent ruler calibration, 6–7
Transverse abdominal muscle, **357,** 360, **361**
Transverse canal, 159, **160**
Transverse fissure, 382, **383**
Transverse muscle, 367
Transverse nerves, 152
Transverse process, **319**
Transverse septum, **297,** 306, 340
Transvs muscle, 321, **322**
Trapezius muscle, **357, 358, 359,** 360
Tree hermit crabs, 237
Trematoda, 85, 153–158, 161
Triatoma, 100
Tribolium confusum, 65–66
Triceps brachii, **321, 323, 358, 359,** 360
Triceps femoris, 321, **321, 322, 323**
Triceps muscles, 320, **357**
Trichinella spiralis, 177, **178**
Trichocysts, 110
Trichonympha, **100,** 101
Trichoptera, 73, 74, 258, 259
Triclads, 150
Tricuspid valve, **374**
Trigeminal nerve, **335, 383,** 384
Triiodothyronine, 364
Trilobita, 225
Trilobite larvae, 222, **223**
Trinomial nomenclature, 77
Triploblastic development, 149, 220
Triradiates, 125, **126**
Trivium, 267
Trochanter
  cat, **352**
  crickets, **252**
  grasshoppers, 245, **245**
  honeybee, **246**
  in spider leg, **224,** 225
Trochlear nerve, **335, 383,** 384
Trophozoites, 92, 101, 106, 107, **107, 108**
True bugs, 74, 261
True coelom, 172, 183, 220
True cornea, 199
True flies, 258
True jellies, 138–140
Truncus arteriosus, 328, 329, **329,** 330
Trunk
  bony fishes, 301
  frogs, 314, 321, **322**
  shark, 296
Trypanosoma, **100,** 100–101
Trypanosomes, 100
Tsetse flies, 100
Tube feet
  brittle stars, 272
  as common feature of echinoderms, 265
  sea cucumbers, 276, **277,** 278
  sea stars, 266, **267,** 268, **268, 269,** 270
  sea urchins, **273,** 274
Tuber cinereum, 382
Tubercles, 274, **274**
Tubeworms, 204, **204**
Tunic (outer covering), 283
Tunicates, 86, 282, 283–286
Turbatrix aceti, 174–176
Turbellaria, **71,** 85, 161
Turbellarians, 150–152, **151,** 161
Turtles, 86, 337–342
Two maternal chromosomes, 30
Two paternal chromosomes, 30
Two-cell stage, 44
Two-chambered heart, 294
Tympanic bulla, **351**
Tympanic membranes, 314, **325**
Tympanum, 245, **245, 335**
Typhlosole, 206, 209, **209, 295**

U

Ulna
  birds, **344,** 345
  cat, **352**
  fetal pig, **352**
  human, **353**
  mammals, 350
  turtles, **339**
Ulnar arteries, 375, **376**
Ulnar nerve, **334, 380,** 380–381
Umbilical arteries, **365,** 367, **369, 371, 376, 377,** 378

Umbilical cord, 350, **365, 371**
Umbilical veins, **365,** 366, **369, 371,** 374, **376, 377**
Umbones, 185, **187**
Undulating membrane, 100, **100**
Unicellular eukaryotes, 89–119. *See also* Protozoans
  Amoebozoa, 89–94 (*See also Amoeba*)
  classification, 113–114
  overview, 89
Unicellular organisms, 70, 77, 83–86. *See also* Protozoans
Uniramous structures, 230
Unrejected hypothesis, 64
Upper mandible, **344,** 345
Ureters
  birds, **346**
  crickets, 254
  fetal pig, **365,** 369, **369, 371**
  frogs, 327, **327**
Urethra, **49,** 369, 370, **371**
Urinary bladder
  bony fishes, **304, 306**
  earthworm, **207,** 208
  fetal pig, **365,** 369, **369, 371**
  frogs, 327, **327**
  parasites of, 157
  tissues, **49**
  turtle, **341**
Urinary papillae, 306
Urinary pores, 306
Urochordata, 86, 282, 283–286
Urogenital arteries, **328,** 330
Urogenital opening, 302, 350, 370
Urogenital papillae
  fetal pig, **371**
  lampreys, **292**
  shark, **297,** 298, 299, **299**
Urogenital pores, 305, **306**
Urogenital sinus, 293, 306, **306,** 371
Urogenital systems
  fetal pigs, 368–372
  frogs, 326–327
  shark, 298
Uropods, 228, **229,** 230, **231, 233**
Urostyle, 319, **320**
Uterine horn, **371**
Uterus
  Ascaris, 19, 174, **175**
  fetal pig, 370
  frogs, 327
  microscopic studies, 19
  shark, **299**
  tapeworms, **159, 160**
  trematodes, **154, 157**
  vinegar eels, **176**

V

Vagina
  Ascaris, **173,** 174
  fetal pig, 370–371, **371**
  Polystoma, **157**
  snails, **193**
  tapeworms, **159, 160**
  tissues, **49**
Vagosympathetic trunk, **380,** 381
  fetal pig, **380,** 381
Vagus nerve
  fetal pig, **380,** 381
  frogs, **335**
  sheep, **383,** 384
Valves, 184
Valvular intestine, 298
Vane of feather, 343
Variable numbers of tandem repeats (VNTRs), 117
Vas deferens
  Ascaris, **173,** 174, **175**
  bony fishes, 306
  crickets, **253,** 254
  crustaceans, **229, 233**
  fetal pig, 369, 370
  planarian, **151**
  squids, **197,** 199
  tapeworms, **159, 160**
  trematodes, **154**
Vas efferens, **154, 159**
Vastus intermedialis, 360, **361**
Vastus lateralis, 360, **361, 362**
Vastus medialis, 360, **361, 362**
Vegetal pole of frog eggs, 44, **45**
Veins
  amphioxus, 288, **289**
  bony fishes, **304**
  fetal pig, 375, 377–378
  frogs, 328–329
  shark, 300
  squids, **197,** 198
  tissues, 54, 57, **57**
  turtles, 342
Velar flaps, 295
Velar tentacles, 288
Velum
  amphioxus, 288
  Gonionemus, 136, **137**
  honeybee, **246,** 247
  lampreys, **294, 295**
Ventral abdominal artery, 232, **233**
Ventral abdominal vein, **328,** 329
Ventral aorta
  amphioxus, 288, 289, **289**
  bony fishes, **304,** 306
  lampreys, 293, 294, **295**
  shark, 300
Ventral branch, 379
Ventral ciliated ridge, **295**
Ventral fins
  amphioxus, 286, **287**
Ventral heart, 286
Ventral median lines, 173
Ventral nerve cord, 174, 207
Ventral ramus, **379**
Ventral root, 379, **379**
Ventral serratus, **358, 359,** 360
Ventral suckers, 46, 153, **157**
Ventral surface of brain, 382–384
Ventral thoracic artery, 232, **233, 376**
Ventral vessel, 206, 209
Ventricles
  bivalves, **187,** 189
  bony fishes, **304,** 306
  frogs, **326,** 328, **329,** 330, **335**
  lampreys, 293, 294, **294**
  shark, **297**
  sheep, **383**
  turtle, **341,** 342
Ventriculus, **253**
Vericrustacea, 228
Vermiform appendix, 367
Vertebrae
  birds, 344
  bony fishes, **303**
  cat, **352**
  fetal pig, **379**
  frogs, 319, **319, 320**
  mammals, 351
  turtles, 338–339, **339**
Vertebral arch, **379**
Vertebral column
  bony fishes, 301, 302, **303**
  cat, **352**
  fetal pigs, **368**
  frogs, 319, **319, 320**
  human, **353**
  mammals, 350
  turtles, **339**
Vertebral ossicles, 271
Vertebrata, 86, 283
Vestigial oviduct, 327, **327**
Vibrissae, 350
Villi, 367
Vinegar eels, 174–176
Viridiplantae, 84, 98–100, **99,** 114
Visceral ganglia, 191
Visceral hump, 193
Visceral mass, **186,** 187
Visceral pericardium, 368, **368**
Visceral peritoneum, 209, 298, 347, 367
Visceral pleura, 368, **368**
Visceral skeleton, **318,** 318–319
Vitgelline membrane, **39**
Viviparous species, 299
VNTRs (variable numbers of tandem repeats), 117
Vocal folds, 384
Vocal sacs, 325
Voluntary muscle, **53**
Volvox, **70,** 98–100, **99**
Volvox globator, 98–100
Vomer, **318,** 319
Vomerine teeth, **318,** 319, 325, **325**
Vorticella, **70,** 112, **112**
Vulvae, 173, 174, **176,** 371

W

Walking legs
  as basic features of insects, 243
  crustaceans, **229,** 230, **231**
  grasshoppers, **245**
  horseshoe crabs, 220–221
  spiders, **224,** 225
Walking sticks, 262
Wasp, 258
Waste disposal. *See* Excretory systems
Water, oxygen availability, 309, 310
Water bug, **74**
Water fleas, 73, 235, **235,** 239–240
Water lilies, 67
Water mites, **72**
Water spider, **72**
Water tubes, 188, **189**
Water-surface habitats, 67
Water-vascular systems
  as basic feature of echinoderms, 265
  sea cucumbers, 276, 278
  sea stars, 266, 267, 270
  sea urchins, 275
Webbed toes, 314
Wheel animals, 72
Wheel organ, 288, **288**
Whitefish, 19, **19**
Whorls, 192
Winged ants, 258
Wingless bugs, 261
Wingless wasps, 260
Wings
  as basic features of insects, 243
  cockroach, 248
  crickets, **252**
  displaying for insects, 256–257
  grasshoppers, 244, **245**
Wishbone (furcula), **344,** 345
Wolffian ducts, 299, 305, **327**
Worker bees, 247
Wuchereria bancrofti, 178

X

Xiphisternum, 319, **320,** 350

Y

Yellow marrow, **51**
Yellow perch, 301–306
Yolk ducts, **154**
Yolk glands
  planarian, **151**
  tapeworms, **159, 160**
  trematodes, **154, 157**
Yolk plug, 46
Yolk sacs, **299**

Z

Zebra mussels, 184
Zoantharia, 144
Zoea, 236
Zonary placenta, 371
Zooids, 99, 135, 137, 285
Zoothamnium, 112, **112**
Zooxanthellae, 142
Zygomatic arch, **351, 352**
Zygomatic bone, **353**
Zygopophyses, 338
Zygote nucleus, 30
Zygotes
  bivalves, 190
  development, 37–38
  hydras, 134–135
  Plasmodium, 106
  schyphozoans, **139**
  Volvox, 99

DEFINITIONS

Regional and Directional References

anterior Pertaining to the front or head end; cephalic; cranial.

posterior Pertaining to the tail or hind end.

dorsal Referring to the back or uppermost side.

ventral Referring to the belly side.

lateral Pertaining to the side of a body; situated to either side of the midline.

medial At or near the middle or midline of a body or organ, in contrast to lateral.

cephalic (cranial) Pertaining to the head end. The opposite of caudal.

caudal Pertaining to the tail end.

proximal Toward or near the point of attachment. The opposite of distal.

distal Away from the base or point of attachment.

longitudinal Pertaining to the long axis of the body; lengthwise.

peripheral Referring to parts away from the center; external to.

superficial On or near the surface.

oral Pertaining to the mouth or the region around the mouth.

aboral Pertaining to a region away from or most distant from the mouth.

Terms That Indicate Planes of Section

sagittal section Longitudinal section through the median vertical plane that bisects the body into right and left halves; or any section parallel to it.

parasagittal section Section to one side of the midline, separating right and left portions of unequal size.

frontal section Longitudinal section made at right angles to a sagittal section. It is parallel to the dorsal and ventral surfaces. Same as coronal section.

transverse (cross) section Any section made through and at right angles to the longitudinal axis.

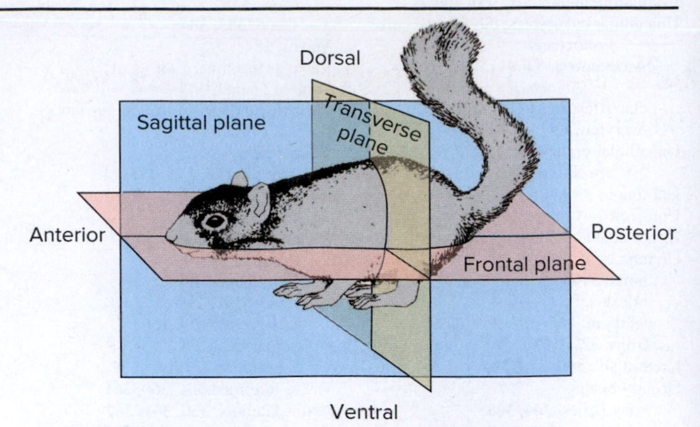

median plane Vertical longitudinal plane that extends from the dorsal to the ventral surface and passes from the anterior end to the posterior end through the middle of the body. This kind of plane produces the sagittal section.

Kinds of Symmetry

symmetry Correspondence in size, shape, and relative position of parts that are on opposite sides of a dividing line or median plane.

spherical symmetry Condition in which any plane passing through the center point will divide the body into like halves. Rare in life; usually associated with floating forms.

radial symmetry Condition in which any plane passing through the longitudinal axis will divide the body into similar halves. Best suited to sessile forms.

bilateral symmetry Condition in which only a sagittal plane (through the longitudinal and dorsoven-tral axes) will divide the body into like halves. The most common type of symmetry.

biradial symmetry Essentially radial, but only two planes through the longitudinal axis create truly similar halves.

Analogy and Homology

analogous Similar in function and superficial structure (e.g., the wings of a bird, bat, and insect are analogous).

homologous Basically similar as a result of similarity in embryonic origin and development (e.g., the forelimbs of a bird, bat, horse, and human are homologous).